Performance of the Jet Transport Airplane

Aerospace Series – Recently Published Titles

Performance of the Jet Transport Airplane: Analysis Methods, Flight Operations, and Regulations	Young	October 2017
Differential Game Theory with Applications to Missiles and Autonomous Systems Guidance	Faruqi	May 2017
Advanced UAV Aerodynamics, Flight Stability and Control: Novel Concepts, Theory and Applications	Marques and Da Ronch	April 2017
Introduction to Nonlinear Aeroelasticity	Dimitriadis	April 2017
Introduction to Aerospace Engineering with a Flight Test Perspective	Corda	March 2017
Aircraft Control Allocation	Durham, Bordignon and Beck	January 2017
Remotely Piloted Aircraft Systems: A Human Systems Integration Perspective	Cooke, Rowe, Bennett. Jr. and Joralmon	October 2016
Adaptive Aeroservoelastic Control	Tewari	March 2016
Theory and Practice of Aircraft Performance	Kundu, Price and Riordan	November 2015
The Global Airline Industry, Second Edition	Belobaba, Odoni and Barnhart	July 2015
Modeling the Effect of Damage in Composite Structures: Simplified Approaches	Kassapoglou	March 2015
Introduction to Aircraft Aeroelasticity and Loads, 2nd Edition	Wright and Cooper	December 2014
Aircraft Aerodynamic Design: Geometry and Optimization	Sóbester and Forrester	October 2014
Theoretical and Computational Aerodynamics	Sengupta	September 2014
Aerospace Propulsion	Lee	October 2013
Aircraft Flight Dynamics and Control	Durham	August 2013
Civil Avionics Systems, 2nd Edition	Moir, Seabridge and Jukes	August 2013
Modelling and Managing Airport Performance	Zografos, Andreatta and Odoni	July 2013
Advanced Aircraft Design: Conceptual Design, Analysis and Optimization of Subsonic Civil Airplanes	Torenbeek	June 2013
Design and Analysis of Composite Structures: With Applications to Aerospace Structures, 2nd Edition	Kassapoglou	April 2013
Aircraft Systems Integration of Air-Launched Weapons	Rigby	April 2013

See more Aerospace Series titles at www.wiley.com

Performance of the Jet Transport Airplane

Analysis Methods, Flight Operations, and Regulations

Trevor M. Young
University of Limerick, Ireland

Registered Offices
John Wiley & Sons, Inc., 111 River Street, Hoboken, NJ 07030, USA
John Wiley & Sons Ltd, The Atrium, Southern Gate, Chichester, West Sussex, PO19 8SQ, UK

Editorial Office
The Atrium, Southern Gate, Chichester, West Sussex, PO19 8SQ, UK

For details of our global editorial offices, customer services, and more information about Wiley products visit us at www.wiley.com.

Wiley also publishes its books in a variety of electronic formats and by print-on-demand. Some content that appears in standard print versions of this book may not be available in other formats.

This publication is intended for educational use and not operational use. While every effort has been made to ensure the correctness of the information herein presented, under no circumstances should this information be used for flight operations. Information has been extracted—and in many cases, for the sake of conciseness, paraphrased from—regulations and specifications issued by the United States Federal Aviation Administration (FAA) and the European Aviation Safety Agency (EASA). For flight operations, it is imperative that the latest revisions of these regulations and specifications are consulted. The data and information contained herein have been compiled by the author, and do not reflect the opinion of any airplane or engine manufacturer or airline.

Library of Congress Cataloging-in-Publication Data

Names: Young, Trevor M., author.
Title: Performance of the jet transport airplane : analysis methods, flight operations and regulations / by Trevor M. Young, University of Limerick, Ireland.
Description: First edition. | Hoboken, NJ, USA : Wiley, 2018. | Includes bibliographical references and index. |
Identifiers: LCCN 2017014995 (print) | LCCN 2017037602 (ebook) | ISBN 9781118534762 (pdf) |
    ISBN 9781118534779 (epub) | ISBN 9781118384862 (pbk.)
Subjects: LCSH: Jet transports–Piloting. | Airplanes–Performance. | Aeronautics, Commercial.
Classification: LCC TL710 (ebook) | LCC TL710 .Y64 2017 (print) | DDC 629.133/349–dc23
LC record available at https://lccn.loc.gov/2017014995

Cover image: Courtesy of Trevor M. Young

Set in 10/12pt WarnockPro by Aptara Inc., New Delhi, India
Printed in Singapore by C.O.S. Printers Pte Ltd

10 9 8 7 6 5 4 3 2 1

Contents

Foreword

The number of textbooks on aircraft performance published since the end of World War II is overwhelming. Most of them treat propeller as well as jet-propelled airplanes, and a few textbooks pay attention to both fixed-wing aircraft and rotorcraft. Dr. Young has made a conscious choice to concentrate on subsonic jet transport airplanes. One of his arguments supporting this choice has been that most books on the market deal with the topic of performance from a pure flight mechanics perspective and pay little attention to FAA/EASA airworthiness regulations and actual flight operations. Also, the majority of the available books have treated the performance of propeller airplanes cruising at speeds up to 700 km per hour as well as jet airplanes cruising at speeds up to approximately 900 km per hour. After digesting Young's book, I came to the conclusion that he also wanted to pay more attention to the effects of compressibility on high-speed performance than is usually done. It is interesting to note that despite his confinement to a single category of aircraft, Dr. Young has produced a comprehensive text of 650 pages.

In my view, the choice of dealing primarily with jet transport airplanes will be an advantage for attracting the maximum audience since, apart from aeronautical engineering students and professionals, there could be considerable interest from the general public. This becomes clear when it is realized how many passengers nowadays travel each day all over the world in long-range, fast, reliable, and safe jetliners. Since the 1950s, when the Boeing 707 and the Douglas DC-8 appeared in the air, and in spite of several oil and financial crises, the growth of air transportation has been nearly uninterrupted with global annual increases of almost 5%. Although many factors have contributed to this trend, there is no doubt that the steadily increasing safety record and the improvement in fuel efficiency (from approximately 10 seat-kilometers per liter of fuel burnt in 1950 to 40 in the year 2000) were dominating factors.

After the appearance of the B-2 stealth bomber in the late 1980s, NASA confronted the American industry with the question: *Is there a renaissance for long-range transport?* The reply was a diagram showing the slow development of the parameter cruise Mach number M times aerodynamic efficiency L/D, which had increased modestly from 13 in 1960 to 15 in 1990. In view of the abovementioned actual fuel efficiency improvement, this apparent contradiction might suggest that the industry had not been creative enough during those years and was eager to start a comprehensive technology development program (to be financed by the U.S. government). But the suggestion that the industry could realize a major technological step forward by the turn of the century has not materialized. And anyone who has digested Young's sophisticated treatment of optimum cruise performance will conclude that the cruise parameter ML/D is a misleading criterion to judge an airplane's cruise efficiency. In fact, the same parameter had been used in 1960 as one of the arguments to start the development of the supersonic British–French Concorde, an airplane that had a fuel consumption per seat-kilometer three times that of contemporary subsonic airliners.

A set of flight safety rules ensuring that a public transport flight is dispatched at an appropriate all-up weight that attains a high level of safety is essential for maintaining the requirements of the commercial air transportation system in modern society. The performance capabilities of a newly developed aircraft type are determined by means of flight testing during the certification process, which is executed according to the stringent regulations of the American Federal Aviation Administration and the European Aviation Safety Agency. Based on these flight tests, operational limitations are prescribed for every flight, so that the risk of unsafe operation is reduced to an acceptable level. In contrast to most textbooks on airplane performance, Young's book pays considerable attention to the close interrelationship between airworthiness and safe operational performance. He has drawn on his knowledge of flying, as a general aviation pilot, to describe the complex procedural and regulatory framework in which flight operations take place.

Another important feature of Young's book is a self-contained, detailed, and scientifically rigorous treatment of topics associated with all aspects of the complete mission of jet transport airplanes. If appropriate data are available to define airplane lift, drag, engine thrust, and fuel consumption, the methods presented will provide solutions with adequate accuracy to be applied in industrial projects. And for professional engineers involved in the early and advanced stages of new airliner design, the performance criteria derived in the chapters on takeoff and landing performance, trip fuel planning, cost considerations, noise, and emissions could serve as design objectives. In particular, the data Items published by IHS ESDU (Engineering Science Data Unit) form a major source of guidance for those associated with flight performance, and many derivations in Young's book agree with the relevant publications. Of interest to pilots and engineers responsible for planning flight operations will be his detailed treatment of field performance, which provides a thorough description of this complex subject.

Comparing the present comprehensive and well-written text with many books on aircraft performance makes it clear that many of them have shortcomings in the sense that they tend to emphasize mathematical treatment at the cost of convincing worked examples and excellent illustrations based on realistic data of existing aircraft. This might explain why preparing this book has cost Dr. Young 12 years. I am sure that the reader will agree with me that the author has spent this time in a very commendable way.

Professor Emeritus Egbert Torenbeek
TU Delft, April 18, 2017

Series Preface

The field of aerospace is multi-disciplinary and wide ranging, covering a large variety of products, disciplines, and domains, not merely in engineering but in many related supporting activities. These combine to enable the aerospace industry to produce innovative and technologically advanced vehicles. The wealth of knowledge and experience that has been gained by expert practitioners in the various aerospace fields needs to be passed on to others working in the industry and also to researchers, teachers, and the student body in universities.

The *Aerospace Series* aims to be a practical, topical, and relevant series of books aimed at people working in the aerospace industry, including engineering professionals and operators, engineers in academia, and those in allied professions such as commercial and legal executives. The range of topics is intended to be wide ranging, covering design and development, manufacture, operation, and support of aircraft, as well as topics such as infrastructure operations and current advances in research and technology.

Aircraft performance concerns the prediction of an airplane's capabilities (range, speed, payload, takeoff, etc.) and evaluation of how well it functions throughout its operation. The topic is inherently multi-disciplinary, requiring not only an understanding of a wide range of individual disciplines such as aerodynamics, flight mechanics, powerplant, loads, etc., but also an appreciation of how these topics interact with each other. Such analyses are essential as a basis for the design of future aircraft.

This book, *Performance of the Jet Transport Airplane: Analysis Methods, Flight Operations, and Regulations*, provides a comprehensive overview of the topics required to carry out performance calculations. It is a welcome addition to the Wiley Series' existing content relating to airplane performance, and will prove to be very useful for undergraduate and graduate aircraft design courses. A complete range of relevant technical topics is examined, complemented by sections on cost, operations, and regulations.

Peter Belobaba, Jonathan Cooper, and Allan Seabridge

Acknowledgments

I wish to pay tribute to my teachers and mentors. It is they who guided my understanding of this wondrous, fascinating subject of flight. It is they who introduced me to the mathematical models and analysis techniques that can be used to predict and study the performance of airplanes. I acknowledge and thank Tony Streather (formerly of the University of the Witwatersrand, Johannesburg); Adrian Marx (posthumous) (formerly of Swissair, Zürich); Martin Eshelby, John Fielding, and Denis Howe (formerly of Cranfield University, UK); and Walt Blake (formerly of Boeing Commercial Airplanes, Seattle).

To everyone who assisted me in the preparation of this book, I wish to express my sincere appreciation. For your many helpful suggestions, comments, and critical reviews, which helped to transform a rough collection of notes into this finished product, I thank you all. In particular, I wish to thank Ruth Bagnell, Roger Bailey, Matthias Bartel, Lance Bays, Peter Belobaba, Clare Bewick, Steve Duffy, Pio Fitzgerald, Ion Fuiorea, Jim Gautrey, Paul Giesman, Martin Hepperle, Tony Khan, Hartmut Koelman, Lars Kornstaedt, Nicholas Lawson, Michael McCarthy, Paul Murray, Andrew Niven, Eoin O'Malley, Paul Peeters, Nicholas Rooney, Nick Stanger, and Egbert Torenbeek for the inspiring discussions, comments, and email exchanges over the past twelve years. I am indebted to Walt Blake, Alastair Cooke, and Robert Telford for their meticulous reviews of early versions of my manuscript. It is acknowledged that any errors and omissions in this work are entirely my own.

I wish to thank my family, Fiona, Michael, Kieran, and Christopher, for their love, support, and understanding.

Trevor M. Young
March 3, 2017

1

Introduction

1.1 Definitions of Performance

The word *performance* has been defined many times—the following two definitions, provided in dictionaries of general word usage, encapsulate much of what is discussed in this book:

(1) Performance describes the "capabilities of a machine or product [in this case, an airplane], especially when observed under particular conditions" [1].
(2) Performance is the "manner in which or the efficiency with which something [in this case, an airplane] reacts or fulfills its intended purpose" [2].

The first definition highlights the *capabilities* of the product. Identifiable attributes of interest that quantify the capabilities of any airplane include the payload that can be carried over a defined distance, the airplane's stall speeds, rates of climb, turn radii, optimum cruise speeds and altitudes, takeoff and landing distances, and so forth. These are all performance attributes that, for a particular set of conditions, can be established by calculation or measurement.

When describing an airplane's performance, the associated conditions under which these attributes were determined is critical information—without which, the data would be meaningless. Frequently, it is the combination of the airplane's weight, the altitude at which it flies, and the ambient air temperature that must be established to describe adequately such performance capabilities. This set of conditions occurs so often in performance discussions, it has its own abbreviation: WAT (for weight, altitude, and temperature). The significance of this combination will become apparent in the ensuing discussions. The barometric pressure of the ambient air around an airplane is a function of altitude, and, through a fundamental law of gas dynamics, air density is linked to pressure and temperature. Consequently, this combination of altitude and air temperature can be seen to influence significantly the performance of an airplane by affecting the engines' thrust and the aerodynamic forces that are generated by the relative motion of the airplane with respect to the surrounding air. An airplane departing from Denver International Airport, Colorado (elevation 5434 ft) on a hot summer day requires a substantially longer takeoff distance compared to that which would be required at Shannon Airport, Ireland (elevation 46 ft) in mid-winter for a similar takeoff weight, for example.

A complete and accurate description of the operating environment is essential when fully describing the capabilities of an airplane, and this goes beyond the WAT conditions. The type of runway surface—for example, textured or smooth—affects braking distances. The presence of standing water or snow on the runway reduces an airplane's acceleration during takeoff, increasing the required takeoff distance. Winds too impact an airplane's performance. Trip times and fuel usage can be significantly affected by jet streams, which are fast flowing, narrow air currents found in the atmosphere at altitudes of about 30 000 to 40 000 ft. Takeoff and landing distances

Performance of the Jet Transport Airplane: Analysis Methods, Flight Operations, and Regulations, First Edition.
Trevor M. Young.

are shortened when operating into a headwind, compared to a nil wind condition. The presence of turbulence is a consideration when a pilot selects an approach speed for landing, and, consequently, this can influence an airplane's landing distance.

The performance capabilities of an airplane naturally depend on the thrust produced by the engines. The thrust of jet engines is not unlimited—it is restricted, or governed, at certain critical flight conditions (e.g., during takeoff) based on such factors as the ambient air temperature, altitude, and airplane speed. The engine's electronic control system governs the engine to ensure that its structural integrity is not compromised when producing the defined, or rated, thrust.

All airplane systems have an indirect impact on an airplane's performance by virtue of their installed weight, which makes up part of the airplane's operating empty weight (OEW). Several systems, however, can also have a direct impact on performance—for example, the cabin pressurization and air conditioning systems extract power from the engines (either as electrical power or as compressed air) at the expense of the available propulsive power. Consequently, the rate of climb for certain airplane types, under demanding conditions, can depend on the air conditioning settings selected by the pilot.

An airplane's performance also depends on its configuration, which, in this context, describes the position or setting of re-configurable parts of the airplane by the flight crew or flight control system—that is, the positions of the flight controls (e.g., flaps, slats, rudder, and ailerons) and undercarriage, and so forth. One important consequence—which all student pilots learn at an early stage of their training—is that an airplane does not have a single stall speed, but rather a range of stall speeds depending on its configuration and the associated WAT values. Interestingly, stall speeds also depend on the pilot's actions prior to the stall (e.g., the airplane's bank angle and the rate at which the airspeed decreases are factors that can influence a stall, albeit to a lesser extent than WAT).

The second general definition of performance, given earlier, extends these ideas by indicating that it is the manner in which, or the efficiency with which, the airplane accomplishes these actions that is a measure of its performance. In other words, performance topics also address the question of how well an airplane accomplishes its task or mission. A definition of performance provided by the United States Federal Aviation Authority is that it is "a measure of the accuracy with which an aircraft, a system, or an element of a system operates compared against specified parameters" [3]. This expands the discussion of performance to the determination of parameters that facilitate comparative assessments of airplane performance to be conducted. An important comparative measure of an airplane's overall performance is its payload–range capability. The standard presentation of this information is a chart that indicates, on one axis, the payload (i.e., the mass or weight of the passengers, baggage, and freight) plotted against, on the other axis, the distance that can be flown for a given set of mission rules (which defines the flight profile and fuel reserves). *Efficiency* is all about achieving the desired result—for example, flying a set distance with a given payload—with the least effort expended or resources consumed. The determination of the optimum conditions, in terms of flight speeds and cruise altitudes, that will minimize the trip fuel or, alternatively, the trip cost are classic airplane performance studies. An extension of studies of this type considers the environmental impact of the flight in terms of the noise generated or the exhaust emissions produced.

Another important attribute that often gets considered under the topic of performance is the manner in which the airplane responds or reacts to control inputs by the pilot (or flight control system) or to external influences such as atmospheric turbulence. The response of the airplane to a sudden vertical gust depends on the airplane's speed and its aerodynamic characteristics (e.g., the lift-curve slope and the wing loading), for example.

The study of the response of an airplane to a set of applied forces (e.g., lift, weight, drag, engine thrust) is known as *flight mechanics*, which is fundamentally based on the application of Newtonian mechanics. The subject of flight mechanics has traditionally covered two separate, but related, topics: airplane performance (which primarily deals with trajectory analysis) and airplane stability and control. This book addresses the former topic, as applied to jet transport airplanes. All elements of a typical flight are considered (i.e., takeoff, climb, cruise, descent, maneuver, approach, and land). For a performance assessment of an airplane to have real value, however, the context and the operating environment need to be clearly defined and understood. For commercial jet transport airplanes, this context is to a large extent imposed by operational and regulatory procedures and constraints. Many of these considerations have traditionally not been addressed in academic textbooks on the subject of airplane performance, but are well known, of course, to those closely involved with the planning and execution of flight operations. In this book, performance discussions have extended many of the traditional, idealized performance representations to include key operational and regulatory aspects. The determination of the maximum takeoff weight that a pilot can safely utilize for a given set of conditions, for example, can be a complex study with many considerations, which include the runway length, slope, and surface condition; the wind and ambient conditions (viz., air temperature and barometric pressure); location of obstacles near to the intended flight path; airplane configuration (e.g., flap setting); and airplane limitations (e.g., brake energy limits, which are important in the event of a rejected takeoff).

Another point that arises from the above-mentioned definitions of performance is that the most important performance characteristics of any product are those that relate most closely to the primary purpose or role of the product. In broad terms, the role of commercial transport airplanes is to transport people and goods efficiently and safely, by air, from one geographical location to another within an aviation infrastructure (comprising airports and related services, air traffic control, ground and satellite navigation systems, and so forth).

1.2 Commercial Air Transportation

Commercial air travel is considered to have started on January 1, 1914, when Tony Jannus—flying a biplane designed to take off and land on water—transported the first recorded fee-paying passenger (across Tampa Bay in Florida, United States) [4]. One hundred years later—a time period that is associated with aviation advancements on a previously unimaginable level—saw 8.6 million passengers transported each day on commercial flights around the world [5]. Nearly 1400 airlines serviced 3800 airports [5]. Passenger and cargo air transportation had grown to become an essential part of modern life and a vital component of global economic activity. In 2014, about 35% of world trade by value was transported by airfreight [5].

Commercial air travel continues to grow. Air traffic—measured in terms of revenue passenger kilometers (RPK)—has experienced global long-term growth rates of approximately 5% per year. Informed forecasts by key stakeholders predict average growth rates of 4.5–5% per year for the forecast period (i.e., until 2035) [6, 7]. In 2015, the number of in-service jet transport airplanes, including freighter aircraft, was 22 510 (made up as follows: widebody types 22.4%, single-aisle types 66.0%, and regional jets 11.6% [7]); to satisfy increasing passenger demand, the global fleet will need to double by about 2035 [7, 8]. As might be expected, air traffic growth is linked to economic growth—both at a regional and global level. Interestingly, air traffic has historically enjoyed growth rates greater than the economic growth rate measured by gross domestic product (GDP) change. Economic cycles coupled with social factors (e.g., pandemics) and political factors (e.g., liberalization of aviation markets and wars) have a significant effect on the demand for air travel, which, consequently, fluctuates over time. One of the key economic factors is the price of crude oil, which impacts airline economics directly through the price of jet fuel and indirectly through its influence on global economic activity. Fluctuations in the global demand for air travel about the long-term trend associated with economic downturns or other significant events (e.g., terror threats or attacks) have historically been short lived, typically lasting about one year, after which the upward trend has continued.

The demand for air travel is also a function of its cost to the consumer (reduced prices stimulate demand), which, in turn, is dependent on the economic factors that impact airlines. The largest single cost element for airlines in recent years has been aviation fuel, which can represent about half of the total operating cost for a long-haul flight. Airplane fuel efficiency is thus a critically important performance metric for the airline industry; efficient airplanes are essential for airline profitability and long-term economic success of the industry.

Two important environmental considerations for the operation of commercial jet airplanes are noise and engine exhaust emissions. Noise limits for airplane certification are issued by aviation regulatory authorities. In addition, local authorities (e.g., airports) frequently impose additional noise limits. Compliance, in certain cases, can necessitate a noise abatement operational procedure that has performance implications (e.g., a reduced climb thrust setting after takeoff). The environmental impact of airplane emissions is of growing concern. For every 1 kg of jet fuel that is burned in an airplane's engine, approximately 3 kg of carbon dioxide is produced [9]. Increasing airplane fuel efficiency is thus a powerful direct means to reduce the environmental impact of air transport.

1.3   Jet Transport Airplanes: A Short History

What is often referred to as the *jet age* is that fascinating period of aviation history since turbojet engines were first installed on passenger transport airplanes (airliners). The first production jet airliner, the DH 106 Comet, manufactured by de Havilland in the United Kingdom (UK), entered commercial service in May 1952. The airplane offered unprecedented performance. In the cabin, it was quieter than its piston engine counterparts. It was also much quicker and was able to fly higher—thus avoiding bad weather—due to the thrust of its four de Havilland Ghost turbojet engines. The Comet 1, however, was grounded in 1954 when its certificate of airworthiness was withdrawn after two aircraft suffered explosive decompression in flight due to metal fatigue (caused by repeated pressurization cycles).

The first generation of successful turbojet commercial airplane types entered service in the late 1950s. In the Soviet Union, the Tupolev Tu-104 began service with Aeroflot in 1956. The design, which was extensively based on the Tu-16 bomber aircraft, had one Mikulin turbojet engine installed in each wing–fuselage junction. In the UK, deliveries from de Havilland of the extensively redesigned Comet commenced in 1958. The Comet 4 had a longer fuselage than its predecessor (the maximum seating capacity was 81) and more powerful Rolls-Royce Avon engines. In the United States, the Boeing 707 and Douglas DC-8 entered service in 1958 and 1959, respectively. Both types featured four turbojet engines mounted under the airplane's wings. Early variants, however, had limited payload–range capability. In France, Sud Aviation developed the first short- to medium-range jet airliner: the SE 210 Caravelle (later, Sud Aviation merged with Nord Aviation to form Aérospatiale). The design featured two Rolls-Royce RA-29 turbojet engines pod-mounted off the rear fuselage—a configuration that has been extensively used since then in the design of regional jet airplanes.

In the 1960s, newly developed low bypass ratio turbofan engines, such as the Pratt & Whitney JT3D, offered much improved airplane performance. Later variants of the highly successful Boeing 707 and DC-8 types offered true intercontinental capability due to the engine's lower fuel consumption. Convair, a division of General Dynamics, produced the medium-range

Convair 880. Despite its high cruise speed, the airplane failed to attract much interest—this was in part due to its poor economic performance compared to the competition, which included the Boeing 720 (a short-range derivative of the Boeing 707). New types that entered service around the world included the three-engine Boeing 727 and the two-engine Douglas DC-9 (this started a four-decade production run of many derivatives, including the MD-80, MD-90, MD-95, and Boeing 717). Three British airliners entered service in the mid-1960s: the short-range British Aircraft Corporation BAC One-Eleven, the short/medium-range Hawker Siddeley HS 121, and the long-range Vickers-Armstrongs VC10 (which featured a pair of rear-mounted engines on each side of the fuselage). In the Soviet Union, the long-range Ilyushin Il-62, which adopted the same engine configuration as the VC10, commenced service in 1967. The highly successful Boeing 737, featuring twin-turbofan engines installed closely coupled below the wing and a six-abreast seating arrangement, began servicing short- and medium-haul routes[1] in 1968 (to date, more Boeing 737 airplanes have been produced than any other jet airliner). To compete with turboprop airplanes in the regional airplane market, Yakovlev (in the Soviet Union) introduced the small Yak-40 (with 24–32 seats) and Fokker (in the Netherlands) introduced the F28 Fellowship (with 60–65 seats) in 1968 and 1969, respectively.

The 1970s will be remembered as the decade that saw the introduction of widebody airliners, that is, with a fuselage wide enough to accommodate two passenger aisles (typically with seven or more seats abreast). The four-engine Boeing 747, three-engine McDonnell Douglas DC-10, and three-engine Lockheed L-1011 TriStar airplane types began servicing long-haul routes worldwide, with significantly increased payload capacity over the Boeing 707, DC-8, and VC10 types. The Boeing 747 (nicknamed *Jumbo Jet*), which entered service as the world's largest passenger jet in 1970, would remain in production for over 45 years and would dominate this market sector for much of this time. The Tu-154, a narrow-body (i.e., single-aisle) airplane with three rear-mounted engines, began service in 1972. This successful medium-range type, which was designed to operate from unpaved or gravel airfields, saw 40 years of service. The A300, the first product of Airbus Industrie,[2] entered service as the first twin-engine widebody airplane in 1974. The decade will also be remembered as a time when airports were extremely noisy places. The first noise regulations for subsonic jet airplanes developed by the International Civil Aviation Organization (ICAO) came into force in 1972. These regulations led to the phasing out of many noisy first-generation jet airplanes, while others were modified with hush kits to reduce their noise during takeoff and landing (these modifications inevitably resulted in performance penalties). Noise regulations also spurred the introduction of higher bypass ratio, quieter turbofan engines, such as the General Electric CF6, Pratt & Whitney JT9D, Rolls-Royce RB211, and CFM International CFM56. These engine types and the many derivatives

---

1 There are no standard definitions of short, medium, or long haul (or range). The terms are used loosely in the aviation industry, with different sectors and organizations defining the categories in their own ways. One definition, used for market categorization, is based on trip duration: short (<3 hr), medium (3–5 hr), and long (>5 hr). Another definition, which is also widely used, defines long haul as a trip time exceeding 6 hr. Ultra-long haul is today taken to mean a flight of more than about 12 hr (although, again, different organizations define it differently). Short haul is sometimes used by airlines to describe domestic or regional routes. When based on distance, short haul is defined by one manufacturer as a trip of less than 3000 miles (2607 nm) and long haul as greater than 4000 miles (3476 nm) [10]. Another definition, which is used in market categorization, defines short haul as a trip distance of less than 2000 nm and long haul as greater than 2000 nm [11].

2 Airbus Industrie, with headquarters in Toulouse, France, was established in 1970 as a consortium of European aerospace companies—a Groupement d'Intérêt Économique (GIE) under French law. Aérospatiale of France, Deutsche Airbus (DASA) of Germany, and Hawker Siddeley of the UK were the initial shareholders. Construcciones Aeronáuticas SA (CASA) of Spain joined soon afterwards. In 2001, following several changes to its structure, the airplane manufacturer became a joint-stock company: Airbus SAS. Airbus's parent company, EADS, was renamed Airbus Group in 2014.

that were subsequently developed based on these designs would power much of the western world's commercial jet airplanes in the decades that followed.

The 1970s also saw the start of supersonic passenger transport services. The Aérospatiale/ BAC Concorde—which was operated in small numbers (seven each by Air France and British Airways)—commenced commercial operations in 1976 (the type was retired in 2003). The Russian Tupolev Tu-144 commenced passenger service in 1977 (the fleet was grounded after 55 scheduled flights). Both designs, which were capable of cruising at speeds greater than Mach 2, employed a tail-less, delta wing configuration with four after-burning turbojet engines. Supersonic cruise, however, came at a considerable aerodynamic penalty (the lift-to-drag ratios of these airplanes were less than half that of a typical subsonic airliner), and the fuel efficiency was significantly poorer (a feature that became increasingly apparent as improved, higher bypass ratio turbofan engines were developed for subsonic airliners). Other issues that plagued both Concorde and TU-144 operations were high levels of takeoff noise and the sonic boom (which essentially limited supersonic flight to overwater sectors).

In the 1980s, first and second-generation narrow-body airliners were withdrawn from service in large numbers and replaced by more fuel-efficient twin-turbofan airplanes—these included the newly developed Boeing 757, Boeing 767 (Boeing's first widebody twinjet), and Airbus A310 types. Boeing also introduced the 737-300/-400/-500 models (the so-called Classics), with CFM56 engines and a redesigned, more aerodynamically efficient wing. In the Soviet Union, the Yakovlev Yak-42 and the Tupolev Tu-204 (both narrow-body designs) and the four-engine widebody Ilyushin Il-86 entered service. The Il-96, a shortened longer-range derivative, followed a few years later. British Aerospace, identifying the need for a STOL (short takeoff and landing) airliner, introduced the high-wing four-engine BAe 146 (later variants were called the Avro RJ). The first member of the very successful Airbus A320 family (which includes the A318, A319, A320, and A321 models) began service on short- to medium-range routes in 1988. The type, which was developed as a direct competitor to the Boeing 737, pioneered the use of digital fly-by-wire flight control systems on airliners. The Fokker 100, a 100-seat regional jet with two rear-mounted engines and 5-abreast seating, also commenced service in 1988. A shortened derivative, the Fokker 70, with 70–80 seats, followed a few years later.

In the 1990s, the first *large twins*—widebody twinjets with a maximum seating capacity exceeding *ca.* 300—were introduced into commercial service. The Airbus A330 and Boeing 777—both featuring two high bypass ratio turbofan engines installed below the wings—were able to compete on long-haul transoceanic routes that, in the West, had previously been the exclusive domain of the four-engine Boeing 747 and the three-engine McDonnell Douglas DC-10 and its derivative, the MD-11. This was made possible by new regulations that became known by the acronym ETOPS (initially defined as Extended Twin Operations). Airbus also introduced the long-range four-engine A340, which shared many common design features and components with the A330. In the single-aisle market, the Next-Generation Boeing 737 airplanes entered service (replacing aging Boeing 737 Classics and DC-9/MD-80 airplanes). In the regional jet market, the first variant of the Bombardier (of Canada) CRJ family commenced service in 1992. Embraer (of Brazil) entered this market with the twin-engine ERJ family in 1996 (models include the -135, -140, and -145). The Dornier 328JET, a 32-seat jet-engine commuter manufactured by the American–German Fairchild Dornier, began commercial operations in 1999.

This decade (i.e., the 1990s) also saw much consolidation and reorganization in the aviation industry with, for example, McDonnell Douglas merging with Boeing, and Fokker ceasing production of its own designs. The Lockheed Corporation, which had withdrawn from the civil aircraft business, merged with Martin Marietta to form the defense-orientated Lockheed Martin. The dissolution of the Soviet Union in 1991 significantly affected civil aircraft production in the region, which declined by 80% within a few years; a substantive downsizing and reorganization of the industry ensued [12]. In Europe, British Aerospace (which, much earlier, had been created by the nationalization and merger of several UK companies, including BAC and Hawker Siddeley) merged in 1999 with Marconi Electronic Systems to form BAE Systems. In the same year, DASA (of Germany), CASA (of Spain), and Aérospatiale-Matra (of France) agreed to merge to create EADS (European Aeronautic Defence and Space Company), which became the majority shareholder of Airbus.

By 2000, Airbus and Boeing effectively shared a duopoly in the large single-aisle and widebody airplane markets (the Tupolev Tu-204/-214 and Ilyushin Il-96 annual production rates were in the low single figures, and the production of the BAe Avro RJ was coming to an end). Following more than 15 years of development effort, Airbus introduced the A380, the world's largest passenger transport airplane (with a seating capacity of 525 passengers in a typical three-class configuration). The A380, a double-deck, widebody, four-engine airliner, began commercial service in 2007. On the other end of the size spectrum, there was much activity in the small single-aisle and regional jet sectors. Embraer, building on their success with the ERJ, introduced the larger E-Jet family in 2004 (models include the E170, E175, E190, and E195). In the Ukraine, Antonov (which had gained worldwide prominence for manufacturing very large military transport aircraft: the An-124 and the one-off giant An-225) developed a new family of high-wing regional jets: the An-148/-158. The An-148 entered service in 2009 after receiving Interstate Aviation Committee (IAC) type certification.

In 2011, the Russian Sukhoi Superjet 100, a low-wing twin-engine regional airliner, began commercial operations with IAC type certification. European certification followed soon afterwards, demonstrating the airplane's compliance with western airworthiness and environmental standards. This opened up international markets for the newly established parent company: United Aircraft Corporation (UAC), a state-owned conglomerate of Russian aerospace companies (including Ilyushin, Irkut, Sukhoi, Tupolev, and Yakovlev). To meet growing international demand for efficient, long-haul operations—and to replace an aging global fleet, which included many four-engine Boeing 747 and Airbus A340 airplanes—both Boeing and Airbus developed new widebody twin-engine airliners: the Boeing 787 entered service in 2011 and the Airbus A350 in 2015. New materials and manufacturing technologies featured strongly in these designs, which made extensive use of carbon fiber composites to reduce airframe weight. In 2016, the Bombardier C Series began commercial operations. This new two-member family of twin-engine medium-range airplanes, comprising the 110-seat CS100 and 135-seat CS300 models, were developed to compete with the Embraer and Sukhoi regional jets and also with the smaller Boeing 737 and Airbus A320 airplane types. In the same year, the Chinese Comac ARJ21-700, a 90-seat single-aisle twinjet with a resemblance to the MD-80, commenced domestic service with Chinese type certification. The next airplane program for this state-owned manufacturer is the single-aisle twin-engine C919. Also competing in this market, on completion of development, will be the UAC Irkut MC-21. Japan's first passenger jet airplane, the 70–90 seat Mitsubishi MRJ, commenced flight testing in 2015, with an entry-into-service target of 2020. The decade (i.e., the 2010s) also saw Airbus and Boeing launching new versions of their single-aisle airplanes—that is, the Airbus A320neo family (first deliveries in 2016) and the Boeing 737-MAX family (first deliveries in 2017). New engine technologies (in the CFM International LEAP and Pratt & Whitney PW1000G turbofan engines) and aerodynamic refinements were key elements of these improved versions of the industry's two best-selling jet transport airplanes.

Although superficially similar in appearance to the jet airliners of the 1960s in terms of their primary geometry—that is, featuring a tubular fuselage with rear empennage, swept wing, and pod-mounted engines installed either under the wing or on the rear fuselage—the latest generation of airliners have many advanced structural, aerodynamic, system, and engine design features that enable them to operate at much improved fuel efficiency levels compared to the early generation jet airliners. Using the Comet 4 as a baseline, these airplanes are more than 70% more fuel efficient per available seat kilometer (ASK) [13]. Noise levels and exhaust emissions have also been significantly reduced over this time. Overall reliability, passenger comfort, and safety[3] have all been dramatically improved.

1.4 Regulatory Framework

The performance of commercial jet transport airplanes needs to be considered within the context of the regulatory framework that applies to the certification and subsequent operation of these aircraft. In the United States (US), these regulations are issued by the Federal Aviation Administration (FAA); equivalent European specifications are issued by the European Aviation Safety Agency (EASA). Additionally, the national aviation authorities of individual countries with significant aviation industries publish their own regulations. In a colloquial setting, these regulations are often referred to as *the rules*. Although substantially similar in many respects, important differences exist between various sets of regulations. Herein, only the US and European regulations are considered. The key documents and the organizations responsible for these measures are described in Chapter 23.

For the purpose of certification and operation, airliners fall within the *transport* category of aircraft. As regards airplane *certification*, key regulations include the US Federal Aviation Regulation Part 25 [15], usually abbreviated as FAR 25, and the European counterpart, EASA Certification Specification 25 (Book 1) [16], abbreviated as CS-25. In many instances, these regulations are identical (the result of many years of effort to harmonize technical details). Common regulations are often written as FAR/CS 25—a practice that has been adopted herein. As regards the *operation* of these airplanes, important regulations are the US Federal Aviation Regulation Parts 91 [17] and 121 [18] and European EASA OPS Part-CAT [19].

The certification of a new airplane involves an extensive series of tests and compliance checks designed to ensure that it meets a minimum set of safety standards. FAR/CS 25 [15, 16] contain specific requirements that pertain to an airplane's performance, which must be demonstrated during the certification process. For example, FAR/CS 25.121(a) deals with an airplane's ability to climb following takeoff with one engine inoperative and with the landing gear extended. It is stipulated that the steady gradient of climb must be positive for two-engine airplanes, not less than 0.3% for three-engine airplanes, and 0.5% for four-engine airplanes, in the critical takeoff configuration. Compliance has to be demonstrated at the appropriate airplane weight, without the benefit of ground effect, with the critical engine (i.e., the engine that most adversely affects the airplane's climb performance) inoperative at a specified speed and thrust setting. This is one of many such performance requirements defined in FAR/CS 25. Airplanes that meet these requirements will have demonstrated a minimum performance capability that is considered appropriate for safe flight operations.[4]

An important output of the certification process is a formal record of the key safety-related performance capabilities of the airplane—this is the Airplane/Aeroplane Flight Manual (AFM). As the AFM is not designed to be used by flight crews, another document, known as the Flight Crew Operations/Operating Manual (FCOM), is produced based on the same performance data, but supplemented by approved manufacturers' data concerning non-safety-critical

---

3 The global accident rate for civil jet airplanes for the five-year period 2010–2014, measured in hull losses per 1 million flights, was 0.45; this is the equivalent of one major accident for every 2.2 million flights [14].
4 Acceptable methods to demonstrate compliance with the primary certification requirements/standards are given in FAA Advisory Circular (AC) 25-7 [20] and EASA CS-25 (Book 2) [16].

topics, such as optimum cruise speeds and all-engine climb performance. The FCOM is the primary source of performance information used by flight crews, who must operate their airplanes within a highly regulated environment, which is a feature of commercial aviation.

The rules that apply to commercial flight operations are many and varied, and these impose yet another set of constraints or limits on the performance that an airplane might achieve in routine flight operations. The International Civil Aviation Organization (ICAO) is responsible for coordinating and regulating international air travel. Central to this role is the Convention on International Civil Aviation [21], which, together with the many standards, policies, and procedures issued by ICAO, provides an internationally agreed framework for the safe operation of aircraft. This includes a set of procedures for operating within controlled airspace. It is often necessary to consider these procedures when analyzing an airplane's performance. Flight altitudes can be restricted and speed restrictions are imposed below 10 000 ft in much of the airspace used for commercial operations worldwide. These factors impact an airplane's achievable performance in service—for example, the previously mentioned restrictions can influence the time that it would take an airliner to reach its initial cruise altitude.

1.5 Performance-Related Activities

1.5.1 Performance Activities Related to the Airplane Life Cycle

Performance analyses are conducted for a variety of reasons, and the techniques that are used vary depending on the nature of the problem—for example, there is the prediction of the performance of a new airplane at the design stage; the reduction of flight-test data from a test airplane; the generation of performance data for the AFM and other key documents; the planning of flight operations taking into account real operational conditions; and the *in situ* calculations and performance monitoring undertaken by the flight crew during flight. Four distinct sets of activities that involve elements of airplane performance are identified in Figure 1.1. The activities have been arranged in the sequence in which they would first be conducted for a new airplane—that is, through the airplane's life cycle. For each activity, the physics does not change as the underlying principles of flight dynamics are the same, but the nature of the work undertaken is different as the purpose—and available data—of each activity is different. Although different calculations are carried out during the different activities, the theoretical basis for the various analyses that are conducted is largely the same.

During the design of a new airplane, engineering analyses are conducted where the performance targets—such as payload–range capability, cruise speed, fuel efficiency, and takeoff and landing capability—have been established, and the airplane's design features and aerodynamic characteristics are to be determined. At the conceptual design stage, it is usual to make a number of simplifying assumptions concerning the behavior of the airplane and regarding the operating environment, which will facilitate analytical solutions to be obtained for many performance problems. The techniques would only need to be accurate to within a few percent. In later stages of the design process, where more accurate performance predictions are required—and more data on the new design are available—more sophisticated techniques, which often involve numerical computation, are likely to be used.

The flight testing and subsequent data analyses of a new airplane type—which establishes the airplane's validated, or demonstrated, performance characteristics—involve a different set of analysis techniques and methodologies. Performance characteristics of the airplane are measured and compared to predicted values. As a range of air temperature and pressure conditions are typically encountered during flight testing, measured performance data are adjusted to

Airplane design (by the manufacturer)

> Activities include performance analyses conducted using projected airplane data (e.g., airplane geometry, weight, aerodynamic and engine characteristics) using idealized mathematical models and historical data in support of design activities associated with the development of a new airplane (or a derivative). Primary considerations include: (1) airworthiness regulations (e.g., FAR/CS 25); (2) operational regulations (e.g., FAR 121, EASA OPS Part-CAT); and (3) customer requirements (e.g., payload–range capability, takeoff and landing distances, fuel economy).

Flight testing and generation of performance documentation (by the manufacturer)

> Activities include the measurement of airplane performance characteristics during a series of standardized tests (described in FAA AC 25-7 and EASA CS-25 Book 2, for example), conducted in actual conditions (not idealized or model conditions) using test airplanes with extensive airborne and ground instrumentation.
>
> Analysis of flight test and supporting performance data is conducted (1) to demonstrate compliance with the regulations (e.g., FAR/CS 25); (2) to produce the Airplane Flight Manual (AFM); and (3) to validate design characteristics and produce data that can be used by the manufacturer and operator.
>
> Data are corrected to standard conditions (e.g., ISA). Correction factors are introduced for flight operations (e.g., to account for anticipated differences in reaction times between test pilots and line pilots in emergency situations, such as engine failure).
>
> Operational documentation (e.g., FCOM) is generated, which must (1) consider likely operational conditions (e.g., off-ISA conditions, airfield limits); (2) introduce conservative correction factors and allowances (e.g., credit for headwind on landing); and (3) establish operational limit speeds (e.g., minimum control speed in the air).

Performance engineering (by the operator)

> Operational flight planning activities are conducted in accordance with (1) the manu-facturer's documentation (e.g., AFM); (2) the requirements of the regulatory authorities (e.g., FAR 121 and EASA OPS Part-CAT); and (3) local and international restrictions (e.g., noise limits). Activities include route planning (for standard and emergency conditions), the analysis of the airplane's performance for all critical phases of the flight, the determination of fuel requirements, weight and balance calculations, and takeoff and landing performance estimations (addressing such issues as noise abatement and reduced thrust, if applicable).

Flight operations (by the flight crew)

> Flight operations are conducted in actual (non-idealized) weather conditions; hence corrections to published performance data may be required.
>
> Preflight activities include the determination of the fuel required for the flight; checking of weight and center of gravity position against airplane limits; route planning (which considers forecast winds, weather, and anticipated delays); and airplane condition (e.g., restrictions due to unserviceable items).
>
> In-flight activities include operating the airplane within the manufacturer's performance limits (e.g., speed, load factor, angle of attack), fuel status and systems monitoring, and the management of routine and emergency situations.

Figure 1.1 Airplane performance-related activities.

represent the data as a function of pressure altitude, which is based on the idealized conditions defined in the International Standard Atmosphere (ISA). Airspeed and altitude instrumentation is calibrated to sea level (datum) conditions of the ISA. The database that is produced through the flight-test program is used to generate the performance values that are recorded in the AFM.

The determination of safety-related performance data during flight testing is a key part of the certification process in which compliance with the relevant airworthiness regulations must be demonstrated for the issue of a *type certificate*. Individual aircraft manufactured to an approved design—that is, a design for which a type certificate has been granted—may be issued a *certificate of airworthiness* by the national aviation authority of the country in which the airplane is registered. Airplanes with a valid certificate of airworthiness (which requires annual renewal) may be legally operated within the regulatory conditions of its issue.

The safe operation of any aircraft depends critically on the relevance and accuracy of the performance data that are available to the operator and flight crew. Performance calculations conducted in support of flight operations are based on manufacturers' performance data—taking account of the forecast weather, prevailing winds, runway conditions and limits, obstacle heights, and so forth. In some respects, the nature of these calculations is the reverse of that conducted during the design of the airplane: here, the airplane's performance attributes have to be determined based on known airplane characteristics; whereas during the design phase, it is the airplane characteristics that have to be determined to meet performance targets.

1.5.2 Performance Engineering and Flight Operations

There is a diverse set of engineering tasks and activities associated with the in-service operation of commercial jet transport airplanes that is customarily addressed under the heading of *airplane performance engineering*. The list of activities undertaken by performance engineers in support of flight operations can, for the sake of convenience, be grouped under five headings, although in reality there are many overlapping aspects. The description given below is not intended to be exhaustive, but rather illustrative of the nature of this work—these descriptions serve to establish a backdrop to the discussions presented in this book.

Performance Activities Associated with Specific Phases of a Flight
These activities include the determination and monitoring of the performance of an airplane during specific phases of the flight, which include takeoff (and rejected takeoff), initial climb after takeoff, takeoff flight path, *en route* climb to cruise altitude, cruise (including step climb), descent, approach (and missed approach), and landing. Key tasks are associated with establishing the airplane's performance during emergencies, such as engine failures at any stage of the flight and the determination of associated operational procedures.

Trip/Mission Performance
These activities consider the trip, or mission, performance of an airplane, which include payload–range assessments, determination of cost index parameters, policies for fuel tankering (fuel transportation), fuel conservation (including policy development and implementation), and considerations associated with exhaust emissions.

Route and Operational Flight Planning
These activities are associated with route analysis and consider such issues as foreign and domestic airspace restrictions, available air traffic tracks and flight levels, air traffic control (ATC) restrictions, suitability of destination and alternate airports, fuel considerations (including fuel planning and fuel usage monitoring), *en route* terrain considerations (including the determination of minimum safe altitudes), in-flight emergency considerations (e.g., oxygen requirements for passengers and crew), and extended-range twin-engine operations.

Weight and Balance

These activities are associated with establishing the airplane's operating empty weight (OEW) and center of gravity (CG) location, developing load sheets and validating payload weight procedures, assessing the CG location with respect to the manufacturer's fore and aft limits for each phase of the flight, and establishing procedures for the correct setting of the stabilizer trim for takeoff.

Operational Support and Organization-Specific Tasks

These activities include the preparation and upkeep of documentation for flight crews and dispatchers; implementation of applicable service bulletins, airworthiness directives and regulations; maintaining dispatch deviation documents (e.g., Master Minimum Equipment List); and a wide range of performance tasks applicable to the individual organization (e.g., monitoring reduced thrust usage in takeoff or climb, and providing input to fleet retirement/renewal decisions).

1.6 Analysis Techniques and Idealizations

The analysis techniques that are used to compute the performance characteristics of an airplane vary considerably depending on the purpose and the availability of data. There are many cases where a quick, simple "back of the envelope" calculation will suffice in order to obtain an approximate answer—for example, to provide a cross check against a result produced by a computer program or data contained in a reference manual. Simple, approximate methods also get used in student exercises, where real airplane data are seldom available. More sophisticated methods are, of course, needed for the computation of performance values that would be used in planning actual flight operations. In this book, a variety of methods of varying degrees of complexity are described (note that certain of these methods are only suitable for rough estimations).

Airplane performance analyses, in most cases, rely on three sets of data—these correspond to descriptions of (1) the characteristic parameters of the airplane (e.g., geometric parameters such as the wing reference area, aerodynamic relationships such as the drag polar, and the airplane's gross weight[5] and center of gravity position); (2) the characteristic parameters of the engine (e.g., net thrust and rate of fuel consumption); and (3) the environment in which the airplane is operating (e.g., ambient air temperature and pressure, wind, and runway features for takeoff and landing calculations). For example, the determination of an airplane's instantaneous rate of climb corresponding to a set of conditions (e.g., airplane gross weight, altitude, air temperature, speed, and acceleration) will require knowledge of a subset of data from the three groups, and—as will be shown later—the rate of climb depends principally on the airplane's thrust and drag.

The task of analyzing the airplane's performance is greatly simplified by the establishment of analytical models. Note that these models are not absolute laws, although many have a theoretical basis. These mathematical models are approximations of measurable parameters, which are usually only valid within specified limits. Some of these relationships can be described by surprisingly simple polynomial functions (such as the parabolic drag polar), which are adequate for approximate calculations. Certain functions, however, do not lend themselves to simple mathematical idealization. The variation of thrust lapse rate with altitude and speed, for example, is far too complex to be modeled by simple functions. Mathematical models can often be extremely useful, as they permit "exact" solutions to be found for complex scenarios, thus enabling

---

5 *Gross weight* is the total airplane weight at any moment during the flight or ground operation (i.e., instantaneous weight).

sensitivity analyses to be completed, thereby identifying the relationship between changes in airplane characteristic parameters and the predicted performance. Quite often, simple mathematical expressions can be used to initially study an idealized problem, and, thereafter, the effect of discrepancies between the model and reality can be addressed by adding smaller, second-order terms or correction factors to refine the calculation.

Idealizations are widely used in airplane performance studies. For most applications, the airplane can be considered to be a rigid body, permitting a classical Newtonian mechanics approach to be used to determine the equations of motion of the airplane. The application of static air loads on the wing will alter the wing's shape as it bends upwards and twists, resulting in a change to the chordwise and spanwise lift distributions, which will change the lift-induced drag. This can, in part, be accounted for in the determination of the airplane's drag polar. Dynamic air loads due to gusts (which tend to be oscillatory) momentarily influence the airplane's trajectory, but have a negligible influence on such performance parameters as climb gradient, range, or endurance.

For certain performance analyses that are associated with point calculations, it is often convenient to assume that a quasi-steady-state condition exists. The airplane's velocity vector is thus assumed not to change in magnitude or direction. This implies a state of equilibrium and a balance between the forces of lift, drag, thrust, and weight, which greatly simplifies the mathematics. Considering, once again, the example of instantaneous rate of climb, such an idealization would ignore the influence of the rate of change of true airspeed with height as well as the influence of wind gradients. In this case, the assumption of quasi-steady state permits the determination of a reasonable approximation. The inclusion of the first acceleration term will provide a small refinement to the calculation; the second acceleration term (i.e., due to a wind gradient) has an even smaller influence on the computed answer for a typical *en route* climb.

Another idealization that is frequently adopted when computing a performance parameter that relies on a time-based integral assumes that the airplane's mass is constant. This is obviously not true for a jet airplane due to the continuous consumption of fuel. However, the time rate of change of the airplane's mass is small, and this permits certain performance analyses to be conducted with the assumption of constant airplane mass. For example, it is commonplace to assume that the airplane's mass does not change during the takeoff run. This assumption simplifies the mathematics and facilitates the development of a closed-form mathematical solution for the takeoff run. Nonetheless, the inclusion of mass change is possible within a sophisticated numerical routine that involves an iterative approach to the determination of the takeoff distance, and this provides a means to refine the computed distance—albeit by a very small amount.

The Earth's curvature is an important consideration for long distance navigation, but, for the most part, the performance of an airplane can be satisfactorily assessed by assuming that the Earth is flat. Additionally, the Earth's rotation does not have a significant influence on most performance parameters and is generally ignored. There are, however, certain times when a high degree of precision is warranted and it is justified to take into account small correction factors. For example, in the analysis of specific fuel consumption data recorded during flight tests, it is possible to take into account the centrifugal influence of the Earth's rotational velocity on the airplane's weight—this correction depends on the airplane's ground speed and direction of flight (or, more precisely, its true track).

A common engineering approach, when assessing the validity of such idealizations, considers the impact that the idealization has on the computed result. For example, when credit is not taken for the reducing airplane mass during takeoff, a conservative result is obtained—that is, a marginally longer takeoff distance is predicted—and this may justify the use of the idealization. Another factor that must always be considered when assessing the merits of including such refinements is the accuracy to which the other parameters in the equation can be established. There is little value in going to considerable computational effort (e.g., by accounting for aircraft mass change in a takeoff analysis) when there is uncertainty associated with another parameter in the equation that has a significantly greater influence on the final result.

References

1 Pearsall, J. (Ed.), *New Oxford dictionary of English*, Oxford University Press, New York, NY, 2001.

2 Stein, J. (Ed.), *The Random House dictionary of the English language*, Random House, New York, NY, 1983.

3 FAA, "Criteria for approval of category I and category II weather minima for approach," Advisory Circular 120-29A, Federal Aviation Administration, Washington, DC, Aug. 12, 2002. Available from www.faa.gov/.

4 IATA, "New year's day 2014 marks 100 years of commercial aviation," International Air Transport Association: Press release number 72, Montréal, Canada, Dec. 31, 2013.

5 ATAG, "Aviation benefits beyond borders," Air Transport Action Group, Geneva, Switzerland, Apr. 2014.

6 Airbus, "Global market forecast: Mapping demand 2016–2035," Airbus S.A.S., Blagnac, France, 2016.

7 Boeing, "Current market outlook 2016–2035," Market Analysis, Boeing Commercial Airplanes, Seattle, WA, July 2016.

8 Airbus, "Global market forecast 2015–2034," Airbus S.A.S., Blagnac, France, 2015.

9 ICAO, "ICAO environmental report 2016: Aviation and climate change," Environment Branch, International Civil Aviation Organization, Montréal, Canada, 2016.

10 Boeing, "Current market outlook 2015–2034," Market Analysis, Boeing Commercial Airplanes, Seattle, WA, 2015.

11 Airbus, "Global market forecast: Delivering the future 2011–2030," Airbus S.A.S., Blagnac, France, 2011.

12 Bragg, L.M., Miller, M.E., Crawford, C.T., *et al.*, "The changing structure of the global large civil aircraft industry and market: Implications for the competitiveness of the U.S. industry," Publication 3143, U.S. International Trade Commission, Washington, DC, Nov. 1998.

13 Peeters, P.M., Middel, J., and Hoolhorst, A., "Fuel efficiency of commercial aircraft: An overview of historical and future trends," NLR-CR-2005-669, National Aerospace Laboratory, the Netherlands, Nov. 2005.

14 IATA, "Annual review 2016," International Air Transport Association, Montréal, Canada, June 2016.

15 FAA, *Airworthiness standards: Transport category airplanes*, Federal Aviation Regulation Part 25, Amdt. 25-143, Federal Aviation Administration, Washington, DC, June 24, 2016. Latest revision available from www.ecfr.gov/ under e-CFR (Electronic Code of Federal Regulations) Title 14.

16 EASA, *Certification specifications and acceptable means of compliance for large aeroplanes*, CS-25, Amdt. 18, European Aviation Safety Agency, Cologne, Germany, June 23, 2016. Latest revision available from www.easa.europa.eu/ under Certification Specification.

17 FAA, *General operating and flight rules*, Federal Aviation Regulation Part 91, Amdt. 91-336A, Federal Aviation Administration, Washington, DC, Mar. 4, 2015. Latest revision available from www.ecfr.gov/ under e-CFR (Electronic Code of Federal Regulations) Title 14.

18 FAA, *Operating requirements: Domestic, flag, and supplemental operations*, Federal Aviation Regulation Part 121, Amdt. 121-374, Federal Aviation Administration, Washington, DC, May 24, 2016. Latest revision available from www.ecfr.gov/ under e-CFR (Electronic Code of Federal Regulations) Title 14.

19 European Commission, *Commercial air transport operations (Part-CAT)*, Annex IV to Commission Regulation (EU) No. 965/2012, Brussels, Belgium, Oct. 5, 2012. Published in *Official Journal of the European Union*, Vol. L 296, Oct. 25, 2012, and reproduced by EASA.

20 FAA, "Flight test guide for certification of transport category airplanes," Advisory Circular 25-7C, Federal Aviation Administration, Washington, DC, Oct. 16, 2012. Available from www.faa.gov/.

21 ICAO, "Convention on international civil aviation," Doc. 7300, 9th ed., International Civil Aviation Organization, Montréal, Canada, 2006.

2

Engineering Fundamentals

2.1 Introduction

This chapter introduces the fundamental engineering parameters, principles, and concepts that underpin the study of the mechanics of flight. An accessible summary of these topics is presented using language and notation that is consistent with subsequent chapters. In essence, this is the foundation upon which the analytical and numerical models that describe the performance of an airplane are constructed.

The topics of notation (representing engineering terms), abbreviations, symbols, and units—applicable to the study of airplane performance—are discussed in Section 2.2. Increasingly, SI units[1] are used in engineering the world over. Nonetheless, the foot–pound–second (FPS) unit system[2] remains widely used in certain English-speaking countries, and many practicing engineers (in the United States, for example) do not have a "feel" for SI units. Furthermore, by international agreement, altitude is measured in feet in most of the world's airspace. For these reasons, equations in this book are presented in a consistent form that will permit either SI units or FPS units to be used (Appendix B contains a comprehensive table of conversion factors).

Converting between units poses little difficulty for most applications; however, in the case of airplane performance analyses, alternative definitions for what could appear to be the same parameter can lead to errors—for example, by the inclusion or exclusion of gravitational acceleration in equations involving fuel flow. Precise definitions of mass, weight, and gravity (see Section 2.3) facilitate a better understanding of affected units and engineering terms.

The equations that describe the motion of an airplane on the ground and in the air rely on the underlying principles of rigid body dynamics and fluid dynamics—a summary of the basic concepts is presented in Sections 2.4 and 2.5, respectively.

---

1 The International System of Units, universally abbreviated as SI from the French *Le Système International d'Unités*, is the modern form of the metric system of measurement [1].

2 United States Customary (USC) units (known also as Standard units) of measurement are very similar—and in many cases identical—to British Imperial units of measurement, as both were historically derived from a common system of measurement, usually called English units, which were used throughout the British Empire and its region of influence. The term foot–pound–second (FPS) is used herein to describe a coherent system of measurement, consistent in all respects with USC units, that considers the pound (lb) as the fundamental unit of force and the slug as the derived unit of mass.

Performance of the Jet Transport Airplane: Analysis Methods, Flight Operations, and Regulations, First Edition.
Trevor M. Young.
© 2018 John Wiley & Sons Ltd. Published 2018 by John Wiley & Sons Ltd.

2.2 Notation, Units, and Conversion Factors

2.2.1 Notation

The subject of flight mechanics is fraught with confusion regarding notation. Various systems have been used over the years by authors, researchers, and practicing engineers around the world. Some of these have been superseded by newer systems, but old texts are still found in libraries (and on the shelves of professors)—a potential source of confusion. The problem is not only restricted to the use of different symbols for the same parameter, but different parameters and non-dimensional groups are sometimes given the same name. This could mean that values for a particular non-dimensional parameter for an airplane could differ by a factor of two (depending on whether an American or British system has been used, for example).

Words used to describe aerodynamic characteristics, such as the breakdown of drag (see Section 7.3), are not universally consistent, and the same word may have a slightly different meaning, depending on the preference of the author or the country of origin of the book or report. A mix of American, British, old and new terms can be found in academic and commercial literature.

Yet another problem encountered in this subject is the inconsistency of what defines the positive sense of a parameter. One example is the stick/yoke force—a positive force will imply that the pilot is pulling back on the stick/yoke in one convention, while another author will define a pushing force as positive.

In this book, an effort has been made to be clear and unambiguous in defining all terms; nonetheless, certain nomenclature will have more than one meaning (an unavoidable situation in this subject).

2.2.2 Abbreviations and Symbols

In any scientific or technical field—and aeronautical engineering is no exception—the use of abbreviations or acronyms as a substitute for cumbersome technical terms is commonplace (e.g., TSFC for thrust specific fuel consumption). Abbreviations facilitate a conciseness in writing and are generally advantageous—readers are more easily able to comprehend the subject matter, as the information is condensed. In this book, abbreviations have been extensively used, and in all cases the abbreviations are defined when they are first mentioned; there is also a list of abbreviations in Appendix G. The abbreviations comprise multiple letters in upper and/or lower case (e.g., SAR, ppm) and are sometimes separated by punctuation marks (e.g., R/C).

The prime purpose of symbols (e.g., V, C_D, α), on the other hand, is to represent technical terms, or parameters, in mathematical formulae. A symbol is best written as a single upper or lower case letter, augmented where necessary by subscripts and/or superscripts (e.g., V_v, δ_e, H^*). The use of abbreviations in mathematical formulae, however, can lead to misunderstanding, and for this reason the practice is discouraged. For example, KE (which can represent kinetic energy) could be understood as the product of K and E. Technical parameters (i.e., variables and constants) that appear in equations are herein assigned symbols, irrespective of whether an abbreviation also exists—for example: TSFC is assigned the symbol c, and this is consistently used in equations. The reverse situation, however, whereby a symbol is used in text, seldom creates a problem (provided that the symbol has been properly defined, of course)—for example, C_L, the widely accepted symbol for lift coefficient, is frequently used both in equations and in text (in academic literature and other technical documentation).

Another convention that has been adopted herein is to set the symbols (representing variables and constants) in italics—this is done in equations and also in the text. Where it is important to identify a vector parameter, this has been done in the customary manner, using either bold font or by placing a small arrow over the symbol.

2.2.3 Base Quantities and Dimensions

All physical quantities used in engineering (e.g., force, entropy) can be expressed in terms of seven mutually independent base quantities—these are length, mass, time, thermodynamic temperature, amount of substance, electric current, and luminous intensity. (Interestingly, it is possible to express almost all quantities that are encountered in mechanics in terms of just three of these quantities: length, mass, and time.) Other quantities—which are called derived quantities—are obtained from equations expressed in the form of a product of powers of the base quantities.

The dimensions of the seven base quantities are usually denoted by the symbols L, M, T, Θ, N, I, and J, respectively (Table 2.1). Dimensions of other quantities are obtained by equation—for example, for an arbitrary quantity Q, the dimension (dim) will be

$$\text{dim}[Q] = \text{L}^{\alpha}\,\text{M}^{\beta}\,\text{T}^{\gamma}\,\Theta^{\delta}\,\text{N}^{\varepsilon}\,\text{I}^{\zeta}\,\text{J}^{\eta} \tag{2.1}$$

where $\alpha, \beta, \gamma, \delta, \varepsilon, \zeta$, and η are the dimensional exponents.

Using this convention, the dimension of energy, for example, is L^2MT^{-2}, which is consistent with the equations for kinetic energy (see Equation 2.42) and potential energy (see Equation 2.44).

Table 2.1 Base quantities

Quantity	Dimension	SI unit	Symbol
length	L	metre *or* meter	m
mass	M	kilogram	kg
time	T	second	s
thermodynamic temperature	Θ	kelvin	K
amount of substance	N	mole	mol
electric current	I	ampere	A
luminous intensity	J	candela	cd

2.2.4 Units and Conversion Factors

Various unit systems have been adopted to express the measurement of engineering quantities. Unit systems that have been defined using the fundamental equations that link quantities are called *coherent*. It is a property of coherent unit systems that the equations of mechanics that describe force, energy (or work), and power in terms of length, mass, and time hold without the introduction of constant factors (see Sections 2.3 and 2.4). The Système International [1, 2], which is widely known as SI, is such a system. Each base quantity has a base unit (see Table 2.1). Each derived quantity has a derived unit, which is obtained from the appropriate equation by replacing the dimension symbols with units. Derived units can thus be expressed in terms of the base units—for example, energy can be expressed in units of $\text{m}^2\,\text{kg}\,\text{s}^{-2}$, which, by definition, is equivalent to a joule (J).

SI units are widely used in aeronautical engineering, but, even then, there are a few exceptions that arise due to customary practices or regulations associated with flight operations. For example, in almost all countries in the world (see Section 5.4 for exceptions), altitude

Table 2.2 Frequently used conversion factors

	US Customary unit	SI unit	Conversion factor
length, height, *or* distance	ft	m	0.3048
distance[a]	nm	m	1852
speed[b]	kt	m/s	0.5144
area	ft^2	m^2	0.09290
volume	US gal *or* USG	dm^3 *or* L	3.785
force[c]	lb	N	4.448
pressure	lb/in^2	Pa	6895
mass	slug	kg	14.59
mass[d]	lbm	kg	0.4536
density	$slug/ft^3$	kg/m^3	515.4
specific energy[e]	Btu/lbm	J/kg	2326
thrust specific fuel consumption[f]	$lb\ lb^{-1}\ h^{-1}$	$mg\ N^{-1}\ s^{-1}$	28.33

Notes:

(a) The unit nm, when taken out of context, can be misinterpreted as the SI unit for nanometers. If confusion is likely, an alternative abbreviation (e.g., NM, nmi, naut.mi.) should be used.
(b) One knot (abbreviated as kt or, alternatively, as kn) is equal to one nautical mile per hour.
(c) The unit abbreviation lb is customarily preferred to the abbreviation lbf for pound-force.
(d) The unit abbreviation lbm is used herein for pound-mass (avoirdupois pound).
(e) Specific energy is energy per unit mass.
(f) The conversion of thrust specific fuel consumption (TSFC) is not dimensionally consistent. The TSFC can be defined as either the *mass* of fuel burned per unit of time divided by the thrust or the *weight* of fuel burned per unit of time divided by the thrust (see Section 8.4.2).

is measured in feet for airplane operations. Furthermore, distance is frequently measured in nautical miles and speed is measured in knots (see Sections 6.2 and 6.3). The acceptance of these non-SI units (i.e., feet, nautical miles, and knots) in flight operations is specifically addressed in ICAO Annex 5 (International Standards and Recommended Practices concerning units of measurement for air and ground operations) [3]—these units have been adopted as the primary measures of height, trip distance, and speed in this book.

The most frequently used conversion factors for this subject (given to four significant figures) are listed in Table 2.2; a more complete listing of conversion factors (with greater precision) is provided in Appendix B.

2.2.5 Temperature Scales

The four commonly used temperature scales are Celsius (unit: °C), kelvin (unit: K), Fahrenheit (unit: °F), and Rankine (unit: °R). Reference temperatures for these temperature scales are given in Table 2.3. Temperature values can be converted between Celsius and Fahrenheit as follows:

$$T_C = \left(\frac{5}{9}\right)(T_F - 32) \tag{2.2a}$$

$$T_F = \left(\frac{9}{5}\right)T_C + 32 \tag{2.2b}$$

where T_C and T_F are temperature values in °C and °F, respectively.

Table 2.3 Reference temperatures for commonly used temperature scales

Temperature scale	Absolute zero	Freezing point of pure water*	Boiling point of pure water*
Celsius	$-273.15\,°C$	$0\,°C$	$100\,°C$
Fahrenheit	$-459.67\,°F$	$32\,°F$	$212\,°F$
Kelvin	$0\,K$	$273.15\,K$	$373.15\,K$
Rankine	$0\,°R$	$491.67\,°R$	$671.67\,°R$

*__Note:__ Measured under standard pressure conditions (see also Table B.4).

The conversion to absolute values can be undertaken using the following expressions:

$$T_K = T_C + 273.15 \tag{2.3a}$$
$$T_R = T_F + 459.67 \tag{2.3b}$$

where T_K and T_R are absolute temperature values in K and °R, respectively.

2.2.6 Scalar and Vector Quantities

Scalar quantities are described by magnitude only, whereas vector quantities are defined by both magnitude and direction. Scalar quantities are traditionally printed in italic typeface (a convention that has been adopted herein). Vectors are printed in bold roman (i.e., upright) typeface; alternatively, they can be identified by an arrow drawn over the italicized symbol. For example, a velocity vector would be written as $\mathbf{V}$ or $\vec{V}$ and its magnitude as V.

2.3 Mass, Momentum, Weight, and Gravity

2.3.1 Mass

Mass is a measure of the amount of matter in a solid or fluid body. Mass is also a measure of a body's resistance to acceleration—a relationship expressed by Newton's second law:

$$F = ma \tag{2.4}$$

where F is the force (typical units: N, lb);
m is the mass (typical units: kg, slug); and
a is the acceleration measured in the direction of the force (typical units: m/s^2, ft/s^2).

The standard units of mass are kilogram (SI unit) and slug (FPS unit). As SI and FPS are coherent unit systems, the following relationships hold:

- 1 kg is the mass of a body that will accelerate at $1\,m/s^2$ when subjected to a force of 1 N.
- 1 slug is the mass of a body that will accelerate at $1\,ft/s^2$ when subjected to a force of 1 lb.

2.3.2 Momentum

Momentum, which is a vector quantity, is defined as the product of a body's mass and its velocity. Standard units are $kg\,m\,s^{-1}$ and $slug\,ft\,s^{-1}$.

2.3.3 Weight

Weight is the gravitational force exerted on a body. When an object is allowed to fall freely in a vacuum under the influence of its own weight, it will accelerate at a rate known as the *gravitational acceleration* (*g*)—which in aviation literature is sometimes written as *gee*. Gravitational acceleration can be considered as the constant of proportionality linking weight and mass:

$$W = mg \tag{2.5}$$

where *W* is the weight (typical units: N, lb); and
   g is the gravitational acceleration, measured in the direction of the weight
      force (typical units: m/s^2, ft/s^2).

The acceleration due to gravity varies with geographical location: it increases in magnitude with increasing latitude (i.e., it is greater at the poles than at the equator), and it reduces with increasing height above the Earth's surface (see Section 2.3.4). To simplify the analysis of air vehicles in the Earth's atmosphere, the standard value of gravitational acceleration[3] (g_0) is widely used, where

$g_0 = 9.80665$ m/s^2 (exactly) in SI units; or
$g_0 = 32.174049$ ft/s^2 (correct to eight significant figures) in FPS units.

This standard value of *g* links the *units* of mass and weight, as follows:

- A body of mass 1 kg weighs 9.80665 N under standard conditions.
- A body of mass 1 slug weighs 32.174049 lb under standard conditions.

Weight-equivalent and mass-equivalent terms, based on g_0, can be defined as follows:

- 1 kgf (1 kilogram-force[4]) is the weight of a body of mass 1 kg under standard gravitational conditions (thus 1 kgf = 9.80665 N).
- 1 lbm (1 pound-mass) is the mass of a body that weighs 1 lb under standard gravitational conditions (thus 1 lbm = 0.03108095 slug).

Note that kilogram-force and pound-mass are non-coherent units; neither are recommended for scientific applications.

2.3.4 Gravitational Acceleration

The weight of a body of fixed mass will depend on its physical location. The law of universal gravitation, formulated by Isaac Newton,[5] states that every particle of matter in the universe attracts

---

3 The standard value of *g*, which is usually identified by the subscript zero, was agreed by the International Bureau of Weights and Measures in 1901 [1]. It is officially defined in SI units; the equivalent value in US Customary (USC) units is obtained by unit conversion.
4 The kilogram-force unit is also known as the kilopond (kp), which, note, is not part of the SI system. Although once widely used as a unit of force, the kilopond is considered an obsolete unit and is today seldom encountered in general engineering work.
5 Isaac Newton (1643–1727), an English physicist and mathematician, published the first mathematical formula of gravity and the basic laws governing the motion of bodies (see Section 2.4.1) in his three-volume work *Principia Mathematica* in 1687.

every other particle with a force proportional to the product of their masses and inversely proportional to the square of the distance between them, that is,

$$F = G\frac{m_1 m_2}{d^2} \tag{2.6}$$

where F is the mutual force of attraction between two bodies (typical units: N, lb);

G is the universal (or Newtonian) gravitational constant (see below);

m_1 and m_2 are the masses of the two bodies (typical units: kg, slug); and

d is the distance between the centers of the two bodies (typical units: m, ft).

The CODATA[6] (2006) recommended value of G in SI units is 6.67428×10^{-11} m³ kg⁻¹ s⁻² with an uncertainty of 0.001×10^{-11} m³ kg⁻¹ s⁻². The converted value of G in FPS units is 3.43978×10^{-8} ft³ slug⁻¹ s⁻².

An expression for the gravitational acceleration at the Earth's surface can be deduced by considering the gravitational force that acts on a body of mass m if the Earth is represented by an idealized, perfectly spherical body. By combining Equations 2.5 and 2.6, it follows that

$$F = mg_{SL} = G\frac{m_E m}{r_E^2}$$

thus $\quad g_{SL} = G\dfrac{m_E}{r_E^2} \tag{2.7}$

where g_{SL} is the gravitational acceleration at sea level (typical units: m/s², ft/s²);

m_E is the mass of the Earth (typical units: kg, slug); and

r_E is an idealized, equivalent radius of the Earth[7] (typical units: m, ft).

The gravitational pull exerted by the Earth on another object (e.g., an airplane) reduces as the object moves away from the Earth's surface. An expression for the gravitational acceleration as a function of height above sea level can be deduced from Equation 2.7, that is,

$$g_z = G\frac{m_E}{(r_E + z)^2} \tag{2.8}$$

where g_z is the gravitational acceleration at height z (typical units: m/s², ft/s²); and

z is the height of the object above sea level (typical units: m, ft).

It is convenient for certain applications to express g_z in terms of the gravitational acceleration at sea level (g_{SL}). By dividing Equation 2.8 by Equation 2.7, it can be shown that

$$g_z = g_{SL} \left(\frac{r_E}{r_E + z}\right)^2 \tag{2.9}$$

Equations 2.7 to 2.9 describe an idealized situation, where the Earth is considered to be perfectly spherical; furthermore, the influence of the Earth's rotation about its axis has been

---

6 The Committee on Data for Science and Technology (CODATA), an interdisciplinary committee of the International Council for Science (ICSU), periodically publishes internationally accepted sets of values of fundamental physical constants [4].

7 For the purpose of defining standard atmospheres (see Section 4.2), an equivalent (nominal) radius of the Earth of 6.356766×10^6 m (2.0855531×10^7 ft) is used [5]. This yields the standard value of gravitational acceleration (g_0) at latitude 45.5425°.

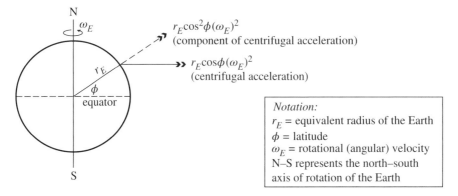

Figure 2.1 Centrifugal acceleration of an object on the Earth's surface due to the Earth's rotation.

ignored. Both of these factors require small corrections to be included in the mathematical description of the Earth's gravitational field if precise calculations involving an object's *weight* are to be conducted. It is known that the gravitational force exerted on an object of fixed mass, at sea level, depends on latitude—it is weaker at the equator than at the poles. There are two reasons for this observation. First, the Earth is not a perfect sphere: it bulges out a little at the equator, effectively increasing the distance from the Earth's center to its surface. Secondly, the Earth's rotational (angular) velocity imparts a centrifugal force on the object, which is perpendicular to the Earth's rotational axis (see Figure 2.1). The upward component (i.e., normal to the Earth's surface) of this centrifugal acceleration opposes gravity, diminishing its apparent effect. The gravitational acceleration at sea level taking into account the Earth's rotation ($g_{SL,\phi}$) is thus

$$g_{SL,\phi} = g_{SL} - r_E \cos^2 \phi \, \omega_E^2 \tag{2.10}$$

where ϕ is the geographic latitude; and

ω_E is the rotational velocity of the Earth (equal to 7.292115×10^{-5} rad/s).

Even though this centrifugal force is not due to the Newtonian gravitational attraction between the Earth and the other body, it is always present, and it is thus customary to include this correction in published data of the Earth's gravitational field.

The Earth, of course, is not a perfect sphere, and, for some applications, Equation 2.10 does not provide sufficiently accurate answers. The Earth's gravitational acceleration is well represented by the international gravity formula, which is used by the World Geodetic System WGS 84 [6]. The formula, which is based on an ellipsoidal representation of the Earth and a rotational velocity of 7.292115×10^{-5} rad/s, expresses gravitational acceleration at sea level ($g_{SL,\phi}$) as a function of latitude (ϕ):

$$g_{SL,\phi} = g_{eq} \left[1 + \left(\frac{b}{a} \frac{g_{po}}{g_{eq}} - 1 \right) \sin^2 \phi \right] \left(\frac{1}{\sqrt{1 - \varepsilon^2 \sin^2 \phi}} \right) \tag{2.11}$$

where g_{eq} is the theoretical gravity at the equator, equal to 9.7803253359 m/s^2;

g_{po} is the theoretical gravity at the poles, equal to 9.8321849378 m/s^2;

a is the semi-major ellipsoidal axis, equal to 6.378137×10^6 m;

b is the semi-minor ellipsoidal axis, equal to 6.3567523142×10^6 m; and

ε is the first ellipsoidal eccentricity, equal to $8.1819190842622 \times 10^{-2}$.

An alternative formulation, which is used for many applications (e.g., to establish standard atmospheres—see Section 4.2), is known as Lambert's equation [5]:

$$g_{SL,\phi} = g_{ref}(1 - 2.6373 \times 10^{-3} \cos 2\phi + 5.9 \times 10^{-6} \cos^2 2\phi) \tag{2.12}$$

where $g_{ref} = 9.80616$ m/s^2 (32.17244 ft/s^2).

The gravitational acceleration corresponding to any given height and latitude ($g_{z,\phi}$) can be obtained by applying a height correction to the corresponding sea level value (given by Equation 2.11 or 2.12). This can be done using Equation 2.9, or, alternatively, using Equation 2.13 (which is a more precise method that accounts for latitude). Equation 2.13 is obtained by including the centrifugal correction in Equation 2.9, and then substituting g_{SL} from Equation 2.10; thus

$$g_{z,\phi} = \left(g_{SL,\phi} + r_E \cos^2 \phi\, \omega_E^2\right) \left(\frac{r_E}{r_E + z}\right)^2 - (r_E + z) \cos^2 \phi\, \omega_E^2 \tag{2.13}$$

2.3.5 Influence of an Aircraft's Motion on Its Apparent Weight

The apparent weight of fast-moving objects, such as aircraft, in the Earth's atmosphere is influenced, albeit by a small amount, by the speed and direction of motion of the object.[8] If an aircraft flies at constant height above the Earth's surface on a fixed heading, it will follow a circular flight path (i.e., curved towards the Earth). Consequently, a centrifugal force will act on the aircraft, which will oppose the gravitation attraction. This is very similar in principle to the centrifugal correction (described in Section 2.3.4) due to the Earth's rotation; however, in this case, the correction depends not only on latitude but also on the aircraft's height (above sea level), ground speed, and true track (defined in Section 6.3.2).

The following equation (which is derived in reference [8]) can be used to estimate the centrifugal correction due to the aircraft's motion (Δg_{cm}):

$$\Delta g_{cm} = -\frac{V_G^2}{r_E + z} - 2\omega_E V_G \cos \phi \sin \chi \tag{2.14}$$

where V_G is the ground speed of the aircraft (typical units: m/s, ft/s); and
       χ is the true track angle of the flight path (unit: degrees, measured clockwise with respect to true north).

The output of Equation 2.14 is added to $g_{z,\phi}$ to give a corrected value of gravitational acceleration ($g_{z,\phi,cm}$):

$$g_{z,\phi,cm} = g_{z,\phi} + \Delta g_{cm} \tag{2.15}$$

2.3.6 Actual Versus Standard Gravitational Acceleration

The use of a single, standard gravitational acceleration (g_0) enables simplified mathematical models involving an object's weight to be used in many engineering tasks, and is sufficiently accurate for most airplane performance applications. It is common practice to use g_0 for flight operations work. For tasks where a high degree of precision regarding the airplane's weight is

---

8 On eastbound flights at the equator, the Concorde cruising at Mach 2 at 60 000 ft would have experienced a 1.5% reduction in apparent weight due to this centrifugal effect [7].

needed, such as cruise flight testing, corrections due to (1) latitude, (2) height, and (3) airplane motion need to be taken into account.

(1) The standard gravitational acceleration (see Section 2.3.3) matches the actual sea level value at about 45° latitude. The discrepancy between g_0 and the actual value of g (at sea level) is thus negligible in mid-latitude regions, with the error increasing to a maximum of 0.27% as the observer moves towards the equator or the poles (as described by Equation 2.11).
(2) The variation of g with height, at a given latitude, within the height range used for commercial airplane operations is small—but not always negligible. It can be shown using Equation 2.13 that g reduces by 0.43% from sea level to a height of 45 000 ft (13 716 m), for example. Precise values of g can be obtained by applying the height correction, given by Equation 2.13 (or by Equation 2.9—which yields almost the same result) to the sea level value predicted by Equation 2.11.
(3) The centrifugal influence on an airplane's weight due to its motion can be estimated using Equation 2.14. The correction depends on the airplane's location (specifically its latitude and height), ground speed, and true track. The influence is greatest for eastbound flights at the equator—for example, at a cruise speed of Mach 0.86 at 40 000 ft, the apparent weight of an aircraft due to its motion is reduced by 0.48%.

2.4 Basics of Rigid Body Dynamics

2.4.1 Newton's Laws of Motion

Before the forces acting on an airplane in flight are considered, it will be useful to review the basic theory of the dynamics of bodies in motion. The starting point of any such discussion is Newton's laws of motion, which, in modern terminology, can be stated as follows:

Law 1: A body remains at rest or continues to move in a straight line at a uniform velocity unless acted upon by an unbalanced force.
Law 2: The acceleration of a body is proportional to the net force applied and acts in the same direction as the force. Alternatively, it can be stated that the time rate of change of a body's momentum vector is equal to the applied force vector.
Law 3: The forces that exist between two interacting bodies are equal in magnitude, collinear, and directed in the opposite sense on each body. Alternatively, it can be stated that forces of action and reaction are equal and opposite.

Newton's laws essentially refer to the motions of *particles*. When applied to the motion of a large body, such as an airplane, a point mass at the airplane's center of gravity is taken to represent the airplane; furthermore, the effects of elastic distortions of the body are ignored (i.e., the airplane is assumed to be a rigid body).

2.4.2 Rectilinear Motion

Rectilinear motion describes the motion of a body in a straight line. Under such a condition, the position of the body (P), at any instant in time t, is given by the coordinate s, which is defined as its distance from a fixed datum (Figure 2.2). For relatively short distances (e.g., runway length), the preferred units of measurement are meters or feet, but for longer distances (e.g., trip length) the preferred units are kilometers (km) or nautical miles (nm), although statute miles (mi) are also used occasionally.

Figure 2.2 Rectilinear motion.

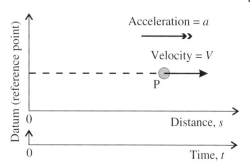

The instantaneous velocity of a body is the rate of change of distance with respect to time at the point of interest (specified by either s or t):

$$V = \frac{\mathrm{d}s}{\mathrm{d}t} \quad \text{or} \quad V\,\mathrm{d}t = \mathrm{d}s \tag{2.16}$$

where V is the velocity (typical units: m/s, ft/s, km/h, kt, mi/h);
$\quad\quad\quad s$ is the distance (typical units: m, ft, km, nm, mi); and
$\quad\quad\quad t$ is the time (typical units: s, h).

The instantaneous acceleration is the rate of change of velocity with respect to time at the point of interest (specified by s, t, or V):

$$a = \frac{\mathrm{d}V}{\mathrm{d}t} = \frac{\mathrm{d}^2 s}{\mathrm{d}t^2} \quad \text{or} \quad a\,\mathrm{d}t = \mathrm{d}V \tag{2.17}$$

where a is the acceleration (typical units: m/s², ft/s²).

A differential equation relating velocity, acceleration, and distance can be obtained from Equations 2.16 and 2.17:

$$a = \frac{\mathrm{d}V}{\mathrm{d}s}\frac{\mathrm{d}s}{\mathrm{d}t} = \frac{\mathrm{d}V}{\mathrm{d}s}V \quad \text{or alternatively} \quad a = \frac{\mathrm{d}}{\mathrm{d}s}\left(\frac{1}{2}V^2\right)$$

Thus $a\,\mathrm{d}s = V\,\mathrm{d}V$ $\tag{2.18}$

Note that velocity V and acceleration a are algebraic quantities that are defined as positive in the positive direction of s.

When the acceleration is constant, the differential Equations 2.17 and 2.18 can be evaluated directly. The initial condition is defined as $s(0) = 0$ and $V(0) = V_0$.

$$a\int_0^t \mathrm{d}t = \int_{V_0}^V \mathrm{d}V \quad \text{or} \quad V = V_0 + at \tag{2.19}$$

$$a\int_0^s \mathrm{d}s = \int_{V_0}^V V\,\mathrm{d}V \quad \text{or} \quad V^2 = V_0^2 + 2as \tag{2.20}$$

Substituting V from Equation 2.19 into Equation 2.16 gives

$$\int_0^s \mathrm{d}s = \int_0^t (V_0 + at)\,\mathrm{d}t \quad \text{or} \quad s = V_0 t + \frac{1}{2}at^2 \tag{2.21}$$

2.4.3 Motion With Respect to a Moving Datum (Reference System)

In Section 2.4.2, the rectilinear motion of bodies is considered with respect to a datum (reference system) that is fixed in space. This is a special case of the more general condition where the

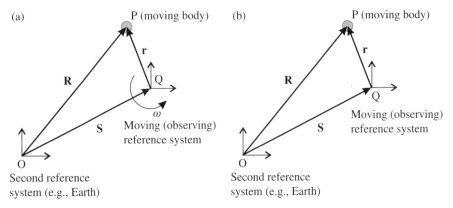

Figure 2.3 Relative motion: (a) Translating and rotating reference system, and (b) Translating reference system.

coordinate system is itself moving with respect to a second reference system (e.g., the Earth). Analyzing the motions of bodies with respect to axes fixed to the Earth is satisfactory for most applications; however, for certain problems, a non-rotating coordinating system (e.g., based on the sun) must be used. There is one particular type of reference system that is convenient for the analysis of the dynamics of bodies: it is called an *inertial frame of reference*. By definition, such a system is either stationary or moving at a constant velocity.

Three cases—representing progressively simpler situations—are described in this section: (1) a moving reference system that translates and rotates; (2) a moving reference system that translates, but does not rotate; and (3) rectilinear motion.

(1) Translating and Rotating Reference System

The origin of the fixed set of axes is designated as O and that of the moving axes as Q (see Figure 2.3a). The vector **r** represents the relative motion of the body (P) with respect to the moving (or observing) axes and the vector **S** represents the motion of Q with respect to O. Furthermore, the moving axes are also turning at a rate ω. The vector **R** describes the position of P with respect to O:

$$\mathbf{R} = \mathbf{S} + \mathbf{r} \tag{2.22}$$

The absolute velocity and acceleration of P (i.e., with respect to O) is given by

$$\dot{\mathbf{R}} = \dot{\mathbf{S}} + \dot{\mathbf{r}} + (\omega \times \mathbf{r}) \tag{2.23}$$

and $\quad \ddot{\mathbf{R}} = \ddot{\mathbf{S}} + \ddot{\mathbf{r}} + (\dot{\omega} \times \mathbf{r}) + \omega \times (\omega \times \mathbf{r}) + 2(\omega \times \dot{\mathbf{r}}) \tag{2.24}$

The last term in Equation 2.24 is known as the Coriolis acceleration.

(2) Translating Reference System

A simpler solution arises when the reference system translates, but does not rotate (Figure 2.3b). The velocity and acceleration of P are obtained by vector summation:

$$\dot{\mathbf{R}} = \dot{\mathbf{S}} + \dot{\mathbf{r}} \tag{2.25}$$

$$\ddot{\mathbf{R}} = \ddot{\mathbf{S}} + \ddot{\mathbf{r}} \tag{2.26}$$

Note that if Q is moving at a constant velocity (as would be the case for an inertial reference system), then $\ddot{\mathbf{S}} = 0$ and the absolute acceleration of P is the same as the observed acceleration (this is a useful property for the determination of inertia forces, for example).

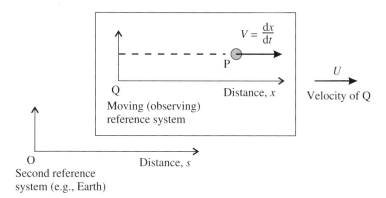

Figure 2.4 Relative motion: rectilinear movement of reference system.

(3) Rectilinear Motion

A further simplification can be introduced when considering rectilinear motion (Figure 2.4). In this case, vector notation is not required as the moving reference system is translating in the same direction as the motion of the body, but not necessarily in the same sense. The velocity and acceleration of the body (P) with respect to O are given by

$$\frac{ds}{dt} = \frac{dx}{dt} + U = V + U \tag{2.27}$$

and

$$\frac{d^2s}{dt^2} = \frac{d^2x}{dt^2} + \frac{dU}{dt} = \frac{dV}{dt} + \frac{dU}{dt} \tag{2.28}$$

where x is the distance (abscissa) of P with respect to the moving axes;
V is the velocity of P with respect to the moving axes;
s is the distance (abscissa) of P with respect to the second axes (e.g., Earth); and
U is the velocity of Q with respect to the second axes.

2.4.4 Curvilinear Motion

Curvilinear motion—which is a special case of the more general three-dimensional motion of a body—describes motion along a curved path that lies on a single plane. Normal (n) and tangential (t) coordinates are usually selected to describe such motions. The n and t coordinates move along the path with the body, such that the positive direction for n is always towards the center of curvature of the path (Figure 2.5). Circular motion is a special case of curvilinear motion where the radius of curvature is constant (Figure 2.6).

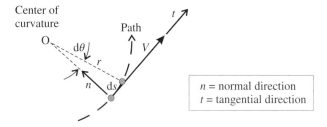

Figure 2.5 Curvilinear motion.

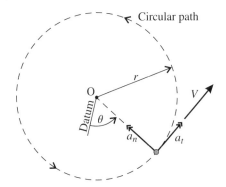

Circular path

Figure 2.6 Circular motion.

In Figure 2.5 the change in direction is shown as $d\theta$, where θ is measured in radians. The magnitude of the velocity, V, can be expressed as follows:

$$V = \frac{ds}{dt} = \frac{ds}{d\theta}\frac{d\theta}{dt} = r\frac{d\theta}{dt} \qquad (2.29)$$

as the differential displacement ds is given by $ds = r\,d\theta$.

Acceleration, by definition, is the time rate of change of velocity. As velocity is a vector, the net acceleration includes both the change in magnitude and the change in direction. The acceleration due to the change in magnitude is given by

$$a_t = \frac{dV}{dt} \qquad (2.30)$$

and the acceleration due to the change in direction is given by

$$a_n = \frac{V^2}{r} = V\frac{d\theta}{dt} = r\left(\frac{d\theta}{dt}\right)^2 \quad \text{(using Equation 2.29)} \qquad (2.31)$$

2.4.5 Newton's Second Law

Newton's second law (see Section 2.4.1) can be used to analyze the motion of a body when subjected to forces applied through the mass center. When Cartesian coordinates (see Section 3.3) are used, the forces are resolved into X, Y, and Z components. The acceleration can be calculated from the following equations:

$$\left.\begin{aligned} \sum F_x &= ma_x \\ \sum F_y &= ma_y \\ \sum F_z &= ma_z \end{aligned}\right\} \qquad (2.32)$$

Similarly, when normal and tangential coordinates are used, the forces are resolved into n and t components and the accelerations calculated:

$$\left.\begin{aligned} \sum F_n &= ma_n \\ \sum F_t &= ma_t \end{aligned}\right\} \qquad (2.33)$$

The net acceleration is given by

$$a = \sqrt{a_x^2 + a_y^2 + a_z^2} \tag{2.34}$$

or $\quad a = \sqrt{a_n^2 + a_t^2}$ (2.35)

2.4.6 Work

Work, which is given the symbol U, is done when the point of application of a force is moved through a finite distance. Work is a scalar quantity; work is positive if the "working" component of the force is in the direction of the displacement and negative if it is in the opposite direction. The work done by force $\mathbf{F}$ during a differential displacement $\mathbf{ds}$ is given by the dot product of the vector $\mathbf{F}$ and the vector displacement $\mathbf{ds}$ (see Figure 2.7).

$$dU = \mathbf{F} \bullet \mathbf{ds} \quad \text{or} \quad U = \int \mathbf{F} \bullet \mathbf{ds} \tag{2.36}$$

Now, as ε is the angle between $\mathbf{F}$ and $\mathbf{ds}$, Equation 2.36 can be written as

$$dU = F \cos \varepsilon \, ds \quad \text{or} \quad U = \int F \cos \varepsilon \, ds \tag{2.37}$$

Equation 2.37 can be interpreted as either the component of force acting in the direction of the displacement multiplied by the displacement, or the force multiplied by the component of displacement measured in the direction of the force. The former interpretation leads to the following expression for work:

$$U = \int \mathbf{F} \bullet \mathbf{ds} = \int F_x \, dx + \int F_y \, dy + \int F_z \, dz \tag{2.38}$$

where F_x, F_y, and F_z are the components of the force $\mathbf{F}$ in the directions X, Y, and Z, respectively.

The standard SI unit of work is the joule (J), where 1 J = 1 N m. In FPS units, work is usually measured in ft lb.

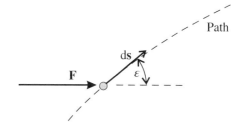

Figure 2.7 Work done by a force that is not collinear with the path.

2.4.7 Power

Power (P) is defined as the time rate of doing work:

$$P = \frac{dU}{dt} \tag{2.39}$$

When a force $\mathbf{F}$ moves through the distance $\mathrm{d}\mathbf{s}$ in the time $\mathrm{d}t$, the power developed is given by

$$P = \frac{\mathrm{d}U}{\mathrm{d}t} = \mathbf{F} \cdot \frac{\mathrm{d}\mathbf{s}}{\mathrm{d}t} = \mathbf{F} \cdot \mathbf{V} \tag{2.40}$$

where P is the power (typical units: W, ft lb s^{-1});

U is the work done (typical units: J, ft lb);

t is time (typical unit: s);

$\mathrm{d}\mathbf{s}$ is the differential displacement (typical units: m, ft);

$\mathbf{F}$ is the force vector (typical units: N, lb); and

$\mathbf{V}$ is the velocity vector (typical units: m/s, ft/s).

Power is a scalar quantity. The standard SI unit of power is the watt (W), where $1\text{ W} = 1\text{ J/s} = 1\text{ N m/s}$. The FPS unit of power is ft lb/s; although in USC units, power is frequently measured in horsepower (hP), where $1\text{ hP} = 550$ ft lb/s.

2.4.8 Kinetic Energy

Kinetic energy, within a conservative system, can be defined as the work that must be done on a body to accelerate it from a state of rest to a given velocity. If the force F results in the mass m being accelerated from zero velocity to velocity V over the distance s, then the work done (U) can be determined from Equations 2.4 and 2.20:

$$U = Fs = m\,a\,s = \frac{1}{2}mV^2 \tag{2.41}$$

The kinetic energy of the body moving at the velocity V is thus

$$E_k = \frac{1}{2}mV^2 \tag{2.42}$$

where E_k is the kinetic energy (typical units: J, ft lb);

m is the mass (typical units: kg, slug); and

V is the velocity (typical units: m/s, ft/s).

Kinetic energy, by definition, is a scalar quantity and is always positive.

2.4.9 Potential Energy

When work is done by moving a body within a force field (such as a gravitational field) or by deforming an elastic member (such as a spring), the body is said to have gained potential energy by virtue of its "elevated" position with respect to an energy baseline or datum.

The gravitational potential energy associated with a body of mass m, when it is raised from a datum elevation h_0 to the height h, is defined as follows:

$$E_p = m \int_{h_0}^{h} g\, \mathrm{d}h \tag{2.43}$$

where E_p is the potential energy (typical units: J, ft lb);

g is the gravitational acceleration (typical units: m/s^2, ft/s^2);

h is height (typical units: m, ft); and

h_0 is the datum height (typical units: m, ft).

If g is considered to be constant, for example if $g = g_0$, then

$$E_p = mg_0(h - h_0) \tag{2.44}$$

2.5 Basics of Fluid Dynamics

2.5.1 Density, Specific Weight, and Specific Gravity

Density, which is it is usually assigned the Greek letter rho (ρ), is defined as mass per unit of volume:

$$\rho = \frac{m}{v} \tag{2.45}$$

where m is the mass (typical units: kg, slug); and
v is the volume (typical units: m^3, ft^3).

The standard units of density are kg/m^3 and $slug/ft^3$ (where 1 slug $ft^{-3} = 1$ lb ft^{-4} s^2).
Specific weight is defined as weight per unit of volume, that is,

$$w = \frac{W}{v} \tag{2.46}$$

where w is the specific weight (typical units: N/m^3, lb/ft^3); and
W is the weight (typical units: N, lb).

The specific weight of a substance relates to its density:

$$w = \rho g \tag{2.47}$$

Specific gravity (relative density[9]) is defined as follows:

$$\sigma_{SG} = \frac{\rho}{\rho_{ref}} \tag{2.48}$$

where σ_{SG} is the specific gravity (dimensionless);
ρ is density (typical units: kg/m^3, $slug/ft^3$); and
ρ_{ref} is a reference density (typical units: kg/m^3, $slug/ft^3$).

The reference density (ρ_{ref}) is the density of pure water at a specified reference temperature, which is normally $4\,°C$ ($39.2\,°F$), although other temperatures, such as $60\,°F$ ($15.6\,°C$) and $68\,°F$ ($20\,°C$), are also used. The reference density of pure water at $4\,°C$ is $1000\,kg/m^3$ ($1.9403\,slug/ft^3$).

2.5.2 Pressure in Fluids At Rest

Pressure (p), by definition, is force per unit of area:

$$p = \frac{F}{A} \tag{2.49}$$

where p is the pressure (typical units: Pa, lb/ft^2, lb/in^2);
F is the force (typical units: N, lb); and
A is the area (typical units: m^2, ft^2).

The standard SI unit of pressure is the pascal (Pa), where $1\,Pa = 1\,N/m^2$. The FPS unit of pressure is lb/ft^2; although in USC units, pressure is often measured in lb/in^2 (psi).

---

9 The usage of the term *specific gravity* in scientific literature has declined in recent decades; many authorities prefer the term *relative density* [9].

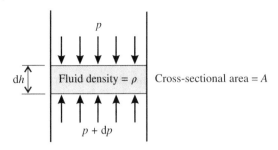

Figure 2.8 Pressure due to the mass of a fluid.

In Figure 2.8 the fluid is stationary under the influence of gravity. The mass (m) of the fluid of the defined element, which has a height dh and an area A, is the product of the density (ρ) and the volume ($A\,dh$). The weight of the element of fluid is mg. As the fluid is in a state of equilibrium, the sum of the forces (F) acting on the element is zero.

$$\sum F = (p + dp)A - pA - mg = 0$$

Thus $(p + dp)A - pA - (\rho A\,dh)g = 0$

and $dp = \rho g\,dh$ (2.50)

Equation 2.50 is an expression of Pascal's principle, or Pascal's law,[10] and is often referred to as the *hydrostatic equation*.

2.5.3 Mass Flow Rate and Continuity of Flow

The mass flow rate (Q) of a fluid moving through a pipe (or stream tube) is equal to the fluid mass that passes a particular station (or location) per unit of time.

$$Q = \frac{\Delta m}{\Delta t} = \rho A V$$ (2.51)

where Δm is the change in mass (typical units: kg, slug);
  Δt is the change in time (typical unit: s);
  ρ is the density (typical units: kg/m^3, slug/ft^3);
  A is the cross-sectional area of the pipe or stream tube (typical units: m^2, ft^2); and
  V is the velocity (typical units: m/s, ft/s).

An alternative—and widely used—designation for mass flow rate is $\dot{m}$, thus

$$Q = \dot{m} = \frac{dm}{dt}$$ (2.52)

The standard units of mass flow rate are kg/s and slug/s (where 1 slug s^{-1} = 1 lb ft^{-1} s).

The continuity equation can be deduced by considering fluid moving uniformly through a horizontal pipe (or stream tube) of changing cross section, as shown in Figure 2.9; it is apparent that the mass of fluid that passes Station 1 every second is equal to the mass of fluid that

---

10 Blaise Pascal (1623–1662), the French mathematician, physicist, and philosopher, defined a set of principles in 1647 that describe the pressure in fluids [10].

Figure 2.9 Flow in a converging pipe (or stream tube).

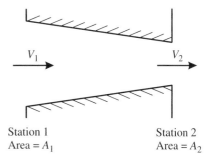

Station 1
Area = A_1

Station 2
Area = A_2

passes Station 2 every second. This principle is known as the *continuity of flow* and is expressed mathematically as

$$\rho_1 A_1 V_1 = Q_1 = \rho_2 A_2 V_2 = Q_2 \tag{2.53}$$

where subscript 1 denotes Station 1 and subscript 2 denotes Station 2.

or $\rho A V = \text{constant}$ (2.54)

The continuity equation can also be written in differential form—from Equation 2.54 it follows that

$$\rho A \, \mathrm{d}V + \rho V \, \mathrm{d}A + AV \, \mathrm{d}\rho = 0 \tag{2.55}$$

Dividing through by $\rho A V$ gives the required equation:

$$\frac{\mathrm{d}V}{V} + \frac{\mathrm{d}A}{A} + \frac{\mathrm{d}\rho}{\rho} = 0 \tag{2.56}$$

Equation 2.56 applies to both liquids and gases. Whereas liquids are considered to be incompressible, gases can easily be compressed. However, when the velocity of the gas is relatively low, the effects of compressibility are small. It is convenient for many low-speed aerodynamic applications to idealize the properties of air and to consider it to be incompressible. In such cases, the density will not change and Equation 2.56 can be simplified as follows:

$$\frac{\mathrm{d}V}{V} + \frac{\mathrm{d}A}{A} = 0 \quad \text{(applicable to incompressible flow)} \tag{2.57}$$

And, it is evident from Equation 2.54 that

$$AV = \text{constant} \tag{2.58}$$

2.5.4   Newton's Second Law Applied to Fluid Flow

According to Newton's second law of motion (see Section 2.4.1), the change in momentum of the object with respect to time is equal to the applied force. Considering unidirectional flow, this can be expressed as

$$F = \frac{\Delta(mV)}{\Delta t} \tag{2.59}$$

where F is the force (typical units: N, lb);
   $\Delta(mV)$ is the change in momentum or impulse (typical units: kg m s^{-1}, slug ft s^{-1}); and
   Δt is the change in time (typical unit: s).

When considering fluid flow, Equation 2.59 is best written in terms of mass flow rates:

$$F = \dot{m}_2 V_2 - \dot{m}_1 V_1 \tag{2.60}$$

where $\dot{m}$ is the mass flow rate (typical units: kg/s, slug/s);
       V is the velocity (typical units: m/s, ft/s);
and where subscript 1 denotes the entry condition and subscript 2 denotes the exit condition.

In a propulsion system, however, it is the reaction of accelerating the fluid—as expressed by Newton's third law of motion—that is of interest. The reaction force, which acts on the power-plant, is equal in magnitude to the accelerative force that acts on the fluid, but opposite in direction.

2.5.5 Bernoulli Equation for Incompressible Flow and Dynamic Pressure

The cross-sectional area of the horizontal frictionless pipe (or stream tube) in Figure 2.10 is shown to reduce in the direction of the flow (x). If the fluid is considered to be incompressible, then Equation 2.58 applies, and it can be concluded that the velocity increases in the direction of the flow; that is, the flow accelerates. By definition, an incompressible flow regime is one in which appreciable changes in fluid density along streamlines do not occur (see Section 3.5.1). By considering the net force that causes this acceleration, a most useful relationship—linking pressure, density, and velocity—can be derived. This is the Bernoulli equation[11] for incompressible flow.

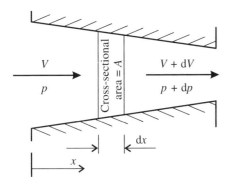

Figure 2.10 Flow velocity and pressure in a converging pipe (or stream tube).

Consider the sum of the forces in direction x acting on the element of fluid, which has a width dx and a mean cross-sectional area A. Now, apply Newton's second law to the element:

$$\sum F = pA - (p + dp)A = m\frac{dV}{dt}$$

or $-A\,dp = (\rho A\,dx)\dfrac{dV}{dt}$

hence $dp = -\rho\dfrac{dV}{dt}dx = -\rho\dfrac{dx}{dt}dV = -\rho V\,dV \tag{2.61}$

and $\displaystyle\int dp = -\int \rho V\,dV \tag{2.62}$

---

11 Named after Daniel Bernoulli (1700–1782), the Swiss mathematician and physicist, who, in 1738, in his acclaimed publication *Hydrodynamica*, established the fundamental relationship between pressure, density, and velocity of fluids.

With the assumption that the flow is incompressible, the density is constant and ρ can be taken out of the integral expression. Evaluation of the integral yields the desired result:

$$p + \frac{1}{2}\rho V^2 = \text{constant} \tag{2.63}$$

Equation 2.63 is the well-known Bernoulli equation for incompressible flow.

The term $(1/2)\rho V^2$, which appears in Equation 2.63, has units of pressure. This term appears frequently in the study of fluid dynamics; it is called the *dynamic pressure* and is often given the symbol q. By definition,

$$q = \frac{1}{2}\rho V^2 \tag{2.64}$$

It is seen from Equation 2.63 that the sum of the pressure p—which in this context is called the *static pressure*—and the dynamic pressure q of the fluid in the pipe (or stream tube) is constant. The sum of these two terms is known as the *total pressure*; note that total pressure will remain constant in the absence of work done on the fluid or losses incurred.

$$p_t = p + q \tag{2.65}$$

where p_t is the total pressure (typical units: Pa, lb/ft^2);
p is the static pressure (typical units: Pa, lb/ft^2); and
q is the dynamic pressure (typical units: Pa, lb/ft^2).

2.5.6 Ideal Gas Law

The ideal (or perfect) gas law is the equation of state of an ideal (or perfect) gas.[12] It describes the state of a gas in terms of the pressure, temperature, volume, and amount of matter. The most useful form of this equation for aerodynamic analyses is

$$p = \rho R T \tag{2.66}$$

where p is the pressure (typical units: Pa, lb/ft^2);
ρ is the density (typical units: kg/m^3, slug/ft^3);
R is the specific gas constant, usually just called the *gas constant*; and
T is the absolute temperature (typical units: K, °R).

The standard values of the gas constant [11] applicable to the International Standard Atmosphere (ISA)[13],[14] are

$R = 287.05287$ m^2 s^{-2} K^{-1} (or N m kg^{-1} K^{-1}) for SI units;
$R = 3089.81138$ ft^2 s^{-2} K^{-1} (or ft lb slug^{-1} K^{-1}) for FPS units, with temperature in K; or
$R = 1716.56188$ ft^2 s^{-2} °R^{-1} (or ft lb slug^{-1} °R^{-1}) for FPS units, with temperature in °R.

---

12 The main constituent gases of dry air, at the temperatures and pressures encountered in the region of the atmosphere used for commercial flight operations, can be treated as ideal gases with reasonable accuracy.
13 The International Standard Atmosphere (ISA) is an idealized model of the properties of the atmosphere; it is widely used in aviation (see Section 4.2).
14 It is common practice, when determining standard atmospheric models, for authorities to define the fundamental constants in SI units and then to convert the computed atmospheric properties to FPS units. If FPS units are used to calculate atmospheric properties, the constants (see Table 4.1) should be expressed with a large number of significant figures (so as to ensure that the computed atmospheric properties are unchanged).

The gas constant for a particular gas is defined as the ratio of the universal gas constant (which is applicable to all gases) to the molar mass of that gas:

$$R = \frac{\bar{R}}{m_0} \qquad (2.67)$$

where $\bar{R}$ is the universal gas constant (typical units: N m K^{-1} kmol^{-1}); and
m_0 is the molar mass of the gas (typical unit: kg/kmol).

The ISO (International Organization for Standardization) value of $\bar{R}$, as derived from the Boltzmann constant, is 8314.4621 N m K^{-1} kmol^{-1} [4]. The determination of atmospheric properties (e.g., pressure, density) in the model atmospheres used in aviation (see Section 4.2), however, is based on a gas constant of 8314.32 N m K^{-1} kmol^{-1} (defined in SI units). Note that this minor disparity from the accepted ISO value, given above, does not represent a reduction in accuracy of the computed atmosphere models.

In a real atmosphere, the composition of the air is not entirely uniform (e.g., it changes due to the presence of water vapor or contaminants), and this influences the molar mass. For the purpose of defining a standard atmosphere, the molar mass of air is based on the properties of a mixture of gases typical of the composition of clean, dry air (see Dubin, *et al.* [12], ISO [13], or ICAO [5] for details). As there is little change in the composition of dry air up to a height of 90 km (295 276 ft), a single value for m_0 is used. In the ISA, the molar mass is taken as

$$m_0 = 28.964420 \text{ kg/kmol}$$

Note that there exists a second, very similar, parameter, which is also known as the gas constant—and this can be confusing as it is sometimes also given the symbol R. Herein, this alternative form of the gas constant is designated as R' and it is defined as

$$R' = \frac{R}{g} \qquad (2.68)$$

It follows that the ideal gas law can thus be written as

$$p = \rho g R' T \qquad (2.69)$$

Equation 2.66, however, is simpler to use as R is valid for all altitudes (the alternative form of the gas constant, R', is not used herein).

2.5.7 Specific Heats of a Gas

The *specific heat* (which is also known as the *specific heat capacity*) of a gas, by definition, is the amount of heat energy needed to raise the temperature of a unit mass of the gas by 1 K (or alternatively by 1 °R). Commonly used SI units are J kg^{-1} K^{-1} (which is equivalent to m^2 s^{-2} K^{-1}). The specific heat of a particular gas, or mixture of gases, depends on the manner in which this operation takes place. It is evident from the ideal gas law (Equation 2.66) that the state of a fixed mass of gas depends on its pressure, volume, and temperature. By holding, in turn, each of the variables constant, three fundamental processes (or operations) can be defined that describe the relationship between the two remaining variables:

(1) constant pressure (isobaric) process;
(2) constant volume (isochoric) process; or
(3) constant temperature (isothermal) process.

The specific heat at constant pressure (C_p) and the specific heat at constant volume (C_v) are parameters of fundamental importance in the study of gas dynamics. The difference between the two values can be shown [14], for an ideal gas, to be equal to the gas constant, that is,

$$C_p - C_v = R \tag{2.70}$$

Moreover, it is the ratio of the two specific heats of air that frequently appears in aerodynamic analyses. This ratio is assigned the Greek letter gamma (γ). As air is assumed to be a mixture of two ideal diatomic gases in the ISA, it has a value of 1.4 (exactly). Thus

$$\gamma = \frac{C_p}{C_v} = 1.4 \qquad \text{(for air in the ISA)} \tag{2.71}$$

2.5.8 Adiabatic Processes

An adiabatic process is one in which no heat is added or removed. The properties of an ideal gas undergoing a reversible adiabatic process can be expressed mathematically by the following equations [14], which can be derived from the first and second laws of thermodynamics:

$$\frac{p}{T^{\left(\frac{\gamma}{\gamma-1}\right)}} = \text{constant} \tag{2.72}$$

$$pv^{\gamma} = \text{constant} \tag{2.73}$$

and $\quad vT^{\left(\frac{1}{\gamma-1}\right)} = \text{constant} \tag{2.74}$

where p is the pressure (typical units: N/m^2, lb/ft^2);
$\qquad$ T is the absolute temperature (typical units: K, °R); and
$\qquad$ v is the volume (typical units: m^3, ft^3).

It is convenient to rewrite these fundamental expressions in terms of density rather than volume. If the fluid mass is constant, Equations 2.73 and 2.74 can be expressed as follows:

$$\frac{p}{\rho^{\gamma}} = \text{constant} \qquad \text{(using Equation 2.45)} \tag{2.75}$$

and $\quad \dfrac{T^{\left(\frac{1}{\gamma-1}\right)}}{\rho} = \text{constant} \qquad \text{(using Equation 2.45)} \tag{2.76}$

2.5.9 Speed of Sound and Mach Number

Gases are highly compressible, and, when disturbed, a pressure wave will result that propagates through the gas; the greater the compressibility of the gas, the lower will be the speed of wave propagation. Sound propagates as a pressure wave in air, and the speed of sound thus depends on the compressibility of the air, which is not constant. The speed of sound in air (a) is given by

$$a = \sqrt{\gamma RT} \tag{2.77}$$

where a is the speed of sound (typical units: m/s, ft/s);
$\qquad$ γ is the ratio of specific heats of air (dimensionless);
$\qquad$ R is the gas constant (defined in Section 2.5.6); and
$\qquad$ T is the absolute temperature (typical units: K, °R).

Since γ and R are both constants in the ISA, the speed of sound (i.e., sonic speed) in air is seen to depend solely on the square root of the absolute temperature. As temperature changes, due to varying atmospheric conditions, for example, so the speed of sound will change. Ambient air temperature usually decreases monotonically with increasing altitude up to the tropopause (see Chapter 4); hence the speed of sound will also decrease with increasing altitude.

The Mach number[15] (M) of a gas, by definition, is the ratio of the speed of the gas (V) to the speed of sound in the gas at the point of interest (a), that is,

$$M = \frac{V}{a} \tag{2.78}$$

When $M < 1$, the flow is subsonic; when $M = 1$, the flow is sonic; and when $M > 1$, the flow is supersonic. Transonic describes a flow field in which there is a mix of subsonic and supersonic flow.

2.5.10 Total Temperature and Total Pressure

When a gas is brought to rest in an adiabatic process, the gas temperature will rise. The total, or stagnation, temperature is given by

$$T_t = T\left(1 + \frac{\gamma - 1}{2}M^2\right) \tag{2.79}$$

where T_t is the total, or stagnation, temperature (absolute); and
T is the static temperature (absolute).

By substituting $\gamma = 1.4$, the total temperature for air, in this idealized condition, is

$$T_t = T(1 + 0.2M^2) \tag{2.80}$$

Instruments designed to measure total temperature (such as the temperature probes used on aircraft) are usually not able to recover the full temperature rise and a recovery factor has to be introduced to account for this.

$$T_{t,i} = T(1 + 0.2k\,M^2) \tag{2.81}$$

where $T_{t,i}$ is the indicated total temperature (absolute); and
k is the recovery factor (typically 0.9 to 1.0, depending on the instrument).

Similarly, the total pressure (p_t) is given by[16]

$$p_t = p\left(1 + \frac{\gamma - 1}{2}M^2\right)^{\left(\frac{\gamma}{\gamma-1}\right)} \tag{2.82}$$

By substituting $\gamma = 1.4$, Equation 2.82 can be written as

$$p_t = p(1 + 0.2\,M^2)^{3.5} \tag{2.83}$$

---

15 Named after Ernst Mach (1838–1916), Austrian-born physicist and philosopher. From *ca.* 1881, Mach conducted pioneering research into supersonic flow, describing and successfully recording the existence of shock waves in 1886 [15]. In honor of his achievements, Jacob Ackeret, the eminent Swiss aeronautical engineer, in 1929 suggested calling the ratio of flow speed to the local speed of sound the *Mach number* [16].
16 The derivation of Equation 2.82 is given in Appendix D, Section D.2.1.

2.5.11 Bernoulli Equation for Compressible Flow

For speeds greater than about Mach 0.3, appreciable changes in density occur due to the compression of the air. Equation 2.63, the incompressible-flow Bernoulli equation, was derived by assuming that the density of the air in the stream tube (illustrated in Figure 2.10) was constant. For higher speeds, the following equation, which takes into account the change in air density, is appropriate:[17]

$$\left(\frac{\gamma}{\gamma-1}\right)\left(\frac{p}{\rho}\right)+\frac{1}{2}V^2 = \text{constant} \tag{2.84}$$

Equation 2.84 is an expression of the Bernoulli equation for compressible flow. The equation can be rearranged and written in several different ways—for example, Mach number can be substituted for velocity, as shown below.

$$\left(\frac{\gamma}{\gamma-1}\right)\left(\frac{p}{\rho}\right)+\frac{1}{2}M^2 a^2 = \text{constant} \qquad \text{(using Equation 2.78)} \tag{2.85}$$

$$\left(\frac{\gamma}{\gamma-1}\right)\left(\frac{p}{\rho}\right)+\frac{1}{2}M^2\gamma RT = \text{constant} \qquad \text{(using Equation 2.77)} \tag{2.86}$$

$$\left(\frac{\gamma}{\gamma-1}\right)\left(\frac{p}{\rho}\right)+\frac{1}{2}M^2\gamma\left(\frac{p}{\rho}\right) = \text{constant} \qquad \text{(using Equation 2.66)} \tag{2.87}$$

or $\quad\left(\frac{\gamma}{\gamma-1}+\frac{\gamma}{2}M^2\right)\left(\frac{p}{\rho}\right) = \text{constant}$ $\tag{2.88}$

2.5.12 Viscosity

The viscous nature of air is evident when observing flow parallel to a stationary body within the boundary layer. The boundary layer (discussed later in Section 3.6) can be described as the layer of air that is slowed down due to the presence of the body. The *no slip* boundary condition implies that the flow has zero velocity at the surface of the body. The flow velocity (u) increases with increasing distance from the surface (y).

When the flow is laminar, the air may be considered to move in discrete layers of finite thickness (see Figure 2.11). The velocity of the flow (which is represented by the length of the arrows) is reduced, or retarded, due to the shear stresses that act between the layers of air. The *dynamic*

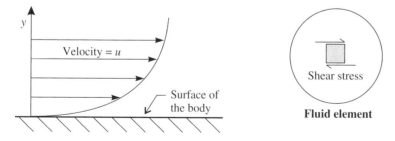

Figure 2.11 Velocity profile in the boundary layer, indicating the shear stresses parallel to the surface acting on a fluid element.

17 The derivation of Equation 2.84 is given in Appendix D, Section D.2.2.

viscosity, which is usually just called the *viscosity*, is assigned the Greek letter mu (μ); it is equal to the constant of proportionality that relates the shear stress (τ) between adjacent layers of the flow to the velocity gradient at that point:

$$\tau = \mu \frac{\mathrm{d}u}{\mathrm{d}y} \tag{2.89}$$

where τ is the shear stress in the fluid (typical units: N/m^2, lb/ft^2);
μ is the viscosity (or dynamic viscosity) of the flow;
u is the flow velocity at a distance y from the surface (typical units: m/s, ft/s); and
y is the distance from the surface (typical units: m, ft).

The standard units of dynamic viscosity are N s m^{-2} (which is equivalent to Pa s or kg s^{-1} m^{-1} in the SI system) and lb s ft^{-2} (which is equivalent to slug s^{-1} ft^{-1}).

Kinematic viscosity, which is assigned the Greek letter nu (v), is defined as the ratio of the dynamic viscosity of a fluid to its density.

$$v = \frac{\mu}{\rho} \tag{2.90}$$

The standard units of kinematic viscosity are m^2 s^{-1} and ft^2 s^{-1}.

The viscosity of air is a function of temperature (T)—this can be fairly accurately described by the Sutherland equation,[18] where the empirical constants β_s and S are determined by experiment:

$$\mu = \frac{\beta_s T^{3/2}}{T + S} \tag{2.91}$$

where $\beta_s = 1.458 \times 10^{-6}$ N s m^{-2} K$^{-1/2}$ for SI units; or
$\beta_s = 3.04509 \times 10^{-10}$ lb s ft^{-2} K$^{-1/2}$ for FPS units; and
$S = 110.4$ K.

It is reported [5, 11] that Equation 2.91 is not valid for very high or very low temperatures or under the atmospheric conditions that exist at altitudes above about 90 km (295 276 ft).

The viscosity of air at the ISA sea-level datum conditions (see Section 4.2) can be determined from Equation 2.91 by setting $T = T_0 = 288.15$ K:

$\mu_0 = 1.7894 \times 10^{-5}$ N s m^{-2} for SI units; or
$\mu_0 = 3.7372 \times 10^{-7}$ lb s ft^{-2} for FPS units.

2.5.13 Reynolds Number

Reynolds number (R_e) is a non-dimensional characteristic number,[19] which is defined as follows:

$$R_e = \frac{\rho u_\infty l}{\mu} \tag{2.92}$$

where u_∞ is the flow velocity (typical units: m/s, ft/s); and
l is a characteristic length (typical units: m, ft).

---

18 The formulation was proposed by William Sutherland (1859–1911), an Australian physicist, in 1893 [17].
19 The characteristic number is named in honor of Osborne Reynolds (1842–1912), the British engineer and physicist, who proposed its use in 1883 for characterizing flow behavior [18].

Alternatively, Reynolds number can be expressed in terms of kinematic viscosity, that is,

$$R_e = \frac{u_\infty l}{\nu} \tag{2.93}$$

The Reynolds number, in a sense, relates inertia effects to viscous effects in the flow. The characteristic length (l) is selected to suit the application—for example, when the object being studied is an airfoil, the characteristic length will usually be the chord. However, when studying the flow at a particular point on an aerodynamic surface (e.g., a wing) the local Reynolds number will be based on a dimension that defines that point, such as the distance from the airfoil leading edge.

Reynolds number is a key parameter in the understanding of the behavior of boundary layers in that it influences the thickness of the boundary layer relative to the body dimensions. For example, with increasing flow velocity and the other parameters (viz., ρ, μ, and l) remaining unchanged, the Reynolds number will increase; there will be reduced rate of diffusion of momentum from outer flow layers resulting in a thinner boundary layer.

2.6 Further Reading

There are many textbooks that contain expanded descriptions of the topics introduced in this chapter—relevant works include:

Hibbeler, R.C., *Engineering mechanics: Statics & dynamics*, 14th ed., Pearson Education, Upper Saddle River, NJ, 2015.

Houghton, E.L., Carpenter, P.W., Collicott, S.H., and Valentine, D.T., *Aerodynamics for engineering students*, 7th ed., Butterworth-Heinemann, Cambridge, MA, 2017.

Kundu, P.K., Cohen, I.M., and Dowling, D.R., *Fluid mechanics*, 5th ed., Academic Press, USA, 2012.

Meriam, J.L. and Kraige, L.G., *Engineering mechanics: Dynamics*, 7th ed., John Wiley & Sons, New York, NY, 2012.

Meriam, J.L. and Kraige, L.G., *Engineering mechanics: Statics*, 7th ed., John Wiley & Sons, New York, NY, 2011.

Munson, B.R., Rothmayer, A.P., Okiishi, T.H., and Huebsch, W.W., *Fundamentals of fluid mechanics*, 7th ed., John Wiley & Sons, New York, NY, 2013.

Rogers, G.F.C. and Mayhew, Y.R., *Engineering thermodynamics: Work and heat transfer*, 4th ed., Pearson Education, Harlow, UK, 1996.

References

1 BIPM, "The international system of units (SI)," 8th ed., Bureau International des Poids et Mesures (BIPM), Paris, France, 2006.

2 ISO, *Quantities and units*, ISO 31, International Organization for Standardization, Geneva, Switzerland, 1992.

3 ICAO, "Units of measurement to be used in air and ground operations," 5th ed., Annex 5 to the Convention on International Civil Aviation, International Civil Aviation Organization, Montréal, Canada, July 2010.

4 Mohr, P.J., Taylor, B.N., and Newell, D.B., "CODATA recommended values of the fundamental physical constants: 2006," *Reviews of Modern Physics*, Vol. 80, Iss. 2, pp. 633–730, 2008.

5 ICAO, "Manual of the ICAO standard atmosphere: Extended to 80 kilometres (262 500 feet)," 3rd ed., Doc. 7488, International Civil Aviation Organization, Montréal, Canada, 1993.

6 NIMA, "Department of defense world geodetic system 1984: Its definition and relationships with local geodetic systems," TR 8350.2, 3rd ed., National Imagery and Mapping Agency, St. Louis, MO, Jan. 3, 2000.

7 Pinsker, W.J.G., "The effect of variations in local gravity and of aircraft speed on the effective weight of aircraft in high performance cruise," R.&M. No. 3680, Aeronautical Research Council, Ministry of Defence, London, UK, 1972.

8 Blake, W. and Performance Training Group, "Jet transport performance methods," D6-1420, Flight Operations Engineering, Boeing Commercial Airplanes, Seattle, WA, Mar. 2009.

9 McNaught, A.D. and Wilkinson, A., *International Union of Pure and Applied Chemistry: Compendium of chemical terminology (Gold Book)*, 2nd ed., Blackwell Science, Oxford, UK, 1997.

10 Adamson, D., *Blaise Pascal: Mathematician, physicist and thinker about God*, Palgrave Macmillan, London, 1995.

11 ESDU, "Equations for calculation of international standard atmosphere and associated off-standard atmospheres," Data item 77022, Amdt. B, IHS ESDU, 133 Houndsditch, London, UK, Feb. 1986.

12 Dubin, M., Hull, A.R., and Champion, K.S.W., "US standard atmosphere 1976," NASA TM-X-74335 / NOAA S/T-76-1562, National Aeronautics and Space Administration, National Oceanic and Atmospheric Administration, United States Air Force, Washington, DC, 1976.

13 ISO, *Standard atmosphere*, ISO 2533, International Organization for Standardization, Geneva, Switzerland, 1975.

14 Rogers, G.F.C. and Mayhew, Y.R., *Engineering thermodynamics: Work and heat transfer*, 4th ed., Pearson Education, Harlow, UK, 1996.

15 Reichenbach, H., "Contributions of Ernst Mach to fluid mechanics," *Annual Review of Fluid Mechanics*, Vol. 15, pp. 1–28, 1983.

16 Rott, N., "Jacob Ackeret and the history of the Mach number," *Annual Review of Fluid Mechanics*, Vol. 17, pp. 1–9, 1985.

17 Sutherland, W., "The viscosity of gases and molecular force," *Philosophical Magazine*, Series 5, Vol. 36, Iss. 223, pp. 507–531, 1893.

18 Jackson, D. and Launder, B., "Osborne Reynolds and the publication of his papers on turbulent flow," *Annual Review of Fluid Mechanics*, Vol. 39, pp. 19–35, 2007.

3

Aerodynamic Fundamentals

3.1 Introduction

This chapter introduces several fundamental aspects of aerodynamics and flight controls, which are key to an understanding of the performance of the jet airplane—these topics include: aerodynamic notation and definitions (Section 3.2); coordinate systems and sign conventions (Section 3.3); aerodynamic forces and moments (Section 3.4); compressibility (Section 3.5); boundary layers (Section 3.6); high lift devices (Section 3.7); and controls for pitch, roll, and yaw (Section 3.8). The purpose of the chapter is to provide background information on these topics in a style and format that is consistent with the treatment of airplane performance as presented in subsequent chapters. There are many excellent texts (as noted in Section 3.9) that provide a wealth of information on the subject of aerodynamics and flight controls—the reader is directed to these works for more information on the topics introduced in this chapter.

3.2 Standard Definitions and Notation

3.2.1 Airfoil Section Definitions

A cross-sectional cut through an airplane's wing defines an *airfoil*, which is also known as an *airfoil section* (frequently shortened to just *section*). The shape definition (profile) of the airfoil section would typically change from the wing root to the wingtip. A great many airfoil sections have been developed and tested, and a considerable amount of data are available regarding the characteristics of these different airfoils. The development of modern, efficient airfoil sections owes a great deal to the work done at NACA[1] in producing the so-called classic NACA airfoils.[2] Although these designs have been around for many years, they are still found in many sectors of the aviation industry and are often used as a benchmark against which modern designs are compared. The main geometric parameters that define an airfoil are shown in Figure 3.1.

Profile Definition
It is convenient to separate the profile (or shape definition) of an airfoil section into its thickness distribution (which has a major influence on the profile drag—see Section 7.3.2) and a zero

---

1 NACA (National Advisory Committee for Aeronautics): precursory organization to NASA (National Aeronautics and Space Administration), which was formed in 1958.
2 References [1] and [2] provide geometrical definitions and experimental results for an extensive range of NACA airfoils.

Performance of the Jet Transport Airplane: Analysis Methods, Flight Operations, and Regulations, First Edition.
Trevor M. Young.
© 2018 John Wiley & Sons Ltd. Published 2018 by John Wiley & Sons Ltd.

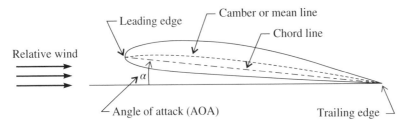

Figure 3.1 Airfoil definitions.

thickness camber line (which has a major influence on lift, pitching moment, and lift-dependent drag). The usual method for constructing an airfoil is from coordinate data of the thickness and camber distributions. The leading edge (or nose) definition is then blended into the upper and lower profiles. For NACA airfoils, the nose section is circular, and the nose radius has its center on a line drawn tangential to the camber line at the leading edge; elliptical or near elliptical nose sections are also used.

Camber Line and Chord Line

The mean camber line—or camber line as it is usually called—is a locus of the midpoints between the upper and lower surfaces measured normal to the camber line itself. A straight line joining the start and end points of the camber line defines the *chord*, which is given the symbol *c*. A measure of the amount of camber of an airfoil is given by the maximum distance from the chord line to the camber line; this is usually expressed as a percentage of the chord. An airfoil with zero camber is symmetrical about the chord line.

Maximum Thickness-to-Chord Ratio

The maximum thickness-to-chord ratio, $(t/c)_{max}$—which is sometimes shortened to thickness ratio—is an important design parameter for an airfoil section. This parameter has a significant influence on the stall and drag characteristics of an airfoil section. Thinner sections have a lower drag coefficient and achieve a delay in the onset of drag rise associated with shock wave formation. Increased structural complexity and weight, as well as reduced wing internal volume (which is required for fuel and, in many cases, the undercarriage), however, limit the extent by which designers can reduce the wing thickness.

Angle of Attack

The angle of attack (AOA) is the angle of inclination of the airfoil section measured from the relative wind (or local airflow) to the chord line (Figure 3.1). The angle of attack is customarily assigned the Greek letter alpha (α) and is defined as positive when the nose of the airfoil is rotated upward.[3]

Two-Dimensional Flow Field

It is convenient for the aerodynamic assessment of a wing to commence with a study of a two-dimensional flow field around an airfoil that represents a small spanwise strip (or segment) of the wing. This is obviously an idealized representation as it considers neither the out-of-plane, or spanwise, component of flow nor the complex three-dimensional effects that arise near the

---

3 In some reference works, principally older British texts, the angle of attack is called the angle of incidence. This term can cause confusion as the angle between the wing centerline chord and the fuselage datum is called the wing incidence angle or simply the incidence angle.

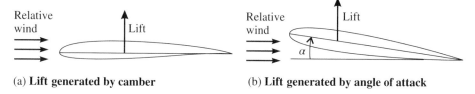

(a) **Lift generated by camber**   (b) **Lift generated by angle of attack**

Figure 3.2 Lift can be generated by (a) camber or (b) angle of attack.

wingtip. Such a flow field can be achieved in a wind tunnel using a zero-sweep, high aspect ratio wing model with a constant airfoil section along its span. With a stream of air flowing over the airfoil, an aerodynamic force acting on the model is generated. This force can be decomposed into *lift* and *drag* components. Lift can be generated by cambered airfoils (Figure 3.2a) or by angle of attack (Figure 3.2b) or—as is frequently the case—by a combination of both.

3.2.2 Wing Geometric Definitions

The wing geometric parameters that are used in aerodynamic analyses are defined on an idealized reference wing planform, which is based on the airplane's actual wing area and shape (see Figure 3.3).

Reference Wing Planform
The reference wing planform is an equivalent, straight-tapered planform that extends to the fuselage centerline. Each wing panel (i.e., left and right wing) is of trapezoidal shape; the leading and trailing edges are straight and the wingtip chords are parallel to the fuselage centerline. A reference wing planform is an idealization of the airplane's actual wing planform, with leading and trailing edge devices (e.g., flaps and slats) retracted. Airplane manufacturers employ their own specific methods to define a reference wing planform; a typical approach is described in ESDU 76003 [3]. A projection of the actual wing shape onto a horizontal plane is initially generated. Features such as engine nacelles, rounding of the wingtips, winglets, fillets at the wing/fuselage intersection, and so forth are then removed. Cranks on the leading and trailing edges are eliminated to generate a straight-tapered wing with the same tip chord and the chord at the wing/fuselage junction adjusted to give the same exposed wing area. The straightened leading edges are extended to the fuselage centerline—a location known as the wing apex. Similarly, the straightened trailing edges are extended to the fuselage centerline to complete the two trapezoidal shapes.

Wing Reference Area
The wing reference area (S) is the area of the reference wing planform. Note that wing reference areas are not necessarily identical to actual wing areas. This may appear to be a problem, but in reality such discrepancies are of no real concern as the wing reference area is only a reference dimension that is primarily used to calculate aerodynamic parameters, such as lift and drag coefficients. It is used to compute the airplane's *wing loading*, which is the ratio of the airplane's mass or weight to the wing reference area. The reference area is a constant dimension for an airplane type and does not change when the actual wing area changes due to the extension of flaps, for example. It is interesting to note that airplane manufacturers will usually use the same reference area for derivative models of an airplane type (e.g., the Boeing 747-400 featured a wingtip extension, but the reference area was the same as earlier versions of the B747).

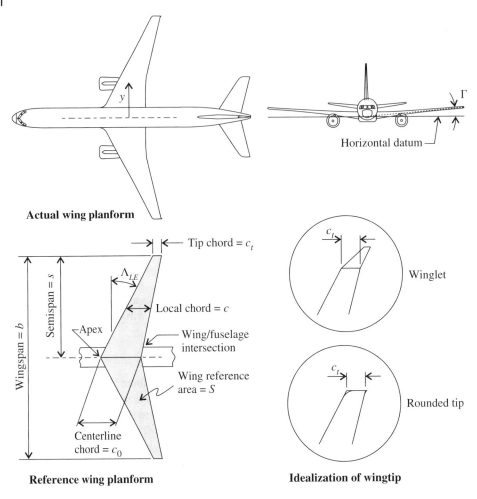

Figure 3.3 Reference wing planform.

Exposed Wing Area

The exposed wing area (S_{exp}) is the portion of the reference area that excludes the part of the wing inside the fuselage; it is measured outwards from the wing/fuselage intersection on the reference wing planform. Typically, S_{exp} is sized to match the area of the actual basic wing planform.

Wingspan

The wingspan (b) is the distance from wingtip to wingtip on the reference planform, measured normal to the airplane's centerline. The semispan distance is usually denoted as s or $b/2$. In constructing a reference wing planform, any rounding of the wingtip is ignored and the leading and trailing edges are extended outwards to define the trapezoidal shape. Out-of-plane wingtip extensions (i.e., winglets) are ignored and a hypothetical wingtip defined at the winglet/wing junction.

Wing Chords

The wing chord (c) is measured from the leading edge to the trailing edge on the reference planform, parallel to the centerline of the fuselage (Figure 3.3). The *centerline chord* (c_0) is unambiguously defined as the chord of the reference planform at the fuselage centerline. Note that certain conventions (e.g., Boeing [4]) use the term *root chord* (c_r) to describe the centerline chord; other conventions (e.g., ESDU [3]) use it to describe the chord at the wing/fuselage junction. The *tip chord* (c_t) is measured at the semispan location on the reference planform.

Wing Taper Ratio

The wing taper ratio, which is frequently shortened to *wing taper*, is assigned the Greek letter lambda (λ); it is defined as the ratio of the tip chord to the centerline chord:

$$\lambda = \frac{c_t}{c_0} \tag{3.1}$$

Taper affects the distribution of lift along the wingspan. The Prandtl wing theory[4] indicates that the minimum induced drag occurs when the lift is distributed in an elliptical fashion. For a wing without twist or sweep, this occurs when the wing planform is itself elliptical. An elliptical wing has superior aerodynamic characteristics than a rectangular wing. By tapering the wing, the lift distribution can be improved—for example, a taper ratio of 0.45 on an unswept wing will approximate the ideal elliptical lift distribution.

Lateral Coordinates

The lateral coordinate (y) measured from the fuselage centerline is frequently non-dimensionalized using the semispan distance. It is usually denoted by the Greek letter eta (η); thus

$$\eta = \frac{y}{s} = \frac{2y}{b} \tag{3.2}$$

Aspect Ratio

The aspect ratio is usually represented by the letter A (note that other conventions exist—for example, the ratio is sometimes written as A_R). For a rectangular wing planform, the aspect ratio is the ratio of the wingspan to the wing chord. For all other wing planform shapes, the aspect ratio can be determined from the reference area (S) and the wingspan (b), as follows:

$$A = \frac{b^2}{S} \tag{3.3}$$

For the reference wing planform (i.e., a straight tapered trapezoidal planform), the centerline chord can be expressed in terms of the wing area and wingspan or, alternatively, the wing area and aspect ratio:

$$c_0 = \frac{2S}{(1 + \lambda)b} = \frac{2}{(1 + \lambda)} \sqrt{\frac{S}{A}} \tag{3.4}$$

Jet transport airplanes have moderately high aspect ratio wings when compared to other classes of aircraft. The aspect ratio has a major influence on the lift-dependent drag (see Section 7.3.4) and also on the lifting ability of the wing at a given angle of attack. A high aspect ratio wing has a steep lift-curve slope (see Section 3.4.7).

---

4 Named after Ludwig Prandtl (1875–1953), the eminent German scientist who conducted pioneering aerodynamics research from 1901 to 1953, initially at the Technical School Hannover (now Technical University Hannover), but mostly at the University of Göttingen [5]. So profound was his influence that he is sometimes called the *father of modern aerodynamics*.

Sweep Angle

The wing sweep angle, or sweepback angle, is measured from a line drawn perpendicular to the fuselage centerline to a reference line on the wing planform, usually taken as the quarter chord line. It is given the upper case Greek letter lambda (Λ). It is common practice to add a subscript to identify the planform reference line to which the angle refers—for example, the leading edge sweep angle is designated as Λ_{LE} or Λ_0. The quarter chord sweep angle (designated as $\Lambda_{1/4}$ or $\Lambda_{0.25}$ or $\Lambda_{0.25c}$) is the sweep angle of a line joining the 25% point of the centerline chord to the corresponding point on the wingtip chord. The 25% chord sweep angle is often selected to characterize a wing planform, although the 50% chord sweep angle may also be used for this purpose. The aerodynamic effect of sweeping the wing back is that it delays the subsonic drag-rise associated with shock-wave formation at transonic speeds and reduces the peak drag coefficient.[5] Wing sweep also improves the lateral (roll) stability of an airplane.

Wing Dihedral Angle

The wing dihedral angle, which is assigned the upper case Greek letter gamma (Γ), is the angle measured from a horizontal datum plane passing through the centerline chord to a reference line (drawn from the wing centerline to the wingtip) when the airplane is viewed from the front—see Figure 3.3. If the wingtips are located above the horizontal datum line, the wing is said to have a positive dihedral angle, whereas if the wingtips are below the datum, the airplane has negative dihedral, or anhedral. The three-dimensional geometry of a modern wing design is complicated, incorporating varying twist, thickness, and camber along the wingspan—and in these cases the establishment of a wing reference line is not trivial. In many cases, the wing is twisted about a reference line (say joining the 25% chord points), and for these aircraft types this line can be used to define the dihedral angle. In flight, dihedral has a strong influence on an airplane's lateral (roll) stability.

Wing Twist

The geometric wing twist (ε_g) is the angle of incidence of the wingtip relative to that of the centerline section. By convention, a positive twist angle is associated with the nose of the wingtip section being rotated upwards. The term *washout* is used to describe a wing where the tip section has been rotated downwards. Washout has a desirable influence on the stalling characteristics of a wing and is widely used in the design of aircraft of many different types. The effect of washout is to reduce locally the angle of attack of the outboard portion of the wing. This results in the wing first stalling at an inboard location, while maintaining good airflow over the outboard portion of the wing. Wing twist also affects the spanwise lift distribution and can be used to reduce the induced drag by tailoring the lift distribution to approximate an elliptical distribution. However, it is not possible to optimize this for all flight conditions. The spanwise lift distribution can also be tailored by varying the airfoil section properties, such as profile and camber, with span—this is known as aerodynamic twist.

Wing Incidence Angle

The wing incidence (or setting) angle is the angle of the wing centerline chord with respect to the fuselage datum line. From a design perspective, it is usually selected to minimize drag at the design operating condition. At the selected flight condition, the wing will be operating at

---

5 The idea of sweeping an airplane's wing back to delay the effects of compressibility was first proposed by Adolf Busemann (1901–1986), a German aerodynamicist, in 1935 [6].

the desired angle of attack for the design lift coefficient and the fuselage will be at the angle of attack for minimum drag; for a typical commercial jet transport airplane this is close to 0°.

Mean Aerodynamic Chord (MAC)

The mean aerodynamic chord ($\bar{\bar{c}}$)—or aerodynamic mean chord, as some authorities prefer—is the chord of a rectangular wing, of span b, that has essentially the same lift and pitching moment characteristics as the actual wing.[6] The MAC of a planar wing (i.e., a wing without dihedral or out-of-plane wingtip extensions) is given by

$$
\bar{\bar{c}} = \frac{\displaystyle\int_0^{b/2} c^2 \, dy}{\displaystyle\int_0^{b/2} c \, dy} \tag{3.5}
$$

If the actual wing is represented by an equivalent straight-tapered planform, the MAC is given by

$$
\bar{\bar{c}} = \frac{2}{3} c_0 \left[1 + \frac{\lambda^2}{1 + \lambda} \right] \tag{3.6}
$$

where c_0 is the centerline chord, which can be expressed in terms of wing area, span, and taper ratio, as indicated by Equation 3.4.

The MAC is used as a reference length for many aerodynamic calculations—for example, it is used to non-dimensionalize pitching moment coefficients (see Section 3.4.6). The position of the airplane's center of gravity (see Section 19.3.1) and aerodynamic center (defined in Section 3.4) are typically expressed relative to $\bar{\bar{c}}$. As the MAC is defined by pitching moment considerations, it follows that it only has a longitudinal position; by convention, $\bar{\bar{c}}$ is shown on the fuselage centerline. For a straight-tapered wing, the leading edge of the MAC (often abbreviated as LEMAC) is located behind the wing apex by a distance of

$$
\bar{\bar{x}}_0 = c_0 \left[1 + \frac{2\lambda}{12} \right] A \tan \Lambda_0 \tag{3.7}
$$

It is convenient, for certain applications, to determine the spanwise position where the local chord is equal to the MAC. For a straight-tapered wing, the lateral coordinate (y) where this occurs is given by

$$
y = c_0 \left[\frac{1 + \lambda}{12} \right] A = \frac{b}{6} \left[\frac{1 + 2\lambda}{1 + \lambda} \right] = \frac{s}{3} \left[\frac{1 + 2\lambda}{1 + \lambda} \right] \tag{3.8}
$$

The fraction of the wing semispan at which $c = \bar{\bar{c}}$ is herein represented by the Greek letter kappa (κ)—thus for a straight-tapered wing,

$$
\kappa = \frac{1}{3} \left[\frac{1 + 2\lambda}{1 + \lambda} \right] \tag{3.9}
$$

A simple graphical construction technique can be used to determine $\bar{\bar{c}}$, $\bar{\bar{x}}_0$, and κ for a straight-tapered wing—this is illustrated in Figure 3.4.

---

6 Cautionary note: The mean aerodynamic chord (MAC) is sometimes given the symbol $\bar{c}$. This practice can be confusing, however, as the symbol $\bar{c}$ is also widely used to represent the geometric mean chord (see Equation 3.10).

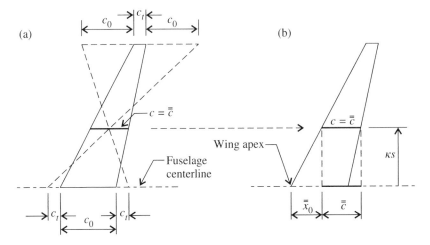

Figure 3.4 Graphical method to determine (a) the length and (b) the centerline location of the mean aerodynamic chord ($\bar{\bar{c}}$) for a straight-tapered wing.

Geometric Mean Chord

The geometric (or standard) mean chord ($\bar{c}$) is the arithmetic mean, or average, chord, which for an arbitrary planform is given by

$$\bar{c} = \frac{\displaystyle\int_0^{b/2} c\,\mathrm{d}y}{\displaystyle\int_0^{b/2} \mathrm{d}y} \tag{3.10}$$

For an equivalent straight-tapered planform, $\bar{c}$ can be determined directly from the wing reference area and wingspan:

$$\bar{c} = \frac{S}{b} \tag{3.11}$$

The geometric mean chord, like the MAC, is used as a reference chord to non-dimensionalize longitudinal parameters and to define longitudinal locations (e.g., of the airplane's CG). By convention, the geometric mean chord is also shown on the fuselage centerline and its longitudinal position established by aligning the quarter-chord location of $\bar{c}$ with the quarter-chord location of $\bar{\bar{c}}$.

3.2.3 Geometric Definitions for the Empennage

The *empennage* is a collective term for the horizontal tailplane (HTP), which is also called the horizontal stabilizer, and the vertical tailplane (VTP), which is also known as the vertical stabilizer or fin. The geometric definitions of the horizontal and vertical stabilizers are basically the same as those given for the wing; exceptions to the definitions given earlier for the wing are described in this section. The subscripts H and V are used to identify the horizontal and vertical stabilizer, respectively.

Reference Areas

The reference area for the horizontal stabilizer (S_H) extends to the airplane's centerline (in an identical manner to that illustrated in Figure 3.3 for the wing). However, this definition is not

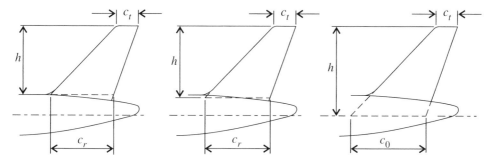

Figure 3.5 Alternative conventions defining the vertical stabilizer reference area.

universally accepted and sometimes only the exposed area is used as the reference area for the horizontal tailplane. The reference area of the vertical stabilizer (S_V) is less well defined and a number of definitions are used (Figure 3.5).

Aspect Ratio

The aspect ratio (A) of the horizontal and vertical stabilizers can be calculated using Equation 3.3. For the vertical stabilizer, it is usual to refer to the height (h) rather than the span (Figure 3.5).

3.3 Coordinate Systems and Conventions

3.3.1 Coordinate Systems

Several right-handed Cartesian coordinate (X, Y, Z) systems are used in the subject of flight dynamics to study airplane motions. In these coordinate systems, translational displacements, velocities, and accelerations are defined to be positive in the positive sense of the axes. Similarly, angular rotations and rotational velocities and accelerations are defined to be positive in the clockwise direction when viewed along the positive direction of the appropriate axis (in compliance with the right-hand rule, as illustrated later in Figure 3.7). Specific terminology and notation is associated with the rotation of one frame (coordinate system) with respect to a reference frame (see also Appendix C).

3.3.2 Ground Axis System

The ground axis system (X_g, Y_g, Z_g) is fixed with respect to the Earth. Its origin, O, is arbitrarily chosen; the X_g–Y_g plane is parallel to an idealized flat surface of the Earth and the Z_g axis points directly downwards towards the Earth (see Figure 3.6a).

3.3.3 Earth Axis System

The Earth axis system (X_e, Y_e, Z_e) has its origin, O, at the airplane's center of gravity (CG) and it moves with the airplane. As with the ground axis system, the X_e–Y_e plane is parallel to an idealized flat surface of the Earth and the Z_e axis points directly downwards towards the Earth (Figure 3.6b). The X_e axis can be defined positive in any convenient direction; however, it is common to align it with either the airplane's longitudinal axis or the direction of flight.

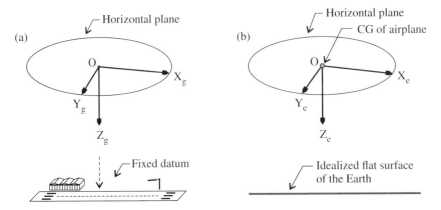

Figure 3.6 (a) Ground axis system; (b) Earth axis system.

3.3.4   Body Axis System

The axes of the body axis system (X_b, Y_b, Z_b) are fixed with respect to the airplane. The origin, O, can be arbitrarily chosen, but it is often convenient to select the airplane's center of gravity. The X axis, which is called the longitudinal axis, is defined positive in the forward direction (typically parallel to the cabin floor). The Y axis, called the lateral or transverse axis, is positive in the starboard direction. The Z axis, called the normal axis, is defined positive in the downward sense (Figure 3.7). The X_e–Z_e plane coincides with the plane of symmetry of the airplane.

The components of linear velocity and force are positive in the positive direction of the corresponding axes. Similarly, components of angular velocity and moment are positive in the conventional right-handed Cartesian sense. The roll angle (ϕ), pitch angle (θ), and yaw angle (ψ) describe rotations about the X_b, Y_b, and Z_b axes, respectively.

Figure 3.7 Body axis coordinate system.

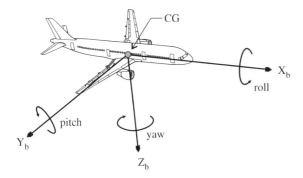

3.3.5   Flight Path (or Velocity or Wind) Axis System

The flight path axis system (X_a, Y_a, Z_a)—which is also called the velocity or wind axis system—has its origin, O, at the body axes origin, but the axes are not fixed with respect to the airplane (Figure 3.8). The X axis coincides with the direction of the velocity vector in that it is defined positive towards the incoming relative wind. The Z axis lies on the plane of symmetry of the airplane and is defined to be positive downwards when the airplane is in a normal attitude. The Y axis is positive in the starboard direction and is aligned with Y_b.

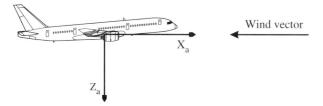

Figure 3.8 Flight path axis coordinate system.

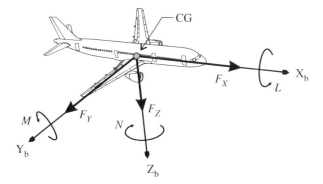

Figure 3.9 Forces and moments with respect to body axes.

3.3.6 Force and Moment Components

The forces that act on the airplane can be resolved in a number of different ways—that is, to suit a particular application. One such system is shown in Figure 3.9, where the forces are resolved with respect to the body axes. For the purposes of airplane performance analyses, however, it is convenient to work with lift and drag forces, which are defined with respect to the flight path axes (see Section 3.4).

The moment about the center of gravity is resolved as follows: rolling moment L (about the X_b axis), pitching moment M (about the Y_b axis), and yawing moment N (about the Z_b axis).

3.4 Aerodynamic Forces and Moments

3.4.1 Airfoil Pressure Distribution

Airfoil pressure distributions—that is, the variations of air pressure with chordwise location around the upper and lower airfoil surfaces—are one of the most important characteristics of the aerodynamic design of any airplane.[7] Figure 3.10 illustrates this variation in air pressure around a cambered airfoil at an angle of attack α. An arrow pointing away from the surface indicates a pressure (p) that is lower than the free-stream ambient pressure (p_∞)—which, in this illustration, is the entire upper airfoil surface with the exception of a very small region at the trailing edge. Almost the entire lower airfoil surface experiences a pressure that is higher than ambient. The length of the arrows indicate pressure differential (i.e., $p - p_\infty$) with respect to ambient conditions.

---

7 Obert [7] has collated pressure distribution data for transonic airfoils.

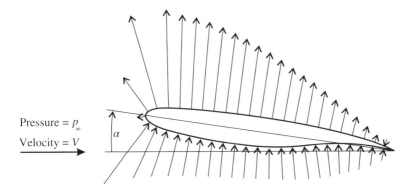

Figure 3.10 Schematic of the distribution of pressure on a cambered airfoil at an angle of attack.

3.4.2 Two-Dimensional Aerodynamic Forces and Coefficients

The net aerodynamic force (R) that acts on an airfoil can be obtained by determining the net effect (or resultant) of the stresses that act on the airfoil's surface. The force R can be computed through a process of integration of both the normal stresses (i.e., pressure) and tangential stresses (i.e., skin friction). This force vector acts through the *center of pressure* (CP), which lies on the airfoil chord. Experimentally, it has been shown that the force R is a function of six independent variables, namely,

$$R = f(V, \rho, S_\pi, \mu, a, \alpha) \tag{3.12}$$

where V is the free-stream velocity (typical units: m/s, ft/s);
ρ is the air density (typical units: kg/m^3, slug/ft^3);
S_π is a characteristic, or reference, area (typical units: m^2, ft^2);
μ is the viscosity of the air (typical units: N s m^{-2}, lb s ft^{-2});
a is the speed of sound in air (typical units: m/s, ft/s); and
α is the angle of attack (typical units: degree, rad).

For many applications, the ease of analysis is enhanced by treating aerodynamic parameters as non-dimensional ratios. The product of dynamic pressure (q) and a selected reference area (S_π) has units of force; this is widely used to non-dimensionalize aerodynamic forces. The resultant aerodynamic force (R) thus becomes a force *coefficient* when divided by qS_π. The reference or characteristic area most commonly used for aerodynamic analyses is the wing reference area (S). Alternative reference areas are, however, also used—for the analysis of the drag of individual components, for example, the frontal area of the component is typically used.

It can be shown by dimensional analysis [8] that the non-dimensional aerodynamic coefficient, is, in fact, dependent only on the angle of attack and two non-dimensional terms—that is,

$$\frac{R}{qS} = f(R_e, M, \alpha) \tag{3.13}$$

where R_e is the Reynolds number (see Section 2.5.13); and
M is the Mach number (see Section 2.5.9).

By convention, the net aerodynamic force (R) is resolved into two components: one normal to the free-stream direction, which, by definition, is the *lift* force and one parallel to the

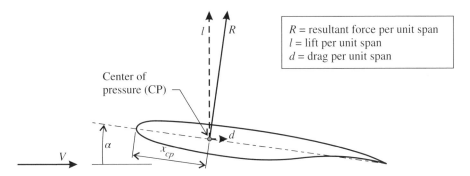

Figure 3.11 Lift and drag forces acting on an airfoil.

free-stream direction, which is the *drag* force (Figure 3.11). It is important to note that the lift vector will not always point directly upwards as it is defined normal to the free-stream vector (only in the case of an airplane operating at constant height with the wings level would the airplane's net lift vector act directly upwards).

The airflow over an airfoil can be treated as two dimensional and the forces acting on the airfoil considered as "per unit of span." The lift per unit of span distance is usually denoted by the lower case letter l and the drag per unit of span distance by the letter d.

3.4.3 Two-Dimensional Lift Coefficient

For most aerodynamic analyses, it is useful to express the lift and drag components as non-dimensional ratios, or coefficients. The two-dimensional lift coefficient (C_l) is defined as follows:

$$C_l = \frac{l}{qc} = \frac{2l}{\rho V^2 c} \qquad (3.14)$$

where l is the lift force per unit of span (typical units: N/m, lb/ft);
q is the dynamic pressure (typical units: N/m^2, lb/ft^2); and
c is the chord length (typical units: m, ft).

The lift coefficient is a function of the angle of attack, the Reynolds number, and the Mach number [1, 7]. Compressibility effects at high Mach numbers arise with transonic and supersonic flight regimes. For most flight conditions, the influence of Mach number on lift coefficient is small and may often be ignored. By keeping the Reynolds number constant, the relationship between lift coefficient and angle of attack can be experimentally determined in a wind tunnel for any airfoil section. It is shown in Figure 3.12a that C_l increases approximately linearly with α up to a critical angle (which is about 12° to 14° for many airfoil sections). After this point, there is a region of non-linearity where the lift decreases rapidly. This is called the separated flow region. When C_l reaches a maximum (i.e., $C_{l_{max}}$), the airfoil is said to have stalled. The air no longer travels smoothly over the upper airfoil surface, and the turbulent airflow produces less lift. The influence of Reynolds number on the C_l versus α graph is largely limited to the stall region for the conditions associated with airplane flight. The measured maximum lift coefficient is thus dependent on the Reynolds number associated with the test conditions.

It can also be seen from Figure 3.12a that there is a difference between a cambered and a symmetrical airfoil section regarding their lift characteristics. The symmetrical section generates no lift at zero angle of attack, whereas a cambered section generates a small amount of lift. The

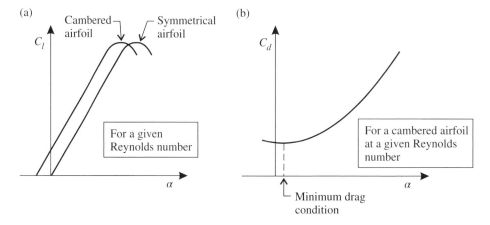

Figure 3.12 (a) Lift coefficient for symmetrical and cambered airfoils; (b) Drag coefficient for cambered airfoil.

angle of attack for zero lift of a cambered airfoil depends on the amount of camber and would typically be between 0° and −5°.

3.4.4 Two-Dimensional Drag Coefficient

The two-dimensional drag coefficient (C_d) is defined as follows:

$$C_d = \frac{d}{qc} = \frac{2d}{\rho V^2 c}$$

(3.15)

where d is the drag force per unit of span (typical units: N/m, lb/ft).

As with the lift coefficient, the drag coefficient depends on the angle of attack, the Reynolds number, and the Mach number. The effects of Mach number and Reynolds number on drag are discussed in Sections 7.3–7.5 (which provide a detailed treatment of the three-dimensional drag characteristics of an airplane). The influence of angle of attack on the drag coefficient of an airfoil is illustrated in Figure 3.12b. In this figure, the drag is shown as a function of α (as would be measured in a wind-tunnel test); however, it is frequently the case that C_d is plotted as a function of C_l. As C_l varies almost linearly with α up to the separated flow region, the shape of the drag curve is unchanged within the linear region. The value of plotting the drag coefficient against the lift coefficient is shown later in Section 7.4.

3.4.5 Two-Dimensional Aerodynamic Center and Pitching Moment

The center of pressure (CP) is located on the chord at a distance x_{cp} from the leading edge (see Figure 3.11). In the absence of compressibility effects, the distance x_{cp} is a function of the Reynolds number and the angle of attack. At a low angle of attack, the center of pressure will be located towards the rear of the airfoil and with an increase in the angle of attack, the center of pressure will move forwards. As the position of the center of pressure moves with changing flow-field conditions, it is not the most convenient reference point for analyzing the loads acting on an airfoil.

It is possible to use any fixed reference point along the chord for the purpose of analyzing the aerodynamic forces. A transformation of the lift and drag forces from the center of pressure to another reference point, located at a distance x from the leading edge, introduces a moment

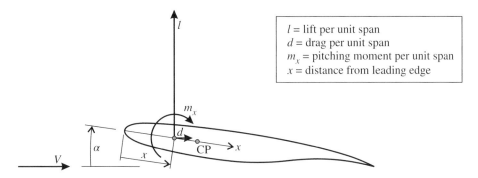

Figure 3.13 Two-dimensional lift force, drag force, and pitching moment.

m_x (shown in Figure 3.13). This is called a pitching moment; by convention, such moments are defined as positive when they act in a "nose-up" orientation (see Figure 3.9).

Taking the sum of the moments about the center of pressure leads to the following equation:

$$\sum Moments_{cp} = m_x - (l\cos\alpha + d\sin\alpha)(x - x_{cp}) = 0 \qquad (3.16)$$

where the subscript *cp* denotes the center of pressure; and
   m_x is the pitching moment at the distance *x*.

This equation can be simplified by noting that (1) the lift is very much greater than the drag; and (2) the angle of attack is generally a small angle and therefore $\cos\alpha \cong 1$. The magnitude of the pitching moment about the selected reference point, at a distance *x* from the leading edge, is thus given by the following simplified equation:

$$m_x = l\,(x - x_{cp}) \qquad (3.17)$$

Alternatively, it can be written in coefficient form:

$$C_m = \frac{m_x}{qc^2} = \left(\frac{2l}{\rho V^2 c}\right)\left(\frac{x}{c} - \frac{x_{cp}}{c}\right) = C_l\left(\frac{x}{c} - \frac{x_{cp}}{c}\right) \qquad (3.18)$$

A surprising—and extremely useful—feature of airfoils is that the moment m_x about one unique reference point on the chord is essentially independent of angle of attack variation. This point is called the *aerodynamic center* (AC). For all other reference points along the chord, the pitching moment will change as the angle of attack (and lift coefficient) varies, but at the aerodynamic center the pitching moment is essentially constant. The aerodynamic center is thus defined as the position that satisfies the following condition:

$$C_{m_{ac}} = C_l\left(\frac{x_{ac}}{c} - \frac{x_{cp}}{c}\right) = \text{constant} \qquad (3.19)$$

where the subscript *ac* denotes the aerodynamic center.

Hence $\left(\dfrac{dC_m}{dC_l}\right)_{ac} = 0$ \qquad (3.20)

As the pitching moment results from the asymmetric distribution of pressure around an airfoil, a symmetrical airfoil will have no pitching moment about its aerodynamic center. For an airfoil with positive camber (see Section 3.2.1), the direction of the pitching moment about the aerodynamic center is always nose-down and is therefore negative by definition.

The aerodynamic center for subsonic flow is located close to one-quarter of the chord from the leading edge. This is supported by thin airfoil theory,[8] which shows that, theoretically, for thin symmetrical airfoils $C_{m_{1/4}} = 0$ [1]. For many conventional, cambered airfoils, the aerodynamic center is located between 26% and 28% of the chord from the leading edge.

It is interesting to work backwards from the definition of the aerodynamic center (Equation 3.19) to an interpretation of the position of the center of pressure. Consider a model of a cambered airfoil in a wind tunnel. At zero angle of attack, the lift will not be zero. It will, however, be very small, and the center of pressure will be located a long way aft of the leading edge. As the angle of attack increases, the lift increases; thus the center of pressure must move forwards, reducing the moment arm to keep $C_{m_{ac}}$ constant. Furthermore, it can be deduced that at the specific angle of attack that produces zero lift (which will be a negative angle for an airfoil with conventional positive camber) the center of pressure will be located at infinity (i.e., $x_{cp} = \infty$), and the moment will be the result of a pure couple.

3.4.6 Three-Dimensional Aerodynamic Coefficients

The three-dimensional aerodynamic characteristics of a wing are a function of the airfoil sections that make up the wing profile and the wing geometric parameters (described in Section 3.2.2). Although a wing can be constructed with the same airfoil section along its span, for most designs, different sections are selected for the root and tip, and the intermediate sections are obtained by smoothly blending the profile of one section into the other.

If the wing is divided into a finite number of spanwise strips or segments, it is possible to quantify the two-dimensional properties of each segment. Associated with each segment will be the lift, drag, and pitching moment resulting from the local flow field that acts around that particular airfoil. An appropriate integration of the distributed forces and moments—taking into account wingtip effects—over the span of the wing will result in the gross (or total) lift (L), drag (D), and pitching moment (M). To distinguish these parameters, which relate to the wing as a whole, from the two-dimensional parameters described earlier (in Sections 3.4.2 to 3.4.5), upper case notation is used.

As with the local aerodynamic forces, the lift (L) and drag (D) acting on a wing of finite span can be expressed in coefficient form—these three-dimensional coefficients are defined as

$$C_L = \frac{L}{qS} = \frac{2L}{\rho V^2 S} \tag{3.21}$$

and $$C_D = \frac{D}{qS} = \frac{2D}{\rho V^2 S} \tag{3.22}$$

where S is the wing reference area (typical units: m^2, ft^2).

It is often required that the aerodynamic forces acting on a body are resolved parallel (axially) and perpendicular (normal) to the longitudinal body axes (see Figure 3.9). The axial force is given the symbol A and the normal force the symbol N (caution: taken out of context, this can be confusing as A is often used for aspect ratio or area and N is also used for yawing moment, as described earlier in Sections 3.2 and 3.3). The *normal force coefficient* is defined as

$$C_N = \frac{N}{qS} = \frac{2N}{\rho V^2 S} \tag{3.23}$$

---

8 Thin airfoil theory was proposed by the German aerodynamicist Michael Max Munk (1890–1986) and further developed by, among others, the British aerodynamicist Hermann Glauert (1892–1934), who published a landmark paper on the subject in 1924 [9].

and can be expressed as a function of C_L, C_D, and the body angle of attack (α):

$$C_N = C_L \cos \alpha + C_D \sin \alpha \tag{3.24}$$

Similarly, the three-dimensional pitching moment coefficient is given by

$$C_M = \frac{M}{qS\bar{\bar{c}}} = \frac{2M}{\rho V^2 S \bar{\bar{c}}} \tag{3.25}$$

where M is the pitching moment (typical units: N m, lb ft); and
$\bar{\bar{c}}$ is the mean aerodynamic chord (typical units: m, ft).

3.4.7   Effects of Finite Span on the Lift Curve

The three-dimensional lift-curve slope ($dC_L/d\alpha$) of a wing of finite span will be smaller than the two-dimensional lift-curve slope ($dC_l/d\alpha$) of the airfoil sections that make up the wing (Figure 3.14). This can be explained by considering the flow field in the vicinity of the wing. There is an upwash (or updraft) ahead of the wing and a downwash (or downdraft) behind the wing. The wing itself can be considered to be operating in a flow field that has a downward direction compared to the undisturbed, free-stream flow. (In this context, the free-stream flow can be taken as the flow field one wingspan ahead of the wing leading edge.) The flow behind a three-dimensional wing (i.e., a wing of finite span) has a notably greater downwash compared to that observed behind a two-dimensional airfoil. Consequently, a three-dimensional wing will require a higher angle of attack to generate the same lift coefficient as the two-dimensional airfoil.

This reduction in lift at a given angle of attack depends significantly on the wing planform geometry, in particular the aspect ratio and the sweep of the wing. The lower the aspect ratio, the lower will be the lift at that particular angle of attack. Horizontal tailplanes, for example, have low aspect ratio planforms and consequently tend to have a low lift-curve slope. The lift curve is also a function of the sweep angle: an increase in sweep will reduce the lift at a given angle of attack.

A two-dimensional airfoil experiences no downwash at the quarter-chord location—consequently, the component of drag known as induced, or vortex, drag (discussed in Section 7.3) is zero. The three-dimensional flow field associated with a finite wing is noticeably different: there is downwash here associated with the trailing wake sheet that is produced.

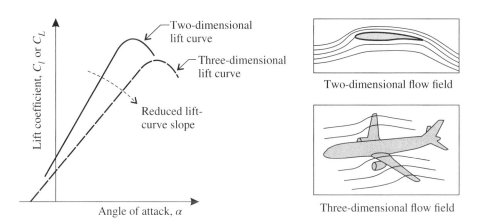

Figure 3.14 Effects of finite span on the lift coefficient versus angle of attack graph.

3.4.8 Aerodynamic Center

The location of the aerodynamic center, with respect to the MAC (defined in Section 3.2.2), is a key parameter for the assessment of the longitudinal trim and stability of an airplane (see Section 19.4).

If it is assumed that the aerodynamic center of the airfoil sections that make up the wing is at 0.25 c (see Section 3.4.5) and the effects of compressibility are ignored, then the aerodynamic center of the wing (excluding the fuselage) will be at 0.25 $\bar{\bar{c}}$. There are other factors that need to be taken into account to determine accurately the aerodynamic center of a jet transport airplane in cruise. By applying the method described in ESDU 70011 [10] to a planar straight-tapered wing equivalent to that found on contemporary airliners (such as the B787 or the B747-8),[9] it can be shown that the aerodynamic center of the wing in cruise is located at approximately 0.41 $\bar{\bar{c}}$. The position shifts forwards to approximately 0.32 $\bar{\bar{c}}$ when the influence of the fuselage is included [11].

3.4.9 Aerodynamic Effect of a Plain Flap

A plain flap is a hinged portion of the trailing edge which, when deflected, locally increases (or decreases) the camber and hence increases (or decreases) the lift at a given angle of attack (Figure 3.15). In simple terms, control surfaces, such as ailerons, elevators and rudders, function as plain flaps. The bulk of the lift increase is achieved by changing the pressure distribution on the forward (fixed) portion of the aerodynamic surface. Flow separation on the flap itself usually occurs at modest angles of flap deflection ($10°-15°$), but the lift coefficient continues to increase until separation occurs on the forward part of the airfoil (typically at a flap deflection of about $40°-50°$).

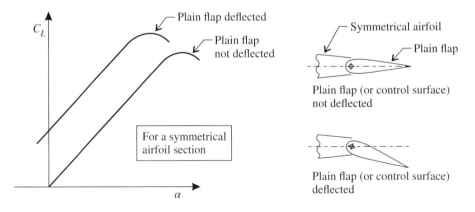

Figure 3.15 Effect on airfoil lift (symmetrical section) of the deflection of a plain flap.

3.4.10 Aerodynamic Effect of a Slot in Multi-Element Airfoils

A slot, which is a gap in an airfoil section, is a common feature of multi-element, high lift wing designs used on jet transport airplanes (Figure 3.16). Gaps separate elements of extended trailing edge flaps (see Section 3.7.2) and certain leading edge slat types (see Section 3.7.3). The

---

9 The relevant wing parameters are $A = 8.20$, $\lambda = 0.19$, and $\Lambda_{0.5c} = 35°$, and the cruise Mach number is 0.85.

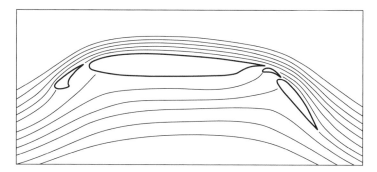

Figure 3.16 Streamlines around an airfoil with a leading edge slat (with a slot) and double-slotted Fowler flap. *Source*: Smith [12]. Reproduced with permission from the American Institute of Aeronautics and Astronautics.

gap permits air to flow from the higher-pressure region below the airfoil to the lower pressure above the airfoil. This movement of air has a profound effect on the pressure distributions seen on both the upstream and downstream airfoil elements (discussed in detail by Smith [12]). In particular, the pressure peak on the downstream element is reduced, which has a favorable influence on flow separation. A gap also means that a fresh boundary layer starts at the leading edge of each airfoil segment; consequently, as thin boundary layers can withstand greater adverse pressure gradients than thick ones, upper surface flow separation is further delayed. This can permit higher angles of attack to be achieved.

3.5 Compressibility

3.5.1 Incompressible Flow

The term *incompressible* is traditionally used in the study of fluid dynamics to describe a flow in which appreciable changes in fluid density do not occur. Air, like all gases, is easily compressed. However, at low Mach numbers, air behaves in a way that mimics an incompressible fluid. What this means, in reality, is that the density changes that occur in airflows at low speed have a negligible influence on the behavior of the airflow. Aerodynamic equations used to describe airflow outside of the boundary layer are dominated by velocity and pressure terms. This observation facilitates accurate mathematical descriptions of low-speed airflow to be achieved with the assumption that the density of the air remains constant along streamlines. The widely used Bernoulli equation for incompressible flow (Equation 2.63) is an example of such an expression.

For Mach numbers of less than 0.3—which corresponds to a velocity of 102 m/s (198 kt or 335 ft/s) at standard sea level conditions—the assumption of incompressibility introduces negligible errors in the determination of flow properties.

3.5.2 Compressible Flow

When considering progressively greater Mach numbers, however, this assumption that air does not undergo significant compression becomes progressively less valid, and a point is soon reached where density changes need to be considered. The Mach number at which this occurs depends on the specific application and on the desired level of accuracy. For example, when comparing the difference between the total pressure predicted using the Bernoulli equation for incompressible flow and that given by Equation 2.82 for Mach 0.5, it is evident that the discrepancy is just less than 1%. For certain applications, this may be acceptable, but not for others (such as airspeed measurement—see Sections 6.6 and 6.7).

The term *compressible flow* is used to describe flow regimes in which density changes are considered. Associated with these density changes are thermodynamic changes, which affect the local air temperature, viscosity, and thermal conductivity.

3.5.3   Shocks and Transonic Flow

The influence of flow compressibility (i.e., density and thermodynamic changes) on the behavior of the airflow as it approaches the speed of sound is so profound that it significantly alters the character of the flow. Disturbances in the flow propagate in a very different way compared to low Mach number flow conditions. Furthermore, the three-dimensional flow fields around an airplane's wing at these speeds are exceedingly complex with non-linear characteristics, and analytical solutions are only possible for very simple, idealized cases [13].

With progressively increasing free-stream Mach number, the local flow will become sonic at some point on the airplane. Modern jet transport airplanes are designed so that this occurs on the forward portion of the upper wing surface. The free-stream Mach number at which localized regions of supersonic flow first appear is known as the *critical Mach number* (M_{cr}). This represents the start of the transonic flow regime. The term *transonic* simply means that there is a mix of subsonic and supersonic flow in the flow field—and for jet transport airplanes this typically occurs at free-stream Mach numbers greater than about 0.65. The critical Mach number depends significantly on the design of the wing (the key design features are the wing sweep, airfoil type, and thickness-to-chord ratio).

A further increase in free-stream Mach number produces a local region of supersonic flow over a portion of the upper surface of the airfoil, as illustrated in Figure 3.17 (and at even greater Mach numbers, a supersonic flow region will also appear on the lower wing surface). As the free-stream flow is subsonic, the airflow has to slow down and become subsonic again before reaching the trailing edge. This deceleration almost always occurs abruptly through a shock, which is accompanied by rapid compression of the air. A feature of a shock, or *shock wave*, is that the flow properties—that is, the air velocity, pressure, density, and temperature—all change abruptly over an exceedingly short distance (sometimes described as a sheet). As the air passes through the shock, it undergoes rapid thermodynamic changes (there is an increase in entropy) and a sudden loss in total pressure. The magnitudes of these changes depend on the nature of the shock (e.g., multiple weak shocks suffer smaller total pressure losses than a single strong shock for comparable flow conditions).

From an airplane performance perspective, an important aspect in the study of transonic flow is the compressibility drag (or *wave drag*) that is associated with shock waves. As the speed increases, the shocks strengthen and the drag increases rapidly—this is described later in

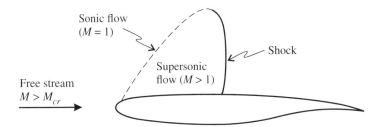

Figure 3.17 Transonic flow around an airfoil, illustrating a region of supersonic flow on the upper surface.

Section 7.4.4. A key feature in the design of modern transonic airplanes is the *supercritical* airfoil (this term refers to the fact that the drag rise is delayed well beyond the critical Mach number). By tailoring the airfoil shape, the position and behavior of the shock can be controlled in a way that minimizes the drag increase, while maintaining the desired pressure distribution over the airfoil.

3.6   Boundary Layers

3.6.1   General Features of the Boundary Layer

From an idealized global perspective, the flow field around a body at high Reynolds number can be described as being made up of two parts: (1) a relatively thin *boundary layer* adjacent to the surface of the body within which the viscosity of the air is very important; and (2) an outer flow field within which the viscosity of the air affects the flow to such a minor extent that it can be assumed to be inviscid (i.e., having zero viscosity).[10] With this idealization, the boundary layer—which tends to behave in a very different way to the outer flow field—can be described by equations and theories that apply specifically to the exceedingly complex flow phenomena that occur within this thin layer of air that surrounds the airplane.[11] The description of the boundary layer over an aerodynamic surface, such as an aircraft's wing, is often presented by considering the wing to be stationary and the flow moving in a time-dependent manner over the surface.

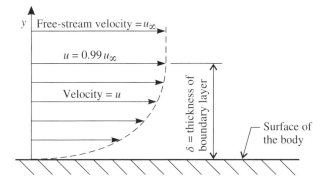

Figure 3.18 Velocity profile in the boundary layer.

For almost all practical aerodynamic conditions (the exception being high altitude flight where the air density is very low), it may be assumed that there is no relative motion between the air molecules that are immediately adjacent to the surface and the surface itself. This is known as the *no-slip* condition, and it has been shown to be an accurate representation of real time-averaged flow conditions. The viscosity of the air (see Section 2.5.12) generates shearing stresses within infinitesimally thin layers of air within the boundary layer, resulting in a velocity profile (illustrated in Figure 3.18) in which the surface velocity is zero, and at some finite

---

10 The idea of dividing the flow field around a body in this way was first proposed by Ludwig Prandtl in 1904 [14], and continues to be used today, albeit with extensive revisions to the equations used to represent the flow. An exception to this subdivision of the global flow field based on viscosity is the downstream wake, the behavior of which is dominated by vorticity, which is the result of the fluid viscosity.

11 A detailed treatment of boundary layer theory is presented by Schlichting and Gersten [15], for example.

distance from the surface, the velocity is equal to the free-stream velocity (u_∞). In practice, the thickness of the boundary layer (δ) is measured from the surface to the point where the velocity reaches 99% of the free-stream velocity.

As the air moves along the surface of the body, the thickness of the boundary layer increases considerably. This increase is influenced by the pressure gradient—an adverse pressure gradient, for example, as would be encountered on the aft portion of an airfoil, would cause a rapid thickening of the boundary layer.

The thickness of the boundary layer (and the presence of viscous wakes) affects the contour around which the inviscid outer flow moves. One way of looking at this is to consider that the boundary layer displaces the inviscid flow outwards away from the body—that is, relative to what would happen in a hypothetical inviscid flow field. This effect, which is called *boundary layer displacement*, changes the effective shape of the body as encountered by the inviscid outer flow. Boundary layer displacement is known to have a big influence on the pressure distribution over airfoils in transonic flow conditions; it also contributes significantly to the drag that acts on the body.

3.6.2 Transition and Turbulent Boundary Layers

Investigations into boundary layer behavior reveal two distinct flow types: laminar and turbulent. A laminar boundary layer is characterized by smooth and uniform flow trajectories, without appreciable mixing between the layers of air. In a turbulent boundary layer, the flow is not uniform: eddies (vortex flow) mix the air and bring relatively high velocity air from the outer regions of the boundary layer towards the surface. This results in higher mean flow velocities near the surface and an increase in the boundary layer thickness.

Experiments conducted on a wing without appreciable leading-edge sweep reveal that the flow is laminar near the leading edge and that the boundary layer thickness (δ) increases with distance along the surface (Figure 3.19). At some point downstream of the leading edge, the flow in the boundary layer usually becomes unstable and a transition from laminar to turbulent flow occurs. The point of transition is influenced by several factors, including the free-stream Mach number, leading-edge-sweep angle, airfoil shape, free-stream turbulence level, and surface roughness. The boundary layer that exists on wings with appreciable leading-edge sweep (as found on most current jet transport airplanes), however, does not feature an initial run of

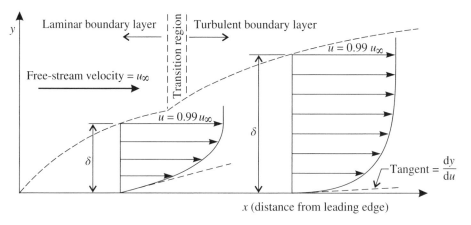

Figure 3.19 Mean (time-averaged) velocity profile in laminar and turbulent boundary layers.

laminar flow; rather, due to flow instabilities that exist at the leading edge associated with the spanwise flow component (known as crossflow), the boundary layer is always turbulent.

3.6.3 Skin Friction

Shear stresses exist in the boundary layer due to the viscosity of the air (μ). The magnitude of the shear stress at a particular location in the flow depends on the rate of change of velocity (i.e., $\mathrm{d}u/\mathrm{d}y$) at that point. The shear stress on the surface (τ_0) can be obtained by setting $y = 0$ in Equation 2.89; thus:

$$\tau_0 = \mu \left(\frac{\mathrm{d}u}{\mathrm{d}y} \right)_{y=0} \tag{3.26}$$

An integration of the shear stresses over the surface of the body is an expression of the *skin friction drag* (see Section 7.3). Laminar boundary layers have a considerably lower skin friction drag compared to turbulent boundary layers; this is illustrated in Figure 3.19, where the velocity gradient at the surface (i.e., $\mathrm{d}u/\mathrm{d}y$) is lower in the case of laminar flow compared to turbulent flow.

3.6.4 Boundary Layer Separation

Boundary layer separation is an important feature of the behavior of a viscous fluid flowing over an aerodynamic surface, such as an aircraft's wing. The air traversing the upper surface of an airfoil initially experiences a decrease in pressure (i.e., a favorable pressure gradient) followed by a region of increasing pressure (i.e., an adverse pressure gradient), a feature known as *pressure recovery*. Due to the viscosity of the air, the boundary layer initially remains attached (i.e., in contact with the airfoil surface) as it moves into this region of rising pressure. The velocity profile within the boundary layer, however, starts to change from the point that an adverse pressure gradient is encountered, with a progressive reduction in the velocity gradient at the surface. Faced with steadily increasing pressure, the boundary layer can detach, creating a separated shear layer away from the airfoil surface. Flow over an airfoil will frequently separate before the trailing edge is reached—the location depending on several factors, including the angle of attack. A turbulent boundary layer can withstand a greater pressure rise than a laminar boundary layer—in other words, laminar boundary layers are less resistant to separation under similar flow conditions than turbulent boundary layers.

Downstream of the point of separation, there will be a region of reverse flow at the surface and pockets of recirculating air near to the surface. Within two-dimensional flow, the point of separation can be defined as the location at which the velocity gradient at the surface is zero (and hence the skin friction is also zero). In three-dimensional flow, the separation phenomenon is more complex, and velocity components in both the streamwise and crossflow directions are considered in defining the line of separation on a wing [13].

3.7 High Lift Devices

3.7.1 Requirement for High Lift

Wings optimized for cruise (with moderate-to-high wing loading and little camber) perform badly at the slow speeds associated with takeoff and landing, where the requirements are for low wing loading and high lift coefficients (which can be obtained by having large amounts of camber). This fundamental incompatibility is the main reason why high lift devices—that is, leading and trailing edge devices—are needed. These devices aid in producing the required lift at slow speeds by reconfiguring the airfoil shape and by introducing slots between the airfoil segments.

There are three principal ways in which high lift devices increase lift (at a given angle of attack); these are:

(1) increasing the wing camber (see Section 3.4.9);
(2) increasing the wing chord, thereby increasing the wing area (since the lift coefficient is referenced to the original wing area, the new lift-curve slope is increased by approximately the ratio of the extended wing area to the original wing area); and
(3) improving the state of the boundary layer, thereby delaying flow separation.

Brief descriptions of trailing and leading edge devices are presented in Sections 3.7.2 and 3.7.3, respectively.[12]

3.7.2 Trailing Edge Flaps

Trailing edge flaps increase the lift coefficient at any given angle of attack. An increase in lift coefficient means that the wing can generate the same lift at slower speeds. Flaps significantly increase the maximum lift coefficient by increasing the wing camber and, in the case of Fowler flaps, by increasing the wing chord (Figure 3.20). Flaps also permit the wing to generate the required lift (to support the airplane's weight) at lower angles of attack. This can be advantageous during landing, as it will reduce the nose-up attitude of the airplane, thereby increasing the pilot's forward visibility (i.e., over the nose of the airplane). The increase in camber, however, results in an increase in nose-down pitching moment, which must be balanced by a down load on the tailplane (achieved by deflecting the elevator and/or the horizontal stabilizer).

A plain flap works by increasing camber and thus increasing lift (as discussed in Section 3.4.9). On jet transport airplanes, flap systems are slotted (that is to say that there is no seal between

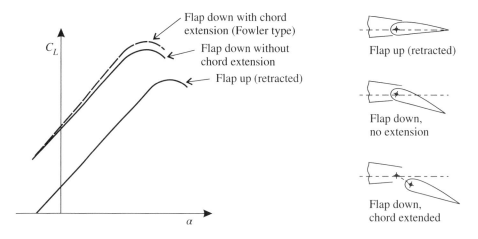

Figure 3.20 Effect of flap extension on lift.

---

12 More detailed reviews of high lift devices, which include descriptions of their aerodynamic and mechanical designs, are presented by Rudolph [16, 17] and van Dam [18], for example.

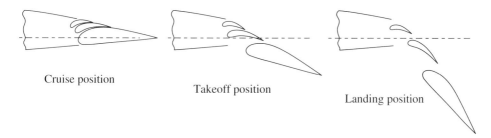

Figure 3.21 Three-segment Fowler flap in cruise, takeoff, and landing positions.

the flap and the main surface), which delays flow separation, increases lift, and reduces drag. Fowler type flaps are slotted flaps that are mechanized to slide rearwards as they are deflected (Figure 3.21). This increases the wing area as well as the camber. To further improve airflow over the flap, double and triple slotted flaps are used on some airplane types. These flaps greatly increase lift, but there is an increase in weight and mechanical complexity.

The principal disadvantage of trailing edge flaps is that they intensify the suction peak on the leading edge of the main wing, thereby increasing the possibility of flow separation. This tendency can be controlled by installing a leading edge device (see Section 3.7.3).

3.7.3 Leading Edge Devices

A leading edge device is a hinged and/or translating portion of the nose section of the wing, which droops downwards to increase the curvature of the upper wing surface and increase the wing chord. On commercial jet transport airplanes, two types of leading edge device are widely used: slats and Krueger (Krüger) flaps.

It is evident from Figure 3.20 that trailing edge flaps, alone, do not increase the stall angle. In fact, they tend to reduce the stall angle by increasing the pressure gradient over the top of the airfoil, which tends to promote flow separation. By incorporating a device that modifies the flow around the leading edge, higher angles of attack and a further increase in the maximum lift coefficient can be achieved (Figure 3.22). Thus, leading edge devices, alone, do little to improve

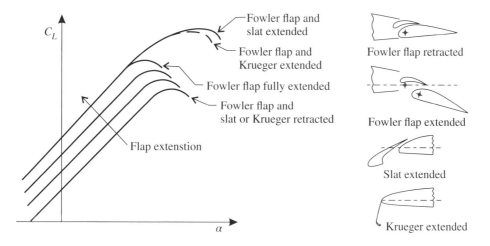

Figure 3.22 Effect of a Fowler flap and alternative leading edge devices on lift coefficient.

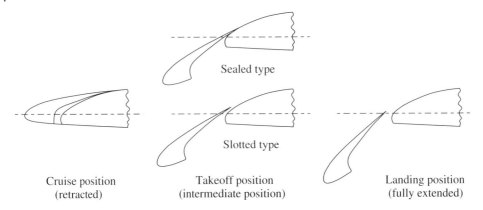

Sealed type

Slotted type

Cruise position   Takeoff position   Landing position
(retracted)   (intermediate position)   (fully extended)

Figure 3.23 Three-position slat (illustrating both sealed and slotted design types in the intermediate position).

lift at takeoff or landing; however, they are very effective when used in combination with a trailing edge flap.

In the retracted position, the slat forms the leading edge of the wing (Figure 3.23). As the slat moves outwards, pushed along a track by an actuator, the wing camber and wing area increase. Although slats are complicated mechanical devices—requiring tracks (rails), rollers, and actuators—they have been used on most jet transport airplanes. Early jet airliners (such as B727, MD-11, and several DC9 models) used two-position slats (retracted for cruise and fully extended for landing), but all newer airplane types use three-position slats (with an intermediate position for takeoff). In the landing position, the slat almost always has a slot, but in the takeoff position, both slotted and sealed designs are used.

There are several aerodynamic changes that arise when a slat is extended. The suction peak on the main wing section is reduced, resulting in a reduction in the adverse pressure gradient and a delay in the onset of stall. The state of the boundary layer is improved; in the case of a slotted design, the boundary layer starts afresh at the leading edge of the main wing. In addition, the lift-curve slope is increased due to an increase in the total wing area (as illustrated in Figure 3.20 for a Fowler flap).

The most common type of Krueger flap is a simple hinged panel that is rotated outwards from a recess in the lower nose portion of the wing (Figure 3.24). The leading edge of the flap can itself incorporate a small retractable nose section. The second type translates forwards and rotates into position by a linkage mechanism, and, in the process, an increase in the flap curvature is achieved (this type of variable-camber Krueger flap is used on the outboard wing panels of the B747). When deployed, an increase in the wing chord and nose curvature results.

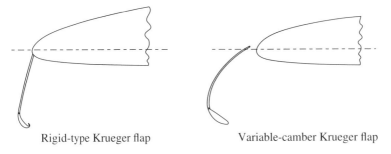

Rigid-type Krueger flap   Variable-camber Krueger flap

Figure 3.24 Two types of Krueger flap used on jet transport airplanes.

Krueger flaps have two positions: retracted for cruise and fully extended for takeoff and landing. Consequently, takeoff performance with a Krueger flap design can be inferior to that with a three-position slat (there are, of course, many other factors to consider during the airplane design process, such as engine thrust). They are, however, lighter, simpler devices compared to slats. Krueger flaps have been used on several airplane types (e.g., on the inboard wing panel of the B707, B727, B737, and B747, and on the outboard wing panel of the B747) [18].

3.7.4   Lift-to-Drag Ratio

The aerodynamic effect of leading and trailing edge high lift devices can be seen by considering the airplane's lift-to-drag ratio (L/D). As the flaps are extended, the lift increases for a given angle of attack (see Figure 3.22). The drag also increases, but at a different rate. At small flap angles, there is an increase in the maximum lift coefficient, which is advantageous for takeoff. At large flap angles, the maximum lift coefficient is further increased, but there is also a significant increase in the drag coefficient. Whereas this is undesirable for takeoff, it is advantageous for landing (as the airplane needs to slow down). The lift-to-drag ratio thus decreases with increasing flap settings at a given reference speed, as illustrated in Figure 3.25 by the line representing flight at 1.3 times the stall speed (V_S).

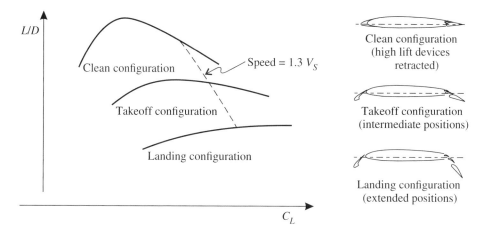

Figure 3.25 Lift-to-drag ratio versus lift coefficient for high lift devices in the retracted (i.e., clean), takeoff, and landing positions.

3.8   Controls for Pitch, Roll, and Yaw

3.8.1   Three-Axes Control

Control of the direction of flight is achieved by the pilot deflecting the airplane's control surfaces (e.g., elevator, aileron, rudder) using the cockpit flight controls (or inceptors). The simplest control surface is essentially a non-extending flap, hinged to the trailing edge of the main surface. By increasing the camber of the airfoil section an increased force is generated on both the hinged and the "fixed" (forward) portion of the aerodynamic surface (e.g., wing, vertical or horizontal tailplane). This force creates a moment about the airplane's center of gravity, which can result in a change in the airplane's flight path.

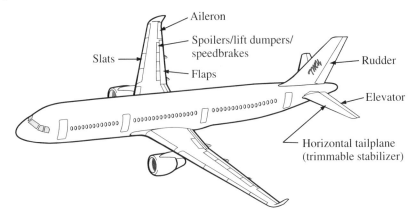

Figure 3.26 Primary flight control surfaces.

The conventional three-axes control system on an airplane has elevators, ailerons, and a rudder, which are primarily responsible for rotating the airplane about the X (pitch), Y (roll), and Z (yaw) axes, respectively (Figure 3.26). The situation is, however, much more complicated than this simple description implies (pitch, roll, and yaw control is described in Sections 3.8.3 to 3.8.5). There exists a significant degree of coupling between the different controls, particularly in roll and yaw. This is illustrated by the following two examples:

(1) When the ailerons are deflected they produce a rolling moment; however, they can also produce a yawing moment, which arises, primarily, from an increase in the induced drag on the upgoing wing as that wing produces more lift.
(2) When the rudder is deflected it produces a yawing moment, but it also produces a rolling moment as the center of pressure of the vertical stabilizer is located high above the centerline of the fuselage.

3.8.2 Pilot's Primary Flight Controls

Deflection of the control surfaces is achieved by means of the pilot's flight controls. Early airplanes had a stick (or joystick) for pitch and roll control. On modern jet airplanes the stick has been replaced by a yoke or a side-stick (a diminutive control stick located on the side console of the cockpit).

Pitch control is achieved by a fore/aft movement of the stick/yoke and roll control by a sideways movement of the stick or by rotating the yoke. Yaw control is achieved by pressing a rudder pedal (pressing the right pedal causes the nose to yaw to the right). Dual controls (which provide a measure of redundancy) are standard on commercial transport airplanes, that is, for the captain (customarily seated on the left), and for the first officer (F/O) or co-pilot (seated on the right).

3.8.3 Pitch Control

Elevator
Pitch control on a conventional airplane is primarily achieved by deflecting the elevators, which are flapped segments of the horizontal tailplane.

Horizontal Stabilizer

On jet transport airplanes, the entire horizontal stabilizer can be rotated nose-up/-down to trim (or balance) the airplane. This is known as a trimmable horizontal stabilizer (THS). Movement of the stabilizer through a relatively small angle, when compared with that required by the elevator, can generate the required pitching moment. This approach will usually result in a reduction in the total drag generated in cruise when compared with use of the elevator alone.

3.8.4   Roll Control

Ailerons

Ailerons—flapped segments on the outboard wing surface—are the primary control surfaces for lateral (roll) control at low speed. Aileron deflection can produce an undesirable yawing moment (see Section 3.8.1). This effect is relatively small and at low angle of attack it is of little consequence; however, at high angle of attack, when the airplane's directional stability is low, it can produce a rapid yawing motion. Differential aileron control—where the upward moving aileron deflects more than the downward aileron—can be employed to counter this problem.

Auxiliary Ailerons

An airplane's rolling ability at high speed, due to aileron deflection, is affected by compressibility and aeroelastic distortion of the wing. Historically, high-speed airplanes have experienced a phenomenon known as aileron reversal, in which the air loads placed on the deflected aileron were so great that the wing twisted to such a degree that the rolling moment induced by the twist exceeded the rolling moment due to the aileron deflection—causing the airplane to roll the wrong way. This is countered in the design of modern aircraft by increasing the wing torsional stiffness or through the use of auxiliary inboard ailerons at high speed.

Spoilers

Spoilers are hinged surfaces located forward of the flaps on the upper wing. Spoilers may be used to augment roll control at low speed—when deflected asymmetrically they cause a large rolling moment as a result of differential loss of lift. They can also be used symmetrically to increase drag as speedbrakes (drag brakes) during the approach, and to reduce lift (as lift dumpers) after touchdown increasing the braking efficiency.

3.8.5   Yaw Control

The rudder is the primary control surface for yaw control. Deflection of the rudder produces camber on the vertical stabilizer, thus generating a side force, which, in combination with the fin, will lead to the development of a sideslip angle. On many large aircraft types, the rudder is split into segments. This can be used to balance the roll/yaw coupling; it also provides control redundancy.

   Note that the rudder is not the prime control for changing the airplane's heading: the ailerons are used to roll the airplane into a banked turn, and, in this way, change the direction of flight (maneuvering flight is discussed in Chapter 15).

3.8.6   Trim Systems

A trim tab is a small secondary control surface hinged to the trailing edge of the primary control surface. On a light aircraft with a manual flight control system, the pitch control force exerted by the pilot on the stick/yoke to hold the airplane in equilibrium depends on the angle of deflection of the elevator trim tab (tab deflections alter the aerodynamic moment about the elevator hinge, and this is counteracted by the pilot applying a force to the stick/yoke). On high-speed aircraft, which have powered flight controls, the trimming procedure for the pilot can be the same—that is, he/she rotates a wheel or uses an electrical switch to deflect the pitch trim control surface. This also changes the control force on the stick/yoke; however, the force felt by the pilot is artificially generated and is not the result of the pilot counteracting the aerodynamic force through a mechanical control system.

On jet transport airplanes, pitch trim, as mentioned earlier, is achieved by changing the pitch angle of the horizontal stabilizer. Control of the stabilizer position is effected by use of an electrical trim switch on the stick/yoke or by rotating the trim wheel. When the stabilizer is moved to the "trimmed" position, the elevator angular deflection will be zero, thereby negating the need for a stick/yoke control input by the pilot. Some modern airplane types have auto-trim functions in which the movement of the stabilizer is triggered automatically by a sustained deflection of the elevator.

3.8.7 Flight Control Systems

On light aircraft, the flight controls are manual and the pilot provides the force to balance the air loads that act on the deflected control surface. A series of mechanical push/pull rods, bellcranks, and pulleys typically connect the control stick/yoke and the pedals to the control surfaces. In such a mechanical flight control system, the pilot directly moves the ailerons, elevators, and rudder via these linkages. The forces felt by the pilot in flying the airplane are thus a function of the gearing of the flight control system (and, of course, the free-stream dynamic pressure).

On high-speed aircraft, the control forces are too great for a human pilot, and hydraulic actuators provide the "muscle" to move the control surfaces. Signals from the pilot's controls can be transmitted mechanically (by cable) or electrically (known as *fly-by-wire*), with the movements of the actuators corresponding to the control inputs of the pilot.

With a fly-by-wire system, there is no direct proportional relationship between the pilot's input and the resulting control surface movement. The input signals are modulated by computer software; this approach can improve the airplane's handling characteristics and provide

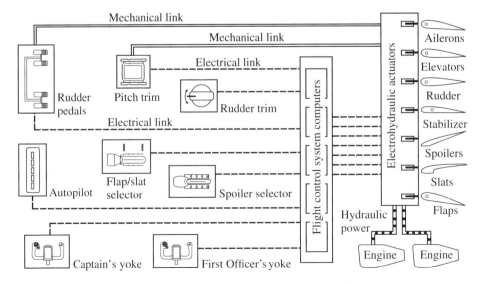

Figure 3.27 Schematic of a typical flight control system for a modern fly-by-wire jet transport airplane.

protection against undesirable flight conditions, such as stall, overspeed, and excessive maneuver loads. The architecture of a typical fly-by-wire system is illustrated in Figure 3.27. Note that in this particular system (selected for illustration) two of the controls have a mechanical backup: the rudder and pitch trim.

On airplanes with powered flight controls, the system is irreversible, and the pilot is given a synthetic, or artificial, feel of the air loads acting on the control surfaces by an artificial feel unit.

3.9 Further Reading

Textbooks that expand on the topics introduced in this chapter include:

Anderson, J.D., *Fundamentals of aerodynamics*, 5th ed., McGraw-Hill, New York, NY, 2010.

Bertin, J.J. and Cummings, R.M., *Aerodynamics for engineers*, 6th ed., Pearson Education, Harlow, UK, 2013.

Clancy, L.J., *Aerodynamics*, John Wiley & Sons, New York, NY, 1975.

Houghton, E.L., Carpenter, P.W., Collicott, S.H., and Valentine, D.T., *Aerodynamics for engineering students*, 7th ed., Butterworth-Heinemann, Cambridge, MA, 2017.

Lan, C.-T.E. and Roskam, J., *Airplane aerodynamics and performance*, DARcorporation, Lawrence, KS, 2010.

McCormick, B.W., *Aerodynamics, aeronautics, and flight mechanics*, 2nd ed., John Wiley & Sons, New York, NY, 1995.

McLean, D., *Understanding aerodynamics: Arguing from the real physics*, John Wiley & Sons, New York, NY, 2012.

Obert, E., *Aerodynamic design of transport aircraft*, IOS Press, Amsterdam, the Netherlands, 2009.

Schlichting, H. and Gersten, K., *Boundary-layer theory*, 8th ed., Springer-Verlag, Heidelberg, Germany, 2014.

Shevell, R.S., *Fundamentals of flight*, 2nd ed., Prentice Hall, Englewood Cliffs, NJ, 1989.

Torenbeek, E. and Wittenberg, H., *Flight physics: Essentials of aeronautical disciplines and technology, with historical notes*, Springer, Heidelberg, Germany, 2009.

References

1 Abbott, I.H.A. and Von Doenhoff, A.E., *Theory of wing sections, including a summary of airfoil data*, Dover Publications, New York, NY, 1959.

2 Abbott, I.H.A., von Doenhoff, A.E., and Stivers, L.S., Jr., "Summary of airfoil data," NACA Report No. 824, National Advisory Committee for Aeronautics, Langley, VA, 1945.

3 ESDU, "Geometrical properties of cranked and straight-tapered wing planforms," Data item 76003, Amdt. C, IHS ESDU, 133 Houndsditch, London, UK, May 2012.

4 Blake, W. and Performance Training Group, "Jet transport performance methods," D6-1420, Flight Operations Engineering, Boeing Commercial Airplanes, Seattle, WA, Mar. 2009.

5 Oswatitsch, K. and Wieghardt, K., "Ludwig Prandtl and his Kaiser-Wilhelm-Institut," *Annual Review of Fluid Mechanics*, Vol. 19, pp. 1–25, 1987.

6 Anderson, J.D., *A history of aerodynamics and its impact on flying machines*, Cambridge University Press, Cambridge, UK, 1997.

7 Obert, E., *Aerodynamic design of transport aircraft*, IOS Press, Amsterdam, the Netherlands, 2009.

8 Anderson, J.D., *Introduction to flight*, 6th ed., McGraw-Hill, New York, NY, 2008.

9 Glauert, H., "A theory of thin aerofoils," Aeronautical Research Committee, Reports & Memoranda No. 910, Royal Aircraft Establishment, London, UK, Feb. 1924.

10 ESDU, "Lift-curve slope and aerodynamic centre position of wings in inviscid subsonic flow," Data item 70011, Amdt. J, IHS ESDU, 133 Houndsditch, London, UK, Mar. 2012.

11 ESDU, "Aerodynamic centre of wing-fuselage combinations," Data item 76015, Amdt. C, IHS ESDU, 133 Houndsditch, London, UK, Sept. 2012.

12 Smith, A.M.O., "High-lift aerodynamics," *Journal of Aircraft*, Vol. 12, Iss. 6, pp. 501–530, 1975.

13 McLean, D., *Understanding aerodynamics: Arguing from the real physics*, John Wiley & Sons, New York, NY, 2012.

14 Prandtl, L., "Über Flüssigkeitsbewegung bei sehr kleiner Reibung," *Third International Congress of Mathematics*, Heidelberg, Germany, Aug. 8–13, 1904. [English translation: "Motion of fluids with very small viscosity," published as NACA TM-452, National Advisory Committee for Aeronautics, USA, 1928.]

15 Schlichting, H. and Gersten, K., *Boundary-layer theory*, 8th ed., Springer-Verlag, Heidelberg, Germany, 2014.

16 Rudolph, P.K.C., "High-lift systems on commercial subsonic airliners," NASA CR-4746, National Aeronautics and Space Administration, USA, 1996.

17 Rudolph, P.K.C., "Mechanical design of high lift systems for high aspect ratio swept wings," NASA CR-1998-196709, National Aeronautics and Space Administration, USA, 1998.

18 van Dam, C.P., "The aerodynamic design of multi-element high-lift systems for transport airplanes," *Progress in Aerospace Sciences*, Vol. 38, Iss. 2, pp. 101–144, 2002.

4

Atmosphere and Weather

4.1 Introduction

The Earth's atmosphere, an envelope of air surrounding our planet, is the domain in which all aircraft operate. It is a dynamic, ever-changing environment that needs to be understood, described, and mathematically modeled, so that accurate and reliable aircraft performance measurements and calculations can be conducted.

The composition and movement of the fluid—that is, the air—around an aircraft affects the aerodynamic forces that are generated and the resulting motions of the aircraft. The normal composition (percentages, by volume) of clean, dry air at sea level is as follows: 78.08% nitrogen, 20.95% oxygen, 0.93% argon, 0.03% carbon dioxide, and trace amounts of several other gases (with a total of less than 0.01%)—these include neon, helium, hydrogen, methane, ozone, nitrogen oxide, and carbon monoxide [1]. Due to the constant mixing of air in the lower atmosphere, these ratios are approximately constant up to about 90 km. Not included in this breakdown are dust and other solid particles, and water vapor. The amount of water vapor is highly variable—it is estimated to be approximately 0.41% of the total volume, on average [1]. Local variations in the composition occur—for example, near to large industrial cities, where larger quantities of carbon monoxide and sulfur dioxide are found, or over deserts, where the water vapor content can be extremely low.

Mathematical models that describe the key properties of the atmosphere are explained in Sections 4.2 and 4.3. These models consider the air to be stationary, with idealized variations of air temperature, pressure, and density with increasing height. The real atmosphere (discussed in Sections 4.4 to 4.6) is dynamic, of course, and pilots encounter widely varying conditions, which can be quite different from those described by the idealized models.

4.2 International Standard Atmosphere

4.2.1 Introduction to Standard Atmospheric Models

Standardized models of the atmosphere have been developed—and periodically revised—since the 1920s.[1] In aviation, reference is frequently made to three models: the International Standard Atmosphere (ISA) [3], the US Standard Atmosphere (1976),[2] and the International Civil

---

1 Reference [2] contains a detailed survey of standard and reference atmospheric models.

2 The US Standard Atmosphere was first published in 1958, and subsequently updated in 1962, 1966, and 1976. The 1976 US Standard Atmosphere, which is fully described in NASA TM-X-74335 / NOAA S/T-76-1562 [1], is identical to the International Standard Atmosphere (ISA) to a height of 50 km.

Performance of the Jet Transport Airplane: Analysis Methods, Flight Operations, and Regulations, First Edition.
Trevor M. Young.

Aviation Organization[3] (ICAO) Standard Atmosphere [4]. The three models are identical up to a height of 32 km, which adequately covers the region of interest for commercial airplane operations. For the sake of brevity and consistency, the ISA (as published by the International Organization for Standardization [3]) is used exclusively in this book.

The ISA is an idealized mathematical model of the Earth's atmosphere, which is universally used for airplane performance analysis and airplane operations. It describes a hypothetical vertical distribution of air temperature, pressure, and density—and this greatly simplifies analytical or numerical analyses of aircraft flight. In very simple terms, the ISA is a linearized approximation of the prevailing conditions that exist at the temperate latitudes of the Northern Hemisphere at about 45°N. The air in the ISA is assumed to be dry (i.e., devoid of water), clean (i.e., devoid of solid particles), and still (i.e., stationary with respect to the Earth).

The ISA model can be used to generate data tables, in which the key properties of the air are described as a function of height (see Appendix A). The model is based on agreed initial values of temperature and pressure at a sea-level datum; density is calculated by assuming that the air is an ideal gas (see Section 2.5.6). The ISA is explicitly defined in SI units. Equivalent values in FPS units are obtained by conversion [5]; published FPS values can thus be subject to small rounding error discrepancies (i.e., at the fifth or sixth significant figure). When calculating atmospheric properties using the governing equations (herein derived) it is advisable to use a large number of significant figures for the constants (recorded later in Tables 4.1 and 4.2).

Commercial aviation takes place in the two lower regions of the atmosphere: the troposphere and the lower stratosphere. In the ISA, these regions are defined as follows:

- The *troposphere* extends from the ISA sea-level datum to the *tropopause*, a boundary that occurs at a defined geopotential height[4] of 11 000 m (36 089.24 ft).
- The lower *stratosphere* is the region above the tropopause up to a geopotential height of 20 000 m (65 616.80 ft).

4.2.2 Principal Features of the ISA

ISA Datum
The ISA sea-level datum is the zero height reference elevation in the ISA. The term *sea level*, in this context, is used in an imprecise manner, as the datum is not a physical datum, but rather a reference elevation defined in terms of a reference pressure (i.e., 1013.25 hPa). This means that the ISA datum at a particular geographical location will seldom coincide exactly with the actual mean sea level (MSL). For most days in the year, the ISA datum will lie either above or below MSL, depending on the prevailing atmospheric conditions at the location of interest.

Temperature
At the ISA sea-level datum, the temperature is specified to be exactly 15 °C (59 °F), which corresponds to an absolute temperature of 288.15 K (518.67 °R). In the troposphere, the temperature gradient (lapse rate) is defined as exactly −6.5 °C per 1000 m (which is equal to −1.9812 °C per 1000 ft or −3.56616 °F per 1000 ft). In the lower stratosphere, the temperature is defined as a constant −56.5 °C (−69.7 °F), which corresponds to an absolute temperature of 216.65 K (389.97 °R). The temperature profile from the ISA sea-level datum to 20 km is illustrated in Figure 4.1.

---

3 The International Civil Aviation Organization (ICAO) is a specialized agency of the United Nations that is charged with coordinating and regulating international air travel (see Section 23.2).
4 Geopotential height is explained in Section 5.2.2.

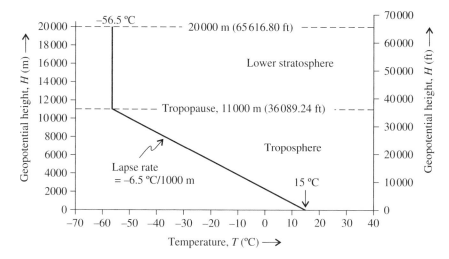

Figure 4.1 Temperature variation in the International Standard Atmosphere (ISA).

Pressure

At the ISA sea-level datum, the atmospheric pressure is 1013.25 hPa in SI units[5] (which is equal to 2116.2166 lb/ft^2 or 29.92126 inHg). The pressure decreases with increasing height in a non-linear manner—for example: at a height of 5000 ft, it falls by about 1 hPa per 32 ft, but at a typical cruising height (for an airliner) of 35 000 ft, the rate is about 1 hPa per 88 ft, and the pressure is approximately 23.5% of the ISA sea-level datum value. Equations that describe the pressure variation in the ISA are derived in Section 4.2.4 and recorded in Table 4.2.

Density

At the ISA sea-level datum, the air density is 1.225 kg/m^3 (0.0023768924 slug/ft^3). The density also decreases in a non-linear manner with increasing height. Equations that describe the density variation in the ISA are also derived in Section 4.2.4 and recorded in Table 4.2.

Speed of Sound in Air

The speed of sound in air (a) is a function of the square root of the absolute temperature of the air and can be computed using Equation 2.77. At the ISA sea-level datum, the speed of sound in air is 340.294 m/s (1116.45 ft/s or 661.479 kt). Within the ISA troposphere, the speed of sound will reduce with increasing height (as the temperature reduces linearly). In the lower stratosphere, the speed of sound is a constant 295.069 m/s (968.076 ft/s or 573.569 kt).

Gravity

The acceleration due to gravity (g) varies with height and geographical location, as described in Section 2.3.4. A reference latitude of 45.5425° is used for the purpose of defining the standard atmosphere—this produces the internationally accepted standard value of g, that is, $g_0 = 9.80665$ m/s^2 (32.174049 ft/s^2). It is assumed for the purpose of defining the ISA that the value of g does not vary with height. The implications of this idealization are considered in Section 4.2.4 and also later in Section 5.2.2 (when geopotential height is discussed).

---

5 The unit millibar (mb) is the preferred unit for measuring atmospheric pressure in some regions of the world; the conversion to SI units is 1 mb = 1 hPa = 100 Pa = 100 N/m^2. In North America, atmospheric pressure is customarily measured in inches of mercury (inHg); the conversion is 1 inHg = 33.86389 hPa. (Unit conversions are provided in Appendix B.)

4.2.3 Relative Temperature, Relative Pressure, and Relative Density

It is customary to define the air temperature (T), pressure (p), and density (ρ)—at any point in any atmosphere—as ratios of the ISA sea-level datum values, which, by convention, are denoted by the subscript zero. This is described below.

Relative Temperature (Temperature Ratio)
The relative temperature, which is assigned the Greek letter theta (θ), is defined as

$$\theta = \frac{T}{T_0} \tag{4.1}$$

where both T (air temperature) and T_0 (ISA sea-level datum temperature) are defined as absolute values (i.e., in kelvin or degrees Rankine).

Relative Pressure (Pressure Ratio)
The relative pressure, which is assigned the Greek letter delta (δ), is defined as

$$\delta = \frac{p}{p_0} \tag{4.2}$$

where p is the air pressure and p_0 is the ISA sea-level datum pressure (any consistent unit of pressure can be used; typical units are N/m^2, hPa, mb, lb/ft^2, inHg).

Relative Density (Density Ratio)
The relative density, which is assigned the Greek letter sigma (σ), is defined as

$$\sigma = \frac{\rho}{\rho_0} \tag{4.3}$$

where ρ is the air density and ρ_0 is the ISA sea-level datum density (any consistent unit of density can be used; typical units are kg/m^3 or $slug/ft^3$).

Ideal Gas Expression
Although Equations 4.1, 4.2, and 4.3 are very useful and can be used independently, it should be remembered that temperature, density, and pressure are not independent. This is evident from the ideal gas law (see Section 2.5.6), which, for arbitrary conditions, can be expressed in the form given by Equation 2.66. Now, if this equation is written out twice, once for the point of interest and once for the ISA sea-level datum, and the former equation is divided by the latter, then the ratios σ, δ, and θ can be combined in a single expression:

$$\frac{\rho}{\rho_0} = \frac{\left(\dfrac{p}{p_0}\right)}{\left(\dfrac{RT}{RT_0}\right)} \quad \text{or} \quad \sigma = \frac{\delta}{\theta} \tag{4.4}$$

where R is the gas constant (see Section 2.5.6).

This is a particularly useful relationship; one application of this equation relates to the determination of atmospheric properties for non-standard conditions (see Section 4.3.4).

Total Temperature and Total Pressure Ratios
Mach number can be used to define two further parameters that relate to the temperature and pressure ratios. The total temperature, as defined by Equation 2.81, can be written for an

arbitrary condition and also for the ISA sea-level datum condition. If the expression for the arbitrary condition is divided by that for the datum condition, the following equation is obtained:

$$\theta_t = \theta(1 + 0.2\,k\,M^2) \tag{4.5}$$

where θ_t is the total temperature ratio;
　　　　M is the Mach number; and
　　　　k is the recovery factor (typically 0.9 to 1, depending on the instrument used to measure the temperature).

Similarly, the total pressure ratio (δ_t) can be deduced from Equation 2.83 and is defined as follows:

$$\delta_t = \delta(1 + 0.2\,M^2)^{3.5} \tag{4.6}$$

4.2.4   Temperature, Pressure, and Density in the ISA

The standard constants of the ISA are recorded in Table 4.1. By defining the sea-level datum conditions and the temperature profile as a function of height (see Figure 4.1), sufficient information is available to determine the pressure and density at any height in the ISA. This can be accomplished by combining the hydrostatic equation (Equation 2.50) with the ideal gas law (Equation 2.66), as follows:

$$\frac{\mathrm{d}p}{\mathrm{d}h} = -\frac{pg}{RT} \tag{4.7}$$

Equation 4.7 is now rearranged in the form of an integral expression:

$$\int \frac{1}{p}\,\mathrm{d}p = -\int \frac{g}{RT}\,\mathrm{d}h \tag{4.8}$$

Table 4.1a ISA constants: fundamental parameters

		Standard values	
Description	Symbol	SI units	Equivalent
Temperature at the sea-level datum	T_0	288.15 K 15 °C	518.67 °R 59 °F
Pressure at the sea-level datum	p_0	101 325 N/m^2 1013.25 hPa	2116.21662 lb/ft^2 29.921255 inHg
Temperature gradient in the troposphere	L	−6.5 K per 1000 m	−1.9812 K per 1000 ft −3.56616 °R per 1000 ft
Temperature gradient in the stratosphere	L	0 K/m	0 K/ft
Height of the tropopause	H^*	11 000 m	36089.24 ft
Gravitational acceleration	g_0	9.80665 m/s^2	32.174049 ft/s^2
Gas constant	R	287.05287 m^2 s^{-2} K^{-1}	3089.81138 ft^2 s^{-2} K^{-1} 1716.56187 ft^2 s^{-2} °R^{-1}
Ratio of specific heats of air	γ	1.40	1.40

Note: The ISA is defined in terms of SI units and corresponding values in FPS units are obtained by conversion, which can lead to small rounding error discrepancies.

Table 4.1b ISA constants: derived values

Description	Symbol	Standard values	
		SI units	Equivalent
Density at the sea-level datum	ρ_0	1.2250 kg/m^3	0.0023768924 slug/ft^3
Speed of sound at the sea-level datum	a_0	340.294 m/s	1116.45 ft/s 661.479 kt
Temperature of the tropopause	T^*	216.65 K −56.5 °C	389.97 °R −69.7 °F
Pressure at the tropopause	p^*	226.320 hPa	472.680 lb/ft^2 6.68324 inHg
Density at the tropopause	ρ^*	0.363918 kg/m^3	0.000706117 slug/ft^3
Speed of sound at the tropopause	a^*	295.069 m/s	968.076 ft/s 573.569 kt

When the temperature variation with height (h) is substituted into Equation 4.8, the variation of pressure with height can be obtained by integration. The temperature in the ISA is described by two functions: one for the troposphere (where temperature reduces linearly with height) and one for the lower stratosphere (where temperature is constant).

The essential difficulty with the evaluation of Equation 4.8 concerns the gravitation term (g), which in a real atmosphere reduces with height; however, in the ISA, gravity is assumed to be constant. The mechanism by which this is done is through the introduction of a height scale called geopotential height. Geopotential height, which is given the symbol H, is the height in a hypothetical uniform gravitational field that would give the same potential energy as the point under consideration in the actual, variable gravitational field. This is discussed further in Section 5.2.2. Equation 4.8 is thus rewritten with gravity set to the standard (i.e., constant) value of g_0 and the height as H.

$$\int \frac{1}{p}\,\mathrm{d}p = -\frac{g_0}{R}\int \frac{1}{T}\,\mathrm{d}H \tag{4.9}$$

This integral expression can be evaluated for the two regions of interest: in the troposphere, the integration is performed from the ISA sea-level datum to an arbitrary height H, and in the stratosphere, the integration is performed from the tropopause to an arbitrary height H.

In the Troposphere

The temperature gradient (lapse rate[6]) in the ISA is herein denoted by the letter L and is defined as

$$L = \frac{\mathrm{d}T}{\mathrm{d}H} \tag{4.10}$$

where L is the ISA temperature gradient (typical units: K/m, K/ft, °R/ft);
T is the absolute temperature (typical units: K, °R); and
H is the geopotential height (typical units: m, ft).

---

6 Caution: A temperature lapse rate in the atmosphere is often defined as positive for a reducing temperature with increasing height; according to this alternative definition, lapse rate $= -\mathrm{d}T/\mathrm{d}H$.

The temperature T at any height H is then given by

$$T = T_0 + LH \tag{4.11}$$

Thus $\theta = 1 + \dfrac{LH}{T_0}$ $\qquad$ (4.12)

Substituting Equation 4.11 into Equation 4.9 gives

$$\int_{p_0}^{p} \frac{1}{p}\, \mathrm{d}p = -\left(\frac{g_0}{R}\right) \int_{0}^{H} \frac{1}{(T_0 + LH)}\, \mathrm{d}H$$

Integrating from the ISA sea-level datum to height H gives

$$\Big[\ln p\Big]_{p_0}^{p} = \frac{-g_0}{RL}\Big[\ln(T_0 + LH)\Big]_{0}^{H}$$

Thus $\ln\left(\dfrac{p}{p_0}\right) = \ln\left[\dfrac{T_0 + LH}{T_0}\right]^{-g_0/RL}$

Finally, the relative pressure (δ) is given by

$$\delta = \frac{p}{p_0} = \left(\frac{T}{T_0}\right)^{-g_0/RL} = \left[1 + \frac{LH}{T_0}\right]^{-g_0/RL} \tag{4.13}$$

Using Equation 4.4, the relative density (σ) is given by

$$\sigma = \frac{\rho}{\rho_0} = \frac{\left(\dfrac{p}{T}\right)}{\left(\dfrac{p_0}{T_0}\right)} = \left(\frac{T}{T_0}\right)^{-g_0/RL - 1}$$

Thus $\sigma = \left[1 + \dfrac{LH}{T_0}\right]^{-g_0/RL - 1}$ $\qquad$ (4.14)

In the Stratosphere

In the stratosphere, the integration of Equation 4.9 is performed from the tropopause, where the conditions are denoted by a superscript asterisk (*), to the height H. As the stratosphere is assumed to be isothermal, the integration is straightforward. Equation 4.9 becomes

$$\int_{p^*}^{p} \frac{1}{p}\, \mathrm{d}p = -\frac{g_0}{RT^*} \int_{H^*}^{H} \mathrm{d}H$$

$$\Big[\ln p\Big]_{p^*}^{p} = \frac{-g_0}{RT^*}\Big[H\Big]_{H^*}^{H}$$

$$\ln\left(\frac{p}{p^*}\right) = \frac{-g_0(H - H^*)}{RT^*}$$

Thus $\dfrac{\delta}{\delta^*} = \dfrac{p}{p^*} = e^{\frac{-g_0(H-H^*)}{RT^*}}$ $\qquad$ (4.15)

Table 4.2 Equations describing the relative properties (i.e., temperature, pressure, and density) in the ISA

Relative temperature	Relative pressure	Relative density
In the troposphere:		
$\theta = 1 + \dfrac{L}{T_0}H$	$\delta = \left[1 + \dfrac{LH}{T_0}\right]^{-g_0/RL}$	$\sigma = \left[1 + \dfrac{LH}{T_0}\right]^{(-g_0/RL)-1}$
	or $\delta = \theta^{5.25588}$	or $\sigma = \theta^{4.25588}$
At the tropopause:		
$\theta^* = 0.751865$	$\delta^* = 0.223361$	$\sigma^* = 0.297076$
In the stratosphere:		
$\theta = \theta^*$	$\delta = \delta^* e^{(-g_0/RT^*)(H-H^*)}$	$\sigma = \sigma^* e^{(-g_0/RT^*)(H-H^*)}$

Using the ideal gas law (see Equation 2.66) for the isothermal case yields

$$\frac{\rho}{\rho^*} = \frac{p}{p^*}$$

Thus $\quad \dfrac{\sigma}{\sigma^*} = \dfrac{\delta}{\delta^*} = e^{\frac{-g_0(H-H^*)}{RT^*}}$ (4.16)

For convenience, Equations 4.12 to 4.16 are repeated in Table 4.2. Using these equations, the temperature, pressure, and density of the air in the ISA can be determined as a function of geopotential height. Relative temperature (θ), relative pressure (δ), and relative density (σ) values, computed using the equations recorded in Table 4.2, are shown in Figure 4.2 for ISA conditions from sea level to 50 000 ft.

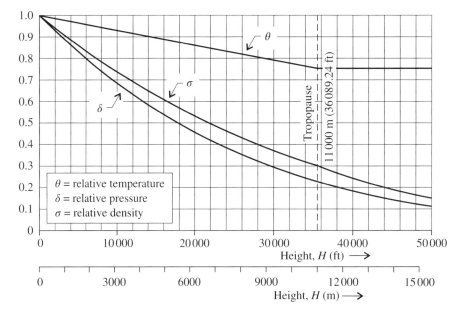

Figure 4.2 Relative temperature (θ), relative pressure (δ), and relative density (σ) in the International Standard Atmosphere (ISA).

Tabulated values of the ISA properties (i.e., temperature, pressure, density, and speed of sound in air) are given in Appendix A as a function of geopotential height measured in feet (which is the unit of measure used for airplane operations in most countries).

4.3 Non-Standard and Off-Standard Atmospheres

4.3.1 Non-Standard Atmospheres

Equations describing the variation of temperature, pressure, and density can, in theory, be fitted to sets of measured atmospheric data—and in the process an almost infinite number of mathematical models of the atmosphere can be generated. The usual approach that is adopted is to describe the temperature profile in terms of layers of constant or zero lapse rate. Then, by combining the hydrostatic equation with the ideal gas law—as explained in Section 4.2.4—the variation of pressure with geopotential height is described for each layer. Finally, the density is described as a function of the temperature and pressure using the ideal gas law (Equation 2.66). These models are generally referred to as *non-standard atmospheres*—they bear little, or no, relationship to the ISA.

Non-standard atmospheric models are useful for meteorological studies or weather forecasting, for example. However, for the purpose of airplane design, certification, or flight operations, they are of limited value. Working models, which more closely match day-to-day conditions than those provided by the ISA, are, however, needed for the aforementioned purposes. The conventional approach is to use what are called *off-standard atmospheres*, which are relatively simple models that are obtained by offsetting the ISA values—this is described in the next section.

4.3.2 Off-Standard Atmospheres

Off-standard atmospheric models are developed by introducing a single change to the governing parameters that describe the ISA—that is, the sea-level datum temperature is increased or decreased. The temperature gradient dT/dH_p, however, is unchanged. The temperature increment/decrement (ΔT) is applied to the ISA model—hence the term *off-standard*. In such cases, it is customary to talk about an ISA + 5 °C atmosphere, for example, where the temperatures are 5 °C higher than those in the ISA model.

An important feature of an off-standard atmosphere is that pressure height (H_p), which is called *altitude* in an operational context (see Section 5.3.1), does not equal geopotential height (H)—the difference between the two height scales is analyzed in Section 4.3.3. Mathematical expressions for temperature, pressure, and density in such atmospheres can easily be defined as functions of pressure height [5].

Temperature

The temperature at each height—where height is specified in terms of pressure height—is obtained by adding a constant value (ΔT) to the temperature defined in the ISA model.

In the troposphere, the temperature T at any height H_p is given by

$$T = T_0 + \Delta T + LH_p \tag{4.17}$$

and $\quad \theta = 1 + \dfrac{\Delta T + LH_p}{T_0} \tag{4.18}$

In the stratosphere, the temperature is given by

$$T = T^* + \Delta T \tag{4.19}$$

$$\text{and} \quad \theta = \theta^* + \frac{\Delta T}{T_0} \tag{4.20}$$

where T_0, T^*, L, and θ^* are defined in Tables 4.1 and 4.2.

For example, the temperature profile of the ISA + 5 °C Off-Standard Atmosphere will start at 20 °C at the sea-level datum and will decrease at a rate of 6.5 °C per 1000 m to the tropopause (which will be at a pressure height of 11 000 m); in the stratosphere, the temperature will be a constant −51.5 °C.

Pressure

The relationships between pressure and *pressure height* in an off-standard atmosphere (for the troposphere and for the stratosphere) are identical to those established between pressure and geopotential height in the ISA. The equations are obtained by replacing H with H_p in Equations 4.13 and 4.16.

In the troposphere, the relative pressure (δ) is given by

$$\delta = \frac{p}{p_0} = \left[1 + \frac{LH_p}{T_0}\right]^{-g_0/RL} \tag{4.21}$$

and in the stratosphere it is given by

$$\frac{\delta}{\delta^*} = \frac{p}{p^*} = e^{\frac{-g_0}{RT^*}\left(H_p - H_p^*\right)} \tag{4.22}$$

where the constants are defined in Tables 4.1 and 4.2.

Density

In an off-standard atmosphere, the density does not match that given in the ISA. The density must be calculated using the ideal gas law (Equation 2.66), where the temperature T is given by Equation 4.17 or Equation 4.19. Alternatively, the relative density can be obtained using Equation 4.4, for the calculated values of δ and θ.

4.3.3 Height Scales in Off-Standard Atmospheres

In the ISA, pressure height (H_p) is, by definition, equal to geopotential height (H), but in off-standard atmospheres this is not the case: there is a difference between H and H_p. However, the sea-level datum pressure is the same for both height scales (i.e., p_0 = 1013.25 hPa), and thus $H = 0$ when $H_p = 0$.

The difference between geopotential height and pressure height depends on the temperature increment/decrement (ΔT) and the pressure height (H_p). The expression that describes this difference can be developed from Equation 4.7. The starting point is to write this equation with geopotential height and the standard value of gravity.

$$\frac{dp}{dH} = -\frac{pg_0}{RT} \quad \text{(from Equation 4.7)} \tag{4.23}$$

In a standard atmosphere, $T = T_{std}$ and $dH = dH_p$, where T_{std} is the temperature in the ISA.

$$\text{Thus} \quad \frac{dp}{dH_p} = -\frac{pg_0}{RT_{std}} \tag{4.24}$$

But in an off-standard atmosphere, $T = T_{std} + \Delta T$ and $\mathrm{d}H \neq \mathrm{d}H_p$.

Thus $\quad \dfrac{\mathrm{d}p}{\mathrm{d}H} = -\dfrac{pg_0}{R\left(T_{std} + \Delta T\right)}$ $\qquad$ (4.25)

Equation 4.24 is now divided by Equation 4.25 (note that p is the same in both equations):

$$\frac{\mathrm{d}H}{\mathrm{d}H_p} = \frac{T_{std} + \Delta T}{T_{std}} \qquad (4.26)$$

The temperature gradient with respect to geopotential height in an off-standard atmosphere is thus

$$\frac{\mathrm{d}T}{\mathrm{d}H} = \frac{\mathrm{d}T}{\mathrm{d}H_p}\left(\frac{T_{std}}{T_{std} + \Delta T}\right) = L\left(\frac{T_{std}}{T_{std} + \Delta T}\right) \qquad (4.27)$$

Equation 4.26 is now integrated from the sea-level datum to the required height, which, in this analysis, is taken as being below the tropopause:

$$\int_0^H \mathrm{d}H = \int_0^{H_p}\left(1 + \frac{\Delta T}{T_{std}}\right)\mathrm{d}H_p$$

$$H = \int_0^{H_p}\left(1 + \frac{\Delta T}{T_0 + LH_p}\right)\mathrm{d}H_p$$

$$H = H_p + \frac{\Delta T}{L}\ln\left(\frac{T_0 + LH_p}{T_0}\right) \qquad (4.28)$$

From Equation 4.21, it is evident that

$$\ln\left(\frac{T_0 + LH_p}{T_0}\right) = \frac{-RL}{g_0}\ln\delta \qquad (4.29)$$

Substituting Equation 4.29 into Equation 4.28 yields the final result:

$$H - H_p = -\frac{R\,\Delta T}{g_0}\ln\delta \qquad (4.30)$$

Although the derivation given above is based on a temperature–height relationship that applies to the troposphere (i.e., constant, non-zero lapse rate), it can be shown by a similar approach that Equation 4.30 also applies to the stratosphere (which has a zero lapse rate). In this case, Equation 4.26 is integrated—for conditions in the stratosphere—from the tropopause to the required height:

$$\int_{H^*}^H \mathrm{d}H = \int_{H_p^*}^{H_p}\left(1 + \frac{\Delta T}{T^*}\right)\mathrm{d}H_p$$

$$H - H^* = H_p - H_p^* + \frac{\Delta T}{T^*}\left(H_p - H_p^*\right)$$

$$H - H_p = \left(H^* - H_p^*\right) + \frac{\Delta T}{T^*}\left(H_p - H_p^*\right)$$

Substituting from Equation 4.30 and Equation 4.15 yields

$$H - H_p = -\frac{R\,\Delta T}{g_0}\ln\delta^* + \frac{\Delta T}{T^*}\left(\frac{-RT^*}{g_0}\right)\ln\left(\frac{\delta}{\delta^*}\right)$$

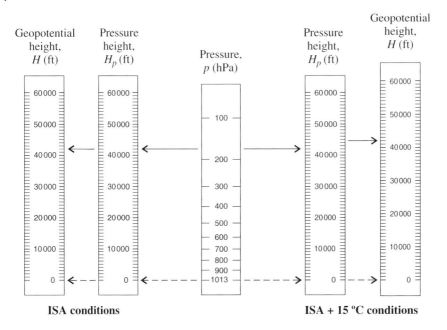

Figure 4.3 Pressure height and geopotential height for standard and off-standard atmospheres.

Thus, as before

$$H - H_p = -\frac{R\,\Delta T}{g_0}\ln\delta \qquad \text{(Equation 4.30)}$$

In conclusion, it is evident that in "warm" conditions (i.e., when ΔT is positive), the pressure height scale can be considered to be compressed in comparison to the geopotential height scale, but—and this is important—the datum is the same. This is illustrated in Figure 4.3. The reverse is true in "cold" conditions (i.e., when ΔT is negative): the pressure height scale is stretched compared to the geopotential height scale.

For the sake of illustration, the difference between H and H_p at a pressure height of 42 000 ft is calculated, using Equation 4.30, for conditions of ISA + 15 °C:

$$H - H_p = \frac{-(287.053)(15)}{9.80665}\ln(0.16812) = 783 \text{ m} = 2570 \text{ ft}$$

It is evident that the difference in height can be considerable when the air temperature varies significantly from ISA conditions. This impacts flight operations. To ensure flight safety—and prevent flight into terrain or collision with another aircraft—a series of standardized operational procedures exist. For example, when pilots climb above the *transition altitude* (defined in Section 5.4.2), they are required to set their altimeters to the same reference pressure, effectively ensuring that all neighboring aircraft are operating on the same height scale, reducing the possibility of an in-flight collision. (The topic of altimetry is discussed in Chapter 5.)

4.3.4 Determination of Density in Non-Standard Conditions

Changes in air density can significantly impact the performance of an aircraft (see Section 5.2.4, for example). Theoretical performance analyses usually determine the density from the ISA; however, in flight tests—and indeed in everyday airplane operations—it is seldom the case that the atmospheric conditions will match those of the ISA model. Although it is possible to determine directly the air density for actual flight conditions, this requires special instrumentation. The usual method—a three-step approach—makes use of the ideal gas law (see Equation 2.66), which links density to pressure and temperature (parameters that are easily determined during testing).

Step 1

The relative temperature (θ) is determined as follows: the ambient temperature (T_{test}) is measured in flight, or estimated from knowledge of the surface (ground) temperature and the lapse rate. The relative temperature is then determined using Equation 4.1.

$$\theta = \frac{T_{test}}{T_0} \tag{4.31}$$

where T_{test} is the ambient (test) temperature (absolute); and
T_0 is the ISA sea-level datum temperature (absolute).

Step 2

The relative pressure (δ) can be determined indirectly using an altimeter. By setting the altimeter reference pressure to 1013 hPa (29.92 inHg), which is the ISA sea-level datum pressure, the pressure height ($H_{p,test}$) is measured. The relative pressure is then read off the ISA tables (see Appendix A) for the pressure height $H_p = H_{p,test}$.

Step 3

The required air density (ρ_{test}) is evaluated from the relative density (σ), which can be determined using Equation 4.4 and the results of steps 1 and 2:

$$\rho_{test} = \sigma \rho_0 = \left(\frac{\delta}{\theta}\right) \rho_0 \tag{4.32}$$

4.4 The Real Atmosphere

4.4.1 Introduction

The ISA is the internationally accepted standard for atmospheric properties, and is the baseline for the determination and subsequent publication of aircraft performance data—to be used, for example, by airplane manufacturers, certification authorities, and airplane operators. The airplane's actual performance, of course, depends on the prevailing atmospheric conditions— and the actual conditions encountered every day by pilots can be vastly different than those described by the ISA. Flight-test data, for example, are routinely adjusted to standard conditions. Conversely, published performance data must be corrected for non-standard conditions for critical flight conditions—in particular, to account for temperature variations.

Another important difference encountered in the real atmosphere is the presence of water vapor. This is not taken into account in the ISA, which assumes that the air is perfectly dry. Humid air is less dense than dry air (this can be deduced from the fact that water vapor is a relatively light gas compared to the main components of air). Thus, an airplane's aerodynamic performance in humid air is affected in a similar manner to an increase in air temperature.

It is important to note that the ISA represents average conditions. These averages, however, are based on measured data that display significant diurnal and seasonal variations. Furthermore, these averages, at best, only represent the temperate latitudes of the Northern Hemisphere at about 45°N, and there exists a large variation in atmospheric conditions for different geographic locations around the world.

4.4.2 Troposphere

The troposphere is the relatively thin layer of the atmosphere adjacent to the Earth in which almost all of the Earth's weather occurs. The nature of the troposphere is dominated by radiative and convective heating of the air by the Earth and by the rotation of the Earth. Air masses, which gain and lose energy within the troposphere, move horizontally and vertically, resulting in substantial mixing.[7] In contrast to the stratosphere, the air in the troposphere is relatively moist and turbulent.

The typical cruise altitude of a commercial jet airplane is in the upper part of the troposphere or the lowest part of the stratosphere (the actual flight level depends on the route flown and the performance capability of the aircraft). In routine flight operations, the airplane must thus climb and descend through much of the troposphere, where air temperature, pressure, and density vary significantly from day to day and from one location to another.

4.4.3 Tropopause

The tropopause, which is the boundary between the troposphere and the stratosphere, is identified by a sharp change in the lapse rate; there is also a slight change in the chemical composition of the air. The meteorological definition of the tropopause is that it is the lowest level at which the rate of change of air temperature is 2 °C/km or less.[8] Above this level, in the stratosphere, temperatures are relatively stable and there is a gradual increase in temperature with increasing height.

Above the equator, the annual average temperature of the tropopause varies from about −75 °C to about −79 °C (variation is evident with longitudinal position, and there is a ± 2 °C change throughout the year). Above the north polar region, the tropopause temperature is about −56 °C, and above the southern polar region, it is about −64 °C [7].

The tropopause acts as a lid on air convection, effectively restricting the Earth's weather systems (e.g., cloud formation, storms, winds) to the troposphere. The tropopause is highest at the equator, where it is forced upwards by strong convective heating, reaching heights of about 52 500 ft (16 km), and it is lowest at the poles, where it is about 29 500 ft (9 km) high [7]. The drop in tropopause height—when moving from the equator to the poles—is not uniform. Between about 25° and 45° latitude (north and south), the drop is relatively steep; this feature is linked to the formation of jet streams (see Section 4.5.6). The positions of these steep gradient features are not stationary and move with the season (migrating a little way north in June–August and south in December–February) and as major weather systems sweep across the Earth. The interface between the upper and lower air masses is also not smooth and the shape can be altered by strong weather systems—for example, "fingers" of stratospheric air can be pulled down into the troposphere in the vicinity of strong, well-developed cold fronts.

---

7 The word troposphere stems from the Greek word *tropos*, which means turning or mixing.
8 Definition used by the World Meteorological Organization (WMO) [6].

4.4.4 Stratosphere

The stratosphere is characterized by layers of "stratified" air, in which there is an absence of mixing. It is heated by ultraviolet radiation from the sun—this produces a lapse rate that is close to zero in the lower levels of the stratosphere; in the upper levels, there is a gradual increase in temperature with increasing height. At about 50 km, near the top of the stratosphere, the temperature reaches 0 °C. This temperature profile—with warmer air located above cooler air—makes the stratosphere very stable and there is no convection or associated vertical air movement. Cruising in the stratosphere is, for the most part, smooth and free of turbulence.

4.5 Weather

4.5.1 Impact on Airplane Performance

The state of the atmosphere with regard to its temperature, barometric pressure, visibility, precipitation, wind, and so forth—that is, the *weather*—can have a substantial influence on the performance, efficiency, and safety of aircraft operations. Moving air masses can increase or decrease an airplane's range, for example. Rising currents of air provide the mechanism for sailplanes to stay aloft for many hours, enabling them to travel hundreds of nautical miles without using a drop of fuel. High-speed tailwinds are utilized each day by eastbound airliners crossing the North Atlantic, for example, to reduce traveling time and to save fuel. Takeoff and landing distances are shortened by operating into a headwind, rather than with a tailwind. These are some of the beneficial aspects of moving air. Conversely, headwinds in cruise extend trip times. Turbulence can be uncomfortable for passengers and the crew, and, in extreme cases, it can result in injuries during flight or fatalities from crashes caused by structural failure or poor flight path control. Violent storms frequently disrupt air travel, and special precautions are needed in icing conditions. Pilots often deviate from planned courses to fly around thunderstorms. These examples illustrate some of the many ways that the weather affects aviation.

4.5.2 Wind Reporting and Forecasts

Winds are designated by their direction and speed (which is reported in knots or kilometers per hour or meters per second depending on the country)—for example, if a wind is described as 225/30KT, this would mean a wind *from* the southwest of 30 knots. Wind speeds and directions are routinely measured and forecasts issued, providing vital information for flight operations. Such forecasts are given with respect to Coordinated Universal Time (UTC),[9] which, by convention, is identified by the letter Z (or by the word Zulu in accordance with the ICAO phonetic alphabet).

Changes in wind direction are described by the terms veering and backing. *Backing* describes a counter-clockwise change in wind direction with time at a given location (e.g., from southerly to southeasterly) or a counter-clockwise change in direction with increasing height. *Veering* has the opposite meaning: it describes a clockwise change in wind direction. Specific meteorological conditions are often associated with these shifts in wind direction (e.g., due to a weather front).

---

9 Coordinated Universal Time (UTC) is essentially synonymous with Greenwich Mean Time (GMT), which is no longer used in a scientific context.

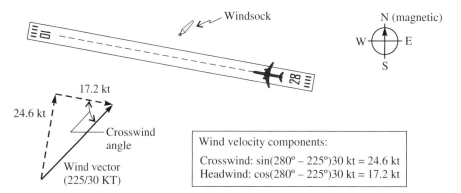

Figure 4.4 Illustration of runway and wind conventions, indicating crosswind and headwind components.

The term *winds aloft* is widely used to describe forecast wind information, generated by numerical weather models, for altitudes greater than 2500 ft elevation above the surface. The direction of these winds is always with respect to true north. These forecasts also provide temperature information (in degrees Celsius).

A *reported surface wind* is the wind reported by an authority (e.g., by air traffic control or via ATIS[10])—for use by flight crew—based on two-minute averages of actual measurements of wind speed and direction at an airfield [8]. By convention, the direction of these reported surface winds is with respect to magnetic north.[11] This is done to facilitate interpretation by pilots, as the orientation (direction) of runways, by convention, is defined with respect to magnetic north (see Section 10.4.2). A trigonometric calculation will yield the crosswind and headwind components—important information for takeoff and landing (Figure 4.4). By definition, the crosswind (which is often abbreviated as "X-wind" by pilots) is the wind component at a 90° angle to the runway. A simple rule of thumb used by some pilots to estimate the magnitude of the crosswind component is based on the position of the numbers on an analog clock face. The crosswind fraction of the reported wind speed is conservatively estimated as one-quarter for a wind 15 degrees "off the runway," one-half for 30 degrees, three-quarters for 45 degrees, and 100 percent for a crosswind angle of 60 degrees or more.

Winds close to the Earth's surface are influenced by the frictional effects of the ground. This tends to reduce the wind speed—and also alter its direction—as measurements are taken closer to the ground. The variation in wind speed below about 400 ft has been shown to approximately follow a power law [9]. A widely used method for adjusting measured wind speeds, which are typically obtained by an anemometer installed on a pole (or tower) at a height of 10 m above the ground at the reporting station (e.g., airfield), is given below.

$$\frac{V_w}{V_{w,an}} = \left(\frac{h}{h_{an}}\right)^{\frac{1}{7}}$$

(4.33)

where V_w is the wind speed at height h above the ground; and
$V_{w,an}$ is the measured wind speed at the height of the anemometer above the ground, that is, at the height h_{an}.

4.5.3 Wind Shear

Wind shear is defined as a significant change in the wind speed and/or direction over a short distance; measurements are given in units of speed per unit distance. The meteorological conditions that give rise to wind shears are many and varied: they include warm and cold fronts, jet streams, mountain waves, thunderstorms, and terrain-induced turbulence. Most are predicable and pose little threat to airplane operations. There is, however, one particular type of wind shear that is a hazard for aircraft operating close to the ground (e.g., during takeoff and landing). This powerful, storm-induced downdraft is called a *microburst*. It is a severe, localized phenomenon that tends to be short-lived (with a duration of approximately 15 minutes). Downdrafts of 4000 ft/min and horizontal winds of 45–100 kt can be experienced in an area of less than 2.5 nm in diameter [10]. The fast moving air descends from a cloud (which may or may not have developed into a thunderstorm), strikes the ground and then spreads out as a shallow layer of air, usually only a few hundred feet thick.

The downdrafts and outflows associated with microbursts have been the cause of serious accidents, where pilots have been unable to maintain their intended flight paths. The problem is not just restricted to the downdraft, which is clearly a serious concern (the speed of the sinking air can exceed an airplane's climb capability), but also to the sudden change in wind direction. This can be illustrated by considering an aircraft on final approach (to land) entering a microburst. The initial outflow will cause an increase in headwind, which will result in an increase in lift, and the airplane will tend to rise above the intended flight path. In this situation, the most likely response of the flight crew will be to reduce thrust. As the airplane moves through the core of the microburst it will experience an intense downdraft, followed by a strong tailwind, resulting in a reduction in lift. The airplane—now in a low-energy (or low-thrust) condition—will rapidly lose height. In extreme cases, if sufficient height cannot be maintained, the airplane will strike the ground.

4.5.4 Gusts

Gusts describe a non-steady wind condition. They are transient, localized phenomena associated with changing vertical or horizontal air movements. Gusts can be a source of discomfort for the crew and passengers, and, over an extended time period, can reduce the fatigue life of an airplane. For aircraft in the takeoff or landing phase, strong gusts—particularly when blowing across the runway—can limit flight operations.

Reporting of surface wind at airfields (see Section 4.5.9) would indicate a gust if the peak wind speed, measured over the preceding 10 minutes, exceeded the mean wind speed by 10 kt or more (based on an ICAO recommendation [11]). For example, a reported surface wind of "135 degrees, 15, gusting 25 knots" (abbreviated as 135/15G25KT) would represent a mean 15 kt wind from the southeast that momentarily increases to 25 kt; the 10 kt increment is the gust.

4.5.5 Temperature Inversions

In the troposphere, air temperature normally decreases with increasing height, but sometimes the temperature gradient is reversed near to the Earth's surface. This is called a temperature inversion. It occurs when a cold frontal system undercuts a warm air mass, pushing it upward. Temperature inversions can also occur at night, when the cold surface of the Earth cools the lower layers of air. The same mechanism is responsible for the temperature inversions that occur over the Arctic and Antarctic regions.

At the top of the inversion layer, the temperature gradient abruptly changes and air temperatures then decrease with increasing height. This change in lapse rate interferes with normal convection—for example, warm, moist air is prevented from rising above this level to the height where condensation would occur, resulting in a cloudless sky, but also trapping pollution and dust, producing hazy flight conditions near to the ground.

4.5.6 Jet Streams

Jet streams are fast moving currents of air that flow from west to east (in both hemispheres), embedded in breaks in the tropopause. The wind speed in the core of these jets is typically between 100 kt and 150 kt, with a reducing speed from the core to the outer boundary (which is arbitrarily set at 50 kt). They form at the interfaces of large air masses—for example, in the Northern Hemisphere, where cold polar air, moving south, meets warm subtropical air at the mid-latitudes. There are two main jet streams—one in each hemisphere—located between 30° and 70° latitude. These are the polar jet streams and occur at a height of approximately 30 000 ft with a depth of about 3000 ft to 7000 ft. There are also two secondary jet streams: the subtropical jet streams, located between 20° and 50° latitude at a height of about 39 000 ft [12].

Knowledge of the jet streams is very important for long-distance flight planning: their locations, altitudes, and speeds are carefully monitored by the world's airlines. For eastbound long-haul flights, it is desirable to be within the core of the jet stream, where the wind speeds are greatest, but when westbound the flight plan must carefully balance the disadvantage of flying a less-direct, longer route against the advantage of flying with a reduced headwind.

The jet streams are neither fixed in location nor in wind speed: their position and speed changes with the seasons and in response to local weather systems. The path of a jet stream tends to meander—in a north–south orientation—and the path itself also moves eastwards. The "driving force" of the jet stream—as is the case with many global weather systems—is the unequal heating of the Earth by the sun. Equatorial regions are hotter than polar regions, and this sets up continuous flows of warm air away from the equator in a northerly direction in the Northern Hemisphere and in a southerly direction in the Southern Hemisphere. Due to the Earth's rotation, however, these air movements are influenced by a Coriolis force—in both hemispheres, the prevailing winds are deflected eastwards as they leave the equatorial region. The result is a global air movement in a predominantly west–east direction. At the boundaries of the major frontal systems, the winds intensify forming jet streams.

The wind speeds in a jet stream depend on the temperature gradient between the equator and the poles, reaching a maximum when the gradient is greatest. For example, during the winter months in the Northern Hemisphere, when the temperature disparity is greatest, the wind speeds in the northern polar jet stream typically reach 200 kt [12].

4.5.7 Precipitation and Icing

Water vapor is a natural component of atmospheric air. The maximum amount of water vapor that air can contain depends on its temperature—for example, at 0 °C, it is about 5 g/m^3, but as the temperature cools to −30 °C it can only support about one-tenth of this amount of water. The limiting condition is called saturation. As air is cooled to the point of saturation, the excess water vapor forms water droplets or ice crystals, which can be visible as clouds. The air can be cooled by being forced to rise, due to the action of a weather system or a wind blowing over a mountain, for example. The air can also be cooled by the ground (at night, for example), leading to low level precipitation (e.g., mist or fog).

In certain, very still, conditions, water droplets may remain liquid at temperatures below 0 °C. These supercooled droplets are unstable and can freeze instantly if disturbed—for example, when hit by the leading edge of an airplane's wing. This, in simple terms, is the mechanism responsible for aircraft icing. Icing can occur at air temperatures of about 0 °C down to −40 °C, but severe icing seldom occurs below −12 °C [13]. At mid-latitudes, this means that severe icing is most likely to occur at altitudes less than about 10 000 ft.

4.5.8 Visual and Instrument Meteorological Conditions

Pilots operate aircraft, at all times, under one of two sets of flight rules (established by the relevant aviation authority). These sets of flight rules, which govern key navigation and safety aspects of the operation, are known by the abbreviations VFR (for Visual Flight Rules) and IFR (for Instrument Flight Rules). When operating under VFR, pilots rely primarily on external visual cues to orientate their aircraft, navigate, and maintain adequate separation of their aircraft from other air traffic, clouds, terrain, buildings, and so forth. IFR is used to conduct flights in conditions where VFR is not safe. IFR flights require reference to cockpit flight instruments, and navigation is conducted under the direction of air traffic control (ATC). IFR operations require the aircraft to be appropriately equipped and the flight crew appropriately trained.

Flights operating under VFR can only take place when the meteorological conditions are adequate—this is known as visual meteorological conditions (VMC). The minimum conditions—known as VMC minima—are defined in terms of requirements for visibility, cloud ceilings (for takeoff and landing), and cloud clearances. These requirements depend on the class of airspace (see Section 5.4.4). The VMC minima specified in ICAO Annex 2 [14] indicate flight visibility requirements of 5 or 8 km depending on altitude and class of airspace (but it can be reduced to 1.5 km in certain situations). Cloud clearance requirements vary from remaining clear of cloud with the surface in sight (applicable to low level flight in certain airspace classes) to maintaining a distance of 1500 m horizontally and 300 m or 1000 ft vertically from cloud during flight. Instrument meteorological conditions (IMC) exist when the VMC minima are not met—and in such conditions flights can only be conducted under IFR.

4.5.9 Weather Information for Flight Operations

There are a number of authenticated sources of weather information for flight operations; the most important ones are described in this section. Recommended procedures for the determination and dissemination of weather information are given in ICAO Annex 3 [11] and World Meteorological Organization (WMO) Technical Regulation 49, Vol. II [15].

ATIS
ATIS is an abbreviation for Automatic Terminal Information Service. It is a continuous broadcast of critical terminal (airfield) information, which includes *current* weather details (e.g., two-minute average wind speed and direction, temperature, and humidity) and other information important for pilots, such as the active runway(s) in use.

METAR
METAR is an abbreviation for Meteorological Aerodrome Report. It is a standard aviation weather report, determined from routine weather observations. The reports usually come from airports and are typically generated once per hour. If conditions change rapidly, a special report, called a SPECI, may be issued (the acronym roughly translates from French as Aviation Selected Special Weather Report). A typical METAR report contains information concerning

the air temperature, dew point, 10-minute average wind speed and direction (in degrees true), precipitation, cloud cover, cloud heights, visibility, and barometric pressure—all presented in a standard coded format.[12] It can also contain information on potentially dangerous weather conditions, such as storms, that would be of interest to flight crews. Barometric pressure is reported in hectopascal or inches of mercury (depending on the authority), temperatures in degrees Celsius (never Fahrenheit), height in feet, and wind speed in knots or meters per second (again, depending on the authority).

TAF

TAF is an abbreviation for Terminal Aerodrome Forecast. It is a complementary report to the METAR, describing the forecasted, rather than the actual, weather. TAFs generally provide a 12 hr or 24 hr forecast, based on numerical weather forecasts, taking into account local effects. TAFs use a similar encoding system to METARs.

SIGMET

SIGMET is an abbreviation for Significant Meteorological Information. It is a weather advisory notice containing safety related information—for example, SIGMETs can describe severe turbulence or icing, or visibility restrictions due to dust, sandstorms, or volcanic ash.

VOLMET

VOLMET is a worldwide network of radio stations that broadcast METAR, TAF, and SIGMET reports, using an automated voice transmission on short-wave frequencies. Pilots listen to these transmissions and typically use them for *en route* weather briefing and flight planning.

4.6 Stability of the Atmosphere

4.6.1 Adiabatic Lapse Rate

From the point of view of thermal activity, it is of interest to study the stability of the atmosphere under various lapse rates. As a starting point, it is convenient to determine the adiabatic lapse rate (ALR). This is obtained by considering a parcel of air at height H_1 under conditions of pressure p_1 and temperature T_1, which is then raised to height H_2, where the pressure is p_2 and the temperature is T_2, as shown in Figure 4.5.

If it is assumed that there is no exchange of heat with the surrounding air, then the air parcel will cool adiabatically from temperature T_1 to temperature T_{2_a} as it rises to H_2. For this rising air, the adiabatic lapse rate can be determined for a reversible adiabatic process. The adiabatic lapse rate L_a can be written as follows:

$$L_a = \left(\frac{\mathrm{d}T}{\mathrm{d}H} \right)_a = \left(\frac{\mathrm{d}T}{\mathrm{d}p} \right)_a \left(\frac{\mathrm{d}p}{\mathrm{d}H} \right) \tag{4.34}$$

where the derivatives $\mathrm{d}T/\mathrm{d}p$ and $\mathrm{d}p/\mathrm{d}H$ are determined separately.

---

12 Details on the coding system are given in ICAO Annex 3 [11] and WMO Technical Regulation 49, Vol. II [15]. The United States adopted the METAR coding system in 1996 (replacing the Surface Aviation Observation code). The METAR coding system used in the US, which is described in the Federal Meteorological Handbook No. 1 [16], has certain national differences compared to the ICAO/WMO system.

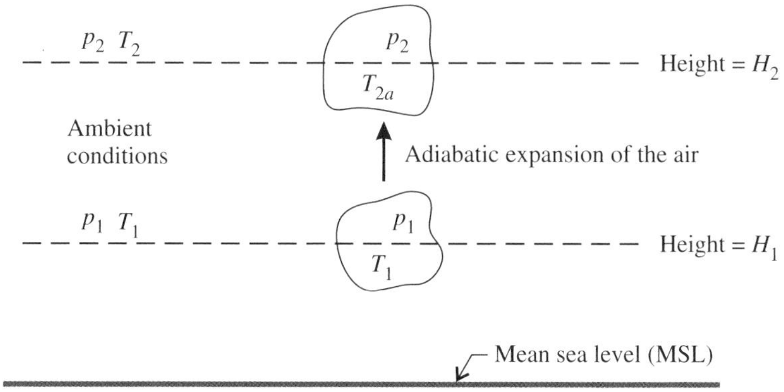

Figure 4.5 Notation for assessing the stability of the atmosphere.

For a reversible adiabatic process, the pressure and temperature are related by Equation 2.72, which is expressed below in the following form:

$$\frac{p}{T^{\left(\frac{\gamma}{\gamma-1}\right)}} = C \tag{4.35}$$

where C is a constant and γ is the ratio of specific heats of air.

Differentiating this equation with respect to T yields

$$\frac{\mathrm{d}p}{\mathrm{d}T} = C\left(\frac{\gamma}{\gamma-1}\right)T^{\left(\frac{\gamma}{\gamma-1}-1\right)}$$

Substitution of C from Equation 2.72 results in the expression

$$\frac{\mathrm{d}p}{\mathrm{d}T} = \left(\frac{\gamma}{\gamma-1}\right)\frac{p}{T}$$

or $$\frac{\mathrm{d}T}{\mathrm{d}p} = \left(\frac{\gamma-1}{\gamma}\right)\frac{T}{p} \tag{4.36}$$

Substituting Equations 4.36 and 4.23 into Equation 4.34 yields the required expression for the adiabatic lapse rate:

$$L_a = \left(\frac{\gamma-1}{\gamma}\right)\left(\frac{T}{p}\right)\left(-\frac{pg_0}{RT}\right) = -\left(\frac{\gamma-1}{\gamma}\right)\frac{g}{R} \tag{4.37}$$

Inserting the values for the ISA (see Table 4.1) produces the following result:

$$L_a = -0.00976 \text{ K/m} \quad \text{or } L_a = -9.76 \text{ K/km}$$

The above value is more accurately called the dry adiabatic lapse rate (DALR) as it does not consider what happens when water vapor in the air condenses to form a cloud of water droplets. In this situation, latent heat is released and this reduces the rate of cooling of the rising air mass. The saturated adiabatic lapse rate (SALR) is thus greater than the dry adiabatic lapse rate.

4.6.2 Stable Atmospheres

The determination of the adiabatic lapse rate leads to an interesting conclusion regarding the stability of the atmosphere. Again, it is useful to refer to Figure 4.5. The temperature T_{2_a} of the rising air mass at height H_2 can be predicted using Equation 4.37. However, the temperature of the surrounding air (T_2) will depend on the actual or environmental lapse rate (ELR) of the prevailing atmospheric conditions and this can be greater than, or less than, the temperature T_{2_a}. If T_{2_a} is less than T_2, then the rising parcel of air will be cooler than the surrounding air; it will also be more dense and will therefore sink back to its original height. This is a stable condition. In mathematical terms, the atmosphere is stable when

$$|ELR| < |ALR| \tag{4.38}$$

The lapse rate for the ISA troposphere is $L = -0.0065$ K/m (see Table 4.1); it can thus be concluded that the ISA troposphere is stable. Furthermore, it can also be concluded that the stratosphere is even more stable, as the lapse rate is zero.

4.6.3 Unstable Atmospheres

On the other hand, if thermal activity results in the rising parcel of air being warmer than the ambient air, it will then be less dense than the surrounding air and will therefore continue to rise. In this situation the atmosphere is unstable—a condition which frequently occurs during cyclonic weather in a meteorological depression. Thus, in such weather conditions, thermal activity is likely to be high, leading to the formation of cumulus cloud at the condensation level, and under very unstable conditions to the formation of cumulonimbus clouds and thunderstorms. Interestingly, the cloud mass in this situation will rise more rapidly than the dry thermal that formed it.

References

1 Dubin, M., Hull, A.R., and Champion, K.S.W., "US standard atmosphere 1976," NASA TM-X-74335 / NOAA S/T-76-1562, National Aeronautics and Space Administration, National Oceanic and Atmospheric Administration, United States Air Force, Washington, DC, 1976.
2 ANSI/AIAA, "Guide to reference and standard atmospheric models," G-003B-2004, American National Standards Institute (ANSI) and American Institute for Aeronautics and Astronautics (AIAA), USA, 2004.
3 ISO, *Standard atmosphere*, ISO 2533, International Organization for Standardization, Geneva, Switzerland, 1975.
4 ICAO, "Manual of the ICAO standard atmosphere: Extended to 80 kilometres (262 500 feet)," 3rd ed., Doc. 7488, International Civil Aviation Organization, Montréal, Canada, 1993.
5 ESDU, "Equations for calculation of international standard atmosphere and associated off-standard atmospheres," Data item 77022, Amdt. B, IHS ESDU, 133 Houndsditch, London, UK, Feb. 1986.
6 WMO, "World Meteorological Organization," Geneva, Switzerland, Retrieved July 29, 2012. Available from http://www.wmo.ch/.
7 Hoinka, K.P., "Temperature, humidity, and wind at the global tropopause," *Monthly Weather Review*, Vol. 127, pp. 2248–2265, 1999.
8 FSF, "FSF ALAR briefing note 8.6—Wind information," ALAR Tool Kit, Flight Safety Foundation, Alexandria, VA, 2009.
9 Thuillier, R.H. and Lappe, U.O., "Wind and temperature profile characteristics from observations on a 1400 ft tower," *Journal of Applied Meteorology*, Vol. 3, pp. 299–306, 1964.
10 Daney, J. and Lesceu, X., "Wind shear: An invisible enemy to pilots?," *Safety first*, Product Safety department, Airbus S.A.S., Blagnac, France, Iss. 19, pp. 38–49, Jan. 2015.

11 ICAO, "Meteorological service for international air navigation," 18th ed., Annex 3 to the Convention on International Civil Aviation, International Civil Aviation Organization, Montréal, Canada, July 2013.

12 Lankford, T.T., *Aviation weather handbook*, McGraw-Hill, New York, NY, 2001.

13 Airbus, "Getting to grips with cold weather operations," Flight Operations Support and Line Assistance, Airbus Industrie, Blagnac, France, Jan. 2000.

14 ICAO, "Rules of the air," 10th ed., Annex 2 to the Convention on International Civil Aviation, International Civil Aviation Organization, Montréal, Canada, July 2005.

15 WMO, "Meteorological service for international air navigation," Technical Regulation No. 49, Vol. II, World Meteorological Organisation, Geneva, Switzerland, 2004.

16 NOAA, "Federal meteorological handbook no. 1: Surface weather observations and reports," FCM-H1-2005, National Oceanic and Atmospheric Administration, Washington, DC, Sept. 2005.

5

Height Scales and Altimetry

5.1 Introduction

The terms height and altitude are often used interchangeably; however, in an aviation context, *height* is understood to be a vertical measure of distance from a defined datum according to a defined scale, but *altitude* is specifically understood to be *pressure height* measured from a defined pressure datum. Pressure height is one of four height scales (see Section 5.2) that are used in airplane performance work—these are geometric height, geopotential height, pressure height, and density height.

Altimetry (discussed in Section 5.3) is the science of measuring altitude using an altimeter, which provides a measure of height by comparing atmospheric pressure to a reference pressure value (or setting). In routine flight operations, three altimeter reference settings—namely, QNH, QFE, and QNE (or STD)—are used. These settings are explained and their use in flight operations described.

The world's navigable airspace is divided into three-dimensional blocks (segments), which generally follow the International Civil Aviation Organization (ICAO) classification system. The rights of pilots to fly at particular heights and the rules associated with the operation of an airplane in *controlled airspace* are highly regulated. Flight levels and flight tracks are defined to ensure safe separation of traffic. These topics are discussed in Section 5.4.

5.2 Height Scales

5.2.1 Geometric Height

Geometric height (h) is the true vertical distance from a datum level, which is usually mean sea level (MSL), to the point of interest. In airplane performance work, geometric height (which is sometimes called tapeline height) is normally confined to the context of the height of flight obstacles, such as aerials and buildings, and the elevation of airports and terrain. Jet transport airplanes are normally equipped with radio altimeters and satellite navigation equipment, which measure geometric height, the output of which is used for terrain proximity warning systems and by autoland systems, for example.

For aircraft performance work, however, it is easier to use *geopotential* height (see Section 5.2.2) rather than geometric height, as altimetry is based on geopotential height.

Performance of the Jet Transport Airplane: Analysis Methods, Flight Operations, and Regulations, First Edition.
Trevor M. Young.
© 2018 John Wiley & Sons Ltd. Published 2018 by John Wiley & Sons Ltd.

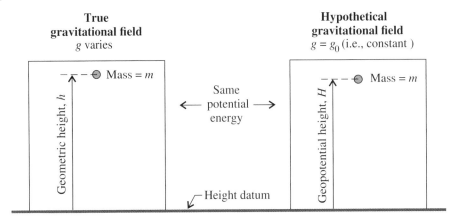

True gravitational field
g varies

Mass = m

Geometric height, h

Same ← potential → energy

Height datum

Hypothetical gravitational field
$g = g_0$ (i.e., constant)

Mass = m

Geopotential height, H

Figure 5.1 Geometric and geopotential height.

5.2.2 Geopotential Height

Geopotential height (H) is a height scale based on potential energy, which takes into account the variation in the Earth's gravitational field. Geopotential height, by definition, is the height in a hypothetical *uniform gravitational field* that would give the same potential energy as the point under consideration in the actual, variable gravitational field (see Section 2.3.4). Thus, if the standard value of gravity—that is, $g_0 = 9.80665$ m/s^2 ($g_0 = 32.174049$ ft/s^2)—is used in conjunction with geopotential height, the variation of gravity with height is automatically taken into account when calculating potential energy. The use of geopotential height greatly simplifies airplane performance analyses.

By equating the work (see Section 2.4.6) needed to raise a body in a real gravitational field to the work required to raise the same body in a hypothetical, constant gravitational field, the geopotential height can be determined as a function of the geometric height (see Figure 5.1):

$$\text{Work performed} = m \int_0^h g \, \mathrm{d}h = mg_0H \tag{5.1}$$

where m is the mass of the body (typical units: kg, slug);
g is the real, variable acceleration due to gravity (typical units: m/s^2, ft/s^2);
h is the geometric height (typical units: m, ft);
g_0 is the standard value of gravity (typical units: m/s^2, ft/s^2); and
H is the geopotential height (typical units: m, ft).

Newton's law of universal gravitation can be used to model the vertical distribution of gravity (see Equations 2.6–2.8). For the purpose of defining the ISA (and for the analysis of aircraft performance based on the ISA), a nominal radius of the Earth (r_E), which is equal to 6 356 766 m (20 855 531 ft), is used. This value of r_E, when used in conjunction with the standard value of gravity (g_0), results in a calculated vertical distribution of gravity that is very close to the true values at latitude 45.5425° [1, 2].

Combining Equations 2.9 and 5.1 results in the following expression for geopotential height:

$$H = \int_0^h \left(\frac{r_E}{r_E + h} \right)^2 \mathrm{d}h = \frac{r_E h}{r_E + h} \tag{5.2}$$

Table 5.1 Geometric and geopotential height scales

Geometric height (h) (ft)	Geopotential height (H) (ft)	Difference (%)
0	0	0.00%
10 000	9995	−0.05%
20 000	19 981	−0.10%
30 000	29 957	−0.14%
40 000	39 923	−0.19%
50 000	49 880	−0.24%
60 000	59 828	−0.29%

Alternatively, Equation 5.2 can be rearranged and the geometric height expressed in terms of the geopotential height:

$$h = \frac{r_E H}{r_E - H} \qquad (5.3)$$

Table 5.1 illustrates the relationship between these two height scales (from MSL to 60 000 ft). The difference between geometric and geopotential height for typical cruise altitudes is small, and for many applications, the discrepancy can be ignored—for example, at 40 000 ft the difference is less than 0.2%. Even for the Concorde, which had a maximum cruise altitude of 60 000 ft, the difference would have been less than 0.3%.

Geopotential Height in the ISA

The integration of Equation 4.8, performed earlier in Section 4.2.4, to give pressure and density variations with height in the ISA, was conducted using the standard value of gravitational acceleration. This means that the resulting equations that describe pressure and density in the ISA (see Table 4.2) are expressed in terms of geopotential height and not geometric height.

5.2.3   Pressure Height

Before introducing the concept of pressure height (H_p), it is useful to review the pressure versus geopotential height equations derived in Section 4.2.4. The relevant expression for the troposphere is

$$p = p_0 \left[1 + \frac{LH}{T_0} \right]^{-g_0/RL} \qquad \text{(Equation 4.13)}$$

and for the stratosphere, it is

$$p = p^* e^{\frac{-g_0(H-H^*)}{RT^*}} \qquad \text{(Equation 4.15)}$$

It is noted that the parameters $p_0, p^*, L, g_0, H^*, T_0, T^*,$ and R are all defined by standard values in the ISA (see Table 4.1). This means that the ISA defines a unique relationship between pressure and geopotential height. In other words, pressure can be uniquely mapped (in a mathematical sense) to geopotential height and vice versa. The height scale in the ISA can thus be considered as a *scale of pressure*—and this is used in the concept of pressure height.

By definition, the *pressure height* (H_p) at a point in any atmosphere (standard, off-standard, or even non-standard) is the height in the Standard Atmosphere (*H*) that has the same pressure as the point of interest. Thus, if the point of interest were in an atmosphere that exactly matches the ISA, the pressure height (H_p) would be identical to the geopotential height (*H*). However, if the point of interest were in an off-standard or non-standard atmosphere (see Section 4.3), then the pressure height would not be the same as the geopotential height.

The concept of pressure height comes from the basic principle behind the operation of an altimeter (see Section 5.3.1), which measures barometric pressure, and, via a calibration function, provides an indication of height. In an operational context, pressure height is called *altitude*.

5.2.4   Density Height

Density height (H_d), by definition, is the height at which the density in the ISA matches the given or ambient air density. Low values of air density, which can be encountered in so-called "hot and high" operations, for example, have a detrimental influence on an airplane's performance. This can be illustrated by considering the lift required to support an airplane's weight during takeoff. The factors that influence lift are given by Equation 3.21. It is evident that, for an airplane to generate sufficient lift to support its weight at a particular angle of attack (which implies a particular value of C_L), the product of ρ and V^2 must have a certain critical value. Thus, in hot and/or high altitude conditions, the true airspeed (*V*) must increase by a squared relationship to compensate for a reduction in air density (ρ).

Density height is a useful concept for understanding airplane performance, particularly when considering hot and high conditions; however, it is rarely, if ever, used in commercial flight operations. Since standard airplane performance data are presented as a function of pressure height with appropriate ambient temperature corrections, density height is not required for safe flight operations.

5.3   Altimetry

5.3.1   The Altimeter

The underlying principle of an airplane's altimeter is essentially the same as that of a barometer, which is an instrument that measures atmospheric pressure. The simplest type of altimeter measures atmospheric pressure based on the size of an expandable, sealed capsule, called an aneroid. An aneroid contains a fixed mass of air and is designed to expand (or contract) when the outside air pressure decreases (or increases). The ambient pressure is conveyed to the instrument via a tube connected to a static port (a small opening on the surface of the airplane). On simple mechanical instruments (as might be found installed in the instrument panel of general aviation aircraft), the expansion and contraction of the aneroid drives a pointer across a scale, which is calibrated to read *height* (not pressure).

An altimeter requires only two inputs to provide a measure of height: the ambient atmospheric pressure and a datum (or reference) pressure. The relationship between pressure and height is based on the equations that were derived in Section 4.2 for the ISA. An altimeter thus provides the pilot with a reading of *pressure height* (and not geometric height). On modern airplanes, altitude is displayed on a vertical "tape" on the Captain's and First Officer's Primary Flight Displays (PFDs), as illustrated in Figure 5.2a, and also on a backup altimeter. A traditional mechanical altimeter is illustrated in Figure 5.2b, for comparison. The reading from a particular

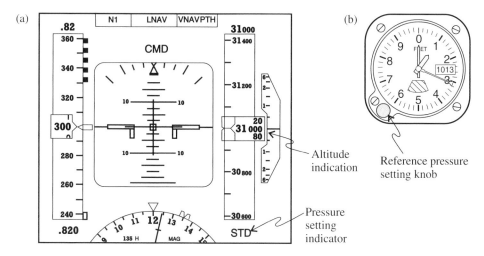

Figure 5.2 (a) Primary Flight Display (PFD); (b) Mechanical altimeter.

instrument is called the *indicated altitude* (altimeter correction factors and errors are discussed later in Sections 5.3.6 to 5.3.8).

For heights up to the tropopause, the height versus pressure relationship is obtained by a rearrangement of Equation 4.13, as follows:

$$H_p = -\frac{T_0}{L}\left[1-\left(\frac{p}{p_0}\right)^{-RL/g_0}\right] \tag{5.4}$$

For an altimeter to be a useful instrument for day-to-day operations, where varying atmospheric (barometric) pressure conditions are encountered, it must be possible to set the zero height (datum) reading. The pilot accomplishes this by rotating a control, or subscale, knob, which on mechanical instruments is located on the front face of the instrument itself (as shown in Figure 5.2b). This, in effect, changes the reference pressure used to determine the height. Replacing p_0 by p_{ref} in Equation 5.4 and inserting the standard values for T_0, L, R, and g_0 (see Table 4.1) results in the following calibration equation:

$$H_p = 145\,442\left[1-\left(\frac{p}{p_{ref}}\right)^{0.190263}\right] \tag{5.5}$$

where H_p is pressure height measured in feet; and
p and p_{ref} are the atmospheric and reference pressures, respectively (as they appear as a ratio, any unit of pressure can be used).

For commercial airplane operations in most countries around the world, altitude is measured in feet—although meters are used in a few countries (see Section 5.4.1). By convention, the reference pressure is specified in hPa (or mb, which is an equivalent unit accepted for use with the SI system) or inches of mercury (inHg). On modern electrical instruments, the option exists to set the reference pressure in units of either hPa or inHg.

Figure 5.3 illustrates the troposphere calibration function, given by Equation 5.5, for two selected reference pressures: 995 hPa and 1013.25 hPa. As expected, both plots indicate zero pressure height at their respective "sea-level" reference pressures. Note that the calibration

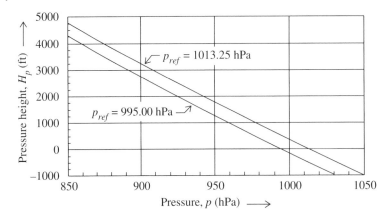

Figure 5.3 Illustration of altimeter reference pressure setting.

curves are seen to be approximately linear over the *ca.* 5000 ft height interval shown (the actual function is curved, as illustrated in Figure 4.2).

In flight operations, the selection of appropriate reference pressures by flight crews follows a well-defined convention. Three reference pressure settings are used:[1]

- QNE setting, which is also known as standard (STD) (see Section 5.3.2);
- QNH setting (see Section 5.3.3); and
- QFE setting (see Section 5.3.4).

5.3.2   QNE or Standard (STD) Setting

The QNE altimeter pressure setting—also known as the standard (STD) setting—corresponds to the ISA sea-level datum value of 1013.25 hPa (which is identical to 1013.25 mb) or 29.92126 inHg (which is used in North America). For flight operations, the values are rounded-off to four significant figures (i.e., 1013 hPa and 29.92 inHg).

By international agreement, all airplane traffic above a certain *transition altitude* (see Section 5.4.2) must utilize the QNE setting. The possibility of a collision occurring at these heights is thus minimal (when correct procedures are followed), as all neighboring aircraft will have their altimeters set to the same reference pressure, irrespective of the actual meteorological conditions. The QNE pressure setting is used to define *flight levels* (described in Section 5.4.1).

5.3.3   QNH Setting

The QNH is a barometric pressure, which, if selected as the reference pressure on the airplane's altimeter, will result in the instrument indicating pressure height above mean sea level (AMSL) for the prevailing atmospheric conditions. The QNH is based on a measured barometric pressure—recorded at a selected location (e.g., on an airfield)—and knowledge of the precise elevation of that location. Using the ISA pressure–height relationship (Equation 5.5),

---

1 QNE, QNH, and QFE are "Q" codes—part of a standardized three-letter coding system that has its origin in maritime communication using Morse code (initially developed by the British government in *ca.* 1909). Although most Q codes are now obsolete, a few remain part of the ICAO standard radiotelephony phraseology [3].

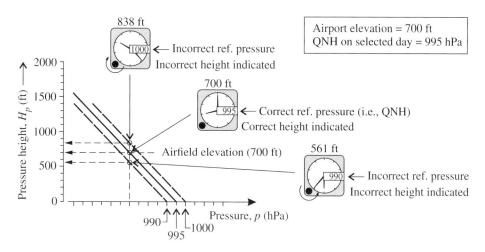

Figure 5.4 Illustration of QNH altimeter setting (with correct and incorrect reference pressure settings).

the pressure corresponding to zero height is computed or otherwise determined using a calibrated instrument—this pressure is called the QNH.

Note that as meteorological conditions change, the QNH for a particular location also changes. Similarly, the QNH will not necessarily be the same for different airports, even those located in the same geographical region. The crew may obtain the QNH applicable to a specific airport or geographical location from the nearest air traffic control (ATC) service or from an automatic air traffic information service. Depending on the location, the QNH will be given in hPa (or mb) or in inHg. Typically, airports will update the QNH every hour, and flight crews of affected air traffic are required to adjust their altimeters accordingly.

Typically, before takeoff, a flight crew would set the QNH for the departure airport as the altimeter reference pressure. The altimeter would then provide a height indication that is equal to the airport's elevation. (As part of normal predeparture checks, the flight crew would routinely confirm that this is correct by referring to the elevation data on their navigation charts or other approved document.) As an illustration, an airplane is assumed to be at an airport that is at an elevation of 700 ft at a time when the QNH is 995 hPa. If the crew correctly adjusts the reference pressure, the altimeter reading would indicate airport elevation, as illustrated in Figure 5.4. Increasing the reference pressure would shift the height–pressure line to the right; reducing the reference pressure shifts the line to the left. In both cases, incorrect heights would be indicated on the altimeter.

In normal operations, a QNH pressure setting is used for takeoff and climb-out to the transition altitude (at which point the pilot switches over to the STD setting), as well as for landing (where the altimeter setting is based on the QNH of the destination airport). For flights below the transition altitude (e.g., by general aviation pilots), QNH is typically used throughout the flight.

5.3.4   QFE Setting

A QFE is a barometric pressure reading, given in hPa (or mb) or inHg, which, if selected as the reference pressure for the altimeter, will result in the altimeter providing the correct pressure height with respect to the *elevation* of the airport or ground station. Thus, when an altimeter has the QFE pressure set, it will indicate the pressure height above the local terrain, or, more specifically, above the datum at the airport (e.g., the runway threshold) or the ground station. If the flight crew sets the local QFE as the reference pressure, the altimeter will indicate zero height (or a value very close to zero) when the airplane is on the runway.

Although QFE is sometimes used for military operations and by general aviation pilots, it is seldom used for commercial aviation (exceptions include some flight operations in Russia and China, for example). Furthermore, QFE information is not available from ATC in all countries.

5.3.5   Illustration of Altimeter Settings

Consider, for the sake of illustration, a particular day for the locality illustrated in Figure 5.5 when the sea-level barometric pressure is 995 hPa. (This, incidentally, would indicate a meteorological low pressure situation.) An aircraft flying at flight level 60 has its altimeter set at 1013 hPa (i.e., STD setting). The reading on the altimeter will thus be 6000 ft, even though the height above mean sea level is only 5500 ft.

The QNH at the local airport, which is at an elevation of 700 ft, is 995 hPa and the QFE is 970 hPa. The altimeter of a light aircraft flying at an altitude of 3000 ft AMSL would indicate this height if the altimeter was set at the QNH pressure of 995 hPa. However, if the pilot were to set the altimeter subscale to the QFE pressure reading of 970 hPa, then the altimeter would indicate 2300 ft (i.e., the pressure height above the airport, neglecting instrument error).

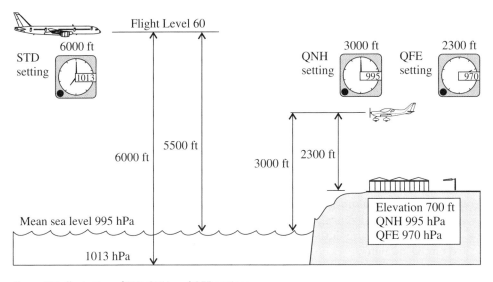

Figure 5.5 Illustration of STD, QNH, and QFE settings.

5.3.6   Altitude Correction: STD to QNH

It is convenient for flight crews to be able to make mental adjustments to the indicated altitude (IA) to account for the local QNH when the altimeter is set to STD. A method is thus needed to facilitate an estimate to be made of the actual height above mean sea level, using the IA and knowledge of the rate of change of pressure with height. The calculation, however, is complicated by the fact that atmospheric pressure decreases in a non-linear fashion with increasing height (see Section 4.2). When the barometric pressure at sea level is 1013 hPa, the pressure falls by 1 hPa over the first 27.3 ft. This value (or the equivalent value when pressure is measured in inHg) can be used to make an approximate adjustment to the IA, using the following equation:

$$H \cong H_{IA} + \left(\frac{dH}{dp}\right)_{H=0} (p_{QNH} - p_{STD}) \tag{5.6}$$

where H is the estimated altitude at the QNH pressure setting (typical unit: ft);
$\quad H_{IA}$ is the indicated altitude at the STD pressure setting (typical unit: ft);
$\quad p_{QNH}$ is the QNH pressure setting (typical units: hPa, inHg);
$\quad p_{STD}$ is the STD pressure setting (i.e., 1013 hPa or 29.92 inHg);

and $\quad \left(\dfrac{dH}{dp}\right)_{H=0} = 27.3$ ft/hPa $\quad$ or $\quad \left(\dfrac{dH}{dp}\right)_{H=0} = 925$ ft/inHg

5.3.7 Altitude Correction for Non-Standard Temperature Conditions

The calibration function used in the design of altimeters is based on the constants defined in the ISA (see Equation 5.4). Although it is possible for a pilot to set the reference pressure— and thus account for varying barometric pressure conditions—it is not possible to adjust the temperature constants (i.e., sea-level datum temperature and lapse rate) that are imbedded in the design of the instrument. This means that a systematic discrepancy is introduced whenever the meteorological conditions differ from those defined in the ISA. On a cold day, when the temperature falls below that defined in the ISA for the particular altitude, the altimeter will "over-read," indicating an altitude that is greater than the actual altitude. The reverse is true for a hot day. Similarly, when the actual lapse rate does not match the ISA lapse rate, such as when there is a temperature inversion, a discrepancy between the indicated height and the actual height arises.

Altimeters provide reference heights for vertically separating traffic and, as all aircraft would be affected in the same way by changes in air temperature, such discrepancies do not create a collision danger; however, for flights in mountainous terrain under conditions of extreme cold, this height discrepancy can create a terrain clearance hazard.

The altimeter calibration function is defined in terms of pressure height (H_p) and not geopotential height (H). The "scale error" that exists between the two measures for off-standard temperature conditions was evaluated in Section 4.3.3—and this provides a means to correct altimeter readings. It follows from Equation 4.30 that

$$H_p - H = \frac{R\,\Delta T}{g_0} \ln \delta \tag{5.7}$$

where R is the gas constant (see Section 2.5.6);
$\quad \Delta T$ is the ISA temperature variation (see Section 4.3.2); and
$\quad \delta$ is the pressure ratio.

The difference between pressure height and geopotential height is illustrated in Figure 5.6 for four selected off-standard atmospheric conditions (determined using Equation 5.7). In an off-standard atmosphere, a constant temperature difference exists at all heights. In an arbitrary, non-standard atmosphere this is not the case. Nonetheless, Equation 5.7 provides a means for correcting for any temperature difference: ΔT is taken as $\Delta T = T_{act} - T_{std}$, where T_{act} is the actual temperature (at a given pressure height) and T_{std} is the ISA temperature (at the same height).

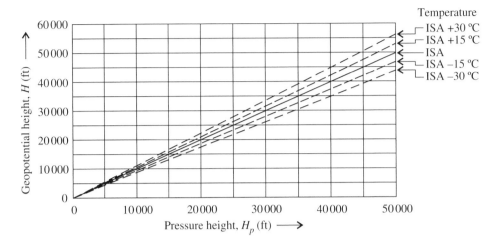

Figure 5.6 Geopotential height versus pressure height for selected off-standard temperature conditions.

In cold conditions, ΔT is negative, and when multiplied by the logarithm of a positive fraction (i.e., ln δ), the product is positive. Hence, in cold conditions, H_p will be greater than H (see Equation 5.7), and the true height of the airplane will be lower than that indicated by the altimeter. This confirms the earlier statement that low temperatures can create a terrain clearance hazard. For this reason, ICAO publishes altimeter correction formulae and tables for cold weather flight operations [4].

It is seen in Figure 5.6 that the lines are almost straight, indicating that the influence of changing pressure ratio (or height) on the error is relatively small—this implies that a simple correction procedure is possible. The following approach provides an approximate height correction for low altitudes (say below 5000 ft). The indicated altitude (which is the pressure height above the datum on either a QNH or QFE setting) is decreased by 4% per 10 °C below ISA (or increased by 4% per 10 °C above ISA). For example: consider an airplane flying over an airport that is at an elevation of 2000 ft, and the measured airport temperature is −10 °C. The ISA temperature for 2000 ft is 11 °C (see Table A.1, Appendix A); thus $\Delta T = -21$ °C. The altimeter will therefore "over-read" by approximately 8.4%. If the indicated altitude on the altimeter is 3000 ft (with the instrument set to the airport's QNH), the corrected altitude will be 2748 ft. This means that the actual height of the airplane above the airport elevation is ~748 ft and not 1000 ft.

Flight crew operating manuals (see Section 23.7.2) usually contain an altitude correction table for low temperature conditions at low altitudes. These tables are often based on the conservative altitude correction data published by ICAO [4]. For *en route* operations, when the altimeter is set to STD, the indicated altitude can be corrected, with reasonable accuracy, by multiplying the height by the ratio of the OAT (outside air temperature) to the standard temperature (for that height).

$$H_{CA} = H_{IA} \left(\frac{T_{act}}{T_{std}} \right) \tag{5.8}$$

where H_{CA} is the corrected altitude (typical units: ft, m);
$\quad\quad H_{IA}$ is the indicated altitude (typical units: ft, m);
$\quad\quad T_{act}$ is the actual OAT at the given height (typical units: K, °R); and
$\quad\quad T_{std}$ is the standard temperature at the given height (typical units: K, °R).

5.3.8   Instrument and Static Pressure Errors

An altimeter indicates altitude based on atmospheric pressure. Its accuracy depends on the accuracy with which the pressure can be determined, the quality of the construction of the instrument, and the precision with which it has been calibrated; collectively, these factors are known as instrument errors. Furthermore, the airplane's static ports are located within a pressure field that is influenced by the airplane itself, and this can lead to small static pressure errors. While it is possible to minimize this influence at a particular flight attitude and speed, static pressure errors can arise if the pressure field changes significantly—for example, due to a change in angle of attack, sideslip, or airplane configuration.

As a consequence of careful design and correct positioning of static ports, it is now common-place to have negligible static pressure errors associated with the primary air data systems on modern jet transport aircraft. Furthermore, the design of modern flight instruments and the tolerances with which they are manufactured have reduced instrument errors to an insignificant level.

5.4   Flight Levels, Tracks, and Airspace

5.4.1   Flight Level

For all airplane operations above a defined height called the *transition altitude* (see Section 5.4.2), altitude is specified in terms of a flight level (FL). Flight levels are usually measured in hundreds of feet and are separated by, at least, 500 ft intervals—for example: FL 55 represents a pressure height of 5500 ft on the STD pressure setting. At typical cruise altitudes, flight levels are separated by 1000 ft intervals. All airplanes operating above the transition altitude use the same reference datum of 1013 hPa (or 29.92 inHg), enabling pilots—and air traffic controllers—to maintain an appropriate vertical separation between aircraft. In a small number of countries,[2] meters, not feet, are used. In metric airspaces, flight levels, which are identified by the word "meters" or the letter "M," are set at 300 m (984.3 ft) intervals. Pilots entering or leaving a metric airspace would usually be required to make a small adjustment to their cruising altitude.

5.4.2   Transition Altitude and Transition Level

The transition altitude (TA) is a defined pressure height above mean sea level, based on the local QNH, at which flight crews are required by ATC to switch to the STD altimeter setting during climb-out (see Figure 5.7). Transition altitudes are set by national or local ATC authorities, and can, for example, vary from 3000 ft (used in some places in Europe[3]) to 18 000 ft (used throughout North America).

Associated with the TA is the transition level (TL), which, by definition, is the lowest usable flight level above the TA. On descent, the change from STD to QNH pressure setting takes place at the TL (see Figure 5.8). The height above MSL of the TA (for a particular airspace) is

---

2 China (but not Hong Kong), Mongolia, North Korea, Russia, and many CIS (Commonwealth of Independent States) countries have traditionally measured altitude in meters (not feet) for flight operations. In 2011, however, Russia and some CIS states switched to using feet above the transition altitude.

3 In the United Kingdom (UK), for example, there are very few mountains above 3000 ft and hence a relatively low TA can be used. In the vicinity of major UK airports the TA is higher (typically between 4000 ft and 6000 ft).

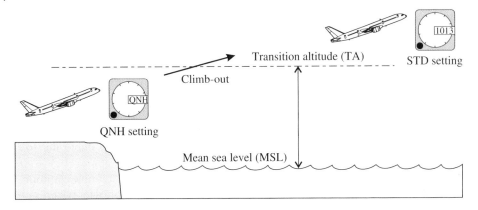

Figure 5.7 Transition altitude (altimeter setting is changed during climb-out).

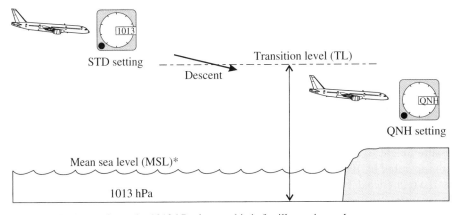

* MSL is shown above the 1013 hPa datum; this is for illustration only.

Figure 5.8 Transition level (altimeter setting is changed during descent).

fixed, but the height above MSL of the TL will change as the meteorological conditions change. For safety reasons, the vertical separation between the TA and the TL must be greater than a specified minimum, usually 500 ft or 1000 ft. Terminal (i.e., airport) information, supplied as a matter of routine to flight crews, will include the TA (and the associated QNH) and the TL for the airspace of interest.

5.4.3 Air Traffic Tracks and Usable Flight Levels

There are defined ATC rules (conventions) that govern the flight levels used for routine flight operations. One such convention, which is used in many parts of the world,[4] is the ICAO semi-circular rule [5]. It requires that IFR (Instrument Flight Rules) traffic, operating in airspace

---

4 In counties where the majority of the air traffic is in a general northerly/southerly direction (e.g., Italy, New Zealand), a semicircular rule is used with an East–West track split (i.e., 090°–269° and 270°–089°).

with usable flight levels that are separated by 1000 ft intervals, adopt the following convention, depending on the direction of the flight:

- Odd thousands of feet (e.g., FL 250, FL 270) are used for tracks 000° to 179°.
- Even thousands of feet (e.g., FL 260, FL 280) are used for tracks 180° to 359°.

Pilots are often taught to remember this convention by the mnemonic ONE, which stands for "Odd for North and East."

Historically, usable flight levels were set at 1000 ft intervals below FL 290 (providing a 2000 ft separation for traffic moving in the same general direction) and at 2000 ft intervals above FL 290. This meant that FL 290 and FL 330, for example, would be used for eastbound traffic and FL 310 and FL 350, for example, for westbound traffic. However, this changed when certain major airspaces[5] reduced the separation interval from 2000 ft to 1000 ft (in effect doubling the airspace capacity), for flight levels from FL 290 to FL 410. Above FL 410, the usable flight levels in these regions are still separated by 2000 ft intervals (giving 4000 ft separation for same-direction traffic).

Table 5.2 ICAO airspace classification

Class	Type of flight[b]	ATC clearance[c]	Separation service provided	Speed limitation[d]
A	IFR	yes	All aircraft are separated from each other	not applicable
B	IFR	yes	All aircraft are separated from each other	not applicable
	VFR	yes	All aircraft are separated from each other	not applicable
C	IFR	yes	IFR traffic are separated from IFR traffic and IFR from VFR traffic	not applicable
	VFR	yes	VFR traffic are separated from IFR traffic	250 kt IAS below 10 000 ft
D	IFR	yes	IFR traffic are separated from IFR traffic	250 kt IAS below 10 000 ft
	VFR	yes	No separation (traffic information)	250 kt IAS below 10 000 ft
E	IFR	yes	IFR traffic are separated from IFR traffic	250 kt IAS below 10 000 ft
	VFR	no	No separation (traffic information as far as practical)	250 kt IAS below 10 000 ft
F	IFR	no	IFR traffic is separated from IFR traffic as far as practical (traffic advisory service)	250 kt IAS below 10 000 ft
	VFR	no	No separation (flight information service)	250 kt IAS below 10 000 ft
G	IFR	no	No separation (flight information service)	250 kt IAS below 10 000 ft
	VFR	no	No separation (flight information service)	250 kt IAS below 10 000 ft

Notes:

(a) *Source*: Based on ICAO Annex 11, Appendix 4 [7].
(b) IFR means Instrument Flight Rules and VFR means Visual Flight Rules (see Section 4.5.8).
(c) Flights subject to an ATC clearance are "controlled."
(d) IAS means indicated airspeed (see Section 6.7). When the height of the transition altitude is lower than 10 000 ft, FL 100 is used in place of the 10 000 ft limit.

---

5 Since 2000, under the guidance of ICAO [6], Reduced Vertical Separation Minimum (RVSM) has been implemented in the airspaces over Europe, North America, South America, Australia, the Middle East, Russia and parts of Asia, and the Atlantic and Pacific Oceans.

In certain geographical regions with high traffic volume, tracks are strictly controlled (organized)—for example: over the North Atlantic, the OTS (Organized Tracking System) applies and over the North Pacific the PACOTS (Pacific Organized Tracking System) applies. The OTS tracks are assigned twice daily, taking into account the position of the jet stream [8].

5.4.4 Airspace Classification

The navigable airspace of the world is divided into three-dimensional segments (zones), each of which is assigned to a defined *class* (or category). Flight operations within these segments are conducted according to the rules that apply to that airspace class.

The airspace classification adopted by ICAO [7] is summarized in Table 5.2. Not all of these airspace categories are used in all countries, and some counties (e.g., China) use their own definitions. Furthermore, despite the gradual harmonization of airspace rules, not all countries have adopted the precise ICAO definitions. Some countries modify the ICAO rules to suit their needs (or to align the rules with those that existed before ICAO standardization was introduced). The definitive rules for each country are documented in that country's Aeronautical Information Publication (see Section 23.4.6).

Note that only IFR operations are permitted in Class A airspace, which is used for the cruise portion of most commercial jet flights. Of particular significance to climb and descent performance analyses is an airspeed limitation: it is required that a speed of 250 kt IAS (indicated airspeed) is not exceeded below 10 000 ft in several airspace classes.

References

1 ESDU, "Equations for calculation of international standard atmosphere and associated off-standard atmospheres," Data item 77022, Amdt. B, IHS ESDU, 133 Houndsditch, London, UK, Feb. 1986.

2 ICAO, "Manual of the ICAO standard atmosphere: Extended to 80 kilometres (262 500 feet)," 3rd ed., Doc. 7488, International Civil Aviation Organization, Montréal, Canada, 1993.

3 ICAO, "Aeronautical telecommunications, Vol. II," Annex 10 to the Convention on International Civil Aviation, 6th ed., International Civil Aviation Organization, Montréal, Canada, Oct. 2001.

4 ICAO, "Procedures for air navigation services: Aircraft operations, Vol. 1, Flight procedures," 5th ed., Doc. 8168, International Civil Aviation Organization, Montréal, Canada, 2006.

5 ICAO, "Rules of the air," 10th ed., Annex 2 to the Convention on International Civil Aviation, International Civil Aviation Organization, Montréal, Canada, July 2005.

6 ICAO, "Manual on implementation of a 300 m (1000 ft) vertical separation minimum between FL 290 and FL 410 inclusive," Doc. 9574, 2nd ed., International Civil Aviation Organization, Montréal, Canada, 2001.

7 ICAO, "Air traffic services," Annex 11 to the Convention on International Civil Aviation, 13th ed., International Civil Aviation Organization, Montréal, Canada, July 2001.

8 Clark, P., *Buying the big jets: Fleet planning for airlines*, 2nd ed., Ashgate, Aldershot, Hampshire, UK, 2007.

6

Distance and Speed

6.1 Introduction

The speed that an airplane travels with respect to the air mass (i.e., the true airspeed) is seldom identical to the speed measured with respect to the ground. This is, of course, due to the influence of wind on the airplane's movement. Similarly, the air distance and the ground distance (see Section 6.2) that an airplane covers are rarely the same.

In flight, the flight crew is provided with an airspeed indication (typically in knots) on their Primary Flight Displays (illustrated in Figure 5.2a). This is called an *indicated airspeed*, and it is generally different from the true airspeed. The flight Mach number, which is also shown on the Primary Flight Displays, is computed from the true airspeed with knowledge of the speed of sound in the surrounding air. The relationship between the indicated airspeed and the true airspeed is a complex one, requiring three separate adjustments (or correction factors) to be applied.

The different speeds that are encountered in airplane performance analysis are described in this chapter, namely, true airspeed and ground speed (Section 6.3), Mach number (Section 6.4), equivalent airspeed (Section 6.5), calibrated airspeed (Section 6.6), and indicated airspeed (Section 6.7). The relationships between these speeds are discussed in Section 6.8.

6.2 Distance

6.2.1 Units of Measurement

The customary units that are used for the measurement of air or ground distance are nautical miles (nm), kilometers (km), and statute miles (mi). The historical definition of a nautical mile, which has a long history of use in sea and air navigation, is that 1 nm is equal to 1 minute of arc measured along a meridian (or line of longitude) of the Earth. This definition, which is today regarded as a good approximation of a nautical mile,[1] is very useful for manual navigation as distances on traditional Mercator projection charts can be estimated using the chart scale (latitude and longitude are indicated in degrees and minutes).

---

1 The current, internationally agreed value of the nautical mile, i.e., 1 nm = 1852 m (exactly), was adopted at the First International Extraordinary Hydrographic Conference (Monaco, 1929) and subsequently accepted by the International Bureau of Weights and Measures (BIPM) [1]. This definition, however, was not universally accepted by all major nations until the 1950s.

Performance of the Jet Transport Airplane: Analysis Methods, Flight Operations, and Regulations, First Edition.
Trevor M. Young.
© 2018 John Wiley & Sons Ltd. Published 2018 by John Wiley & Sons Ltd.

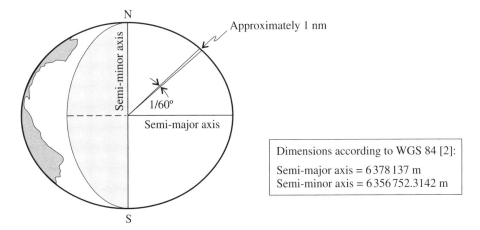

Figure 6.1 Oblate ellipsoid (exaggerated) representation of the Earth, indicating the historical definition of one nautical mile.

There is, of course, a problem with this historical definition: the Earth is not perfectly spherical. Due to its oblate shape, the precise length of an arc formed on a meridian on the Earth's surface by sweeping through an angle of 1 minute of a degree at the Earth's center varies, depending on latitude (Figure 6.1).

6.2.2 Air Distance

The *air distance* can be defined as the distance traveled with respect to the surrounding (or ambient) air mass. The best way to visualize this is to view the airplane as traveling in a large air mass, while the air mass itself is moving with respect to the ground. In theoretical performance analyses, the air distance is often called the *still air distance* to make it clear that the effects of wind have been ignored.

In navigation, the term *equivalent still air distance* (ESAD) is used to describe the air distance that is equivalent to a particular ground distance under a prevailing wind condition (which can be defined by speed and direction with respect to the ground track). The ESAD can be determined by vector algebra, as discussed later in Section 6.3.2.

6.2.3 Ground Distance

The *ground distance* that an airplane covers is the length of the airplane's track, which is the projected path of the airplane onto the Earth's surface. In other words, it is the air distance corrected for the influence of the wind. The units are the same as those used for air distance.

6.2.4 Great Circle Distance

A *great circle* of a sphere is the intersection of the sphere with a plane that passes through the center of the sphere. In mathematics, a great circle or a portion of one is known as an orthodrome. Any two points on the surface of the sphere can be joined by an arc of a great circle—the center of which coincides with the center of the sphere. The significance of a great circle arc is that the length of the arc is the shortest distance of any route along the surface between the two points. The same concept can be applied to an ellipsoid representation of the Earth; this, of course, is a very useful navigation technique for long distance travel.

The *great circle distance* (GCD) between two points on the Earth's surface can be approximated by the following expression, which is derived for a sphere:[2]

$$d_{gc} = r_E \cos^{-1}[\sin\phi_1 \sin\phi_2 + \cos\phi_1 \cos\phi_2 \cos(\psi_2 - \psi_1)] \qquad (6.1)$$

where the arccosine angle is expressed in radians, and

where d_{gc} is the great circle distance (units are the same as those used for r_E);

r_E is a mean (or equivalent) radius of the Earth, which can be taken as 6371.01 km (3440.07 nm) [2];

ϕ is the geographic latitude (north is positive, south is negative);

ψ is the geographic longitude (east is positive, west is negative); and

subscript 1 denotes the starting point and subscript 2 denotes the end point.

The longest possible direct route that can be flown is between antipodes—that is, points on the Earth's surface that are diametrically opposite each other (i.e., a straight line connecting the two points will theoretically pass through the center of the Earth). The great circle distance between Madrid, Spain and Wellington, New Zealand (two cities that are nearly antipodal), for example, is approximately 10 700 nm. Long-range airplanes, such as the A340-500 and B777-200LR, can cover such distances, but without a viable payload. A shorter route that has been flown with an economically viable payload is Singapore to New York[3]—a great circle distance of almost 8300 nm.

The disadvantage of great circle navigation is that the true bearing is not constant, and must continuously be adjusted—this is not the case for rhumb navigation (discussed below).

6.2.5   Rhumb-Line Distance

A rhumb-line (also called a loxodrome) is a navigation term that describes a course (or path) that crosses all lines of longitude at the same angle. Such a course maintains a constant bearing with respect to true north (which is not the case with a great circle). Rhumb navigation has been used for centuries as the course appears as a straight line on a Mercator projection navigation chart (or map). Only under specific conditions—that is, a course between locations that have the same longitude or a course between locations on the equator—will the rhumb-line distance (d_{rl}) equal the great circle distance (d_{gc}); for all other courses on the Earth's surface, rhumb-line distances are longer than comparable great circle distances. The deviation is greatest for west–east (or east–west) courses and increases with increasing latitude (the influence is greatest near the poles). For short distance travel (other than in the vicinity of the poles) the deviation is negligible. From New York (40°45′N/73°58′W) to London (51°32′N/0°10′W), for example,

---

2 Equation 6.1 is an application of the spherical law of cosines. The formula provides satisfactory estimates of great circle trip distances; however, it is ill-conditioned (in a mathematical sense) when the two points on the Earth's surface are in close proximity—an alternative computational approach, which resolves this issue, is based on a haversine formulation [3]. As the flattening of the Earth's shape is quite small, errors introduced by the spherical idealization are also small. Increased accuracy is achieved by representing the Earth by an ellipsoid; in this case, however, the calculation requires iteration (a popular method was developed by Vincenty [4]).

3 In June 2004, Singapore Airlines, using A340-500 aircraft, introduced a non-stop scheduled service between these two cities—a record at that time [5]. The service was discontinued in 2013 when the airline stopped operating the airplane type. The flight time was approximately 18 hr 30 min, depending on the winds.

$d_{gc} = 3004$ nm and $d_{rl} = 3133$ nm (the deviation is 4.3%); whereas, from New York to Beijing (39°55′N/116°23′E), $d_{gc} = 5950$ nm and $d_{rl} = 7765$ nm (the deviation is 30.5%) [6].

6.3 True Airspeed, Ground Speed, and Navigation

6.3.1 True Airspeed

An airplane's true airspeed (TAS), which is denoted by the letter V, is the speed of the airplane relative to the surrounding (or ambient) air mass.

For navigation purposes, the true airspeed is associated with the airplane's heading (often abbreviated as HDG). The heading is the compass direction of the airplane's longitudinal axis with respect to either true north or magnetic north. It is measured in degrees from 0° (i.e., north) in a clockwise direction. By convention, it is expressed in three digits, using preliminary zeros if needed (e.g., 090° denotes an eastward heading). A true heading is in relation to the lines of longitude (meridians). A magnetic heading, which customarily has the capital letter M appended to the heading, is measured with respect to local magnetic north (e.g., 045° M means that the nose of the airplane is pointing north-east with respect to magnetic north).

When an airplane is being flown such that it does not have a component of velocity in the Y direction (see Section 3.3.4 and Figure 3.7), it has no sideslip and is said to be in "balance." This is the normal situation in flight, and it simply means that the correct rudder deflection is being provided by the pilot (or autopilot). In the case of an airplane being held in balance, the heading will also be the direction of the airplane's motion with respect to the air mass in which it is flying. The combination of the airplane's heading and its true airspeed is its velocity vector ($\vec{V}$) with respect to the surrounding air.

The standard operational unit for true airspeed is the nautical mile per hour (i.e., knot, which is abbreviated as kt or kn) or kilometer per hour (km/h), although statute miles per hour (mi/h) are also used. For engineering analysis, however, it is frequently necessary to convert airspeeds to m/s or ft/s.

6.3.2 Ground Speed and Navigation

An airplane's TAS will differ from its speed over the ground (i.e., the ground speed) in the presence of wind. The wind speed (V_w) and wind direction make up the wind vector ($\vec{V}_w$). By meteorological convention, the wind direction (which is often abbreviated as WD) is defined as the direction from which the wind *comes*. The airplane's track (usually abbreviated as TR) is the path of the airplane's center of gravity over the ground. The ground speed (V_G) and the track make up the ground speed vector ($\vec{V}_G$). These vectors, which are important for aerial navigation, are summarized in Table 6.1. The angle of the track, measured clockwise with respect to true north (called the true track), is customarily denoted by the Greek letter chi (χ). The angular variation between an airplane's heading and its track is known as *drift*.

The ground speed (GS) and the airplane's track can be determined from a vector summation of the true airspeed and the wind speed:

$$\vec{V}_G = \vec{V} + \vec{V}_w \tag{6.2}$$

where $\vec{V}_G$ is the ground velocity vector;
$\vec{V}$ is the TAS vector; and
$\vec{V}_w$ is the wind vector.

Table 6.1 Speed and direction for navigation

	Magnitude	Direction	Vector
True airspeed	V	Heading (HDG)	$\vec{V}$
Wind	V_w	Wind direction* (WD)	$\vec{V}_w$
Ground speed	V_G	Track (TR)	$\vec{V}_G$

*__Note:__ The wind direction is the direction from which the wind comes (meteorological convention).

This is illustrated in Figure 6.2. The ground speed (V_G) is given by

$$V_G = V \cos \varepsilon + V_w \cos \upsilon \tag{6.3}$$

where ε is the flight path drift angle (Figure 6.2); and
υ is the wind direction with respect to the track (Figure 6.2).

A typical navigation computation involves the determination of the ground speed, ESAD, and trip time given the following information: ground distance, intended track, airspeed, and wind speed and direction.

The average drift angle during cruise is usually small and to a good approximation $\cos \varepsilon \cong 1$; hence Equation 6.3 can be written as

$$V_G \cong V + V_w \cos \upsilon \tag{6.4}$$

where $V_w \cos \upsilon$ is the speed of the headwind/tailwind component.

Note that when $\upsilon = 0°$ there is a direct tailwind and $V_G = V + V_w$, and when $\upsilon = 180°$ there is a direct headwind and $V_G = V - V_w$. It is important to note also that Equation 6.4 is an approximation that is only valid when the drift angle is small, as would occur when $V \gg V_w$. The

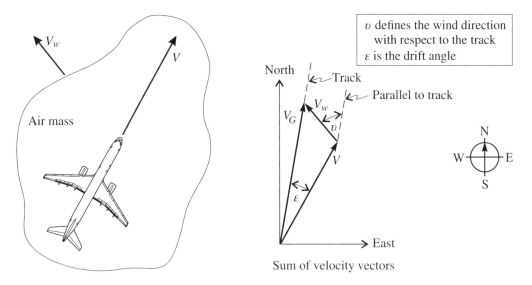

Figure 6.2 Sum of velocity vectors: true airspeed plus wind.

approximation should not be used to compute an airplane's ground speed when flying at slower speeds, as would occur in takeoff, climb-out, approach, and landing, for example.

6.3.3 Speed Measurement

The primary measure of airspeed on an airplane comes from the airplane's air data system (the basic concept is described in Sections 6.6 and 6.7). Modern jet transport airplanes are also equipped with a suite of inertial reference and satellite navigation systems,[4] which provide real-time positional data, ground speed, and ground track information. By combining this information with the aircraft's airspeed and heading, the navigation system can compute, using vector geometry, the wind speed and wind direction. The computed wind information is indicated on the flight crews' navigation displays.

6.4 Speed of Sound and Mach Number

6.4.1 Speed of Sound

The speed of sound (a) in air is described in Section 2.5.9; it can be calculated as follows:

$$a = \sqrt{\gamma R T} \qquad \text{(Equation 2.77)}$$

where γ is the ratio of specific heats (see Section 2.5.7);
R is the gas constant (see Section 2.5.6); and
T is the absolute temperature (see Section 2.2.5).

Alternatively, this equation can be written as a function of the temperature ratio (θ) of the air. To derive this relationship, Equation 2.77 is initially written for the ISA sea-level datum conditions, that is,

$$a_0 = \sqrt{\gamma R T_0} \tag{6.5}$$

where a_0 is the speed of sound at the ISA sea-level datum; and
T_0 is the absolute temperature at the ISA sea-level datum.

Equation 2.77 is then divided by Equation 6.5 to yield the required expression:

$$\frac{a}{a_0} = \sqrt{\frac{T}{T_0}} \qquad \text{or} \quad a = a_0 \sqrt{\theta} \tag{6.6}$$

The speed of sound in air under ISA sea-level datum conditions is 340.29 m/s (or 661.48 kt or 1116.5 ft/s); this reduces to 295.07 m/s (or 573.57 kt or 968.08 ft/s) at the ISA tropopause and remains constant throughout the ISA stratosphere. A consequence of this reduction in the speed of sound with increasing height is that aircraft encounter compressibility effects at lower true airspeeds in the upper troposphere and stratosphere compared to flight near sea level.

---

4 Satellite navigation is based on global networks of satellites in medium-Earth orbit, which transmit radio signals to onboard receivers. The best-known systems are GPS (Global Positioning System), GLONASS (Global Navigation Satellite System), and the more recently developed Galileo system.

6.4.2   Mach Number

Mach number was introduced in Section 2.5.9; the defining equation (i.e., Equation 2.78) can be used to describe the speed of an airplane:

$$M = \frac{V}{a} \tag{6.7}$$

where M is the flight Mach number (dimensionless);
$\qquad$ V is the true airspeed of the airplane (typical unit: kt); and
$\qquad$ a is the speed of sound in the ambient air mass (typical unit: kt).

For example, a cruise Mach number of 0.84 (which, incidentally, is the design cruise speed of the B777) means that the airplane is cruising at 84% of the speed of sound determined for the atmospheric conditions of the air surrounding the airplane (specifically, the outside air temperature).

Combining Equations 6.7 and 6.6 enables the true airspeed to be written in a convenient form:

$$V = Ma = Ma_0 \sqrt{\theta} \tag{6.8}$$

This equation leads to an interesting deduction regarding the velocity–Mach number relationship during a climb. As the air temperature decreases linearly with increasing altitude from sea level to the tropopause in the ISA, it is evident from Equation 2.77 that the speed of sound will also decrease, although not linearly but as a function of $\sqrt{\theta}$. In the lower ISA stratosphere, temperature is constant and hence the speed of sound is also constant. A hypothetical constant TAS climb from sea level would thus mean that the Mach number would progressively *increase* as the airplane climbs to the tropopause, and thereafter, for flight in the stratosphere, the Mach number would be *constant* (climb speeds are discussed later in Section 17.2).

6.5   Dynamic Pressure and Equivalent Airspeed

6.5.1   Dynamic Pressure

The dynamic pressure of moving fluids is described in Section 2.5.5. The free-stream dynamic pressure, which is associated with the airplane's flight speed (i.e., TAS), is given by the following expression:

$$q = \frac{1}{2}\rho V^2 \tag{6.9}$$

where q is the free-stream dynamic pressure (typical units: Pa, lb/ft^2); and
$\qquad$ ρ is the density of the surrounding (ambient) air (typical units: kg/m^3, slug/ft^3).

Now, for high-speed operations, it is Mach number, and not TAS, that is the preferred measure of speed. For this reason, it is convenient for many applications to express the dynamic pressure in terms of Mach number (and air pressure ratio), rather than in terms of TAS. From Equations 6.9 and 4.3, it follows that

$$q = \frac{1}{2}\rho V^2 = \frac{1}{2}\rho_0 \sigma V^2 \tag{6.10}$$

Substituting for V using Equation 6.8 leads to the following equation for dynamic pressure:

$$q = \frac{1}{2}\rho a^2 M^2 = \frac{1}{2}\sigma \rho_0 \theta a_0^2 M^2 \tag{6.11}$$

And substituting for θ using Equation 4.4 gives

$$q = \frac{1}{2}\rho_0 a_0^2 \delta M^2 \tag{6.12}$$

6.5.2 Equivalent Airspeed

Equivalent airspeed (V_e), by definition, is the equivalent speed that an airplane would have at the ISA sea-level datum (where the air density is ρ_0) if it developed the same *dynamic pressure* as it does moving at its true airspeed at the altitude concerned (where the air density is ρ). Mathematically, this can be written as follows:

$$q = \frac{1}{2}\rho_0 V_e^2 = \frac{1}{2}\rho V^2 \tag{6.13}$$

Hence $V_e = \sqrt{\dfrac{\rho}{\rho_0}}\, V = \sqrt{\sigma}\, V$ \hfill (6.14)

It is convenient for several applications to be able to relate the equivalent airspeed (EAS) directly to Mach number:

$$V_e = \sqrt{\sigma}\, V = \sqrt{\sigma}\, Ma = \sqrt{\sigma}\, Ma_0 \sqrt{\theta} = \sqrt{\delta}\, a_0 M \tag{6.15}$$

   Equivalent airspeed is an extremely useful parameter for aircraft engineering analyses. Aerodynamic load calculations for airframe structural design, for example, are performed using EAS, rather than TAS. It is also a very useful speed for the analysis of airplane performance—as will be seen in subsequent chapters. However, for normal flight operations, EAS is not used and reference is made instead to calibrated airspeed (see Section 6.6) or indicated airspeed (see Section 6.7).

6.6 Calibrated Airspeed

6.6.1 Definition

Calibrated airspeed[5] (V_c), by definition, is the airspeed reading on a calibrated airspeed indicator (ASI) that is connected to Pitot and static ports that are assumed to be entirely free of error (these errors are discussed in Section 6.7.2). An ASI measures the difference between the total (or stagnation) pressure (p_t) and the static pressure (p) of the air. To explain further this definition of calibrated airspeed (CAS), it is necessary to describe first the operation of the Pitot–static system.

---

5 Historically, the term *rectified airspeed* was used for calibrated airspeed in certain parts of the world (and is still occasionally encountered, for example, in some pilot training literature).

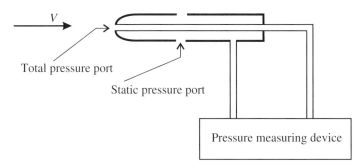

Figure 6.3 Schematic of Pitot–static system.

6.6.2   Pitot–Static System

A Pitot–static system[6] essentially utilizes the Bernoulli equation (see Sections 2.5.5 and 2.5.11) to measure airspeed by determining the difference between the total air pressure and the static (or ambient) air pressure. The Bernoulli equation (see Equation 2.63) simply states that in steady, incompressible, streamline flow, the sum of the static pressure and dynamic pressure is constant, thus:

$$p_t = p_s + \frac{1}{2}\rho V^2 = \text{constant} \tag{6.16}$$

where p_t is the total air pressure; and
p_s is the static air pressure.

Figure 6.3 is a schematic of a Pitot–static system. There are two pressure openings: the *total pressure port* at the front of the tube and the *static pressure port* at the side of the tube. In flight, the tube is aligned approximately in the direction of the incoming air. The air directly in line with the central opening (port) is slowed down and momentarily comes to rest, before proceeding around the tube. The central port thus senses the *total pressure* of the air. The flow accelerates around the nose of the tube and essentially reaches its original speed when it gets to the second port, which senses the *static air pressure*. A mechanical measuring device, usually incorporating a diaphragm or bellows, is used to measure the difference in air pressure between the two ports. This, by the application of Bernoulli's equation, is a measure of the *dynamic pressure*, and, providing air density is known, can be used to indicate the true airspeed for incompressible flow.

For airspeeds above about Mach 0.3, compressibility effects are important and the form of the Bernoulli equation given by Equation 6.16 is no longer satisfactory. This is evident from Equation 2.84, which is a form of the Bernoulli equation that is applicable to compressible flow (see Sections 2.5.11 and 3.5.2).

The effect of compressibility means that the airspeed indicator (ASI) requires an airspeed correction for moderate to high speeds. This calibration of the instrument is possible for all Mach numbers at sea level; however, flight at altitude introduces an error, which requires correction (see Section 6.6.3).

---

6 Named after Henri Pitot (1695–1711), the French hydraulic engineer who, in *ca*. 1732, invented a device, now called a Pitot tube, to measure the speed of water flowing in rivers and canals [7].

6.6.3 Compressibility Correction Factor

For subsonic speeds, it can be shown (see Appendix D, Section D.2.3) for compressible isentropic flow that the difference between the total and static pressures is equal to the product of the dynamic pressure (see Equation 6.9) and a function of Mach number, which can be written in the form of a binomial expansion series:

$$\left(p_t - p\right)_{compressible} = \frac{1}{2}\rho V^2 f(M) \tag{6.17}$$

$$\text{where } f(M) = \left[1 + \frac{M^2}{4} + \frac{M^4}{40} + \frac{M^6}{1600} + \cdots\right]$$

An airspeed indicator is essentially an accurate differential pressure gauge that is designed to produce a reading of velocity. It is evident from Equation 6.17 that, in general, the true airspeed (V) can only be accurately determined if the density and the Mach number are correctly taken into account; however, as both of these factors depend on altitude, this is not a simple correlation that can be built into a mechanical device. A different airspeed scale is required for each pressure altitude, and this is not possible for a relatively simple instrument. If a single atmospheric condition were used to determine the calibration factors for the instrument, then an error would be introduced for flight operations at other altitudes—and this is indeed what is done in practice. The ASI bases speed on ISA sea-level datum conditions.

Equation 6.17 leads to a useful deduction regarding flights at relatively low speeds: for speeds less than about Mach 0.3, the effect of compressibility can be neglected as $f(M) \cong 1.0$. At any altitude, the ASI will then display the equivalent airspeed as

$$\left(p_t - p\right)_{incompressible} = \frac{1}{2}\rho V^2 = \frac{1}{2}\rho_0 V_e^2 \tag{6.18}$$

In general, for speeds above about Mach 0.3, this approximation is not acceptable and, as the ASI will account for *f(M)* based on ISA sea-level datum conditions, a difference between EAS and CAS results—the CAS will be greater than the EAS by an amount that increases with Mach number and altitude. This difference is known as the compressibility correction factor (or scale-altitude correction) and is designated as ΔV_c. A commonly used convention defines ΔV_c as a negative quantity[7] by the following equation [9, 10]:

$$\Delta V_c = V_e - V_c \tag{6.19}$$

6.6.4 Expressions for Calibrated Airspeed

A better representation of the operation of an airspeed indicator is obtained from Equation 2.82, which is the Bernoulli equation applied to compressible airflow:

$$\left(\frac{p_t}{p}\right)_{compressible} = \left(1 + \frac{\gamma - 1}{2}M^2\right)^{\frac{\gamma}{\gamma - 1}} \qquad \text{(Equation 2.82)}$$

---

7 The alternative convention, which is used by several organizations (e.g., Boeing [8]), defines ΔV_c as a positive quantity where $\Delta V_c = V_c - V_e$.

As a Pitot–static system provides a measurement of $(p_t - p)$, this term will be isolated on the left-hand side of the equation:

$$\left(\frac{p_t - p}{p}\right) = \left(1 + \frac{\gamma - 1}{2} M^2\right)^{\frac{\gamma}{\gamma - 1}} - 1 \tag{6.20}$$

Noting that $\gamma = 1.4$ for air and that $M = V/a$, this equation becomes

$$(p_t - p) = p \left\{ \left[1 + 0.2 \left(\frac{V}{a}\right)^2 \right]^{3.5} - 1 \right\} \tag{6.21}$$

Now, Equation 6.21 can be turned into a useful expression for determining airspeed. This is done by *selecting* ISA sea-level datum values for the required parameters on the right-hand side of Equation 6.21 and by *defining* the resulting velocity as *calibrated airspeed* (V_c). For airspeed measurement, the left-hand side of Equation 6.21 is provided by the Pitot–static system.

$$(p_t - p) = p_0 \left\{ \left[1 + 0.2 \left(\frac{V_c}{a_0}\right)^2 \right]^{3.5} - 1 \right\} \tag{6.22}$$

Equation 6.22 is the standard calibration equation for an airspeed indicator, where the airspeed is based on the difference between the total and the static pressures, as sensed by the aircraft's Pitot–static system.

It is evident from Equation 6.22 that an airspeed indicator can be correctly calibrated for any Mach number at sea level, where CAS will always equal EAS. However, due to the use of ISA sea-level datum values for pressure and speed of sound in Equation 6.22, CAS will not equal EAS at any point above the ISA sea-level datum. This means that a systematic discrepancy, which increases with altitude, exists between CAS and EAS. The difference can be obtained by combining Equations 6.21 and 6.22:

$$p \left\{ \left(1 + 0.2M^2\right)^{3.5} - 1 \right\} = p_0 \left\{ \left[1 + 0.2 \left(\frac{V_c}{a_0}\right)^2 \right]^{3.5} - 1 \right\} \tag{6.23}$$

thus $\left[1 + 0.2 \left(\frac{V_c}{a_0}\right)^2 \right]^{3.5} = \delta \left\{ \left(1 + 0.2M^2\right)^{3.5} - 1 \right\} + 1$

and $\quad V_c = a_0 \sqrt{ 5 \left\{ \left[\delta \left\{ \left(1 + 0.2M^2\right)^{3.5} - 1 \right\} + 1 \right]^{\frac{1}{3.5}} - 1 \right\} } \tag{6.24}$

An expression for the compressibility correction factor (ΔV_c), defined in Equation 6.19, can now be determined:

$$\Delta V_c = V_e - V_c = M a_0 \sqrt{\delta} - V_c \tag{6.25}$$

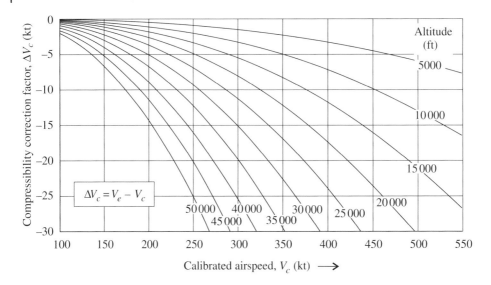

Figure 6.4 Compressibility correction factor.

From Equations 6.24 and 6.25, it follows that

$$\Delta V_c = a_0 \left\{ M\sqrt{\delta} - \sqrt{5\left\{ \left[\delta\left\{ \left(1 + 0.2M^2\right)^{3.5} - 1 \right\} + 1 \right]^{\frac{1}{3.5}} - 1 \right\}} \right\}$$

(6.26)

Figure 6.4 illustrates the variation of the compressibility correction factor with CAS and altitude. It is evident that the compressibility correction factor is negligibly small for operations below about 10 000 ft at speeds less than 200 kt CAS—for these conditions (which are associated with takeoff and landing, for example), the correction is usually ignored.

It is convenient for many applications to be able to express an airplane's Mach number in terms of CAS; this is achieved by rearranging Equation 6.23, as follows:

$$M = \frac{V}{a} = \sqrt{5\left\{ \left[\frac{1}{\delta}\left\{ \left[1 + 0.2\left(\frac{V_c}{a_0}\right)^2 \right]^{3.5} - 1 \right\} + 1 \right]^{\frac{1}{3.5}} - 1 \right\}}$$

(6.27)

The Mach number versus CAS function, as given by Equation 6.27, is illustrated in Figure 6.5. The Mach number is seen to increase in an almost linear fashion with increasing CAS.

A similar relationship between EAS and CAS can be determined using Equation 6.24:

$$V_e = a_0 \sqrt{5\delta\left\{ \left[\frac{1}{\delta}\left\{ \left[1 + 0.2\left(\frac{V_c}{a_0}\right)^2 \right]^{3.5} - 1 \right\} + 1 \right]^{\frac{1}{3.5}} - 1 \right\}}$$

(6.28)

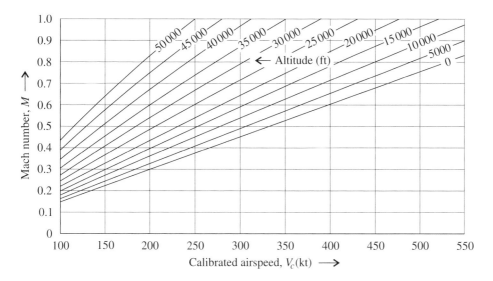

Figure 6.5 Mach number versus CAS for selected altitudes.

The compressibility correction factor can be described in terms of the CAS:

$$\Delta V_c = a_0 \sqrt{5\delta \left\{ \left[\frac{1}{\delta} \left\{ \left[1 + 0.2 \left(\frac{V_c}{a_0} \right)^2 \right]^{3.5} - 1 \right\} + 1 \right]^{\frac{1}{3.5}} - 1 \right\} - V_c} \tag{6.29}$$

Equation 6.29 is a particularly useful expression for operational analysis. The magnitude of the compressibility correction factor can be determined directly from the CAS, as illustrated in Figure 6.4 for selected altitudes.

6.7   Indicated Airspeed

6.7.1   Instrument Indicated Airspeed

The *instrument indicated airspeed* (V_I) is the reading of an actual air airspeed indicator (ASI). An ASI can have a small error (ΔV_I) associated with the individual instrument. Errors associated with mechanical instruments are principally due to internal friction and hysteresis effects. In general, such errors can be corrected using data supplied by the component manufacturer. Modern instruments, however, are very accurate, and individual instrument errors are thus negligibly small.

6.7.2   Indicated Airspeed

Correcting the instrument indicated airspeed (V_I) for instrument error (ΔV_I) results in the *indicated airspeed* (IAS), which is customarily denoted as V_i.

$$V_i = V_I + \Delta V_I \tag{6.30}$$

In general, the indicated airspeed will differ from the calibrated airspeed due to position error (ΔV_p), which is defined as follows:

$$V_c = V_I + \Delta V_p \tag{6.31}$$

The *position error* essentially arises because of an inability to measure precisely the true static pressure in the immediate vicinity of an airplane. It is usually not possible to locate a single point on the airplane that will remain at ambient static pressure for all flight conditions. The static port thus measures pressure in a flow field that is disturbed by the presence of the airplane. The wing, by its design, is largely responsible for this disturbance. An error will, therefore, arise due to the physical location or *position* of the static port. The selected position must be in an area where the pressure is close to that of the ambient atmosphere, but also at a location that minimizes the effect of angle of attack variation, which alters the pressure field around the wing.

In addition to the static pressure error, there could also be an error in the *total pressure recovery*. This is a consequence of locating the Pitot tube in a region of reduced energy. (It is for this reason that the Pitot tube is attached to the side of the airplane by means of a small strut or mount, designed to place the tube outside of the boundary layer.) A pressure recovery error will also result if the Pitot tube is not properly aligned with the incoming air. For correctly designed Pitot installations, the total pressure recovery error is negligible; consequently, it is only the static pressure errors that need to be considered.

The position error can be calibrated at different speeds by various experimental methods [11], which are beyond the scope of this book. It should also be noted that the position error would change as the pressure field around the airplane changes during flight. For example, at low speed with the flaps deployed, the correction is likely to be different than that required at high speed. When the airplane is flown near to the ground, during takeoff or landing, the presence of the ground will change the flow pattern around the airplane by restricting the downwash from the wing. The position error correction for the flaps-down configuration will usually include the influence of ground effect. On modern airliners, the correction is automatically taken into account by the Air Data Computer (ADC), and for these airplanes $V_I = V_c$.

In a modern cockpit, IAS is displayed on the left-hand vertical "tape" on the Captain's and First Officer's Primary Flight Displays (illustrated in Figure 5.2a). As the speed changes, so the tape scrolls up or down to present the correct speed in the central window. The tape will also indentify key reference speeds (such as those used in takeoff—see Section 10.2) by means of markers or identification codes.

6.8 Relationship Between Airplane Speeds

6.8.1 Summary of Airplane Speeds

Table 6.2 contains a summary of the speeds described in this chapter, with their corresponding correction/conversion factors.

6.8.2 Speed Variation During a Constant CAS Climb

It is instructive to study the effect of a hypothetical constant CAS climb on the various speeds presented in this chapter—this is illustrated in Figure 6.6. CAS, EAS, and TAS are shown on a common scale (abscissa). The Mach number, which is based on ISA temperature at

Table 6.2 Airplane speed correction/conversion factors

Name	Symbol	Correction/conversion
Instrument indicated airspeed (i.e., ASI reading)	V_I	
Indicated airspeed (IAS)	V_i	$V_i = V_I + \Delta V_I$ where ΔV_I is the instrument error correction
Calibrated airspeed (CAS)	V_c	$V_c = V_i + \Delta V_p$ where ΔV_p is the position error correction
Equivalent airspeed (EAS)	V_e	$V_e = V_c + \Delta V_c$ where ΔV_c is the compressibility correction
True airspeed (TAS)	V	$V = V_e / \sqrt{\sigma}$ where σ is the relative density
Ground speed (GS)	V_G	$\vec{V}_G = \vec{V} + \vec{V}_w$ where $\vec{V}_G$ is the ground velocity vector, $\vec{V}$ is the TAS vector, and $\vec{V}_w$ is the wind vector
Mach number	M	$M = V/(a_0 \sqrt{\theta})$ where a_0 is the speed of sound at the ISA sea-level datum and θ is the relative temperature

the corresponding altitude, is shown on an independent scale. Several observations can be made:

- The compressibility correction factor is zero at sea level, but its magnitude increases during the climb (see Figure 6.4).
- The TAS will increase as the air density drops (this can be seen from Equation 6.14). As there is a change in the density function at the tropopause (see Table 4.2), this will alter the rate of change of true airspeed with height.
- The Mach number will increase as the TAS increases and also as the speed of sound decreases as a function of $\sqrt{\theta}$ (see Section 6.4.1).

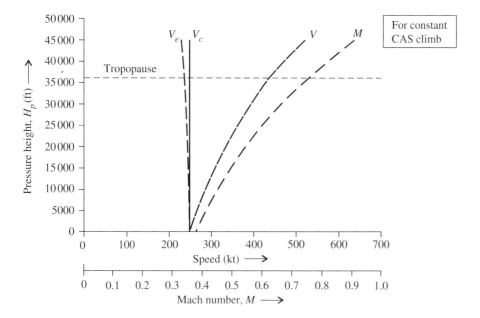

Figure 6.6 Variation in EAS, TAS, and Mach number for a hypothetical constant CAS climb from sea level into the stratosphere.

Techniques for analyzing climb performance are presented in Section 12.4, and climb–speed schedules used in flight operations are discussed in Section 17.2.

References

1 BIPM, "The international system of units (SI)," 8th ed., Bureau International des Poids et Mesures (BIPM), Paris, France, 2006.

2 NIMA, "Department of defense world geodetic system 1984: Its definition and relationships with local geodetic systems," TR 8350.2, 3rd ed., National Imagery and Mapping Agency, St. Louis, MO, Jan. 3, 2000.

3 Sinnott, R.W., "Virtues of the haversine," *Sky and Telescope*, Vol. 68, Iss. 2, pp. 158, 1984.

4 Vincenty, T., "Direct and inverse solutions of geodesics on the ellipsoid with application of nested equations," *Survey Review*, Vol. XXIII, Iss. 176, pp. 88–93, 1975.

5 Clark, A., "Record longest flight flies in the face of its critics," *The Guardian*, Guardian News and Media Ltd., London, UK, June 29, 2004.

6 Alexander, J., "Loxodromes: A rhumb way to go," *Mathematics Magazine*, Vol. 77, Iss. 5, pp. 349–356, 2004.

7 Anderson, J.D., *A history of aerodynamics and its impact on flying machines*, Cambridge University Press, Cambridge, UK, 1997.

8 Blake, W. and Performance Training Group, "Jet transport performance methods," D6–1420, Flight Operations Engineering, Boeing Commercial Airplanes, Seattle, WA, Mar. 2009.

9 Lawford, J.A. and Nippress, K.R., "Calibration of air-data systems and flow direction sensors," AGARD Flight Test Techniques Series Vol. 1, AGARDograph No. 300, Advisory Group for Aerospace Research and Development, North Atlantic Treaty Organization, Neuilly sur Seine, France, Sept. 1983.

10 ESDU, "Airspeed data for performance calculations," Data item 69026, Amdt. B, IHS ESDU, 133 Houndsditch, London, UK, Oct. 1992.

11 Gallagher, G.L., Higgins, L.B., Khinoo, L.A., and Pierce, P.W., "Fixed wing performance: Flight test manual," USNTPS-FTM-No. 108, United States Naval Test Pilot School, Patuxent River, MD, Sept. 30, 1992.

7

Lift and Drag

7.1 Introduction

This chapter introduces the topics of lift and drag and the relationship between these two aerodynamic forces, which act on an airplane in flight. The lift and drag *coefficients*, which are non-dimensional ratios, are fundamental parameters that are used in the analysis of an airplane's performance. Although their definitions are straightforward (given by Equations 7.1 and 7.5) and the equation that is often used to link the two coefficients is relatively simple (see Equation 7.14), the subject is deceptively complex.

An overview of airplane lift and the factors that influence the lift coefficient (e.g., airplane weight, load factor, and configuration[1]) are discussed in Section 7.2. The airplane's drag, which has a critical influence on the overall efficiency of the airplane, is discussed in Section 7.3. There are several alternative ways in which the drag that acts on an airplane can be sub-divided into constituent elements. An introduction to the non-trivial topic of drag decomposition is given in this section.

Central to the analysis of airplane performance is the drag–lift relationship or *drag polar* as it is usually called. Several models that are used to represent this relationship mathematically, such as the parabolic drag polar (Equation 7.14), are presented in Section 7.4. This is followed by a brief discussion of drag correction factors, needed to account for such issues as aircraft configuration and ground effect (Section 7.5). The simple parabolic drag representation enables several interesting mathematical expressions to be developed, which can be used as a basis to assess the flight characteristics of an airplane. The following conditions are analyzed: maximum lift-to-drag ratio (Section 7.6), minimum drag (Section 7.7), minimum drag power (or required power) (Section 7.8), and minimum drag-to-speed ratio (Section 7.9). Concluding remarks and a summary table indicating the relationships between these flight conditions are presented at the end of the chapter (Section 7.10).

---

1 An airplane's *configuration*, in this context, describes the position of reconfigurable parts of the airplane (i.e., the position of the landing gear, flaps, slats, and so forth). The influence of configuration changes on lift is discussed in Section 7.2.7 and on drag in Section 7.5.2.

Performance of the Jet Transport Airplane: Analysis Methods, Flight Operations, and Regulations, First Edition.
Trevor M. Young.
© 2018 John Wiley & Sons Ltd. Published 2018 by John Wiley & Sons Ltd.

7.2 Airplane Lift

7.2.1 Airplane Lift and Lift Coefficient

The airplane's lift is that component of the net aerodynamic force that acts normal to the path of flight (i.e., perpendicular to the relative wind) in the plane of symmetry of the airplane. The airplane's *lift coefficient* (C_L) is a dimensionless ratio of its net lift to the product of the free-stream dynamic pressure and the wing reference area:

$$C_L = \frac{L}{qS} \tag{7.1}$$

where L is the net lift acting on the airplane (typical units: N, lb);
q is the free-stream dynamic pressure (typical units: N/m^2, lb/ft^2); and
S is the wing reference area (typical units: m^2, ft^2).

The dynamic pressure can be expressed in several different ways, depending on the application. For reference purposes, the dynamic pressure (see Section 6.5) is expressed below in terms of the airplane's true airspeed, equivalent airspeed, and, for high-speed operations, Mach number:

$$q = \frac{1}{2}\rho V^2 = \frac{1}{2}\rho_0 \sigma V^2 \qquad \text{(Equation 6.10)}$$

$$q = \frac{1}{2}\rho_0 V_e^2 \qquad \text{(Equation 6.13)}$$

$$q = \frac{1}{2}\rho_0 a_0^2 \delta M^2 \qquad \text{(Equation 6.12)}$$

where ρ is the ambient air density (typical units: kg/m^3, slug/ft^3);
V is the true airspeed of the airplane (typical units: m/s, ft/s);
ρ_0 is the air density at the ISA sea-level datum (typical units: kg/m^3, slug/ft^3);
σ is the relative air density (dimensionless);
V_e is the equivalent airspeed of the airplane (typical units: m/s, ft/s);
a_0 is the speed of sound in air at the ISA sea-level datum (typical units: m/s, ft/s);
δ is the relative air pressure (dimensionless); and
M is the flight Mach number of the airplane (dimensionless).

7.2.2 Lift Coefficient Versus Angle of Attack

The *angle of attack* is customarily assigned the Greek letter alpha (α); in fact, the convention is so widely used that the term *alpha* is now synonymous with angle of attack in the aviation industry. Alpha is measured from the relative wind (i.e., the direction of the free-stream airflow) to a datum (or reference line). For an *airfoil*, it is measured with respect to the chord line (see Figure 3.1). For an airplane, different conventions are used, depending on the specific application. For general aerodynamic work, the datum that is often selected is the aircraft's longitudinal axis (i.e., the X axis, as defined by the body axis system in Section 3.3.4). It is also common to define alpha with respect to a reference wing chord. For airplane stability analysis, it is often convenient to measure the angle of attack with respect to a zero lift condition (in which case $C_L = 0$ when $\alpha = 0$). It is customary to distinguish the various conventions by adding a subscript to α—for example, *b* for body (i.e., X-axis datum), *w* for wing, and *zl* for zero lift. Where confusion is unlikely to arise, the subscript is routinely omitted.

(a)

(b)

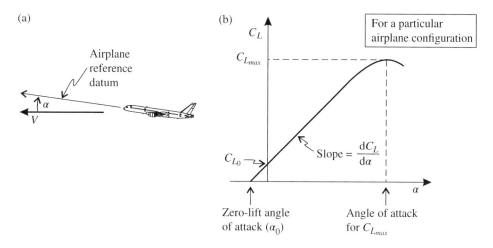

Figure 7.1 (a) Airplane angle of attack; (b) Lift coefficient versus angle of attack.

As illustrated in Figure 7.1, C_L increases in a nearly linear manner with increasing angle of attack up to a limiting value; it then enters a region of significant non-linearity, which displays a clear maximum (when $C_L = C_{L_{max}}$) associated with the wing stalling (stall is discussed in more detail in Section 20.2). The C_L versus α graph is usually called the *lift curve* and the derivative $\mathrm{d}C_L/\mathrm{d}\alpha$ the *lift-curve slope*.

7.2.3 Lift Coefficient in Straight, Level (i.e., Constant Height) Flight

In straight, level flight—that is, when the wings are not banked and the airplane is neither climbing nor descending—the airplane's net lift exactly balances its weight, as shown in Figure 7.2. This is an example of 1-*g* flight (*g*-loads and maneuvers are discussed later in Section 7.2.5). For this flight condition, the lift coefficient can be written as follows:

$$C_{L_1} = \frac{W}{qS} = \frac{\left(\dfrac{W}{S}\right)}{q} \tag{7.2}$$

where C_{L_1} is the 1-*g* lift coefficient (i.e., $L = W$) (dimensionless);
W is the airplane weight (typical units: N, lb); and
W/S is known as the wing loading (typical units: N/m², lb/ft²).

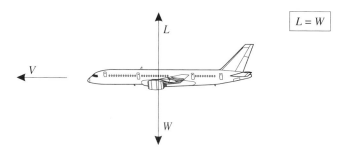

Figure 7.2 Lift and weight in straight, level (i.e., constant height) flight.

7.2.4   Influence of CG on the Lift Curve

The airplane's center of gravity (CG) position has a small, but not insignificant, influence on the C_L versus α relationship. The airplane's CG location on takeoff depends on several factors, including the distribution of passengers and freight and the location of the fuel. It may also move during the flight as fuel is consumed. There are limits for each phase of the flight as to how far forward and far rearwards the CG can move from its nominal reference position (see Section 19.3.3).

The explanation as to how the CG location influences the lift curve is based on the concept of a *trimmed* flight condition (which is discussed in Section 19.4), where the sum of the pitching moments about the airplane's CG is zero. In a trimmed, level-flight condition, the horizontal tailplane produces an aerodynamic force that acts downwards to balance the nose-down aerodynamic moments about the CG generated by the wing and fuselage. When the CG is located at an aft position, the tailplane down-force, needed to keep the airplane in trim, would be smaller compared to a forward CG position. This means that the required wing lift—which is needed to balance the sum of the airplane's weight and the downward-acting tailplane force—is less. Corresponding to the reduced wing lift would be a reduced angle of attack.

In a trimmed level-flight condition, the net lift (L) must equal the airplane's weight (W). Consequently, if the airplane weight (and speed and altitude) do not change, then C_L will not change even if the CG position moves. Consequently, to generate the same total C_L for steady 1-g flight, the lift curve for an aft CG must lie above that for a forward CG (see Figure 7.3). The angle of attack for a typical jet transport airplane is reported to be about 2° less in cruise when the CG is at the aft limit compared to when it is at the forward limit [1].

The conventional approach adopted for the generation of performance data (for use by flight crews, for example) is based on a reference CG location (e.g., 25% mean aerodynamic chord is often used); data to correct these values for other CG locations are provided, where necessary.

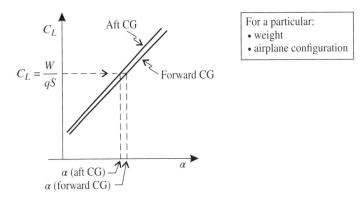

Figure 7.3 Influence of CG on lift coefficient.

7.2.5   Lift Coefficient in Maneuvering Flight

The mechanics of maneuvering flight, which involve the airplane changing its flight path through a rotation about the pitch, roll, or yaw axes, are discussed in Chapter 15. When an airplane performs a pitching maneuver (as illustrated in Figure 7.4a) or a rolling maneuver (see

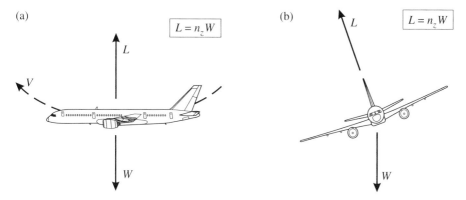

Figure 7.4 Airplane lift and weight during (a) pull-up maneuver, or (b) banked turn.

Figure 7.4b), the airplane's lift will generally not equal its weight. To describe an airplane's lift during maneuvering flight, it is convenient to introduce a load factor.

By definition, a *load factor* is an aerodynamic force (load), non-dimensionalized by dividing by the airplane's weight (this is discussed further in Section 15.2.3). A load factor is thus a vector; its direction is the same as that of the force upon which it is based. In the study of maneuvering flight, it is the ratio of the airplane's lift to its weight that is of interest:

$$n_z = \frac{L}{W} \tag{7.3}$$

where n_z is the load factor normal to the flight path (dimensionless).

Using this definition, it is possible to express the lift coefficient in terms of the airplane weight and load factor—in several alternative, but equivalent, ways—for a general flight condition.

$$C_L = \frac{L}{qS} = \frac{2n_z W}{\rho V^2 S} = \frac{2n_z W}{\rho_0 \sigma V^2 S} = \frac{2n_z W}{\rho_0 V_e^2 S} = \frac{2n_z \left(\frac{W}{\delta} \right)}{\rho_0 a_0^2 M^2 S} \tag{7.4}$$

The reason for expressing the weight parameter as W/δ (as done in the last form of the lift coefficient in Equation 7.4 above) is that engine thrust is often written in this way (see Section 8.2.9). Expressing both weight and thrust in an identical manner simplifies the analysis process in certain situations.

7.2.6   Lift Coefficient in Cruise

For long periods during cruise, jet transport airplanes fly at a constant (or nearly constant) height, with constant Mach number and with the wings level—thus, during these times the lift will be equal to the weight. A typical lift coefficient during cruise for a conventional airliner is 0.5 (this value is frequently used for preliminary analyses).

During cruise, fuel is consumed; the airplane thus becomes lighter and the required lift decreases. Furthermore, the reduction in lift will result in a reduction in total drag. If the speed and height remain unchanged, then the lift coefficient will progressively reduce (in other words, the angle of attack will decrease). The lift coefficient versus airplane weight relationship, under these conditions, is linear. This is illustrated in Figure 7.5 for a constant-speed, constant-height cruise. It is, however, commonplace during long-haul flights for the pilot to execute one or more

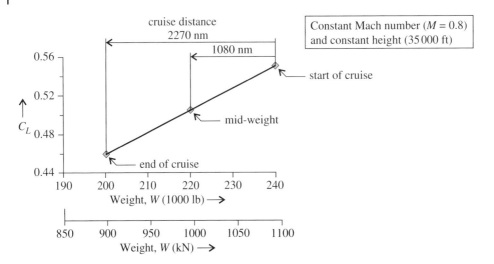

Figure 7.5 Typical C_L variation for a mid-size, twin-engine airliner during a constant-speed, constant-height cruise.

step climbs in cruise (this is discussed in Section 13.6.5); in such cases, the lift coefficient versus weight relationship will display a step change, associated with the lower air density at the higher altitude.

The lift coefficient reduces linearly with airplane weight in Figure 7.5. The rate of change of lift coefficient with *time* during cruise, however, would not be exactly constant. The time rate of change of an airplane's weight is equal to the rate that fuel is consumed (other changes to an airliner's weight are considered negligible). The fuel flow can be expressed as the product of the thrust specific fuel consumption (TSFC) and the net thrust (F_N), as described by Equation 8.30 (see Table 8.2) in Section 8.4.2. Now, if the airplane is to maintain a constant speed and height, the thrust must be reduced as the airplane gets lighter and the drag decreases. The thrust specific fuel consumption may be assumed to be approximately constant in cruise (although in reality, small non-linear changes do occur). This means that the fuel flow will decrease during the cruise, and hence the time rate of change of lift coefficient will not be exactly constant.

7.2.7 Lift Coefficient in Takeoff and Landing

The extension of high lift devices (see Section 3.7) enables the airplane to be flown at a significantly greater lift coefficient than that required for cruise. The maximum C_L value that can be achieved depends on several wing design features (e.g., airfoil type, wing sweep, and type of leading and trailing edge devices) and on the flap setting (which governs the deployment angle and the increase in chord). With the flaps fully deployed, the greatest C_L value that is possible for the particular airplane type is usually achieved. Roughly speaking, $C_{L_{max}}$ values for jet transport airplanes vary between about 2.4 to more than 3.3 (reported for the BAe 146).

A critical phase of any flight is the approach to landing. The combination of the lift coefficient in the approach ($C_{L_{app}}$) and the airplane's wing loading directly determines the approach speed (this is evident from Equation 7.4). Figure 7.6 illustrates this trend with data for airplanes with various flap types. Low approach speeds (which would allow a shorter landing distance, for example) are associated with airplanes with high $C_{L_{app}}$ values and/or low wing loading.

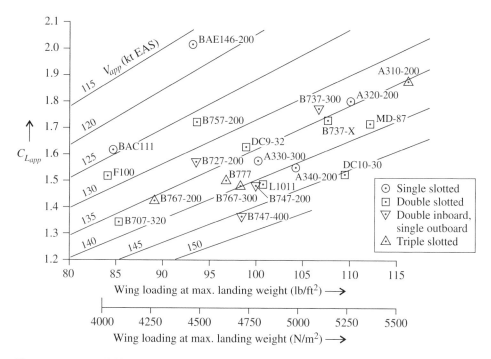

Figure 7.6 Approach lift coefficient versus approach wing loading for various high lift devices. *Source*: Adapted from [2].

7.3 Airplane Drag

7.3.1 Drag Coefficient

The airplane's drag is that component of aerodynamic force which acts aftwards, parallel to the direction of flight (i.e., in the direction of the relative wind), such as to impede the forward motion of the airplane. The ratio of the airplane's net drag (D) to the product of the free-stream dynamic pressure and the wing reference area is defined as the drag coefficient of the airplane:

$$C_D = \frac{D}{qS} \tag{7.5}$$

The dynamic pressure can be expressed in terms of the airplane's true airspeed, equivalent airspeed, or Mach number—as described earlier in Section 7.2.1.

The drag coefficient is a dimensionless ratio. There is, however, a tradition of using units of "drag counts" to account for minute drag increments/decrements associated with changes to the airframe or to the airflow around the airplane. The evaluation of drag changes associated with excrescences (see Section 7.5.7) would be a typical application. By definition, one drag count is equal to $1/10\,000^{\text{th}}$ of a drag coefficient. When using drag counts, it must be remembered, though, that a drag count is a relative—not an absolute—measure of drag (i.e., it is associated with a specific wing reference area and dynamic pressure).

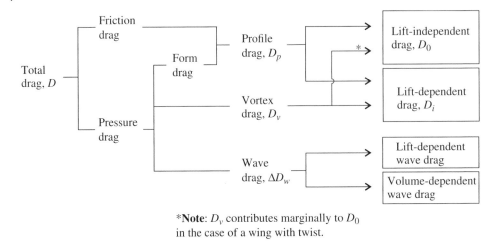

*Note: D_v contributes marginally to D_0 in the case of a wing with twist.

Figure 7.7 Typical drag breakdown.

7.3.2 Drag Breakdown

The decomposition of an airplane's drag into its constituent elements is an exceedingly complex problem—and one that has been the subject of much study and debate since Otto Lilienthal's pioneering research on lift and drag in the 1880s.[2] An understanding of what causes drag—and how it might be minimized—is of fundamental importance to the design and operation of any modern airplane. Drag elements associated with various flow phenomena have been identified and described using a combination of analytical and empirical expressions, which are functions of the airplane's geometry, weight, and operating conditions (e.g., Mach number, angle of attack, atmospheric properties, Reynolds number). This approach can be used to establish a framework for the categorization of drag elements.

There are a few points, however, that should be noted in regard to drag breakdown models: (1) there is no single set of drag accounting procedures (or "book-keeping" rules) that is universally adopted; (2) the simple "textbook" drag models that lend themselves to the decomposition of an airplane's drag are imprecise; and (3) drag breakdown schemes are required for several different purposes, such as airplane design, flight testing, and operations engineering. It is not surprising, then, that many different drag categorization schemes exist. This warrants a note of caution: within the various schemes, specific drag elements are sometimes assigned to different categories, different names are sometimes given to the same thing, and different things are sometimes given the same name.[3] Probably, the biggest cause of confusion is that some people use terminology for the wing drag that others use for the whole airplane, and vice versa.

The subdivision shown in Figure 7.7 is typical and provides a framework to introduce the drag components that are important for airplane performance studies. A short description of the key

---

2 Otto Lilienthal (1848–1896), German civil engineer and inventor, resolved the net aerodynamic force acting on a wing into lift and drag components, innovatively expressing them as coefficients. Lilienthal is remembered as the person who first demonstrated the viability of heavier-than-air human flight (he conducted over 2000 flights in gliders, of his design, from 1891 until his death in a gliding crash) [3].

3 There have been a few attempts to standardize terminology—for example, the British Aeronautical Research Council (ARC), in 1958, issued a technical report [4] on the subject. Unambiguous, descriptive terms, however, can be cumbersome. This fact, coupled with a reluctance by people and organizations to change their customary practices, means that there is a lack of widespread agreement on the subject of drag terminology.

drag components follow. In all cases, the drag component can be expressed as a coefficient by dividing by the product of free-stream dynamic pressure and wing reference area.

Friction Drag and Pressure Drag

The total drag that acts on an airplane can be divided, entirely, into two parts: *friction drag* (which is often called skin friction drag or surface friction drag) and *pressure drag*. This is evident, as the sum of the tangential (i.e., friction) and normal (i.e., pressure) stresses on any surface make up the resultant (i.e., total) stress acting on that surface. By considering the tangential and normal stresses on elements of the airplane's surface and integrating those stresses over the entire wetted area of the airplane, the total aerodynamic force is obtained; the component of that force that acts parallel with the flight direction is, by definition, the total drag. Friction drag is thus caused by the tangential, shearing stresses (due to the viscous effects in the boundary layer) acting on the aircraft's surface by the air (see Section 3.6.3). Its magnitude depends on the exposed surface area, the smoothness of the surface, and the nature of the boundary layer (i.e., laminar or turbulent). The friction drag coefficient, which tends to decrease with increasing Mach number, accounts for just less than half the total drag acting on a jet transport airplane in cruise [5]. Pressure drag is due to the normal stresses acting on the surface of the aircraft (i.e., at right angles to the surface) by the air. The pressure drag can be divided into several components, based on the drag-inducing flow phenomena (e.g., viscous effects, lift-induced, shocks).

Profile Drag

Profile drag (or boundary layer drag) is the drag arising due to the presence of a boundary layer on the airplane's surface as it moves through the air. It comprises friction drag and *form drag* (boundary layer normal pressure drag). The form drag depends on the shape (or form) of the body and hence on the boundary layer displacement effect, which changes the effective aerodynamic shape of the body due to the thickness of the boundary layer (see Section 3.6.1). In a detailed drag breakdown—which would be developed during the design of an airplane—the profile drag contribution of individual airplane components would be ascertained and summed (Hoerner [6], for example, provides extensive data on empirical drag estimation). Added to this total would be the interference drag, which is caused by the coalescing of the pressure fields and merging of the boundary layers associated with the individual airframe components. Profile drag, as indicated in the breakdown in Figure 7.7, also includes powerplant-related drag influences (see Section 7.5.4) and excrescence drag (see Section 7.5.7). An important observation is that the profile drag element associated with boundary layer displacement varies with changes in angle of attack, and, for the sake of convenience, can thus be categorized as being lift dependent. The bulk of the profile drag, that is, the skin friction drag, is essentially independent of angle of attack (and hence lift) for attached flow conditions; however, once separation occurs the skin friction reduces (see Section 3.6.4).

Vortex Drag

Vortex drag (also called induced drag[4] or lift-induced drag) is the drag produced on a wing (or other lifting surface such as the horizontal stabilizer) due to the trailing vortex system formed when any segment of the wing produces a lift force (i.e., an aerodynamic force that is normal

---

4 The term *induced drag* appears to have been coined by Michael Max Munk (1890–1986), an eminent German aerodynamicist and student of Ludwig Prandtl. It is the subject of his classic paper "The minimum induced drag of aerofoils" [7]. The term is arguably best used to describe the vortex drag of wings and not the complete lift-dependent drag of an airplane.

to the free-stream). Wing vortex drag (described in Section 7.3.3) accounts for approximately 40% of the total drag of a jet transport airplane in cruise.

Wave Drag
Wave drag (or compressibility drag) is the drag associated with shock waves and is only encountered in high-speed flight (i.e., at speeds greater than the critical Mach number, M_{cr}). The drag (described in Section 7.4.4) is due to a reduction in total pressure, which occurs as the air moves through the shock, and the interaction of the shock with the boundary layer. It has a lift-dependent component and a volume-dependent component (which is independent of angle of attack). The wave drag is usually measured as an increment with respect to the airplane's drag at a defined reference Mach number (typically Mach 0.5 or Mach 0.6).

Lift-Independent (Zero-Lift) and Lift-Dependent Drag Components
For airplane performance analysis, it is convenient to separate those drag elements that depend on angle of attack—and can hence be considered to be *lift dependent*—from those that do not. In this sub-division of the airplane's total drag, the lift-dependent drag should not be considered as being *caused* by the lift, but rather that this drag component can be mathematically expressed as a function of lift. This simple sub-division means that the drag coefficient in the low-speed regime (i.e., $M < M_{dd}$) can be expressed as

$$C_D = C_{D_0} + C_{D_i} \tag{7.6}$$

and in the high-speed regime (i.e., $M \geq M_{dd}$) as

$$C_D = C_{D_0} + C_{D_i} + \Delta C_{D_w} \tag{7.7}$$

where C_{D_0} is the zero-lift drag coefficient (i.e., C_D when $C_L = 0$);
$\quad\quad C_{D_i}$ is the lift-dependent drag coefficient; and
$\quad\quad \Delta C_{D_w}$ is the wave drag (or compressibility drag) increment.

The term *parasite drag* (or parasitic drag) is widely used in this context, but not always in a consistent manner. For example, it has been used to describe all drag contributions that are not associated with lift or compressibility [8]. Alternatively, some authors will define it such that it includes wave drag [9]. Under other drag decomposition schemes, profile drag is defined as the parasite drag of an airfoil [10], and, sometimes, it is used to describe the drag of non-lifting airplane components (i.e., fuselage, engine nacelle, vertical tail) [11].

The lift-dependent drag (see Section 7.3.4) of an airplane is often called the *induced drag* (or lift-induced drag). This usage, however, can be confusing as the term induced drag is also widely used (in current literature and historically) as an equivalent to vortex drag—and the vortex drag is only a part, albeit the overwhelmingly dominant part, of the lift-dependent drag.

7.3.3 Wing Vortex Drag

The vortex drag (or induced drag) of the wing is the main component of the airplane's lift-dependent drag. The vortex sheet can be considered as a row of closely spaced co-rotating vortex "filaments" that originate at the trailing edge of a lift-generating surface. The "filaments" formed behind each wing are attracted towards each other and roll-up over a short distance into a pair of counter-rotating vortices, which trail behind the airplane. These vortices are often called *wingtip vortices*. This term, however, is imprecise and its usage tends to be discouraged for two reasons: (1) each vortex is generated by the half-span of the wing and not just the wingtip;

and (2) the vortices are actually formed inboard of the wingtip (for an elliptical lift distribution, the distance between the centers of the two vortices is theoretically $\pi/4$ times the wingspan).

From a design perspective, the vortex drag associated with a given airplane weight depends on several wing design parameters—the most important being the wingspan and the shape of the spanwise lift distribution. The vortex drag of a wing without appreciable dihedral or out-of-plane wingtip extensions or winglets (i.e., a planar wing) is a minimum when the spanwise lift distribution has an elliptical shape—in such cases, the vortex drag coefficient is directly proportional to the square of the lift coefficient. For an airliner without winglets, operating at its design cruise condition, the lift distribution approaches the ideal elliptical distribution [12].

A lower limit of the vortex drag coefficient can be predicted theoretically for planar wings with elliptic lift distributions. This is given by the following expression, which was first derived by Ludwig Prandtl using what is now generally known as the Prandtl lifting line theory:[5]

$$C_{D_V} = \frac{1}{\pi A}\, C_L^2 \tag{7.8}$$

where C_{D_V} is the wing vortex drag coefficient; and
$\quad\quad A$ is the wing aspect ratio.

The vortex drag for a real planar wing, however, is always greater than that given by Equation 7.8. This can be taken into account by introducing a correction factor.

$$C_{D_V} = \frac{1+\delta}{\pi A}\, C_L^2 \tag{7.9}$$

where δ is a correction factor that accounts for deviations from the ideal, elliptical spanwise lift distribution.

The correction factor is not exactly constant during flight, as the lift distribution will change as the flight condition (i.e., speed, air density, airplane weight) changes. One of the reasons for this is that the geometric twist[6] of the wing is optimized for the design cruise condition (where, typically, C_L is about 0.5) and operation at other flight conditions results in unfavorable changes to the shape of the lift distribution.

Non-planar wings, however, can have lower vortex drag than comparable planar wings. Several non-planar wing design concepts have been devised (e.g., C-wing, box-wing, ring-wing) [15, 16], but most are considered impractical. Of interest for the design of jet transport airplanes are winglets (i.e., vertical or near-vertical wing-like aerodynamic surfaces mounted at the wing-tips). In the presence of a winglet, the trailing vortex sheet is extended vertically, leading to a reduction in the downwash and vortex strength—and consequently, the vortex drag is reduced. When evaluating the potential benefits of winglets, the conditions of the comparison need to be carefully considered, as misleading conclusions are easily reached. The addition of a winglet reduces the vortex drag, but it also increases the viscous drag (due to the increased wetted surface area) and it increases the wing weight (due to an increase in the wing bending moment). In the case where the wingspan is constrained as a design parameter (e.g., limited

---

5 Ludwig Prandtl (1876–1953), the renowned German scientist, established many foundation principles of aerodynamics [13]. The lifting line theory is sometimes referred to as the Lanchester-Prandtl wing theory in honor of the British physicist Frederick W. Lanchester (1868–1946), who carried out similar work independently. The vortex drag expression (Equation 7.8) was published in a translation from German in 1920 [14].

6 The wing geometry usually incorporates a twist, such that the angle of incidence at the wingtip is less than that at the wing root. This is done to improve stall characteristics (i.e., move the point of stall initiation inboard) and to modify the spanwise lift distribution. The B747-100, for example, has an incidence angle of $-1.5°$ at the wingtip and $2°$ at the wing root, which gives a 3.5° twist [8].

by taxiing or parking considerations), a well-designed winglet will result in a significant drag reduction. As a retrofit to an existing design, for example, a vortex drag reduction in cruise of 3–8% is possible [8, 17]. The situation, however, is not so clear when designers have the freedom to choose either a winglet or an increase in wingspan. When a trade study between these two concepts is conducted where the wing weight increase is held constant, the drag reduction—that is, with respect to the reference wing—is seen to be practically identical for the two designs [17].

In Equation 7.9 (which describes a planar wing), the aspect ratio is given by the ratio of the wingspan squared to the wing reference area (see Equation 3.3). This will be unchanged with the addition of winglets if neither the wingspan nor the reference area changes. An accepted means of taking into account the reduction in vortex drag for a wing with winglets is to define an effective aspect ratio based on an equivalent span and to redefine Equation 7.9 as follows:

$$C_{D_V} = \frac{1 + \delta}{\pi A_{eff}} C_L^2 \tag{7.10}$$

where A_{eff} is the effective aspect ratio (which takes into account the influence of non-planar wingtip devices, such as winglets).

Studies comparing the vortex drag of wings with winglets to that of planar wings, reveal that optimally designed winglets yield a similar drag benefit (not considering structural issues, such as changes to the wing bending moment) to a span extension on a comparable wing where the semispan has been increased by approximately 55% of the height of the winglet [17, 18].

7.3.4 Lift-Dependent Drag

There are several smaller drag elements that need to be added to the wing vortex drag to give the total lift-dependent drag coefficient of the airplane. First, the profile drag is largely, but not entirely, independent of angle of attack and, hence, lift. A portion of the friction and pressure drag associated with the airframe components (and some of the interference drag) will change when the angle of attack (and hence lift) changes—but not necessarily as a quadratic function of lift. Secondly, the vortex drag described by Equation 7.9 (or Equation 7.10) only applies to the wings, and does not include trim drag (see Section 7.5.3), which is also lift dependent. If these smaller drag terms are lumped together with the wing vortex drag and it is assumed that, collectively, they vary as a function of C_L^2, then the following approximate expression for the lift-dependent drag coefficient of the airplane is obtained:

$$C_{D_i} = \frac{1}{\pi Ae} C_L^2 \tag{7.11}$$

or $\quad C_{D_i} = K\, C_L^2 \tag{7.12}$

where e is known as the Oswald factor or Oswald efficiency factor;[7] and
K is the lift-dependent drag factor.

The Oswald factor is typically about 0.83 for a conventional airliner. From a design perspective, it is seen that an aerodynamically efficient airplane requires a combination of a high aspect ratio wing and a high value of the Oswald factor. In general terms, the Oswald factor is linked to the

---

7 Named after William Bailey Oswald (1906–1996), who used this formulation in his aircraft performance studies at the California Institute of Technology, Pasadena, CA [19]. The term is sometimes called the span efficiency factor. This practice, however, can be misleading as e (as defined in Equation 7.11) also includes lift-dependent profile drag elements and is not solely dependent on the spanwise lift distribution.

spanwise lift distribution; as the lift distribution approaches the ideal elliptical shape, the value to e tends towards unity.

It is instructive, when assessing the design attributes of a jet transport airplane that make it aerodynamically efficient, to consider the dimensional form of Equation 7.11. Using Equations 7.3 and 3.3, the lift-dependent drag (D_i) can be written as follows:

$$D_i = qS \left(\frac{1}{\pi Ae} \right) \left(\frac{n_z W}{qS} \right)^2 = \frac{n_z}{q\pi e} \left(\frac{W}{b} \right)^2 \tag{7.13}$$

Equation 7.13 is an interesting result as it indicates that for a given load factor (in level flight, n_Z equals 1), the lift-dependent drag depends on the square of the airplane's span loading (W/b). It is evident from this equation that it is the wingspan, rather than the aspect ratio *per se*, that is important. Vortex drag scales in a different way to the other drag terms, which are roughly proportional to the wing area. The process of non-dimensionalizing all drag terms in a consistent way (i.e., by dividing the force by the product of dynamic pressure and wing reference area) disguises the true nature of vortex drag and makes it appear that vortex drag depends on wing area or aspect ratio, when, in fact, it is a function of wingspan.

7.4 Drag Polar

7.4.1 Introduction

The relationship between drag and lift—or lift and drag—has been the focus of much aerodynamic research for over a century. The traditional representation is a plot of lift coefficient (ordinate) against drag coefficient (abscissa)—this is universally known as a *drag polar*.[8] Today, the term is applied in a loose manner to describe both graphical and mathematical representations of the relationship between the lift and drag coefficients.

7.4.2 Typical Drag Versus Lift Relationship

A typical series of drag polars for a jet transport airplane in the clean configuration are shown in Figure 7.8. Two trends are immediately apparent:

(1) The drag coefficient increases monotonically with increasing lift coefficient.
(2) Mach number has a significant influence on the drag–lift relationship above a certain value: the critical Mach number (M_{cr}). For the particular data set shown in Figure 7.8, M_{cr} is seen to be between Mach 0.6 and Mach 0.7.

The second observation provides a convenient simplification for the description of airplane drag. The drag polars are divided into two groups:

(1) The *low-speed* (incompressible flow) drag polars are those where Mach number has little or no influence. For many applications, a single curve can be used to represent the drag–lift relationship at these speeds.

---

8 A more correct term would be drag polar graph or drag polar diagram, as polar coordinates can be used to represent this relationship (this is seldom done, though). The first use of this representation is attributed to Otto Lilienthal (1848–1896), who presented such graphs in his landmark work, *Der Vogelflug als Grundlage der Fliegekunst* (Bird flight as the basis of aviation) in 1889 [20]. Gustave Eiffel (1832–1923), in 1911, called them *courbes polaires* (polar curves).

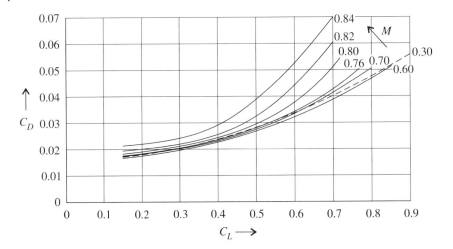

Figure 7.8 Typical C_D versus C_L relationship for a jet transport airplane.

(2) The *high-speed* (transonic) drag polars are those above M_{cr}. Here it is necessary to take account of compressibility effects and the resulting drag increase (see Section 7.4.4). In the high-speed regime, there is a unique drag polar for each Mach number.

A number of alternative mathematical formulations have been devised to describe the C_D versus C_L relationship, taking into account the influence of Mach number—this is discussed in Sections 7.4.3–7.4.8. In Section 7.5, corrections to the basic drag polar, for example to account for changes to the airplane's configuration or proximity to the ground, are discussed.

7.4.3 Low-Speed Drag Polars

In this context, *low speed* describes flight Mach numbers below the critical Mach number. At these speeds, a single function is frequently used to describe the drag–lift relationship for the airplane in the clean configuration. If the lift-dependent drag term is expressed by Equation 7.12, the total drag coefficient of the airplane can be approximated by a parabolic function of the lift coefficient, that is,

$$C_D = C_{D_0} + KC_L^2 \tag{7.14}$$

The validity of this expression can be assessed by redrawing the drag curves shown in Figure 7.8 with C_D plotted against C_L^2 (Figure 7.9). It is seen in Figure 7.9 that for Mach numbers less than 0.6, the drag functions are indeed almost exactly linear, indicating that Equation 7.14 provides an acceptable model for this data set. However, above Mach 0.6, there is a progressive departure from the parabolic relationship.

The value of K in Equation 7.14 can be found by determining the slope of the line within the linear region and the value of C_{D_0} by extrapolating the line to $C_L = 0$. Close inspection of the low-speed data reveals that the lines are approximately parallel, indicating that K does not change appreciably with Mach number, but there are small changes to the values of C_{D_0}, with the lowest value occurring at about Mach 0.5 (for this particular drag data).

It should be noted that Figures 7.8 and 7.9 describe the drag of a single airplane type, and significant differences are evident between airplanes types (in this illustration, the drag polars are based on those of a mid-size, single-aisle, twin-engine jet airliner). For example, the Mach

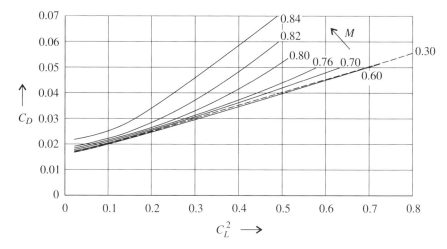

Figure 7.9 Typical C_D versus C_L^2 relationship for a jet transport airplane.

number that separates the drag data into low and high-speed regimes (i.e., M_{cr}) depends significantly on the wing design (the primary wing design parameters that influence M_{cr} are the sweepback angle, airfoil type, and thickness-to-chord ratio). Airplanes designed with a high cruise Mach number will have a relatively high critical Mach number.

7.4.4   High-Speed (Transonic) Drag Polars

The critical Mach number (M_{cr}) is the free-stream Mach number at which localized regions of supersonic flow first appear (see Section 3.5.3). As the free-stream Mach number increases past M_{cr}, the drag coefficient starts to increase rapidly (i.e., diverge) due to compressibility effects associated with the shocks that are formed on the wing surface. This drag increase is known as *wave drag* or compressibility drag. The drag coefficient can thus be expressed as follows:

$$C_D = (C_D)_{low\,speed} + \Delta C_{D_w} \tag{7.15}$$

where $(C_D)_{low\,speed}$ is the low-speed drag coefficient (for the given C_L); and
ΔC_{D_w} is the wave drag increment (for the given C_L and Mach number).

The drag increase is the result of several factors. A feature of shocks is that the air undergoes a "jump" in pressure as it moves through the shock, and this manifests as an increase in pressure drag. The increase in pressure through the shock also results in a thickening of the boundary layer and an increase in its rate of growth—and this directly increases the pressure drag due to the boundary layer displacement effect [21], described earlier in Section 3.6.1. With an increase in free-stream Mach number, the shocks strengthen and tend to move towards the trailing edge; strong shocks can cause separation of the boundary layer resulting in further, substantial increases in drag.

The drag divergence Mach number (M_{dd}), which is always greater than the critical Mach number (i.e., $M_{dd} > M_{cr}$), is defined as the Mach number at which there is a significant increase in drag coefficient (Figure 7.10a). For subsonic jet transport airplanes, the drag increases so rapidly that operating at speeds much greater than M_{dd} is usually impractical and uneconomical. There are several interpretations regarding what constitutes a significant drag

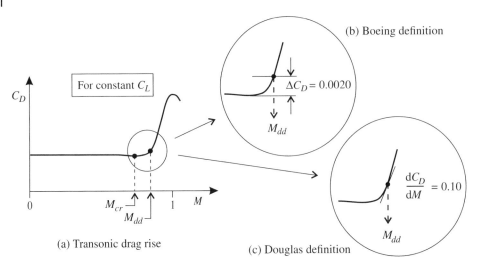

Figure 7.10 (a) Drag coefficient versus Mach number, illustrating the drag divergence Mach number (M_{dd}); (b) Boeing definition of M_{dd}; (c) Douglas definition of M_{dd}.

increase.[9] The so-called Boeing definition of M_{dd} corresponds to the speed where the drag increases by 20 drag counts above the baseline condition (Figure 7.10b); the so-called Douglas definition is based on a rate of change of C_D with Mach number equal to 0.10 (i.e., $dC_D/dM = 0.10$) (Figure 7.10c) [8, 22]. The two definitions, not unexpectedly, yield slightly different values of M_{dd} in most cases.

The drag divergence Mach number for an airplane depends on the design of its wing. The key parameters are the wing sweep, thickness-to-chord ratio, and airfoil section. A convenient method for estimating M_{dd} for an airfoil section is the Korn equation,[10] which has been extended by Mason [24] to represent a finite wing incorporating sweep:

$$M_{dd} = \frac{\kappa_A}{\cos \Lambda} - \frac{1}{\cos^2 \Lambda} \left(\frac{t}{c} \right) - \frac{C_L}{10 \cos^3 \Lambda} \qquad (7.16)$$

where κ_A is an airfoil technology factor;
Λ is the wing sweep angle;
t/c is the thickness-to-chord ratio of the wing; and
C_L is the lift coefficient.

The airfoil technology factor takes into account the airfoil section—for example, the performance of NACA 6-series airfoils can be approximated by a value of κ_A of 0.87, whereas supercritical sections are best represented by a value of 0.95 [24, 25]. Equation 7.16 represents a linear relationship of M_{dd} with C_L. This expression was validated, by the author, using the drag data given in Figure 7.8 for the Douglas definition of M_{dd}. The result is shown in Figure 7.11 for a range of cruise lift coefficients.

---

9 The two definitions of M_{dd} described herein correspond to conventions historically adopted by Boeing Commercial Airplanes and by the Douglas Aircraft Company [8, 22]. Other conventions also exist—for example, some authors/organizations do not use the term drag divergence and instead refer to the speed at which an appreciable drag rise is observed as the critical Mach number.

10 A two-dimensional empirical equation to estimate M_{dd} for an airfoil was devised by David Korn at the NYU Courant Institute in the early 1970s [23].

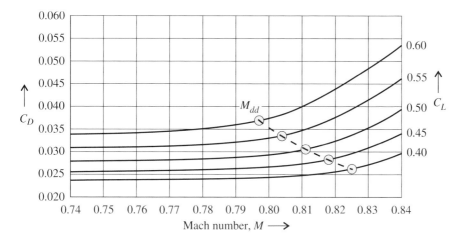

Figure 7.11 Drag divergence Mach number (Douglas definition) for the drag data given in Figure 7.8.

Several empirically derived equations have been proposed to represent the shape of the drag curve, as it rises with increasing Mach number. One common approach assumes that the wave drag can be modeled as a power law, that is,

$$C_{Dwave} = p(M - M_{cr})^n \quad \text{and} \quad M > M_{cr} \tag{7.17}$$

where p and n are empirical constants selected to best represent the data.

The original formulation of this expression is attributed to Lock [26], who evaluated airfoil wave drag with an exponent $n = 4$. The value of p depends on the aerofoil geometry, the critical Mach number, and the angle of attack.

However, Equation 7.17 does not always provide a particularly accurate representation of the drag rise of modern wing designs. Another approach, which produces a better approximation to the manner in which the drag increases at typical cruise lift coefficients, is presented by Torenbeek [27]. This is based on the M_{dd} and takes the following form:

$$\Delta C_{D_w} = 0.002 \left(1 + n\frac{M_{dd} - M}{\Delta M}\right)^{-1} \quad \text{for} \quad M \leq M_{dd} \tag{7.18}$$

and

$$\Delta C_{D_w} = 0.002 \left(1 + \frac{M - M_{dd}}{\Delta M}\right)^n \quad \text{for} \quad M > M_{dd} \tag{7.19}$$

where n and ΔM are empirical constants selected to best represent the data.

Note that when $M = M_{dd}$, $\Delta C_{D_w} = 0.002$, which matches the Boeing definition of the drag rise Mach number. Torenbeek [27] illustrated the application of this method with $n = 2.5$ and $\Delta M = 0.05$. Acceptable agreement to the drag curve shown in Figure 7.8 for $C_L = 0.5$ was obtained with constants $n = 3$ and $\Delta M = 0.05$.

7.4.5 Effect of Camber and Wing Twist on the Drag Polar

Many analytical problems can be solved using the parabolic drag approximation given as Equation 7.14. Actual airplane drag polars, however, do not exactly follow this form, particularly at high and low angles of attack. The departure from the idealization at low angles of attack is principally due to airfoil camber and wing twist. As shown in Figure 7.12a, the idealized parabolic

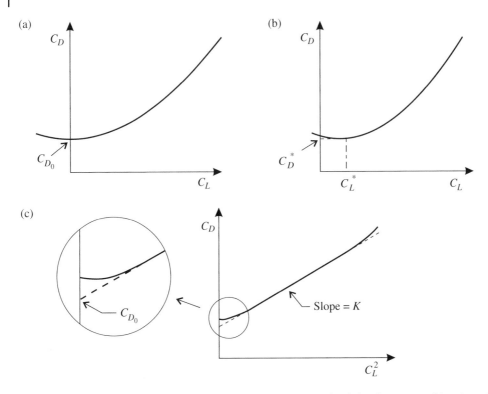

Figure 7.12 Drag polars constructed for (a) a wing with a symmetrical airfoil and zero twist; (b) a wing with a cambered airfoil and/or twist; (c) a wing with a cambered airfoil and/or twist, plotted as a function of C_L^2.

drag relationship is symmetrical about the ordinate, with the lowest drag value corresponding to $C_L = 0$. This feature is true for wings without camber (i.e., symmetrical airfoil sections). Camber or wing twist shifts the curve slightly to the right. The lowest point on the drag curve (denoted as C_D^*) now corresponds to a small positive value of C_L (denoted as C_L^*), as shown in Figure 7.12b. Such drag polars can be represented by the following equation:

$$C_D = C_D^* + K^*(C_L - C_L^*)^2 \tag{7.20}$$

or $$C_D = \left\{ C_D^* + K^*(C_L^*)^2 \right\} - (2K^*C_L^*)C_L + K^*C_L^2 \tag{7.21}$$

where C_D^* is the minimum value of the drag coefficient (see Figure 7.12b);
    K^* is a lift-dependent drag factor (similar to K in Equation 7.14); and
    C_L^* is the lift coefficient corresponding to C_D^* (see Figure 7.12b).

Although Equation 7.20 provides a more accurate representation of the drag acting on a conventional airplane at low C_L (in the low-speed regime), it has, in fact, little real value for general airplane performance analyses. This is because the region where an appreciable discrepancy between the two functions (i.e., Equations 7.14 and 7.20) exists corresponds to lift coefficients that are lower than those associated with the angles of attack typical of normal steady-state flight. The minimum value of C_L in steady-state flight corresponds to a condition of maximum speed and minimum weight (see Equation 7.4). For conventional jet transport airplanes, operating at a load factor of one (or more), C_L is unlikely to ever be less than 0.2 (lift coefficients less than this are only achieved during maneuvers that produce load factors less than one). Drag data for very low values of C_L are obtained from computational analysis and/or wind tunnel tests conducted at angles of attack that do not correspond to steady, level flight.

A pragmatic approach that is widely adopted ignores this small departure from reality and determines a fictitious value of C_{D_0} by extrapolating to zero lift a best-fit line on a graph of C_D versus C_L^2, taken within the linear region of the graph (as illustrated in Figure 7.12c). This enables the simpler equation given by Equation 7.14 to be used, where K is given by the slope of the line.

7.4.6 Drag Polar Represented by the Sum of Two Parabolic Segments

A refinement of the drag representation given by Equation 7.14 assumes that an airplane's drag—for any given Mach number—can be represented by two parabolic segments. This approach, which can be used to model the drag in the high-speed regime, is described in ESDU 66031 [28] (albeit, for a plane symmetrical section wing). The approach is best seen on a plot of C_D versus C_L^2, as shown in Figure 7.13. Up to a certain value of the lift coefficient (C_{L_K}), the variation of C_D with C_L^2 is linear and can be represented by the slope k_1. Above C_{L_K}, the slope is given by k_2, where $k_2 > k_1$.

A limitation of this method is that the slope of the C_D versus C_L^2 line does not, in reality, change abruptly at C_{L_K}, but instead a gradual change is evident. This limitation is eliminated by the approach described in Section 7.4.7, which introduces higher-order terms into the drag polar.

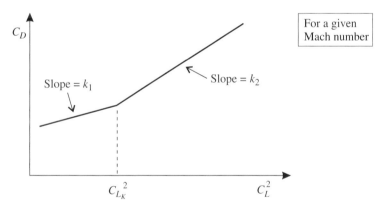

Figure 7.13 Drag approximation based on two parabolic segments.

7.4.7 Drag Polars with Higher-Order Terms

The drag polars shown in Figure 7.8 can be represented, rather accurately, by the following polynomial expression:

$$C_D = C_{D_0} + K_1\, C_L^2 + K_2\, C_L^n \tag{7.22}$$

where

    (1) for low-speed drag polars, C_{D_0} and K_1 do not vary significantly with Mach number and K_2 is very small or equal to zero; and
    (2) for high-speed drag polars, C_{D_0}, K_1, and K_2 vary with Mach number.

The exponent n in Equation 7.22 can be found from curve fitting to experimental data. ESDU 66031 [28] indicates that, in some cases, a term of $n = 4$ "is not negligible" in drag representation. In a study of the drag of a fighter airplane with maneuver flaps extended, it was concluded that the exponent $n = 4$ provided an effective representation of flight-test data [29].

Curve fitting of drag data for a conventional airliner (see Figure 7.8), undertaken by the author, indicated a good fit, across the entire operating Mach range, using the exponent $n = 6$. In the low-speed regime, constant values of C_{D_0} and K_1 were employed (with K_2 set equal to zero), whereas in the high-speed regime, C_{D_0}, K_1, and K_2 were observed to change as smoothly varying functions of Mach number.

Yet another approach is to include additional terms in the drag polar, in an effort to increase the accuracy of the representation. For example, in reference [30] it is maintained that a better fit to experimental data of elliptic and crescent wing planforms than that possible with the parabolic drag polar can be obtained using the following drag representation:

$$C_D = C_{D_0} + K_1 C_L + K_2 C_L^2 + K_3 C_L^3 + K_4 C_L^4 \tag{7.23}$$

This expression, however, introduces additional terms (compared to Equation 7.22) that have to be determined by curve fitting to experimental data, without appearing to offer any real advantage to the representation given by Equation 7.22.

7.4.8 Drag Polars: Concluding Remarks

The more complex drag representations described in Sections 7.4.5 to 7.4.7 provide alternative means to model experimental data. These formulations, however, do not readily lend themselves to closed-form analytical analyses of the performance of an airplane (such as those described in Chapters 9 to 15, for example). For this reason, the simple parabolic drag polar, given by Equation 7.14, is widely used to establish approximate descriptions of the behavior of an airplane in flight. The parabolic representation, although not wholly realistic, facilitates the generation of many closed-form solutions, thereby promoting a fundamental understanding of the flight characteristics of an airplane. It is thus a very useful analytical tool. It must be kept in mind, though, that Equation 7.14 is an approximation—not a law—and that the parameters C_{D_0} and K depend on several factors (such as the airplane's configuration and Mach number) and are therefore not constant across the operating envelope of the airplane.

7.5 Drag Polar Corrections

7.5.1 Correction Factors

The drag polar for a specific airplane or airplane type can be represented using the graphical or mathematical methods outlined in Section 7.4. This is often referred to as the basic or reference drag polar and it describes the drag–lift relationship of the airplane, at the Mach number of interest, for a defined set of reference (baseline) conditions. To analyze the performance of the airplane for many practical situations, however, it is necessary to apply one or more correction factors to the basic drag polar to account for changes from the reference conditions. The following factors are described in this chapter:

- configuration of the airplane (Section 7.5.2);
- center of gravity (Section 7.5.3);
- engine thrust (Section 7.5.4);
- inoperative engine (Section 7.5.5);

- ground effect (Section 7.5.6);
- airframe condition with respect to excrescence drag (Section 7.5.7);
- atmospheric conditions (Section 7.5.8); and
- aeroelastic wing shape change (Section 7.5.9).

The last two correction factors are very small (typically less than 1%) and are only relevant to performance analyses conducted for actual flight operations, where very accurate drag polars have been established by the manufacturer. Historically (prior to the 1980s), drag polars could only be established with an accuracy of 1–2% and hence these smaller corrections were not considered.

7.5.2 Influence of Aircraft Configuration on the Drag Polar

The drag characteristics of an airplane will be altered when its configuration changes. In this context, the *configuration* is understood to be the position of any reconfigurable part of the airframe—on instruction by the flight crew or flight control system—that will have an influence on the aerodynamic characteristics of the airplane. In other words, it describes the position of the leading and trailing edge devices, spoilers, undercarriage (landing gear), flight control surfaces (e.g., rudder), and so forth. The term "clean" is often used to describe the basic configuration, that is, without flaps or undercarriage, and so forth, extended. The rather odd-sounding description of "dirty" is occasionally used to describe any configuration with flaps or undercarriage, and so forth, extended.

The extension (or deployment) of any of these items will alter the clean, basic aircraft drag coefficient. For a given lift coefficient, the drag coefficient can be written as follows:

$$C_D = (C_D)_{clean} + \Delta C_{D_{slats}} + \Delta C_{D_{flaps}} + \Delta C_{D_{spoilers}} + \Delta C_{D_{uc}} + \Delta C_{D_{controls}} + \Delta C_{D_{other}} \quad (7.24)$$

where $(C_D)_{clean}$ is the clean, basic drag coefficient at the given C_L; and
ΔC_D represents the drag change due to the slats, flaps, spoilers, undercarriage (uc), flight controls, etc., at the appropriate C_L.

These drag changes (usually increments) depend on the geometry of the airplane (e.g., flap position) and on the associated flight conditions (such as speed, angle of attack, and load factor). The magnitudes of the changes, in almost all cases, will be a function of the airplane's lift coefficient. The usual approach, then, is to determine $(C_D)_{clean}$ from an appropriate drag polar for the relevant C_L and then to apply a series of drag corrections, as per Equation 7.24, where each term is found using a lookup table or is read off a graph of drag change versus C_L.

An alternative approach considers the aerodynamic impact of the configuration changes as factors that alter the clean (basic) drag polar. In other words, a new drag polar is generated for the particular aircraft configuration of interest (this could be, for example: undercarriage extended, spoilers retracted, slats and flaps set for takeoff). This technique is useful when conducting analytical investigations across a range of C_L values (e.g., during climb-out after lift off). If the drag polar is represented by the parabolic drag model (Equation 7.14), corrections ΔC_{D_0} and ΔC_{D_i} can be introduced as follows for a particular configuration:

$$C_D = \left(C_{D_0}\right)_{clean} + \Delta C_{D_0} + K C_L^2 + \Delta C_{D_i} \quad (7.25)$$

where $(C_{D_0})_{clean}$ is the zero-lift drag coefficient of the clean airplane;
ΔC_{D_0} is the change in the zero-lift drag coefficient;
K is the lift-dependent drag factor (clean airplane); and
ΔC_{D_i} is the change in the lift-dependent drag coefficient.

The simplifying assumption is often made that ΔC_{D_i} varies as a function of C_L^2:

$$\Delta C_{D_i} = \phi\, C_L^2 \tag{7.26}$$

thus $\quad C_D = \left(C_{D_0}\right)_{clean} + \Delta C_{D_0} + (K + \phi)C_L^2 \tag{7.27}$

where ϕ represents the change to the lift-dependent drag factor due to the changed configuration of the airplane.

In the case of leading and trailing edge devices, their deployment will result in an increase in zero-lift drag. Furthermore, there will be an increase in lift-dependent drag. An airplane's wings are optimized for cruise; thus, when the flaps are extended, the spanwise lift distribution will be altered in an unfavorable manner. With more lift being generated over the flapped, inboard sections of the wing, the lift distribution will be further away from the ideal, low-drag elliptical distribution, and this means that ϕ is a positive value. In other words, the Oswald factor reduces with the deployment of slats and flaps. Flap deployment may also result in another drag increment. When the flaps are extended, a nose-down pitching moment is generated that must be balanced, or trimmed out, by a deflection of either the horizontal stabilizer or the elevator. In either situation, an increase in what is known as *trim drag* results (see Section 7.5.3).

In the case of the undercarriage, however, it can be expected that the extension of the undercarriage will result in a significant increase in zero-lift drag with little or no change to the lift-dependent drag component.

Deflection of the rudder, elevator, or ailerons are required to maneuver the airplane—in normal flight operations, these are momentary, resulting in short-term drag increments (which can be ignored for the purpose of climb, cruise, or descent analysis, for example). During certain emergencies, however, control inputs are required for extended time periods. An engine failure on a twinjet, for example, would require constant deflections of the rudder (to balance the yawing moment caused by the asymmetric thrust) and ailerons (to balance the rolling moment caused by the rudder deflection). In such cases, the drag increase due to the control surface deflection should be taken into account.

7.5.3 Longitudinal Trim Drag and Center of Gravity Location

An associated topic to the drag caused by the deflection of a flight control is *longitudinal trim drag*. In normal, level flight (i.e., $n_z = 1$), the horizontal stabilizer (tail) produces a downward aerodynamic force, which is needed to balance the airplane's nose-down pitching moment (generated, primarily, by the camber of the wing). The tail lift-force produces vortex drag. Furthermore, an increase in wing lift—and thus additional wing vortex drag—is needed to counter the downward direction of the tail lift-force. These factors are all taken into account in the determination of the basic drag polar for the airplane; however—and this is the salient point—it is done so for a nominal or reference center of gravity (CG) position. If the airplane's CG were ahead of this reference position, for example, a greater deflection of the horizontal stabilizer would be needed to trim the airplane for steady flight (see Section 19.4).

Trim drag increments or decrements are thus needed to account for the airplane's actual CG position. The A330/340 family of airplanes, for example, which have a CG reference position of 28% mean aerodynamic chord (MAC), can be safely operated with the CG between 20% MAC and 37% MAC. The magnitude of the drag increment or decrement (ΔC_{D_0}) depends on the airplane type (certain types, e.g., A320, are relatively insensitive to CG movement) and also on the airplane's weight and flight altitude. A drag increase (or decrease) of up to 2% with respect to the reference condition is possible [31]. To minimize drag—and thus fuel burn—it is thus desirable to operate the airplane with the CG near to the certified aft limit (see also Sections 19.3 to 19.5).

7.5.4 Powerplant Considerations

The thrust produced by the engines affects the airplane's lift and drag in several ways. First, the thrust vector is offset from the flight path vector (Figure 7.14). The component of the thrust that acts parallel to the flight path is the *net thrust* (F_N)—this opposes the drag. The thrust offset angle (ϕ_T), which is the angle between the fuselage datum (X_b) and the thrust vector, is very small. In cruise, the angle of attack (α) is also small (typically 2–3°)—consequently, the thrust offset has a relatively minor influence in cruise; however, during flight phases that require higher angles of attack, the influence is much greater. The component of thrust normal to the flight path augments the lift, effectively reducing the required lift that has to be generated by the wing (and, consequently, reducing the vortex drag). Secondly, as the net thrust vector does not, in the general case, act through the center of gravity, a pitching moment is generated that has to be balanced by the horizontal tailplane—and this in turn affects the trim drag (see Section 7.5.3). Thirdly, the engine inlet and exhaust stream-tubes alter the flow field around the aircraft. A change to the airflow around the wing and flaps, for example, will change the lift and drag.

The accurate prediction of an airplane's performance requires a consistent set of definitions as to what constitutes propulsion system thrust and what constitutes propulsion system drag. The categorization of these forces (and their secondary effects), however, is a complex task. For example, during cruise, the high velocity flow over the inlet lip of the engine nacelle results in an intense low pressure region, part of which acts on forward-facing surfaces producing a forward component of force. It can be argued that this aerodynamic force reduces the drag on the nacelle and should thus be regarded as a negative drag contribution; alternatively, it can be argued that this force is part of the engine thrust. Airplane manufacturers apply bespoke thrust/drag accounting definitions to categorize such forces. In simple terms, the fore and aft components of force (parallel to the flight path vector) whose magnitude depends directly on the thrust lever position are usually considered as increments or decrements of thrust, rather than drag changes. Furthermore, the component of thrust normal to the flight path vector is considered a lift increment.

Most powerplant considerations are taken into account in the determination of the basic drag polar for the airplane. There are, however, certain flight conditions where a correction to the drag polar is necessary. Air spillage around the engine inlet is one such case. If a flight crew were

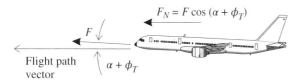

$F = $ installed thrust of all engines
$F_N = $ net thrust (defined parallel to the flight path)
$\alpha = $ angle of attack (i.e., angle between the flight path vector and the fuselage datum)
$\phi_T = $ thrust offset angle (i.e., angle between the fuselage datum and the thrust vector)

Figure 7.14 Illustration of thrust offset and net thrust.

to move the thrust levers to the in-flight idle position during cruise, for example, the engines' demand for air would be significantly reduced and air would be deflected, or spilled, around the engine inlet, generating what is called *spillage drag*. Performance calculations for a low-thrust descent, for example, should include this drag increase to the basic drag polar.

7.5.5 Drag Increments Due to Inoperative Engine(s)

Another condition that requires a correction to the basic drag polar is flight with one or more inoperative engine. The drag increase typically has three components: (1) windmilling drag, (2) spillage drag, and (3) yaw (or control) drag—these factors are discussed in this section.

Moving air entering an inoperative engine will cause the rotating assemblies (described in Section 8.2.2) to spin. The energy needed to produce this effect, can be viewed as an effective drag force acting on the engine—this is known as *windmilling drag*. Furthermore, the amount of air that can pass through an engine in this condition will be substantially less than what would normally occur in a fully functioning engine at the associated flight speed, causing air to spill around the nacelle—this results in spillage drag. Both of these factors can be expressed in coefficient form and included in a description of the airplane's drag polar.

There can also be a secondary effect on the airplane's drag associated with the pilot's actions in controlling the aircraft. If the inoperative engine is not on the centerline of the airplane, the resulting asymmetric thrust produced by the "good" engine(s) will introduce a significant yawing moment, which will have to be balanced (or countered) by the pilot deflecting the flight controls. Figure 7.15 illustrates this effect for a twin-engine airplane with the right (starboard) engine inoperative. To maintain a constant heading, the pilot would counter the yawing moment by moving the rudder (to the left in this case). As the line of action of the aerodynamic force due to the deflected rudder is above the fuselage, a rolling moment—which, by definition, is positive (i.e., right-wing down) in this example—is produced. To achieve a flight condition in which the aerodynamic forces are balanced, it would also be necessary for the pilot to provide an aileron input and to utilize some sideslip (the airplane's longitudinal axis would not be aligned with the direction of flight). The deflected flight controls and the yawed condition of the fuselage relative to the flight direction generates additional drag—this is usually called *yaw drag* (or control drag).

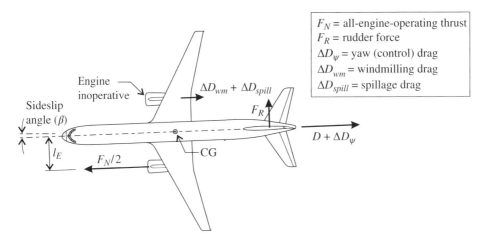

Figure 7.15 One engine inoperative condition (for a twin).

The yaw drag, which can be expressed as an increment to the basic airplane drag coefficient as ΔC_{D_ψ}, is a function of the yawing moment coefficient (C_N) produced by the inoperative engine(s). A plot of ΔC_{D_ψ} versus C_N, applicable to the airplane type can be used to determine the associated drag increment. In the case of the twinjet illustrated in Figure 7.15, the yawing moment coefficient produced by the propulsion system is described by the following equation:

$$C_N = \frac{l_E \left(\dfrac{F_N}{2} + \Delta D_{wm} + \Delta D_{spill} \right)}{qSb} \tag{7.28}$$

where C_N is the yawing moment coefficient (dimensionless);
$\quad l_E$ is the moment arm, as defined in Figure 7.15 (typical units: m, ft);
$\quad \Delta D_{wm}$ is the windmilling drag increment (typical units: N, lb);
$\quad \Delta D_{spill}$ is the spillage drag increment (typical units: N, lb); and
$\quad b$ is the wingspan (typical units: m, ft).

The drag acting on an airplane with one or more inoperative engines can be written in coefficient form taking into account the influence of the windmilling effect ($\Delta C_{D_{wm}}$), the engine spillage effect ($\Delta C_{D_{spill}}$), and the yaw drag (ΔC_{D_ψ}):

$$C_D = (C_D)_{clean} + \Delta C_{D_{wm}} + \Delta C_{D_{spill}} + \Delta C_{D_\psi} \tag{7.29}$$

7.5.6 Ground Effect and Its Influence on the Drag Polar

The operation of the airplane in close proximity to the ground (or to water) results in a change in the nature of the trailing vortex sheet, which, in turn, produces a small increase in lift and a significant reduction in drag. This phenomenon is called *ground effect* (even when referring to flight near water). The principal influence of the close proximity of the ground to the aircraft is a reduction in the vortex drag (and consequently an increase in the lift-to-drag ratio)—compared to when the airplane is operating "out of ground effect." Theoretical studies and wind tunnel experiments reveal that the ground effect increases the lift-curve slope and the angle of attack for zero lift, and decreases the nose-down (i.e., negative) pitching moment of the wing [32, 33]. This reduction in pitching moment results in a reduction in trim drag (see Section 7.5.3). The magnitude of these changes diminish as the airplane moves away from the ground, and when the airplane's wings are at a height approximately equal to one wingspan, the effects are negligible.

The ratio of the height of the lifting surface (i.e., the wing) above the ground (h) to the airplane's wingspan (b) is seen as the critical parameter influencing the magnitude of the observed aerodynamic changes, and this provides a convenient parameter to characterize—and mathematically model—the phenomenon. The drag polar of the airplane operating in ground effect (IGE) is compared to that of the airplane operating out of ground effect (OGE) for the same configuration. In this way, ground effect factors, which are a function of h/b, can be used to correct the basic OGE drag polar for the purpose of analyzing the airplane's performance near to or on the ground—for example, during takeoff or landing.

If the parabolic drag polar (given by Equation 7.14) is used to represent the OGE drag polar, then the IGE drag polar requires additional terms (in a similar format to that given by Equation 7.25). The change to the zero-lift drag coefficient is small and can usually be ignored. The effect of the ground on the lift-dependent drag coefficient can be expressed in terms of a drag reduction factor, herein assigned the Greek letter lambda (λ), which is defined as follows:

$$\lambda = \frac{\left(C_{D_i}\right)_{IGE}}{\left(C_{D_i}\right)_{OGE}} \tag{7.30}$$

This enables a simplified drag expression to be used to describe the drag coefficient of the airplane in ground effect (C_{D_g}):

$$C_{D_g} = C_{D_0} + \lambda K C_L^2 \tag{7.31}$$

where C_{D_0} is the zero-lift drag coefficient of the airplane (OGE);
λ is the drag reduction factor due to ground effect; and
K is the lift-dependent drag factor (OGE).

Flight in close proximity to the ground has a similar effect to a hypothetical increase in the wing's aspect ratio. Using Equation 7.11, it is possible to relate the ground effect factor to the wing aspect ratio (A) and a hypothetical effective aspect ratio (A_{eff}):

$$\lambda = \frac{A}{A_{eff}} \tag{7.32}$$

The classic, theoretical technique used to investigate this phenomenon introduces a mirror vortex system below the wing, such that the vertical flow components cancel each other out at the mid-plane, which represents the ground. This involves placing an inverted replica (mirror image) of the wing at a distance $2h$ below the wing. Using this model, the upper (real) wing can be considered to be flying in the upwash of the inverted mirror image. Thus, the lift as well as the lift-curve slope (i.e., lift change due to the angle of attack) is increased, which is a similar result to a change of aspect ratio, as indicated by Equation 7.32. Wieselsberger [32], using Prandtl's lifting line theory (with an elliptical lift distribution) to analyze the resulting twin wing model, determined what he called a ground-effect influence coefficient. When this influence coefficient is written as a drag reduction factor, as defined by Equation 7.30, it takes the following form:

$$\lambda = \frac{0.05 + 8.72\left(\dfrac{h}{b}\right)}{1.05 + 7.4\left(\dfrac{h}{b}\right)} \tag{7.33}$$

where h is the vertical distance from the ground to the wing aerodynamic center; and
b is the wingspan.

Although Wieselsberger [32] suggested limits of applicability of $0.033 < h/b < 0.25$, other researchers have extrapolated the use of this expression to greater ratios of h/b.

An empirical model proposed by Hoerner and Borst [34] produces comparable results to those given by Equation 7.33. This is based on the interpolation of test data; it is presented in the following form:

$$\lambda = \frac{33\left(\dfrac{h}{b}\right)^{3/2}}{1 + 33\left(\dfrac{h}{b}\right)^{3/2}} \tag{7.34}$$

Equations 7.33 and 7.34 are simple expressions that do not take into account the full complexity of the trailing vortex system that develops around an actual aircraft wing (particularly with the deployment of high lift devices). Furthermore, the expressions were determined for a wing-alone condition and, as such, do not account for changes in trim drag. In conclusion, then, their value is limited to preliminary performance analyses in the absence of more substantive data.

7.5.7 Excrescence Drag and Drag "Growth"

The term *excrescence drag*, in general, describes drag caused by physical deviations (or imperfections) from an ideal, perfectly smooth, sealed object. In this context, it describes the drag caused by small items that protrude into the airflow (e.g., probes, lights, antennae, static-discharge wicks); drains and vents; air leakage through damaged or incorrectly fitted seals; gaps around access panels and doors; alignment mismatches (e.g., around doors, hatches, or fairings); surface imperfections (e.g., caused by poor paint finish, non-flush fasteners, dirt); structural repairs and dents in the skin (e.g., from hail, birdstrike, or other such causes); misrigging[11] of control surfaces; and so forth.

Several of these elements are unavoidable and are associated with the design and manufacture of the airplane—these drag contributions are included in the basic drag polars determined for a new airplane. The rest must be taken into account by considering the condition of the particular airplane. Good maintenance practice can minimize excrescence drag increments during the life of an airplane. Nonetheless, it is almost inevitable that there will be an increase in drag as the airplane ages—this is sometimes called *drag growth*, but it is probably better termed aerodynamic deterioration. A drag increase for a new airplane of 2% over the first five years after entry into service is considered typical [35], with a subsequent reduction in the rate of change thereafter.

7.5.8 Atmospheric Conditions (Reynolds Number Correction)

The reference drag polar determined by the airplane manufacturer for a particular airplane type is based on a reference temperature condition (e.g., ISA or ISA + 5 °C). Operation of the airplane at any time when the ambient temperature differs from the reference condition results in a small discrepancy with regard to the predicted drag based on the reference drag polar. This necessitates a drag correction, which is usually called a *Reynolds number correction* as it is linked to a change in the viscosity of the air. It is evident from the Sutherland equation (Equation 2.91) that viscosity increases with temperature, and this in turn increases the friction drag (see Equation 3.26). An increase in air temperature (above the reference condition) thus results in a positive Reynolds number correction and an increase in the airplane's drag coefficient.

The exact manner in which this correction is undertaken is complex [8, 36], requiring empirical airplane data. The magnitude of the drag correction is relatively small (less than 1% for a 20 °C temperature difference in cruise, for example), and it can thus be ignored for most theoretical performance studies. The correction factor, however, should be taken into account when very accurate reference drag polars are available—such as those contained within the computer databases used for the operational flight planning of modern transport airplanes.

---

11 Misrigging is a colloquial term used to describe poor alignment of a control surface—for example, a slat or spoiler that does not retract flush with the wing profile. Misrigging is often the largest single cause of avoidable excrescence drag on in-service aircraft.

7.5.9 Aeroelastic Wing Shape Change

Another small correction that may get applied to the reference drag polar accounts for the change in the shape of the wing during flight. The aeroelastic distortion of the wing under high loading can produce a change in the wing twist and dihedral, resulting in a change in drag—this is due to a combination of upward-acting lift forces and downward-acting inertia forces. This effect is likely to become more important as more flexible wings are developed for future generations of aircraft. The airplane's speed, weight, and the distribution of fuel in the wing tanks are all factors that influence this drag correction.

7.6 Lift-to-Drag Ratio

7.6.1 Definition

The ratio of an airplane's lift to its drag[12] is herein denoted by the letter E. By definition,

$$E = \frac{L}{D} = \frac{C_L}{C_D}$$

(7.35)

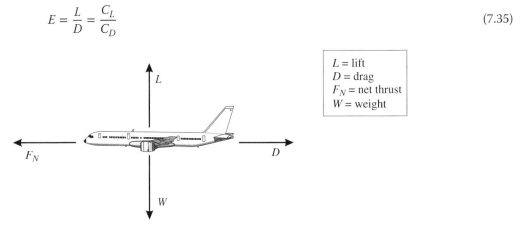

L = lift
D = drag
F_N = net thrust
W = weight

Figure 7.16 Steady, level flight (i.e., at constant speed and constant height).

The lift-to-drag ratio, which is often called the *aerodynamic efficiency*,[13] is a key aircraft design parameter and features in many performance calculations. As a simple illustration of its value, the forces acting on an airplane in steady, level flight (i.e., at constant speed and constant height) are considered. The forces acting on the airplane (see Figure 7.16) are summed—both horizontally and vertically—to give the following expressions:

$$F_N = D$$

(7.36)

and $\quad W = L$

(7.37)

---

12 An alternative convention to the one adopted herein characterizes the aerodynamic performance of an airfoil or wing in terms of its ratio of drag to lift (i.e., the reciprocal of E). It should be noted that there is no overarching advantage or disadvantage for either convention; it is a matter of custom or personal preference.

13 Note that L/D is not a measure of efficiency in the strict scientific sense, which considers efficiency as the ratio of output to input of a system (e.g., energy). Customary usage of the term, however, has been adopted herein, by which *efficiency* is understood as the ability (of a system) to do something with the least waste—of time, materials, energy, or money, for example.

Equations 7.36 and 7.37 are now combined to yield the following relationship, which applies to steady, level flight:

$$\frac{F_N}{W} = \frac{D}{L} = \frac{1}{E} \tag{7.38}$$

or $\quad F_N = \frac{D}{L} W = \frac{C_D}{C_L} W = \frac{W}{E} \tag{7.39}$

Equation 7.39 shows that the thrust (F_N) required to sustain steady, level flight for a given airplane weight, depends on the ratio C_D/C_L. From an airplane design perspective, the influence of the lift-to-drag ratio on airplane performance is immediately apparent: an increase in E reduces the required thrust to maintain steady, level flight at a particular airplane weight.

7.6.2   Maximum Lift-to-Drag Ratio

The lift-to-drag ratio (E) will change during flight, as the lift coefficient and/or Mach number changes. For a given Mach number, an airplane will have a maximum lift-to-drag value (E_{max}), which it cannot exceed (see Figure 7.17). The value of E_{max} for a selected Mach number can be determined using a graphical technique, which involves drawing a line tangent to the drag polar through the origin, as illustrated in Figure 7.17. The lift coefficient that corresponds to E_{max} is the minimum-drag lift coefficient, $C_{L_{md}}$.

Mach number has a significant influence on the lift-to-drag ratio, as illustrated in Figure 7.18 (the chart was generated using the drag polars shown in Figure 7.8). The general trend illustrates an increasing value of the E with C_L, at a given Mach number, which reaches a maximum (at the minimum-drag lift coefficient, $C_{L_{md}}$) and then reduces. It is evident from Figure 7.18 that $C_{L_{md}}$ reduces with increasing Mach number.

Figure 7.18 appears to indicate that the greatest value of E is obtained, for this airplane, at Mach 0.60. At higher Mach numbers, the value of E_{max} is lower. What is not particularly clear on this chart is that at lower Mach numbers, E_{max} is also lower. As the line for Mach 0.30 overlaps higher Mach number data, it is difficult to represent accurately this characteristic on a single chart of this type.

An alternative representation of the lift-to-drag ratio data for an airplane is illustrated in Figure 7.19, for a selected range of E values. Isolines (contours) of equal lift-to-drag ratio have

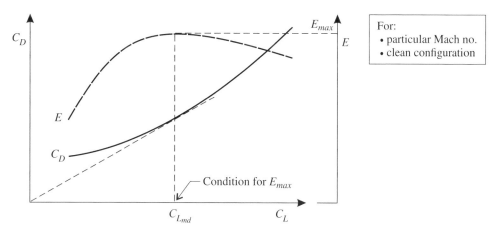

Figure 7.17 Drag coefficient and lift-to-drag ratio versus lift coefficient, indicating the condition for E_{max}.

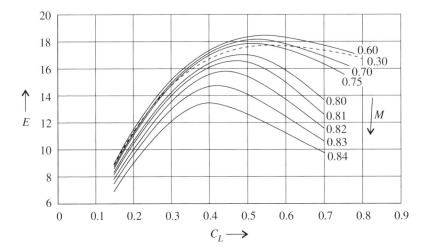

Figure 7.18 Lift-to-drag ratio versus lift coefficient (generated using the drag polars shown in Figure 7.8).

been generated as a function of two independent variables: lift coefficient and Mach number. An advantage of this format is that constant E lines do not intersect on a plane of C_L versus M, yielding an unambiguous representation of the lift-to-drag ratio data over a region of interest. Superimposed on the constant E isolines are two dashed lines, identified as I and II—these lines represent partially constrained optima. Line I is the locus of maximum E for a given Mach number (i.e., a plot of $C_{L_{md}}$ versus Mach number) and line II is the locus of maximum E for a given lift coefficient. The position of these lines on the chart can be visualized by interpreting the isolines as elevation contours and by intersecting the virtual three-dimensional shape with normal planes parallel to the C_L axis (for line I) and parallel to the M axis (for line II). The point where the two lines intersect is the unconstrained optimum for E—in other words, the point where E is greatest considering all possible combinations of C_L and M. This representation of

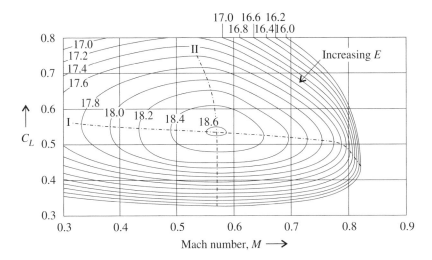

Figure 7.19 Constant lift-to-drag ratio (E) lines drawn as a function of lift coefficient and Mach number (generated using the drag polars shown in Figure 7.8).

an airplane's aerodynamic characteristics is particularly useful for cruise optimization studies (see Section 13.5).

7.6.3   Maximum Lift-to-Drag Ratio Based on the Parabolic Drag Polar

In this section, the discussion is restricted to the Mach number range in which the parabolic drag polar provides a reasonable approximation (see Section 7.4). Using Equation 7.14, the lift-to-drag ratio can be written as follows:

$$E = \frac{C_L}{C_D} = \frac{C_L}{C_{D_0} + KC_L{}^2} \tag{7.40}$$

The conditions that will yield a maximum lift-to-drag ratio are readily determined by considering the reciprocal of E (i.e., the drag-to-lift ratio):

$$\frac{1}{E} = \frac{C_D}{C_L} = \frac{C_{D_0} + KC_L{}^2}{C_L} = \frac{C_{D_0}}{C_L} + KC_L \tag{7.41}$$

The two components of $1/E$ are plotted against C_L in Figure 7.20. The zero-lift drag term is seen to reduce as a hyperbolic function of C_L and the lift-dependent term increases as a linear function of C_L. The sum of the two terms displays a clear minimum value, that is, $(1/E)_{min}$ at the corresponding lift coefficient, $C_{L_{md}}$.

Analytically, the minimum value of the drag-to-lift ratio can be obtained by differentiating Equation 7.41 with respect to C_L and setting the result equal to zero:

$$\frac{d}{dC_L}\left(\frac{1}{E}\right) = -\frac{C_{D_0}}{C_L^2} + K = 0 \qquad \text{(for a minimum value)}$$

Hence $C_{L_{md}} = \sqrt{\dfrac{C_{D_0}}{K}}$ $\tag{7.42}$

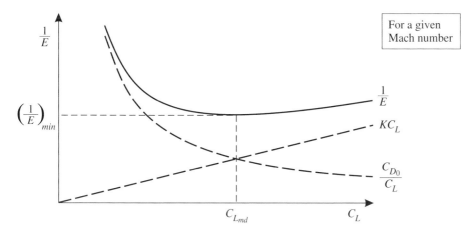

Figure 7.20 Drag-to-lift ratio (1/E) versus lift coefficient.

The maximum value of E is obtained by substituting $C_{L_{md}}$ into Equation 7.41, that is,

$$\left(\frac{1}{E}\right)_{min} = C_{D_0}\sqrt{\frac{K}{C_{D_0}}} + K\sqrt{\frac{C_{D_0}}{K}} = 2\sqrt{KC_{D_0}}$$

Thus $\quad E_{max} = \frac{1}{2}\sqrt{\frac{1}{KC_{D_0}}}$ \hfill (7.43)

It can be seen in this derivation that there are equal contributions to $(1/E)_{min}$ from the two drag components given in Equation 7.41, namely, C_{D_0}/C_L and KC_L. In other words, at $C_{L_{md}}$, the zero-lift and lift-dependent contributions to the total drag of the airplane are equal. An interesting observation that is evident from this simple analysis based on the parabolic drag polar is that the value of E_{max} depends on C_{D_0} and K, which are airplane design—not flight—parameters. The associated lift coefficient, $C_{L_{md}}$, however, links this optimum to a flight condition (the lift coefficient, as indicated by Equation 7.4, depends on W, δ, and M).

7.7 Minimum Drag Condition

7.7.1 Definition

The minimum drag flight condition is achieved at the minimum-drag lift coefficient ($C_{L_{md}}$), which produces the highest lift-to-drag ratio (see Section 7.6.2). Operation of the airplane at $C_{L_{md}}$ is aerodynamically efficient, resulting in the lowest thrust requirement for sustained level flight.

7.7.2 Minimum Drag Based on the Parabolic Drag Polar

The expression for the minimum drag (D_{min})—which corresponds to operating the airplane at the lift coefficient $C_{L_{md}}$—can be determined directly from Equation 7.43.

$$D_{min} = \left(\frac{1}{E}\right)_{min} n_z W = 2n_z W\sqrt{KC_{D_0}} \qquad \text{(and in level flight } n_z = 1) \hfill (7.44)$$

This equation assumes that the drag can be approximated by the parabolic drag polar (i.e., Equation 7.14). The airplane's speed associated with the minimum drag condition can be expressed in terms of true airspeed (V_{md}), equivalent airspeed ($V_{e_{md}}$), or Mach number (M_{md}). These speeds can be obtained from the definition of the lift coefficient (see Equation 7.4) in conjunction with Equation 7.42.

$$C_{L_{md}} = \frac{2n_z W}{\rho V_{md}^2 S} = \frac{2n_z W}{\rho_0 V_{e_{md}}^2 S} = \frac{2n_z W}{\rho_0 \delta a_0^2 M_{md}^2 S} \qquad \text{(and in level flight } n_z = 1)$$

Thus $\quad V_{md} = \left[\frac{2W}{\rho S C_{L_{md}}}\right]^{1/2} = \left[\frac{2W}{\rho S}\sqrt{\frac{K}{C_{D_0}}}\right]^{1/2}$ \hfill (7.45) and $\quad V_{e_{md}} = \left[\dfrac{2W}{\rho_0 S C_{L_{md}}} \right]^{1/2} = \left[\dfrac{2W}{\rho_0 S} \sqrt{\dfrac{K}{C_{D_0}}} \right]^{1/2}$ (7.46)

and $\quad M_{md} = \left[\dfrac{2W}{\rho_0 a_0^2 S \delta} \sqrt{\dfrac{K}{C_{D_0}}} \right]^{1/2}$ (7.47)

As with many of these results, Equations 7.45 to 7.47 can be derived from alternative starting points. It is instructive to consider one such alternative route, as it introduces a methodology that is useful for several other applications. This approach requires the drag (D) to be expressed directly as a function of equivalent airspeed. This is achieved using the definition of the parabolic drag polar (Equation 7.14) and Equation 7.5:

$$D = \dfrac{\rho_0 V_e^2 S}{2} \left[C_{D_0} + K C_L^2 \right]$$

Thus $\quad D = \left[\dfrac{C_{D_0} \rho_0 S}{2} \right] V_e^2 + \left[\dfrac{2KW^2}{\rho_0 S} \right] \dfrac{1}{V_e^2}$ (7.48)

To simplify the ensuring mathematics, the following substitutions are introduced:

$$A_1 = \dfrac{C_{D_0} \rho_0 S}{2} \quad \text{and} \quad B_1 = \dfrac{2KW^2}{\rho_0 S}$$

Thus $\quad D = A_1 V_e^2 + B_1 V_e^{-2}$ (7.49)

Note that A_1 and B_1 are not standard nomenclature and are merely introduced here to simplify the mathematics; furthermore, A_1 and B_1 will change if the airplane's configuration or weight changes.

Using Equation 7.49, the drag D is plotted against V_e^2 in Figure 7.21. It is evident from Equation 7.49 that there are two contributions to the drag of the airplane: the first term, the zero-lift drag (D_0), is proportional to V_e^2; whereas, the second term, the lift-dependent drag (D_i), is inversely proportional to V_e^2. These relationships indicate that at low speeds, D_i is the dominant part of the total drag, whereas at high speeds D_0 is dominant.

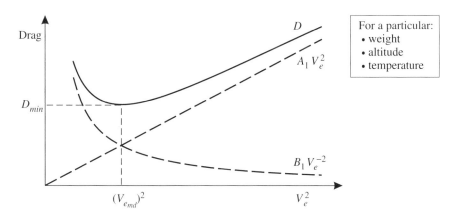

Figure 7.21 Minimum drag condition (parabolic representation of low-speed drag polar).

The drag function given by Equation 7.49 is now differentiated with respect to V_e and the result set equal to zero to determine the minimum drag speed ($V_{e_{md}}$), that is,

$$\frac{dD}{dV_e} = 2A_1 V_e - 2B_1 V_e^{-3} = 0 \quad \text{(for a minimum value)}$$

Thus $\quad V_{e_{md}} = \left[\frac{B_1}{A_1}\right]^{1/4}$ \hfill (7.50)

or $\quad V_{e_{md}} = \left[\frac{2W}{\rho_0 S}\sqrt{\frac{K}{C_{D_0}}}\right]^{1/2}$ \qquad (Equation 7.46)

This confirms the earlier result (given by Equation 7.46). Equation 7.50 is now substituted into Equation 7.49 to give the required expression for minimum drag:

$$D_{min} = A_1\sqrt{\frac{B_1}{A_1}} + B_1\sqrt{\frac{A_1}{B_1}} = 2\sqrt{A_1 B_1}$$

or $\quad D_{min} = 2W\sqrt{KC_{D_0}}$ \qquad (see Equation 7.44, with $n_z = 1$)

This result validates the equation derived earlier as Equation 7.44, and once again shows that there are equal contributions to the drag from the zero-lift drag term (C_{D_0}) and lift-dependent drag term (C_{D_i}) at this flight condition. In other words, at the minimum drag speed, $C_{D_i} = C_{D_0}$ and $C_D = 2C_{D_0}$.

There are several situations where it is advantageous to operate the airplane at a speed where the drag is a minimum (e.g., to stay airborne for the greatest length of time). This will be discussed in the performance chapters that follow. At this juncture, it is useful to consider the impact on the airplane's fuel consumption when operating at the minimum drag speed. The net fuel flow (mass flow rate) to all engines (Q) can be expressed as the product of the thrust specific fuel consumption (c) and the net thrust (F_N), as described in Section 8.4.2. During steady, level flight the net thrust (F_N) equals the drag (D). If the idealization is made that TSFC is constant across the speed range of interest (in reality, small changes occur), then the fuel flow will be a minimum for flight at the minimum drag speed, and it can be approximated by the following expression:

$$Q_{min} = cD_{min} = 2cW\sqrt{KC_{D_0}} \qquad (D_{min} \text{ is given by Equation 7.44})$$ \hfill (7.51)

where Q_{min} is the minimum fuel flow; and
$\qquad c$ is the thrust specific fuel consumption (mass flow basis).

7.8 Minimum Drag Power (Required Power) Condition

7.8.1 Definition

Drag power (P_D), by definition, is the product of the airplane's total drag (D) and its true airspeed (V).

$$P_D = DV$$ \hfill (7.52)

As the thrust produced by the airplane's engines is required to overcome the drag, the drag power can be considered as the power required for flight. In fact, the term drag power is not universally adopted and many authors prefer to call this parameter the *required power* (or *power required*).

A related concept to that of drag power is *thrust power* (P_T), which, by definition, is the product of the airplane's net thrust and its true airspeed. In steady, level flight, where the thrust will equal the drag, the thrust power will equal the drag power. However, in climbing, descending, or accelerated flight, this will generally not be true (these flight conditions are discussed later in Chapter 12).

7.8.2 Minimum Drag Power Based on the Parabolic Drag Polar

In Section 7.7, it was shown that the drag acting on the airplane in steady, level flight varies as a function of airspeed and that it has a clear minimum value (see Figure 7.21), which, when expressed in terms of equivalent airspeed, is written as $V_{e_{md}}$. The drag power has a similar, but not identical, variation with airspeed. Drag power is a minimum at the minimum drag power speed, which, when expressed in terms of equivalent airspeed, is written as $V_{e_{mp}}$. The associated drag is denoted as D_{mp}, the lift coefficient as $C_{L_{mp}}$ and the lift-to-drag ratio as E_{mp}. The relationship between these parameters and the corresponding parameters for the minimum drag condition are derived in this section.

There are several alternative approaches to study the flight condition corresponding to minimum drag power. The most direct approach is to express the drag power (see Equation 7.52) in terms of the equivalent airspeed and to substitute for drag using Equation 7.49:

$$P_D = DV = D\frac{V_e}{\sqrt{\sigma}} = \frac{V_e}{\sqrt{\sigma}}\left(A_1 V_e^2 + B_1 V_e^{-2}\right)$$

or $\quad P_D = \frac{1}{\sqrt{\sigma}}\left(A_1 V_e^3 + B_1 V_e^{-1}\right)$ (7.53)

where A_1 and B_1 are defined in Equation 7.49.

For a given altitude (the air density ratio is thus constant), the minimum value of P_D can be obtained by differentiating Equation 7.53 with respect to V_e and setting the result equal to zero.

$$\frac{dP_D}{dV_e} = \frac{3A_1}{\sqrt{\sigma}}V_e^2 - \frac{B_1}{\sqrt{\sigma}}V_e^{-2} = 0 \quad \text{(for a minimum value)}$$

hence $V_{e_{mp}} = \sqrt[4]{\frac{B_1}{3A_1}} = \frac{1}{\sqrt[4]{3}}V_{e_{md}} = 0.760 V_{e_{md}}$ (using Equation 7.50) (7.54)

The drag at the minimum power speed can be obtained from Equation 7.49, as follows:

$$D_{mp} = A_1 V_{e_{mp}}^2 + B_1 V_{e_{mp}}^{-2}$$

hence $D_{mp} = A_1 \sqrt{\frac{B_1}{3A_1}} + B_1 \sqrt{\frac{3A_1}{B_1}} = \frac{4}{\sqrt{3}}\sqrt{A_1 B_1}$ (7.55)

or $\quad D_{mp} = \frac{2}{\sqrt{3}}D_{min} = 1.15 D_{min}$ (7.56)

Also, the value of C_L at the minimum power speed is given by

$$\frac{C_{L_{mp}}}{C_{L_{md}}} = \frac{\left\{ \dfrac{2n_z W}{\rho_0 V_{e_{mp}}^2 S} \right\}}{\left\{ \dfrac{2n_z W}{\rho_0 V_{e_{md}}^2 S} \right\}} = \frac{V_{e_{md}}^2}{V_{e_{mp}}^2} = \sqrt{3} \qquad \text{(for straight, level flight } n_z = 1) \qquad (7.57)$$

or $\quad C_{L_{mp}} = \sqrt{\dfrac{3\,C_{D_0}}{K}} \qquad$ (from Equation 7.42) $\hfill (7.58)$

In a similar way that $C_{L_{md}}$ describes the flight condition where C_L/C_D is a maximum, $C_{L_{mp}}$ describes the flight condition where $C_L^{3/2}/C_D$ is a maximum.

There is an interesting observation regarding the relative magnitudes of the lift-independent and the lift-dependent drag contributions in this flight condition. In Section 7.7 it was shown that at the minimum drag speed, these components are equal in magnitude. However, at the minimum power speed, the lift-dependent term is proportionally greater, and $C_{D_i} = 3\,C_{D_0}$. This can be shown by noting that C_{D_i} is proportional to C_L^2, and, hence,

$$\frac{\left(C_{D_i}\right)_{mp}}{\left(C_{D_i}\right)_{md}} = \left(\frac{C_{L_{mp}}}{C_{L_{md}}}\right)^2$$

Thus $\quad \left(C_{D_i}\right)_{mp} = \left(\dfrac{C_{L_{mp}}}{C_{L_{md}}}\right)^2 C_{D_0} = 3C_{D_0} \qquad$ (from Equation 7.57) $\hfill (7.59)$

Hence, at the minimum power speed, $C_D = 4\,C_{D_0}$.

The relationship between the lift-to-drag ratios E_{mp} and E_{max} can be derived as follows:

$$E_{mp} = \frac{C_{L_{mp}}}{C_{D_{mp}}} = \frac{\sqrt{3}\,C_{L_{md}}}{4\,C_{D_0}} = \frac{\sqrt{3}}{2}\left(\frac{C_{L_{md}}}{C_{D_{md}}}\right) \qquad \text{(using Equation 7.57)}$$

hence $E_{mp} = \dfrac{\sqrt{3}}{2} E_{max} = 0.866 E_{max} \hfill (7.60)$

7.9 Minimum Drag-to-Speed Ratio Condition

7.9.1 Definition

The ratio of the airplane's drag to its true airspeed (i.e., D/V) is a useful parameter that has a direct bearing on the distance (range) that a jet airplane can fly on a given quantity of fuel. From an economic standpoint, it is desirable to maximize the distance flown per unit of fuel mass consumed, which is equivalent to minimizing the fuel consumed per unit of distance flown. In other words, the objective is to determine the conditions that will produce a minimum value of the parameter $-\mathrm{d}m_f/\mathrm{d}x$, where m_f is the onboard fuel mass and x is the still air distance. This parameter is the reciprocal of what is known as the *specific air range* (which is defined later in Section 13.2.2). The reason for the minus sign is that the change of fuel mass is negative for increasing distance, and specific air range (SAR) is defined as a positive quantity.

During steady (i.e., constant speed), level cruising flight, when net thrust equals drag, the change in the onboard fuel mass per unit distance will be given by

$$\left(-\frac{dm_f}{dx}\right) = \frac{\left(-\dfrac{dm_f}{dt}\right)}{\left(\dfrac{dx}{dt}\right)} = \frac{Q}{V} = \frac{cF_N}{V} = c\left(\frac{D}{V}\right) \tag{7.61}$$

where m_f is the onboard fuel mass;
  x is the still air distance;
  t is time;
  Q is the net fuel flow (mass flow), given by $-dm_f/dt$;
  V is the true airspeed, given by $V = dx/dt$; and
  F_N is the net thrust.

It can be concluded from Equation 7.61 that if the thrust specific fuel consumption (c) is approximately constant for the cruise speeds of interest, then the ratio D/V must be minimized in order to maximize the range.

7.9.2 Minimum Value of the Drag-to-Speed Ratio

The conditions corresponding to a minimum value of D/V can be determined by constructing a line through the origin tangent to the drag curve on a graph of drag versus true airspeed (TAS), as illustrated in Figure 7.22, below.

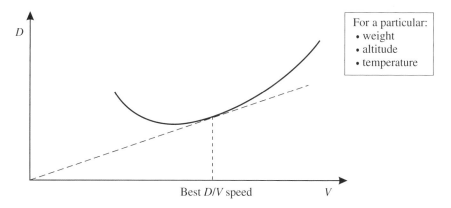

Figure 7.22 Drag versus TAS indicating best D/V speed.

7.9.3 Minimum Value of the Drag-to-Speed Ratio Based on the Parabolic Drag Polar

The conditions associated with a minimum value of the drag-to-speed ratio can be determined for a constant-height cruise using the drag expression given by Equation 7.49:

$$\frac{D}{V} = \frac{D}{V_e}\sqrt{\sigma} = A_1 V_e \sqrt{\sigma} + B_1 V_e^{-3}\sqrt{\sigma} \tag{7.62}$$

where A_1 and B_1 are defined in Equation 7.49.

The equivalent airspeed that gives a minimum value of D/V can be found by differentiating Equation 7.62 with respect to V_e and setting the result equal to zero. If the altitude is constant, σ is fixed and the required speed is given by

$$\frac{\mathrm{d}}{\mathrm{d}V_e}\left(\frac{D}{V}\right) = A_1\sqrt{\sigma} - 3B_1 V_e^{-4}\sqrt{\sigma} = 0 \qquad \text{(for a minimum value)}$$

$$V_e = (3)^{1/4}\left(\frac{B_1}{A_1}\right)^{1/4} = (3)^{1/4}\left\{\left(\frac{2W^2 K}{\rho_0 S}\right)\left(\frac{2}{C_{D_0}\rho_0 S}\right)\right\}^{1/4}$$

or $\qquad V_e = (3)^{1/4}\left\{\frac{2W}{\rho_0 S}\sqrt{\frac{K}{C_{D_0}}}\right\}^{1/2} \qquad$ (for minimum drag-to-speed ratio) $\qquad$ (7.63)

This result can be compared to the expression for the minimum drag, as given by Equation 7.46:

$$V_e = (3)^{1/4} V_{e_{md}} = 1.32 V_{e_{md}} \tag{7.64}$$

The lift coefficient corresponding to this speed can also be determined. Writing the lift coefficient for the required condition and dividing by the lift coefficient for minimum drag yields the following relationship:

$$\frac{C_L}{C_{L_{md}}} = \frac{\left\{\dfrac{2n_z W}{\rho_0 V_e^2 S}\right\}}{\left\{\dfrac{2n_z W}{\rho_0 V_{e_{md}}^2 S}\right\}} = \frac{V_{e_{md}}^2}{V_e^2} = \frac{1}{\sqrt{3}} \qquad \text{(for straight, level flight } n_z = 1\text{)}$$

thus $\quad C_L = \dfrac{1}{\sqrt{3}} C_{L_{md}} = 0.577 C_{L_{md}} \tag{7.65}$

or $\qquad C_L = \sqrt{\dfrac{C_{D_0}}{3K}} \tag{7.66}$

And, in a similar way that $C_{L_{mp}}$ describes the flight condition where $C_L^{3/2}/C_D$ is a maximum, the lift coefficient defined by Equation 7.66 describes the flight condition where $C_L^{1/2}/C_D$ is a maximum.

The ratio of the lift-dependent to the lift-independent drag contributions can be determined using the same approach adopted in Section 7.8.2. It can be shown that in this flight condition, $C_{D_i} = (1/3)C_{D_0}$ and hence $C_D = (4/3)C_{D_0}$.

The lift-to-drag ratio, in this flight condition, can similarly be expressed in terms of E_{max}.

$$E = \frac{C_L}{C_D} = \frac{\left(\dfrac{1}{\sqrt{3}} C_{L_{md}}\right)}{\left(\dfrac{4}{3} C_{D_0}\right)} = \frac{\sqrt{3}}{2}\left(\frac{C_{L_{md}}}{2C_{D_0}}\right) = \frac{\sqrt{3}}{2} E_{max} \tag{7.67}$$

Note that these derivations assumed a constant altitude—the results given by Equations 7.64 to 7.67 are thus only valid for this constraint.

7.9.4 Concluding Remarks—Minimum Value of the Drag-to-Speed Ratio

The use of the parabolic drag polar to determine expressions for the airplane's speed (Equations 7.63 and 7.64), lift coefficient (Equations 7.65 and 7.66), and lift-to-drag ratio (Equation 7.67) for the minimum drag-to-speed ratio condition provides interesting trend information. However, it should be noted that Section 7.9.3 contains a highly simplified—and potentially misleading—explanation of the flight condition necessary to maximize the range of a jet airplane.

It can be correctly deduced from Equation 7.64 that to achieve the greatest possible range, an airplane must operate at a speed faster than $V_{e_{md}}$. It is also correct to deduce that the corresponding lift-to-drag ratio in cruise (for maximum range) is less than E_{max}. However, the optimum cruise speed—required to maximize range—for an actual jet airplane is not exactly 32% faster than the minimum drag speed (as predicted by Equation 7.64). The reason for this apparent discrepancy is that the analyses presented in Sections 7.7.2 and 7.9.3 assumed that the drag across the entire operating speed range could be represented by a single drag–lift expression (i.e., the parabolic drag polar with constant values of C_{D_0} and K), and this is clearly a gross over-simplification that ignores the influence of Mach number on the drag polar. Furthermore, the simplified analysis presented in this section did not consider the rate at which fuel is consumed.

The conclusion, which is often presented in the general literature, that an optimum cruise speed for an airplane is a fixed multiple of its minimum drag speed is not valid for aircraft designed to cruise at high subsonic speeds. This topic is discussed further in Sections 13.5 and 13.6, where a more complete analysis of the flight conditions necessary to maximize an airplane's range is presented.

7.10 Summary of Expressions Based on the Parabolic Drag Polar

Figure 7.23 illustrates the relative magnitude of the three speeds derived in Sections 7.7–7.9: the minimum drag speed (V_{md}), the minimum power speed (V_{mp}), and the best D/V speed. Using

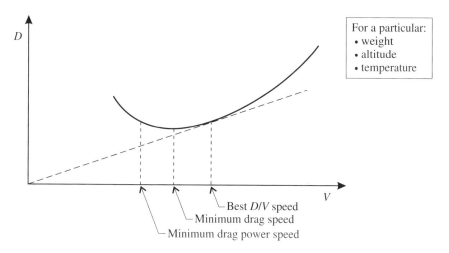

Figure 7.23 Airplane drag versus speed, indicating the speed for minimum drag power, minimum drag, and best drag-to-speed ratio.

Table 7.1(a) Summary of performance parameters based on the parabolic drag polar

Drag	Drag power
Function = D, where	Function = P_D, where $P_D = DV$
$$D = \left[\frac{C_{D_0}\rho_0 S}{2}\right]V_e^2 + \left[\frac{2KW^2}{\rho_0 S}\right]\frac{1}{V_e^2}$$	$$P_D = \left[\frac{C_{D_0}\rho_0 S}{2\sqrt{\sigma}}\right]V_e^3 + \left[\frac{2KW^2}{\sqrt{\sigma}\rho_0 S}\right]\frac{1}{V_e}$$
At the minimum drag condition:	**At the minimum drag power condition:**
which corresponds to $\left(\dfrac{C_L}{C_D}\right)_{max}$	which corresponds to $\left(\dfrac{C_L^{3/2}}{C_D}\right)_{max}$
$$V_{e_{md}} = \left[\frac{2W}{\rho_0 S}\sqrt{\frac{K}{C_{D_0}}}\right]^{1/2}$$	$$V_{e_{mp}} = \frac{1}{\sqrt[4]{3}}V_{e_{md}} = 0.760 V_{e_{md}}$$ $$V_{e_{mp}} = \left[\frac{2W}{\rho_0 S}\sqrt{\frac{K}{3C_{D_0}}}\right]^{1/2}$$
$$C_{L_{md}} = \sqrt{\frac{C_{D_0}}{K}}$$	$$C_{L_{mp}} = \sqrt{3}\,C_{L_{md}} = 1.73 C_{L_{md}}$$ $$C_{L_{mp}} = \sqrt{\frac{3C_{D_0}}{K}}$$
$$C_{D_i} = C_{D_0}$$ hence $C_D = 2C_{D_0}$	$$C_{D_i} = 3C_{D_0}$$ hence $C_D = 4C_{D_0}$
$$D_{min} = 2W\sqrt{KC_{D_0}}$$ $$D_{min} = \left(\frac{1}{E_{max}}\right)W$$	$$D_{mp} = \frac{2}{\sqrt{3}}D_{min} = 1.15 D_{min}$$ $$D_{mp} = \frac{4}{\sqrt{3}}W\sqrt{KC_{D_0}}$$
$$E_{max} = \frac{1}{2}\sqrt{\frac{1}{KC_{D_0}}}$$	$$E_{mp} = \frac{\sqrt{3}}{2}E_{max} = 0.866\,E_{max}$$ $$E_{mp} = \frac{1}{4}\sqrt{\frac{3}{KC_{D_0}}}$$

Note: The parabolic representation of an airplane's drag provides reasonable accuracy over a limited Mach range (see Figure 7.9).

the parabolic drag polar, it was shown that it is possible to uniquely describe these speeds in terms of the airplane design parameters C_{D_0} and K, and the airplane's weight and flight altitude. Associated with each of these speeds is a unique value of C_L, and as the drag polar is known, the corresponding values of C_D and hence E could be determined. For convenience, these expressions are summarized in Table 7.1.

One advantage of the use of the parameters E_{max}, E_{mp}, $C_{L_{md}}$, and $C_{L_{mp}}$ in preliminary airplane performance work (and aircraft conceptual design) is that these non-dimensional parameters are independent of the airplane's speed, weight, or altitude (neglecting Reynolds number and Mach number effects). They are dependent on the airplane's geometric design characteristics—which, from an aircraft performance perspective, is rather useful, as they can usually be estimated from knowledge of the airplane's geometry [27, 37].

Table 7.1(b) Summary of performance parameters based on the parabolic drag polar

Drag/velocity ratio

Function $= D/V$, where

$$\frac{D}{V} = \left[\frac{C_{D_0} \rho_0 \sqrt{\sigma} S}{2}\right] V_e + \left[\frac{2KW^2}{\rho_0 S}\right] \frac{1}{V_e^3}$$

At the minimum D/V condition:

which corresponds to $\left(\dfrac{C_L^{1/2}}{C_D}\right)_{max}$

$$V_e = (3)^{1/4} V_{e_{md}} = 1.32 V_{e_{md}}$$

$$V_e = (3)^{1/4} \left[\frac{2W}{\rho_0 S} \sqrt{\frac{K}{C_{D_0}}}\right]^{1/2}$$

$$C_L = \frac{1}{\sqrt{3}} C_{L_{md}} = 0.577 C_{L_{md}}$$

$$C_L = \sqrt{\frac{C_{D_0}}{3K}}$$

$$C_{D_i} = \frac{1}{3} C_{D_0}$$

hence $C_D = \dfrac{4}{3} C_{D_0}$

$$D = \frac{2}{\sqrt{3}} D_{min} = 1.15 D_{min}$$

$$D = \frac{4}{\sqrt{3}} W \sqrt{K C_{D_0}}$$

$$E = \frac{\sqrt{3}}{2} E_{max} = 0.866 E_{max}$$

$$E = \frac{1}{4} \sqrt{\frac{3}{K C_{D_0}}}$$

Note: The results given for the minimum D/V condition warrant careful interpretation (see Section 7.9.4).

References

1 Cashman, J.E., Kelly, B.D., and Nield, B.N., "Operational use of angle of attack," *Aero*, The Boeing Company, Seattle, WA, Vol. 12, pp. 10–22, Oct. 2000.
2 Rudolph, P.K.C., "High-lift systems on commercial subsonic airliners," NASA CR-4746, National Aeronautics and Space Administration, USA, 1996.
3 Otto Lilienthal Museum, "Otto Lilienthal," Retrieved Feb. 20, 2012. Available from www.lilienthal-museum.de.
4 ARC, "Report of the Definitions Panel on definitions to be used in the description and analysis on drag," C.P. No. 369, Aeronautical Research Council (ARC), HMSO, London, UK, 1958.

5 Marec, J.-P., "Drag reduction: A major task for research," *CEAS / DragNet European Drag Reduction Conference*, Potsdam, Germany, June 19–21, 2000, in *Notes on numerical fluid mechanics*, Vol. 76, Thiede, P. (Ed.), Springer Verlag, Heidelberg, Germany, pp. 17–28, 2001.

6 Hoerner, S.F., *Fluid-dynamic drag: Practical information on aerodynamic drag and hydrodynamic resistance*, Hoerner Fluid Dynamics, Albuquerque, NM, 1965.

7 Munk, M.M., "The minimum induced drag of aerofoils," NACA Report No. 121, National Advisory Committee for Aeronautics, USA, 1921.

8 Blake, W. and Performance Training Group, "Jet transport performance methods," D6-1420, Flight Operations Engineering, Boeing Commercial Airplanes, Seattle, WA, Mar. 2009.

9 Anderson, J.D., *Introduction to flight*, 6th ed., McGraw-Hill, New York, NY, 2008.

10 Torenbeek, E., *Advanced aircraft design: Conceptual design, analysis and optimization of subsonic civil airplanes*, John Wiley & Sons, Chichester, UK, 2013.

11 Torenbeek, E. and Wittenberg, H., *Flight physics: Essentials of aeronautical disciplines and technology, with historical notes*, Springer, Heidelberg, Germany, 2009.

12 Bowes, G.M., "Aircraft lift and drag prediction and measurement," AGARD Lecture Series 67: Prediction methods for aircraft aerodynamic characteristics, AD-780-608, Advisory Group for Aerospace Research and Development, North Atlantic Treaty Organization, Neuilly sur Seine, France, May 1974.

13 Oswatitsch, K. and Wieghardt, K., "Ludwig Prandtl and his Kaiser-Wilhelm-Institut," *Annual Review of Fluid Mechanics*, Vol. 19, pp. 1–25, 1987.

14 Prandtl, L., "Theory of lifting surfaces, Part 1," NACA TN-9, National Advisory Committee for Aeronautics, July 1920. [Translated from German.]

15 Kroo, I., "Nonplanar wing concepts for increased aircraft efficiency," in *Innovative Configurations and Advanced Concepts for Future Civil Aircraft*, Torenbeek, E. and Deconinck, H. (Eds.), von Karman Institute, June 6–10, 2005.

16 Kroo, I., "Drag due to lift: Concepts for prediction and reduction," *Annual Review of Fluid Mechanics*, Vol. 33, pp. 587–617, 2001.

17 McLean, D., "Wingtip devices: What they do and how they do it," *Boeing Performance and Flight Operations Engineering Conference*, Seattle, WA, Sept. 2005, Article 4.

18 Jones, R.T. and Lasinski, T.A., "Effect of winglets on the induced drag of ideal wing shapes," NASA TM-81230, Ames Research Center, National Aeronautics and Space Administration, Moffett Field, CA, Sept. 1980.

19 Oswald, W.B., "General formulas and charts for the calculation of airplane performance," NACA TR-408, National Advisory Committee for Aeronautics, USA, 1932.

20 Lilienthal, O., *Der Vogelflug als Grundlage der Fliegekunst ein Beitrag zur Systematik der Flugtechnik [Birdflight as the basis of aviation]*, R. Gaertner Verlagsbuchhandlung, Berlin, Germany, 1889.

21 McLean, D., *Understanding aerodynamics: Arguing from the real physics*, John Wiley & Sons, New York, NY, 2012.

22 Lan, C.-T.E. and Roskam, J., *Airplane aerodynamics and performance*, DARcorporation, Lawrence, KS, 2010.

23 Boppe, C.W., "CFD drag prediction for aerodynamic design," Technical status review on drag prediction and analysis from computational fluid dynamics: State of the art, AGARD-AR-256, Advisory Group for Aerospace Research and Development, North Atlantic Treaty Organization, Neuilly sur Seine, France, June 1989.

24 Mason, W.H., "Analytic models for technology integration in aircraft design," *Aircraft Design, Systems and Operations Conference*, Dayton, OH, Sept. 17–19, 1990.

25 Malone, B. and Mason, W.H., "Multidisciplinary optimization in aircraft design using analytic technology models," *Journal of Aircraft*, Vol. 32, Iss. 2, pp. 431–438, 1995.

26 Lock, C.N.H., "The ideal drag due to a shock wave: Parts I and II," ARC Technical Report R&M No. 2512, Aeronautical Research Council, London, UK, Feb. 19, 1945.

27 Torenbeek, E., *Synthesis of subsonic airplane design*, Delft University Press, Delft, the Netherlands, 1982.

28 ESDU, "Introductory sheet on subcritical lift-dependent drag of wings," Data item 66031, Amdt. C, IHS ESDU, 133 Houndsditch, London, UK, Nov. 1995.

29 Yajnik, K.S. and Subbaiah, M.V., "Representation of the drag polar of a fighter aircraft," *Journal of Aircraft*, Vol. 13, Iss. 2, pp. 155–156, 1976.

30 Ardonceau, P.L., "Aerodynamic properties of crescent wing planforms," *Journal of Aircraft*, Vol. 31, Iss. 2, pp. 462–465, 1994.

31 Airbus, "Getting to grips with fuel economy," Iss. 4, Flight Operations Support and Line Assistance, Airbus S.A.S., Blagnac, France, Oct. 2004.

32 Wieselsberger, C., "Wing resistance near the ground," NACA TM-77, National Advisory Committee for Aeronautics, USA, 1922. [Translated from German.]

33 Carter, A.W., "Effects of ground proximity on the longitudinal aerodynamic characteristics of an unswept aspect-ratio-10 wing," NASA TN-D-5662, National Aeronautics and Space Administration, Washington, DC, 1970.

34 Hoerner, S.F. and Borst, H.V., *Fluid-dynamic lift: Practical information on aerodynamic and hydrodynamic lift*, 2nd ed., Hoerner Fluid Dynamics, Albuquerque, NM, 1985.

35 Speyer, J.-J., "Getting hands-on experience with aerodynamic deterioration," *FAST*, Airbus Industrie, Blagnac, France, Vol. 21, pp. 15–24, 1997.

36 Boeing, "Jet transport performance methods," D6-1420, The Boeing Company, Seattle, WA, May 1989.

37 Howe, D., *Aircraft conceptual design synthesis*, Professional Engineering Publishing, London, UK, 2000.

8

Propulsion

8.1 Introduction

The type of powerplant used on modern jet transport airplanes is a high bypass ratio gas turbine engine, usually called a *turbofan* engine. Due to the vastly superior fuel efficiency and lower noise of turbofans, this engine type has completely replaced turbojets (which are gas turbine engines that do not incorporate bypass airflow) on commercial jet transport airplanes. High bypass ratio turbofans, such as those installed on airliners, are often called *civil turbofans*, a name that distinguishes these engine types from low bypass ratio turbofans and turbojets installed on many military aircraft types. A characteristic feature of this powerplant type is its large diameter multi-bladed propeller—known as a *fan*—located at the front of the engine. The fan, which is mechanically driven by the low pressure turbine via a shaft, compresses and accelerates the air that enters the engine through the air intake. A central air stream then travels through the engine core (which comprises the basic jet engine components: compressor, combustion system, and turbine), while the remainder—the bypass flow—is ducted around the engine core. Typically, more than three-quarters of a turbofan's thrust is due to the acceleration of the bypass air.

A brief description of the turbofan engine is presented in Section 8.2 of this chapter (more detailed treatments of the subject are presented in the reference works listed in Section 8.9). The emphasis is on issues that relate to the performance of the installed powerplant. The net thrust (F_N) that is generated and the factors that influence the thrust are then discussed (Section 8.3). An overview of key engine design limits and the thrust ratings that are used in commercial flight operations are presented in Section 8.5.

Turbine engines are capable of burning a wide range of fuel products. Jet transport airplanes predominantly burn kerosene (these fuels are required to meet stringent specifications, as described in Section 22.7.1). As the cost of fuel consumed in flight constitutes a significant portion of the total cost of any trip, the rate at which fuel is burned is thus an important performance parameter for any commercial flight operation. The *thrust specific fuel consumption* (TSFC)—which, by definition, is the mass or weight of fuel burned per unit of time divided by the net engine thrust—is a measure of the efficiency with which a jet engine can convert fuel into thrust. The topics of fuel flow and TSFC are addressed in Section 8.4.

The factors that influence the two key performance parameters of the engine—that is, the net thrust and the TSFC—are described in Sections 8.6 and 8.7, respectively. Also presented in these sections are a series of simple algebraic functions that are sometimes used to model these parameters. The chapter closes with a description of installation losses and in-service engine deterioration (Section 8.8).

Performance of the Jet Transport Airplane: Analysis Methods, Flight Operations, and Regulations, First Edition. Trevor M. Young.
© 2018 John Wiley & Sons Ltd. Published 2018 by John Wiley & Sons Ltd.

8.2 Basic Description of the Turbofan Engine

8.2.1 Basic Principle of Operation

In very simple terms, the operation of a jet engine is as follows:

(1) Air enters into the engine through the intake (inlet).
(2) The air is compressed in the compressors.
(3) Fuel is added, the mixture is ignited, and the gas is heated in the combustion chambers.
(4) The high-energy gas exiting the combustor spins the turbines.
(5) The turbines drive the compressors through connecting shafts.
(6) The gas is accelerated through a propelling nozzle, exiting at the exhaust.

The earliest jet engines had a single shaft—known as a single-spool design. This meant that all stages of the compressor rotated at the same speed, which is not an efficient design. Consequently, multiple spool designs soon replaced the original single-spool concept. The introduction of a second shaft—rotating concentrically within the first shaft—in a two-spool design meant that an engine can have compressors that operate at different rotational speeds, powered by two independent turbines. In this way, improved efficiency through better matching of the compressor rotation speed to the flow conditions is achieved. In a twin-spool design, the low pressure compressor (which is the first phase of compression) is driven by the second turbine—this is called the low-speed spool. The high pressure compressor is driven by the first turbine—this is the high-speed spool. The majority of contemporary turbofan engines are of a twin-spool design, but there are also several very successful triple-spool designs[1] that have three independent turbine–compressor units rotating on three concentric shafts. A schematic of a generic triple-spool jet engine is given in Figure 8.1.

Air, on entering the inlet duct (or intake), is drawn through the fan. It then splits into two separate airflows: the core stream, or *hot stream*, and the bypass stream, or *cold stream*. The bypass stream flows in a concentric duct around the engine core (which is often referred to as the *gas generator*). The core stream passes through the compressors into the combustor (burner). Here, fuel is burned (at approximately constant pressure), the temperature increases, and the hot gas then passes through the turbines, where power is extracted to drive the compressors. The hot gas leaving the turbine is accelerated through a nozzle, after which it is mixed with the slower moving bypass air. It is the acceleration of this large mass flow of air—through the engine core and bypass duct—that produces the thrust. The bypass air may join the hot gases at the nozzle exit—this is known as a *long duct* (or *mixed flow*) design. The alternative concept is a *short duct* system, where the bypass air emerges from the engine immediately after the fan (as illustrated in the schematic in Figure 8.1). As the core and bypass gas streams are not actively mixed within the engine, this is called an *un-mixed* design.

8.2.2 Main Engine Components and Systems

Air Intake
The type of intake (or inlet) that is most suitable for subsonic speeds is the Pitot-type inlet. The circular, or nearly circular, inlet lip (leading edge) is aligned approximately normal to the local airflow. Inside the intake is a divergent duct, which is designed to deliver the air to the compressor without flow distortions and with minimal loss of energy across a broad operating

---

1 The only manufacturer of triple-spool turbofan engines that are compliant with the FAA's airworthiness standard for engines FAR 33 [1] and EASA's certification specification CS-E (engines) [2] is Rolls-Royce plc (UK).

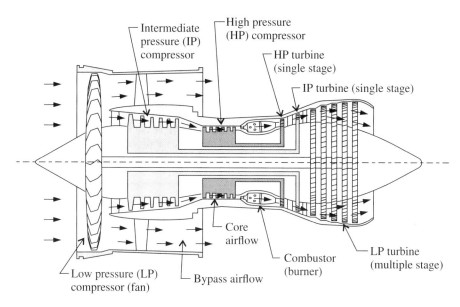

Figure 8.1 Schematic of a generic triple-spool turbofan engine (the number of stages shown is for illustration only).

range, including high angles of attack and large sideslip angles. The intake lip can be heated to melt preformed ice or to prevent the formation of ice.

Spool

A spool is a rotating assembly in the engine—the key components are a compressor, shaft, and turbine. All turbofan engines are of a multiple-spool design: two or three spools are used. The concentric shafts operate at different speeds, enabling each compressor to operate at or near to its optimum condition. For a three-spool design, the low pressure (LP) compressor (i.e., the fan), intermediate pressure (IP) compressor, and high pressure (HP) compressor are each driven by their own turbine, designated as the LP turbine, IP turbine, and HP turbine, respectively. For a two-spool design, the HP compressor is coupled with the HP turbine, and the fan is coupled with the LP turbine. In some designs, where additional compression is required, a so-called *booster*—consisting of one or more compressor stage(s) driven by the LP shaft—is installed between the fan and the HP compressor.

Fan

The fan blades, which have an airfoil cross section, incorporate a root attachment that secures the blade to the fan disk (this assembly is the LP rotor). The disk is a critical structural component that is designed to withstand the very high centrifugal loads of the spinning fan blades (as well as loads that might arise from a blade failure). The disk is coupled directly to the LP shaft in the vast majority of turbofan engines, rotating at the same speed as the LP turbine. This is not the case in a *geared turbofan*, however, where a planetary gearbox is employed to reduce the rotation speed of the fan. The incorporation of a gearbox adds complexity and weight to the design, but, on the positive side, such an arrangement means that the fan speed can be independently optimized for the envisaged mass flow and the LP turbine design simplified.

Compressors

The turbofan engines that power modern jet transport airplanes utilize compressors that are of an axial flow design. The compression is usually spread over several stages. A *stage* comprises a rotating disk with a row of airfoil-shaped blades (the rotor) and a row of stationary blades or vanes (the stator) that are fixed to the engine casing. Several modern rotor designs feature *blisks*—a single component comprising both disk and blades (this concept reduces the overall weight of the compressor and improves its efficiency by eliminating the need for disk slots and blade root attachments). The air is accelerated by the rotor blades and swept rearwards into the passages between the stators, where it decelerates (diffuses); the kinetic energy imparted by the rotors, in effect, is translated into an increase in pressure and temperature. Multiple-stage compressors are very efficient, achieving the required pressure increase with minimal losses—a significant factor that influences the engine's overall efficiency. A two-spool engine has two compressor units (LP and HP), whereas a triple-spool engine has an additional intermediate pressure (IP) compressor.

Combustion System

The fuel is vaporized before entering the combustor or it is injected as an atomized spray. Combustion can only take place at relatively slow air speeds at air/fuel ratios of approximately 15:1 [3], and, as the airflow leaving the compressor will exceed this ratio, the air is diffused and only a relatively small amount of the core airflow is used in the primary combustion zone; the rest is introduced downstream to cool the combustion chamber. Engine thermal efficiency increases with gas temperature; the development trend over the past decades has therefore been to increase the temperature of the gas stream entering the turbine—this is known as the *turbine entry temperature* (TET).

Turbines

A turbine consists of one or more stage(s)—each stage has a row of stationary vanes called nozzle guide vanes (NGVs) and a row of rotating airfoil-shaped blades. The blades are attached to a disk, which is fixed to a shaft. The number of stages in a turbine depends on the gas flow properties, the speed that the shaft must rotate the compressor, and the power needed for the accessory units (i.e., power off-take). The turbines extract energy from the hot gas stream to provide the shaft power to drive the fan, compressors, and engine accessories.

Engine Support Structures

The engine case, or casing, is essentially a pressurized tube, usually circular in cross section, which contains the hot gas flow. Bearing housings that support the rotating assemblies are joined to the casing by a number of radial struts or vanes. Mounting lugs to attach the engine to the airframe are typically installed on the exterior of the engine casing. The fan case (or LP case), which surrounds the fan, is designed to contain a blade failure without releasing blade or casing fragments. On certain engine types, the fan case also has attachment lugs as part of the engine mounting system. The fan case attaches to the engine nacelle. On the internal wall of the air intake, an acoustic liner (comprising a perforated face-sheet, honeycomb core, and back-skin) is installed to attenuate fan noise.

Exhaust System

The hot gas leaves the engine through a propelling nozzle, which has the effect of increasing the gas velocity. On long duct engines, the hot and cold gas mixes at the exit or just before the exit; on short duct designs the two gas streams mix externally, with the bypass stream being ejected at a station ahead of the core nozzle.

Accessory Units

Accessory units mounted on the engine provide the electrical, pneumatic, and hydraulic power for the airplane. The drive for these units is taken from a compressor shaft, typically via an internal gearbox, through a radial driveshaft, to an external gearbox. The external gearbox provides the drives for accessory units (which provide power for the airplane systems), and it also provides power for the engine pumps and engine control systems. The gearbox and associated drives are used to start the engine by rotating the compressor shaft.

Bleed Air System

High pressure air is extracted (bled) from the compressors for use in several engine and airplane systems. Within the engine, it can be used for internal cooling and also for oil system sealing, for example. Bleed air is also used by several airplane systems that are external to the engine—these include cabin pressurization (see Section 22.3), air conditioning (see Section 22.4), and airframe de-icing (see Section 22.5). Bleed air can also be used for "cross-bleed" engine starting if air from the auxiliary power unit (described in Section 22.6) or a ground cart is not available. The amount of air extracted from the engines can be automatically controlled or manually controlled by the flight crew by adjusting the position of bleed air valves. Note that the extraction of bleed air from the compressor negatively influences an engine's performance (as discussed later in Section 8.8.3). A departure from this traditional system architecture has been adopted on certain modern engines, where the amount of bleed air taken from the engine is significantly reduced.[2]

Thrust Reversal System

The thrust reverser—which operates by turning, or redirecting, the exhaust gas streams through an angle of approximately 120°—is used to decelerate the airplane on the ground. Reverse thrust can only be selected by the crew when the engine is operating at a low power setting and thereafter the level of reverse thrust can be increased. The system augments the airplane's conventional wheel brakes and is particularly effective on slippery runways. A number of different systems have been devised to redirect the exhaust gas streams. One system—which was widely used on early jet engines—relies on bucket-shaped doors being moved into the path of the hot gas stream, redirecting the flow such that it has a forward directional component. On modern turbofan engines, it is usually only the bypass air that is redirected. A cascade-type thrust reverser system utilizes internal blocker doors to obstruct the path of the bypass gas stream, forcing the air outwards through rows of cascade vanes located on the side of the engine (the vanes are not visible in flight as they are covered by a translating cowl). Another popular system utilizes large pivoting doors (typically four per engine), which are flush with the side of the nacelle when retracted. When actuated, the rear portion of the door rotates into the bypass duct, blocking the gas stream and redirecting the air outwards. The forward portion of the door, which is rotated into the external airflow, further directs the exiting bypass air.

8.2.3 Station Identification

Key points in the gas flow paths are assigned station numbers. SAE Aerospace Standard AS755,[3] which is widely used throughout the gas turbine industry for all types of turbine engines, provides a framework for the identification of engine locations that are significant in terms

---

2 For the B787, Boeing adopted a "no-bleed" engine design concept; electrical systems replaced most of the pneumatic systems in the design of the airplane (bleed air is only used for engine cowl ice protection and pressurization of hydraulic reservoirs). Benefits indicated by Boeing include improved fuel efficiency, reduced maintenance costs, and improved systems reliability [4].
3 SAE Aerospace Standard AS755 [5] superseded SAE Aerospace Recommended Practice ARP755 [6], which, through its widespread adoption since it was first issued in 1962, had become a *de facto* standard in the industry.

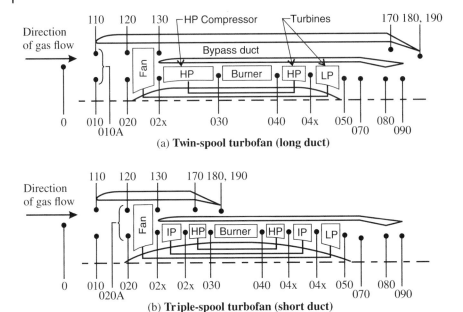

Figure 8.2 Station numbers for (a) twin-spool long duct and (b) triple-spool short duct turbofan engines. *Source*: Adapted from [5].

of propulsion system performance. The precise manner in which this framework gets implemented varies, and differences can thus exist between numbering schemes used by different manufacturers. Furthermore, the rules for numbering the stations have historically undergone several revisions. The description presented herein is based on the new (alternate) numbering system[4] described in SAE AS755 [5]. This is illustrated in Figure 8.2 for two turbofan designs: (a) a twin-spool engine with a long duct (mixed flow) design, and (b) a triple-spool engine with a short duct (un-mixed flow) design.

The numeral 0 (zero) is used in the context of engine performance to identify the free-stream, ambient conditions ahead of the engine, as per SAE AS755 [5]. This convention, however, can be a source of confusion for airplane performance, as the numeral zero is customarily used to denote ISA sea-level atmospheric conditions (as described in Section 4.2). To avoid ambiguity, ambient conditions in the free stream are often identified by the letters "amb" or, alternatively, by the symbol ∞ (the latter notation has been used herein).

In compliance with SAE AS755 [5] recommendations, the primary (core) gas stream stations have a leading 0, as shown in Figure 8.2. This is convenient for computer coding. It is, however, commonplace in the industry that the zero is omitted; this policy of dropping the leading zero in station identification has been adopted herein. The main station numbers for the primary (core) gas flow are allocated as follows: the entrance to the intake is identified as 10, the entrance to the first mechanical compression stage (i.e., the fan) is 20, the last compression stage discharge or burner entrance is 30, the burner discharge or first turbine stage entrance is 40, and the last

---

4 The traditional numbering system, as first defined in ARP755 [6], used a single digit (ranging from 0 to 9) to identify the main stations in the primary gas stream and a second digit following a decimal point to denote in-between stations. This meant that the fan entrance would be identified as station 2 and the entrance to the HP compressor could be 2.5, for example. The main stations in the bypass stream had only two digits—i.e., without the trailing 0 (zero) shown in Figure 8.2.

turbine stage discharge is 50. Station number 60 applies to engines with afterburners; station 70 is at the entrance to the expansion nozzle and station 80 is the nozzle throat. Some nozzles have a divergent section aft of the throat, the exit of which would be station 90. In-between stations are allocated numbers between the main station numbers—for example, for a twin-spool design, the discharge from the fan or entrance to the HP compressor, which is identified as 02x in Figure 8.2a, is usually assigned the station number 25 (or 025 if the leading zeros are retained) in this system.

Station identification for the bypass gas flow follows the same rules, but with a leading numeral 1, as illustrated in Figure 8.2. The letter A, when appended to a station number, denotes average properties of two or more gas streams—for example, 20A would describe average properties of the core and bypass streams across the fan entrance.

8.2.4 Thermodynamic Cycle

Gas turbine engines employ the Brayton (Joule) thermodynamic cycle.[5] This is an open cycle in which the working fluid (air) is continuously refreshed. The characteristic properties of the air—that is, total temperature (see Equation 2.79), total pressure (Equation 2.82), and entropy—as it progresses through the cycle phases of compression, heating, and expansion essentially determine the performance of the engine in terms of its power output and its efficiency.

The station identification system (Section 8.2.3) is used in the description of the gas properties at the various cycle points. By convention, the station number is added, usually as a subscript, to the symbol that denotes the parameter of interest—for example, $T_{t,20}$ would be the total temperature of the core gas stream at the fan face.

8.2.5 Gas Temperatures

The ambient air temperature, which in an operational context is usually called the *outside air temperature* (OAT), is a critically important operational parameter for any jet engine, as it can directly impact the thrust generated by the engine. As the air flows through the engine components, it undergoes rapid changes in temperature, reaching a peak at the combustion chamber exit (i.e., station 40). The total temperature of the gas able to do work at the first turbine rotor—known as the *turbine entry temperature*[6] (TET)—is a thermodynamic cycle parameter that has a powerful affect on the engine's overall performance. The turbine entry total temperature is herein identified as $T_{t,41}$. This is consistent with industry practice, which differentiates the conditions at the combustor outlet and the turbine inlet, which has a slightly lower temperature due to the cooling airflow that surrounds the combustors. Moreover, it is the ratio $T_{t,41}/T_{t,20}$ that is important. Increasing the TET (for engine conditions that will hold the pressure ratio constant) results in an increase in thermal efficiency and net power. Temperatures exceeding 1550 °C are common for modern engines, requiring sophisticated internal and external cooling concepts as these temperatures exceed the melting point of the nickel-based alloys from which the HP turbine blades are cast [3]. During a typical mission, the ratio $T_{t,41}/T_{t,20}$ will change; the highest values are usually reached at the top of climb.

---

5 The cycle is named in honor of George Bailey Brayton (1830–1892), the American mechanical engineer who, in 1872, patented an internal combustion engine that utilized a constant pressure cycle [7]. He designed the first continuous ignition combustion engine to be commercially manufactured (Brayton's Ready Motor).
6 The turbine entry temperature (TET) is also known variously as the *stator outlet temperature* (SOT), *turbine inlet temperature* (TIT), or *rotor inlet temperature* (RIT).

Temperature is monitored at various locations in the engine during normal operation. Such measurements are used as system health monitoring parameters, for example to prevent over-heating of the turbine. The time that an engine can operate with temperatures approaching the design limits is restricted—this sets operating limits for the engine (see Section 8.5). Note that the TET is too high to measure in service, but several measurements are taken aft of the HP turbine using special high temperature thermocouples.

8.2.6 Gas Pressures

The gas pressure, which is linked to the gas temperature and density through the ideal gas law (Equation 2.66), also undergoes massive changes as the gas flows through the engine. Pressure ratios, taken as the ratio of the total gas pressures at two defined points (stations) in the engine, are similarly defined as explained above for the total temperature ratios.

A key performance parameter influencing the engine cycle of a turbojet is the compressor pressure ratio; for a turbofan, it is the fan pressure ratio and the core pressure ratio that matter. For the analysis of engine performance, an overall *engine pressure ratio* (EPR) is taken as the total pressure of the gas (p_t) at the nozzle (station 90) divided by the total pressure at the fan face (station 20). A more convenient ratio for monitoring or setting the engine thrust, however, is the ratio of the total pressure at station 70 to that at station 20, that is,

$$\text{EPR} = \frac{p_{t,70}}{p_{t,20}} \tag{8.1}$$

This dimensionless ratio is often called the *hot* EPR as a *cold* EPR can be defined as the ratio of total pressure in the bypass flow at station 170 to that at station 120.

8.2.7 Mass Flow and Bypass Ratio

The mass flow of the air entering the engine ($\dot{m}_a$) splits into two parts: the mass flow of the air passing through the bypass duct ($\dot{m}_{a,b}$) and the mass flow of the air passing through the engine core ($\dot{m}_{a,c}$). It thus follows that

$$\dot{m}_a = \dot{m}_{a,b} + \dot{m}_{a,c} \tag{8.2}$$

where mass flow is defined by Equation 2.51 (typical units: kg/s or slug/s).

The *bypass ratio* (BPR), herein denoted by the Greek letter lambda (λ), is defined as

$$\lambda = \frac{\dot{m}_{a,b}}{\dot{m}_{a,c}} \tag{8.3}$$

There has been a progressive increase in the BPR used in engine designs since the first jet airliners were introduced in the 1950s due to the general improvement in fuel efficiency and reduced noise that result from the use of higher bypass ratios. The turbofan engines installed on most jet transport aircraft currently in service have a BPR of between 4 and 9, with some modern engines featuring BPRs as high as 12.5 [8]. Ultra high bypass ratio engines, under development, will have BPRs of about 15.

The bypass ratio of an engine is not exactly constant: it tends to increase slightly with increasing airplane speed and reduce slightly with increasing altitude. Quoted values of BPR (provided by engine manufacturers, for example) are given for a reference condition.

8.2.8   Engine Rotor Speeds

The rotational speeds of the engine rotors, which are usually measured in revolutions per minute (RPM), are key parameters influencing the performance of the engine. The speed of the LP shaft is denoted as N_1 (or alternatively as *N1* or *NL*). The rotation speed N_1 is used on many engine types to control the engine's thrust (see Section 8.5). The HP shaft speed of a two-spool engine is denoted as N_2 (or alternatively as *N2* or *NH*). In the case of a three-spool engine, N_2 (or *N2* or *NI*) denotes the speed of the IP shaft and N_3 (or *N3* or *NH*) the speed of the HP shaft.

Typical rotational speeds on a modern three-spool engine at takeoff condition are 3000 RPM for the LP rotors, 6000 RPM for the IP rotors, and 10 000 RPM for the HP rotors [3]. Very high stresses are induced in the blades and disks of the compressors and turbines at these speeds due to high centrifugal forces; rotational speeds are monitored to ensure that the engine design limits are not exceeded in flight.

8.2.9   Corrected Engine Parameters

Engine data are frequently presented as *corrected* data (also known as *referred* data), rather than as absolute values. Various engine parameters are treated in this way, including gas temperatures and pressures, shaft rotational speeds, mass flow, net thrust (defined in Section 8.3.1), and fuel flow. Non-dimensional values for the parameter of interest are obtained by dividing the actual value by a defined reference value (e.g., the gas temperature at a particular station can be divided by the ambient air temperature).

Corrected parameters have also been defined based on dimensional analysis. Two important parameters are the corrected net thrust, F_N/δ, where δ is the pressure ratio of the ambient air (see Equation 4.2), and the corrected LP shaft speed, $N_1/\sqrt{\theta_{t,20}}$, where $\theta_{t,20}$ is the total temperature ratio of the air at station 20 (see Equations 4.1 and 2.79). The value of presenting the information in this way is that it allows the data to be generalized, permitting a single chart to be used for a range of conditions (e.g., different altitudes), rather than having a separate chart for each set of operating conditions. The technique works particularly well for initial studies (i.e., first-order calculations) as a large number of graphs can be collapsed onto a single plot. Measured engine data or those resulting from sophisticated numerical analysis techniques, however, do not always generalize perfectly. Corrected parameters are also extremely useful when carrying out installed engine tests following maintenance or inspection as it facilitates a check, by maintenance engineers, to determine if the engine is performing as designed.

The actual net thrust (F_N) can be obtained by multiplying the corrected value by the pressure ratio of the ambient air, which is given by

$$\delta = \frac{p_\infty}{p_0} \tag{8.4}$$

where p_∞ is the ambient air (static) pressure (typical units: N/m^2, lb/ft^2); and
p_0 is the ISA sea-level datum air pressure (typical units: N/m^2, lb/ft^2).

Similarly, the actual LP shaft speed (N_1) can be obtained by multiplying the corrected value by $\sqrt{\theta_{t,20}}$, where

$$\theta_{t,20} = \theta \left(1 + \frac{\gamma - 1}{2} M^2 \right) = \frac{T_\infty}{T_0} \left(1 + \frac{\gamma - 1}{2} M^2 \right) \tag{8.5}$$

where $\theta_{t,20}$ is the total temperature ratio of the air at station 20 (dimensionless);
θ is the temperature ratio of the ambient air (dimensionless);

T_∞ is the ambient air temperature (typical units: K, °R);
T_0 is the ISA sea-level datum air temperature (typical units: K, °R);
γ is the ratio of specific heats of air (dimensionless); and
M is the free-stream (or flight) Mach number (dimensionless).

8.2.10 Indicated Engine Parameters

Several engine parameters are measured and displayed, as part of the engine instrumentation, in the cockpit. These indications enable the flight crew to monitor the status of the airplane's engines and operate the engines within set limits. The key parameters that are indicated for engine performance monitoring are engine gas temperature (with units of °C or °F), engine pressure ratio (dimensionless), rotor speeds (which are indicated as percentages of nominal maximum rotational speeds), and fuel flow (with units of kg/h or lb/h). The indicated gas temperature can have different names depending on the location of the sensors—for example: *turbine gas temperature* (TGT), which has sensors located between turbine units, and *exhaust gas temperature* (EGT), which has sensors located downstream of the last turbine unit. Similarly, there are several different ways to determine an engine pressure ratio (EPR), which is used on many engine types as a thrust control parameter (see Section 8.5.3).

8.3 Engine Thrust

8.3.1 Basic Thrust Equation for a Turbojet Engine

The basic theory regarding the generation of thrust by a jet engine is introduced in this section. In simple terms, the thrust is the result of two separate mechanisms: (1) the momentum change of the gas stream, and (2) the pressure differential that arises across the engine. These factors are described below.

(1) The "momentum" force is the *reaction* (which is to be understood in the context of Newton's third law) of accelerating an air mass through the engine. By the application of Newton's second law (see Section 2.5.4) the force is equal to the time rate of change of momentum of the gas stream:

$$F_m = (\dot{m}_a + \dot{m}_f)U_e - \dot{m}_a U_\infty \qquad (8.6)$$

where F_m is the engine momentum force (typical units: N, lb);
$\dot{m}_a$ is the gas mass flow (typical units: kg/s, slug/s);
$\dot{m}_f$ is the mass flow of the fuel (typical units: kg/s, slug/s); and
U is the speed of the gas stream (typical units: m/s, ft/s), and where e
denotes exit and ∞ denotes free stream (Figure 8.3).

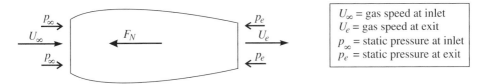

Figure 8.3 Idealized turbojet engine.

(2) The "pressure" force component arises due to the pressure differential that acts across the engine (shown in Figure 8.3)—this can be expressed as follows:

$$F_p = (p_e - p_\infty)A_e \tag{8.7}$$

where F_p is the engine pressure force (typical units: N, lb);

p_e is the exit pressure (typical units: N/m^2, lb/ft^2);

p_∞ is the free-stream (i.e., ambient) pressure (typical units: N/m^2, lb/ft^2); and

A_e is the exit area (typical units: m^2, ft^2).

The total thrust (F) is given by the sum of F_m and F_p:

$$F = (\dot{m}_a + \dot{m}_f)U_e - \dot{m}_a U_\infty + (p_e - p_\infty)A_e \tag{8.8}$$

Equation 8.8 can be expressed in terms of the fuel–air ratio (f), which is defined as

$$f = \frac{\dot{m}_f}{\dot{m}_a} \tag{8.9}$$

Thus $\quad F = \dot{m}_a(1+f)U_e - \dot{m}_a U_\infty + (p_e - p_\infty)A_e \tag{8.10}$

These force components act on the fixed and rotating parts of the engine to produce a propelling thrust. Two simplifications are now considered:

(1) As the fuel mass flow is much less than the gas stream mass flow (a typical fuel/air ratio is roughly 0.008–0.02), it is reasonable to assume that $1 + f \cong 1$.
(2) As the momentum force is much greater than the pressure force, the smaller term is ignored (the nozzle of a turbofan engine is usually designed such that the exit pressure is very close to the ambient pressure).

Hence $F \cong \dot{m}_a U_e - \dot{m}_a U_\infty \tag{8.11}$

The first term in Equation 8.11 is called the *gross thrust* (or *gross momentum thrust*); the second term is called the *momentum drag* (or *ram drag*). The difference between the gross thrust and the momentum drag is the *net thrust*—and, by convention, it is written as F_N. Equation 8.11 can be written as

$$F_N \cong \dot{m}_a(U_e - U_\infty) \tag{8.12}$$

This simple analysis yields an interesting result: the net thrust, for a given increase in gas flow velocity, is approximately proportional to the mass flow of the air passing through the engine. The parameters that influence the mass flow are thus important; this can be explored by developing Equation 8.12 as follows:

$$F_N \cong \rho_\infty A_0 U_\infty (U_e - U_\infty) \tag{8.13}$$

where ρ_∞ is the free-stream air density (typical units: kg/m^3, slug/ft^3); and

A_0 is the free-stream intake capture area (typical units: m^2, ft^2).

The factors that influence air density are ambient pressure and temperature—as described by Equation 2.66 (the influence of humidity is usually negligible). These factors directly affect the thrust that can be generated by an engine—as a general trend, the net thrust will reduce with increasing pressure altitude and also with increasing ambient air temperature (these trends are explored further in Sections 8.5 and 8.6).

8.3.2 Basic Thrust Equation for a Turbofan Engine

For a bypass engine, the total mass flow ($\dot{m}_a$) has two components: bypass and core, as described by Equation 8.2. The thrust of a turbofan engine can thus be expressed in terms of the components of flow:

$$F_N \cong \dot{m}_{a,c}(U_{90} - U_\infty) + \dot{m}_{a,b}(U_{190} - U_\infty) \tag{8.14}$$

where U is the speed of the gas stream (typical units: m/s, ft/s); the station numbers are defined in Figure 8.2.

Recalling the definition of the bypass ratio (Equation 8.3) permits Equation 8.14 to be written as

$$F_N \cong \dot{m}_{a,c}(U_{90} + \lambda U_{190}) - \dot{m}_a U_\infty \tag{8.15}$$

8.3.3 Influence of Airplane Speed

The influence of the airplane's forward speed on the net thrust is twofold. As shown by Equation 8.11, there is a linear *increase* in momentum drag with increasing airplane speed. This reduces the net thrust. There is also an *increase* in gross thrust with increasing speed due to the increased mass flow—this is known as the *ram effect*. The incoming air undergoes a compression in the air intake. By definition, the *ram pressure ratio* is the ratio of the total air pressure at the fan face to the static air pressure at the entry to the air intake (i.e., $p_{t,20}/p_\infty$). This pressure ratio, which increases with flight Mach number, is an important design parameter—an efficient air intake tends to compensate, to some extent, for the loss of thrust due to the momentum drag associated with the airplane's speed.

In the case of a turbojet engine, the increase in gross thrust tends largely to compensate for the increasing momentum drag across a range of subsonic Mach numbers. A turbofan engine, however, has a much higher mass flow ($\dot{m}_a$) than a comparable turbojet (for the same thrust, under the same conditions). Consequently, for a turbofan engine the momentum drag increases more quickly with Mach number and there is a progressive decrease in net thrust with increasing Mach number for high BPR engines. These trends are illustrated in Figure 8.4.

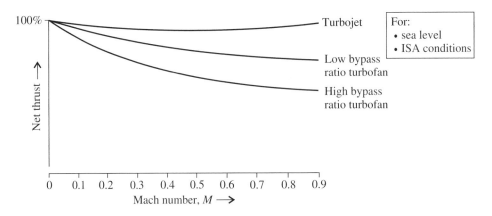

Figure 8.4 Net thrust as a function of Mach number for turbojet and turbofan engines.

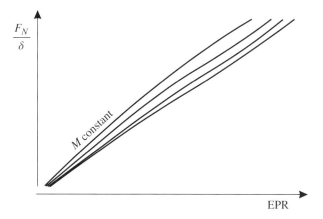

Figure 8.5 Corrected thrust as a function of EPR.

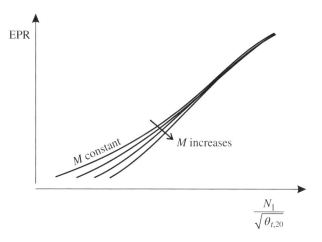

Figure 8.6 EPR as a function of corrected N_1.

8.3.4 Generalized Thrust Functions

For an engine of a particular design, the net thrust (F_N) will be proportional to the EPR when operating at a given airplane speed, altitude, temperature, and thrust lever position [9]. The EPR can thus be used as an index for thrust setting and monitoring. Generalized thrust functions are illustrated in Figure 8.5 for varying Mach number. The EPR can be mapped to the engine speed as a function of flight Mach number (as illustrated in Figure 8.6). This relationship provides an alternative means for setting and monitoring thrust—that is, through the use of N_1 (this is discussed further in Section 8.5).

8.3.5 Overall Engine Efficiency

A turbofan engine is a heat engine that converts fuel energy (or heat energy) into mechanical energy and then converts the mechanical energy into useful work, which is done on the airplane. Recall that heat and work are both forms of energy, and are interchangeable, albeit often with losses.

The *overall efficiency* (η_0) of a turbofan engine can be defined as by the product of the thermal efficiency (η_t) and the propulsive efficiency (η_p):

$$\eta_0 = \eta_t \eta_p \tag{8.16}$$

where $\eta_t = \dfrac{\text{Rate that mechanical work is done}}{\text{Rate that heat energy (fuel energy) is delivered}}$ (8.17)

and $\eta_p = \dfrac{\text{Rate that useful work is done on the airplane}}{\text{Rate that mechanical work is done}}$ (8.18)

The rate at which fuel energy is supplied to the engine is the product of the fuel mass flow and the net heating value (or net calorific value measured at constant pressure) of the fuel.[7] The fuel energy is converted, with losses (notably the unusable enthalpy), into mechanical work, which is imparted on the gas stream, resulting in a change in kinetic energy. The thermal efficiency can thus be expressed as

$$\eta_t = \frac{\frac{1}{2}\dot{m}_a U_e^2 - \frac{1}{2}\dot{m}_a U_\infty^2}{\dot{m}_f H_f} = \frac{\dot{m}_a \left(U_e^2 - U_\infty^2\right)}{2\dot{m}_f H_f} \tag{8.19}$$

where H_f is the net heating value of the fuel, i.e., energy content per unit mass (typical units: MJ/kg, Btu/lbm).

The propulsive efficiency (η_p) is the ratio of the propulsive energy to the sum of the propulsive energy and the unused kinetic energy, which is the kinetic energy of the exhaust gas with respect to a stationary reference (e.g., the Earth). The time rate that propulsive energy is used is the rate that useful work is done by the engine—this is called the thrust power (P_T). Thrust power is the product of the net thrust and the forward speed of the airplane, which is equal in magnitude to the free-stream velocity (U_∞); thus

$$P_T = F_N U_\infty = \dot{m}_a (U_e - U_\infty) U_\infty \qquad \text{(from Equation 8.12)} \tag{8.20}$$

The time rate of unused kinetic energy in the exhaust gas is

$$\dot{E}_k = \frac{1}{2}\dot{m}_a (U_e - U_\infty)^2 \tag{8.21}$$

An expression for the propulsive efficiency can be deduced from Equations 8.20 and 8.21:

$$\eta_p = \frac{P_T}{P_T + \dot{E}_k} = \frac{\dot{m}_a(U_e - U_\infty)U_\infty}{\dot{m}_a(U_e - U_\infty)U_\infty + \frac{1}{2}\dot{m}_a(U_e - U_\infty)^2} = \frac{2}{1 + \dfrac{U_e}{U_\infty}} \tag{8.22}$$

---

7 The *net heating value*—which, in this context, is often called the *lower heating value* (LHV) or, alternatively, the *lower calorific value* (LCV)—is a measure of the energy content of the fuel per unit mass (see Section 22.7.2). The properties of jet engine aviation fuel can vary slightly depending on the precise chemical makeup of the fuel.

The propulsive efficiency, as expressed by Equation 8.22, is known as the *Froude efficiency*.[8] Two interesting observations can be made by comparing Equation 8.22 with Equation 8.12:

(1) From Equation 8.12, it is seen that F_N is a maximum when $U_\infty = 0$, but from Equation 8.22, it is seen that $\eta_p = 0$ at this condition.
(2) From Equation 8.22, it is seen that η_p is a maximum when $U_e = U_\infty$, but from Equation 8.12, it is seen that $F_N = 0$ at this condition.

This is, however, a simplistic analysis and it does not fully represent actual engine performance; nonetheless, it does provide a useful observation: U_e must be greater than U_∞ to generate thrust, but the difference should not be too great, otherwise the efficiency is impaired.

It should be borne in mind that the Froude efficiency is only one of two terms that describes the overall efficiency, which is obtained by combining Equation 8.16 with Equations 8.19 and 8.22:

$$\eta_0 = \left\{ \frac{\dot{m}_a \left(U_e^2 - U_\infty^2 \right)}{2\, \dot{m}_f H_f} \right\} \frac{2}{1 + \dfrac{U_e}{U_\infty}} = \frac{\dot{m}_a U_\infty (U_e - U_\infty)}{\dot{m}_f H_f} \tag{8.23}$$

Equation 8.23 can also be written in terms of the thrust using Equation 8.12:

$$\eta_0 = \frac{F_N U_\infty}{\dot{m}_f H_f} \tag{8.24}$$

The thermal efficiency of the engine increases with an increase in TET; this, in turn, increases the jet exit velocity (U_e), resulting in an increase in thrust (see Equation 8.12). The inherent problem for a jet engine is that an increase in exit velocity reduces the propulsive efficiency. A turbofan engine overcomes this problem by accelerating a greater total mass flow through the engine by utilizing a bypass duct, while reducing the mean exit velocity—resulting in an increase in the propulsive efficiency. The bypass principle thus allows a greater amount of useful work to be done on the airplane (resulting in greater thrust at a given speed) for a given amount of heat energy contained in the fuel.

8.3.6 Specific Thrust

The *specific thrust*, which is a good overall figure of merit for the performance of a jet engine, is defined as the thrust divided by the gas *mass flow*.

$$F_s = \frac{F_N}{\dot{m}_a} \tag{8.25a}$$

where F_s is the specific thrust (typical units: N kg^{-1} s, m/s, lb slug^{-1} s, ft/s);
  F_N is the engine net thrust (typical units: N, lb); and
  $\dot{m}_a$ is the gas mass flow (typical units: kg/s, slug/s).

---

8 The efficiency expression is named after William Froude (1810–1879), the British engineer and hydrodynamicist who first used it. Froude, an eminent naval architect, also established a formula, now called the Froude number, by which the results of small-scale experiments can be used to predict the behavior of full-sized vessels.

Alternatively, it can be defined as the thrust divided by the *weight flow* rate. To avoid a potential misunderstanding, the specific thrust, when based on weight flow rate, will herein be given the symbol F_s'.

Thus $$F_s' = \frac{F_N}{\dot{W}_a} \qquad\qquad (8.25b)$$

where F_s' is the specific thrust (typical unit: s); and
$\dot{W}_a$ is the gas weight flow (typical units: N/s, lb/s).

The takeoff specific thrust for modern turbofan engines is approximately 0.25–0.35 kN kg^{-1} s. This means that these engines produce ~0.3 kN for every 1 kg/s of air that enters the intake.

The significance of the specific thrust is that it provides an indication of the relative size of an engine—for example: an engine with a comparatively high specific thrust will have a relatively low mass flow for a given thrust, and this would imply a relatively small inlet area and low engine weight.

8.4 Fuel Flow and Thrust Specific Fuel Consumption

8.4.1 Fuel Flow Definitions

In the context of airplane performance, *fuel flow* can be defined as the fuel consumed (or combusted) per unit of time—it can be expressed as

(1) the *mass* of the fuel consumed per unit of time ($\dot{m}_f$), which, for the purpose of airplane performance, is often given the symbol Q; or alternatively as
(2) the *weight* of the fuel consumed per unit of time ($\dot{W}_f$), which herein is given the symbol Q'.

The definitions and corresponding units for the two conventions are summarized in Table 8.1. The choice of convention (i.e., mass based or weight based) is a matter of personal preference, and for almost all applications either convention will yield satisfactory results. It should be noted, however, that the two approaches are not exactly equivalent, as the energy supplied to the engines depends on certain fuel properties—namely, the net heating value and density (see Section 22.7.2). For a set mass flow, the energy supplied will not change as the gravitational acceleration (g) varies (e.g., with changes in height or geographical location), but this will not be true for a set weight flow rate. For precise energy-based calculations, the variation of g must be taken into account when the fuel flow is based on weight. For all other calculations, it is acceptable to use the standard sea-level value (g_0) in converting weight flow to mass flow (see Equation 8.27).

Table 8.1 Fuel flow definitions and units

	Mass flow basis		Weight flow basis	
Fuel flow definitions:	$Q = \dot{m}_f$	(8.26a)	$Q' = \dot{W}_f$	(8.26b)
Typical SI units:	mg/s, kg/h		N/s, N/h	
Typical FPS units:	slug/h		lb/h	
Conversion:		$Q' = Qg$	(8.27)	

Note: It is customary to use the mass flow definition when using SI units and the weight flow definition when using FPS units.

For the purpose of airplane performance analysis, it is usual that the term *fuel flow* is understood as the net fuel flow to all engines. Depending on the application, the net fuel flow may also include fuel supplied to the auxiliary power unit (APU).

8.4.2 Thrust Specific Fuel Consumption

Thrust specific fuel consumption (TSFC)—which is often shortened to *specific fuel consumption* (SFC)—is a key performance parameter for a jet engine. The TSFC can be defined in two equally correct ways—a potential source of confusion:

(1) By definition, TSFC is the *mass* of fuel burned per unit of time (Q), divided by the engine thrust (F_N). It is given the symbol c.
(2) Alternatively, TSFC can be defined as the *weight* of fuel burned per unit of time (Q'), divided by the engine thrust. To avoid a potential misunderstanding, the TSFC when based on weight is herein given the symbol c'.

The definitions and corresponding units of TSFC are given in Table 8.2. A typical TSFC for a modern turbofan engine in cruise is 16–19 mg $N^{-1}s^{-1}$ (0.57–0.67 lb $lb^{-1}h^{-1}$).

The TSFC is inversely proportional to the overall engine efficiency (defined in Section 8.3.5) for a given free-stream velocity and fuel type—note that a high efficiency implies a low value of TSFC. The two measures are linked by the following relationship:

$$c = \frac{1}{\eta_0} \left(\frac{U_\infty}{H_f} \right) \qquad (8.28)$$

The TSFC also relates to another measure of engine efficiency—specific thrust:

$$c = \frac{f}{F_s} \qquad (8.29)$$

where f is the fuel/air mass flow ratio (dimensionless); and
F_s is the specific thrust (typical units: N kg^{-1} s, lb $slug^{-1}$ s).

Note that a high specific thrust corresponds to a low TSFC.

Table 8.2 Thrust specific fuel consumption (TSFC) definitions and units

	Mass flow basis		Weight flow basis	
TSFC definitions:	$c = \dfrac{Q}{F_N}$	(8.30a)	$c' = \dfrac{Q'}{F_N}$	(8.30b)
Typical SI units:	mg $N^{-1}s^{-1}$, kg $N^{-1}h^{-1}$		N $N^{-1}s^{-1}$, N $N^{-1}h^{-1}$	
Typical FPS units:	slug $lb^{-1}h^{-1}$		lb $lb^{-1}h^{-1}$	
Conversion:		$c' = cg$	(8.31)	

Notes:

(a) It is customary to use the weight flow definition when expressing TSFC in FPS units; however, both mass and weight flow definitions are used with SI units.

(b) The weight-based units of TSFC are effectively "per unit of time." It is, however, customary to include a unit of force in both numerator and denominator. Several authorities recommend that the solidus (forward slash) not be used more than once in an unit expression without brackets to avoid ambiguity [10]. The tradition of writing TSFC units as lb/lb/h (or lb/h/lb) or mg/N/s is, however, widespread in the aviation community.

8.4.3 Cruise TSFC Variation

A very important efficiency metric for an airliner is the engine's TSFC at the range of thrust values required for cruise. The relationship of TSFC versus net installed thrust is illustrated in Figure 8.7, for varying cruise Mach number. Such graphs are sometimes called *TSFC loops* due to their characteristic shape. The required thrust produced by the installed engines—which equals the airplane's drag for a constant-speed, constant-altitude cruise—will be at or near the bottom of the loop at the design cruise Mach number for properly matched airframe–engine combinations.

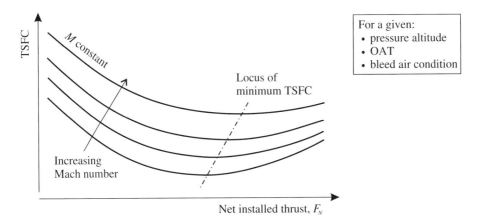

Figure 8.7 Thrust specific fuel consumption (TSFC) for a turbofan engine in cruise as a function of net installed thrust.

8.4.4 Corrected Fuel Flow

Fuel flow data (like several other engine parameters—see Section 8.2.9) are often presented in a corrected form. The *corrected fuel flow* (Q_{cor}) is usually defined as

$$Q_{cor} = \frac{Q}{\delta_{t,20} \sqrt{\theta_{t,20}}}$$

(8.32)

where $\delta_{t,20}$ is the total pressure ratio of the air at station 20; and
$\theta_{t,20}$ is the total temperature ratio of the air at station 20.

The group of terms used to correct the fuel flow has its origin in propulsion theory dimensional analysis. The benefit of such data presentation is that it is possible to estimate an engine's performance (in this case to determine the fuel flow) at different Mach numbers, pressure altitudes, and ambient temperature conditions when the engine is operating at a particular non-dimensional or corrected condition (e.g., described by EPR or $N_1/\sqrt{\theta_{t,20}}$).

The total temperature ratio is given by Equation 8.5. The total pressure ratio used in Equation 8.32 can be deduced from Equation 2.82:

$$\delta_{t,20} = \delta \left[1 + \frac{\gamma - 1}{\gamma} M^2 \right]^{\left(\frac{\gamma}{\gamma - 1} \right)}$$

(8.33)

Equations 8.5 and 8.33 can be substituted into Equation 8.32 to give an expression for the true fuel flow, which can now be determined from knowledge of Q_{cor}, M, and the ambient conditions:

$$Q = Q_{cor}\delta\sqrt{\theta}\left[1 + \frac{\gamma - 1}{\gamma}M^2\right]^{\left(\frac{1.5\gamma - 0.5}{\gamma - 1}\right)} \tag{8.34}$$

Actual engine data seldom generalize completely—in order words, a single chart cannot be used to represent accurately all the data under all conditions. In such cases, it is usual that a chart is prepared for each pressure altitude of interest. Furthermore, if the fuel flow characteristics do not align perfectly with the theoretical model, then Equation 8.32 is modified by introducing a correction factor (k), thus:

$$Q_{cor} = \frac{Q}{k\delta_{t,20}\sqrt{\theta_{t,20}}} \tag{8.35}$$

Since k is a function of the total temperature, it can be combined with the temperature ratio, as follows: $k\sqrt{\theta_{t,20}} = \theta_{t,20}^x$. The exponent x has to be determined specifically for each engine. Typical values may vary from 0.50 (i.e., the value determined from theoretical analysis) to 0.67 [9]. The general form of the corrected fuel flow is thus given as

$$Q_{cor} = \frac{Q}{\delta_{t,20}\theta_{t,20}^x} \tag{8.36}$$

And taking $\gamma = 1.4$, the actual fuel flow will be given as

$$Q = Q_{cor}\delta\theta^x(1 + 0.2M^2)^{3.5+x} \tag{8.37}$$

A plot of Q_{cor} as a function of F_N/δ for four selected cruise Mach numbers (at 35 000 ft) is shown in Figure 8.8 for a typical turbofan engine. Using these data (and Equations 8.37

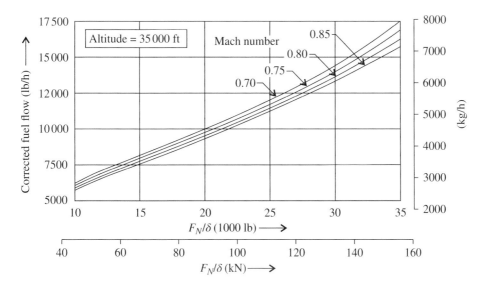

Figure 8.8 Corrected fuel flow versus corrected thrust for selected Mach numbers at a particular cruise altitude for a typical turbofan engine.

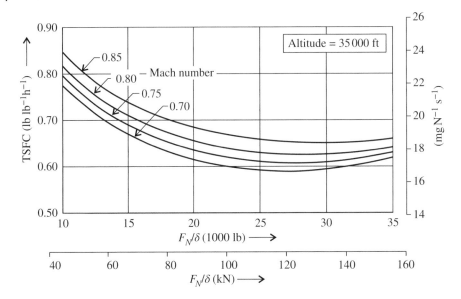

Figure 8.9 TSFC versus corrected thrust for selected Mach numbers at a particular cruise altitude for a typical turbofan engine.

and 8.30), TSFC values have been computed and plotted in Figure 8.9 (the abscissa is the same in both figures).

8.5 Thrust Control, Engine Design Limits, and Ratings

8.5.1 FADEC

FADEC is an acronym for Full Authority Digital Engine Control (it is also defined as Full Authority Digital Electronic Control by certain organizations). It is an engine control system in which the primary functions are provided electronically and the electronic control unit has full-range authority over the engine power or thrust [11]. FADEC has replaced the traditional hydromechanical engine control and interface system; it is employed on all modern turbine-powered aircraft. The airworthiness authorities require that FADEC systems are designed with either two identical channels that can provide full-operational capability after failure of one channel or with a hydromechanical backup that can provide an alternate operating mode.

FADEC systems control and monitor all critical engine performance parameters, using digital computers—called Electronic Engine Controllers (EECs) or Electronic Control Units (ECUs), depending on the manufacturer—and related accessory equipment and instrumentation. A FADEC system can support the following functions: manual thrust selection, autothrottle (or autothrust), reverser control, engine start-up, engine limit protection, power management, fuel flow computation, engine data acquisition (for cockpit indication), and engine condition monitoring. A key function of a FADEC system is the manual thrust lever interface.

8.5.2 Thrust Control

The *thrust levers*—one for each engine installed on the airplane—are located in the cockpit center console, between the flight crew seats. An engine's thrust is controlled, in manual operation, by the flight crew advancing (i.e., pushing forward) or retarding (i.e., pulling back) the thrust lever. The thrust levers on an airplane can be moved in unison with one hand or individually. The positions of the thrust levers, which are designed to rotate through a segment of an arc, are sensed electrically on FADEC-equipped aircraft. When a thrust lever is advanced, a signal is sent to the fuel pump to supply more fuel to the combustors. The increased gas temperature causes the speed of the turbines to increase and this, in turn, causes an increase in fan and compressor speeds and an increase in gas pressure. The resulting increase in mass flow passing through the core and bypass duct produces more thrust. This is, of course, a very simple explanation of an exceedingly complex engineering process. Flight path control can thus be achieved, in manual operations, by the flight crew manipulating the flight controls.

The *autothrottle*[9] (automatic throttle) system, which is a key component of an airplane's automatic flight control system, enables the pilot to operate in a supervisory capacity, rather than being part of the control loop. The primary interface for the flight crew would be a multifunctional Control Display Unit (CDU), which would communicate with the Flight Management Computer (FMC). Each CDU (there is one installed for the captain and one for the first officer) incorporates a keyboard or touchscreen and a small display screen. The flight crew can indirectly control the engine's thrust by commanding a desired flight parameter via the CDU—such as airspeed—rather than by manually positioning the thrust levers. In this mode, the FMC will aim to hold the targeted indicated airspeed (IAS) within a defined tolerance. Manufacturers have adopted different design philosophies regarding the pilot interface—for example, a Boeing autothrottle system has servos that move the thrust levers, during operation, in response to signals sent by the FMC, whereas the thrust levers remain in the last set position on an Airbus *autothrust* system.

On many airplane types, reverse thrust is commanded by the flight crew through the rearward movement of a second set of levers, which are attached to the forward thrust levers. The *reverse thrust* levers, which are retracted in flight, can only be deployed when the forward thrust levers are in the idle position. An alternative system—employed on the A320 family, for example [13]—has only one thrust lever per engine. Reverse thrust is selected by engaging latching levers at the idle stop position and moving the thrust levers into the reverse thrust range.

8.5.3 Thrust Setting Parameters

The thrust of a jet engine is neither directly measured nor directly controlled—in fact, pilots do not know explicitly how much thrust is being produced by the engines. Instead, the thrust of a jet engine is controlled via a *thrust setting parameter* (TSP) that has a close relationship to thrust, and which can be reliably obtained by measurement. Engine manufacturers use several alternative engine parameters to indirectly control an engine's thrust via the EEC/ECU.

The first method (which has been employed on many Rolls-Royce, Pratt & Whitney, and IAE[10] engines, for example) uses an engine pressure ratio based on actual pressure measurements. The EPR selected to control thrust is usually the *hot EPR*, as described by Equation 8.1. On certain older turbofan engines, the *cold EPR* has been used, while on other engine types, such as the Rolls-Royce RB211, an EPR (known as an integrated EPR) has been used that is the average of the total pressures measured in the core and bypass exhausts divided by the total pressure at station 20.

---

9 The Airbus system is known as *autothrust*, rather than an autothrottle [12].
10 International Aero Engines (IAE) is a joint venture of Pratt & Whitney (USA), Japanese Aero Engine Corporation (Japan), and MTU Aero Engines (Germany). Rolls-Royce (UK) was also a shareholder until 2012.

The second method (which has been used by several engine manufacturers including General Electric Aviation and CFM International,[11] for example) uses the speed of the LP shaft (N_1) to control thrust. The control parameter is expressed as %N_1, that is, a percentage of a nominal reference N_1 speed, which is usually taken as a typical rotational speed of the LP shaft at maximum takeoff thrust (defined in Section 8.5.5).

The third method (which has only been used on a series of Rolls-Royce three-spool engines) employs a turbofan power ratio (TPR) to control thrust. The TPR is a function of corrected measurements of the burner inlet total pressure and a total turbine gas temperature (TGT).

The choice of thrust setting parameter—which is a decision that is made at the beginning of every engine development program—has a major impact on the software logic in the EEC/ECU and also on the instrumentation that will be installed to measure the required engine parameters during normal operation (viz., temperatures, pressures, or rotational speeds).

8.5.4 Engine Design Limitations

The maximum thrust that an engine can produce under various operating conditions—such as flight Mach number and ambient air temperature and pressure—depends fundamentally on the structural limitations of critical engine components. These components are designed to meet a complex set of operating conditions for prolonged periods of time. The main engine design parameters that can theoretically limit the engine thrust for a given set of operating conditions are:

(1) the turbine entry temperature (exceeding the design temperatures can cause damage to the turbine blades, for example);
(2) the pressure differential across the engine case (exceedingly high pressures at station 30 compared to the ambient pressure can cause structural damage); and
(3) the N_1 shaft speed (excessive rotational speeds can cause damage to the fan blades or to the disk due to high centrifugal forces).

The first condition (temperature limit) impacts the engine's allowable operating range in that the EPR must decrease as the ambient air temperature increases beyond an acceptable value. This is driven by the increase in gas temperature that is needed to maintain a given thrust level with increasing ambient air temperature. The second condition (pressure limit) imposes an EPR limit that is dependent on the ambient air pressure. An increase in pressure altitude is associated with a reduction in ambient pressure, which increases the pressure differential for the same internal pressure. The third condition (N_1 limit) is rarely more restrictive than the other two conditions, but it can be a factor limiting takeoff thrust at high altitude.

8.5.5 Standard Thrust Ratings

To avoid exceeding one or more engine design limitations, the thrust output is restricted, or governed, at key flight conditions, based on such factors as the ambient air temperature, pressure altitude, and flight speed. The engine control system governs the engine to ensure that the structural integrity of the engine is not compromised when producing a defined thrust. These thrust levels are known as *thrust ratings*. Demonstration of an engine's rated performance under prescribed conditions is a critical part of the engine certification process.[12]

---

11 CFM International (CFMI) is a 50:50 joint-owned company of Snecma (France) and GE Aviation (USA).
12 Certification requirements for establishing engine ratings are described in FAR 33.7 [1] and CS-E 40 [2]. The definitive record of engine certification is known as a Type Certificate Data Sheet (TCDS).

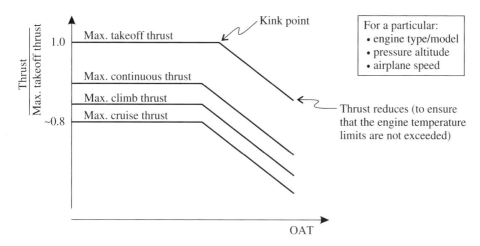

Figure 8.10 Thrust versus outside air temperature (OAT) for selected thrust ratings.

There are five standard thrust ratings commonly used in the operation of turbofan engines installed in jet transport airplanes. Data for three of these ratings are published in the Airplane Flight Manual (AFM)—these are the maximum takeoff thrust, maximum go-around thrust, and maximum continuous thrust. These are *certified ratings*. As they are published in the AFM, they have a legal standing, and compliance with the conditions of these ratings by operators is mandatory. The other two thrust ratings are the maximum climb thrust and the maximum cruise thrust (the distinction is that no airplane performance data presented in the AFM are based on either of these ratings).

Figure 8.10 illustrates the influence of ambient temperature on the maximum thrust levels for selected ratings. Note that there is a constant level of thrust—that is, the thrust output is *flat rated*—up to a critical ambient temperature, which is known as the *kink point* (or break point) temperature, after which the thrust reduces with increasing OAT (this is discussed further in Section 8.5.6).

Rated thrust values for any particular engine type (and model) are the minimum thrust values that are guaranteed by the engine manufacturer for operation under the specified conditions. All engines of that type (and model) should be capable of delivering at least the rated thrust levels during their service life. Due to the many tiny variations in the as-manufactured engine components and due to the calibration tolerances associated with the installed instrumentation used to measure the thrust setting parameter, not all engines of a particular type and model will deliver precisely the same thrust under the same operating conditions. A conservatism is thus implemented in the design of the engine ratings, which results in an *average* engine typically producing about 1–2% more than the documented rated thrust values.

Maximum Takeoff (T/O) Thrust Rating

This thrust rating produces the greatest thrust level for the engine. It is time limited to a maximum of 5 minutes. This limit can be extended to 10 minutes for one-engine-inoperative takeoff conditions (see Section 9.2.4) where there is an obstacle in the takeoff flight path, provided that the operator has purchased a special AFM appendix [14]. The 10 minute EPR (or $\%N_1$) limit values are published in the AFM corresponding to pressure altitude and ambient temperature for various bleed air options. Published rated thrust values apply to low speeds (e.g., less than 80 kt).

Go-Around (GA) Thrust Rating

This thrust rating corresponds to the maximum thrust available during a go-around (missed approach); the time limit is 5 minutes. This is sometimes called the *in-flight takeoff thrust*. The thrust is normally the same as the takeoff thrust (in some cases, it is slightly lower). The engine settings, however, are adjusted for air compression in the intake, and the corresponding EPR (or N_1) is different from that of the takeoff thrust. The presentation in the AFM is the same as for the maximum takeoff thrust. The combined takeoff/go-around thrust is abbreviated as TOGA or, alternatively, as TO/GA.

Maximum Continuous Thrust (MCT) Rating

This is the greatest thrust rating that may be used continuously by the pilot. It is exclusively reserved for emergencies (e.g., one-engine-inoperative situations). It may be used for an unrestricted time; however, the higher operating temperatures will reduce the life of the engine. The presentation of the data in the AFM is different from that given for takeoff and go around as the EPR (or $\%N_1$) limits also depend on the airplane's speed. For airplane certification, the MCT rating is used to demonstrate compliance with the final segment climb gradient requirement[13] (see Section 10.6.1). In flight operations, it can be used to clear distant obstacles following an engine failure once the time limit for the T/O thrust has expired.

Maximum Climb Thrust (MCLT) Rating

This is the thrust rating that is used for the normal *en route* climb to cruising altitude. There is no time limit for this rating. It would also apply to a step climb in cruise (i.e., when the airplane climbs to a higher cruise flight level). For certain engine types, the manufacturer's operating instructions makes no distinction between the maximum continuous thrust and the maximum climb thrust ratings. EPR (or $\%N_1$) limits are recorded in such publications as the Flight Planning and Performance Manual (see Section 23.7.5)—but not in the AFM—for varying pressure altitude, ambient temperature, and airplane speed, with corrections for bleed air extraction (for air conditioning and anti-icing).

Maximum Cruise Thrust (MCRT) Rating

This thrust rating produces the maximum thrust that is available for continuous, normal (i.e., non-emergency) cruise conditions and is typically about 95% of the MCLT rating. It is essentially defined by engine life considerations. EPR (or $\%N_1$) limits are recorded in a similar manner to that of the maximum climb thrust rating. As with the MCLT rating, there is no time limit. In most day-to-day situations, however, the thrust setting used by the pilot in cruise will be lower than MCRT (as there is usually no good reason to fly any faster).

8.5.6 Design of Engine Rating Structure

Engine manufacturers program their engines to deliver the rated thrust levels through settings in the EEC/ECU, taking into account the ambient temperature, pressure altitude, and flight Mach number. The design of the rating structure of an engine (which is shown in simplified form in Figure 8.10) is undertaken to meet several key thrust values on the operating envelope of the engine, while preserving the structural integrity of the engine and conserving engine life. (These key thrust values are usually part of a customer contractual agreement.)

---

13 Minimum climb gradients that must be demonstrated with the critical engine inoperative are stipulated in FAR/CS 25.121 [15, 16].

Various methods are used to define the rating structure for an engine. The most basic feature considers the thrust as constant (i.e., flat rated) with increasing OAT until the kink point is reached. This means that the engine, at a given pressure altitude, will deliver the same thrust over a range of ambient temperature conditions up to the *kink point temperature* (T_{KP}), which is normally defined as a fixed increment above ISA. The most commonly used T_{KP} value for maximum takeoff thrust ratings is ISA + 15 °C, which is adequate for a large proportion of commercial flight operations worldwide (e.g., it covers over 95% of all commercial flights in the Northern Hemisphere).

The design of the basic rating structure for takeoff thrust is linked to a fundamental feature of jet engines: the corrected thrust parameter F_N/δ increases, approximately linearly, with increasing total temperature ratio $T_{t,41}/T_{t,20}$. The impact of changing OAT is now considered with the assumption that Mach number is constant. If the OAT were to increase, this would result in an increase in $T_{t,20}$. Over the flat rated region, it is thus apparent that $T_{t,41}$ must increase as the OAT increases to maintain a constant value of F_N/δ. The limiting condition is reached when the OAT equals T_{KP}. The most commonly used (i.e., traditional) takeoff rating structure for ambient temperatures greater than T_{KP} is based on a constant TET limit. This results in a decrease in thrust with increasing OAT. The available takeoff thrust at these high temperatures would thus be lower than that which is available at ambient temperatures less than T_{KP}. This method of defining the takeoff rating structure can be extended to the full range of takeoff altitudes from which the airplane is expected to operate, by maintaining the same constant TET limit. With increasing altitude, T_{KP} (defined as a fixed increment above ISA) would reduce, as illustrated in Figure 8.11a.

The traditional approach to the design of a rating structure, which is described above, considers the TET as the key parameter that limits, or governs, the engine's performance capability. This approach, however, does not necessarily lead to an optimized solution for all applications. Alternative approaches—which, in addition to TET, consider shaft rotational speeds and other engine temperatures (e.g., $T_{t,30}$)—are used by engine manufacturers to develop rating structures that allow the engine to perform optimally across the required flight envelope. In such cases, the constant TET line is replaced by two or more lines of varying slope, as illustrated in Figure 8.11b.

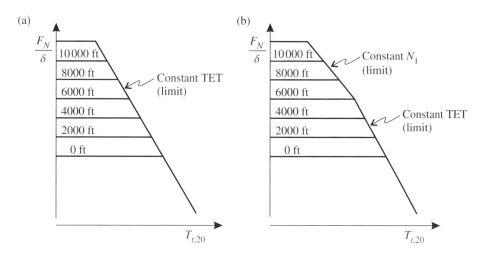

Figure 8.11 Maximum takeoff thrust rating structures: (a) traditional approach with constant TET; (b) optimized approach with other considerations (e.g., constant N_1).

The final consideration, when defining the basic takeoff rating structure, is the influence of flight Mach number. Again, there are various ways in which this can be done. The traditional takeoff rating structure holds TET constant with increasing takeoff Mach number—this sets the manner in which the thrust will reduce with forward speed during takeoff.

8.5.7 Engine Setting Parameters for Rated Thrust

The two widely adopted methods that are used to control engine thrust are based on the use of EPR or %N_1 as the thrust setting parameter (as mentioned earlier in Section 8.5.3). Over the flat rated region, the EPR is constant, delivering a constant thrust irrespective of the OAT up to the limiting temperature T_{KP}. The shaft speed (%N_1) will, however, not be constant as the engine speed must increase to maintain the same EPR with increasing OAT. At temperatures greater than T_{KP}, both EPR and %N_1 decrease with increasing OAT—and the thrust delivered will also decrease. This is illustrated in Figure 8.12 for the maximum takeoff thrust rating.

8.5.8 Takeoff Thrust Bump

In specific cases, engine manufacturers provide a rated takeoff thrust that has an increased performance capability over the standard maximum takeoff thrust rating—this is called a *thrust bump*. The enhanced performance can take several forms—for example, there can be an increase to the flat rated thrust level or the TET-limited thrust or both. The increased thrust is usually applicable to takeoff operations within a defined pressure altitude range. This would be of benefit to an operator that utilizes a specific high altitude airport (e.g., Denver, Johannesburg, Mexico City). Details of bump ratings are described in an AFM appendix and would be applicable to specified engines only.

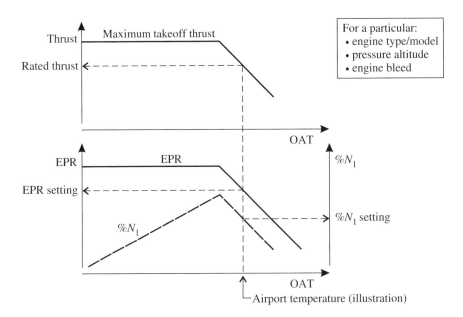

Figure 8.12 Setting takeoff thrust using EPR or %N_1 for various OAT.

8.5.9 Reduced and Derated Takeoff Thrust

To increase the life of key engine components, operators are provided with a means to take off safely with a lower thrust than the rated maximum takeoff thrust. This procedure reduces the stresses on the engine components, which, consequently, reduces maintenance costs and increases the life of the engine. It also improves engine reliability, which can result in fewer in-flight shutdowns. Takeoff procedures using an approved lower thrust[14] are described in Section 10.8. There are two approved techniques.

The first technique is the *assumed temperature method* (ATM). This involves the pilot selecting a hypothetical OAT (i.e., an assumed OAT) that is higher than the actual OAT. The assumed OAT must be greater than T_{KP}. The assumed OAT is then entered, via a keypad, into the airplane's digital control system by the flight crew as part of the takeoff preparation. This results in a lower thrust during takeoff than that which would be produced if the actual OAT were selected.

The second technique employs a *derated takeoff thrust* rating, which is often called a *derate*. The selection of a derated thrust, from a performance perspective, essentially equates—for that takeoff—to having less powerful engines installed on the airplane. Performance charts for each approved derated thrust setting are published in the AFM, separate to the maximum takeoff rating (but using the same format). Derates are certified ratings; when selected by the flight crew, the lower (i.e., derated) thrust becomes the new takeoff limit. Derates can be defined in different ways—for example, they can be defined as a fixed percentage reduction below the maximum takeoff thrust of the engine (e.g., 10% or 20% lower). Alternatively, a derate can be defined as a nominal lower thrust value—for example, the CFM56-7B26 engine, installed on the B737-800, has a maximum rated takeoff thrust of 26 000 lb and two derated thrust options: 24K (for 24 000 lb) and 22K (for 22 000 lb) [19].

8.5.10 Idle Thrust

Several considerations are taken into account by the engine manufacturer when determining the engine settings for the *idle thrust* condition. This complex scenario typically results in different engine settings—and, consequently, different thrust levels—for ground and flight operations. These settings ensure that the engine remains in a stable operating condition when it produces an acceptably low thrust (and low fuel burn), while simultaneously satisfying the aircraft systems' demand for bleed air and power off-take.

During descent (from cruise), idle thrust is normally selected. Very little thrust is generated when the thrust levers are moved to the idle position. At high Mach numbers, the aerodynamic drag due to the spillage of air around the engine nacelle can be of a similar magnitude to the flight idle thrust, resulting in zero or even negative net thrust. These factors need to be taken into account when computing an airplane's rate of descent (see Section 12.5.2). Idle fuel flow, while affected by altitude, tends not to be influenced much by Mach number variation.

Another important consideration concerns the approach (to landing). The regulatory requirement (given in FAR/CS 25.119 [15, 16]) for the engine to spool up within 8 seconds to the go-around thrust setting, in the event of an aborted landing, can result in a dedicated *approach idle*

---

14 Note on terminology: EASA and the FAA use the term *reduced takeoff thrust* to mean the *assumed temperature method* [16, 17]. Airbus and several other organizations refer to a reduced thrust takeoff as a *flexible* takeoff; the assumed temperature is known as the *flexible temperature* (flextemp) and the corresponding thrust the *flexible* (FLEX or FLX) thrust [12]. Generally speaking, Boeing have used the term *reduced takeoff* thrust to mean either *assumed temperature* thrust or *derated* thrust [18].

setting for certain aircraft types. The *ground idle* engine settings are required to ensure that the fuel burn is a minimum, that the taxi thrust is not too high (resulting in excessive brake wear), and that bleed air and power off-take are adequate for the aircraft systems' needs when on the ground.

The idle thrust and associated fuel flow data are not recorded in the AFM, but, rather, in such documents as the Performance Engineer's Manual (see Section 23.7.6). This information is required for several performance calculations: for example, the ground idle thrust—which is about 3% of the maximum takeoff thrust—is taken into account when computing required braking distances for landing (see Section 11.4) or following a rejected takeoff (see Section 9.7).

8.6 Thrust Variation

8.6.1 Functional Relationship and Performance Trends

The net thrust of a turbofan engine, when expressed in corrected form (i.e., F_N/δ), can be described by the following functional relationship in terms of the dominant parameters [20]:

$$\frac{F_N}{\delta} = f\left(\frac{N}{\sqrt{\theta}}, M\right) \tag{8.38}$$

where N is the rotational speed of the engine, which, by convention, is taken as N_1 in the case of multiple shaft engines; and
θ is the temperature ratio of the ambient air.

Figure 8.13 provides a graphical illustration of this relationship for selected Mach numbers, where $\theta_{t,20}$ is the total temperature ratio for the engine at station 20.

The effect of flight Mach number and altitude on maximum climb thrust is illustrated in Figure 8.14. The thrust of a turbofan engine tends to reduce with increasing flight speed; however, at cruise altitudes this effect can be negligible. These relationships depend significantly on the manner in which the engine rating has been defined (see Section 8.5.6).

Due to the inherent complexity of turbofan engines, general expressions for the function f (in Equation 8.38) do not exist. It is possible, however, to approximate the engine thrust using simple algebraic expressions, which would apply over limited ranges of Mach number, engine

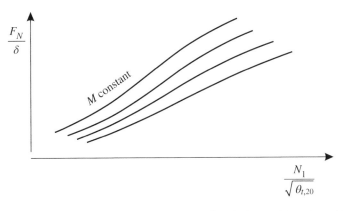

Figure 8.13 Corrected thrust versus corrected LP shaft speed for selected Mach numbers.

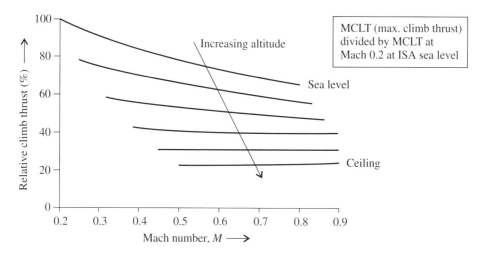

Figure 8.14 Typical maximum climb thrust variation with Mach number and altitude.

speed, or ambient conditions. A number of empirical models have been proposed (by different authors) to describe the variation of thrust with operating conditions. Several idealizations and models are presented in this section.

8.6.2 Effect of Ambient Pressure

The thrust produced by a jet engine is fundamentally influenced by changes to the inlet pressure at the fan face (p_{t_2}). Factors that change this pressure can change the thrust produced, depending on how the engine is controlled. A decrease in ambient air pressure—associated with an increase in pressure altitude, for example—will reduce the thrust. An increase in ram pressure, associated with an increase in airplane speed, will increase the thrust. Changes in inlet pressure result in changes to air density, and this alters the mass flow. For a given engine rotational speed, the volume capacity of the compressor is fixed—thus if the air density decreases, the mass flow will decrease, and this, in turn, will directly reduce the engine thrust (see Section 8.3).

Takeoff thrust, when expressed as F_N/δ (as illustrated in Figure 8.11), is seen to increase with increasing pressure altitude over the flat rated region of the chart. The actual thrust, F_N, however, will reduce with increasing pressure altitude.

A power law function of air density is often used to estimate thrust changes for a constant Mach number for limited changes in altitude. The general form of this relationship is given below:

$$\frac{F_N}{F_N^*} = \left(\frac{\sigma}{\sigma^*} \right)^n \tag{8.39}$$

where F_N^* is the thrust at the reference altitude (typical units: N, lb);
   σ is the relative density of the air (dimensionless);
   σ^* is the relative density of the air at the reference altitude (dimensionless); and
   n is a positive empirical constant, selected to best match the engine's performance.

The index n is typically in the range 0.60–0.85. Equation 8.39 has been shown to provide a reasonable approximation of takeoff thrust variation from ISA datum conditions (in which case $\sigma^* = 1$).

Equation 8.39 can also be used to account for changes in pressure altitude in cruise (for a given Mach number and engine speed). Note that in the stratosphere, where temperature is constant, the ratio of relative air densities that appears in Equation 8.39 can be replaced by a ratio of relative air pressures, that is,

$$\frac{F_N}{F_N{}^*} = \left(\frac{\delta}{\delta^*}\right)^n \tag{8.40}$$

where the superscript asterisk denotes the reference condition.

8.6.3   Effect of Ambient Temperature

Turbofan engines used in jet transport airplanes are flat rated (see Section 8.5.6). This means that ambient air temperature has no influence on the engine thrust for a range of temperatures up to the kink point temperature (T_{KP}). At temperatures greater than T_{KP}, the available thrust will decrease with increasing OAT.

Raymer [21] suggests that for every 1 °C increase, the thrust will reduce by 0.75%. Data evaluated by the author indicated that the thrust reduction can be between 0.75% and 1% for every 1 °C increase in OAT above T_{KP} (for a constant TET engine rating).

8.6.4   Effect of Airplane Speed on Takeoff Thrust

The influence of airplane speed on thrust is complex and is influenced by the primary design characteristics of the engine, such as the bypass ratio. The exact manner in which the thrust reduces with increasing speed during takeoff on a modern turbofan engine, however, depends fundamentally on the method used by the engine manufacturer to control the takeoff thrust rating. For example, it is common that TET is held constant (as mentioned in Section 8.5.6), which will set the thrust–speed relationship. More complex methods, however, are also used, which rely on other parameters being controlled by the EEC/ECU.

For the purpose of calculating an airplane's acceleration during takeoff (see Section 9.4), it is usually possible to adequately represent the engine thrust by linear or quadratic functions of Mach number or true airspeed—several such approaches are discussed below.

Linear Function
The simplest way to model the takeoff thrust is to assume that it reduces linearly with Mach number (or, alternatively, with true airspeed). This method—which starts to lose accuracy at about 150 kt—can provide an acceptable approximation for preliminary takeoff calculations.

$$\frac{F_N}{F_{N_0}} = 1 - kM \tag{8.41}$$

where F_{N_0} is the static thrust (typical units: N, lb); and
k (dimensionless) is a positive empirical constant, selected to best match the engine's performance.

The value of k varies from approximately 0.8 to 1.1 depending on the bypass ratio; a greater reduction in thrust with speed is associated with higher BPR engines (see Section 8.3.3).

Quadratic Function

A better representation of the reduction in takeoff thrust with forward speed than that given by 8.41 expresses the relative thrust (i.e., F_N/F_{N_0}) as a quadratic function of Mach number (or true airspeed):

$$\frac{F_N}{F_{N_0}} = 1 - k_1 M + k_2 M^2 \tag{8.42}$$

where k_1 (dimensionless) and k_2 (dimensionless) are positive empirical constants, selected to best match the engine's performance.

Torenbeek [22] investigated the use of this expression, indicating that it is accurate to approximately Mach 0.3.

Torenbeek [22]

Torenbeek [22] related the k_1 and k_2 factors (in Equation 8.42) to engine component efficiencies, the bypass ratio (λ), and to a dimensionless gas generator function, defined as G. Using typical (historical) efficiency values, he expressed the constants as follows:

$$k_1 = \frac{0.454\,(1 + \lambda)}{\sqrt{(1 + 0.75\lambda)\,G}} \tag{8.43}$$

$$k_2 = 0.6 + \frac{0.11\lambda}{G} \qquad \text{(see footnote[15])} \tag{8.44}$$

where $G \cong 0.9$ for low BPR engines and $G \cong 1.1$ for high BPR engines.

Bartel and Young [23]

Bartel and Young [23] followed a similar approach to that of Torenbeek [22]. In their study, TET was considered to be constant and invariant with Mach number. Using data for several two-shaft engines, they concluded that the following empirical equations provided a better representation of available engine data than the original versions:

$$k_1 = \frac{0.377\,(1 + \lambda)}{\sqrt{(1 + 0.82\lambda)\,G}} \tag{8.45}$$

$$k_2 = 0.23 + 0.19\sqrt{\lambda} \tag{8.46}$$

Comparing the thrust values for the reference engines with the thrust values given by Equations 8.42, 8.45, and 8.46, they demonstrated an accuracy of $\pm 1\%$ for airplane speeds up to Mach 0.4. In their study, G was evaluated based on known engine parameters. A correlation between G and the bypass ratio was also produced [23]—the resulting empirical equation, which is expected to be roughly valid for two-shaft engines, is

$$G = 0.06\lambda + 0.64 \tag{8.47}$$

Example of Takeoff Thrust Variation with Speed and Altitude

To illustrate the takeoff thrust variation, Figure 8.15 was constructed using Equation 8.42 (expressed in true airspeed) with the empirical constants $k_1 = 1.43 \times 10^{-3}$ kt^{-1} and

---

[15] There is some uncertainty regarding the value of 0.11 in Equation 8.44, as the same source [22] also gives a value of 0.13. This inconsistency was seen to have little impact on the results obtained, however.

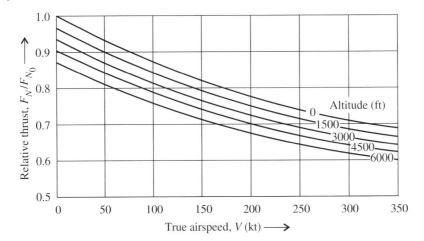

Figure 8.15 Relative takeoff thrust (F_N/F_{N_0}) variation with true airspeed (TAS) and with pressure altitude (ISA conditions assumed).

$k_2 = 1.54 \times 10^{-6}$ kt^{-2}. The effect of altitude was taken into account using Equation 8.39, with an exponent n of 0.77.

8.6.5 Effect of Altitude and Speed on Climb Thrust

There are several factors that need to be taken into account when assessing the manner in which the thrust of an airplane's engines reduce during the climb to the cruise altitude. Due to the reducing air density, the engines' thrust (at a fixed rotational speed) will decrease monotonically with increasing pressure altitude. The rate at which the thrust reduces (i.e., the *thrust lapse rate*) tends to increase slightly when the airplane enters the stratosphere. In the troposphere, the ambient temperature reduces with increasing altitude; this has a beneficial effect in terms of the engines' thrust output as it tends to offset—to some extent—the effect of reducing air pressure. This benefit does not occur when the airplane is climbing in the stratosphere, as the air temperature is constant. Figure 8.16a illustrates the thrust reduction with increasing altitude (for an engine with a BPR of ~4.3).

The second factor that needs to be considered is the airplane's speed, which directly influences the thrust produced by the engine (this is a complex relationship, which is influenced by several engine design parameters, as mentioned in Section 8.3.3). Jet transport airplanes, in normal flight operations, adhere to a climb–speed schedule, which would be indicated in the relevant Standard Operating Procedures (SOP) for the airplane (see Section 17.2.1). After takeoff, the airplane will accelerate to its initial climb speed, which is often restricted to a calibrated airspeed (CAS) of 250 kt until a defined altitude is reached (as defined by air traffic control restrictions). Following a short acceleration, the airplane will again climb at a steady CAS (note that climbing at constant CAS results in a steady increase in Mach number). Above a certain altitude (typically about 30 000 ft), the airplane will usually be flown at a constant Mach number (which is close to the cruise Mach number) until it reaches the top of the climb. Figure 8.16b shows how Mach number will change for two selected climb–speed schedules (i.e., 250 kt/290 kt/M 0.78 and 250 kt/310 kt/M 0.8). The engine thrust at a particular altitude could be obtained from Figure 8.16a for the associated Mach number.

Another important factor to bear in mind when modeling the thrust lapse of a particular engine is the engine settings implemented by the manufacturer. Modern engines can be

(a)

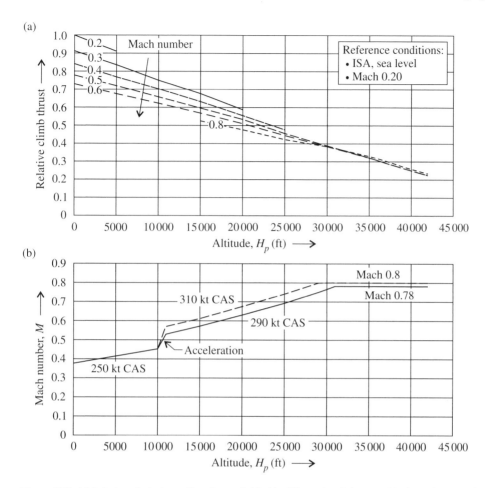

(b)

Figure 8.16 (a) Relative climb thrust (i.e., thrust divided by ISA sea-level thrust at Mach 0.20) versus altitude; (b) Mach number versus altitude for two typical climb–speed schedules (in ISA conditions).

customized to suit the requirements of the customer (i.e., the airplane manufacturer). Limits are set on the TET at each altitude (generally speaking, TET will increase with increasing altitude). A TET limit versus altitude profile can be selected to achieve a customer-specified rate of climb at a particular airplane speed at the top of climb. This means that variant models of the same basic engine—fitted to different airplane types, for example—can have different thrust lapse rates, depending on the selected engine settings.

Simple algebraic functions are not capable of accurately representing an engine's thrust for this complex situation of changing atmospheric conditions and changing flight Mach number. The traditional approach—used for initial performance calculations—either treats the Mach number as constant or considers it as a function of altitude and includes the Mach effects in the altitude terms. Four empirical models, which have been devised in an attempt to approximate the thrust lapse with increasing altitude, are discussed below.

Power Law Function

This model is the same as that given by Equation 8.39, but with the reference conditions corresponding to the ISA sea-level datum. It assumes that the engine's thrust reduces as a power law function of the air density for any constant Mach number and engine speed.

$$\frac{F_N}{F_{N,0}} = \sigma^n \tag{8.48}$$

where $F_{N,0}$ is a sea-level reference thrust; and
   n is a positive empirical constant selected to best match the engine's performance; it is typically in the range 0.6–0.85.

Equation 8.48 can provide an acceptable approximation of the engine thrust lapse over a finite altitude interval, for a fixed Mach number (and thrust setting). Note that even for the same engine, different values of n are required to model the engine data for different Mach numbers.

The reference thrust ($F_{N,0}$) is the climb thrust at sea level at standard ISA conditions (not the maximum takeoff thrust). To fit the data best at high Mach numbers (say above ~Mach 0.6) for the upper part of the climb, a hypothetical sea-level reference thrust may be used. This approach, in essence, treats both n and $F_{N,0}$ as empirical constants, selected to fit the data.

Mair and Birdsall [24] demonstrate that Equation 8.48 provides a good correlation to data provided by Rolls-Royce for the RB-211-535E4 engine for Mach 0.7, and they suggest an exponent n of 0.6. They report that a "further investigation has shown that this result is also true for other civil turbofans." Other authors [25, 26] suggest an exponent of 0.7. A curve fitting exercise conducted by the author using thrust data for an engine similar to that studied by Mair and Birdsall [24] indicated that a range of n values from 0.6 to 0.8 were required to model the maximum climb thrust at different Mach numbers.

Linear Function of Altitude

The relative thrust (i.e., $F_N/F_{N,0}$), for this model, is assumed to reduce linearly with increasing altitude:

$$\frac{F_N}{F_{N,0}} = 1 - C_1 H \tag{8.49}$$

where C_1 is a positive empirical constant, selected to best match the engine's performance (typical unit: ft^{-1}); and
   H is altitude (typical unit: ft).

Raymer [21] comments on this model: he reports that "as a rough but reasonable" approximation for standard atmospheric conditions below 40 000 ft, the thrust relationship may be represented as a straight line from 100% thrust at sea level to zero thrust at 55 000 ft. An investigation by the author showed that extrapolated trend lines on a chart of relative climb thrust versus altitude for a particular engine (BPR of ~4.3) intersects the altitude axis (corresponding to zero thrust) at various altitudes between 55 000 ft and 65 000 ft, depending on the Mach number (see Figure 8.16a). Evaluating the constant C_1 (defined by Equation 8.49) for this observation yields a range of values for C_1 between 1.8×10^{-5} ft^{-1} and 1.5×10^{-5} ft^{-1}.

Quadratic Function of Altitude

Another approach that has been used represents thrust as a quadratic function of altitude:

$$\frac{F_N}{F_{N,0}} = 1 - C_2 H + C_3 H^2 \tag{8.50}$$

where C_2 (typical unit: ft^{-1}) and C_3 (typical unit: ft^{-2}) are constants selected to best match the engine's performance.

This engine model was used by EUROCONTROL [27], for example, to establish the BADA[16] database for airplane performance modeling. The empirical constants C_2 and C_3 are fitted to engine data for a defined (or assumed) climb–speed schedule; in other words, Mach number is not necessarily considered constant, but can change as a function of altitude, H (see Section 17.2.1).

Howe [28]

Howe [28] proposed the following version of the power law model for subsonic flight for engines with different bypass ratios, which accounts for air density and Mach number:

$$\frac{F_N}{F_{N,0}} = \{K_1 + K_2\lambda + (K_3 + K_4\lambda)M\}\sigma^n \tag{8.51}$$

where $n = 0.7$ (troposphere) or $n = 1.0$ (stratosphere); and
K_1, K_2, K_3, and K_4 are empirical constants (see Table 8.3).

Table 8.3 Thrust parameters for Equation 8.51

Bypass ratio	Mach no.	K_1	K_2	K_3	K_4
3–6	0–0.4	1.0	0	−0.60	−0.04
3–6	0.4–0.9	0.88	−0.016	−0.30	0
8	0–0.4	1.0	0	−0.595	−0.03
8	0.4–0.9	0.89	−0.014	−0.30	0.005

Source: Data from [28].

8.7 Fuel Flow and TSFC Variation

8.7.1 Functional Relationships and General Performance Trends

The following functional relationship describes the corrected fuel flow (i.e., $Q/\delta\sqrt{\theta}$) in terms of the dominant parameters [20]:

$$\frac{Q}{\delta\sqrt{\theta}} = f\left(\frac{N}{\sqrt{\theta}}, M\right) \tag{8.52}$$

Using Equations 8.30 and 8.38, the TSFC (c) can be expressed as

$$c = \frac{Q}{F_N} = \sqrt{\theta} f\left(\frac{N}{\sqrt{\theta}}, M\right) \tag{8.53}$$

Equation 8.53 indicates that the TSFC depends fundamentally on ambient temperature (and hence altitude) and Mach number, for a given engine speed. For an airplane cruising in the stratosphere, the TSFC will depend only on Mach number (for a given engine speed). As with

---

16 BADA (Base of Aircraft Data) is an aircraft performance model and corresponding database, developed by EUROCONTROL (Brétigny-sur-Orge, France). The main application of BADA is airplane trajectory simulation and prediction within the domain of air traffic control (ATC).

the engine thrust, it is usually not possible to obtain general expressions for the functions identified in Equations 8.52 and 8.53.

The effect of flight Mach number and altitude on the fuel flow and TSFC are complex relationships. In the absence of any atmospheric changes (i.e., at constant height), the fuel flow will increase with increasing Mach number. The TSFC—which directly depends on F_N and Q—will thus increase with increasing airspeed, in this situation. Thrust and fuel flow both decrease with increasing pressure altitude at a given Mach number. The TSFC will vary depending on the relative rate of change of these two parameters—and this is a function of the engine type. For the purpose of approximate airplane performance calculations, simple algebraic expressions are often used that allow Q or c to be approximated over limited ranges of Mach number, ambient conditions, and engine speed. Several empirical models are described in this section.

8.7.2   Fuel Flow Data and Models

The rate that fuel is consumed is a critically important parameter for several airplane performance analyses, such as the determination of an airplane's range (see Chapter 13) or its endurance (see Chapter 14). The fuel flow (Q) corresponding to the net thrust (F_N) must thus be determined for the flight conditions of interest. There is one particular flight condition that requires special mention: a steady-state or quasi-steady-state condition where the airplane is neither accelerating nor decelerating. This is a typical flight condition during cruise, for example. In this situation, the net thrust (of all engines) will be equal to the airplane's drag. The *thrust required* (or required thrust) can thus be determined from knowledge of the airplane's drag characteristics and operating conditions (this is discussed in Section 12.3.2).

The best approach to determine the fuel flow for a particular thrust value depends on the available data and the precise application. If data of the form illustrated in Figure 8.8 are available, then Q can be determined from Q_{cor} (see Section 8.4.4). Alternatively, if TSFC data of the form illustrated in Figure 8.9 are available, then Q can be obtained by multiplying TSFC by F_N. Generalized data can also be used; this requires the evaluation of a generalized engine parameter (i.e., EPR or $N_1/\sqrt{\theta_{t,20}}$) corresponding to the net thrust, from which Q_{cor} can be established.

In the absence of such substantive engine data—or when only partial data are available—algebraic models of Q can be used. One such model takes the form of a polynomial function of F_N/δ. ESDU 73019 [29] reports that for a range of turbojet and turbofan engines, manufacturers' fuel flow data, at a given Mach number, can be satisfactorily represented by the following relationship:

$$\frac{Q}{\delta\sqrt{\theta}} = k_1 + k_2\left(\frac{F_N}{\delta}\right) + k_3\left(\frac{F_N}{\delta}\right)^2 \tag{8.54}$$

where k_1, k_2, and k_3 are constants for a particular engine type and must be determined for each value of M.

8.7.3   TSFC Models and Idealizations

Constant (Mean) TSFC
The simplest TSFC model assumes that TSFC does not change over the flight segment of interest. For example, the variation of TSFC during cruise segments—when Mach number and altitude are held nearly constant—is small, and the use of a mean value can yield satisfactory results for many applications. The shape of the TSFC curve is relatively flat about the minimum point

(as shown in Figures 8.7 and 8.9); small deviations about the minimum point—as might occur with changes in drag during cruise—would result in very small changes in TSFC. The derivation of the widely used Bréguet range equation (see Section 13.3.3), for example, assumes a constant TSFC.

If there is an appreciable change to one or more of the governing parameters (i.e., Mach number, altitude, or thrust setting), however, a more complex fuel flow or TSFC model is required.

Power Law Function

A widely used model of TSFC is based on a power law function of Mach number, that is,

$$c = c_1 \sqrt{\theta} M^n \tag{8.55}$$

where c_1 (typical units: mg $N^{-1}s^{-1}$, lb $lb^{-1}h^{-1}$) and n (dimensionless) are suitably selected constants that best represent the particular engine.

According to ESDU 73019 [29], the model "may be applied over limited ranges of Mach number but is, strictly speaking, valid at only one value of $N/\sqrt{\theta}$." With suitably chosen values for the constants (viz., c_1 and n), this expression can provide an accurate approximation of measured TSFC values within the range of typical conditions associated with subsonic cruising flight.

Mair and Birdsall [24] report that for the Rolls-Royce RB-211-535E engine, this model provides a "reasonably good approximation to the variations of c due to changes of both Mach number and height." They concluded that the best results were achieved by setting the exponent n equal to 0.48 for a cruise altitude of 19 700 ft (6 000 m) and 0.45 for 29 500 ft (9 000 m). An independent study conducted by the author determined that n is best set equal to 0.40 for cruise at 35 000 ft, based on data for a similar engine.

If the TSFC is known for a given reference condition, then Equation 8.55 can be used to describe the TSFC over a range of speeds and/or altitudes:

$$\frac{c}{c^*} = \left(\frac{\theta}{\theta^*} \right)^{0.5} \left(\frac{M}{M^*} \right)^n \tag{8.56}$$

where n is the constant given in Equation 8.55 and the superscript asterisk designates the reference condition.

Howe [28]

A TSFC model that is appropriate for preliminary analyses associated with conceptual airplane design and which accounts for varying BPR is presented by Howe [28]:

$$c = c_2(1 - 0.15\lambda^{0.65}) \{1 + 0.28(1 + 0.063\lambda^2)M\}\sigma^{0.08} \tag{8.57}$$

where c_2 is a suitably selected constant that best represents the specific engine (typical units: mg $N^{-1}s^{-1}$, lb $lb^{-1}h^{-1}$).

"Very approximate" values for the uninstalled TSFC can be determined using the following values for c_2, according to Howe [28]:

$$c_2 = 24 \, \text{mg} \, N^{-1}s^{-1} \qquad \text{for an engine with a low BPR;}$$

or $\quad c_2 = 20 \, \text{mg} \, N^{-1}s^{-1} \qquad$ for a large turbofan with a high BPR.

Linear Function of Mach Number

An alternative model that attempts to take into account the variation of TSFC with Mach number at a fixed altitude is

$$c = c_3 + c_4 M \tag{8.58}$$

where c_3 (typical units: mg $N^{-1}s^{-1}$, lb $lb^{-1}h^{-1}$) and c_4 (typical units: mg $N^{-1}s^{-1}$, lb $lb^{-1}h^{-1}$) are suitably selected constants that best represent the specific engine.

It is reported [29] that Equation 8.58 usually provides a satisfactory approximation to manufacturers' data (at constant altitude and engine speed) over a considerably greater range of values of Mach number than is the case for Equation 8.55. Several authors (e.g., references [24] and [30]) prefer to write Equation 8.58 in an alternative, but equivalent, form:

$$c = c_5(1 + kM) \tag{8.59}$$

where c_5 (typical units: mg $N^{-1}s^{-1}$, lb $lb^{-1}h^{-1}$) and k (dimensionless) are suitably selected constants that best represent the specific engine.

8.7.4 TSFC Variation with Engine Rotational Speed

The best fuel efficiency, that is, lowest TSFC, is achieved for a given set of conditions (i.e., ambient pressure and temperature and Mach number) when the engine is operating at a specific rotational speed (N_1). This is evident from the TSFC "loops" illustrated in Figures 8.7 and 8.9; the lowest point on each curve corresponds to an optimum thrust condition, denoted herein as $F_{N,opt}$. The corresponding TSFC is denoted as c_{opt}. It is important, particularly for the cruise segment of a flight, that the airplane is operated near this optimum point to reduce its trip fuel consumption.

The *thrust required* in cruise (see Section 12.3.2) depends on the airplane's drag coefficient (C_D), which is a function of the lift coefficient (C_L)—and, as illustrated in Figure 7.5 for the cruise condition, the lift coefficient depends on the airplane's weight. Now, the airplane's weight will reduce as the flight progresses due to the consumption of fuel. The lift coefficient will thus reduce and so will the drag coefficient. If the airplane's speed and altitude are unchanged, the thrust required will steadily reduce during a typical cruise segment. Consequently, the TSFC will change. This is illustrated in Figure 8.17. The thrust and TSFC for an airliner in the cruise condition (i.e., at a specific Mach number and altitude) have been normalized with respect to the lowest TSFC (c_{opt}), and plotted against the ratio of thrust to the corresponding reference or optimum thrust ($F_{N,opt}$). The variation in TSFC about the optimum condition is seen to be small (less than 3%) for thrust changes of up to 20% above and below the optimum condition.

8.8 Installation Losses and Engine Deterioration

8.8.1 Installation Effects

The thrust produced by an engine is affected by the installation of the engine in the airplane. Thus, when the performance of a particular engine/airframe combination is being considered, it is necessary to take into account the associated installation factors—these include: (1) the intake

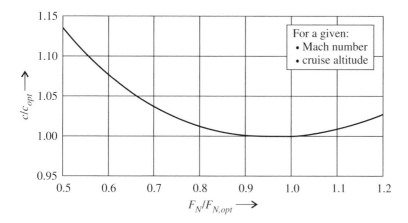

Figure 8.17 Typical cruise TSFC versus thrust relationship normalized with respect to the optimum (or lowest TSFC) condition.

pressure recovery; (2) the engine bleed air and shaft power off-take; (3) engine nacelle drag, including the nozzle; and (4) secondary losses due to local flow distortions and engine/airframe interactions. Thus

$$F_{N, inst} = k_{pr} k_{ba} k_{ot} F_{N, uninst} - \sum D_{inst} \tag{8.60}$$

where $F_{N,inst}$ is the installed net thrust (typical units: N, lb);
$\quad F_{N,uninst}$ is the uninstalled net thrust (typical units: N, lb);
$\quad k_{pr}$ is the intake pressure recovery thrust correction factor;
$\quad k_{ba}$ is the bleed air thrust correction factor;
$\quad k_{ot}$ is the shaft power off-take thrust correction factor; and
$\quad \sum D_{inst}$ is the sum of the installation drag terms (typical units: N, lb).

The uninstalled engine test data are determined by ground testing in a facility that incorporates an ideal, bell-mouth-shaped inlet, which has known pressure recovery characteristics (see Section 8.8.2). The baseline performance data—as determined by the engine manufacturer by testing and/or computer modeling—would thus need to be corrected for the installation losses associated with the less-than-ideal efficiency of actual engine intakes on the airplane.

The performance data would also need to be corrected for bleed air and shaft power taken from the engine (see Sections 8.8.3 and 8.8.4) to provide energy for both engine and airplane system needs (engine manufacturers refer to the latter category as the *customer* bleed and off-take requirements).

Finally, the uninstalled thrust is corrected for all drag contributions allocated to the propulsion system. This is a non-trivial task. In theory, a component of drag may be considered as a reduction in thrust. The thrust–drag accounting system adopted by an airplane manufacturer requires a strict set of definitions of what is thrust and what is drag. An overriding principle that gets used considers all drag components whose magnitudes depend primarily on the position of the thrust levers as negative thrust.

For the purpose of *airplane* performance analysis it is usual that the term *engine thrust* is taken to mean *net installed engine thrust*. This will be the understanding used in this book.

The installed TSFC will also differ from the uninstalled value. If the engine manufacturer uses the uninstalled thrust as the reference parameter to determine the TSFC, then a correction is needed for airplane performance analyses:

$$c_{inst} = \frac{F_{N,uninst}}{F_{N,inst}} c_{uninst} \tag{8.61}$$

where c_{inst} is the installed thrust specific fuel consumption (TSFC); and
c_{uninst} is the uninstalled thrust specific fuel consumption (TSFC).

8.8.2 Intake (Inlet) Total Pressure Loss

The *intake pressure recovery* is the average total pressure of the gas at the fan face divided by the total pressure in the free stream:

$$\text{Intake pressure recovery } = \frac{p_{t,20}}{p_{t,\infty}} \tag{8.62}$$

The pressure recovery is a measure of the efficiency of the intake; it depends on the design of the intake and is a function of the free-stream (flight) Mach number. The intake *total pressure loss*, which is frequently expressed as a percentage, is defined as

$$\text{Intake total pressure loss } = 1 - \frac{p_{t,20}}{p_{t,\infty}} \tag{8.63}$$

A well-designed nacelle will have a very small total pressure loss. Walsh and Fletcher [31] indicate that in cruise, a pod-mounted turbofan engine can have an intake total pressure loss as low as 0.5% (giving a pressure recovery of 0.995).

The thrust loss due to the intake total pressure loss can be estimated using the following equation [21]:

$$\frac{F_N}{F_{N,ref}} = 1 - c_{ram} \left[\left(\frac{p_{t,20}}{p_{t,\infty}} \right)_{ref} - \left(\frac{p_{t,20}}{p_{t,\infty}} \right)_{actual} \right] \tag{8.64}$$

where F_N is the engine thrust with actual pressure recovery (typical units: N, lb);
$F_{N,ref}$ is the reference engine thrust (typical units: N, lb); and
c_{ram} is the ram recovery factor (dimensionless).

Raymer [21] indicates that, typically, c_{ram} ranges from 1.2 to 1.5 and is approximately 1.35 for subsonic flight.

Walsh and Fletcher [31] provide typical *exchange rate* data that can be used to assess the impact of a small change to a leading engine parameter on the engine's performance (when operating at its design point). Such sensitivity data enable estimates to be obtained for installation effects. For example, they indicate that in cruise (at Mach 0.80 at 36 089 ft) an intake total pressure loss of 0.5% would result in an increase in TSFC of approximately 0.9% for an engine with a BPR of 4.5 and 2.4% for an engine with a BPR of 10.5 [31].

8.8.3 Bleed Air Losses

The extraction of hot, high pressure air from the engine for use by the aircraft's systems (e.g., air conditioning, cabin pressurization, and de-icing)—which is known as external or customer bleed air—negatively impacts the engine's performance, as this mass flow is no longer available to do work in the engine. The effect that a given mass flow of bleed air, which is usually extracted from the HP compressor, will have on the thrust and TSFC depends on the engine design (key parameters are the bypass ratio and pressure ratio) and on the operating conditions (i.e., Mach number, altitude, and OAT). It also depends on the extraction station: bleed air is more "expensive" the farther downstream it is extracted from the compressor.

Generally speaking, the impact of an increase in external bleed air on an engine's performance can be considered under two scenarios: (1) when TET is limited or unchanged, and (2) when TET can increase. If the bleed air demand increases and the TET remains unchanged, the thrust will reduce due to the reduced core mass flow. This is of particular importance for takeoff when the TET is limited. On the other hand, if the TET is not limited, for example in cruise, then the same thrust can be delivered by the engine, but at a higher TET. In both cases, the engine would be operating in a less efficient state and the TSFC would increase.

An evaluation by Bartel and Young [23] into the effect of bleed air extraction from the HP compressor of two-shaft engines on takeoff thrust indicated that Equation 8.42 can be modified, as follows, to take into account external bleed air:

$$\frac{F_N}{F_{N_0}} = A - k_1 M + k_2 M^2 \tag{8.65}$$

where A is a fraction, the magnitude of which depends on the percentage of the core airflow extracted, the extraction station, and on the engine design.

8.8.4 Shaft Power Losses

Direct shaft power off-take for the accessories gearbox results in a reduction in available engine thrust; this has a comparatively smaller impact on the engine's performance than that which results from bleed air extraction. The impact of shaft power off-take on the engine's thrust and TSFC depend on the engine design characteristics and on the operating conditions (i.e., Mach number, altitude, and OAT).

To illustrate the impact of an increase in power off-take, a cruise condition is considered where the TET is not limited. The required engine thrust can be maintained by operating at a higher TET. The TSFC, however, will be negatively impacted. This is shown in Figure 8.18, which illustrates how thrust (F_N) and TSFC (c) vary as functions of TET. Consider, for example, a cruise flight condition that requires a certain thrust level; the baseline power off-take

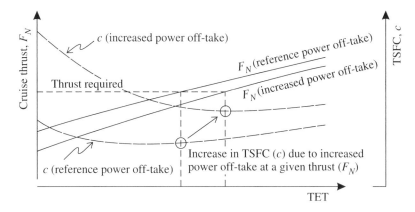

Figure 8.18 Cruise thrust (F_N) and TSFC (c) versus TET for two levels of shaft power off-take—"reference" and "increased." *Source*: Adapted from [32].

is indicated by the "reference power off-take" curve. If the shaft power off-take changed to the "increased power off-take" level, the TET and fuel flow would, theoretically, increase to maintain the same thrust. The TSFC for the new condition would be at the higher TET value, corresponding to the "increased power off-take" curve. The TSFC penalty (when expressed as a percentage), for a given increase in power off-take, is not constant—the penalty (i.e., per-cent change in TSFC) increases approximately linearly with decreasing engine thrust (as would occur during cruise).

In a study conducted by the author [32], the TSFC penalty corresponding to an increased power off-take was investigated using an engine-modeling computer program,[17] utilizing the Rolls-Royce RB211-535 and Trent 772 engines as references. The TSFC change, at a given cruise thrust level, was seen to increase almost linearly with increasing power off-take. This observa-tion can be represented by the following empirical relationship:

$$\frac{\Delta c}{c} = k_{ot} \left(\frac{\Delta P_{ot}}{F_N} \right) \tag{8.66}$$

where Δc is the TSFC change (typical units: mg $N^{-1}s^{-1}$, lb $lb^{-1}h^{-1}$);
 c is the reference TSFC (typical units: mg $N^{-1}s^{-1}$, lb $lb^{-1}h^{-1}$);
 k_{ot} is an empirical correlation factor (typical units: kN/kW, lb/hp);
 ΔP_{ot} is the change to the power off-take (typical units: kW, hp); and
 F_N is the reference engine thrust (typical units: kN, lb).

The value of k_{ot} was seen to vary between 0.0020 kN/kW and 0.0025 kN/kW for the two refer-ence engines, across a wide operating thrust range at typical cruise altitudes. Based on a review of published and unpublished data of TSFC penalties due to power off-take, Scholz, *et al.* [33] established an average value for k_{ot} of 0.0023 kN/kW.

8.8.5 Engine Deterioration

Jet engine components operate in an exceedingly demanding environment. During normal ser-vice, turbofan engines suffer slow, but noticeable, performance degradation—this is caused pri-marily by mechanical wear, the erosion of engine parts, and the accumulation of dirt. Washing the engine core is an effective, standard procedure to address the performance degradation associated with dirt in the engine. The deterioration can be seen by monitoring the EGT. As an engine's condition deteriorates, it will need to operate at a higher temperature to produce the same thrust—this is illustrated in Figure 8.19 for the maximum takeoff thrust. As a consequence of this deterioration, the TSFC will increase.

The *EGT margin* is the difference between the EGT limit (called the red line) and the EGT required to achieve the rated thrust (usually the maximum takeoff thrust). The EGT margin, which is typically tracked as an indicator of an engine's condition, reduces with engine usage. When the EGT limit is reached, the engine needs to be removed from the airplane and sent to a maintenance facility (i.e., a shop visit). Refurbishment of the engine core involves replacement or restoration of worn parts. A key parameter is the gap between the blade tips and the inner wall of the engine casing. Restoring these critical dimensions minimizes leaks around the blade tips, increasing the EGT margin. Typically, about 70% of a new engine EGT margin can be regained following an engine's first shop visit [34].

---

17 The investigation was conducted using Turbomatch, a gas turbine performance simulation computer program developed at Cranfield University, Bedfordshire, UK.

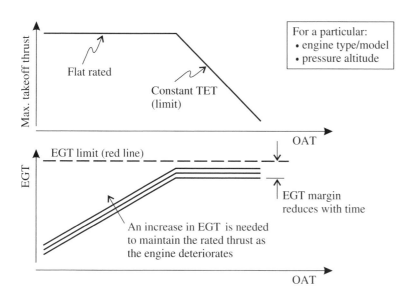

Figure 8.19 Impact of engine deterioration on EGT for maximum (max.) takeoff thrust.

The reduction of the EGT margin in service is primarily a function of *engine flight hours* (EFH) and *average flight cycle time* (short flight times imply a greater number of takeoffs). The rate of EGT margin reduction depends significantly on the operating conditions (e.g., hot, dusty conditions accelerate engine wear) and whether reduced/derated thrust is used on takeoff. A typical reduction in EGT margin per 1000 EFH is 3–5 °C for an engine installed on a single-aisle airplane [34]. Engines on widebody airplanes, used for longer haul operations, would normally suffer a slower rate of deterioration. The rate is, however, not constant and tends to be greater for a new engine or one that has recently been overhauled.

It should be noted that while being critically important as a health monitoring parameter, the EGT margin is by no means the only factor used to schedule engine removal for maintenance. Turbofan engines have many life-limited parts. In fact, modern engines—which tend to have greater EGT margins than earlier generation engines—operating in benign conditions will frequently be removed due to a life limit being reached on a component before the EGT limit is reached.

8.9 Further Reading

Textbooks that provide a more comprehensive description of turbine engines than that presented in this chapter include:

Cumpsty, N., *Jet propulsion: A simple guide to the aerodynamic and thermodynamic design and performance of jet engines*, 2nd ed., Cambridge University Press, Cambridge, UK, 2003.
El-Sayed, A.F., *Aircraft propulsion and gas turbine engines*, CRC Press, Boca Raton, FL, 2008.
Hünecke, K., *Jet engines: Fundamentals of theory, design and operations*, Airlife Publishing, Shrewsbury, UK, 1997.
Linke-Diesinger, A., *Systems of commercial turbofan engines: An introduction to systems functions*, Springer-Verlag, Berlin, Germany, 2008.

Mattingly, J.D., *Elements of gas turbine propulsion*, American Institute of Aeronautics and Astronautics, Reston, VA, 2005.

Mattingly, J.D., Heiser, W.H., and Pratt, D.T., *Aircraft engine design*, 2nd ed., American Institute of Aeronautics and Astronautics, Reston, VA, 2002.

Saravanamuttoo, H.I.H., Rogers, G.F.C., Cohen, H., and Straznicky, P.V., *Gas turbine theory*, 6th ed., Pearson Education, Harlow, UK, 2009.

References

1 FAA, *Airworthiness standards: Aircraft engines*, Federal Aviation Regulation Part 33, Amdt. 33-34, Federal Aviation Administration, Washington, DC, Nov. 4, 2014. Latest revision available from www.ecfr.gov/ under e-CFR (Electronic Code of Federal Regulations) Title 14.

2 EASA, *Certification specifications and acceptable means of compliance for engines*, CS-E, Amdt. 4, European Aviation Safety Agency, Cologne, Germany, Mar. 12, 2015. Latest revision available from www.easa.europa.eu/ under Certification Specification.

3 Rolls-Royce, *The jet engine*, 5th ed., Rolls-Royce, London, UK, 2005.

4 Sinnett, M., "787 No-bleed systems: Saving fuel and enhancing operational efficiencies," *Aero*, The Boeing Company, Seattle, WA, Vol. Q 04, pp. 6–11, 2007.

5 SAE, *Aircraft propulsion system performance station designation*, AS755 Rev. F, SAE International, Warrendale, PA, Dec. 2014.

6 SAE, *Gas turbine engine performance station identification and nomenclature*, ARP755A, Society of Automotive Engineers, Warrendale, PA, 1974.

7 Brayton, G.B., Improvement in gas engines, Patent number US125166 A, USA, Apr. 2, 1872.

8 Pratt & Whitney, "PurePower PW1000G-JM Engine," Document S16307-G.07.14, Pratt & Whitney, East Hartford, CT, 2014.

9 Boeing, "Jet transport performance methods," D6-1420, The Boeing Company, Seattle, WA, May 1989.

10 BIPM, "The international system of units (SI)," 8th ed., Bureau International des Poids et Mesures (BIPM), Paris, France, 2006.

11 FAA, "Compliance criteria for 14 CFR §33.28, aircraft engines, electrical and electronic engine control systems," Advisory Circular 33.28-1, Federal Aviation Administration, Washington DC, June 29, 2001.

12 Airbus, "Getting to grips with aircraft performance," Flight Operations Support and Line Assistance, Airbus S.A.S., Blagnac, France, Jan. 2002.

13 Airbus, "A318/A319/A320/A321 flight deck and systems briefing for pilots," STL 945.7136/97, Iss. 4, Airbus S.A.S., Blagnac, France, 2007.

14 Boeing, "Propulsion," Training course notes, Flight Operations Engineering, Boeing Commercial Airplanes, Seattle, WA, undated.

15 FAA, *Airworthiness standards: Transport category airplanes*, Federal Aviation Regulation Part 25, Amdt. 25-143, Federal Aviation Administration, Washington, DC, June 24, 2016. Latest revision available from www.ecfr.gov/ under e-CFR (Electronic Code of Federal Regulations) Title 14.

16 EASA, *Certification specifications and acceptable means of compliance for large aeroplanes*, CS-25, Amdt. 18, European Aviation Safety Agency, Cologne, Germany, June 23, 2016. Latest revision available from www.easa.europa.eu/ under Certification Specification.

17 FAA, "Reduced and derated takeoff thrust (power) procedures," Advisory Circular 25-13, Federal Aviation Administration, Washington, DC, May 4, 1988. Available from www.faa.gov/.

18 Blake, W. and Performance Training Group, "Jet transport performance methods," D6-1420, Flight Operations Engineering, Boeing Commercial Airplanes, Seattle, WA, Mar. 2009.

19 Boeing, "Flight planning and performance manual: 737-800 CFM56-7B26," Rev. 03, Flight Operations Engineering, Boeing Commercial Airplane Group, Seattle, WA, Nov. 1, 2005.

20 ESDU, "Non-dimensional approach to engine thrust and airframe drag for the analysis of measured performance data: Aircraft with turbo-jet and turbo-fan engines," Data item 70020, IHS ESDU, 133 Houndsditch, London, UK, Sept. 1970.

21 Raymer, D.P., *Aircraft design: A conceptual approach*, AIAA Education Series, 3rd ed., American Institute of Aeronautics and Astronautics, Reston, VA, 1999.

22 Torenbeek, E., *Synthesis of subsonic airplane design*, Delft University Press, Delft, the Netherlands, 1982.

23 Bartel, M. and Young, T.M., "Simplified thrust and fuel consumption models for modern two-shaft turbofan engines," *Journal of Aircraft*, Vol. 45, Iss. 4, pp. 1450–1456, 2008.

24 Mair, W.A. and Birdsall, D.L., *Aircraft performance*, Aerospace Series, Cambridge University Press, Cambridge, UK, 1992.

25 Eshelby, M.E., *Aircraft performance: Theory and practice*, Arnold, London, UK, 2000.

26 Ojha, S.K., *Flight performance of aircraft*, AIAA Education Series, American Institute of Aeronautics and Astronautics, Reston, VA, 1995.

27 EUROCONTROL, "User manual for the base of aircraft data (BADA)," Rev. 3.6, EUROCONTROL Experimental Center, Brétigny-sur-Orge, France, July 2004.

28 Howe, D., *Aircraft conceptual design synthesis*, Professional Engineering Publishing, London, UK, 2000.

29 ESDU, "Approximate methods for estimation of cruise range and endurance: Aeroplanes with turbo-jet and turbo-fan engines," Data item 73019, Amdt. C, IHS ESDU, 133 Houndsditch, London, UK, May 1982.

30 Anderson, J.D., *Aircraft performance and design*, McGraw-Hill, New York, 1999.

31 Walsh, P.P. and Fletcher, P., *Gas turbine performance*, 2nd ed., Blackwell Science, Oxford, UK, 2004.

32 Young, T.M., *Investigations into the operational effectiveness of hybrid laminar flow control aircraft* [PhD thesis], College of Aeronautics, Cranfield University, Bedfordshire, UK, 2002.

33 Scholz, D., Seresinhe, R., Staack, I., and Lawson, C., "Fuel consumption due to shaft power off-takes from the engine," *4th International Workshop on Aircraft System Technologies AST 2013*, Hamburg, Germany, Apr. 23–24, 2013.

34 Anon, "CFM56-3 maintenance analysis & budget," *Aircraft Commerce*, Iss. 45, pp. 18–28, Apr./May 2006.

9

Takeoff Performance

9.1 Introduction

The takeoff phase of a flight is considered to extend from the point of brake release at the start of the takeoff run to a point 1500 ft above the height of the runway or to the point at which the airplane has fully transitioned to the *en route* climb configuration and reached the V_{FTO} (final takeoff) speed (defined later in Section 10.2.2), whichever point is higher.

When a commercial airplane takes off, the flight crew is expected to follow a carefully constructed set of procedures. During the takeoff and landing phases of the flight it is even more important than is the case for other phases of the flight that the techniques employed by the crew are acceptably close to the nominal procedures established for the operation of the airplane due to the proximity of the airplane to the ground and to other traffic. These procedures have been meticulously considered to provide an acceptable margin of safety for the operation of the airplane. Underpinning these activities is a range of regulations/specifications[1] that apply to the certification and subsequent operation of commercial transport airplanes. The takeoff and landing of commercial flights are highly controlled flight activities, with many conditions imposed on these operations by the regulatory authorities. An understanding of the regulatory requirements is thus paramount for the determination of the takeoff and landing performance characteristics of any airplane.

During the certification testing of a new airplane type, takeoff distances are measured, from which key takeoff performance parameters are determined and used to establish the takeoff distances recorded by the manufacturer in the Airplane Flight Manual (AFM). These test situations, however, are not fully representative of typical in-service airline operations. The reactions of a test pilot during a simulated emergency—who would be anticipating a failure event, for example—will typically be much quicker than could reasonably be expected of an average airline pilot when a non-normal situation, such as engine failure, is encountered in an everyday line-operation takeoff. For this reason, the measured takeoff distances are increased to produce the standard takeoff data that are recorded in the AFM and used for routine flight operations. The test data are adjusted by introducing time delays and by multiplying the resulting distances by safety factors. Furthermore, there are specific requirements regarding the use of thrust reversers, the position of the airplane's center of gravity, and the condition of the

---

1 The US regulations are issued by the Federal Aviation Administration (FAA); equivalent European specifications are issued by the European Aviation Safety Agency (EASA). Details on aviation (regulatory) authorities are provided in Section 23.3; several key regulations/specifications applicable to jet transport aircraft are identified in Section 23.4.

Performance of the Jet Transport Airplane: Analysis Methods, Flight Operations, and Regulations, First Edition.
Trevor M. Young.
© 2018 John Wiley & Sons Ltd. Published 2018 by John Wiley & Sons Ltd.

brakes, which must be adhered to for certification testing; in all cases, a conservative approach is adopted, which provides additional margins of safety.

An introduction to the takeoff performance of commercial transport airplanes is described in this chapter, with an emphasis on the determination of takeoff distances under normal and emergency conditions (as outlined in Section 9.2). A review of the forces that act on the airplane during the ground run is presented in Section 9.3; this is followed by a description of several analytical and empirical methods that can be used to estimate the ground run (Sections 9.4 and 9.6), the rotation and flare to a 35 ft reference height (Section 9.5), and the rejected (i.e., aborted) takeoff distance (Section 9.7). Section 9.8 describes braking forces due to the action of wheel brakes during a rejected takeoff (RTO). The complex subject of taking off on icy or slush-covered runways is introduced in Section 9.9.

In Chapter 10, the takeoff discussion is extended to consider the key operational factors that are vital for safe flight operations. This includes a review of relevant FAA and EASA regulations (including the definitions of takeoff speeds) and techniques for the determination of an airplane's maximum takeoff weight for a given set of operational conditions (such as runway length limitations). The requirements for takeoff climb-out after liftoff are also discussed in Chapter 10.

9.2 Takeoff Distances

9.2.1 Runway Surface Conditions

The surface condition of a runway is an important consideration in determining the runway distance required for a safe takeoff. For the purpose of flight operations, runways are described as being either (1) dry, or (2) wet, or (3) contaminated. Note that the precise definitions provided by the FAA [1] and EASA [2] for wet and contaminated runways are almost the same, but not identical in all respects. The term *contaminated*, in this context, is used to describe a runway where more than 25% of the runway surface area within the required length and width being used is covered by:

(1) surface water to a depth of more than 3 mm (EASA rules) or 0.125 inch (FAA rules); or
(2) loose snow or slush equivalent to more than 3 mm (or 0.125 inch) of water; or
(3) ice or compacted snow.

A *wet* runway is a runway that is covered with surface water not exceeding 3 mm (EASA rules [2]) or 0.125 inch (FAA rules [1]) in depth, such as to cause it to appear reflective [2]. A *damp* runway is associated with light precipitation where the water has penetrated the macrostructure of the runway surface, but it has not yet reached saturation. The runway surface will be observably darker in color than a dry runway, but will not have a shiny appearance. From a performance standpoint, a damp runway is considered equivalent to a dry runway [2]. Dry and wet runway operations are described below in Sections 9.2.2 to 9.2.7. The special case of flight operations from contaminated runways is discussed later in Section 9.9.

9.2.2 Required Runway Distances for Takeoff on a Dry or Wet Surface

The determination of the required runway distance for a given set of takeoff conditions (i.e., aircraft weight, flap setting, runway altitude, ambient temperature, wind strength and direction, and so forth) is a critical part of any preflight takeoff planning exercise. The airworthiness regulations contain specific requirements in this respect, as described in FAR/CS 25.109 [3, 4] and FAR/CS 25.113 [3, 4]. The flight planning is based on a conservative set of assumptions that requires the determination of the longest runway distance needed for several probable scenarios—these distances for a *dry runway* are:

(1) the factored all-engines-operating (AEO) dry runway takeoff distance (Section 9.2.3);
(2) the dry runway one-engine-inoperative (OEI) takeoff distance (Section 9.2.4); and
(3) the dry runway rejected takeoff (RTO) distance following an engine failure (Section 9.2.5) or other takeoff abort event (Section 9.2.6).

For a *wet runway*, the distance must be the greater of the dry runway distances (described in Sections 9.2.3. to 9.2.6) and the wet runway distances determined according to the appropriate regulations[2] (described in Section 9.2.7).

In an operational situation, the available runway length will, of course, be known and the analysis procedure is reversed: the airplane's allowable takeoff weight is determined considering the different limiting factors (e.g., runway dimensions, climb obstacles, and brake energy limit)—this is discussed in Chapter 10.

9.2.3 All-Engines-Operating Takeoff Distance on a Dry Runway

During takeoff, the airplane accelerates from rest, which marks the start of the takeoff distance, to a predetermined speed called V_R, the rotation speed. It then begins to rotate (i.e., raise the nosewheel off the runway), while continuing to accelerate down the runway. When the speed and body attitude are such that the wing generates an upward force greater than the airplane's weight, the airplane lifts off the runway. The liftoff speed is known as V_{LOF}. The airplane then establishes its initial climb path.

The point at which the airplane reaches a defined height (as specified in the regulations) has a particular significance: that height is called the *screen height* and the corresponding point on the initial climb path is considered, by convention, to mark the end of the takeoff distance. The screen height (h_{sc}), for normal commercial airplane operations on a dry runway, is defined as 35 ft (10.7 m) in FAR 25.113 [3, 4] and as 11 m in CS 25.113 [4]. For the sake of brevity, the dry runway screen height will herein be referred to as the 35 ft height for both sets of regulations.

The distance from the point of brake release to the point where the airplane is 35 ft above the runway surface—determined by measurement or calculation when all engines are operating normally—is called the unfactored all-engines-operating (AEO) takeoff distance. During certification tests, the distance is measured horizontally from a point on the main landing gear (undercarriage) when the brakes are released to the same point on the landing gear when the lowest part of the airplane (which is usually the main landing gear) is 35 ft above the runway surface. An operational margin of 15% is then added to this distance to give the *factored takeoff distance*, which may be used to determine the required runway length for operational planning purposes.

During a takeoff, the airplane will be rotated to its liftoff attitude at a rotation speed (V_R) determined by the pilots. The V_R speed is computed such that the airplane can achieve an airspeed known as V_2 (i.e., the takeoff safety speed) when reaching a height of 35 ft with one engine inoperative. The V_2 speed is defined in the regulations in a manner that ensures that

---

2 The regulations for wet and contaminated runway operations have changed several times over the past few decades. For example, prior to 1998, the FAA did not require that manufacturers certify wet runway takeoff and landing performance, but it was a feature of European regulations at that time. With the introduction of FAR 25 Amdt. 25-92 (1998), this also became part of the US regulations. (Note that only current regulations (2016) are described herein.)

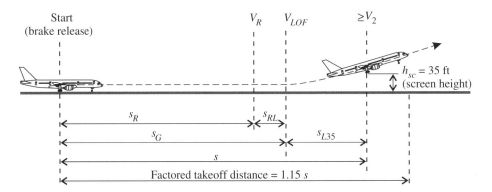

Figure 9.1 All-engines-operating (AEO) takeoff distance.

there is a sufficient margin from stall and also a sufficient margin from the minimum speed at which the airplane is controllable with the critical engine inoperative (further details are given in Section 10.2.2).

As shown in Figure 9.1, the takeoff distance (s) consists of a ground segment (s_G) and an air segment (s_{L35}). The ground segment, which is also called the *ground run*, is further divided into two parts: the distance taken from rest to the point where the airplane rotates (s_R) and the distance from the point of rotation to liftoff (s_{RL}). At the speed V_R the airplane starts to rotate, increasing the angle of attack of the wing and the amount of lift that is generated. Soon afterwards, liftoff occurs at the speed V_{LOF}. In normal circumstances, the airplane will continue to accelerate, clearing the hypothetical 35 ft screen at a safe speed.

The takeoff segments (i.e., s_R, s_{RL}, and s_{L35}) are typically analyzed separately and summed to give the total takeoff distance (s)—as illustrated in Figure 9.1.

9.2.4 One-Engine-Inoperative Takeoff Distance on a Dry Runway

Modern jet engines are extremely reliable and the probability of an engine failing during takeoff is very small. It is, nonetheless, one of the key safety-related directives within the airworthiness regulations [2–5] that routine flight operations consider the possibility of an engine failure—which may theoretically occur at any time—during the takeoff. Multiple engine failures during takeoff, however, are considered to be extremely improbable[3]—such an extreme scenario is thus not considered for takeoff weight planning (see Section 10.3).

The standard operational approach considers the possibility of a complete failure of the so-called *critical engine* (i.e., the engine that would most adversely affect the airplane's performance during takeoff) at the speed V_{EF} and that the engine would remain inoperative for the rest of the takeoff. If the engine failure occurs shortly after the start of the takeoff run, the obvious—and safest—choice for the crew would be to abort the takeoff; this is known as a *rejected takeoff* (RTO). However, if the engine failure occurs much later during the takeoff run—say, as the airplane's speed approaches the rotation speed—the safest option, under normal circumstances, would be to continue with the takeoff, albeit with less thrust available.

For the purpose of analyzing takeoff distances, as presented in this chapter, the assumed speed at which the pilot takes the first action to slow the airplane (e.g., brake application) in the event

---

3 The reliability requirements for safety-critical systems, as specified by the FAA and EASA, are outlined in Section 22.2.

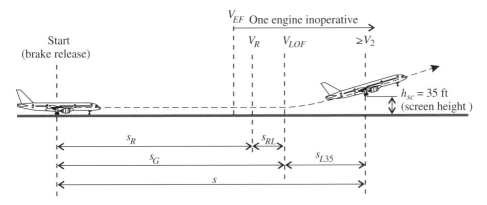

Figure 9.2 One-engine-inoperative (OEI) takeoff distance for a dry runway.

of a decision to abort a takeoff is called the V_1 speed and it is assumed to occur one second after V_{EF}. Regulatory details on V_1 are given in Section 10.2.3, and a more complete explanation on how V_1 speeds are determined for takeoff weight planning for flight operations is presented later in Section 10.5.

The OEI takeoff distance (Figure 9.2), which is also called the *accelerate–go distance*, is analyzed in much the same way as the normal all-engines-operating condition (Section 9.2.3), but taking into account the loss of thrust and the increase in drag associated with the inoperative engine. The operational margin of 15%, which was added to the AEO takeoff distance, does not apply in this case.

9.2.5 Rejected Takeoff Distance Following an Engine Failure on a Dry Runway

Following the recognition, by the flight crew, of an engine failure and a decision to abort the takeoff, the crew would be expected to take several actions in quick succession—these include applying brakes, retarding the thrust levers to idle, selecting reverse thrust, and deploying the ground spoilers (speedbrakes). As V_1 is defined as the airspeed at which the crew takes the first corrective action, the speed at which the brakes become effective, for theoretical calculations, can be assumed to be equal to V_1. During certification tests—that is, when the failure is anticipated—a test pilot would typically complete the actions of applying brakes, retarding the thrust levers, and deploying ground spoilers within two seconds. For the determination of AFM data—which are intended for use by line pilots during routine flight operations—additional allowances are made (see Section 9.7). As a means of introducing a further safety margin (i.e., an operational conservatism), the regulatory authorities do not permit credit to be taken for the effect of reverse thrust. This measure, which applies to operations from dry runways, is designed to account for the possible reduction of friction on operational runway surfaces when compared to that of the runway used for certification testing [6].

The RTO distance, which is usually called the *accelerate–stop distance*, comprises the following three segments, which must be determined and summed (see Figure 9.3):

(1) the acceleration distance (s_{EF}), from brake release to the point of engine failure;
(2) the distance required for the brakes to be applied and to become fully effective and for the ground spoilers to be deployed (s_{EB}); and
(3) the braking distance (s_B).

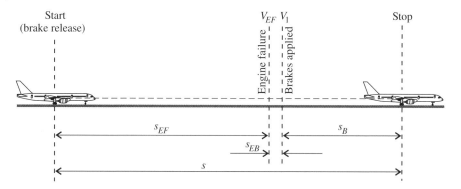

Figure 9.3 One-engine-inoperative (OEI) accelerate–stop distance for rejected takeoff.

9.2.6 All-Engines-Operating Rejected Takeoff Distance on a Dry Runway

There are several alternative reasons why a takeoff may be aborted other than for an engine failure—for example, a major system failure, fire warning, burst tire, airplane structural damage, or an emergency associated with an event on the airfield. The planning for such cases assumes that the thrust levers are retarded simultaneously to the idle position. As the engines spin down (or spool down) they will produce a residual thrust—the magnitude of which decays rapidly to the idle setting. In this scenario, an abort *event speed* (V_{EV}) replaces the engine failure speed (V_{EF}) in the takeoff sequence, but in all other respects the procedure for determining the accelerate–stop distance is the same as that described in Section 9.2.5. The accelerate–stop distance, in this case, will typically be a little longer than that required for a comparable engine failure scenario, because of the greater amount of idle thrust during the deceleration. There are, however, exceptions to this observation; therefore, both RTO distances (i.e., for an engine failure and for another abort event) need to be evaluated and the most conservative scenario—that is, the one that produces the longest takeoff distance—used for flight planning [7].

9.2.7 One-Engine-Inoperative Takeoff and Rejected Takeoff Distances on a Wet Runway

The wet runway OEI takeoff distance is determined in the same way as the dry runway OEI distance, but with a reduced screen height of 15 ft. The takeoff distance is measured from the start of the takeoff to the point that the airplane is 15 ft above the runway surface, achieved in a manner that would enable the V_2 speed to be reached before the airplane crosses the 35 ft screen height (Figure 9.4).

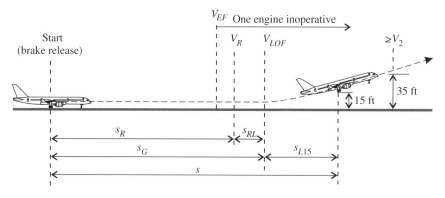

Figure 9.4 One-engine-inoperative (OEI) takeoff distance for a wet runway.

The determination of the RTO distance on a wet runway is determined in the same way as that for a dry runway, with one notable difference: credit for the use of reverse thrust is permitted.

9.3 Forces Acting on the Airplane During the Ground Run

9.3.1 Vector Diagram and Overview of the Forces Acting on an Airplane

The forces that can act on an airplane during takeoff are shown in Figure 9.5. The runway is assumed to have an uphill slope[4] of angle γ_G and to be clear of standing water or snow or icy slush that would impede the airplane's acceleration. (The special case of airplanes operating from contaminated runways is described in Section 9.9.) Resistance to forward motion also comes from the aerodynamic drag, rolling friction of the tires on the runway, and the component of the airplane's weight acting parallel to the runway. Opposing these forces is the thrust of the engines (F_N), which will cause the airplane to accelerate along the runway.

Before a general equation for the takeoff run can be determined, it is necessary to first identify and describe the forces that act on the airplane. The following factors are important: wind (Section 9.3.2), engine thrust (Section 9.3.3), lift and drag forces (Section 9.3.4), airplane configuration (Section 9.3.5), ground effect (Section 9.3.6), and rolling friction (Section 9.3.7).

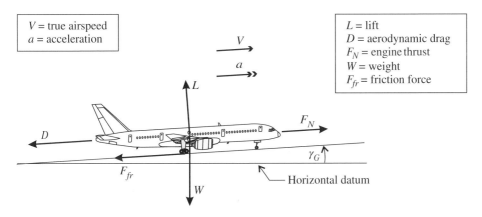

Figure 9.5 Forces acting on an airplane during the ground run.

9.3.2 Influence of Wind

It is usual that the takeoff direction is into the prevailing wind (i.e., giving a headwind rather than a tailwind) unless the winds are slack and are of little consequence. It is customary to resolve the prevailing wind vector (at the height of the airplane) into crosswind and headwind/tailwind components (see Section 4.5.2). The crosswind component has a negligible influence on the takeoff distance (despite the fact that the pilot may have to use the rudder and ailerons to counter the tendency for the airplane to turn into the wind). The component of wind acting along the runway is herein designated as V_w, which is defined as positive for a headwind and negative for a tailwind.[5]

---

4 By convention, the runway slope angle is defined as positive for uphill.
5 The convention of defining a headwind as positive for takeoff or landing is opposite to that which is customarily adopted for *en route* performance analyses—for climb, cruise, or descent (see Chapters 12 and 13) a tailwind is

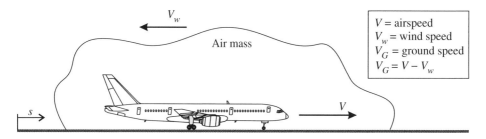

Figure 9.6 Effect of wind on ground speed.

The ground speed (V_G), which is the time rate of change of ground distance (s), is illustrated in Figure 9.6. It is defined as follows:

$$V_G = \frac{ds}{dt} = V - V_w \tag{9.1}$$

where V is the true airspeed (typical units: m/s, ft/s); and
$\quad\quad$ V_W is the wind speed along the runway, defined as positive for a headwind
$\quad\quad\quad$ (typical units: m/s, ft/s).

9.3.3 Engine Thrust

In general, an engine's thrust depends on the engine setting, bleed air demand, atmospheric conditions (air temperature and pressure), and airspeed (see Section 8.3). For a given thrust setting (e.g., maximum takeoff thrust) and atmospheric conditions, the thrust will decrease during the ground run as the airspeed increases; this thrust–speed relationship depends on the powerplant type. The following expression, which follows directly from Equation 8.42, has been shown [8] to provide a reasonably accurate representation of jet engine thrust at low speeds:

$$\frac{F_N}{F_{N_0}} = 1 - k_1 V + k_2 V^2 \tag{9.2}$$

where F_N is the thrust (typical units: N, lb);
$\quad\quad$ F_{N_0} is the static thrust (typical units: N, lb); and
$\quad\quad$ k_1 and k_2 are positive empirical constants selected to match the engine
$\quad\quad\quad$ data.

Although Equation 9.2 provides an acceptable representation of the takeoff thrust, it, unfortunately, leads to a mathematical description of the takeoff ground run for which there is no closed-form analytical solution. To facilitate the development of analytical expressions for the ground run, it is necessary that the thrust be described by simpler expressions, as discussed later in Section 9.4.

---

defined as positive. The reason for this inconsistency is that headwinds are beneficial for takeoff and landing and are thus viewed as positive or advantageous; whereas, for the *en route* performance, it is a tailwind that is usually advantageous.

9.3.4 Lift and Drag Forces

From the definitions of the lift and drag coefficients, given by Equations 7.1 and 7.5, respectively, the lift and drag forces can be expressed as follows:

$$L = qSC_L = \frac{1}{2}\rho SC_L V^2 \tag{9.3}$$

$$D = qSC_{D_g} = \frac{1}{2}\rho SC_{D_g} V^2 \tag{9.4}$$

where L and D are the lift and drag forces, respectively (typical units: N, lb);
q is the dynamic pressure (typical units: N/m^2, lb/ft^2);
C_L and C_{D_g} are the lift and drag coefficients, respectively (in ground effect);
ρ is the ambient air density (typical units: kg/m^3, slug/ft^3); and
S is the wing reference area (typical units: m^2, ft^2).

When the airplane accelerates down the runway during takeoff, it does so at a nearly constant angle of attack up to the point of rotation. Changes in the angle of attack can only result from differential extension or contraction of the nose and main landing gear legs. The landing gear will extend a little as the lift builds up over the wing and this may result in an angle of attack change, but the magnitude of this change is negligible. For this reason, C_L can, for all practical purposes, be considered constant during the takeoff, up to the point of rotation.

The drag coefficient will also be essentially constant. This is evident from Equation 7.14: C_D is a function of C_L (which is constant), and neither C_{D_0} nor K will change during a typical ground run (as the airplane's configuration does not change). During the ground run, it is generally advantageous to have a low angle of attack for as long as possible. Although the resistance due to the rolling coefficient of friction would decrease if the angle of attack were increased (generating more lift), this effect is outweighed by the increase in lift-induced drag.

The lift and drag forces acting on the airplane thus increase as linear functions of airspeed-squared during the ground run, up until the point of rotation. From the rotation speed (V_R) to the liftoff speed (V_{LOF}) the angle of attack will change, and hence C_L and C_D will also change—both C_L and C_D will increase during this short segment of the ground run.

9.3.5 Airplane Configuration

It is important that the drag polar used to evaluate the takeoff run includes appropriate correction factors to account for the airplane's configuration (see Section 7.5.2). With the slats and flaps set for takeoff and the landing gear extended, the zero-lift drag coefficient can be determined from the following expression:

$$C_{D_0} = \left(C_{D_0}\right)_{clean} + \left(\Delta C_{D_0}\right)_{slats} + \left(\Delta C_{D_0}\right)_{flaps} + \left(\Delta C_{D_0}\right)_{gear} \tag{9.5}$$

where C_{D_0} is the zero-lift drag coefficient;
$(C_{D_0})_{clean}$ is the clean airplane zero-lift drag coefficient; and
ΔC_{D_0} describes the drag coefficient increments due to the extension of the slats, flaps, and landing gear.

Furthermore, the deployment of the flaps will change the spanwise lift distribution and this will alter the lift-dependent drag coefficient.

9.3.6 Ground Effect

During the ground run, the airplane's lift-dependent drag is reduced due to ground effect (discussed in Section 7.5.6). In the absence of substantive test or wind-tunnel data, a ground effect factor (λ) can be applied to the out-of-ground-effect (OGE) drag polar to establish an approximate drag polar for the airplane on the ground. The magnitude of the ground effect factor depends on the airplane type (see also Equations 7.33 and 7.34). The drag coefficient in ground effect (IGE) is given by Equation 7.31, which can be expressed as follows:

$$C_{D_g} = C_{D_0} + \lambda K C_L^2 \tag{9.6}$$

where C_{D_g} is the drag coefficient of the airplane in ground effect;
$\quad\quad C_{D_0}$ is the zero-lift drag coefficient (OGE);
$\quad\quad \lambda$ is the ground effect factor; and
$\quad\quad K$ is the lift-dependent drag factor (OGE).

9.3.7 Rolling Friction

The coefficient of friction (or friction coefficient), which is usually assigned the Greek letter mu (μ), is defined as the ratio of the tire friction force (which acts in a direction parallel to the runway surface) to the force acting on the tire normal to the runway surface (i.e., the reaction force):

$$\mu = \frac{F_{fr}}{R} \tag{9.7}$$

where μ is the coefficient of friction (dimensionless);
$\quad\quad F_{fr}$ is the friction force (typical units: N, lb); and
$\quad\quad R$ is the reaction force acting on the tire (typical units: N, lb).

The total friction force that acts on the airplane can be expressed as follows:

$$F_{fr} = \mu R_{MW} + \mu R_{NW} \tag{9.8}$$

where the subscripts MW and NW designate the main landing gear and nose gear, respectively.

A very good approximation of the total friction force acting on all the tires (nose and main landing gear), up to the point of rotation on an inclined runway (see Figure 9.5), is given by the following expression:

$$F_{fr} = \mu(W \cos \gamma_G - L) \tag{9.9}$$

where γ_G is the "slope" angle of the runway (see Figure 9.5), defined as positive for an uphill runway (typical units: degrees, radians).

The *coefficient of friction* depends on the nature of the tire–runway surface interface and, critically, on the amount of slip that occurs between the tires and the runway. Slip describes the relative motion between a tire and the runway surface; a slip condition exists when a tire's rotational speed is less than or greater than the free-rolling speed. During takeoff, when the aircraft's tires are rolling freely, they will experience zero slip, but, obviously, this is not the case when the brakes are applied (wheel braking is discussed in Section 9.8).

The coefficient of friction for a free rolling tire is usually taken as a measure of the total resistance to the rolling motion of the wheel; this includes the internal friction of the bearings in the wheel and the energy associated with the lateral flexing of the walls of the tire under the airplane's weight. The tire type and internal pressure thus influence the frictional force. The value of μ changes a little during the takeoff as the speed increases; however, this has a negligible influence, and for the purpose of computing the takeoff distance, a mean value can be used. The coefficient of friction is approximately 0.015 for a jet transport airplane under rolling conditions on a dry, concrete runway [9].

To accurately measure the effective coefficient of friction of a rolling airplane tire is a surprisingly difficult task. A widely adopted approach utilizes airplane acceleration data—measured during takeoff under known conditions—with knowledge of the other forces that act on the airplane to determine the coefficient of friction. In other words, the friction coefficient is treated as a fudge factor rather than an independently evaluated parameter (such as ambient temperature), and adjusted within a mathematical model to give the measured distance. The resulting model is then used to determine the takeoff distances for other conditions.

9.3.8 Net Acceleration Force

For the sake of brevity, the net acceleration force acting on the airplane in the direction of motion is herein denoted as $\sum F$, where

$$\sum F = F_N - D - F_{fr} - W \sin \gamma_G \tag{9.10}$$

or $\quad \sum F = F_N - D - \mu(W \cos \gamma_G - L) - W \sin \gamma_G \quad$ (from Equation 9.9) $\tag{9.11}$

As the runway gradient is a very small quantity,[6] the small angle approximations $\cos \gamma_G \cong 1$ and $\sin \gamma_G \cong \gamma_G$ (for γ_G measured in radians) can be used without loss of accuracy.

Thus $\quad \sum F = F_N - D - \mu(W - L) - W\gamma_G \tag{9.12}$

Substituting Equations 9.3 and 9.4 into Equation 9.12 results in the required expression for the net acceleration force:

$$\sum F = F_N - \frac{1}{2}\rho V^2 S\left(C_{D_g} - \mu C_L\right) - (\mu + \gamma_G)W \tag{9.13}$$

where C_{D_g} can be approximated by Equation 9.6.

The group of terms $C_{D_g} - \mu C_L$ does not change appreciably during the ground run and a constant value can be used to evaluate the takeoff distance. This is usually determined from tests conducted for a particular airplane type for each required flap setting.

With knowledge of the net force acting on the airplane (in the direction of motion), the acceleration can be determined for a given airplane weight, which can then be used to calculate the ground run.

---

6 It is rare that a runway slope (gradient) will exceed 2%, which is a slope angle of 1.146° (0.020 rad). Runways with slopes greater than 2% are considered special cases for flight operations (see Section 10.4.3).

9.4 Evaluation of the Takeoff Distance from Brake Release to Rotation

9.4.1 Basic Equation for the Takeoff Distance s_R

The distance (s) that the airplane will travel (see Figure 9.6) in a given period of time depends on its ground speed (V_G) and its acceleration (a). The airplane's acceleration (with respect to the ground) can be expressed as follows:

$$a = \frac{d}{dt}\left(\frac{ds}{dt}\right) = \frac{d}{dt}(V - V_w) = \frac{dV}{dt} \qquad (\text{if } dV_w/dt = 0) \tag{9.14}$$

In the development of Equation 9.14, the wind speed is considered invariant with time. Equation 9.14 can now be written in a manner that facilitates the development of an integral expression for the ground distance (s), that is,

$$a = \frac{dV}{dt} = \frac{dV}{ds}\frac{ds}{dt} = \frac{dV}{ds}(V - V_w) \tag{9.15}$$

In the analysis that follows, the ground distance (s) is expressed as a function of *airspeed* (not ground speed). At the start of the ground run (i.e., brake release), the airplane is stationary, but the presence of wind means that the airspeed is not zero, but is equal in magnitude to the headwind V_w. This sets the lower limit to the integral expression. The upper limit is the rotation speed, which is independent of the wind speed (the lift required for takeoff depends on the airspeed and not the ground speed). The ground distance measured from the start of the takeoff to the point of rotation (s_R) can be determined from the following expression:

$$s_R = \int_{start}^{end} ds = \int_{V_w}^{V_R} \frac{V - V_w}{a}\, dV \tag{9.16}$$

At any instant during the ground run, the airplane's acceleration can be obtained from Newton's second law (see Section 2.4.5):

$$a = \frac{\sum F}{m} = \frac{g \sum F}{W} \tag{9.17}$$

where a is the instantaneous acceleration (typical units: m/s^2, ft/s^2);
  $\sum F$ is the net acceleration force, given by Equation 9.13 (typical units: N, lb);
  m is the airplane's mass (typical units: kg, slug);
  g is the acceleration due to gravity (typical units: m/s^2, ft/s^2); and
  W is the airplane's weight (typical units: N, lb).

Substituting Equation 9.17 into Equation 9.16 yields the required equation:

$$s_R = \int_{V_w}^{V_R} \frac{W(V - V_w)}{g \sum F}\, dV = \int_{V_w}^{V_R} \frac{WV}{g \sum F}\, dV - \int_{V_w}^{V_R} \frac{WV_w}{g \sum F}\, dV \tag{9.18}$$

where the net accelerating force is given by Equation 9.13 (and where C_{D_g} can be approximated by Equation 9.6).

In order to integrate Equation 9.18, it is necessary to describe the variations of thrust, lift, drag, weight, and friction with speed. The airplane's weight changes very little during the short time taken for the takeoff; thus, for "hand" calculations (i.e., analytical solutions), weight may be regarded as constant. In detailed computational evaluations (as conducted by aircraft manufacturers, for example), the very small influence of reducing airplane weight due to fuel burn is taken into account. The rolling coefficient of friction can also be regarded as constant (see Section 9.3.7). The values of C_L and C_{D_g} are dictated by the airframe geometry, and therefore the lift and drag forces can be expressed as functions of V^2. Both lift and drag are also functions of air density—hence altitude and ambient temperature have a major effect on the takeoff.

In general, the thrust will vary as a function of velocity, air temperature, and pressure (see Section 9.3.3). During the ground run, the thrust will reduce as the speed builds up. The exact manner by which the thrust will reduce depends on the powerplant type. The integration of Equation 9.18 can be performed analytically for simple functions of F_N. In Section 9.4.2, this is undertaken by assuming that F_N is constant. The resulting expression is useful for obtaining analytical estimates of the takeoff distance. The solution is also useful for developing an understanding of the impact of the various parameters on the takeoff distance (or for conducting sensitivity analyses, for example).

9.4.2 Determination of the Distance s_R Assuming F_N is Constant

In this section, a mathematical expression for the distance s_R is determined for the general case of an airplane taking off on a dry or wet (i.e., non-contaminated) runway with a gradient γ_G and with a headwind V_w. The problem is simplified by assuming that the engine thrust is equal to $\bar{F}_N$, a mean constant value. This value is not the arithmetic mean of the thrust values at the start and end of the ground run; instead, it is evaluated at a particular speed selected to give an accurate approximation of the takeoff distance. The most frequently used approach assumes that acceleration varies linearly with V^2 (this is discussed further in Section 9.4.3). The mean thrust ($\bar{F}_N$) is calculated at a reference speed, herein denoted as V^*, which is determined as follows:

$$(V^*)^2 = \frac{1}{2}(V_R - V_w)^2 \tag{9.19}$$

or $\quad V^* = \frac{1}{\sqrt{2}}(V_R - V_w) = 0.707(V_R - V_w) \tag{9.20}$

An expression for the net accelerating force ($\sum F$) can be obtained by substituting Equation 9.6 into Equation 9.13, that is,

$$\sum F = \bar{F}_N - \frac{1}{2}\rho V^2 S\left(C_{D_0} + \lambda K C_L^2 - \mu C_L\right) - (\mu + \gamma_G)W \tag{9.21}$$

The required distance (s_R) is obtained by a summation of the two integrals that are given in Equation 9.18. For the sake of convenience, the two integral expressions will be evaluated separately, where

$$\int_1 = \int_{V_w}^{V_R} \frac{WV}{g\sum F}\,dV \quad \text{and} \quad \int_2 = \int_{V_w}^{V_R} \frac{WV_w}{g\sum F}\,dV \tag{9.22}$$

For the first integral, the following variable substitution is made:

$$f = V^2 \tag{9.23}$$

thus $\quad df = 2V\,dV \quad$ and $\quad dV = \frac{1}{2V}\,df$

hence $\int_1 = \int_{V_w}^{V_R} \frac{WV}{g\sum F}\,dV = \int_{\sqrt{f_w}}^{\sqrt{f_R}} \frac{WV}{g\sum F}\left(\frac{1}{2V}\right)df = \frac{1}{2g}\int_{\sqrt{f_w}}^{\sqrt{f_R}} \frac{W}{\sum F}\,df \tag{9.24}$

Now substitute $\sum F$ from Equation 9.21 into Equation 9.24:

$$\int_1 = \frac{1}{2g} \int_{\sqrt{f_w}}^{\sqrt{f_R}} \frac{1}{\left(\dfrac{\bar{F}_N}{W} - \mu - \gamma_G\right) - \dfrac{\rho S}{2W}\left(C_{D_0} + \lambda K C_L^2 - \mu C_L\right)f}\, df \tag{9.25}$$

To simplify the analysis, arbitrary constants A_2, B_2, and C_2 are now introduced, where

$$A_2 = \frac{\bar{F}_N}{W} - \mu - \gamma_G \quad \text{and} \quad B_2 = \frac{\rho S}{2W}\left(C_{D_0} + \lambda K C_L^2 - \mu C_L\right)$$

and $\quad C_2 = \sqrt{\dfrac{A_2}{B_2}}$ \tag{9.26}

Thus $\quad \int_1 = \dfrac{1}{2g}\displaystyle\int_{\sqrt{f_w}}^{\sqrt{f_R}} \dfrac{1}{A_2 - B_2 f}\, df = \dfrac{-1}{2gB_2}\left[\ln(A_2 - B_2 f)\right]_{\sqrt{f_w}}^{\sqrt{f_R}}$

$$\int_1 = \frac{-1}{2gB_2}\left[\ln(A_2 - B_2 V^2)\right]_{V_w}^{V_R} = \frac{1}{2gB_2}\ln\left(\frac{A_2 - B_2 V_w^2}{A_2 - B_2 V_R^2}\right) \tag{9.27}$$

The second integral is now considered:

$$\int_2 = \frac{V_w}{g}\int_{V_w}^{V_R}\frac{1}{A_2 - B_2 V^2}\, dV = \frac{V_W}{gB_2}\int_{V_w}^{V_R}\frac{1}{C_2^2 - V^2}\, dV$$

$$\int_2 = \frac{V_w}{2gB_2 C_2}\left[\ln\left(\frac{C_2 + V}{C_2 - V}\right)\right]_{V_w}^{V_R} = \frac{V_w}{2gB_2 C_2}\ln\left\{\frac{(C_2 + V_R)(C_2 - V_w)}{(C_2 - V_R)(C_2 + V_w)}\right\} \tag{9.28}$$

The two results are now added to give the distance s_R:

$$s_R = \frac{1}{2gB_2}\left[\ln\left(\frac{A_2 - B_2 V_w^2}{A_2 - B_2 V_R^2}\right) - \frac{V_w}{C_2}\ln\left\{\frac{(C_2 + V_R)(C_2 - V_w)}{(C_2 - V_R)(C_2 + V_w)}\right\}\right] \tag{9.29}$$

where A_2, B_2, and C_2 are defined in Equation 9.26.

By selecting an appropriate speed at which the mean thrust is calculated (see Equation 9.20), an accurate estimate of the required distance can be obtained by this method.

It is evident that an analytical solution can be found for the ground distance s_R if a constant thrust value is assumed. Equation 8.42 (see Section 8.6.4) provides a reasonably accurate representation of a turbofan engine's thrust variation with speed. Unfortunately, it is not possible to obtain a closed-form solution for the ground distance s_R using Equation 8.42, and a numerical technique—such as the one described later in Section 9.4.4—must be employed if greater accuracy is needed.

9.4.3 Determination of the Distance s_R Using a Mean Acceleration

A popular, and relatively simple, method to estimate the ground distance s_R is based on the use of a mean acceleration ($\bar{a}$). The estimate for s_R is then given by the equation for

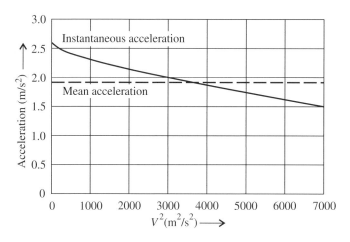

Figure 9.7 Instantaneous and mean accelerations for a medium-sized twinjet plotted as a function of airspeed squared (note: the same airplane data were used to generate Figures 9.8 and 9.9).

uniform acceleration:

$$s_R = \frac{1}{\bar{a}} \int_{V_w}^{V_R} (V - V_w)\, \mathrm{d}V = \frac{1}{\bar{a}} \left[\frac{V^2}{2} - V V_w \right]_{V_w}^{V_R} = \frac{1}{2\bar{a}} (V_R - V_w)^2 \tag{9.30}$$

The influence of the wind can be easily seen from Equation 9.30: a headwind will directly reduce the takeoff ground run and a tailwind will increase the distance.

The acceleration of a jet airplane during the ground run reduces in an almost linear manner with V^2—this is illustrated in Figure 9.7. The mean value $\bar{a}$ can therefore be calculated at the reference speed given by Equation 9.20.

The method for calculating the takeoff distance s_R based on a mean acceleration is summarized below:

(1) Determine the reference speed V^* using Equation 9.20.
(2) Determine the thrust $\bar{F}_N$ at V^* from thrust versus velocity data or from an appropriate empirical model, such as Equation 9.2.
(3) Calculate $\bar{a}$ for $V = V^*$ and $F_N = \bar{F}_N$ using Equations 9.17 and 9.13.
(4) Calculate the takeoff distance s_R using Equation 9.30.

9.4.4 Numerical Evaluation of the Distance s_R

It was shown in Section 9.4.2 that an analytical integration of Equation 9.18, to yield the distance s_R, is possible if the thrust is expressed as a mean constant value. An alternative approach is to conduct a numerical integration of the governing equation (this is preferential when actual engine thrust data are available). By substituting the acceleration term in Equation 9.16 using Equation 9.17, the following integral expression is obtained:

$$s_R = \int_{V_w}^{V_R} \frac{V - V_w}{a}\, \mathrm{d}V = \frac{W}{g} \int_{V_w}^{V_R} \frac{V - V_w}{\sum F}\, \mathrm{d}V \tag{9.31}$$

where $\sum F$ is given by Equation 9.13.

Table 9.1 Example of takeoff distance calculation

V_G (kt)	V_G (m/s)	V (kt)	V (m/s)	q (Pa)	F_N (kN)	$(C_{D_g} - \mu C_L)qS$ (kN)	μW (kN)	$\gamma_G W$ (kN)	a (m/s^2)	Δs (m)
0	0.0	10	5.1	16.2	312	0.24	17.6	10.7	2.60	
20	10.3	30	15.4	146	303	2.12	17.6	10.7	2.50	20.8
40	20.6	50	25.7	405	294	5.88	17.6	10.7	2.39	64.9
60	30.9	70	36.0	794	286	11.5	17.6	10.7	2.26	114
80	41.2	90	46.3	1313	279	19.0	17.6	10.7	2.12	169
100	51.4	110	56.6	1961	271	28.4	17.6	10.7	1.97	233
120	61.7	130	66.9	2739	265	39.7	17.6	10.7	1.81	308
140	72.0	150	77.2	3647	258	52.9	17.6	10.7	1.63	401
150	77.2	160	82.3	4150	255	60.2	17.6	10.7	1.53	243

Total $= 1552$

Notes:
(a) Airplane: medium-size twinjet, with a takeoff weight of 1067 kN (240 000 lb).
(b) Conditions: 1% runway gradient (uphill), 10 kt headwind, dry runway surface ($\mu = 0.0165$), and ISA conditions.

To integrate Equation 9.31 numerically, it is possible to divide the takeoff distance s_R into n segments (or velocity intervals) and to use the trapezoidal rule for incremental values of V:

$$s_R = \frac{W}{2g} \sum_{i=0}^{n} \left(\frac{(V)_{i+1} - V_w}{(\sum F)_{i+1}} + \frac{(V)_i - V_w}{(\sum F)_i} \right) \Delta V \tag{9.32}$$

The intervals do not have to be very small for an acceptable result to be obtained (10 kt intervals, for example, are adequate for most applications).

The distance s_R can also be calculated by a more direct approach using a spreadsheet—this is illustrated in Table 9.1 for a medium-size twinjet. For the sake of illustration, 20 kt speed intervals have been used. It is convenient to start the calculation with the ground speed (columns 1 and 2) and then determine the true airspeed (V) and dynamic pressure (q). When evaluating the terms on the right-hand side of Equation 9.13, note that the thrust (F_N) is a function of V not V_G, the term $C_{D_g} - \mu C_L$ can be assumed constant, and that the resisting forces due to friction and runway slope are both functions of the airplane's weight. The instantaneous acceleration (a), given in column 10, is determined using Equations 9.17 and 9.13. Each entry (or spreadsheet cell) in the table for columns 3 to 10 contains point (or instantaneous) values evaluated for the corresponding ground speed. The interval distance (Δs), given in column 11, is calculated using an *average* acceleration for the speed interval. The calculation is repeated for each speed interval until the rotation speed is reached. Finally, the interval distances are summed.

Figures 9.8 and 9.9 illustrate the takeoff variables for the medium-size twinjet considered in the previous paragraph; the takeoff conditions are unchanged. In Figure 9.8, the components of force given by Equation 9.12 are plotted against the ground speed. The resultant acceleration, which is shown on the second Y axis, was calculated using the numerical approach illustrated in Table 9.1. In Figure 9.9, the resulting acceleration is repeated (again on the second Y axis), but this time it is plotted against the takeoff distance. The airspeed is seen to increase rapidly at

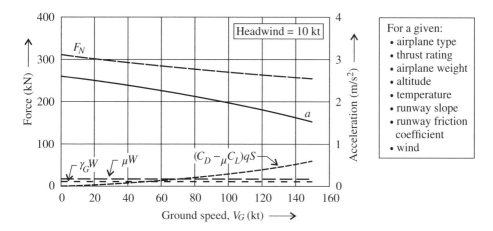

Figure 9.8 Example of takeoff calculations: components of force and airplane acceleration plotted against ground speed.

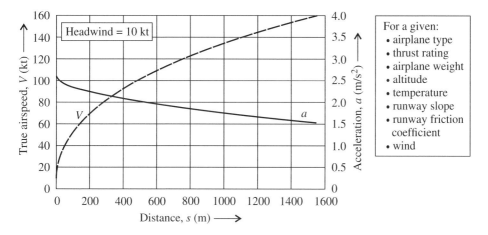

Figure 9.9 Example of takeoff calculations: airspeed and acceleration plotted against takeoff distance.

the start of the takeoff when the acceleration is greatest. The rotation speed (TAS) was assumed to be 160 kt.

9.4.5   Estimation of the Time to Reach V_R

The time that the airplane takes to accelerate to the rotation speed (V_R) may be determined using similar approaches to those presented for the distance calculation (Sections 9.4.1 to 9.4.4). The time–velocity relationship is given by

$$dt = \frac{1}{a} dV \quad \text{(based on Equation 9.15)}$$

Thus $\quad t = \int_{start}^{end} dt = \int_{V_w}^{V_R} \frac{1}{a} dV$ \hfill (9.33)

where the acceleration (a) is given by Equation 9.17.

The development of an analytical solution to the integral expression of Equation 9.33 is complex. A simplified approach—which assumes a mean, constant acceleration—will usually provide a satisfactory result, though. With a constant acceleration, it follows that the time for the ground run is

$$t_R = \frac{1}{\bar{a}} \int_{V_w}^{V_R} dV = \frac{V_R - V_w}{\bar{a}} \tag{9.34}$$

In Section 9.4.3, the distance s_R was calculated using a mean acceleration determined at a reference velocity V^* given by Equation 9.20. The rationale behind this approach is that the acceleration is seen to be almost a linear function of V^2; however, the time expression is a function of both V and V^2. It is reported by reference [9] that a better estimate of the time for the ground run can be obtained by determining a mean acceleration at the reference speed $V^* = 0.6V_R$.

The method can be summarized as follows:

(1) Determine the reference speed by assuming that $V^* = 0.6V_R$.
(2) Determine the thrust $\bar{F}_N$ at $V = V^*$ from thrust versus velocity data.
(3) Calculate $\bar{a}$ for $V = V^*$ and $F_N = \bar{F}_N$ from Equation 9.17.
(4) Calculate the time for the ground run (t_R) from Equation 9.34.

9.5 Rotation and Climb-Out to Clear the Screen Height

9.5.1 Rotation and Flare

At the point of rotation, the pilot will pull the stick/yoke back, which will have the following effect: the elevator will be deflected upwards (creating a negatively cambered airfoil shape), which will produce an aerodynamic load acting downwards on the horizontal tail and this, in turn, will raise the nose of the airplane as it pivots around the main wheels. The angle of attack will thus increase, resulting in a progressive increase in lift. The time that it takes the airplane to rotate (typically about 2 to 4 seconds) depends on several factors, including the airplane type, CG position, speed, airplane weight, and the rate that the pilot pulls the stick/yoke back. The typical rate of rotation for jet transport airplanes is reported to be 2.5 to 3.0 degrees per second [10].

If a pilot executes an incorrect takeoff technique by rotating too early and/or by pulling the stick/yoke back too quickly, this can result in too high a rate of rotation, which can cause the rear part of the fuselage to strike the runway. This is called a *tailstrike* and can occur on airplanes that are geometrically limited.[7] Such events occur infrequently; they tend to be costly mistakes resulting in damage to the aft fuselage.

After the liftoff, there will be a short phase in which the flight path is curved—called the flare[8]—when the lift is greater than the weight. During this phase the airplane will continue to

---

7 An airplane is said to be *geometrically limited* if it cannot achieve its maximum angle of attack—corresponding to the maximum lift coefficient in the takeoff configuration—when the main landing gear is still in contact with the ground as the lower rear fuselage would strike the runway. Most commercial transport airplanes are geometrically limited [11].

8 Certain authors and organizations consider the flare to start at the point of rotation and not the point of liftoff, as described herein.

accelerate, and there may be a small change in C_L. After the flare, the airplane will climb steadily at a climb angle (γ) that is approximately constant. The screen height, as mentioned earlier in Section 9.2, is 35 ft for a dry runway and 15 ft for a wet runway. The point at which the screen height is reached may be either before or after the end of the flare.

The distance from the point of rotation to the point where the airplane clears the screen is difficult to calculate accurately using a mechanics approach due to the variation of the governing parameters and the influence of varying pilot technique. The most straightforward approach is based on experimentally determined average times for the rotation and flare segments.

9.5.2 Estimation of the Rotation Distance

The distance from rotation to liftoff (s_{RL}) is difficult to estimate accurately—this is largely due to the changes in C_L. The distance, however, is small in comparison to s_R. Flight-test data are usually used in performance analyses. An estimate of the rotation distance may be obtained by multiplying a typical rotation time by the average ground speed:

$$s_{RL} = \Delta t_{RL} \left(\frac{V_R + V_{LOF}}{2} - V_w \right) \tag{9.35}$$

where s_{RL} is the distance from rotation to liftoff (typical units: m, ft);
$\quad\quad \Delta t_{RL}$ is the time from rotation to liftoff (typical unit: s);
$\quad\quad V_R$ is the rotation speed (typical units: m/s, ft/s);
$\quad\quad V_{LOF}$ is the liftoff speed (typical units: m/s, ft/s); and
$\quad\quad V_w$ is the headwind speed (typical units: m/s, ft/s).

An estimate of the liftoff speed can be obtained by noting that the airworthiness regulations FAR/CS 25.105 [3, 4] require V_{LOF} to be greater than 1.1 times the minimum unstick speed, V_{MU} (defined in Section 10.2.2).

Once the rotation distance is known, the total ground run (s_g) can be determined by summation of the two ground segments:

$$s_g = s_R + s_{RL} \tag{9.36}$$

9.5.3 Energy Method to Estimate the Distance from Liftoff to 35 ft

The takeoff distance from liftoff to clear the 35 ft screen can be determined by considering the work done in accelerating the airplane from its liftoff speed to the speed at the 35 ft screen and simultaneously raising it 35 ft above the runway surface. This approximation ignores the fact that the flight path distance is slightly greater than the runway distance and also assumes zero runway slope and zero wind.

The work done is given by the following approximate expression:

$$U \cong \int (F_N - D)\,\mathrm{d}s = (F_N - D)_{avg}\, s_{L35} \tag{9.37}$$

where U is the work done (typical units: J, ft lb);
$\quad\quad (F_N - D)_{avg}$ is the average excess thrust (typical units: N, lb); and
$\quad\quad s_{L35}$ is the runway distance from liftoff to the 35 ft screen (typical units: m, ft).

The work done can be equated to the sum of the changes in kinetic and potential energies (see Equations 2.42 and 2.44), that is,

$$U = \frac{W}{2g} \left(V_{35}^2 - V_{LOF}^2 \right) + h_{sc} W \qquad (9.38)$$

where V_{35} is the speed at the 35 ft screen (typical units: m/s, ft/s); and
$\quad\quad h_{sc}$ is the screen height (i.e., 35 ft or 10.7 m).

By combining Equations 9.37 and 9.38, the following expression for the distance from liftoff to the 35 ft point is obtained (for zero wind):

$$s_{L35} = \left(\frac{W}{(F_N - D)_{avg}} \right) \left(\frac{V_{35}^2 - V_{LOF}^2}{2g} + h_{sc} \right) \qquad (9.39)$$

9.5.4   Time-Based Method to Estimate the Distance from Liftoff to 35 ft

An alternative—and relatively straightforward—method to estimate the distance from liftoff to the 35 ft point is based on the product of the average time and the average ground speed for this segment of the takeoff:

$$s_{L35} = \Delta t_{L35} \left(\frac{V_{LOF} + V_{35}}{2} - V_W \right) \qquad (9.40)$$

where Δt_{L35} is the time from liftoff to clear the 35 ft screen (unit: s).

The time Δt_{L35} is largely a function of the thrust-to-weight ratio of the airplane (the influence of the thrust-to-weight ratio can be seen in Equation 9.39). The time Δt_{L35}, which is best determined from experimental data, typically varies between 2 and 5 seconds (see Figure 9.10).

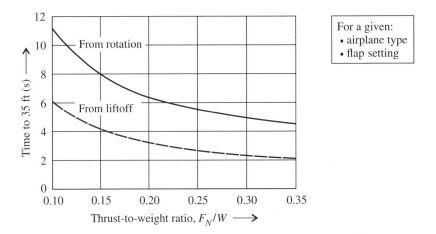

Figure 9.10 Illustration of how the time to clear the 35 ft threshold from rotation or liftoff varies with thrust-to-weight ratio. *Source*: Adapted from [7].

9.6   Empirical Estimation of Takeoff Distances

9.6.1   Introduction

A simple, graphical method[9] to estimate the effect of thrust setting, weight, and air density on the takeoff distance—from brake release to the 35 ft screen height—is presented in this section. The method, which is based on the use of a takeoff parameter (TOP), only applies to nil wind operations from a runway with zero slope. This parameter is used to characterize the actual measured performance (from flight test) so that the effects of changes in thrust or weight on the takeoff distance, for example, can then be approximated. A number of different takeoff parameters can be found in the literature, although they all have a similar format. In Section 9.6.2, an all-engines-operating (AEO) parameter (TOP_{AEO}) is derived; the development is repeated in Section 9.6.3 for one-engine-inoperative (OEI) conditions. Note that the correlation method relies on the assumption that the TOP is proportional to the takeoff distance.

9.6.2   Takeoff Parameter for All-Engines-Operating Condition

This takeoff parameter (TOP_{AEO}) is derived assuming that the airplane has a constant acceleration ($\bar{a}$). This is a mean acceleration, which would result in the same distance as that obtained during the actual conditions where the acceleration reduces as airspeed increases. From Equation 9.30, the takeoff distance can be expressed as

$$s_{35} = \frac{V_{35}^2}{2\bar{a}} \tag{9.41}$$

where s_{35} is the distance from brake release ($V = 0$) to the 35 ft screen height
(typical units: m, ft); and
V_{35} is the TAS at 35 ft (equal to the ground speed as $V_w = 0$) (typical units: m/s, ft/s).

As the greatest part of the takeoff distance will be the ground run, Equation 9.21 can be used to derive a relationship for the mean acceleration. Assuming zero slope (i.e., $\gamma_G = 0$), it follows that

$$\bar{a} = \frac{\sum F}{m} = \frac{g}{W}\left[\bar{F}_N - \frac{1}{2}\rho V^2 S\left(C_D - \mu_R C_L\right) - \mu_R W\right] \tag{9.42}$$

Equation 9.42 can be simplified by noting that the acceleration depends primarily on the mean engine thrust ($\bar{F}_N$) and airplane weight (W). The rolling friction and aerodynamic drag terms together account for approximately 10% to 20% of the net acceleration force applied on the airplane. If this fraction is denoted by a fixed constant k, then the equation can be written as

$$\bar{a} = \frac{g\bar{F}_N}{W}(1 - k) \tag{9.43}$$

Implicit in this derivation is the assumption that, like acceleration, a constant mean thrust can be used to replace the actual thrust (which reduces with airspeed). It was shown earlier (see Sections 9.4.2 and 9.4.3) that the appropriate reference speed at which to calculate $\bar{F}_N$ for the ground run (in nil wind) is $0.707\,V_R$. Extending the estimated distance to the 35 ft screen height requires account to be taken of the small increase in potential energy and the

---

9 The method is based on ESDU 76011 [12].

fact that the airplane must accelerate to V_{35}. The assumption is made that these influences are negligible. The mean thrust has been determined at 70% of V_{35}, which is consistent with the approach taken in ESDU 76011 [12]. (The precise value is not critical as it is only used to determine a correlation parameter.) Substituting Equation 9.43 into Equation 9.41 yields a simplified expression for the takeoff distance:

$$s_{35} = \frac{V_{35}^2 \, W}{2g(1-k)F_{N_{0.7}}} \tag{9.44}$$

where $F_{N_{0.7}}$ is the thrust evaluated at 0.7 V_{35} (typical units: N, lb).

The speed term in Equation 9.44 is now eliminated using the lift coefficient at the screen height ($C_{L_{35}}$), that is,

$$C_{L_{35}} = \frac{2L}{\sigma \rho_0 V_{35}^2 S} = \frac{2n_z W}{\sigma \rho_0 V_{35}^2 S} \tag{9.45}$$

where σ is the relative air density (dimensionless);
ρ_0 is the ISA sea-level datum air density (typical units: kg/m^3, slug/ft^3);
V_{35} is the TAS at 35 ft (it can be assumed that this is equal to the takeoff safety speed, V_2) (typical units: m/s, ft/s);
S is the wing reference area (typical units: m^2, ft^2);
n_z is the load factor normal to the flight path (dimensionless); and
W is the airplane weight (typical units: N, lb).

Thus $\quad s_{35} = \dfrac{n_z}{g\rho_0(1-k)} \left[\dfrac{W^2}{\sigma C_{L_{35}} SF_{N_{0.7}}} \right] \tag{9.46}$

The TOP can be defined, as follows, by assuming that the load factor (n_z) is constant:

$$s_{35} \propto \frac{W^2}{\sigma C_{L_{35}} SF_{N_{0.7}}} = \text{TOP}_{\text{AEO}} \tag{9.47}$$

where TOP_{AEO} is the all-engine-operating takeoff parameter (typical units: N/m^2, lb/ft^2).

If the distance to the screen height (s_{35}) is measured for a series of actual takeoffs at different combinations of thrust, weight, and ambient conditions, then a correlation curve can be established by plotting s_{35} against TOP_{AEO}. Subsequently, s_{35} can be estimated for an untested combination of thrust, weight, and ambient conditions, simply by calculating the appropriate value for the takeoff parameter and then reading off the corresponding distance from the chart.

9.6.3 Takeoff Parameter for One Engine Inoperative

As it is a requirement to consider the possibility of an engine failure when planning a "regulated" takeoff (see Section 9.2.4), it is necessary to obtain correlations of s_{35} for this case. In deriving a suitable takeoff parameter, it is assumed that the aircraft accelerates normally from rest to V_1 (see Section 10.2.3), while covering a distance $s_{0,1}$. The aircraft then suffers an instantaneous total failure of one engine, but continues the takeoff with the remaining engine(s) operating normally, and clears the 35 ft screen height after covering a farther distance $s_{1,35}$. The takeoff distance is the sum of two segments, that is,

$$s_{35} = s_{0,1} + s_{1,35} \tag{9.48}$$

As before (see Equation 9.41), mean acceleration is used to determine the takeoff distance segments. This time, however, two values are needed:

$$s_{35} = \frac{V_1^2}{2\bar{a}_{0,1}} + \frac{V_{35}^2 - V_1^2}{2\bar{a}_{1,35}} = \frac{1}{2}\left[\frac{V_1^2}{\bar{a}_{0,1}} + \frac{V_{35}^2 - V_1^2}{\bar{a}_{1,35}}\right] \tag{9.49}$$

The mean accelerations are now related to a mean thrust, $F_{N_{0.7}}$, which is evaluated at 0.7 V_{35}. The reduction in mean acceleration following the engine failure is assumed to be solely a function of the number of live engines (i.e., $N - 1$); no account is taken of the increase in drag from engine windmilling and/or yaw correction. The acceleration for the two segments has a similar form to that of Equation 9.43, that is,

$$\bar{a}_{0,1} = \frac{gF_{N_{0.7}}(1 - k)}{W} \tag{9.50}$$

and $$\bar{a}_{1,35} = \frac{gF_{N_{0.7}}(1 - k)}{W}\left(\frac{N - 1}{N}\right) \tag{9.51}$$

where N is the number of engines installed.

Substituting Equations 9.50 and 9.51 into Equation 9.49 yields the following expression:

$$s_{35} = \frac{WV_1^2}{2gF_{N_{0.7}}(1 - k)} + \frac{W\left(V_{35}^2 - V_1^2\right)}{2g\,F_{N_{0.7}}(1 - k)}\left(\frac{N}{N - 1}\right)$$

$$= \left(\frac{WV_{35}^2}{2gF_{N_{0.7}}(1 - k)}\right)\left\{N - \left(\frac{V_1}{V_{35}}\right)^2\right\}\left(\frac{1}{N - 1}\right) \tag{9.52}$$

The lift coefficient, as expressed by Equation 9.45, is now introduced:

$$s_{35} = \frac{n_z W^2}{g\rho_0 \sigma C_{L_{35}} S F_{N_{0.7}}(1 - k)}\left\{N - \left(\frac{V_1}{V_{35}}\right)^2\right\}\left(\frac{1}{N - 1}\right)$$

or $$s_{35} = \frac{n_z}{g\rho_0(1 - k)}\left(\frac{W^2}{\sigma C_{L_{35}} S F_{N_{0.7}}}\right)\left\{N - \left(\frac{V_1}{V_{35}}\right)^2\right\}\left(\frac{1}{N - 1}\right) \tag{9.53}$$

Thus $$s_{35} \propto \left(\frac{W^2}{\sigma C_{L_{35}} S F_{N_{0.7}}}\right)\left\{N - \left(\frac{V_1}{V_{35}}\right)^2\right\}\left(\frac{1}{N - 1}\right) = \text{TOP}_{\text{OEI}} \tag{9.54}$$

where TOP_{OEI} is the one-engine-inoperative takeoff parameter (typical units: N/m², lb/ft²).

This approach to estimate the takeoff distance is illustrated in Figure 9.11. A trend line has been drawn through data taken from ESDU 76011 [12]. The takeoff distance to 35 ft, following a single engine failure at V_1, is shown as a function of TOP_{OEI}. In this case, the speed at screen height has been taken as the takeoff safety speed, V_2 (see Section 10.2.2), and the mean thrust evaluated at 0.7 V_2.

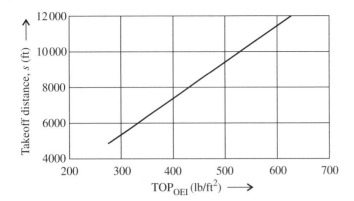

Figure 9.11 Approximate takeoff distance to 35 ft versus TOP_{OEI} following an engine failure at V_1.
Source: Adapted from [12].

9.7 Evaluation of Rejected Takeoff Runway Distances

9.7.1 Regulations Regarding the Rejected Takeoff (RTO)

The certification regulations for the determination of the accelerate–stop (i.e., rejected takeoff) distance have undergone several revisions (amendments) over the past few decades.[10] Consequently, the rules that applied at the time of certification need to be considered in evaluating the RTO performance of a particular airplane type. The description presented herein is in accordance with FAR 25 Amendment 25-92.

Accelerate–stop distances are defined in FAR/CS 25.109 [3, 4] for both dry and wet runways. As mentioned in Section 9.2.5, the additional deceleration available from the use of reverse thrust is not included when determining the required accelerate–stop distance for dry runway operations. For wet runway operations, however, a different procedure is adopted as the runway surface friction is addressed directly within the calculations. Reverse thrust, in this case, can be included provided it is safe, reliable, and its selection does not require exceptional skill to operate the airplane [6].

9.7.2 Requirements for the Accelerate–Stop Distance on a Dry Runway

The required accelerate–stop distance is the greater of the one-engine-inoperative (OEI) RTO distance and the all-engines-operating (AEO) RTO distance. The one-engine-inoperative RTO distance on a dry runway is the sum of the distances necessary to:

(1) accelerate the airplane from a standing start with all engines operating to V_{EF};
(2) allow the airplane to accelerate from V_{EF} to the highest speed reached during the rejected takeoff, assuming the critical engine fails at V_{EF} and the pilot takes the first action to reject the takeoff at V_1; and

---

10 There have been several changes over the years to the FAR 25 rules for the determination of the accelerate–stop (i.e., rejected takeoff) distance. Under Amendment 25-92 rules (applicable to the B737-600/700/800/900, 757-300, 767-400, A321, A330 and A340 types, for example), the distance calculated for the two second "delay" is at constant speed (as illustrated in Figure 9.12), whereas, under the older Amendment 25-42 rules (applicable to the B777, A320 and MD-11 types, for example), the distance is calculated with the assumption that the airplane continues to accelerate [10, 13].

(3) come to a full stop; plus

(4) a distance equivalent to two seconds at the V_1 speed.

The time allowed for the first action by the pilot following the engine failure (e.g., to apply the brakes) is the demonstrated time during testing or one second, whichever is the greater [4, 14]. The time required for the second and third actions (e.g., to retard the thrust levers and extend the ground spoilers) is the demonstrated time during testing, which is typically less than one second. (On most modern transport airplanes, retarding the thrust levers in this situation engages the RTO autobrake function, which is described in Section 11.3.8.)

The principle for calculating one-engine-failure RTO distances, as required for standard flight operations, is based on an expansion of certification test data—this is explained in Section 9.7.4. The all-engines-operating RTO distance is determined in an identical manner. The only differences are that all engines operate until the crew takes the first action at the V_1 speed to reject the takeoff, and that the airplane must come to a full stop with all engines still operating.

9.7.3 Requirements for the Accelerate–Stop Distance on a Wet Runway

The accelerate–stop distance on a wet runway is determined in accordance with the procedures for the dry runway, except that:

(1) the wet runway values of V_{EF} and V_1 are used;

(2) the wet runway braking coefficient of friction (as defined in FAR/CS 25.109 [3, 4]) applies;

(3) the maximum tire-to-ground wet runway braking coefficient of friction must be adjusted to take into account the efficiency of the anti-skid system on a wet runway; and

(4) available thrust reversers can be used (i.e., for the engine failure RTO, it must be assumed that the corresponding thrust reverser is inoperative).

9.7.4 Evaluation of the Accelerate–Stop Distance

Figure 9.12 illustrates the process for determining an acceptable accelerate–stop distance for flight operations based on an expansion of certification test data, considering either (1) a failure of the critical engine, or (2) another, unspecified, abort event with AEO. The accelerate–stop distance for an RTO is the sum of the following distances (which are defined in Figure 9.12):

$$s = s_{AEO} + s_{E1} + s_{V1} + s_T + s_B \tag{9.55}$$

The distance s_{AEO} corresponds to the AEO acceleration from a standing start to V_{EF}, the speed at which the critical engine is assumed to fail or, alternatively, to V_{EV}, the speed at which the abort event is assumed to occur. The distance can be determined using the techniques described in Section 9.4—in particular, the numerical approach (see Section 9.4.4) is appropriate. The distance s_{E1} in the case of the engine-failure RTO is the OEI acceleration from V_{EF} to V_1. The minimum time interval between engine failure (at V_{EF}) and initiation of the first pilot action to stop the airplane (at V_1), permitted by the regulatory authorities, is one second [4, 14]. In the case of the event-abort RTO, the distance s_{E1} is the AEO acceleration distance from V_{EV} to V_1. For either case, the distance can be determined using the method described in Section 9.4.4, taking into account the appropriate engine thrust and drag changes. The distance s_{V1} is determined by multiplying the speed V_1 by the mandatory two second time allowance. The distance s_T, usually called the transition distance, is the distance covered after the first action of the pilot until the airplane is fully configured for braking (i.e., wheel brakes fully efficient and ground spoilers fully deployed). The corresponding transition time depends on the airplane type; data for six airplane types indicated transition times between 1.3 and 3.6 seconds [13]. The distance s_B is the

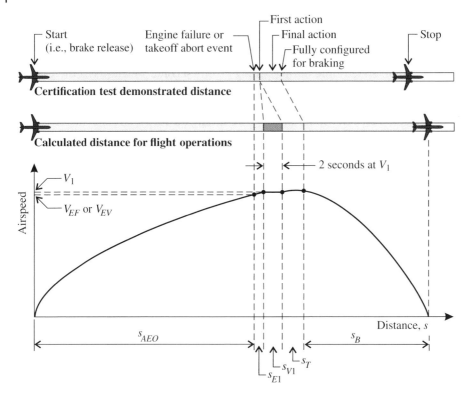

Figure 9.12 Determination of accelerate–stop (rejected takeoff) distance.

braking distance. The analytical evaluation of this distance is non-trivial due to the complexities associated with analyzing the wheel braking force—this is discussed in Section 9.8.

Accurate determination of the RTO distances requires knowledge of the residual thrust generated by the airplane's engines as they spin down (spool down). In the case of a failed engine, a conservative assumption can be made that the thrust decays to zero in a few seconds in a manner that mimics a sudden stoppage of the fuel supply. However, when the pilot moves the thrust levers to the idle position (sometimes called a "throttle chop"), the residual thrust from a "good" engine will decay to the idle-thrust level over a slightly longer period (typically about six seconds). It can thus be assumed for theoretical analyses that a "good" engine will produce a little more thrust during the spool-down than a failed engine, and that the idle thrust will be maintained for the entire deceleration period.

The time-dependent nature of the events after V_{EF} or V_{EV} requires a different approach for the numerical evaluation of the accelerate–stop distance to that described earlier for the accelerate–go distance (see Section 9.4.4). To determine s_{AEO}, a numerical method based on *velocity* increments (as described earlier) can be used. Following an engine failure or closure of the thrust levers, however, it is necessary to determine the distance traveled based on *time* increments (of, say, one second). The procedure for the first time interval (i.e., $t = 0$ s to $t = 1$ s) is as follows:

(1) At $t = 0$ s (corresponding to an engine failure or closure of the thrust levers), calculate the instantaneous acceleration ($a_{t=0}$) based on the net accelerating force (see Section 9.3.8).
(2) At $t = 1$ s, calculate the airplane speed ($V_{t=1}$) based on the assumption that the acceleration determined at the start of the interval (i.e., $a_{t=0}$) applies for the entire interval.

(3) At the speed $V_{t=1}$, calculate the instantaneous acceleration ($a_{t=1}$).
(4) Determine the interval mean acceleration, i.e., $\bar{a}_1 = (a_{t=0} + a_{t=1})/2$.
(5) In a repeat of step 2, calculate a revised estimate of the airplane speed ($V_{t=1}$) at $t = 1$ s based on the interval mean acceleration ($\bar{a}_1$).
(6) Determine the distance covered from $t = 0$ s to $t = 1$ s (i.e., s_1) based on the interval mean acceleration ($\bar{a}_1$).

Having established the distance covered during the first time interval and the end-of-interval speed, it is now possible to move on to the second time interval, which requires steps 1–6 to be repeated. The process is then repeated for each subsequent time interval; the interval distances are evaluated and summed.

9.8 Wheel Braking

9.8.1 Brake System Design

Braking that is accomplished by brakes fitted to the airplane's wheels is often called *mechanical braking* (to distinguish this feature from aerodynamic braking or reverse thrust). A brake unit, which is fitted to each wheel on the main landing gear, has an array of disks that comprise alternating rotor disks (which rotate with the wheel) and stator disks (which are stationary with respect to the landing gear). The stack of rotor disks is securely attached to the wheel such that their axes of rotation are aligned with the wheel axle. Stator disks, which separate each rotor disk, are installed such that when hydraulic pressure[11] within the brake system increases—as controlled by the pilot pressing the brake pedal or commanded by the autobrake system—pistons press the stator disks against the rotor disks. The friction between the disks produces a braking torque that acts on the wheel, reducing its rotational velocity. The friction wears the contacting surfaces down, necessitating periodic replacement of worn disks.

The heat that is generated by the friction between the rotor and stator disks can affect the performance of steel brakes. At extreme temperatures, the steel disks start to melt, reducing the braking force that can be generated—this is known as *brake fade*. The standard method for quantifying brake fade is to conduct a series of high-speed tests with increasing kinetic energy at the point of brake application (the brakes are allowed to cool after each test). The maximum braking force that can be achieved is seen to be initially independent of kinetic energy; however, as the kinetic energy increases beyond a limiting value, the maximum braking force tends to reduce. Modern brake disks manufactured from a carbon-fiber-ceramic composite material, however, have a much higher melting temperature and do not fade in this manner. (Brake energy limits, which apply to an RTO, are discussed in Section 10.3.7.)

9.8.2 Airplane Braking Coefficient

The coefficient of friction for a wheel (μ) moving over a hard surface is defined by Equation 9.7 as the ratio of the tire friction force to the force acting on the tire normal to the runway surface (i.e., the reaction force on the tire). When braking takes place, the friction force increases considerably compared to the rolling friction case (described in Section 9.3.7).

---

11 An electrically activated braking system (which replaces the traditional hydraulically activated system) was approved for the B787 in 2011; this was the first application of an electric brake system on a jet transport airplane.

It is customary for the evaluation of airplane stopping distances to use an airplane *braking coefficient*, which is defined as follows:

$$\mu_B = \frac{F_{br}}{W \cos \gamma_G - \bar{L}}$$

(9.56)

where μ_B is the airplane braking coefficient (dimensionless);
 F_{br} is the braking force (typical units: N, lb);
 W is the airplane weight (typical units: N, lb);
 γ_G is the runway slope angle (uphill is positive) (typical unit: rad); and
 $\bar{L}$ is the mean lift (typical units: N, lb).

The braking coefficient (μ_B), as defined by Equation 9.56, is not exactly the same as a coefficient of friction (μ). This braking coefficient is based on the airplane's entire weight and not just the reaction force on the main landing gear (where the brakes are installed). Furthermore, braking coefficients are usually determined by experiment and based on a mean reaction force corresponding to the test duration. In such evaluations, the lift is taken as a mean value.

Unlike the rolling coefficient of friction, the braking coefficient depends significantly on the airplane's speed. As the airplane slows down, the braking coefficient increases. When the runway is dry, the increase is small—permitting a constant value to be used to determine stopping distances. When the runway is wet, however, there is a large non-linear increase as the airplane slows down. Other factors that influence the braking coefficient include:

(1) the amount of slip that occurs between the tire and the runway (see Section 9.8.3);
(2) the maximum brake torque (see Section 9.8.4); and
(3) the nature of the tire–runway-surface interface, which depends on (i) whether the surface is dry, wet, or contaminated; (ii) the type of tire and its pressure and condition (i.e., new or worn); and (iii) the runway surface profile and condition (see Sections 9.8.5 and 9.8.6).

9.8.3 Slip and Anti-Skid Systems

Slip is the difference between the airplane's ground speed and the tire speed; it is usually expressed as a percentage of the airplane's ground speed—for example: an airplane traveling at 120 kt, with a tire speed of 90 kt (based on the rotational speed of the tire), experiences a slip of 30 kt or 25%. A free rolling tire (i.e., zero slip) has the same speed as the airplane, and provides no braking force—just a rolling resistance to the motion. The friction force increases as the amount of slip increases; it reaches a maximum at about 12% slip [15] and then starts to decrease (Figure 9.13). In the extreme condition, 100% slip (i.e., a locked wheel)

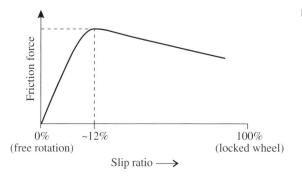

Figure 9.13 Friction force versus slip ratio.

will produce a skid, resulting in a substantially reduced friction force (and a danger of a loss of directional control).

Anti-skid systems—employed on all modern jet airplanes—work by comparing the speed of each main gear wheel (obtained from tachometer readings) with the airplane's forward speed. When the wheel speed drops below a predetermined percentage of the airplane's speed (say ~87%), the system instructs the brakes to momentarily release, thus maintaining the optimum brake slip ratio (and protecting the wheels from locking).

The braking coefficient (μ_B) can be expressed as the product of an anti-skid braking efficiency factor (η_B) and the maximum tire-to-ground braking coefficient (μ_{max}):

$$\mu_B = \eta_B \, \mu_{max} \tag{9.57}$$

The value of η_B depends on the type of anti-skid system installed and is best determined experimentally for the airplane type. The following data are provided in FAR/CS 25.109 [3, 4] for three different systems: for an *on-off* system, $\eta_B = 0.3$; for a *quasi-modulating* system, $\eta_B = 0.5$; and for a *fully modulating* system, $\eta_B = 0.8$.

9.8.4 Maximum Brake Torque Limit

The determination of the braking force due to the wheel brakes is further complicated by the fact that the theoretical braking force determined by the product of the braking coefficient and the reaction force (see Equation 9.56) may not be achievable in practice because of aircraft brake system limitations. The "internal" braking torque, which is due to the friction that is generated between the stator and rotor disks, opposes the "external" torque, which is the product of the tire friction force (which acts at the tire–runway-surface interface) and the tire radius.

Accelerate–stop tests are conducted with varying airplane weight and configuration. The CG is required to be in the most forward position allowed for takeoff (this conservatism is introduced as it results in the lowest percentage of the airplane's weight being carried by the tires on the main landing gear). The test data are used to establish a range of braking coefficient values for different flap settings and runway conditions. The maximum braking force values achieved in such tests can be used to generate a graph of the form illustrated in Figure 9.14. The indicated limits establish a region of viable braking conditions. At low airplane weights, the anti-skid system limits the braking force—the slope of this line represents the maximum braking coefficient that the system can deliver. At higher weights, a limit is imposed by the maximum torque that the brake system can produce. In the latter case, the anti-skid system plays no part in limiting the braking action.

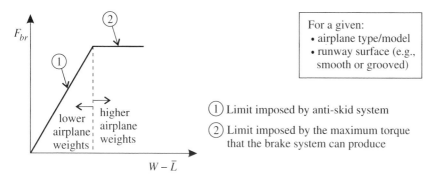

Figure 9.14 Braking force versus mean reaction force ($W - \bar{L}$) for a dry runway.

9.8.5   Runway Pavement Construction

The braking performance of an airplane depends on the micro- and macrotexture of the runway pavement surface. Microtexture, which is important at low speeds, is the fine-scale roughness that would be apparent to the touch. Macrotexture, which is important at high speeds when the runway is wet, is a larger-scale visible roughness of the pavement surface. The purpose of the macrotexture is to provide paths for water to be removed from beneath the airplane's tires, and thus improve braking under wet conditions.

There are two main types of runway surface treatment currently used to produce the desired macrotexture: grooved and Porous Friction Course (PFC). Grooved runways that comply with FAA AC 150-5320-12 [16] have transverse grooves of 6 mm (0.25 inch) deep by 6 mm (0.25 inch) wide at 38 mm (1.5 inch) spacing, which are created either by forming the concrete before it is fully hardened or by saw-cutting the surface. A PFC construction has a top layer, known as a course, of an open-graded hot-mix asphalt matrix, which is 19–38 mm (0.75–1.5 inch) thick [16]. Grooved and PFC runway surfaces can achieve airplane braking coefficients under wet conditions in excess of 70% of that which is achievable under dry conditions.

Another aspect that can affect braking is runway maintenance. Localized deposits of rubber—left on the surface when the tires on landing airplanes spin-up on touchdown—reduce braking effectiveness and have to be removed during routine runway maintenance.

9.8.6   Wet Runway Conditions

Water on the runway leads to a loss of contact between the tire and the runway surface, resulting in a reduction in friction compared to a dry runway. As the tire rolls over the surface, the water tends to be squeezed out from beneath the tire; however, this is not always fully achieved. At high speeds, there is insufficient time for the water to be completely squeezed out, resulting in reduced friction—the braking coefficient thus reduces as the airplane's speed increases.

The braking coefficient on wet runways at high speed is relatively low, but it increases substantially with decreasing speed. FAR/CS 25.109 [3, 4] provides data on the maximum braking coefficients that can be used for braking evaluations on wet runways.[12] This information is provided for two categories of runway surface: smooth runways (Figure 9.15) and runways that have a grooved or PFC surface (Figure 9.16). It is evident that at high speed the maximum tire-to-ground braking coefficient can be as low as about 0.15, increasing to about 0.6 to 0.9 (depending on the tire pressure) as the airplane's ground speed drops. These maximum tire-to-ground braking coefficient values (μ_{max}) must be multiplied by the braking efficiency factor (η_B) to account for the anti-skid system installed in the airplane (see Section 9.8.3) to give an appropriate airplane braking coefficient.

Further information on airplane braking coefficients under dry and wet conditions on various runway surfaces is provided in CS-25 Book 2 (section AMC 25.109(c)(2)) [4], FAA AC 25.7 [14], ESDU 71025 [18], and ESDU 71026 [17].

9.8.7   Determination of the Braking Force

As described earlier in this chapter (see Sections 9.8.2–9.8.6), it is possible—albeit with some difficulty—to determine the braking force (F_{br}) that can be achieved by the wheel brakes based

---

12 This method is sometimes called the "ESDU method" as it is based on ESDU 71026 [17].

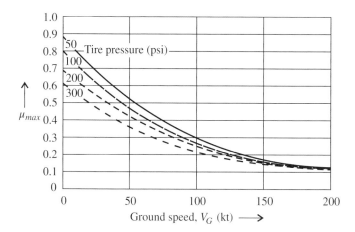

Figure 9.15 Maximum braking coefficient versus airplane ground speed for a wet, smooth runway. *Source*: Data from FAR/CS 25.109 [3, 4].

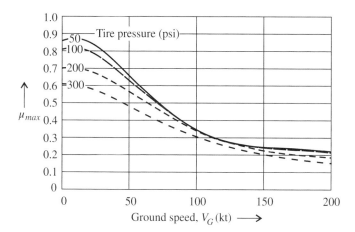

Figure 9.16 Maximum braking coefficient versus airplane ground speed for a wet, grooved or PFC runway. *Source*: Data from FAR/CS 25.109 [3, 4].

on an airplane braking coefficient. The force can be evaluated as follows:

$$F_{br} = \mu_B(W \cos \gamma_G - L) \tag{9.58}$$

where L is the instantaneous lift (typical units: N, lb).

The small angle approximation $\cos \gamma_G \cong 1$ is now applied to Equation 9.58, and the lift is expressed in terms of the airplane's lift coefficient (using Equation 9.3):

$$F_{br} = \mu_B W - \frac{1}{2}\rho V^2 S \mu_B C_L \tag{9.59}$$

For a dry runway, a single value of μ_B can be used to determine the stopping distance. This means that the stopping distance can be calculated using a closed-form solution in the manner described in Section 9.4.2 for the takeoff distance. In the absence of actual airplane data, the

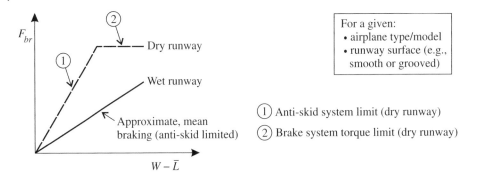

Figure 9.17 Wet runway braking (approximately 50% of dry runway braking).

braking coefficient can be estimated from Equation 9.57, with a value of μ_{max} taken from Figure 9.15 for zero speed. For initial performance calculations, a value of μ_B of between 0.38 and 0.41 can be used for dry, hard runway surfaces (this would represent a pilot "standing on" the brakes, with the anti-skid system fully operational).

For a wet runway, the braking coefficient varies significantly as a function of speed. The evaluation of the stopping distance should ideally be undertaken using a numerical technique in which braking coefficient values are selected to match the airplane's speed as it slows down. It is possible, however, to obtain a reasonably accurate estimate for the wet runway stopping distance using a constant braking coefficient that is 50% of the corresponding dry value (as illustrated in Figure 9.17). This simplified approach, in fact, has been used in the past for the determination of actual AFM stopping distance data [9].

It is important to note that under normal operations, these maximum braking forces are seldom required. Only under emergency conditions would the braking forces approach the maximum braking level.

9.9 Takeoff on Contaminated Runways

9.9.1 Contaminated Runway Conditions

In an operational context, runways are considered to be *contaminated* if more than 25% of the runway surface is covered by water, slush, snow, or ice, with the proviso that runways covered with water (or slush) to a depth of less than 3 mm (EASA rules) or 0.125 inch (FAA rules) are classified as *wet*, but not contaminated.

A description of the various contaminants applicable to aircraft operations is given in Table 9.2. The terminology used varies across the industry; the descriptions provided herein are typical, and are based on details provided in CS-25 Book 2 (section AMC 25.1591) [4]. Typical braking coefficient (μ_B) values for the various contaminated runway surfaces are given. Contaminant drag (see Section 9.9.2)—which applies to water, slush, or snow, but not compacted snow or ice—needs to be taken into account when analyzing the airplane's acceleration or deceleration if the contaminant depth exceeds a critical value (i.e., 3 mm for water or slush, 5 mm for wet snow, and 10 mm for dry snow). Specific gravity (σ_{SG}) data are provided in the table for this calculation. According to the regulations, a takeoff should not be attempted on a runway where the contaminant depth exceeds 15 mm (EASA rules) or 12.5 mm (FAA rules) [4, 19].

Table 9.2 Runway contaminant data

Contaminant	Typical μ_B Specific gravity (σ_{SG})	Depths to be considered	Contaminant drag applies?
Standing water (i.e., flooded runway of water-depth greater than 3 mm)	μ_B (see note b) $\sigma_{SG} = 1.0$	3–15 mm (0.12–0.59 inch)	yes
Slush (i.e., partly melted snow or ice with a high water content, from which water can readily flow)	μ_B (see note b) $\sigma_{SG} = 0.85$	3–15 mm (0.12–0.59 inch)	yes
Wet snow (i.e., snow that will stick together when compressed, but will not readily allow water to flow from it when squeezed)	$\mu_B = 0.17$ $\sigma_{SG} = 0.5$	< 5 mm (0.20 inch)	no
		5–30 mm (0.2–1.18 inch)	yes
Dry snow (i.e., fresh snow that can be blown, or, if compacted by hand, will fall apart upon release; also called *loose snow*)	$\mu_B = 0.17$ $\sigma_{SG} = 0.2$	< 10 mm (0.39 inch)	no
		10–130 mm (0.39–5.12 inch)	yes
Compacted snow (i.e., solid mass of snow compressed such that airplane wheels do not cause significant rutting)	$\mu_B = 0.20$	0 mm (0 inch)	no
Ice (i.e., frozen water or compacted snow that has transitioned to a "polished" ice surface)	$\mu_B = 0.05$	0 mm (0 inch)	no

Notes:

(a) *Source*: Based on EASA CS-25 Book 2 [4].

(b) The braking coefficient for standing water and slush is determined as follows:

For $V_G < V_{G,hp}$, $\mu_B = 0.3485 - 0.0632 \left(\dfrac{V_G}{100}\right)^3 + 0.2683 \left(\dfrac{V_G}{100}\right)^2 - 0.4321 \left(\dfrac{V_G}{100}\right)$

and for $V_G \geq V_{G,hp}$, $\mu_B = 0.05$

where V_G is the airplane's ground speed (unit: kt); and
$V_{G,hp}$ is the airplane's hydroplaning speed (see Section 9.9.3) (unit: kt).

9.9.2 Contaminant Drag

Fluid contaminants (e.g., standing water or slush) and loose contaminants (e.g., snow that is not compact) on a runway create additional drag on the airplane—this is known as *contaminant drag* or *slush drag*. It results from two distinct mechanisms: the fluid in the path of the tires gets displaced, resulting in a retarding force on the tires; and secondly, the contaminant material gets sprayed up from the tires, impinging on the landing gear and airframe, thus slowing the airplane down.

A simple model to predict the retarding force for both the displacement and impingement force components, herein denoted as F_{con}, is based on an experimentally derived coefficient:[13]

$$F_{con} = \frac{1}{2}\rho_{con}V_G^2 A_{gear}C_{D_{con}} \tag{9.60}$$

where ρ_{con} is the contaminant density (typical units: kg/m^3, slug/ft^3);
V_G is the ground speed (typical units: m/s, ft/s);
A_{gear} is a reference area based on a frontal projection of the impinged tire and landing gear components (typical units: m^2, ft^2); and
$C_{D_{con}}$ is the contaminant drag coefficient (dimensionless).

---

13 Tests conducted by the FAA and NACA using a Convair 880, which generated full-scale experimental data on this effect, are described by Sommers, *et al.* [20].

Details on approved methods for the determination of displacement drag and impingement drag can be found in CS-25 Book 2 (section AMC 25.1591) [4]; more detailed information is given in references [20–24].

9.9.3 Dynamic Hydroplaning

Fluid contaminants on a runway can lead to hydroplaning (also called aquaplaning). This condition exists, at relatively high speeds, when the water cannot be fully squeezed out from beneath the tires and, to a large extent, the tires are separated from the runway surface by a thin fluid layer. Dynamic hydroplaning is a limiting condition in which the tire is supported clear of the runway surface by fluid pressure generated principally by inertial effects [25]. Tire traction is significantly reduced; the friction force is almost zero, and nosewheel steering is virtually ineffective.

The *hydroplaning speed* is defined as the speed at and above which dynamic hydroplaning occurs. The hydroplaning speed depends critically on the tire pressure: high inflation pressure tires can remove surface water more efficiently from the tire–pavement footprint than low inflation pressure tires. The following semi-empirical relationship—which is accepted by EASA [4]—is widely used to estimate the hydroplaning speed for an *unbraked rolling tire* where the water (or slush or snow) depth is greater than the tire tread depth:[14]

$$V_{G,hp} = 9\sqrt{p_{tire}} \tag{9.61}$$

where $V_{G,hp}$ is the hydroplaning ground speed (unit: kt); and

p_{tire} is the tire pressure (unit: lb/in^2).

An alternative formulation, which is also widely used, includes the density of the surface contaminant [7]:

$$V_{G,hp} = 8.63\sqrt{\frac{p_{tire}}{\sigma_{SG}}} \tag{9.62}$$

where σ_{SG} is the specific gravity of the contaminant (see Table 9.2).

9.9.4 Estimating Takeoff and RTO Distances on Contaminated Runways

Contaminant drag should always be taken into account when analyzing the takeoff (and landing) distances on runways with loose or fluid contaminants, as this has a significant impact on the airplane's acceleration. Reference [28] reports that the OEI takeoff acceleration capability of Boeing 737/747/757/767 airplanes at 130 kt is reduced by 15% to 50% (depending on the airplane type) on runways with only 6 mm (0.25 inch) of slush.

The takeoff distance on a contaminated runway can be evaluated using the step-wise numerical approach described in Section 9.4.4. In such analyses, the contaminant drag needs to be included in the equation that describes the net accelerating force (see Section 9.3.8); Equation 9.10 can thus be written as:

$$\sum F = F_N - D - F_{fr} - W\sin\gamma_G - F_{con} \tag{9.63}$$

The contaminant drag increases in magnitude as a square of the airplane's ground speed. Once the ground speed has reached the hydroplaning speed, the contaminant drag will cease increasing at this rate: it will then remain approximately constant before starting to reduce. At the

---

14 Equation 9.61, proposed by Horne [26, 27], was initially based on testing conducted by NASA in the 1960s.

rotation speed, the nosewheels will be raised out of the water (or slush or snow), and the contaminant drag associated with the nose landing gear will abruptly disappear. Thereafter, the contaminant drag will rapidly reduce to zero as the airplane lifts off the runway. The screen height is 35 ft for the AEO takeoff case and 15 ft for the OEI takeoff case (which is the same as for a wet runway—see Section 9.2.7).

The determination of the RTO distance for a contaminated runway can be undertaken using the method described earlier in Section 9.7 for a wet runway, and the same time delays apply. The airplane will have a reduced wheel braking capability, but the presence of the contaminant drag will act to slow the airplane down. Credit for reverse thrust is permitted when determining accelerate–stop distances.

References

1 FAA, "Reduced and derated takeoff thrust (power) procedures," Advisory Circular 25-13, Federal Aviation Administration, Washington, DC, May 4, 1988. Available from www.faa.gov/.

2 European Commission, *Commercial air transport operations (Part-CAT)*, Annex IV to Commission Regulation (EU) No. 965/2012, Brussels, Belgium, Oct. 5, 2012. Published in *Official Journal of the European Union*, Vol. L 296, Oct. 25, 2012, and reproduced by EASA.

3 FAA, *Airworthiness standards: Transport category airplanes*, Federal Aviation Regulation Part 25, Amdt. 25-143, Federal Aviation Administration, Washington, DC, June 24, 2016. Latest revision available from www.ecfr.gov/ under e-CFR (Electronic Code of Federal Regulations) Title 14.

4 EASA, *Certification specifications and acceptable means of compliance for large aeroplanes*, CS-25, Amdt. 18, European Aviation Safety Agency, Cologne, Germany, June 23, 2016. Latest revision available from www.easa.europa.eu/ under Certification Specification.

5 FAA, *Operating requirements: Domestic, flag, and supplemental operations*, Federal Aviation Regulation Part 121, Amdt. 121-374, Federal Aviation Administration, Washington, DC, May 24, 2016. Latest revision available from www.ecfr.gov/ under e-CFR (Electronic Code of Federal Regulations) Title 14.

6 IFALPA, "Runway safety manual," Aerodrome & Ground Environment Committee, International Federation of Air Line Pilots' Associations, Montréal, Canada, 2009.

7 Blake, W. and Performance Training Group, "Jet transport performance methods," D6-1420, Flight Operations Engineering, Boeing Commercial Airplanes, Seattle, WA, Mar. 2009.

8 Bartel, M. and Young, T.M., "Simplified thrust and fuel consumption models for modern two-shaft turbofan engines," *Journal of Aircraft*, Vol. 45, Iss. 4, pp. 1450–1456, 2008.

9 Boeing, "Jet transport performance methods," D6-1420, The Boeing Company, Seattle, WA, May 1989.

10 Boeing, "Takeoff performance," Training course notes, Flight Operations Engineering, The Boeing Company, Seattle, WA, 2009.

11 Airbus, "Getting to grips with aircraft performance," Flight Operations Support and Line Assistance, Airbus S.A.S., Blagnac, France, Jan. 2002.

12 ESDU, "First approximation to take-off field length of multi-engined transport aeroplanes," Data item 76011, IHS ESDU, 133 Houndsditch, London, UK, May 1985.

13 Airbus, "Takeoff safety training aid," Iss. 2, Flight Operations Support and Line Assistance, Airbus S.A.S., Blagnac, France, Nov. 2001.

14 FAA, "Flight test guide for certification of transport category airplanes," Advisory Circular 25-7C, Federal Aviation Administration, Washington, DC, Oct. 16, 2012. Available from www.faa.gov/.

15 Airbus, "Getting to grips with cold weather operations," Flight Operations Support and Line Assistance, Airbus Industrie, Blagnac, France, Jan. 2000.

16 FAA, "Measurement, construction, and maintenance of skid-resistant airport pavement surfaces," Advisory Circular 150-5320-12C, Federal Aviation Administration, Washington DC, Mar. 18, 1997.

17 ESDU, "Frictional and retarding forces on aircraft tyres – Part II: Estimation of braking force," Data item 71026, Amdt. D, IHS ESDU, 133 Houndsditch, London, UK, June 1995.

18 ESDU, "Frictional and retarding forces on aircraft tyres – Part I: Introduction," Data item 71025, Amdt. D, IHS ESDU, 133 Houndsditch, London, UK, Apr. 1995.

19 FAA, "Water, slush and snow on the runway," Advisory Circular 91-6A, Federal Aviation Administration, Washington DC, May 24, 1978.

20 Sommers, D.E., Marcy, J.F., Klueg, E.P., and Conley, D.W., "Runway slush effects on the takeoff of a jet transport," Project No. 308-3X, National Aviation Facilities Experimental Center, Atlantic City, NJ, May 1962.

21 ESDU, "Estimation of spray patterns generated from the sides of aircraft tyres running in water or slush," Data item 83042, Amdt. A, IHS ESDU, 133 Houndsditch, London, UK, Apr. 1998.

22 ESDU, "Estimation of airframe skin-friction drag due to impingement of tyre spray," Data item 98001, IHS ESDU, 133 Houndsditch, London, UK, Apr. 1998.

23 ESDU, "Frictional and retarding forces on aircraft tyres – Part V: Estimation of fluid drag forces," Data item 90035, Amdt. A, IHS ESDU, 133 Houndsditch, London, UK, Oct. 1992.

24 ESDU, "Operations on surfaces covered with slush," Memorandum No. 96, IHS ESDU, 133 Houndsditch, London, UK, Feb. 1998.

25 ESDU, "Frictional and retarding forces on aircraft tyres – Part III: Planing," Data item 72008, IHS ESDU, 133 Houndsditch, London, UK, May 1972.

26 Horne, W.B. and Dreher, R.C., "Phenomena of pneumatic tire hydroplaning," NASA TN-D-2056, National Aeronautics and Space Administration, Langley Research Center, Hampton, VA, 1963.

27 Horne, W.B., "Wet runways," NASA TM-X-72650, National Aeronautics and Space Administration, Langley Research Center, Hampton, VA, Apr. 1975.

28 Boeing, "Takeoff/landing on wet, contaminated and slippery runways," Training course notes, Flight Operations Engineering, Boeing Commercial Airplanes, Seattle, WA, Sept. 2009.

10

Takeoff Field Length and Takeoff Climb Considerations

10.1 Introduction

Takeoff is a critical part of any flight. To ensure that a high level of safety is consistently achieved for the thousands of flight operations that take place each day around the world (transporting several million people), a series of regulations have evolved and will likely continue to evolve as new information, technologies, and transportation demands emerge. These regulations, which are issued by the aviation authorities, fall into two categories: (1) airplane certification and (2) flight operation. *Certification*—as described in FAR/CS 25 [1, 2], for example—is essentially a series of tests and compliance checks that ensure that a candidate airplane meets a minimum set of safety standards. The airplane's proven performance during certification demonstrates a performance potential that has been deemed to be adequate for the safe operation of the airplane. The *operational procedures*—as described in FAR 121 [3] and EASA OPS Part-CAT [4], for example—ensure that the airplane is then utilized in a safe and efficient manner. An overview of the key regulatory topics that directly impact the takeoff performance of jet transport airplanes is presented in this chapter.

It is important to note that airworthiness regulations are usually not retroactive. This means that the rules that currently apply to the operation of a particular in-service airplane are the regulations under which certification of the airplane type took place (and not the current rules). This means that slightly different definitions and procedures are in force for older generation airplane types compared to newer types.[1]

A summary of the key reference speeds, relevant to the takeoff, is provided in Section 10.2. Section 10.3 introduces the topic of takeoff limitations and considers the various factors that can limit an airplane's takeoff weight. This is followed by a description of the conventions used to describe runway distances (Section 10.4) and runway-limited takeoff weights (Section 10.5). The topics of takeoff climb performance and obstacle clearance are discussed in Sections 10.6 and 10.7, respectively. The widely used practices of derated thrust and reduced thrust takeoff are discussed in Section 10.8.

---

1 Occasionally these changes are beneficial to a particular airplane type—for example, Airbus re-certified the A320 to take advantage of a significant revision to the regulations.

Performance of the Jet Transport Airplane: Analysis Methods, Flight Operations, and Regulations, First Edition.
Trevor M. Young.
© 2018 John Wiley & Sons Ltd. Published 2018 by John Wiley & Sons Ltd.

10.2 Takeoff Reference Speeds

10.2.1 Overview

The airworthiness regulations/specifications FAR 25 [1] and CS-25 [2] define a number of reference speeds for the operation of multiple engine transport airplanes—the key reference speeds applicable to takeoff are described in this section. Note that the precise definition of some of these speeds has changed in the past (as the regulations/specifications have evolved).

There is a subtle point that should be noted in regard to these reference speeds. The speeds are selected by the airplane manufacturer for the purpose of certification of the airplane. The manner in which they are defined—and validated—corresponds to a controlled test environment, which is not representative of a typical operational scenario. The operating procedures, as published in the airplane's Operations/Operating Manual (OM), are based on these reference speeds, but take into account realistic operational conditions.

Furthermore, in many cases, conservative restrictions are placed on the certification test procedures, which would not apply in normal operations—for example: during the evaluation of V_{MCG} (minimum control speed on the ground), nosewheel steering is not permitted, but it would normally be available—and helpful—to control the airplane following an actual in-service engine failure.

The reference speeds are described in FAR/CS 25.107 and FAR/CS 25.149 [1, 2]; supporting information relating to certification testing is provided in FAA AC 25-7 [5] and CS-25 Book 2 [2]. The descriptions that follow in Section 10.2.2 are based on the regulations/specifications; however, for the sake of brevity, certain of the conditions relating to some of the speeds have been omitted (the reader should consult the regulations/specifications for a complete description).

The reference speeds are defined in the regulations/specifications as *calibrated* airspeed. It should be noted, however, that the same notation (i.e., V_R for rotation speed and V_{LOF} for lift-off speed) is used in the preceding chapter, where the determination of the takeoff distance is based on the use of true airspeed (and not calibrated airspeed).

10.2.2 Regulatory Definitions of Key Reference Speeds

Minimum Unstick Speed, V_{MU} (FAR/CS 25.107 [1, 2])
V_{MU} is the airspeed at and above which the airplane can safely lift off the ground and continue the takeoff. V_{MU} is determined at the greatest pitch (longitudinal) angle that an airplane can achieve on the ground (i.e., in a "tail-dragging" condition). V_{MU} speeds must correspond to the range of thrust-to-weight ratios to be certificated.

Minimum Control Speed—Ground, V_{MCG} (FAR/CS 25.149 [1, 2])
V_{MCG} is the slowest airspeed at which, when the critical engine is suddenly made inoperative, it is possible to maintain control of the airplane using the rudder control alone (i.e., without the use of nosewheel steering), as limited by 150 lb (667 N) force, and the lateral control to the extent of keeping the wings level to enable the takeoff to be safely continued using normal piloting skill. Furthermore, assuming that the airplane's initial path is along the centerline of the runway, it may not deviate more than 30 ft laterally from the centerline. V_{MCG} is established with the most critical takeoff configuration, with maximum available takeoff thrust on the operating engines.

Minimum Control Speed, V_{MC} (FAR/CS 25.149 [1, 2])

V_{MC} is the airspeed at which, when the critical engine is suddenly made inoperative (in the most critical mode of failure with respect to controllability), it is possible to maintain control of the airplane and maintain straight flight with an angle of bank of not more than 5°. V_{MC} may not exceed 1.13 times the reference stall speed V_{SR} (see Section 20.5.2) with the airplane in the most critical takeoff configuration existing along the flight path, except with the landing gear retracted, at maximum available takeoff thrust and maximum sea-level takeoff weight, and with negligible ground effect. The rudder forces required to maintain control at V_{MC} may not exceed 150 lb (667 N) nor may it be necessary to reduce thrust of the operative engines. During recovery, the airplane may not assume any dangerous attitude or require exceptional piloting skill, alertness, or strength to prevent a heading change of more than 20°.

Engine Failure Speed, V_{EF} (FAR/CS 25.107 [1, 2])

V_{EF} is the airspeed at which the critical engine is assumed to fail. V_{EF} may not be less than V_{MCG}.

Liftoff Speed, V_{LOF} (FAR/CS 25.107 [1, 2])

V_{LOF} is the airspeed at which the airplane first becomes airborne.

Rotation Speed, V_R (FAR/CS 25.107 [1, 2])

V_R is the speed at which rotation commences. V_R may not be less than:

(1) V_1;
(2) 105% of V_{MC};
(3) the speed that allows reaching V_2 at or before reaching a height of 35 ft (or 11 m) above the takeoff surface; or
(4) a speed that, if the airplane is rotated at its maximum practicable rate, will result in a V_{LOF} of not less than:
  (i) 110% of V_{MU} in the all-engines-operating condition, and 105% of V_{MU} determined at the thrust-to-weight ratio corresponding to the one-engine-inoperative condition; or
  (ii) if the V_{MU} attitude is limited by the geometry of the airplane (i.e., tail contact with the runway), 108% of V_{MU} in the all-engines-operating condition, and 104% of V_{MU} determined at the thrust-to-weight ratio corresponding to the one-engine-inoperative condition.

Takeoff Safety Speed, V_2 (FAR/CS 25.107 [1, 2])

V_2 is the airspeed that provides at least the gradient of climb required by FAR/CS 25.121(b) [1, 2] (details are provided later in Section 10.6, Table 10.1), but may not be less than:

(1) V_{2MIN};
(2) V_R plus the speed increment attained before reaching a height of 35 ft (or 11 m) above the takeoff surface; or
(3) a speed that provides the maneuvering capability specified in FAR/CS 25.143(h) [1, 2].[2]

Minimum Takeoff Safety Speed, V_{2MIN} (FAR/CS 25.107 [1, 2])

V_{2MIN} may not be less than:

(1) 1.08 V_{SR} for airplanes with provisions for obtaining a significant reduction in the one-engine-inoperative power-on stall speed;

---

2 This requirement is for the airplane, at the V_2 speed, to be free of stall warning or other characteristics that might interfere with normal maneuvering, with asymmetric thrust, during a coordinated turn with a 30° bank angle.

(2) 1.13 V_{SR} for airplanes without provisions for obtaining a significant reduction in the one-engine-inoperative power-on stall speed; or

(3) 1.10 times V_{MC}.

Final Takeoff Speed, V_{FTO} (FAR/CS 25.107 [1, 2])

V_{FTO} is the airspeed that will provide at least the gradient of climb required by FAR/CS 25.121(c) [1, 2] (see Table 10.1 in Section 10.6), but may not be less than:

(1) 1.18 V_{SR}; or

(2) a speed that provides the maneuvering capability specified in FAR/CS 25.143(h) [1, 2].[3]

10.2.3 V_1 Speed

The V_1 speed is described in FAR/CS 25.107 [1, 2] as follows:[4] V_1 may not be less than V_{EF} plus the speed gained with the critical engine inoperative during the time interval between the instant at which the critical engine fails and the instant at which the pilot recognizes and reacts to the engine failure, as indicated by the pilot's initiation of the first action (e.g., applying brakes or reducing thrust) to stop the airplane during accelerate–stop tests.

V_1 is a much more complex takeoff reference speed than those described earlier in Section 10.2.2. It is also quite different from the reference speeds V_R and V_2, for example. When the weight, altitude, ambient temperature,[5] and flap setting are known for a particular take-off, V_R and V_2 can be uniquely determined. The values are specific to those takeoff conditions and do not vary; they depend only on the airplane's thrust and aerodynamic characteristics. V_1 too depends on weight, altitude, temperature, and flap setting, but it also depends on the length characteristics of the runway to be used—specifically the TORA, TODA, and ASDA (these runway distances are defined later in Section 10.4). Furthermore, except in the most limiting case, V_1 can vary within a range of speeds, while still satisfying all of the regulatory takeoff requirements. This feature can sometimes be exploited to increase an airplane's allowable takeoff weight. The relationship between V_1 and the takeoff weight, on the one hand, and the *accelerate–go* and *accelerate–stop* distances, on the other hand, is discussed later in this chapter when the takeoff web chart is introduced in Section 10.5.

For the purposes of determining the takeoff speeds and distances that are published in the Airplane Flight Manual (AFM), V_1 is simply the speed at which the first action is taken—usually brake application—to abort the takeoff. For the flight crew, the V_1 speed—which can be ascertained from the Quick Reference Handbook (see Section 23.7.3) or a computer program, for example—is the primary factor influencing the so-called *go/no-go* decision. A good way for a pilot to think about V_1 is that if he/she cannot be "standing" on the brakes by V_1 then it is too late to abort. The flight crew would usually set a marker (commonly called a bug) for V_1 on the airspeed indicators, before takeoff. On the vertical speed scale of a typical Primary Flight Display (PFD) (illustrated in Figure 5.2a), V_1 is identified by the numeral "1" located adjacent to the selected speed.

---

3 This requirement is for the airplane, at the V_{FTO} speed, to be free of stall warning or other characteristics that might interfere with normal maneuvering, with asymmetric thrust, during a coordinated turn with a 40° bank angle.
4 The definition of V_1 contained in the regulations has changed several times in the past; the most recent change by the FAA was in 1998. Prior to this amendment of FAR 25, V_1 was called the *takeoff decision speed*—a term that is no longer considered appropriate (as the go/no-go decision must already have been made by the pilot by the time V_1 is reached).
5 In flight operations, the ambient temperature is usually called the *outside air temperature* (OAT).

10.3 Takeoff Weight Limitations

10.3.1 Overview

The overarching regulatory requirement concerning takeoff weight limitations is given by the FAA and EASA in FAR 121.189(a) [3] and OPS Part CAT.POL.A.205(a) [4], respectively. It is simply stated that no person may take off an airplane at a weight greater than that listed in the Airplane Flight Manual (AFM) for the elevation of the airport and for the ambient temperature existing at takeoff. The determination of an airplane's maximum takeoff weight for a given set of operational conditions—which will naturally dictate the maximum payload that can be used for a particular mission—is a vital part of the operational flight planning process. The various considerations that could potentially impose a limit on an airplane's takeoff weight are discussed in this chapter.

10.3.2 Structural-Limited Weights and Certified Weights

The maximum taxi weight (MTW), maximum takeoff weight (MTOW), and maximum landing weight (MLW) are absolute weight limits that are dictated by the structural design of the airplane—these are referred to as *structural-limited weights* (see Section 19.2.3). The maximum airplane weights that an operator can legally use are those recorded in the AFM—these are referred to as *certified weights* (see Section 19.2.4). An operator may elect, for financial or operational reasons, to have certified weights for a particular airplane that are less than the corresponding structural-limited weights that apply to the airplane type or model. For example, airport landing fees are sometimes based on an airplane's certified maximum landing weight—thus, if an airline can accept a lower certified landing weight it can save money.

10.3.3 Runway-Limited Takeoff Weights

The physical characteristics of the runway (e.g., length, slope, runway surface) and the size and location of obstacles near the intended flight path can restrict the takeoff weight that can safely be used. The key runway parameters, including definitions of the available takeoff distances, are described later in Section 10.4.

The takeoff requirements, which consider runway limitations, for modern airplanes are described in FAR 121.189(c) [3] and OPS Part CAT.POL.A.205(b) [4]. The airplane weight may not be greater than that at which compliance with the following can be shown:

(1) The *accelerate–stop distance* must not exceed the length of the runway plus the length of any stopway (i.e., the ASDA, as defined in Section 10.4.6).
(2) The *takeoff distance* must not exceed the length of the runway plus the length of any *clearway* (defined in Section 10.4.5), except that the length of any clearway included must not be greater than one-half the length of the runway.[6]
(3) The *takeoff run* must not be greater than the length of the runway (i.e., the TORA, as defined in Section 10.4.4).

In determining maximum takeoff weights, minimum distances, and flight paths, corrections must be made for the runway to be used, the elevation of the airport, the effective runway

---

6 There are operational limits to how much of the clearway associated with a particular runway can be included in a takeoff calculation—these limits are described in Section 10.5.4.

gradient (slope), the ambient temperature and wind component at the time of takeoff, and the runway surface condition (dry or wet). The standard techniques to show compliance with these regulations—and to determine *runway-limited takeoff weights*—are discussed in Section 10.5.

10.3.4 Climb-Gradient-Limited Weights

The takeoff portion of the flight does not end at the 35 ft screen height, but extends to a point at which the airplane is 1500 ft above the takeoff surface, or at which the transition from the takeoff to the *en route* configuration is completed and the speed V_{FTO} is reached, whichever point is higher. It is customary to break up the flight path for this critical portion of the flight into four parts, known as *segments*. The FAA and EASA stipulate minimum performance requirements for the segments in FAR/CS 25.121 [1, 2]. These requirements take the form of OEI still air climb gradients that have to be met with the airplane in a defined configuration at a defined speed. Takeoff weight limits that are established to comply with these requirements are known as *climb-gradient-limited weights*. These requirements are designed to ensure that compliant airplanes operating at the correct weight have a minimum acceptable performance potential, which can be used to climb or accelerate (or both). This is discussed later in Section 10.6.

10.3.5 Obstacle-Clearance-Limited Weights

Obstacle clearance requirements are defined by the FAA in FAR 121.189 [3] and EASA in OPS Part CAT.POL.A.210 [4]. It is a requirement that no person takes off an airplane at a weight greater than that which allows a *net takeoff flight path* to clear all obstacles by a height of at least 35 ft vertically within a defined horizontal corridor along the intended takeoff flight path. Due to the inherent safety concerns of the takeoff climb, which must consider an engine failure, a conservatism is introduced through the *net flight path* concept. The net flight path is determined by theoretically reducing the airplane's one-engine-inoperative climb gradient by a fixed percentage (as stipulated in FAR/CS 25.115 [1, 2]). The limiting airplane weight that is constrained by obstacle clearance requirements, as determined for a particular runway and operating conditions (including forecast winds), is known as an *obstacle-clearance-limited weight*. This is discussed in Section 10.7.

10.3.6 Tire Speed Limit

Airplane tires have a maximum rated speed, which for safety considerations must not be exceeded during takeoff or landing. Tires experience enormous internal stresses due to the weight of the airplane and the centrifugal forces that are induced at high rotational velocities [6]. The tires also get hot due to the repeated flexing of the rubber as the tires roll over the runway (or taxiway) surface. The combination of high stresses and high temperatures makes tires vulnerable to catastrophic failure if the design rated speed is exceeded or if the tire is damaged or excessively worn. Consequences can be very serious with possible damage occurring to the airplane from impacting pieces of the fractured tire or a possible reduction or loss of directional control.

The tires used on modern commercial airplanes are designed to accommodate maximum speeds of 225–235 mph (195.5–204.2 kt), with tires fitted to some older airplane types and to regional jets limited to 210 mph (182.5 kt) [7]. It is critical that the airplane's true ground speed at liftoff does not exceed the rated tire speed, and this, in certain circumstances, can limit the airplane's takeoff weight. In addition to airplane weight, the takeoff speed depends on the selected flap setting and the pressure altitude and ambient temperature at the airfield. It is the so-called "hot and high" airfield conditions that are likely to result in takeoff weight limits due to tire speed. Another factor that has to be considered is wind, as tailwinds increase the airplane's ground speed for a given liftoff speed, V_{LOF} (airspeed).

Tire-speed-limit takeoff weights are determined for each aircraft type and published in the AFM for combinations of pressure altitude, ambient temperature, and flap setting, with corrections for head/tailwinds.

10.3.7 Brake Energy Limit

An airplane's takeoff weight can also be limited by the maximum amount of kinetic energy that the brakes can absorb during a rejected takeoff (RTO). The wheel brakes essentially convert the airplane's kinetic energy, at the time of brake initiation, to heat. There is a limit to the amount of heat that brakes can absorb; this can place a limit on the airplane's takeoff weight.

The determination of the brake energy limit is accomplished through accelerate–stop testing for the most critical combination of takeoff weight and speed, as described in FAR/CS 25.735 [1, 2]. There are several specific requirements for these tests; FAR AC 25-7C [5] includes the following details:

(1) An accelerate–stop test, which should be conducted at a weight not less than the maximum takeoff weight, is to be preceded by a taxi of at least 3 miles (4.8 km) and which features three full stops using normal braking.
(2) If the brakes are not in a fully worn state at the beginning of the test, the accelerate–stop distance should be corrected to represent the stopping capability of fully worn brakes. The calculation of maximum *brake-energy-limited takeoff weights* and speeds, for presentation in the AFM, should be based on the most critical wear range of the brakes.
(3) The maximum airplane brake energy allowed for dispatch should not exceed the value for which a satisfactory after-stop condition exists. This condition is defined as one in which fires are confined to tires, wheels, and brakes, such that progressive engulfment of the rest of the airplane would not occur during the time of passenger and crew evacuation. The application of fire fighting means or artificial coolants should not be required for a period of 5 minutes following the stop.

The results of these RTO tests are used to calculate the airplane's *maximum brake energy speeds*, which are then recorded in the AFM. The maximum brake energy speed (V_{MBE}) is the maximum takeoff speed, for a given airplane mass, at which the brakes may be applied in the event of a rejected takeoff without exceeding the brake energy absorption limits [8]. Note that the V_{MBE} speeds are derived, initially, as true ground speeds—in which case, they depend only on the established brake energy limit, the airplane mass, and the runway slope (it requires more energy to stop an airplane on a downhill slope). For flight operations, however, it is convenient to express the V_{MBE} speeds as calibrated airspeed. This means that runway altitude, ambient temperature, and wind must also be taken into account.

The calculation of V_{MBE} is based on the premise that the pilot applies the maximum manual brake pressure, which would consequently produce the greatest possible braking effort. The V_{MBE} speed thus sets an upper bound on the range of V_1 speeds (see Section 10.5.6) that can be used—and in this way it can restrict the airplane's takeoff weight in certain conditions.

It is interesting to note that there is no allowance in the V_{MBE} values published in the AFM for any residual heat built-up in the brakes from previous takeoffs or landings. Airlines that have short flight and/or ground times need to be cognizant of this fact, as they have an increased exposure to the possibility of brake fire in an RTO. This possibility can be reduced by extending the ground cooling time or through the use of supplemental brake cooling, designed to force air over the brakes on the ground (electric fans mounted within the wheel axles have been offered by manufacturers as optional equipment).

10.3.8 Forward Center of Gravity Limit

The analyses conducted to determine the maximum takeoff weight corresponding to defined runway and obstacle clearance criteria are always based on the assumption that the airplane's center of gravity (CG) is located at the certified forward limit (see Section 19.3 for details on CG limits). This is a conservative assumption as a forward CG location is punitive compared to a more aft location in respect of takeoff performance. For certain operators who do not utilize the forward extremes of the CG envelope, this introduces the possibility of increasing the takeoff weight by restricting the CG to an *alternate forward takeoff limit*. This is discussed further in Section 19.3.4.

Essentially, the use of an approved alternate forward CG limit—which is located aft of the normal forward CG takeoff limit—results in lower takeoff speeds for comparable conditions. For a given runway and set of operating conditions, this will directly increase the runway-limited-takeoff weight and the tire-speed-limited takeoff weight. Depending on the airplane type and flap setting, the use of an alternate forward CG limit may or may not offer an improvement in climb performance; hence, the climb-gradient-limited weight may not always increase [8]. Similarly, the obstacle-clearance-limited weight may or may not increase—in this case the limiting weight depends not only on the change to the climb gradient, but also on the distance to the obstacle as the reference *zero point* (defined later in Figure 10.7) would be farther from the obstacle.

10.3.9 Prevailing Wind

Accounting for the wind can be imprecise due to the variable nature of wind (see Sections 4.5.2–4.5.4). A conservative policy is used to take into account the effect of the wind for the purpose of publishing takeoff distances. The following operational correction factors apply for the determination of the effective headwind or tailwind component acting along the runway: (1) for a headwind, not more than 50% of the nominal wind speed component is used, and (2) for a tailwind, not less than 150% of the nominal wind speed component is used (FAR/CS 25.105 [1, 2]). In other words, when the wind direction is advantageous, only partial credit is taken for the wind's influence, whereas when it is punitive, the penalty is increased.

The AFM will also indicate operational limits and advisory information for takeoff with respect to winds. A tailwind limit (typically about 10–15 kt) is usually indicated; it is normal that no headwind limit is indicated in the AFM, although operators may impose their own specific limits. The crosswind information, provided by the manufacturer in the AFM, is indicated as a maximum demonstrated crosswind speed (30 kt is typical, but some airplanes have higher values); operators can impose their own limits in this respect based on their particular circumstances. Note that an airplane's crosswind performance, as described in the AFM, is a maximum *demonstrated* crosswind and it is not considered to be an airplane performance limit, in the strict sense. The value merely indicates the strongest crosswind in which the manufacturer demonstrated the airplane's crosswind takeoff and landing capabilities to the satisfaction of the regulatory authorities during certification testing.[7] However, in the event of certain non-normal airplane operations—for example, a dispatch with certain system malfunctions (in compliance

---

7 As part of the airplane's certification, the manufacturer will seek out airports having very strong crosswinds for these tests—for example, Boeing flew a B787 to Keflavik, Iceland in 2010 to conduct crosswind testing.

with the provisions of the Minimum Equipment List)—there can be specific crosswind limits imposed by the manufacturer—and these would be considered limit values.

10.4 Runway Limitations and Data

10.4.1 Approved Runway Data

Information on runway features at a particular airport is given in the aerodrome (airport) section of the Aeronautical Information Publication (AIP) of the relevant state (see Section 23.4.6). It is the responsibility of the national aviation authority of that state to record—and to ensure the accuracy of—the data that are published in the document.

10.4.2 Runway Identification

Runways are identified by their orientation (or direction). The runway *number* is usually determined from its magnetic bearing divided by ten and rounded to the nearest integer.[8] For example, a runway designated as Runway 28 will be oriented such that it has a direction of between 275° and 285° with respect to magnetic north (as illustrated in Figure 4.4).[9] The same runway, when used in the opposite direction, will be called Runway 10 (i.e., corresponding to a direction of 100° ± 5° magnetic). It is interesting to note that, for certain runways, the designation will occasionally change as a consequence of the slow migration of the magnetic poles. Parallel runways are identified by appending a letter to the runway number: L (for left), C (for center), and R (for right)—for example, the shorter of the two runways at London Gatwick is denoted as 08L in one direction and 26R in the opposite direction.

10.4.3 Runway Slopes (Gradients)

Runway code numbers (1 to 4) are used by ICAO (International Civil Aviation Organization) in runway design specifications [9]. The *longitudinal slope* (gradient) of a runway—defined as positive for an uphill slope—is computed by dividing the difference between the maximum and minimum elevation along the runway centerline by the corresponding runway distance. The limit for the longitudinal slope along any portion of a runway is 2% for code numbers 1 and 2, 1.5% for code number 3 (0.8% for the first and last quarter of a precision approach runway category II or III), and 1.25% for code number 4 (0.8% for the first and last quarter).

The airworthiness regulations FAR/CS 25.105 [1, 2] indicate that the effective runway slope (gradient) must be taken into account for the purpose of establishing AFM takeoff data. These data are prepared for the range of runway slopes that the airplane is expected to operate from. Published runway slope data are then used by operators, in a conservative manner, in the estimation of required takeoff distances. Changes in slope over the length of a runway are commonplace. In such cases, operators are required to establish an average or equivalent slope that is conservative (i.e., the method of estimation needs to pay greatest attention to that portion of the runway where the performance is more critical).

---

8 An exception to this convention applies to runways in polar regions, where magnetic compass indications can be erratic. In Canada's Northern Domestic Airspace, for example, runway headings are defined with respect to True North.

9 Pilots are trained to check the runway number against the airplane's compass heading when "lining up" on the runway prior to takeoff.

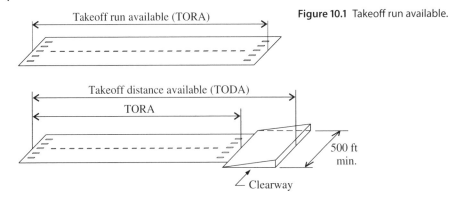

Takeoff run available (TORA)

Figure 10.1 Takeoff run available.

Takeoff distance available (TODA)

TORA

500 ft min.

Clearway

Figure 10.2 Runway with clearway.

10.4.4 Declared Runway Distances

The lengths of the runways that are available for takeoff and landing at approved (certified) airports are measured and documented by the appropriate airworthiness authority. These *declared* distances must comply with established criteria, as described in FAR 1.1 [10] and CS-Definitions [11].

The takeoff run available (TORA) is the length of the runway that is available for the takeoff ground run. It is equal to the total runway length (Figure 10.1) or the distance from the runway entry point (i.e., the point where the taxiway intersects the runway) to the end of the runway.

10.4.5 Clearway

A clearway is a rectangular area beyond the runway (ground or water) under the control of the airport authority over which an airplane may make a portion of its initial climb. The clearway may not be less than 500 ft (152 m) wide, and must be centrally located about the extended centerline of the runway (Figure 10.2). No object (other than threshold lights) or terrain may protrude above a plane that extends from the end of the runway with an upward slope of 1.25%. Although the declared takeoff distance available (TODA) can include the clearway distance to a maximum of one-half the length of the runway, there is an operational limit to how much of the clearway can be included in a takeoff calculation (see Section 10.5.4).

10.4.6 Stopway

A stopway is a hard surfaced area at the end of the runway that may be used for braking (Figure 10.3). The stopway, which may not be narrower than the runway, must be capable

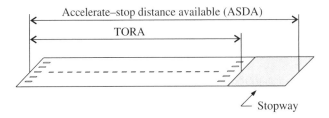

Accelerate–stop distance available (ASDA)

TORA

Stopway

Figure 10.3 Runway with stopway.

of supporting the airplane without causing damage to the airplane. The additional braking distance can be taken into account to determine the accelerate–stop distance available (ASDA).

10.4.7   Clearways and Stopways—Concluding Remarks

Under certain conditions, a runway may have a clearway that is not usable as stopway (e.g., a clearway over water past the departure end of the runway), and under other conditions there may be a stopway that is not usable as clearway (e.g., a stopway that is less than 500 ft wide). For many runways, a clearway and a stopway coexist; however, on any given runway, the presence of clearway should never be assumed to indicate the presence of stopway, or vice versa. Under these conditions, the OEI takeoff distance available and the accelerate–stop distance available are not equal.

10.4.8   Line-Up Corrections (Allowances)

In many cases, it will not be possible for the pilot to position the airplane at the start of the takeoff with the main landing gear at the beginning of the runway. According to ICAO Annex 6, Part 1 [12], account must be taken of the loss, if any, of runway length due to alignment (positioning) of the airplane prior to commencing the takeoff. The "lost" runway distance is known as a *line-up correction* or *line-up allowance*. Runways with a displaced takeoff threshold or a large turning apron may not require a line-up correction.

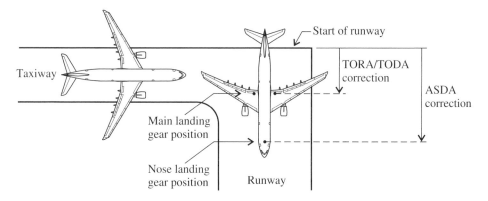

Figure 10.4   Line-up corrections for 90° taxiway–runway intersection.

The corrections are specific to a particular airplane type (and model) as they are a function of the airplane's geometry and landing gear design. Standard corrections are calculated (and published in such documents as the Aircraft Characteristics for Airport Planning Manual) for a 90° turn onto a runway (see Figure 10.4) and also for a backtrack[10] followed by a 180° turn. The line-up correction should be subtracted from the TORA, TODA, and ASDA when planning the takeoff (this is regulated by EASA, for example, in OPS Part CAT.POL.A.205 [4], with supporting information given in AMC1 CAT.POL.A.205 [13]).

---

10 *Backtracking* describes an airplane taxiing in the opposite direction on the runway that is intended for takeoff. This procedure is standard practice at airports that do not have suitable taxiways that extend to the start of the runway. Pilots are required to backtrack and then execute a 180° turn, before commencing the takeoff run.

10.4.9   Runway (Pavement) Loading Limits

The Aircraft Classification Number/Pavement Classification Number (ACN/PCN) system is the most widely used standard for the assessment of the bearing strength of a runway for use by a particular airplane. The system, which came into use in 1981, has been adopted by ICAO [14] and is used to determine whether a particular airplane can operate from a given runway.

The system works as follows: Aircraft are rated in terms of an ACN, which is a relative measure of the stress that the aircraft will introduce in the pavement (based on the aircraft's weight, landing gear geometry, number of wheels, tire pressure, and pavement type). Runways are rated in terms of a PCN, which is a relative measure of the bearing strength of the pavement (based on the pavement type, i.e., rigid or flexible, and construction). If the ACN is less than or equal to the PCN, the aircraft can use the runway in question without risk that it will cause stresses in the pavement that, over time, will cause damage to the pavement. ACN values are calculated and published by the airplane manufacturer; whereas, PCN values are determined by the relevant aviation or airport authority (based on an assessment of the expected cumulative damage over, say, a 20 year life of the runway).

The bearing strength of the pavement can thus impose a takeoff weight (TOW) limit on a particular aircraft type. The PCN should, however, not be regarded as an absolute limit for airplane operations and overweight waivers can—in certain circumstances—be granted to an operator by an airport authority. For the sake of illustration, consider the A380-800 operating from Shannon Airport, Ireland. The main runway at Shannon (i.e., 06/24) has a rigid pavement (subgrade category C). The PCN is 75 [15], which thus limits the aircraft's gross weight to 510 000 kg, based on data provided in the A380 Aircraft Characteristics—Airport and Maintenance Planning document [16].

10.5   Operational Field Length and Runway-Limited Takeoff Weight

10.5.1   Field-Length-Limited TOW

In Section 10.3, it was indicated that there are several considerations that need to be taken into account when determining the maximum TOW for a given runway, prevailing conditions (altitude, ambient temperature, and so on), and airplane performance characteristics (including tire speed limits and CG location). This results in the need to determine the maximum TOW for each consideration—as described in Sections 10.3.2 to 10.3.9—and thereafter to establish which one is truly limiting for the particular operation. The most complex of these calculations involves the determination of the *field-length-limited* TOW (runway-limited TOW). The procedure for computing this TOW for a given runway (as described by the TORA, TODA, and ASDA) is outlined in this section.

10.5.2   Overview of Requirements

The *operational field length* for a particular airplane gross weight—at a given airport elevation, ambient temperature, and wind—is equal to the longest of the following three distances:

(1) 115% of the all-engines-operating (AEO) takeoff distance (see Section 9.2.3);
(2) the one-engine-inoperative (OEI) takeoff distance (i.e., accelerate–go distance), which is the distance required to reach a height of 35 ft for a dry runway (15 ft for a wet runway) following an engine failure that is assumed to occur one second before the V_1 speed is reached (see Sections 9.2.4 and 9.2.7); and

(3) the accelerate–stop (i.e., rejected takeoff) distance, which is the distance resulting from an all-engine acceleration, followed by a deceleration to bring the airplane to a complete stop following an engine failure or other takeoff abort event, where the first RTO action by the crew is assumed to take place at the V_1 speed (see Sections 9.2.5 and 9.2.6).

The determination of the field-length-limited weight for the AEO case (number 1 above) is relatively straightforward. The second and third cases, however, are much more complex as the field lengths are dependent on the V_1 speed. The complexity is due to the fact that the V_1 speed is a variable that can be selected by the operator to suit the operating conditions best. The relationships between V_1 and the OEI takeoff distance (case 2) and between V_1 and the accelerate–stop distance (case 3), taking into account takeoff weight, are explored in Sections 10.5.3 to 10.5.6, which follow.

10.5.3 Balanced and Unbalanced Field Lengths

By considering a range of progressively increasing V_1 speeds (for a given set of takeoff conditions), it can be concluded that the corresponding OEI takeoff distance will decrease. This can be deduced by noting that an increased V_1 speed means that more of the takeoff will be conducted with all engines operating; hence, the OEI takeoff distance will reduce. The accelerate–stop distance, however, will increase as an increased V_1 speed implies that the distance covered before the brakes are applied is greater and braking distance is also greater due to the higher V_1 speed.

This is illustrated in Figure 10.5. It is also evident that there is a unique V_1 speed where the two distances are equal. This concept is used in the determination of a *balanced field length* (for a given airplane weight, airport altitude, ambient temperature, and wind); whereby the selected V_1 speed ensures that the OEI takeoff distance and the accelerate–stop distance are equal. This concept of a balanced V_1 speed is very useful for routine flight operations.

A balanced V_1 speed, for a given set of conditions, thus yields equal OEI takeoff and accelerate–stop distances. It is also apparent from the figure that different distances corresponding to the OEI takeoff and the accelerate–stop situation can be accommodated by adjusting the V_1 speed downwards or upwards from the balanced value. If the runway has a clearway and/or a stopway, then it may be possible to increase the airplane's takeoff weight above that which would be possible with a balanced V_1 (for a given TORA and operating conditions). In such a case, the takeoff will be planned using an *unbalanced field length*, and the OEI takeoff and accelerate–stop distances will not be equal.

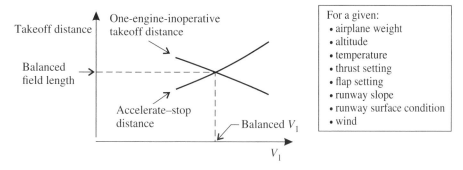

Figure 10.5 Balanced field length.

10.5.4 Operations from a Runway with a Clearway

When the runway has a clearway, the pilot may be permitted, in some cases, to lift off farther down the runway and use the clearway to climb to the 35 ft screen height. In order to retain the same accelerate–stop distance with an increased takeoff weight, the V_1 speed must be reduced.

The *takeoff run* on a dry runway with a clearway is defined as the distance from the start of the takeoff to a point equidistant between the point at which V_{LOF} is reached and the point at which the airplane is 35 ft above the takeoff surface (as described in FAR/CS 25.113 [1, 2]). This requirement effectively sets an upper limit on the allowable clearway that credit can be taken for—that is, a distance equal to one-half of the airplane's expected takeoff distance from liftoff to 35 ft. In other words, the first half of the flare distance to 35 ft must be over the runway. As flare distances for jet transport airplanes are typically between 800 and 1500 ft, the maximum usable clearway is effectively limited to distances of about 400 to 750 ft.

For a wet runway, the same clearway allowance—as described for the dry runway—is permitted for the all-engines-operating case (i.e., one-half of the flare distance to 35 ft); however, for the one-engine-inoperative case, no clearway credit is allowed.[11]

10.5.5 Operations from a Runway with a Stopway

When there is stopway at the end of a runway (see Figure 10.3), the additional braking distance can be utilized in the case of an RTO. The required accelerate–stop distance cannot exceed the length of the runway plus the length of the stopway (i.e., the ASDA). In the case of a runway with stopway but no clearway, the potential exists to increase the allowable takeoff weight by increasing V_1 above its balanced value.

10.5.6 Permissible Range of V_1 and TOW for a Given Field Length (Takeoff Web Chart)

It is apparent from the discussion presented in Section 10.5.3 that there is a relationship between, on the one hand, the TOW and V_1 speed and, on the other hand, the resulting (or required) OEI takeoff and accelerate–stop distances. This relationship will now be explored by constructing a chart of the OEI takeoff distance (ordinate) plotted against the accelerate–stop distance (abscissa) for a range of takeoff weights and V_1 speeds (other factors, i.e., altitude, ambient temperature, flap setting, wind, runway slope, and runway surface are not considered to vary). This is illustrated in Figure 10.6. The balanced V_1 condition is indicated by a dashed line, where the two distances are equal. The area to the left of the dashed line on the chart indicates combinations of TOW and V_1 that result in the OEI takeoff distance being greater than the accelerate–stop distance; the reverse is true for the area to the right of the dashed line.

It is convenient to consider initially a single V_1 speed. If the TOW were to increase, it is apparent from the preceding discussions that both the OEI takeoff and the accelerate–stop distances would increase. For the illustration given in Figure 10.6, such lines have been constructed for the speed range of 120 kt to 150 kt in 5 kt intervals. Similarly, if a single TOW were considered and the V_1 speed were to increase, then the accelerate–stop distance would increase, but the OEI takeoff distance would decrease. By considering a range of takeoff weights, a family of such lines can be generated—as shown in the figure. Due to the resulting hatched appearance of this

---

11 These allowances apply to airplanes certified to current regulations (the rules that apply to airplane types that were certified under earlier regulations are those rules that were applicable at the time of certification).

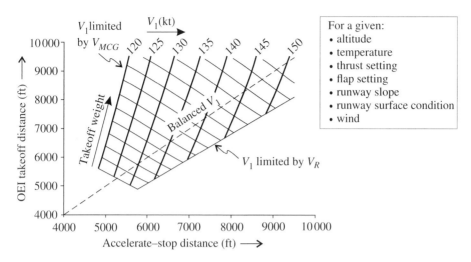

Figure 10.6 Illustrative takeoff web chart indicating permissible combinations of V_1 speed and TOW for OEI takeoff and accelerate–stop distances.

chart, it is usually called a *web chart*. The value of such charts is that permissible combinations of V_1 and TOW can readily be determined for a given set of runway dimensions.

In Figure 10.6, it is indicated that the left-hand boundary of allowable V_1 speeds—which represents the slowest, or minimum, V_1 speed that is permissible—is "limited by V_{MCG}." The exact relationship between the minimum allowable V_1 speed and V_{MCG} is governed by the regulatory requirement that V_{EF} may not be less than V_{MCG}. Now, as V_1 is at least one-second-acceleration faster than V_{EF}, this enables the minimum allowable V_1 to be computed for a particular airplane based on V_{MCG}. Certain manufacturers (e.g., Boeing) call this speed V_{1MCG} and interpret the regulations by stating that V_1 must not be less than V_{1MCG} [8]. On the right-hand boundary, the allowable range of V_1 speeds are bounded on the chart by the requirements that V_1 may not be greater than V_R (note that with increasing TOW, the V_R speed increases). Furthermore, V_1 cannot exceed V_{MBE} (as mentioned earlier in Section 10.3.7).

It can be seen on the chart that for any given set of OEI takeoff and accelerate–stop distances (which need not be equal), corresponding unique values of TOW and V_1 are defined. This is an important observation, as it provides the mechanism to determine the *field-length-limited weight* (and associated V_1) for the given runway. It must be remembered, though, that this calculation only considers the requirements for the OEI takeoff (i.e., case 2 in Section 10.5.2) and the engine failure/abort event RTO (i.e., case 3)—the result thus needs to be compared to the field-length-limited weight that comes from the AEO takeoff condition (case 1), and the most conservative value used.

The web chart can be an extremely useful tool for any takeoff in which the TOW will be less than the field-length-limited weight. In any such situation, there will be a range of acceptable V_1 speeds, and any V_1 speed within that range will yield acceptable takeoff distances. By analyzing the takeoff at different V_1 speeds within that acceptable range, it may be possible to take off with an increased weight, depending on the nature of the limiting condition. Consider, for example, the situation—at a particular airport—where the TOW is not limited by the runway parameters but rather by an obstacle located below the planned climb-out flight path (see also Section 10.7). In this case, there will be a range of V_1 speeds that can be used, which are associated with the particular obstacle-limited TOW. For each of those V_1 speeds, there will be a unique climb path. By considering increased V_1 speeds, it is evident that the OEI takeoff distance will be reduced.

The airplane will thus become airborne earlier, resulting in a higher flight path over the obstacle. As the clearance margin is now greater than the minimum value required for the operation, it may thus be possible to increase the TOW for this situation. Note that by increasing V_1, the accelerate–stop distance is increased, but is still within the runway limits because of the way in which the web chart is defined.

The V_1 range concept can be utilized by operators to solve a number of problems, including takeoffs in which the V_1 is greater than V_{MBE} or V_R, or less than V_{1MCG}, or the obstacle-limited case mentioned above.

10.6 Takeoff Climb Gradient Requirements

10.6.1 Takeoff Climb Gradient Requirements (FAR/CS 25.121)

The takeoff flight path, after liftoff, can be divided into four segments. For each segment, a minimum still air climb gradient requirement is specified in FAR/CS 25.121 [1, 2]. Compliance with each gradient requirement has to be demonstrated at the appropriate weight with the airplane at a defined speed and thrust setting, and with the airplane in a specified configuration. Furthermore, it is stated that the minimum climb gradients must be achieved without the benefit of ground effect (the airplane can be considered to be out of ground effect when it reaches a height equal to its wingspan) and with the critical engine inoperative (i.e., that engine which most adversely affects the airplane's climb performance). The required climb gradients and the associated conditions are summarized in Table 10.1.

There is an important, but subtle, point associated with these requirements. The requirements are not intended to replicate an operational scenario and are not directly related to obstacle clearance *per se*. Rather, these requirements are designed to ensure that all certified airplanes have an acceptable *performance potential*. It is argued that if an airplane can meet these climb requirements, then it will have exhibited a minimum performance capability that is deemed necessary for commercial airplane flight operations. By stipulating a minimum climb gradient, the regulatory authorities, in effect, are ensuring that all compliant airplanes have a minimum potential to climb or accelerate or execute an accelerated climb with the critical engine inoperative after liftoff.

The first segment is from liftoff to the point of complete gear retraction (which can only commence once the airplane is airborne). The airplane will thus have the gear partially extended, the flaps (and slats) set for takeoff, and the engine thrust set for takeoff. In practice, this segment does not impose any limitations on the takeoff weight.

The second segment starts at the point of complete retraction of the gear and ends at a height no less than 400 ft (122 m) above the height of the runway. In this segment the airplane climbs at a speed not less than V_2 with the gear retracted and the flaps (and slats) set for takeoff. The engine thrust remains set for takeoff.

The third segment, which is often called the transition segment, is flown at constant height. It extends from the end of the second segment to the point where the airplane has accelerated to an appropriate flaps-up speed and fully retracted the flaps (and slats). The thrust will initially be set for takeoff, but will reduce to maximum continuous thrust (MCT) at the end of the segment.

The final segment, which begins at the end of the third segment, ends at a height of 1500 ft (457 m) above the runway surface or at the height when the transition from takeoff to *en route* configuration is complete and V_{FTO} is reached, whichever is higher. V_{FTO} is the final takeoff speed, which may not be less than 1.18 times the reference stall speed, V_{SR}.

Table 10.1 Takeoff segment characteristics and gradient requirements

	First segment	Second segment	Third segment (transition)	Final segment
Minimum still air climb gradient with N-1 engines				
$N = 2$ (twin):	>0%	>2.4%	–	>1.2%
$N = 3$:	>0.3%	>2.7%	–	>1.5%
$N = 4$ (quad):	>0.5%	>3.0%	–	>1.7%
Airplane speed and segment characteristics				
Starts when:	V_{LOF} is reached	gear is fully retracted	acceleration height is reached (min. 400 ft)	*en route* configuration is achieved
Applicable weight:	at start of gear retraction	when gear is fully retracted	at start of acceleration segment	at end of acceleration segment
Speed:	V_{LOF}	V_2	accelerating	V_{FTO}
Engine thrust:	T/O	T/O	T/O	MCT
Slats/flaps configuration:	takeoff	takeoff	slats/flaps retracting	clean
Gear position:	extended	retracted	retracted	retracted

Notes:

(a) *Source*: FAR/CS 25.121 [1, 2].
(b) N is the number of engines on the airplane.
(c) T/O is the takeoff thrust setting and MCT is the maximum continuous thrust setting (see Section 8.5.5).
(d) Compliance with the minimum still air gradient requirements are to be demonstrated with the critical engine inoperative at the applicable weight and indicated conditions without the benefit of ground effect.
(e) Takeoff thrust is usually only available for 5 minutes. For certain airplane types, its use can be extended to a maximum of 10 minutes in the case of engine failure; consequently, the end of the third segment must be achieved within 10 minutes after takeoff [8, 17].

10.6.2   Takeoff Path Requirements (FAR/CS 25.111)

Specific climb performance requirements are stipulated in FAR/CS 25.111 [1, 2], which apply to the takeoff flight path following the failure of the critical engine during the takeoff run. At each point along the takeoff path, starting at the point at which the airplane reaches 400 ft (122 m) above the takeoff surface, the available gradient of climb may not be less than 1.2% for two-engine airplanes, 1.5% for three-engine airplanes, and 1.7% for four-engine airplanes. The *available gradient*, in this context, describes an available potential that can also be used to accelerate the airplane. This interpretation is important, as this requirement applies to all climb segments and that includes the third (or transition) segment, which has a level flight path.

10.6.3   Climb-Limited Takeoff Weight

Methods for analyzing an airplane's climb performance are presented later in Section 12.4. An airplane's climb gradient is seen to depend on its weight, the lift-to-drag ratio (L/D), and the available thrust (see Equations 12.23 and 12.24). The lift-to-drag ratio depends primarily on the flap setting (i.e., L/D reduces as the flaps are extended). As the scenario under consideration has one engine inoperative, yaw and windmilling drag corrections (see Section 7.5.5) need to be included. Furthermore, L/D and thrust depend on the airplane's speed. This complicates the analysis as the V_2 speed is itself a function of weight. Consequently, an iterative approach is required to determine the limiting weight.

For each of the climb segments, it is possible to calculate the maximum weight at which the stipulated one-engine-inoperative climb gradient can be met at the specified speed, airplane configuration, and thrust setting. The lowest of these weights is the critical *climb-limited takeoff weight*. In many, but not all, cases it is the second segment requirement that is most limiting [8]. As an airplane's climb performance is also a function of the altitude and ambient temperature for a given flap setting, the climb-limited takeoff weight is not a single value, but rather a range of values that will vary depending on the operating conditions. The AFM presents the data in a manner that permits operators to determine the critical climb-limited takeoff weight for a given airport altitude and temperature.

10.7 Takeoff Climb Obstacle Clearance

10.7.1 Obstacle Clearance Requirements

In addition to the climb gradient requirements (described earlier in Section 10.6), all commercial airplanes must be operated within a set of safety regulations concerning obstacle clearance that applies to actual flight conditions. The FAA requirements are described in FAR 121.189 [3] and the EASA requirements in OPS Part CAT.POL.A.210 [4]. Tall obstacles near the airport—such as buildings, trees, and communication towers—are potential dangers for an airplane that suffers an engine failure during takeoff.

Furthermore, it has to be considered that the airplane's flight path in such an emergency may be lower than the predicted OEI flight path due to a variety of operational factors (such as a change in wind direction or extreme cold weather). A conservatism is thus introduced by reducing (i.e., degrading) the airplane's predicted climb gradient through the *net flight path* concept. The predicted flight path with the critical engine inoperative—which is called the *gross flight path* (or *actual flight path* in the regulations)—is computed using the airplane's thrust and drag data taking into account envisaged flight conditions, including forecast winds. (The influence of wind on climb performance is discussed later in Sections 12.4.13 and 12.4.14.) The net flight path is a conservative definition that requires the computed gross OEI climb gradient to be reduced by a fixed percentage; the amount is 0.8% for two-engine airplanes, 0.9% for three-engine airplanes, and 1.0% for four-engine airplanes (FAR/CS 25.115 [1, 2]).

When an operator plans to take off from a specific airport, it has to be shown that the airplane's net flight path can clear all obstacles by at least 35 ft (10.7 m) vertically and 200 ft (61 m) horizontally; the horizontal clearance distance is increased to 300 ft (91.4 m) after the airplane passes the airport boundary (FAR 121.189 [3]). The EASA [4] vertical requirement is identical, but the horizontal requirements are defined in a slightly different way. The vertical clearance requirements are illustrated in Figure 10.7. Determination of the net flight path must be based on the airplane's predicted weight, with corrections made for the runway to be used, the elevation of the airport, the effective runway gradient, and the ambient temperature and wind component at the time of takeoff.

This policy is designed to ensure that variations in pilot technique or minor errors in loading, optimistic approximations in performance predictions, or changes in wind speed or direction, will not result in a catastrophe should an engine failure occur during takeoff.

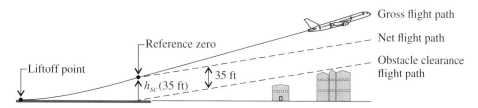

Figure 10.7 Takeoff climb gross and net flight paths (one engine inoperative).

10.7.2 Demonstrating Compliance with the Obstacle Clearance Requirements

The takeoff flight path is considered to begin 35 ft above the takeoff surface at the end of the takeoff distance (FAR/CS 25.115 [1, 2]). It is common practice in takeoff obstacle clearance analysis to refer to this point as *reference zero*. Note that the end of the *takeoff distance* will not necessarily coincide with the end of the runway. For example, it may be before the end of the runway when the takeoff weight is less than the field-length-limited takeoff weight or it may be beyond the end of the runway if the takeoff weight is based on a distance that includes a clearway (see Section 10.5).

The FAA's Advisory Circular 120-91 [18] and EASA's Acceptable Means of Compliance and Guidance Material to Part CAT [13] provide detailed information to show compliance with the regulations in regard to obstacle clearance (note that the FAA and EASA requirements differ in several respects). The FAA [18] provides two methods to ensure clearance of critical obstacles: the Area Analysis Method (described below) and the Flight Track Analysis Method (this is a more complex procedure that can allow for a smaller affected area through the use of modern navigation equipment).

The Area Analysis Method defines an area centered on the intended takeoff departure flight path. This area, known as the *obstacle accountability area* (OAA), must be free of all obstacles that cannot be cleared vertically. The OAA is acceptable for use without accounting for operational factors that may affect the actual flight track relative to the intended track, such as wind and available course guidance. Obstacles lying outside the OAA need not be considered. Rules for defining the OAA for both straight-out departures and turning departures are provided by the FAA [18] and EASA [13] (although EASA do not use the same terminology). For the sake of illustration, the OAA based on FAA AC 120-91 [18] for a straight-out departure is illustrated in Figure 10.8.

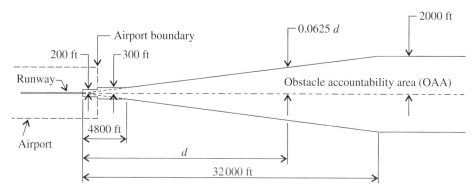

Figure 10.8 Obstacle accountability area (OAA) for straight-out departure. *Source*: Adapted from [18].

10.7.3 Impact of Turns on Obstacle Clearance

Obstacle clearance requirements need to be carefully considered for flight paths that incorporate a turn (turning performance is discussed in Chapter 15). The FAA (in FAR 121.189 [3]) and EASA (in OPS Part CAT.POL.A.210 [4]) provide specific details on allowable bank angles for various heights during the climb—for example, a 15° banked turn should not be started below 50 ft, or one-half of the wingspan, whichever is greater. A consequence of the airplane executing a banked turn is that the stall speed will be increased in comparison to the wings-level stall case (see Section 20.2.4). To maintain an adequate margin between the V_2 speed (in the second climb segment) and the stall speed, it will be necessary for the climb speed to be increased above the V_2 speed for bank angles greater than 15°.

The radius of a level turn is easily computed as it depends solely on the TAS and selected bank angle (see Equation 15.11) and does not depend directly on the airplane's weight or performance characteristics. Note that air density has an influence on the radius for a given CAS (the radius will increase with decreasing air density). Note also that the derived equation, strictly speaking, applies to a coordinated turn. To maintain the desired heading with a failed engine that is not located on the airplane's centerline requires rudder and aileron control inputs by the pilot to counteract the asymmetric thrust—this means that turns will not be coordinated. Consequently, when an airplane turns in the direction of a failed wing-mounted engine, the resulting turn radius will be less than the value predicted by this equation; turning away from the failed engine will result in a radius that is greater than the predicted value [8].

The determination of the climb gradient needs to take into account the increase in drag associated with the failed engine (see Section 7.5.5). Furthermore, the climb gradient in a turn is less than that which would occur for straight (i.e., non-turning) flight under comparable conditions—this is a consequence of the increase in drag associated with the turn (see Section 15.3.4). The magnitude of the *gradient decrement* for a given angle of bank depends on the airplane type and flap setting.

Winds complicate the task of determining an obstacle clearance departure path that incorporates a turn. As the path is defined with respect to the ground, winds have the effect of demanding that the pilot increase or decrease the bank angle to compensate for the wind and achieve the desired turn radius.

10.7.4 Obstacle-Limited Takeoff Weight

The *obstacle-limited takeoff weight* is that weight which would yield a climb gradient equal to the minimum acceptable value that satisfies the net flight requirements in terms of obstacle clearance. The determination of this limiting weight for operations out of airports with neighboring obstacles is nontrivial when taking into account all relevant operating conditions, which include airport altitude and ambient temperature, runway type and surface condition, and forecast winds. References [13] and [18] provide supporting information and should be consulted for acceptable means of compliance with the regulatory requirements.

What makes the analysis particularly demanding is that there can be several solutions (corresponding to different speeds, airplane configuration, and scheduled climb profile) for a given set of operating conditions. The following factors are relevant as they will impact the flight profile and can be used to avoid specific obstacles:

(1) the V_1 speed (as described in Section 10.5.6, a range of V_1 speeds may be possible);
(2) the flap setting (a smaller flap setting will increase the climb gradient but at the expense of a longer takeoff distance);

(3) the location and radii of planned turns on the intended track (if the bank angle is increased, the turn radius will reduce, but the climb gradient will also reduce);

(4) the height of the third climb segment (the regulations only require that it be at least 400 ft above the runway surface); and

(5) the duration of the acceleration period in the third climb segment (certain airplanes are approved to use takeoff thrust for 10 minutes—not 5 minutes—thus extending the acceleration time).

By analyzing a number of viable solutions, it is possible to optimize the takeoff climb procedure in order to achieve the greatest takeoff weight considering all operational conditions and obstacle clearance requirements.

10.8 Derated Thrust and Reduced Thrust Takeoff

10.8.1 Introduction

It is apparent from the preceding sections that the takeoff weight that can be used for a particular operation can be limited by any one of several factors (e.g., structural limits, brake energy, tire speed, runway characteristics, regulatory climb gradient requirements, and obstacle clearance). For a particular set of operating conditions (which includes altitude, ambient temperature, wind, and runway surface conditions), one of these factors will be limiting, thus determining the maximum allowable takeoff weight—this limiting weight is the maximum *regulatory takeoff weight* (RTOW).

It is frequently the case, however, that the airplane's actual weight is lower than the RTOW. This could be the result of a light payload or a light fuel load (for a short trip) or the consequence of operating from a sea level airport with dry, long runways and no obstacles, for example. As the RTOW is determined with the assumption of maximum takeoff thrust, whenever the takeoff weight is less than the RTOW, the option exists to utilize a lower thrust setting for takeoff, while still meeting all regulatory requirements. And, this is routinely done. There are two methods that are acceptable to the authorities to safely decrease the engine thrust for these situations (described in FAA AC 25-13 [19] and CS-25 Book 2 [2]). The first method utilizes an approved *derated thrust*, which is specified in the AFM and is lower than the rated takeoff thrust (described in Section 10.8.2); the second is based on the selection of a hypothetical higher-than-actual ambient temperature, which results in the engines producing a *reduced thrust* (described in Sections 10.8.3 and 10.8.4).

There is a substantial financial benefit for operators to utilize a lower thrust for takeoff. As the engines are not required to "work" as hard, maintenance and operating costs are reduced. Takeoff margins are reduced through the use of a lower thrust, but engine reliability is improved.

10.8.2 Derated Thrust

A derated thrust is an approved takeoff engine rating (see Section 8.5.9), which is published in the AFM (separate to the maximum takeoff thrust rating), and, as such, carries the same legal status as any other rating. The AFM provides full details to calculate the airplane's takeoff performance for each derated thrust rating, which is not done—nor needed—for the *assumed temperature method* (described in Section 10.8.4). The available thrust at the derated thrust rating must exceed the thrust required for the takeoff at the associated ambient temperature, as illustrated in Figure 10.9.

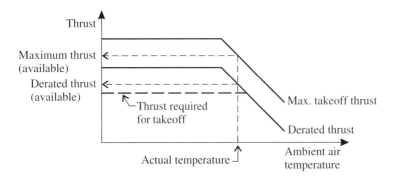

Figure 10.9 Derated thrust.

10.8.3 Regulatory Basis for Reduced Thrust Operations

The rules that govern the use of reduced takeoff thrust are outlined in FAA AC 25-13 [19] and CS-25 Book 2 [2]. Three key aspects are herein described. A reduced takeoff thrust setting may be used if:

(1) it does not result in loss of systems or functions that are normally operative for takeoff (e.g., automatic spoilers);
(2) it enables compliance with the applicable engine operating and airplane controllability requirements in the event that takeoff thrust (or derated takeoff thrust if such is the performance basis) is applied at any point in the takeoff path; and
(3) it is at least 75% of the takeoff thrust (or derated takeoff thrust if such is the performance basis) for the existing ambient conditions.[12]

Reduced thrust may not be used on runways contaminated with standing water, snow, slush, or ice; however, it may be used on wet runways (provided that the increased stopping distance on the wet surface is taken into account). Two important considerations for the flight crew are indicated: (1) application of reduced takeoff thrust in service is always at the discretion of the pilot; and (2) when conducting a takeoff using reduced thrust, takeoff thrust (i.e., maximum takeoff thrust) may be selected at any time during the takeoff operation. In respect of the second point, it is important to note that by virtue of the way that reduced takeoff thrust settings are computed, it is not necessary for the execution of a safe OEI takeoff to advance the thrust levers in the event of an engine failure. In fact, certain flight crew training procedures recommend against changing the thrust setting following an engine failure during a reduced thrust takeoff because of controllability issues associated with changing thrust and crew distractions that may result.

10.8.4 Reduced Takeoff Thrust: Assumed Temperature Method

The *assumed temperature method* (ATM) is an approved technique that provides flight crews with the means to conduct the takeoff with a reduced thrust, which is adequate to meet all

---

12 EASA rules [2] differ in this respect: A reduced takeoff thrust setting may be used provided that it is at least 60% of the maximum takeoff thrust (no derate) for the existing ambient conditions. Thus, when reduced thrust is combined with derated thrust, the overall thrust reduction must be such that the thrust is at least 60% of the maximum takeoff thrust.

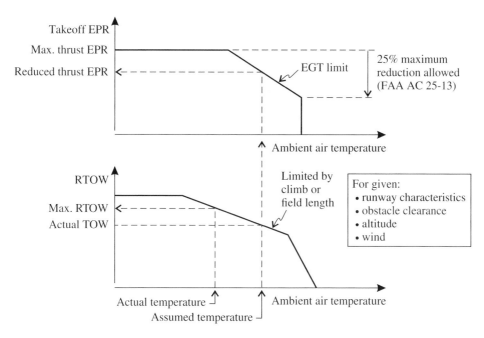

Figure 10.10 Assumed temperature method (reduced takeoff thrust) with EPR thrust setting parameter.

regulatory requirements for the particular takeoff being considered, by the selection of a hypothetical higher-than-actual ambient temperature. The selected temperature is called the *assumed temperature.*

The process can be explained by referring to Figure 10.10. The lower chart defines an operating envelope of regulatory takeoff weight (RTOW) as a function of the ambient air temperature. The boundaries of the envelope, which applies to a particular set of operating conditions, are established by considering the various takeoff weight limiting factors (e.g., structural, field length, climb gradient, obstacle clearance). It is apparent that the usable takeoff weight reduces as the temperature increases. Corresponding to the *actual temperature* will be the maximum RTOW. And, corresponding to the actual TOW is a higher ambient temperature—this is called the *assumed temperature.* The value of using an assumed temperature in this way to set the engine thrust is that all limiting takeoff constraints are automatically met.

The procedure for determining the engine thrust setting that corresponds to the assumed temperature depends on the engine type. For engines that use EPR as the thrust setting parameter, the process is straightforward: the thrust can be obtained directly from EPR–temperature data, as illustrated in the upper chart in Figure 10.10. For engine types that use N_1 as the thrust setting parameter, the process is more complicated as it requires several corrections to be implemented (as described by reference [8], for example). Manufacturers thus publish specific tables or charts so that N_1 can be determined for an assumed temperature.

The assumed temperature method is inherently conservative [8]. This stems from two facts. First, setting a takeoff EPR corresponding to a higher (i.e., assumed) temperature, but doing so at the actual ambient temperature, yields greater engine thrust than would be the case if the temperature were actually equal to the higher (i.e., assumed) value. Secondly, the takeoff indicated airspeed at the actual temperature for an ATM takeoff is the same as that which would be used for takeoff at the higher assumed temperature—and this means that the *true airspeed* will be lower for the ATM takeoff compared to actually taking off at that higher temperature. The combination of higher thrust and lower true takeoff airspeeds, when making an ATM takeoff, means that the airplane's takeoff performance will be better for that flight than it would be for a takeoff at the higher (i.e., assumed) temperature.

10.8.5   Derated Thrust and Reduced Thrust Takeoff Speeds

When a lower (i.e., derated or reduced) takeoff thrust is used, as described above, it will be necessary to recalculate the takeoff speeds. There are specific requirements that address this issue in the regulations (see references [19] and [2]), which have a bearing on the way in which the minimum V_1 and V_R speeds are determined. The issue relates to the need for the pilot to be able to maintain lateral control of the airplane—at the derated or reduced thrust—following the failure of the critical engine when (1) the airplane is on the ground, and (2) the airplane is in the air. In the first case, this is ensured by the requirement that V_1 may not be less than V_{MCG}. In the second case, V_R may not be less than V_1 nor less than 105% of V_{MC}. (The reference speeds V_{MCG} and V_{MC} are defined in Section 10.2.2.) The V_2 speed is similarly affected.

When utilizing derated thrust, the minimum V_1 speed is obtained from the AFM for the derated thrust setting. This speed will be lower than the minimum V_1 at the same ambient temperature at full thrust (this is acceptable as the lower thrust will produce a smaller yawing moment). When the takeoff is conducted using reduced thrust (i.e., assumed temperature method), the minimum V_1 speed that can be used should correspond to the actual, and not the assumed, temperature [8].

References

1 FAA, *Airworthiness standards: Transport category airplanes*, Federal Aviation Regulation Part 25, Amdt. 25-143, Federal Aviation Administration, Washington, DC, June 24, 2016. Latest revision available from www.ecfr.gov/ under e-CFR (Electronic Code of Federal Regulations) Title 14.

2 EASA, *Certification specifications and acceptable means of compliance for large aeroplanes*, CS-25, Amdt. 18, European Aviation Safety Agency, Cologne, Germany, June 23, 2016. Latest revision available from www.easa.europa.eu/ under Certification Specification.

3 FAA, *Operating requirements: Domestic, flag, and supplemental operations*, Federal Aviation Regulation Part 121, Amdt. 121-374, Federal Aviation Administration, Washington, DC, May 24, 2016. Latest revision available from www.ecfr.gov/ under e-CFR (Electronic Code of Federal Regulations) Title 14.

4 European Commission, *Commercial air transport operations (Part-CAT)*, Annex IV to Commission Regulation (EU) No. 965/2012, Brussels, Belgium, Oct. 5, 2012. Published in *Official Journal of the European Union*, Vol. L 296, Oct. 25, 2012, and reproduced by EASA.

5 FAA, "Flight test guide for certification of transport category airplanes," Advisory Circular 25-7C, Federal Aviation Administration, Washington, DC, Oct. 16, 2012. Available from www.faa.gov/.

6 Goodyear, "Aircraft tire care & maintenance," Rev. 1/11, The Goodyear Tire & Rubber Company, Akron, OH, Jan. 2011.

7 Goodyear, "Aircraft tire data book," Rev. 10/02, The Goodyear Tire & Rubber Company, Akron, OH, Oct. 2002.

8 Blake, W. and Performance Training Group, "Jet transport performance methods," D6-1420, Flight Operations Engineering, Boeing Commercial Airplanes, Seattle, WA, Mar. 2009.

9 ICAO, "Aerodrome design manual: Runways," 3rd ed., Doc. 9157 Part 1, International Civil Aviation Organization, Montréal, Canada, 2006.

10 FAA, *Definitions and abbreviations*, Federal Aviation Regulation Part 1, Amdt. 1-10, Federal Aviation Administration, Washington DC, Mar. 29, 1966. Latest revision available from www.ecfr.gov/ under e-CFR (Electronic Code of Federal Regulations) Title 14.

11 EASA, *Definitions and abbreviations used in certification specifications for products, parts and appliances*, CS-Definitions, Amdt. 2, European Aviation Safety Agency, Cologne, Germany, Dec. 23, 2010. Latest revision available from www.easa.europa.eu/ under Certification Specification.

12 ICAO, "Operation of aircraft: Part 1, International commercial air transport – aeroplanes," Annex 6 to the Convention on International Civil Aviation, 9th ed., International Civil Aviation Organization, Montréal, Canada, July 2010.

13 EASA, "Acceptable means of compliance (AMC) and guidance material (GM) to Part-CAT," Iss. 2, Amdt. 5, European Aviation Safety Agency, Cologne, Germany, Jan. 27, 2016. Available from www.easa.europa.eu/.

14 ICAO, "Aerodromes: Vol. 1, aerodrome design and operations," Annex 14 to the Convention on International Civil Aviation, 5th ed., International Civil Aviation Organization, Montréal, Canada, July 2009.

15 IAA, "Aeronautical Information Publication (AIP): Ireland," Aeronautical Information Service, Irish Aviation Authority, Shannon, Ireland, 2013. Latest revision available from www.iaa.ie/ as Integrated Aeronautical Information Package.

16 Airbus, "A380 airplane characteristics for airport planning (AC)," Iss. Mar. 30/05, Rev. Nov. 01/10, Customer Services, Airbus S.A.S., Blagnac, France, 2010.

17 Airbus, "Getting to grips with aircraft performance," Flight Operations Support and Line Assistance, Airbus S.A.S., Blagnac, France, Jan. 2002.

18 FAA, "Airport obstacle analysis," Advisory Circular 120-91, Federal Aviation Administration, Washington, DC, May 5, 2006.

19 FAA, "Reduced and derated takeoff thrust (power) procedures," Advisory Circular 25-13, Federal Aviation Administration, Washington, DC, May 4, 1988. Available from www.faa.gov/.

11

Approach and Landing

11.1   Introduction

The *approach* is that phase of a flight immediately prior to the airplane crossing the runway threshold with the intention of landing. Following the descent from cruise, for normal flight operations, the airplane will be flown along a flight path that has defined "gates" located at specified heights and distances from the airport. The configuration (i.e., position of the flaps, slats, speedbrakes, and undercarriage) and the speed of the airplane as it passes through each "gate" are key aspects of the flight crew's Standard Operating Procedures (SOP) for approach and landing. For the purpose of certain performance analyses (e.g., trip fuel calculation), a height of 1500 ft AFE (above field elevation) is often selected as a reference point for the start of the approach. Note that the 1500 ft height is not defined in any regulation or manual pertaining to operational procedures—it is merely a convenient reference height that is often used to separate the approach from the descent phase of the flight for certain airplane performance assessments.

Typically, the airplane will be flown such that its ground track is closely aligned with the extended centerline of the runway. The vertical profile of a typical flight path is approximately straight with a shallow descent angle of about 3° to the horizontal (slightly steeper approaches are sometimes flown, as explained later). This will be followed by a short curved flight segment (when viewed in profile), known as a *flare*, during which time the airplane's pitch attitude is changed with the angle of attack being progressively increased as the pilot pulls back on the stick/yoke. Following touchdown, with the main landing gear wheels contacting the runway, the nose landing gear wheels will then be lowered onto the runway surface. After this, the airplane will be slowed down to a full stop or to a safe speed for taxiing—this is called the *rollout*.

As with the takeoff (described in Chapters 9 and 10), the landing of a jet transport airplane in commercial flight operations is a highly controlled activity, which is undertaken within a complex regulatory and procedural framework. The analysis of the landing performance of an airplane thus requires an understanding of the relevant certification and operational regulations—these include FAR 25 [1], CS-25 [2], FAR 91 [3], FAR 121 [4], and OPS Part-CAT [5]. The characteristics of the runway (e.g., landing distance available, runway slope, and braking coefficient), the height and position of obstacles located near to the approach flight path, ambient conditions (i.e., temperature and pressure), and the prevailing weather (e.g., precipitation and wind) are all factors that need to be considered in the performance analysis of the approach and landing.

An overview of the various elements of the approach and landing phase of a flight is described in Section 11.2. The forces that act on the airplane during the ground run are essentially the same as those described in Section 9.3 for the takeoff—this means that many of the performance analysis methods described in the earlier chapter are also applicable here—this is discussed in Section 11.3. A description of methods that may be used to estimate the landing distance

Performance of the Jet Transport Airplane: Analysis Methods, Flight Operations, and Regulations, First Edition. Trevor M. Young.
© 2018 John Wiley & Sons Ltd. Published 2018 by John Wiley & Sons Ltd.

from a 50 ft reference height (or screen height) is given in Sections 11.4 and 11.5. The complex topic of landing on runways contaminated by standing water or snow or icy slush is introduced in Section 11.6. This is followed, in Section 11.7, by a review of several important operational issues applicable to the approach and landing and, finally, by a discussion on the rejected landing (go-around maneuver) in Section 11.8.

11.2 Procedure for Approach and Landing

11.2.1 Approach

The approach profile angle leading to touchdown is known as the *glide path* angle or *glideslope* angle (although, strictly speaking, the glideslope is an element of the Instrument Landing System, as described later in Section 11.7.1). The glide path angle is typically 3° from the horizontal for commercial flight operations, although greater angles of up to about 5.5° are used in special circumstances.[1] During the approach, the flight crew will slow the airplane down and configure it for landing. This is a multi-step process, which involves—among many tasks—the incremental extension of the flaps according to a defined flap-position–speed schedule, the extension of the slats, and lowering of the landing gear. On a 3° glide path, a jet transport airplane with idle thrust selected and landing flaps and gear extended will decelerate at about 20–30 kt per nautical mile [7]. Speedbrakes may be used to increase the deceleration (depending on the airplane type). The airspeed is then held approximately constant during the last portion of the approach (i.e., below about 1000 ft AFE). The rate of descent[2] is a key parameter that would be monitored by the flight crew. Rates exceeding 1000 ft/min are regarded as excessive and outside of normal operating procedures. For example, the rate of descent on a 3° glide path would be about 740 ft/min for a typical approach speed of 140 kt (ground speed). During the final part of the approach, with landing flaps extended, the thrust is usually set above idle to maintain the target approach speed (see Section 11.7.2 for details on stabilized approaches).

Specific conditions exist in the airworthiness regulations with regard to the airplane's speed when crossing the reference or screen height (h_{sc}). This height is defined as 50 ft (15.2 m) above the landing surface in FAR 25.125 [1] and 15 m in CS-25.125 [2]. It is required that the final approach speed at this point (V_{APP}) not be less than V_{REF}, the reference landing speed. To establish a target approach speed, the flight crew would add a correction (or increment) to V_{REF}, based on certain operational factors—such as wind strength and direction, turbulence, and icing conditions—as defined in the operator's SOP. For calm-wind conditions, a typical target final approach speed is $V_{REF} + 5$ kt [8]. (The influence of wind is discussed later in Section 11.7.7.)

Reference landing speeds are determined by the manufacturer and are based on the airplane's stall speed, as defined in FAR/CS 25.125 [1, 2]:

$$V_{REF} \geq 1.23\, V_{SR_0} \qquad (11.1)$$

where V_{REF} is the reference landing speed; and
V_{SR_0} is the reference stall speed in the landing configuration (see Section 20.5.3).

---

1 The regulatory authorities consider a glideslope angle equal to or greater than 4.5° a *steep approach* for which special conditions apply [2, 6]. Steep approaches are usually flown to clear obstacles, such as buildings surrounding a city center airport (e.g., London City Airport has a 5.5° glideslope), or to reduce the airplane's noise footprint (see Section 21.4).
2 The rate of descent (i.e., the time rate of change of height), in the absence of vertical up- or downdrafts, is equal to the vertical component of the true airspeed.

As an airplane's stall speed depends on its weight and flap position (see Section 20.2), V_{REF} is thus also a function of airplane weight and flap position. To determine the target approach speed for flight operations, the flight crew would ascertain V_{REF} using tables or charts—contained in, for example, the Flight Crew Operating Manual (FCOM)—based on the airplane's landing weight and the selected landing flap position, and then add appropriate corrections (or increments), if required [8].

11.2.2 Landing Flare

In normal operations, the rate of descent (or sink rate) will be approximately constant as the airplane approaches the runway. The objective of the flare is to reduce the vertical speed to an acceptably low value at the time when contact is made with the ground. The rate of descent at touchdown can vary from a value close to zero, when the pilot "greases" the landing, to a hard landing[3] of four or more feet per second. Typically, the rate of descent on touchdown is 2 to 3 ft/s.

The flare usually starts after the 50 ft screen is crossed; however, with steep approaches, it may start earlier. The maneuver involves the pilot pulling the stick/yoke back to increase the angle of attack and generate more lift. As the airplane's lift—for the few seconds before touchdown—is greater than its weight, the load factor $n_z > 1$ and the rate of descent will reduce. Typically, the airplane will slow down a little in the flare and the touchdown speed will be about 3 to 5 kt less than the speed at the screen height [6, 11].

As the touchdown point is approached, the increasing proximity of the ground (i.e., the ground effect) will lead to a reduction in the drag coefficient (C_D) and an increase in the lift coefficient (C_L) for a given angle of attack, so that the glide path tends to become less steep. For this reason, it is found that for certain airplane types a flare is initiated naturally by the influence of the ground. Pilot technique is an important consideration when analyzing the landing flare. In some cases, there will be a period of *float*, as the pilot holds the airplane a few feet off the runway, allowing the speed to decrease until the airplane settles onto the runway. Such landings, however, are frowned upon in flight operations as they come at the expense of hundreds—or even thousands—of feet of additional runway distance (a well-used teaching mantra is that a good touchdown is a firm touchdown).

As the manner in which the approach and flare are executed can significantly affect the landing distance, it is usual, therefore, that published performance data (e.g., in the FCOM) are based on a prescribed landing technique that in-service pilots are expected to use for every landing. For the purpose of theoretical analysis, it is often assumed that there is no float and that the touchdown occurs with a zero vertical component of velocity.

11.2.3 Transition

At the touchdown point, the nosewheel will still be well above the runway, and it will be necessary for the pilot to rotate the airplane by moving the stick/yoke forwards (thus reducing the angle of attack) and bringing the nosewheel down onto the runway. Following touchdown, a time interval of approximately 2 seconds is required for the airplane to be configured by the crew for braking, and the speed will reduce by approximately 2–3%. During this time, the wheel

---

3 There is no universal definition of what constitutes a *hard landing* [9]. For airplane certification, FAR/CS 25.473 [1, 2] requires that commercial airplanes are designed for a rate of descent of 6 ft/s (1.83 m/s) at the design takeoff weight and 10 ft/s (3.05 m/s) at the maximum design landing weight. A structural inspection is warranted if these values are approached or exceeded. Boeing [10] reports that "service experience indicates that most flight crews report a hard landing when the sink rate exceeds approximately 4 ft/s."

brakes will be activated, and the ground spoilers and thrust reversers, if operational, will be deployed. Spoilers, on many airplane types, function as speedbrakes in flight and in a ground spoiler mode during the landing roll (or during a rejected takeoff), when they are extended to their maximum deflection angle. The purpose of the ground spoilers is to increase wheel braking by "spoiling" the lift and by providing additional aerodynamic drag. On modern commercial airplanes, the flight crew can set the ground spoilers and wheel brakes to activate automatically after touchdown. Automatic activation of these systems requires that several conditions are met, such as compression on the main landing gear being registered by installed sensors or switches, the thrust levers being in the idle position, and the wheels having "spun-up" (typically to a speed exceeding 60 kt). The precise logic required to trigger these automatic systems depends on the airplane type.

The transition segment extends from the point of touchdown to the point where all deceleration devices (e.g., wheel brakes, ground spoilers, thrust reversers) are operative. The transition time (Δt_{TR}), which is needed for the airplane to transition into its full braking configuration, depends on the airplane type, whether or not auto-spoilers and autobrakes are selected and, of course, on how quickly the pilot performs the necessary actions.

11.2.4 Braking

In the final phase of the landing, the airplane will be slowed down by use of the wheel brakes (discussed earlier in Section 9.8), assisted by the ground spoilers and the thrust reversers, if operational. Note that the thrust reversers are not used throughout the rollout, but only down to a relatively slow speed, at which time the thrust levers are moved to a reverse idle position and thereafter the thrust reversers are stowed. The pilot will use the rudder (which provides aerodynamic steering) and nosewheel steering (which is used at low speeds) to control the airplane's direction. The *stopping distance* (i.e., the distance associated with braking) is measured from the point at which the airplane is fully configured for braking to the point at which the airplane is stationary.

11.2.5 Landing Distance

The *landing distance* (see Figure 11.1) is measured from the ground location where the lowest point on the airplane (usually taken as the main landing gear) is at the screen height (h_{sc}) to the location when the airplane comes to a complete stop. This distance (s) can be broken

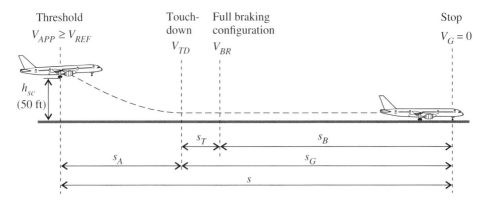

Figure 11.1 Landing distance.

into two segments: the air distance (s_A), which is the final part of the approach and the flare, and the ground distance (s_G), which is also known as the ground run. The ground distance is further broken into the transition distance (s_T), which follows touchdown, and the stopping distance (s_B).

Landing distances that are determined through flight testing and simulation for airplane certification are called *unfactored distances*. These distances are determined strictly according to a defined set of procedures that stipulate the airplane's approach speed, the rate of descent at touchdown, the time allowed to configure the airplane for braking, and so forth [6]. It is important to note that the manner in which these landings are conducted is not typical of routine flight operations. Such tests are conducted to establish the performance capabilities of the airplane. The unfactored landing distances are then multiplied by field length factors (see Section 11.7.6) to give the *factored distances* that are used for operational flight planning. It is the factored distances that are used to check if an airplane can legally land on a runway within the available landing distance at the anticipated landing weight and forecast conditions.

11.3   Forces Acting on the Airplane During the Ground Run

11.3.1   Vector Diagram and Forces Acting on the Airplane

The forces acting on an airplane during the braking segment are shown in Figure 11.2. The runway is assumed to have an uphill slope of angle γ_G. The engine thrust (F_N) is defined as positive in the direction of the velocity (which is the same convention used for the takeoff)—idle thrust will thus be positive, but if thrust reversers are used, then the thrust will be negative. The resistance to forward motion is also due to aerodynamic drag, tire friction, and, for an inclined (uphill) runway, the component of the airplane's weight that acts parallel to the runway. Landing on a runway contaminated by slush or snow introduces an additional contaminant drag force (this is discussed later in Section 11.6).

The same approach adopted for the takeoff analysis (Chapter 9) is used in this chapter: a general description of the forces acting on the airplane is first given, before the equations that can be used to determine the takeoff run are presented. For this analysis, the following factors are important: wind (Section 11.3.2), thrust and reverse thrust (Section 11.3.3), lift and drag forces (Section 11.3.4), airplane configuration (Section 11.3.5), ground effect (Section 11.3.6), rolling friction (Section 11.3.7), and wheel braking (Section 11.3.8).

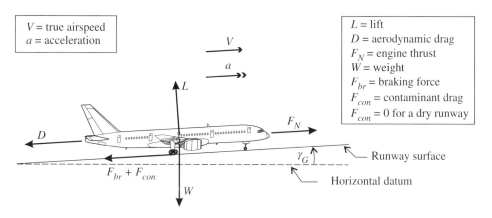

Figure 11.2 Forces acting on the airplane during the landing run.

11.3.2 Influence of the Wind

It is desirable—and, indeed, part of routine operational procedures if there is an appreciable wind speed—that airplanes land "into wind" (i.e., the direction of the component of wind along the runway will be towards the aircraft). The wind speed along the runway (V_w) is, by convention, defined as positive for a headwind and negative for a tailwind (see Section 4.5.2). A headwind results in a lower ground speed at touchdown and reduced kinetic energy, which under comparable braking conditions will mean that the stopping distance will be shorter. (Operational considerations for windy conditions are discussed later in Section 11.7.7.)

11.3.3 Thrust and Reverse Thrust

On touchdown, the engines will produce a small amount of forward thrust—usually corresponding to the idle setting. If thrust reversers are selected by the crew, the direction of the engine thrust is reversed (the operation of thrust reversers is described in Section 8.2.2). The time that it takes for the thrust reversers to deploy and for the engines to spin-up depends on the engine type, but is typically about 5 to 8 seconds [12]. The reverse thrust is achieved by redirecting the air leaving the engine, such that it has a forward component; this produces a powerful braking force, which is particularly important for operations on runways covered with water, ice, or snow. Reverse thrust contributes a greater percentage of the total retarding force acting on the airplane on a wet or contaminated runway compared to a dry runway, due to the reduction in wheel braking effectiveness under these conditions.

During normal flight operations, reverse thrust is discontinued at about 60 kt (the actual speed depends on the airplane and engine types). As the airplane slows down, the thrust reversers are stowed—this is done to prevent the ingestion of engine exhaust gases (which can result in engine surging) or foreign objects. In an emergency, such as a rejected takeoff, thrust reversers can be used all the way to a full stop. Reverse thrust contributes a relatively small amount to the braking effort at low speeds; hence, the impact on the braking distance of stowing the thrust reversers during the final stage of the ground run is small.

11.3.4 Lift and Drag Forces

The magnitude of the lift and drag forces depends on the dynamic pressure and the corresponding lift or drag coefficients, as described by Equations 9.3 and 9.4 (repeated below):

$$L = qSC_L = \frac{1}{2}\rho SC_L V^2 \qquad \text{(Equation 9.3)}$$

$$D = qSC_{D_g} = \frac{1}{2}\rho SC_{D_g} V^2 \qquad \text{(Equation 9.4)}$$

where L and D are the lift and drag forces, respectively (typical units: N, lb);
  q is the dynamic pressure (typical units: N/m^2, lb/ft^2);
  C_L and C_{D_g} are the lift and drag coefficients, respectively (in ground effect);
  ρ is the ambient air density (typical units: kg/m^3, slug/ft^3);
  S is the wing reference area (typical units: m^2, ft^2); and
  V is the true airspeed (typical units: m/s, ft/s).

The lift and drag forces depend on the square of the airspeed; the forces will thus reduce as the airplane slows down. The influence of air density is also apparent in these equations: low air density—associated with hot temperature and/or high altitude operations—reduces the aerodynamic forces. The influence of changes in air density can also be seen on the approach speed. If the airplane is considered to approach at the same calibrated airspeed (CAS), the reduced air density will result in a higher TAS and therefore a higher ground speed for a given headwind strength.

After the nosewheel touches the ground, the angle of attack will be approximately constant for the remainder of the ground run and will thus not induce changes to the lift and drag coefficients. The lift coefficient, however, can be reduced to zero or even negative by the deployment of ground spoilers. Negative lift means that a load greater than the weight of the airplane is placed on the landing gear; this will significantly increase the wheel braking forces (described later in Section 11.3.8). The landing configuration is one of high drag due to the position of the flaps and the extension of the landing gear and ground spoilers (see Section 11.3.5, below). The drag can be described by a nearly constant drag coefficient, which will be much higher than that associated with the takeoff.

11.3.5 Airplane Configuration

The drag coefficient corresponding to the landing configuration must include appropriate corrections and increments to that which describes the "clean" airplane. With the flaps/slats set for landing (the position of the flaps will not be the same as for takeoff) and the undercarriage extended, the zero-lift drag coefficient will naturally increase—as illustrated by Equation 11.2, below. Of particular importance to the analysis of the braking segment is the deployment of ground spoilers, which provide a significant increase in drag.

$$C_{D_0} = \left(C_{D_0}\right)_{clean} + \left(\Delta C_{D_0}\right)_{slats} + \left(\Delta C_{D_0}\right)_{flaps} + \left(\Delta C_{D_0}\right)_{spoilers} + \left(\Delta C_{D_0}\right)_{gear} \quad (11.2)$$

where C_{D_0} is the zero-lift drag coefficient;

$\left(C_{D_0}\right)_{clean}$ is the clean airplane drag coefficient; and
ΔC_{D_0} describes the drag coefficient increments due to the extension of the slats, flaps, ground spoilers, landing gear, and so forth.

11.3.6 Ground Effect

The influence of *ground effect* on the lift and drag is described in Section 9.3.6 for the takeoff run—this also applies to the landing. The ground effect factor (λ) can be used in the same way in the drag polar to account for the airplane's proximity to the ground. This is described by Equation 9.6, which is repeated below:

$$C_{D_g} = C_{D_0} + \lambda K C_L^2 \quad \text{(Equation 9.6)}$$

where λ is the ground effect factor (dimensionless); and
K is the lift-dependent drag factor (dimensionless).

11.3.7 Rolling Friction

After touchdown, the rolling friction (described in Section 9.3.7) will contribute to the aerodynamic and mechanical forces that act to slow the airplane down. A typical value for the rolling coefficient of friction (μ), which is needed to compute the landing distance, is 0.015 [11]. From the point where the wheel brakes are applied, the rolling resistance can be replaced in the mathematical models by a much larger *braking force*.

11.3.8 Wheel Braking

The wheel braking force (F_{br}) can be determined from an appropriate braking coefficient, as described by Equation 9.56 in Section 9.8.2:

$$F_{br} = \mu_B(W - \bar{L}) \tag{11.3}$$

where F_{br} is the wheel braking force (typical units: N, lb);
μ_B is the airplane braking coefficient (dimensionless);
W is the airplane weight (typical units: N, lb); and
$\bar{L}$ is the mean lift (typical units: N, lb).

Wheel braking is significantly influenced by the interface between the tire and the runway surface. The presence of water or surface contaminants (see Section 11.6) can dramatically reduce the braking force.

An autobrake (automatic braking) system, if installed in the airplane, controls the stopping distance by varying the brake pressure to achieve a set (or target) deceleration. Prior to landing, the flight crew—if they elect to use this function—would engage the autobrake system by selecting, typically on a rotary switch, one of several braking options. Each setting corresponds to a target deceleration (e.g., setting 1 may correspond to 4 ft/s^2, setting 2 to 6 ft/s^2, and so forth).[4] The brake pressure is automatically adjusted to account for the influence of reverse thrust and the condition of the runway surface. The system can be disengaged at any time by the crew pressing the foot brakes. Autobrake systems typically take between 2 and 4 seconds after touchdown to reach the target brake pressure. From a computational viewpoint, the estimation of the stopping distance, when an autobrake system is used, is straightforward as the deceleration is approximately constant.

11.3.9 Net Deceleration Force

The discussion presented in Section 9.3, concerning the forces acting on the airplane during the takeoff run, also applies to the landing. It is common practice to use the same conventions— that is, regarding what is positive and what is negative—for forces, velocities, and accelerations for both takeoff and landing. This means that the engine thrust is defined as positive in the direction of motion (even when thrust reversers are applied) and that the acceleration (a) will be a negative quantity when the airplane is slowing down.

The net force acting on the airplane in the direction of motion ($\sum F$) is given by Equation 9.10 (derived for the takeoff)—the only difference is that a braking force replaces the rolling friction force:

$$\sum F = F_N - D - F_{br} - W \sin \gamma_G \tag{11.4}$$

where F_N is the engine thrust (typical units: N, lb); and
γ_G is the runway gradient, positive for uphill (typical unit: rad).

---

4 The Brake-To-Vacate (BTV) system [13], introduced by Airbus on the A380, is an extension of the standard autobrake system. The flight crew selects their desired runway exit point (i.e., taxiway) and arm the autobrake system in the normal way. The BTV system then utilizes GPS with the airplane's onboard airport database and flight control systems to manage the brake pressure such that the selected runway exit is reached at the desired runway vacate speed. A complementary function, known as Runway Overrun Protection, computes the landing distance in real time and alerts the flight crew if the remaining available runway distance becomes too short. The system compensates for such factors as wind, ambient temperature, runway elevation, and surface condition.

The same substitutions and small angle simplifications, as introduced for the takeoff, apply for the landing. Equation 9.13, which describes the net accelerating force for the takeoff, takes the following form for the landing:

$$\sum F = F_N - \frac{1}{2}\rho V^2 S\left(C_{D_g} - \mu_B C_L\right) - (\mu_B + \gamma_G)W \tag{11.5}$$

Once the net force acting on the airplane has been quantified, the deceleration can be calculated, thus permitting the ground run to be evaluated (as described in Section 11.4, below).

11.4   Landing Distance Estimation

11.4.1   Total Landing Distance

As shown in Figure 11.1, the total landing distance (s) is given by

$$s = s_A + s_T + s_B \tag{11.6}$$

The distance for each segment will be determined separately; the total distance is then given by the sum of the air segment, s_A (Section 11.4.2), the transition segment, s_T (Section 11.4.3), and the braking segment, s_B (Sections 11.4.4 and 11.4.5).

11.4.2   Air Segment

There are several methods that can be used to calculate the air distance (s_A).

Method 1: Circular Arc

This method, which assumes a condition of zero wind, idealizes the flight path as two parts (see Figure 11.3):

(1) a linear element, which is assumed to be a steady-state glide at constant airspeed (V), with the airplane in the landing configuration; and
(2) a circular arc element representing the flare, where the touchdown rate of descent is assumed to be zero.

The *flight path angle*, which in the absence of wind is identical to the climb angle (γ), is measured with respect to a horizontal datum; it is defined as positive for a climb and negative for a descent (see Section 12.2). The flight path angle on approach is a small quantity, permitting small angle approximations to be used in this case without loss of accuracy (i.e., $\cos\gamma \cong 1$ and $\sin\gamma \cong \tan\gamma \cong \gamma$ for γ in radians).

The analysis is similar to that used for calculating the distance of the air segment in a takeoff. As indicated in Figure 11.3, the air distance is the sum of two components:

$$s_A = s_{A,1} + s_{A,2} \tag{11.7}$$

The first term can be expressed as a function of the screen height (h_{sc}) and the flight path angle, and the second term can be expressed in terms of the radius of the arc (r) and the flight path angle:

$$s_{A,1} = \frac{h_{sc}}{\tan(-\gamma)} \cong \frac{h_{sc}}{(-\gamma)} \qquad \text{(for } \gamma \text{ in radians)} \tag{11.8}$$

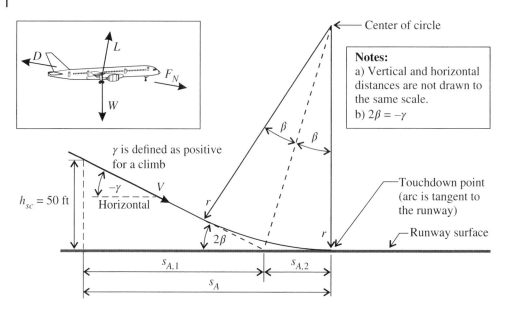

Figure 11.3 Estimating the air segment.

and $\quad s_{A,2} \cong \dfrac{(-\gamma)r}{2}$ (11.9)

Thus $\quad s_A = \dfrac{h_{sc}}{(-\gamma)} + \dfrac{(-\gamma)r}{2}$ (11.10)

Referring again to Figure 11.3, the forces acting on the airplane can be resolved parallel to and normal to the flight path, that is,

$$D - F_N + W \sin\gamma = 0$$ (11.11)

and $\quad L - W \cos\gamma = 0$ (11.12)

From Equations 11.11 and 11.12, the flight path angle can be expressed as follows:

$$\gamma \cong \frac{F_N}{W} - \frac{D}{W} \cong \frac{F_N}{W} - \frac{D}{L}\cos\gamma \cong \frac{F_N}{W} - \frac{1}{E}$$ (11.13)

where E is the lift-to-drag ratio (dimensionless).

In the approach, the lift coefficient is given by

$$C_{L_{app}} = \frac{L}{qS} = \frac{2W\cos\gamma}{\rho V^2 S} \cong \frac{2W}{\rho V^2 S}$$ (11.14)

Thus $\quad V^2 = \dfrac{2W}{\rho S C_{L_{app}}}$ (11.15)

In the flare, the lift will increase as the pilot pulls the stick/yoke back, increasing the angle of attack. The load factor (n_z), as defined in Equation 7.3, will be greater than 1 and the airplane will accelerate towards the center of the circle (radius r). The normal (centripetal) acceleration can be determined from Newton's second law, as it is proportional to the normal force. The sum of the forces on the airplane normal to the flight path in the flare can be approximated by the following expression:

$$\sum F_n = L_{flare} - W \cos \gamma = n_z W - W \cos \gamma \cong W(n_z - 1) \tag{11.16}$$

where L_{flare} is the lift in the flare; and
n_z is the load factor normal to the flight path.

And the normal acceleration (a_n) follows from Newton's second law:

$$a_n = \frac{g}{W} \sum F_n = (n_z - 1)g \qquad \text{(from Equation 2.33)} \tag{11.17}$$

where g is the acceleration due to gravity (typical units: m/s^2, ft/s^2).

The radius (r) of the circle can now be determined:

$$r = \frac{V^2}{a_n} \qquad \text{(from Equation 2.31)} \tag{11.18}$$

Substituting from Equations 11.17 and 11.15 into Equation 11.18 yields the required equation for the radius:

$$r = \frac{V^2}{(n_z - 1)g} = \frac{2W}{(n_z - 1)g\rho SC_{L_{app}}} \tag{11.19}$$

The final result for the landing distance covered in the air is obtained by substituting Equations 11.13 and 11.19 into Equation 11.10, that is,

$$s_A = \frac{h_{sc}}{\left(\dfrac{1}{E} - \dfrac{F_N}{W}\right)} + \frac{\left(\dfrac{W}{S}\right)\left(\dfrac{1}{E} - \dfrac{F_N}{W}\right)}{(n_z - 1)g\rho C_{L_{app}}} \tag{11.20}$$

The value of C_L in the approach can be determined from the approach speed (see Section 11.2.1). The load factor depends on the rate at which the stick is pulled back.

A limitation that exists for this method is that the airplane is assumed to touch down with zero rate of descent (i.e., the flight path is tangent to the runway surface). This is seldom the case. The air distance would be reduced for a landing with a high touchdown rate of descent (vertical impact velocity).

Method 2: Glideslope and Flare Load Factor

If the glideslope and flare load factor are known (or can be estimated), then it possible to simplify the technique described above for the determination of the zero-wind air distance (s_A). It is assumed that the flight path matches the glideslope and that the flare is executed with a constant load factor. An expression for the air segment can be obtained directly from Equations 11.10 and 11.19, that is,

$$s_A = \frac{h_{sc}}{(-\gamma)} + \frac{(-\gamma)V^2}{2(n_z - 1)g} \tag{11.21}$$

where γ is the flight path angle (defined as positive for a climb; unit: rad); and
V is the mean airspeed for the segment (typical units: m/s, ft/s).

If a standard glideslope of 3° is assumed (i.e., the flight path angle is −3°), then the first term in Equation 11.21 is equal to 954 ft (290.8 m). This value, it should be noted, is only a little shorter than the unfactored air distance of 1000 ft that is published in Boeing advisory material, such as in the Quick Reference Handbook, for several airplane types (i.e., B737, B757, B767, B777) [14], and which is supported by airplane certification flight tests. (A distance of 1200 ft was used for the B747 [14].) The small difference between 1000 ft and 954 ft reflects the fact that landing tests for certification can be conducted with minimal flare and a high touchdown rate of descent of up to 6 ft/s (1.83 m/s).

With knowledge of the load factor and airplane's speed, the air distance (s_A) can be estimated using Equation 11.21. A "nominal" load factor of 1.2 for operational performance calculations is suggested in reference [11]. For preliminary evaluations, it may be assumed that airspeed "bleeds off" during the flare and that the mean speed for the air segment is 1% to 1.5% slower than V_{REF}.

Method 3: Empirical Data

FAA AC 25-7 [6] provides data in the form of an upper bound to measured air distances obtained during past FAR Part 25 certification campaigns under zero wind conditions—this is described by the following empirical equation:

$$s_A = 1.55(V_{REF} - 80)^{1.35} + 800 \tag{11.22}$$

where s_A is the landing air distance (unit: ft); and
$\quad\quad V_{REF}$ is the reference landing speed (unit: kt).

Method 4: Mean Speed

A simple method that is widely used to determine the air distance (s_A) is based on the product of the average speed of the airplane and an average time for the air segment, as determined from flight testing or simulation. In the expression given below, the influence of wind has been included.

$$s_A = \left(\frac{V_{APP} + V_{TD}}{2} - V_w \right) \Delta t_{APP} \tag{11.23}$$

where V_{APP} is the approach airspeed at 50 ft (typical units: m/s, ft/s);
$\quad\quad V_{TD}$ is the airplane's airspeed at touchdown (typical units: m/s, ft/s);
$\quad\quad V_w$ is the wind speed, defined as positive for a headwind (typical units: m/s, ft/s); and
$\quad\quad \Delta t_{APP}$ is the time from the 50 ft screen to touchdown (unit: s).

Reference [14] reports that descent rates of 4–6 ft/s are "targeted" for touchdown during certification tests (in zero wind conditions), and this results in flight times of 5–6 seconds.

11.4.3 Transition Segment

To calculate the transition distance (s_T), the average ground speed is multiplied by the transition time, that is,

$$s_T = \left(\frac{V_{TD} + V_{BR}}{2} - V_w \right) \Delta t_{TR} \tag{11.24}$$

where V_{TD} is the airplane's airspeed at touchdown (typical units: m/s, ft/s);
$\quad\quad V_{BR}$ is the airspeed at which the airplane is fully configured for braking (typical units: m/s, ft/s); and
$\quad\quad \Delta t_{TR}$ is the transition time (typical unit: s).

The transition time will vary depending on the airplane type and the response times of the flight crew; it is typically about 2 seconds [11]. For certification purposes, specific time allowances are stipulated: the time for each manual action must be the demonstrated test time or one second, whichever is longer [6]. The airplane's speed (without brake application and at idle thrust) will reduce by approximately 1.3% for each second after touchdown.

11.4.4 Braking Segment: Governing Equation

The governing equation of motion to be used for calculating the portion of the ground run after the brakes become effective (i.e., the stopping distance) is essentially the same as that used for the takeoff. The instantaneous acceleration can be determined by the application of Newton's second law (see Section 2.4.5) in which the net accelerating force is described by Equation 11.5. The following expression describes the acceleration (a) of an airplane on a runway that is free of surface contamination:

$$a = \frac{\sum F}{m} = \frac{g \sum F}{W} = g \left[\frac{F_N}{W} - \frac{\rho V^2 S}{2W}(C_{D_g} - \mu_B C_L) - (\mu_B + \gamma_G) \right] \tag{11.25}$$

Note that the acceleration is defined as positive in the direction of motion. During the braking segment, a will thus be negative, which implies a *deceleration*. The thrust F_N will be positive for idle thrust and negative for reverse thrust. In cases where reverse thrust can be used, it is generally applied after the ground spoilers and brakes become effective. The negative value of F_N due to reverse thrust makes a significant contribution to the deceleration (particularly when the runway is wet or slippery).

The stopping distance (s_B) can be found by integration of the following expression:

$$s_B = \int_{V_{BR}}^{V_w} \frac{V - V_w}{-a} \, dV \qquad \text{(from Equation 9.16)} \tag{11.26}$$

where V_{BR} is the airspeed at which the airplane is fully configured for braking (typical units: m/s, ft/s); and
V_w is the wind speed (positive for a headwind) (typical units: m/s, ft/s).

Solutions to the governing equation are described in Section 11.4.5, below.

11.4.5 Braking Segment: Stopping Distance Estimation

Method 1: Closed-Form Solution

It is possible to determine a closed-form mathematical solution to Equation 11.26 where the acceleration is described by Equation 11.25; in fact, the solution has the identical form to that which was derived in Section 9.4.2 for the takeoff. The usefulness of such a solution, however, can be limited due to the idealizations that must be introduced to describe several of the key parameters. First, the braking coefficient μ_B (see Section 9.8.2) must be represented by a mean constant value. For a dry runway surface, this is usually acceptable; however, for a wet surface, this assumption is fundamentally less accurate as the braking coefficient increases in a non-linear fashion with reducing speed. Secondly, the force produced by thrust reversers is difficult to model in a simple manner, which is needed in order to evaluate the integral (i.e., Equation 11.26). Thirdly, the deployment of ground spoilers introduces additional drag and a significant reduction in lift—both factors need to be accounted for. Finally, as the airplane's configuration changes during the braking sector (e.g., the spoilers and thrust reversers are deployed and then retracted), it is necessary to consider the stopping distance as comprising several smaller distances—and to evaluate each distance separately.

Method 2: Mean Acceleration
If the acceleration is constant or assumed to be constant, then it is possible to use the following equation (derived in Section 9.4.3) to compute the stopping distance:

$$s_B = \frac{-1}{2\bar{a}}(V_{BR} - V_w)^2 \quad \text{(from Equation 9.30)} \tag{11.27}$$

Equation 11.27 can be used to obtain an initial estimate of the stopping distance (s_B) using a mean acceleration ($\bar{a}$) that is calculated for a reference speed given by $V^* = 0.707V_{BR}$. This is the same approach as that described in Section 9.4.3 for the takeoff. Furthermore, Equation 11.27 provides an acceptable method for calculating the stopping distance when autobrakes are used because an autobrake system controls the brake system pressure so as to produce a preselected—and nearly constant—deceleration (see Section 11.3.8).

Method 3: Numerical Evaluation
A superior method to those described above involves evaluating the integral expression (i.e., Equation 11.26) using a numerical technique, which can be implemented on a spreadsheet. The process is almost identical to that described in Section 9.4.4 for the takeoff and is based on a step-by-step computation of the distance covered during selected speed intervals. This technique permits the inclusion of realistic speed-dependent braking coefficients and the appropriate application of reverse thrust.

An example of a stopping distance calculation is presented in Table 11.1 for a medium-size twinjet. It has been assumed that braking commences at an airspeed of 130 kt. For the sake of illustration, 20 kt speed intervals have been used. The calculation starts with the ground speed (V_G), which is used to compute the true airspeed (V) and dynamic pressure (q). The forces acting on the airplane (see Equation 11.5) are then calculated, as indicated by the results given

Table 11.1 Example of stopping distance calculation

V_G (kt)	V_G (ft/s)	V (kt)	V (ft/s)	q (lb/ft^2)	F_N (lb)	$\left(C_{D_g} - \mu_B C_L\right)qS$ (lb)	$\mu_B W$ (lb)	$\gamma_G W$ (lb)	a (ft/s^2)	Δs (ft)
120	203	130	219	57.2	4454	19 990	76 000	2000	−15.0	
100	169	110	186	41.0	4872	14 313	76 000	2000	−14.1	431
80	135	90	152	27.4	5353	9581	76 000	2000	−13.2	376
60	101	70	118	16.6	5897	5796	76 000	2000	−12.5	310
40	67.5	50	84.4	8.46	6505	2957	76 000	2000	−12.0	232
20	33.8	30	50.6	3.05	7177	1065	76 000	2000	−11.6	145
0	0	10	16.9	0.34	7913	118.3	76 000	2000	−11.3	50
									Total =	1543

Notes:
(a) Airplane: medium-size twinjet, with $W = 200\,000$ lb, $S = 1951$ ft^2, $C_L = 0.134$, and $C_{D_g} = 0.230$.
(b) Conditions: 1% runway gradient (uphill), 10 kt headwind, dry runway surface, and ISA conditions.
(c) The braking is accomplished with "good" wheel braking ($\mu_B = 0.38$) and idle thrust (no thrust reversers are used).

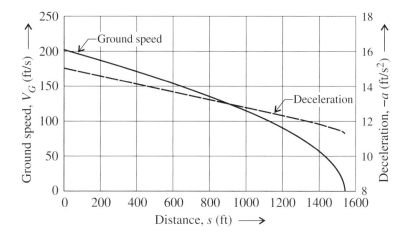

Figure 11.4 Ground speed and deceleration (i.e., $-a$) plotted as a function of runway distance.

in columns 6 to 9. With these results, the instantaneous acceleration (a) is computed using Equation 11.25 (column 10). An average acceleration for the speed interval is then found, from which the interval distance (Δs) is determined. The calculation is then repeated for the subsequent speed intervals and the distances summed. The computed ground speed and deceleration (i.e., $-a$) are indicated as functions of runway distance in Figure 11.4.

11.5   Empirical Estimation of the Landing Distance

11.5.1   Introduction

A simple, approximate method is presented to estimate the landing distance—from the 50 ft screen height to standstill—for airplanes landing on runways of zero slope. The method relies on the correlation of the landing distance with the square of the airplane's approach speed (measured at the screen height), for different braking conditions. The underlying principle for the method is that the energy that must be dissipated by the airplane's braking system depends significantly on the square of the ground speed on approach.

11.5.2   Basic Relationship

The stopping distance, given by Equation 11.27, is substituted into Equation 11.6 to give the following expression for the landing distance:

$$s = s_A + s_T - \frac{1}{2\bar{a}_B}(V_{BR} - V_w)^2 = (s_A + s_T) + \frac{1}{2g_0\left(\dfrac{-\bar{a}_B}{g}\right)}(V_{G,BR})^2 \qquad (11.28)$$

where s_A, s_T, and s_B are the air, transition, and stopping distances, respectively (typical units: m, ft);

$\bar{a}_B$ is the mean braking acceleration (defined as positive in the direction of the motion) (typical units: m/s², ft/s²); and

$V_{G,BR}$ is the ground speed at brake application (typical units: m/s, ft/s).

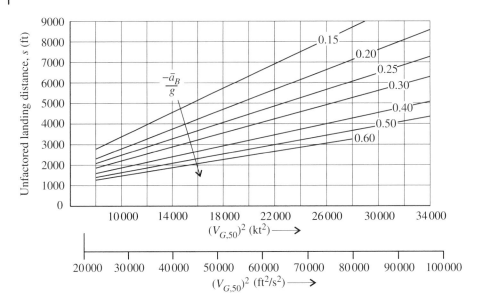

Figure 11.5 Landing distance versus approach speed squared. *Source*: Adapted from ESDU 84040 [15].

The braking speed is, however, not a particularly useful reference speed to correlate landing performance data. It is preferable to work with the approach speed at the 50 ft screen height (V_{50}), which is either known or can be deduced (see Section 11.7.2). It is assumed that the airplane will slow down by about 5% from the 50 ft screen to the point of brake application, thus, $V_{G,BR} = kV_{G,50}$, where $k \cong 0.95$. This permits Equation 11.28 to be written as follows:

$$s = (s_A + s_T) + \frac{k^2}{2g_0 \left(\dfrac{-\bar{a}_B}{g}\right)}(V_{G,50})^2 \tag{11.29}$$

Equation 11.29 provides the underlying relationship for this empirical estimation technique. A graph of landing distance (s) versus $(V_{G,50})^2$ for selected values of $(-\bar{a}_B/g)$ is prepared—as illustrated in Figure 11.5. The figure may be supplemented by the addition of actual airplane data, which are plotted for different landing conditions (e.g., runway surface condition, thrust reversers operative/inoperative).

11.6 Landing on Contaminated Runways

11.6.1 Runway Contaminants

The impact of runway surface contamination on takeoff is discussed in Section 9.9; much of that discussion also applies directly to the landing. Table 11.2 categorizes surface contamination and its effects on the airplane during landing. Due to their impact on an airplane's performance during takeoff or landing, it is necessary to distinguish between (1) loose or fluid contaminants, and (2) hard contaminants. Both categories will reduce the tire–surface friction. Loose or fluid contaminants, however, will also slow the airplane down due to the combined effects of the tires having to displace the surface contaminant and from the impingement of snow, water, or slush

Table 11.2 Runway surface conditions and braking implications

Runway condition	Description	Implications
Dry runway	Absence of water on the runway	• Good wheel braking
Wet runway	Standing water shallower than 3 mm (0.12 inch)	• Moderate wheel braking
Contaminated runway with loose or fluid contaminants	Standing water or slush more than 3 mm (0.12 inch) deep; wet or dry (loose) snow	• Poor wheel braking • Contaminant drag likely • Possibility of hydroplaning
Contaminated runway with solid or compact contaminants	Ice; compact snow	• Poor wheel braking

on the landing gear and airframe. Loose or fluid contaminants on the runway create additional drag on the airplane, as described in Section 9.9.2.

11.6.2 Runway Condition Reporting

Information on the condition of slippery runways can be obtained from three sources:

(1) pilot reports (PIREPs) on observed braking action during landing;
(2) an observation of the runway condition (e.g., using descriptive terms, such as those recorded in Table 9.2); and
(3) friction values determined by a special test device or test vehicle.

Air traffic control (ATC) has traditionally furnished pilots with information on the expected quality of braking action, based on information supplied by other pilots and/or airport management. The braking action on various sections of the runway has been characterized using the terms *good*, *medium* (or *fair*), *poor*, and *nil* or a combination of these terms. The following approximate correlations of pilot reports on braking action to airplane braking coefficients are suggested [16]:

• *good* ($\mu_B = 0.2$), typically describing a wet runway (to a depth of less than 3 mm);
• *medium* or *fair* ($\mu_B = 0.1$), typically describing a runway with dry snow or ice or slush; and
• *poor* ($\mu_B = 0.05$), typically describing a runway covered by melting ice.

These traditional PIREP descriptors are being replaced by a numerical rating scale, with a range of 0 to 6, which is allocated to each one-third portion of the landing surface.[5] The Runway Condition Assessment Matrix (RCAM) correlates the new codes with the traditional pilot reported braking action as follows: 0 (nil braking), 1 (poor braking), 2 (medium to poor braking), 3 (medium braking), 4 (medium to good braking), 5 (good braking) and 6 (dry surface) [18, 19].

Reported mu ($\mu_{reported}$) values, obtained through the use of a ground-friction measuring device, provide a useful comparative measure of the state of a runway. Two problems, however, arise when attempting to use measured reported mu data for airplane performance analysis. First, a variety of testing devices/vehicles are used (by different authorities), and the results can

---

5 These changes resulted from an in-depth policy review by the FAA's Takeoff and Landing Performance Assessment (TALPA) Aviation Rulemaking Committee (ARC), which was established in 2005, following a B737-700 landing overrun at Chicago Midway International airport [17].

thus differ. A common friction index for all ground friction measuring devices does not exist at present [2]. Secondly, the correlation between measurements made with such testing devices and actual aircraft is not always satisfactory. There are good reasons for this: the airplane friction coefficient depends not only on the condition of the runway surface, but also on the tire characteristics (e.g., tire type, wear, and pressure), the efficiency of the anti-skid system, and the speed and weight of the vehicle. (It has been said that the only really useful vehicle for determining the braking force applicable to a jet airplane is another jet airplane.)

11.6.3   Determination of the Landing Distance

The determination of the landing distance for a contaminated runway follows the same process described earlier for a wet runway, and the same time delays apply.

The air distance (see Section 11.4.2) can be estimated by multiplying the mean ground speed by an assumed time from the 50 ft screen to touchdown. For contaminated runways, EASA AMC 25-1591 [2] introduces a conservatism by stipulating that this time interval should be 7 seconds, and that, in the absence of flight-test data, the touchdown speed should be assumed to be 93% of the threshold speed.

Dynamic hydroplaning (see Section 9.9.3), if it occurs, will take place at the beginning of the ground run following touchdown and will cease when the airplane's ground speed falls below the hydroplaning speed (given by Equation 9.61 or 9.62).

The numerical step integration technique described in Section 11.4.5 can be used to estimate the stopping distance taking into account the contaminant drag (F_{con}), which can be approximated by Equation 9.60, and a braking force determined from braking coefficient values that change as the airplane slows down.

11.7   Flight Operations

11.7.1   Instrument Approach Systems

Several different navigation systems are used to conduct instrument-assisted approaches. Widely used, throughout the world, is the Instrument Landing System (ILS); the principle of operation of this radio-navigation system is described below. Systems based on radar and microwave are also used (as described in references [20] and [21], for example). Modern satellite navigation systems, with augmentation through ground-based receivers to improve accuracy, are becoming increasing important for autoland capability.

ILS provides lateral and vertical guidance information along the approach path to the runway. The system, which has both ground-based and airplane-installed components, comprises the following main elements: (1) a *localizer*, which provides horizontal (left/right) guidance; (2) a *glideslope*, which provides vertical (up/down) guidance; (3) marker beacons, which provide range information; and (4) approach lights, which assist in the transition from instrument to visual flight [21].

The localizer (LOC) is a ground-based antenna array, which transmits a narrow radio signal of about 5° in width along the extended runway centerline (Figure 11.6). The companion signal is the glideslope (GS), which is a very thin sloping field pattern (typically 1.4° in depth) transmitted from a ground station located near the approach end of the runway. The glideslope projection angle is normally set at 3° to the horizontal, but angles of 2.5° to 5.5° are sometimes used. The glide path is the straight, sloping line on the center of the glideslope, along which the airplane needs to fly to get to the runway touchdown zone. The localizer and glideslope signals

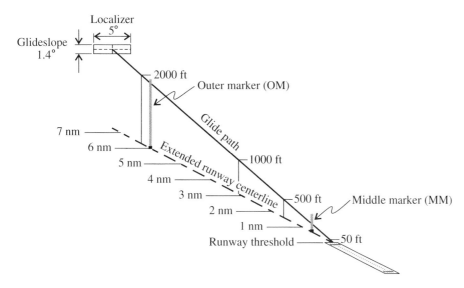

Figure 11.6 Typical Instrument Landing System (ILS) installation with 3° glideslope.

are received by onboard navigation equipment and displayed as vertical and horizontal needles on the flight crew's Primary Flight Displays (PFDs). The airplane's horizontal position with respect to the extended runway centerline and vertical position with respect to the glideslope can easily be seen by the flight crew.

Two or more ground-based VHF beacons, installed on the extended runway centerline, transmit narrow vertical signals. An indication on the PFDs occurs when the airplane overflies the markers, providing the flight crew with confirmation of their location. The outer marker (OM), which is located on the centerline of the localizer course about 3.5 to 6 nm from the runway threshold, indicates the position at which an airplane, at the appropriate altitude (for that airport), will intercept the glide path. The middle marker (MM) is located about 3500 ft (1067 m) from the runway threshold—this roughly corresponds to a position where the glideslope is 200 ft above the elevation of the touchdown zone.

11.7.2   Approach Requirements

A *stabilized approach* is one in which all of the following elements are achieved at the applicable stabilization height [22]:

(1) the airplane is in the desired landing configuration;
(2) the thrust is stabilized, usually above idle, to maintain the target approach speed along the desired final approach path;
(3) vertical and lateral deviations from the desired flight path are within limits;
(4) only small changes in heading and pitch are required to maintain the flight path; and
(5) any deviation of the key flight parameters—that is, airspeed, vertical speed, pitch attitude, and bank angle—is within defined limits (e.g., airspeed limits could be set as $V_{APP} - 5$ kt and $V_{APP} + 10$ kt in the SOP).

The recommended minimum stabilization height is 1000 ft AFE for IMC (instrument meteorological conditions—see Section 4.5.8) and 500 ft AFE for VMC (visual meteorological conditions) [23]. If the airplane is not stabilized on the approach path at the minimum stabilization height, the approach should, in almost all cases, be discontinued and a go-around executed[6] (see Section 11.8).

In determining certified landing distances (see Section 11.7.5), a stabilized approach with the airplane in the landing configuration and with an airspeed not less than V_{REF} must be maintained down to a height of 50 ft (or 15 m) above the runway surface. According to FAR/CS 25.125 [1, 2], in non-icing conditions V_{REF} may not be less than:

(1) 1.23 V_{SR_0}, the reference stall speed in the landing configuration (see Section 20.5.3);
(2) V_{MCL}, the minimum control speed during approach and landing (see Section 20.5.3); or
(3) a speed that provides a minimum maneuvering capability, as defined in FAR/CS 25.143(h) [1, 2]).[7]

In addition, specific requirements exist for flights in icing conditions—these are described in FAR/CS 25 paragraphs 125(a)(2) and 125(b)(2)(ii) [1, 2].

11.7.3 Runway Data

In Section 10.4, the runway data relevant to takeoff operations are described; in this section, additional information relevant to the landing is given.

Published data for commercial airports—determined in accordance with ICAO Annex 14 [25] recommendations—indicate the *landing distance available* (LDA) for each runway.[8] When there are no obstacles below the flight path, within a defined approach corridor, the LDA is the actual runway length (Figure 11.7). However, if there are obstructions that infringe upon the flight path (e.g., buildings, towers, antennae), a "displaced threshold" is defined with respect to the most penalizing obstacle, to provide adequate clearance for landing aircraft. The displayed threshold is defined by considering a 2° plane over the top of the critical obstacle plus a 60 m margin (Figure 11.8). A stopway (see Section 10.4.6), if present, cannot be included in the determination of the LDA.

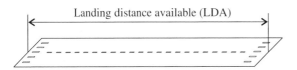

Figure 11.7 Landing distance available (LDA)—no obstacles under flight path.

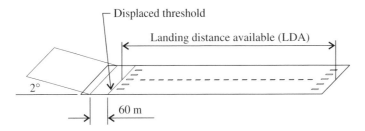

Figure 11.8 Landing distance available (LDA)—with displaced threshold.

6 Unstabilized approaches are relatively rare, accounting for only 3% of landings of commercial airplanes worldwide [24]. Non-compliance with best industry practice (and the airline Standard Operating Procedures) by which pilots continue the approach when it is not stabilized is the leading cause of runway excursions and has been a causal factor in 40% of approach and landing accidents [22, 24].
7 This requirement is for the airplane, at the V_{REF} speed, to be free of stall warning or other characteristics that might interfere with normal maneuvering at a −3° flight path angle, during a coordinated turn with a 40° bank angle.
8 Published in the Aeronautical Information Publication (AIP) of the relevant state (see Section 23.4.6).

11.7.4 Regulatory Requirements for Landing

Landing requirements for airplanes powered by turbine engines are set out in FAR 121.195 [4] and OPS Part-CAT.POL.A.230 [5]. It is stated that no person may take off an airplane at such a weight that—allowing for normal consumption of fuel and oil in flight to the destination or alternate airport—the weight of the airplane on arrival would exceed the landing weight given in the Airplane Flight Manual (AFM) for the elevation of the airport and the ambient temperature anticipated at the time of landing. Furthermore, the landing weight at the intended destination or alternate airport may not be greater than that at which compliance with the following can be shown:

(1) From a point 50 ft above the intersection of the obstruction clearance plane and the runway, it must be possible to execute a full stop landing within 60% of the effective length of the runway. The probable wind speed and direction and the ground handling characteristics of the airplane are to be considered.
(2) For wet or slippery runways, it must be possible to execute a full stop landing within a distance of at least 115% of the required runway length determined under paragraph (1) above (unless, based on actual operating landing techniques on wet runways, a shorter landing distance has been approved by the authorities).

Compliance with these requirements is accomplished by factoring (i.e., multiplying by the appropriate factor) the airplane's certified landing distances.

11.7.5 Certified Landing Distances

As part of an airplane's certification program, the manufacturer undertakes a series of landing tests; the specifications for these tests are given in FAR/CS 25.125 [1, 2] and supporting information on acceptable experimental procedures that would satisfy the requirements are given in FAA AC 25-7 [6] and EASA AMC 25.125 [2].

The measured landing distance is taken as the horizontal distance from the position of the main landing gear when the main landing gear is 50 ft above the landing surface (which can be treated as a horizontal plane through the touchdown point) to the position of the nose landing gear[9] when the airplane is at a full stop. Traditionally, these tests were conducted with a high approach angle and a very high touchdown rate of descent (of up to 8 ft/s). Today, manufacturers can adjust measured data acquired with touchdown rates of 2 to 6 ft/s, in a parametric process, to equivalent values consistent with the traditional experimental technique.

It is specified that the tests be conducted on a level, smooth, dry, hard-surface runway in a manner that does not require exceptional piloting skill. Following touchdown, braking is accomplished by means of the wheel brakes only; reverse thrust is not used. A minimum time delay of 1 second applies between touchdown and the pilot's activation of the first deceleration device and also between the activation of subsequent devices. If, however, automatic devices (e.g., autobrakes, auto-spoilers) are approved then this 1 second delay does not apply. Distances are measured for a series of landing tests covering a range of landing weights using the same set of tires and brakes (thus generating data for the AFM on the stopping distances that can be achieved with worn brakes).

---

9 The policy of referencing the end of the landing distance to the *nose landing gear* when the airplane has stopped was introduced in FAA AC 25-7 rev. C (Oct. 2012) [6].

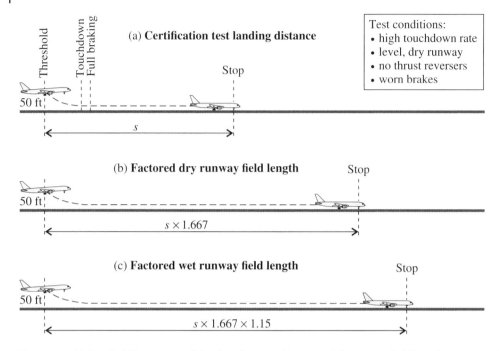

Figure 11.9 (a) Certified (demonstrated) landing distance; (b) Factored dry runway field length; (c) Factored wet runway field length.

11.7.6 Factored Landing Field Lengths

As described in Section 11.7.4, it is a requirement for standard dry runway operations that the airplane is capable of landing from the 50 ft screen height—at the given weight, altitude, and temperature—using no more than 60% of the landing distance available. The remaining 40% can be considered as an operational reserve or margin.

The minimum *required landing distance* (or landing field length) for flight operations can be obtained by multiplying the appropriate certified landing distance by a *field length factor*. For dry runways, the field length factor is 1.667 (i.e., 1/0.6) and for wet or slippery runways, it is 1.917 (i.e., 1.667 × 1.15). The process is illustrated in Figure 11.9. In the case of runways contaminated by ice or snow (see Section 11.6), additional allowances are made for the reduced braking capability of the airplane.

The policy of not using thrust reversers for the determination of the certified landing distance means that the airplane would stop in an even shorter distance if they were used. This provides a further level of safety (or conservatism) for actual airplane operations, and would accommodate the extra stopping distance that would be needed if an engine failure occurred (or if the thrust reversers failed to deploy). The policy also accommodates the longer stopping distances that could occur in practice from lower friction coefficient values than those experienced during testing.

11.7.7 Wind Considerations

Wind can have a big impact on landing performance—for example, headwinds reduce landing distances (which is desirable). Strong winds blowing across the runway, however, can make landings unsafe. As the direction and speed of the wind can change suddenly, conservative policies are adopted to determine procedures and limits for flight operations.

One example of this conservatism is seen in the airworthiness regulations FAR/CS 25.125 [1, 2], where the following operational correction factors must be applied for the determination of the effective headwind or tailwind component acting along the runway: (1) for a headwind, not more than 50% of the nominal wind speed component is used, and (2) for a tailwind, not less than 150% of the nominal wind speed component is used. Thus, when the wind direction is advantageous, only partial credit is taken for the wind's influence, whereas when it is punitive, the penalty is increased.

The regulatory authorities in FAR/CS 25.237 [1, 2] describe the certification requirements for crosswind takeoffs and landings. It is indicated that a 90° cross component of wind velocity of at least 20 kt or $0.2\,V_{SR_0}$, whichever is greater, except that it need not exceed 25 kt, must be demonstrated to be safe for takeoff and landing on dry runways. The *demonstrated crosswind* landing capability is recorded in the AFM. Note that is not necessarily equal to the maximum crosswind that the airplane is able to cope with. Rather, as mentioned in the discussion on takeoff in Section 10.3.9, the recorded value is equal to the strongest crosswind experienced in flight testing during which the manufacturer demonstrated the airplane's capabilities to the satisfaction of the regulatory authorities.

Operators establish their own limits based on AFM data and other operational factors. When an airplane is in close proximity to the ground and the winds are strong, the crab angle and/or airplane drift crosswise to the runway can create difficulties for the pilot to transition from the approach flight path to rollout on the runway. The runway surface condition also affects the limiting speeds at which an airplane can be safely landed. A poor runway friction coefficient can impose a crosswind limit—for example, an airplane with a 30 kt crosswind limit on surfaces with braking action reported as "good" (see Section 11.6.2), could typically have a crosswind limit of 15–20 kt for runways reported as "medium/poor" [26]. Landings become progressively more difficult for a pilot to execute when the crosswind strength increases—this is particularly true in gusty conditions. When deciding if it is safe to land, it is usual for pilots to add half of the gust speed to the steady crosswind speed to give an equivalent (representative) wind speed. This value will then be compared to the operational limit for the airplane for the particular runway surface conditions.

Strong winds also influence the speed that pilots select for the approach. The recommended *wind correction* to the final approach speed (see Section 11.2.1) depends on the airplane type and operator. One such policy requires that pilots add to V_{REF} one-half of the steady headwind component plus the full intensity of the gust, with the total wind correction being limited to a maximum value (usually 20 kt) [8]. The reason for the increase in approach speed is to provide an additional margin of safety. Increasing the airspeed on the approach has the effect of increasing the ground speed. This higher inertial speed will prevent the airspeed from getting dangerously low should the wind speed abruptly reduce during the flare. In this way, the probability of a hard landing is reduced.

11.7.8   Brake Heating Considerations

High-speed braking during landing results in a considerable build-up of heat in the brakes; this is especially true for heavy landing weights. Dissipation of the heat from the brakes causes the temperature of the entire wheel assembly to increase, and this, in turn, causes the gases within the tires to expand. If a tire gets hot enough and the gas is not released, it will explode. As a safety measure, fuse plugs (also called thermal plugs) are installed in the landing gear wheels (in compliance with FAR/CS 25.731 [1, 2]). The plugs, which are made of a low-melting-temperature metal, are inserted into small holes drilled in each wheel. The purpose of the plugs is to melt—releasing the gases within the tires—before the wheels get too hot and the tires burst.

There are two safety-related aspects regarding fuse plugs that should be noted. First, the heat transfer from the brakes through the wheel assembly takes place rather slowly and this means that there is a considerable time lag between the braking event and a fuse plug reaching its peak temperature. Blake, *et al.* [16] indicate that melting of a fuse plug could occur "as much as forty-five minutes to an hour after brake application." Secondly, the heat build-up is cumulative. If there is inadequate time for the brakes to cool completely between braking events, there can be residual heat in the wheel assembly from the previous landing when the airplane touches down. Short-haul operators have adopted operational techniques to cool the brakes for short turnaround operations—for example, by delaying the retraction of the landing gear after takeoff (brakes cool much quicker in flight when the landing gear is extended than when retracted).

If a tire deflates due to a fuse plug melting when the airplane is taxiing or when it is stationary, this would not pose a safely concern (it will obviously have an operational consequence). However, at high speed during takeoff, the airframe could be damaged from impacting pieces of tire thrown off the wheel. To reduce the likelihood of such an event, most modern airplane types are fitted with tire pressure indicators in the cockpit. (On older types, however, pilots would not necessarily be aware of a deflated tire on the airplane.)

Manufacturers publish in the airplane's AFM a landing weight limit called the *maximum quick turnaround weight* (which is a function of the flap setting, pressure altitude, OAT, runway slope and headwind component). This is a certified limit. Whenever an airplane lands at a weight exceeding this value, it must remain on the ground for a specified time interval (which can exceed 1 hour) before leaving the gate for the next flight. The waiting time is intended to ensure that if the tires were to deflate, then this would happen in a safe manner at the ramp and not afterwards when the airplane has left the gate for the next takeoff. Alternative procedures that reduce the waiting time through monitoring the brake temperature are available on certain airplanes.

11.7.9 Runway Excursion

The term *runway excursion* describes an event in which an airplane veers off the side of the runway or overruns off the end of the runway. The vast majority of runway excursions occur on landing, rather than on takeoff. In recent years, landing overruns have been the subject of several safety studies and initiatives, with the expressed goal of mitigating associated risks and further reducing the incidence rate.[10] Addressing this topic is FAA Advisory Circular 91-79A [18], which identifies the following factors (based on FAA and NTSB[11] data) as increasing the risk of a runway overrun: (1) an unstabilized approach; (2) high airport elevation or high-density altitude, resulting in increased ground speed; (3) excess airspeed over the runway threshold; (4) excess airplane landing weight; (5) landing beyond the touchdown point; (6) downhill runway slope; (7) excessive height over the runway threshold; (8) delayed use of deceleration devices; (9) landing with a tailwind; and (10) a wet or contaminated runway.

The precise manner in which these factors influence the landing distance depends on the airplane type, landing conditions, and, of course, the actions of the flight crew. Several

---

10 A review of landing accidents worldwide (1970–2004) for passenger and cargo aircraft with a takeoff mass of 5500 kg or greater indicated a landing overrun accident rate of about 0.5 per million landings [27].
11 The National Transportation Safety Board (NTSB), based in Washington DC, is an independent Federal agency charged by the US Congress with investigating civil aviation accidents in the United States (see Section 23.5) [28].

general conclusions are given in the Flight Safety Foundation's ALAR Briefing Note 8.3 on landing distances [29], including the following observations:

(1) A 5% increase in final approach speed increases the landing distance by 10%, when a normal flare and touchdown occur—that is, without the airplane "floating" to reduce speed. However, if the flare is extended and the touchdown delayed with a period of float to reduce, or bleed-off, the airspeed, then a 5% increase in final approach speed can increase the landing distance by 30%.
(2) Crossing the runway threshold at 100 ft (i.e., 50 ft higher than recommended) increases the landing distance by about 1000 ft (305 m), regardless of runway condition and airplane type.

11.8 Rejected Landing

11.8.1 Discontinued Approach and Go-Around

There are several reasons why a pilot might discontinue an approach—these include:

(1) when the approach is not stabilized (see Section 11.7.2);
(2) when complying with an air traffic control request (e.g., if there were another airplane or a vehicle on the runway);
(3) following a significant unexpected event (e.g., an airplane failure warning or a wind shear encounter); or
(4) when the landing cannot be made at an appropriate speed within the touchdown zone (which is defined as the first 3000 ft or the first third of the runway, whichever is shorter) [24].

Following the decision to discontinue the landing attempt, the flight crew will undertake a series of actions, which will include applying go-around thrust (see Section 8.5.5), raising the nose of the airplane to an appropriate climb attitude while maintaining an appropriate airspeed, raising the landing gear, and retracting the flaps in compliance with standard operating procedures. These operations on modern airplanes can also be performed by the autopilot following flight crew selection and initiation. This transition from an approach to a stabilized climb is known as a *go-around* (Figure 11.10). The term *missed approach* is typically used to describe a discontinued approach and initiation of a go-around following an instrument approach procedure. Specific requirements are given in the regulations regarding the performance of the airplane during this critical flight phase (see Section 11.8.2). Typically, the airplane will climb to an appropriate, safe altitude of at least 1500 ft AFE and then re-enter the traffic pattern for another landing attempt.

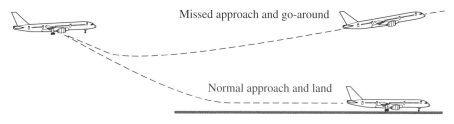

Missed approach and go-around

Normal approach and land

Figure 11.10 Missed approach and go-around.

11.8.2   Climb Requirements Following a Rejected Landing

A situation that requires careful consideration during flight planning is a discontinued approach with one engine inoperative. With the airplane configured for a one-engine-inoperative go-around, the airplane's climb performance must be such that it can safely clear all obstacles. This consideration can limit an airplane's landing weight—thus restricting payload—under certain conditions (e.g., when landing at high-elevation airports on hot days).

Two conditions are described in the regulations. First, it is required that the still air climb gradient, with one engine inoperative, exceeds a minimum specified value as stipulated in FAR/CS 25.121(d) [1, 2]. The airplane's configuration for the *approach-climb* case is landing gear retracted and flaps/slats set at the normal approach position. The required minimum climb gradient, with the critical engine inoperative and the remaining engines at the go-around thrust setting, is 2.1% for two engine airplanes, 2.4% for three engine airplanes, and 2.7% for four engine airplanes. Compliance has to be possible at an airspeed not greater than 1.4 V_{SR}. (V_{SR} is the reference stall speed—see Section 20.5.2.)

The second requirement applies to rejected landings with all engines operating (AEO). The AEO *landing-climb* requirement is described in FAR/CS 25.119 [1, 2]. In the landing configuration—that is, landing gear extended, flaps/slats set for landing—all airplane types (irrespective of the number of engines) must be capable of maintaining a climb gradient of 3.2% with the thrust that is available 8 seconds after moving the thrust levers to the go-around thrust setting.[12] This must be possible at an airspeed no greater than V_{REF}.

The climb gradients for these two cases are computed using such information as the drag polar and the available engine go-around thrust (climb performance is discussed later in Section 12.4). The maximum airplane weights at which the climb gradient requirements can be met are established and published in the AFM charts of *climb-limited landing weight*. The data are provided for a range of altitudes and temperatures, with appropriate corrections for the use of air conditioning packs and engine anti-ice (as both systems extract power from the engine, reducing its output and, hence, the airplane's climb performance).

References

1   FAA, *Airworthiness standards: Transport category airplanes*, Federal Aviation Regulation Part 25, Amdt. 25-143, Federal Aviation Administration, Washington, DC, June 24, 2016. Latest revision available from www.ecfr.gov/ under e-CFR (Electronic Code of Federal Regulations) Title 14.

2   EASA, *Certification specifications and acceptable means of compliance for large aeroplanes*, CS-25, Amdt. 18, European Aviation Safety Agency, Cologne, Germany, June 23, 2016. Latest revision available from www.easa.europa.eu/ under Certification Specification.

3   FAA, *General operating and flight rules*, Federal Aviation Regulation Part 91, Amdt. 91-336A, Federal Aviation Administration, Washington, DC, Mar. 4, 2015. Latest revision available from www.ecfr.gov/ under e-CFR (Electronic Code of Federal Regulations) Title 14.

4   FAA, *Operating requirements: Domestic, flag, and supplemental operations*, Federal Aviation Regulation Part 121, Amdt. 121-374, Federal Aviation Administration, Washington, DC,

---

12 FAR 33.73 [30] and EASA CS-E 745 [31] describe specific engine response criteria that must be demonstrated during engine certification testing (one such condition is that the engine must enable an increase from 15% of the rated takeoff thrust to 95% rated thrust in less than 5 seconds).

May 24, 2016. Latest revision available from www.ecfr.gov/ under e-CFR (Electronic Code of Federal Regulations) Title 14.

5 European Commission, *Commercial air transport operations (Part-CAT)*, Annex IV to Commission Regulation (EU) No. 965/2012, Brussels, Belgium, Oct. 5, 2012. Published in *Official Journal of the European Union*, Vol. L 296, Oct. 25, 2012, and reproduced by EASA.

6 FAA, "Flight test guide for certification of transport category airplanes," Advisory Circular 25-7C, Federal Aviation Administration, Washington, DC, Oct. 16, 2012. Available from www.faa.gov/.

7 Airbus, "Approach techniques: Aircraft energy management during approach," Flight operations briefing notes, FLT_OPS–APPR–SEQ03–REV02, Customer Services, Airbus S.A.S., Blagnac, France, Oct. 2005.

8 FSF, "FSF ALAR briefing note 8.2—The final approach speed," *Flight Safety Digest*, Flight Safety Foundation, Alexandria, VA, Vol. Aug.–Nov., pp. 163–166, 2000.

9 FSF Editorial Staff, "Stabilized approach and flare are keys to avoiding hard landings," *Flight Safety Digest*, Flight Safety Foundation, Alexandria, VA, pp. 1–25, Aug. 2004.

10 Garber, R.M. and van Kirk, L., "Conditional maintenance," *Aero*, The Boeing Company, Seattle, WA, Vol. 14, Apr. 2001.

11 Boeing, "Jet transport performance methods," D6-1420, The Boeing Company, Seattle, WA, May 1989.

12 Schmid, P., "Performance margins," *Boeing Performance and Flight Operations Engineering Conference*, Seattle, WA, Sept. 2007, Article 10.

13 Villaumé, F., "Brake-to-vacate system," *FAST*, Airbus S.A.S., Blagnac, France, Vol. 44, pp. 17–25, July 2009.

14 Boeing, "Landing," Training course notes, Flight Operations Engineering, The Boeing Company, Seattle, WA, 2009.

15 ESDU, "First approximation to landing field length for civil transport aeroplanes (50 ft, 15.24 m screen)," Data item 84040, IHS ESDU, 133 Houndsditch, London, UK, Dec. 1984.

16 Blake, W. and Performance Training Group, "Jet transport performance methods," D6-1420, Flight Operations Engineering, Boeing Commercial Airplanes, Seattle, WA, Mar. 2009.

17 Chiles, P., "Unveiling the matrix: A new tool for assessing and reporting runway condition," *AeroSafety World*, Flight Safety Foundation, Alexandria, VA, Vol. Nov., pp. 33–36, 2010.

18 FAA, "Mitigating the risks of a runway overrun upon landing," Advisory Circular 91-79A, Federal Aviation Administration, Washington DC, Sept. 17, 2014.

19 Lignee, R. and Kornstaedt, L., "Landing on contaminated runways," *Safety first*, Product Safety Department, Airbus S.A.S., Blagnac, France, Iss. 19, pp. 12–25, Jan. 2015.

20 FAA, "Instrument procedures handbook," FAA-H-8083-16, Flight Procedures Standards Branch, Federal Aviation Administration, Oklahoma City, OK, 2014.

21 FAA, "Instrument flying handbook," FAA-H-8083-15B, Airman Testing Standards Branch, Federal Aviation Administration, Oklahoma City, OK, 2012.

22 Airbus, "Approach techniques: Flying stabilized approaches," Flight operations briefing notes, FLT_OPS-APPR-SEQ 01-REV 02, Customer Services, Airbus S.A.S., Blagnac, France, Oct. 2006.

23 FSF, "FSF ALAR briefing note 6.1—Being prepared to go around," *Flight Safety Digest*, Flight Safety Foundation, Alexandria, VA, Vol. Aug.–Nov., pp. 117–120, 2000.

24 Coker, M., "Why and when to perform a go-around maneuver," *Aero*, The Boeing Company, Seattle, WA, Vol. Q 02, pp. 5–11, 2014.

25 ICAO, "Aerodromes: Vol. 1, aerodrome design and operations," Annex 14 to the Convention on International Civil Aviation, 5th ed., International Civil Aviation Organization, Montréal, Canada, July 2009.

26 Airbus, "Landing techniques: Crosswind landings," Flight operations briefing notes, FLT_OPS-LAND-SEQ 05-REV 03, Customer Services, Airbus S.A.S., Blagnac, France, Mar. 2008.

27 van Es, G.W.H., "Running out of runway: Analysis of 35 years of landing overrun accidents," NLR-TP-2005-498, National Aerospace Laboratory, Amsterdam, the Netherlands, Sept. 2005.

28 National Transportation Safety Board, "About the National Transportation Safety Board," Washington, DC, Retrieved Aug. 2015. Available from www.ntsb.gov/.

29 FSF, "FSF ALAR briefing note 8.3—Landing distances," *Flight Safety Digest*, Flight Safety Foundation, Alexandria, VA, Vol. Aug.–Nov., pp. 167–171, 2000.

30 FAA, *Airworthiness standards: Aircraft engines*, Federal Aviation Regulation Part 33, Amdt. 33-34, Federal Aviation Administration, Washington, DC, Nov. 4, 2014. Latest revision available from www.ecfr.gov/ under e-CFR (Electronic Code of Federal Regulations) Title 14.

31 EASA, *Certification specifications and acceptable means of compliance for engines*, CS-E, Amdt. 4, European Aviation Safety Agency, Cologne, Germany, Mar. 12, 2015. Latest revision available from www.easa.europa.eu/ under Certification Specification.

12

Mechanics of Level, Climbing, and Descending Flight

12.1 Introduction

In this chapter, the mechanics of *level, climbing*, and *descending* flight, outside of the influence of the ground, will be discussed. The airplane will be restricted to a *symmetric flight condition*— that is, one in which (1) the true airspeed vector lies in the vertical plane of symmetry of the airplane (this implies zero sideslip), and (2) the vertical plane of symmetry is perpendicular to a horizontal datum (this implies that the wings are level). Maneuvers involving the airplane rotating about the roll or yaw axis will be considered later in Chapter 15. General equations will be developed for accelerated flight, where the associated unbalanced forces will cause the airplane to increase or reduce speed. Simplified equations of motion will then be developed for the airplane in a *steady* (i.e., non-accelerating) condition. In all cases, the airplane will be treated as a rigid body, thus permitting classic techniques of the dynamics of bodies in motion to be applied.

The motion of the airplane will be described with respect to the surrounding air mass within which it moves. In other words, the air mass will be treated as the inertial frame of reference (see Section 2.4.3). Airplane motions with respect to the ground can be established by mathematical transformation taking into account the relative motion of the air with respect to the ground. The motion of the airplane's center of gravity (CG) at an instant in time—determined with respect to coordinates that move with the air mass—can be described by the velocity vector $\vec{V}$, the magnitude of which is V, the true airspeed (TAS). The angle measured in the vertical plane from a horizontal reference datum to $\vec{V}$ is assigned the Greek letter gamma (γ); this is known as the *still air climb angle.* When the air mass is stationary, γ is equal to the airplane's *flight path angle*—that is, the angle between a horizontal datum fixed with respect to the ground and the instantaneous flight path.

The approach adopted in this chapter has been to consider initially the forces acting on an airplane moving along a curvilinear flight path, at an instant in time, and to derive the basic equations of motion through the application of Newton's second law—this is presented in Section 12.2. The resulting equations are then used to describe an airplane in level flight (i.e., $\gamma = 0$) in Section 12.3, climbing flight (i.e., $\gamma > 0$) in Section 12.4, and descending flight (i.e., $\gamma < 0$) in Section 12.5. Still air conditions are initially assumed; the effects of wind and up- or downdrafts are discussed in Sections 12.4.13 to 12.4.15.

Mechanics of flight is a topic that has been previously addressed many times by different authors, and there is a wide range of publications dealing with the performance of fixed-wing aircraft—examples of textbooks on the subject are listed in Section 12.6 (Further Reading).

Performance of the Jet Transport Airplane: Analysis Methods, Flight Operations, and Regulations, First Edition.
Trevor M. Young.
© 2018 John Wiley & Sons Ltd. Published 2018 by John Wiley & Sons Ltd.

12.2 Basic Equations of Motion

12.2.1 Curvilinear Motion Applied to Airplane Flight

The schematic given in Figure 12.1 shows an airplane flying along a curved path of radius r in the vertical plane at constant speed. Still air conditions are assumed. The direction of the velocity vector $\vec{V}$ defines the instantaneous flight path. Gamma (γ) is shown as positive in this figure, indicating a climb.

Newton's second law of motion can be applied to a body that is moving along a plane curvilinear path (see Sections 2.4.1 and 2.4.4). Normal (n) and tangential (t) coordinates will be used to resolve the forces acting on the airplane. Taking the sum of the forces in the tangential direction (see Equations 2.30 and 2.33) yields the expression:

$$\sum F_t = ma_t = m\frac{\mathrm{d}V}{\mathrm{d}t} = \frac{W}{g}\frac{\mathrm{d}V}{\mathrm{d}t} \tag{12.1}$$

and taking the sum of the forces in the normal direction (see Equations 2.31 and 2.33) gives

$$\sum F_n = ma_n = mV\frac{\mathrm{d}\gamma}{\mathrm{d}t} = \frac{W}{g}V\frac{\mathrm{d}\gamma}{\mathrm{d}t} \tag{12.2}$$

where m is the instantaneous airplane mass (typical units: kg, slug);

a_t is the acceleration tangential to the curved flight path (typical units: m/s², ft/s²);

V is the true airspeed, measured in the direction of the flight path (typical units: m/s, ft/s);

t is time (typical unit: s);

W is the instantaneous airplane weight (typical units: N, lb);

g is the acceleration due to gravity (typical units: m/s², ft/s²);

a_n is the acceleration normal to the curved flight path, i.e., centripetal acceleration (typical units: m/s², ft/s²); and

γ is the still air climb angle, defined as positive upwards (typical unit: rad).

It is important to note that the summation of forces must be consistent with the corresponding sense of the acceleration terms. The centripetal acceleration is defined as positive towards the center of the circle. An airplane flying as indicated in Figure 12.1 will have an increasing flight path angle with time, and hence $\mathrm{d}\gamma/\mathrm{d}t$ will be positive. The centripetal acceleration can

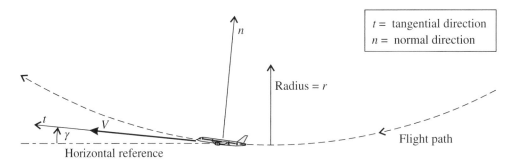

Figure 12.1 Curvilinear motion with the airplane's flight path in the vertical plane (in still air).

be expressed in terms of the true airspeed (V) and the radius (r). Using the notation defined in Figure 12.1, Equation 2.29 can be written as

$$V = r\frac{d\gamma}{dt} \tag{12.3}$$

Thus $a_n = V\frac{d\gamma}{dt} = r\left(\frac{d\gamma}{dt}\right)^2 = \frac{V^2}{r}$ (12.4)

where r is the radius of the circle (typical units: m, ft).

12.2.2   General Equations for Climb, Descent, and Level Flight

The discussion of plane curvilinear motion is now elaborated taking into account the forces that act on the airplane, which are shown in Figure 12.2. As noted earlier, the airplane will be treated as a rigid body, not rotating about the roll or yaw axis (i.e., a symmetrical flight condition). Several terms will now be defined before the basic equations of motion are derived.

A longitudinal datum line, fixed with respect to the airframe and passing through the CG of the airplane parallel with the cabin floor, is used to define the body axes (see Section 3.3.4). The body axes coordinate system is fixed with regard to the airplane (i.e., the axes *move* with the airplane). The origin of the axes is at the airplane's reference CG position; the X axis is defined forward along the datum line; the Y axis is positive along the starboard wing, and the Z axis is positive downwards.

The angle between the X axis and the flight path is defined as the *body angle of attack* (or reference angle of attack) of the airplane—herein, it is denoted by the Greek letter alpha[1] (α). The angle of attack (AOA) is defined positive as shown in Figure 12.2. Although angle of attack

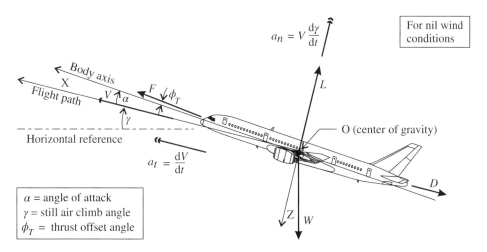

Figure 12.2 Definition of angles, forces, and velocities for symmetrical flight (in still air).

---

1 The use of the symbol α for *angle of attack* is commonplace. For an airplane, the angle can be measured with respect to the body longitudinal axis (i.e., body datum)—as defined above—or with respect to the wing chord (i.e., wing datum). The subscripts "b" and "w" are customarily used to make the distinction. Herein, for the sake of brevity, the subscript "b" has been omitted.

is measured—and used to provide stall margin indication on airspeed indicators and to trigger stall warnings (e.g., stick shakers)—dedicated AOA indicators are not widely used on jet transport airplanes. What the flight crew see is the pitch angle—or pitch *attitude*, as it is often called—which is prominently displayed on the attitude indicator of the Primary Flight Display (a Primary Flight Display is illustrated in Figure 5.2a). The pitch angle (θ) is the angle between the horizontal reference (datum) and the X axis (see also Appendix C.2). The pitch angle, angle of attack, and flight path angle are linked, as follows:

$$\theta = \alpha + \gamma \tag{12.5}$$

As per convention, the lift (L) and the drag (D) are defined normal and parallel to the flight path. The lift vector is the net force normal to the flight path, resulting from the aerodynamic pressure acting on the complete airplane. A change in the angle of attack will change the lift and the drag (see Sections 7.2 and 7.3).

In Figure 12.2, the letter F represents the combined thrust of the engines. The direction of the thrust vector is typically offset by a small amount from the body X axis; the included angle is herein given the symbol ϕ_T and is arbitrarily defined as positive in the "nose-up" orientation.

Equation 12.1 is now expanded by considering the forces acting parallel to the flight path:

$$\sum F_t = F\cos(\alpha + \phi_T) - D - W\sin\gamma = \frac{W}{g}\frac{dV}{dt} \tag{12.6}$$

Similarly, Equation 12.2 is expanded by considering the forces normal to the flight path:

$$\sum F_n = F\sin(\alpha + \phi_T) + L - W\cos\gamma = \frac{W}{g}V\frac{d\gamma}{dt} \tag{12.7}$$

Two simplifications applicable to these equations are now introduced.

(1) Centripetal Acceleration Simplification

In this chapter, maneuvering flight resulting in high pitch rates is not considered; instead, the discussion is limited to typical *climb* or *descent* maneuvers associated with routine operations. Under these flight conditions, the airplane's flight path will not be exactly straight; it will have a slight curve. However, the curvature is very small and the rate of change of climb angle with respect to time (i.e., $d\gamma/dt$) is therefore a very small quantity. The centripetal acceleration (i.e., normal acceleration) is thus negligible (see Equation 12.2) and to a very good approximation can be assumed to be zero.

(2) Thrust Simplification

For most problems involving level, climbing, or descending flight, the angle between the thrust line and the flight path is relatively small (flight at high angles of attack is not considered in this chapter). By ignoring this angular discrepancy, an apparent error in the magnitude of the thrust is introduced, which is a function of the angle of attack. This error has a magnitude of $F\{1 - \cos(\alpha + \phi_T)\}$. Now there exists an elegant way for taking into account this angular discrepancy in thrust for the purpose of airplane performance analysis. This involves *defining* the net installed thrust (F_N) parallel to the flight path, which means that the thrust discrepancy is taken into account in the thrust/drag accounting of the airplane (see Section 7.5.4). In other words, the associated drag polar, which is used for such performance analyses, is defined in a way that takes into account the thrust "re-alignment." Note that installed thrust cannot be measured directly during flight testing—it is derived from models of the drag established by flight testing under steady, level flight conditions and correlated against theoretical engine performance predictions.

Implementing these simplifications enables Equations 12.6 and 12.7 to be written as follows:

$$F_N - D - W \sin \gamma = \frac{W}{g} \frac{dV}{dt} \tag{12.8}$$

$$L - W \cos \gamma = 0 \tag{12.9}$$

Equations 12.8 and 12.9 will now be applied to level flight (Section 12.3), climbing flight (Section 12.4), and descending flight (Section 12.5).

12.3   Performance in Level Flight

12.3.1   Level (Constant Height) Accelerated Flight

The discussion presented in this section applies to an airplane that is flying with the wings level at constant geopotential height. The forces acting on the airplane can be simplified as shown in Figure 12.3. As $\gamma = 0$, Equations 12.8 and 12.9 can be written as follows:

$$F_N - D = \frac{W}{g} \frac{dV}{dt} \tag{12.10}$$

$$L = W \tag{12.11}$$

Figure 12.3 Forces acting on the airplane in level flight.

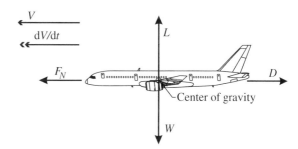

12.3.2   Level (Constant Height) Unaccelerated Flight

For long periods during cruise, airplanes operate at constant true airspeed and at constant or nearly constant geopotential height. In this flight condition, the installed net thrust (F_N) is thus equal in magnitude to the airplane's drag (D), that is,

$$F_N = D \tag{12.12}$$

This is the *thrust required* to maintain level (i.e., constant height) unaccelerated flight. The drag in Equation 12.12 can be expanded in terms of the airplane's dynamic pressure (q), wing reference area (S), and drag coefficient (C_D) using Equation 7.5. The dynamic pressure, as noted in Section 6.5.1, can be expressed in several equivalent ways. For high-speed flight, it is typical that Mach number (M) is used as the speed measure. Using Equations 7.5 and 6.12, the thrust required to maintain this steady-state flight condition can be written as

$$F_N = qSC_D = \frac{1}{2}\rho_0 a_0^2 \delta M^2 S C_D \tag{12.13}$$

where q is the dynamic pressure (typical units: N/m^2, lb/ft^2);

S is the wing reference area (typical units: m^2, ft^2);

ρ_0 is the air density at the ISA sea-level datum (typical units: kg/m^3, slug/ft^3);

δ is the relative air pressure (dimensionless); and

a_0 is the speed of sound in air at the ISA sea-level datum (typical units: m/s, ft/s).

It is frequently the case, when analyzing the cruise performance of an airplane, that reference is made to engine data where the engine thrust is expressed in corrected form (i.e., as F_N/δ)—for example, on tables of corrected fuel flow versus corrected thrust. For such studies, it is convenient to re-arrange Equation 12.13 as follows:

$$\frac{F_N}{\delta} = \frac{1}{2}\rho_0 a_0^2 M^2 S C_D \tag{12.14}$$

The drag coefficient can be expressed as a function of the lift coefficient for a given Mach number (see Section 7.4). If the lift equals the airplane's weight (as indicated by Equation 12.11), the load factor (n_z) is unity and Equation 7.4 can be used to express the lift coefficient (C_L) in a convenient form:

$$C_L = \frac{L}{qS} = \frac{2W}{\rho_0 a_0^2 \delta M^2 S} \tag{12.15}$$

Once the lift coefficient is known, the drag coefficient can be established from an appropriate drag polar. (Drag polars can be recorded in tabular or graphical form or expressed as an empirical model, as described in Section 7.4.)

In this hypothetical cruise analysis, it is evident that the determination of the lift coefficient, as described by Equation 12.15, requires knowledge of the airplane's weight, the flight Mach number, and the relative pressure of the air (which depends on the cruise altitude and the ambient air temperature). The lift coefficient, together with an appropriate drag polar, can then be used to establish the required thrust, and, if appropriate engine data were available, the cruise fuel flow could then be determined, for example.

12.3.3 Graphical Representation of Steady-State Flight Performance

A useful technique for visualizing an airplane's performance potential involves superimposing the available thrust (corresponding to a particular engine rating) on a graph of drag versus speed for a given set of flight conditions. These conditions include airplane weight, altitude, and temperature (usually abbreviated as WAT). For the sake of illustration, the drag (D) at sea level is shown in Figure 12.4 as a function of equivalent airspeed (EAS), rather than true airspeed (TAS). The advantage of working with EAS is that the relative air density (σ) is eliminated from idealized models of the drag polar (as illustrated by Equation 7.48). The drag can thus be represented by two variables: W and V_e. As the rate of change of weight of an airliner with time is small in normal flight operations (due solely to the burning of fuel) point performance calculations can be carried out that ignore changes in aircraft weight. The drag acting on an airplane at a selected weight can thus be modeled as a single function of V_e, which will be valid for a range of altitudes.

In general, the net installed thrust of a turbofan engine tends to reduce with increasing flight speed (see Section 8.6). This thrust lapse with speed is more apparent at sea level than at cruise altitudes. Engine thrust also reduces with increasing altitude and may furthermore be limited, or governed, at high ambient air temperatures. The precise manner in which the thrust varies with speed and height depends on the powerplant type and the approach adopted by the engine manufacturer in defining the engine ratings (see Section 8.5.6). The thrust (F_N) must be

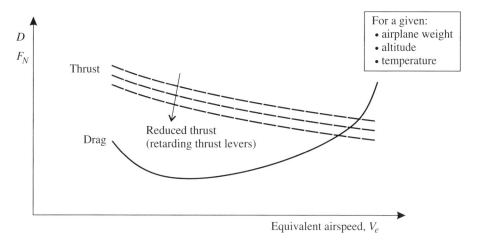

Figure 12.4 Superposition of thrust and drag relationships expressed as a function of EAS.

expressed in the same way as the drag—that is, as a function of EAS. In the illustration presented in Figure 12.4, a family of F_N curves have been constructed for the engine operating at various thrust lever positions and superimposed on the drag graph. The intersection points of the F_N and D curves represent a series of steady-state, level flight conditions. This technique can thus be used to graphically determine an airplane's steady-state speed for actual or modeled drag and thrust data.

12.3.4 Speed Stability

The concept of speed stability can be assessed by considering an airplane's thrust and drag relationships—both expressed as functions of equivalent airspeed (EAS), that is,

$$D = f(V_e) \quad \text{and} \quad F_N = f(V_e) \tag{12.16}$$

Figure 12.5 shows the thrust, for a selected thrust setting, superimposed on the drag relationship (for a particular flight condition). If the airplane is flying straight and level at a speed V_{e_1},

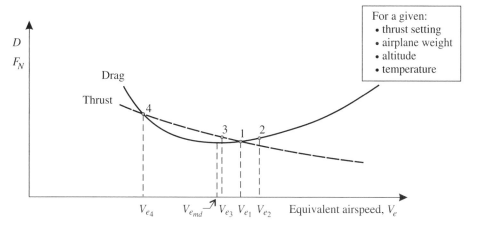

Figure 12.5 Illustration of speed stability concept.

it will be operating at a condition of *equilibrium* (point 1) as $F_N(V_{e_1}) = D(V_{e_1})$. By considering off-equilibrium conditions to the left or right of point 1, the concept of *speed stability* can be illustrated.

Consider what happens when the airplane speeds up a little—say, due to atmospheric turbulence—without an input from the flight crew. If the speed increases from V_{e_1} to a speed V_{e_2}, then $F_N(V_{e_2})$ will be less than $D(V_{e_2})$ and the airplane will tend to slow down to the original speed, that is, V_{e_1}. By a similar argument, if the speed were to decrease to V_{e_3}, this would cause the airplane to accelerate, as the drag would decrease and $F_N(V_{e_3})$ would be greater than $D(V_{e_3})$.

Consider now the slow-speed equilibrium point V_{e_4}. In this situation, if the airplane were to slow down, the drag would *increase*, causing the airplane to slow down further. Without input from the flight crew (or autothrottle/autothrust system), the airplane's speed would move away from its equilibrium point. Clearly, this is not a stable condition. It can be concluded that an airplane is inherently statically unstable with regard to speed changes for a range of slow flight speeds. Note that the point at which an airplane with the thrust–speed relationship illustrated in Figure 12.5 becomes speed unstable is not the minimum drag speed ($V_{e_{md}}$), but a point slower than the minimum drag speed. This region is often called the "back end" of the drag curve.

It is interesting to pursue this discussion a little further. If the airplane slows down from the minimum drag speed, it will slowly move up the back end of the drag curve. Associated with this decrease in speed will be an increase in drag, which will now be greater than the thrust. The immediate consequence of this *thrust deficit* will be that the airplane cannot maintain level flight, and a corrective action will be needed by the flight crew (or autopilot). In this high drag situation, with the airplane sinking, a considerable increase in thrust is needed to maintain constant height and accelerate the airplane.

The incorporation of autothrottle/autothrust systems—described in Section 8.5.2—in modern jet airplanes means that this speed-unstable feature is not apparent to flight crews in the same way as it would have been in earlier generations of jet airplanes. Furthermore, the thrust–speed relationship of turbojet or low bypass ratio turbofan engines displays little change in thrust with changes in speed, irrespective of altitude. This meant that the speed unstable region of old jet airliners was comparatively larger than that of modern airplanes and did, in fact, extend up to or close to $V_{e_{md}}$.

In the landing configuration—with landing gear extended and flaps and slats deployed—the minimum drag speed is reduced (this is evident from Equation 7.46). The impact of these configuration changes is beneficial in this situation as it lowers the speed at which speed instability first occurs when slowing an airplane down for landing.

The situation in cruise is different as thrust is significantly affected by altitude changes. The maximum thrust available reduces with increasing altitude (see Section 8.6) and the thrust–speed relationship, in many cases, will be flatter than that illustrated in Figure 12.5. As an airplane approaches its ceiling,[2] the slow-speed equilibrium point V_{e_4} associated with the maximum thrust available can get relatively close to $V_{e_{md}}$. Operation at speeds below the minimum drag speed in cruise is generally not recommended. Airbus note that if flight slightly below the minimum drag speed is required for some reason during cruise, then "vigilant monitoring is necessary to ensure that further uncommanded speed reductions are immediately checked and recovered from" [1].

---

2 The maximum altitude that an airplane can reach or maintain under defined conditions is known as a *ceiling*—this is discussed later in Sections 17.3.1 and 20.4.1.

12.4 Performance in Climbing Flight

12.4.1 Climb–Speed Schedules

During any climb, potential energy increases; the kinetic energy, however, may reduce, remain the same, or even increase depending on the climb–speed schedule—as discussed below. By considering how the airplane's potential and kinetic energies change, it is possible to describe any climb as one of three types:

(1) The first type of climb involves a reduction in TAS; kinetic energy is thus traded for potential energy. The technique of building up speed and then using the kinetic energy to gain height is often called a *zoom maneuver*. Zoom climbs are not a feature of a typical commercial airplane flight operation, but are often used by aerobatic or fighter pilots. Climbing at constant Mach number in the upper troposphere, however, is routinely done and this implies a reducing TAS, which is due to the fact that the speed of sound reduces with increasing altitude (below the tropopause).

(2) The second type of climb is a *steady-speed* climb. The airplane climbs at a constant TAS and is in a state of equilibrium at all points along the flight path. Potential energy increases, but kinetic energy is constant. The sum of the forces acting on the airplane in the direction of the flight path is zero, and, from an analytical point of view, the climb is relatively easy to analyze. Climbing at constant Mach number in the stratosphere is an example of a steady-speed climb as the speed of sound is constant. If the rate of change of speed with respect to time is small—and in many climb situations this is indeed the case—then it can be convenient to *assume* that the airplane is in a steady flight condition. The assumption of quasi-steady state is useful as it greatly simplifies the climb analysis.

(3) The third alternative involves an increase in both potential and kinetic energies. The TAS increases as the airplane accelerates along the flight path. This is typical for airliners that follow climb–speed schedules that have constant calibrated airspeed (CAS) climb segments. In this case, the TAS will increase as the air density drops.

Climb–speed schedules used in routine commercial flight operations typically comprise constant calibrated airspeed and constant Mach number segments (see Section 17.2).

12.4.2 Angle of Climb and Climb Gradient (Still Air Conditions)

The forces acting on an airplane climbing at a true airspeed V in still air are shown in Figure 12.6. The net thrust (F_N) is defined in the direction of the flight path (as described in

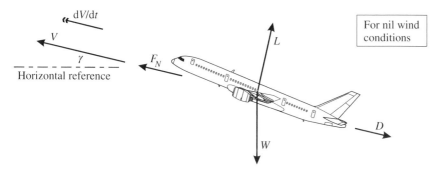

Figure 12.6 Climbing flight (in still air).

Section 12.2.2), and the airplane is shown accelerating along the flight path. The still air climb angle (γ) is measured from a horizontal reference to the instantaneous flight path; the still air climb gradient is $\tan \gamma$. In the absence of wind or up- or downdrafts, the climb gradient represents the ratio of the gain in height to the horizontal distance flown. This ratio is frequently expressed as a percentage—for example, a climb gradient of 8.7% represents a climb angle of 5° (i.e., $\tan 5° = 0.087$).

Equations 12.8 and 12.9, derived in Section 12.2, are used as the starting point for this analysis. Equation 12.8 is re-arranged to give an expression for the angle of climb in which the derivative dV/dt is expanded to introduce the time rate of change of geopotential height (dH/dt):

$$\sin \gamma = \frac{F_N - D}{W} - \frac{1}{g}\frac{dV}{dt} = \frac{F_N - D}{W} - \frac{1}{g}\frac{dV}{dH}\frac{dH}{dt} \tag{12.17}$$

In the absence of up- or downdrafts (gusts), the vertical component of the true airspeed vector is equal to the change in height with respect to time (vertical speed is discussed later in Section 12.4.4), that is,

$$V \sin \gamma = \frac{dH}{dt} \tag{12.18}$$

The forces and accelerations that apply to the airplane in flight and the resulting motions are defined relative to a fixed inertial datum. To be absolutely correct, the height measure should thus be geometric height. However, the difference between geometric height (h) and geopotential height (H) in this situation is very small (as shown in Section 5.2.2). It is thus acceptable to equate the airplane's true vertical speed to the time rate of change of geopotential height.

Substituting Equation 12.18 into 12.17 yields the following expression:

$$\sin \gamma = \frac{F_N - D}{W} - \frac{1}{g}\frac{dV}{dH}V \sin \gamma$$

which can be re-arranged to provide the required equation for the climb angle:

$$\sin \gamma = \frac{\dfrac{F_N}{W} - \dfrac{D}{W}}{1 + \dfrac{V}{g}\dfrac{dV}{dH}} \tag{12.19}$$

The usual way of evaluating Equation 12.19 is to introduce an *acceleration factor*,[3] which is herein defined as

$$f_{acc} = \frac{V}{g}\frac{dV}{dH} \tag{12.20}$$

$$\text{thus} \quad \sin \gamma = \frac{\dfrac{F_N}{W} - \dfrac{D}{W}}{1 + f_{acc}} \tag{12.21}$$

Climb–speed schedules for airline operations (see Section 17.2.1) would typically involve several segments—for example: a lower segment conducted at a CAS not exceeding 250 kt below 10 000 feet, a middle section at an increased CAS that is appropriate to the airplane

---

3 Caution: There are two alternative conventions for the *acceleration factor* described herein, which can also be used to simplify climb/descent equations, such as Equations 12.19 and 12.26. ESDU [2] defines the climb/descent acceleration factor as $1 + (V/g)(dV/dH)$. Blake, *et al.* [3] also describe this alternative convention; an earlier Boeing publication [4], however, utilizes the definition given by Equation 12.20. Another, less widely used convention, which is described by Shevell [5], defines a "kinetic energy correction factor" as $[1 + (V/g)(dV/dH)]^{-1}$.

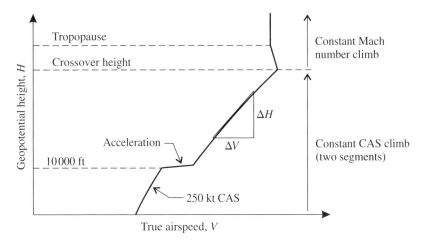

Figure 12.7 Typical climb–speed schedule: two constant CAS climb segments, followed by a constant Mach number climb to the initial cruise altitude (assumed to be in the stratosphere).

type (typically about 300 kt), and an upper segment conducted at constant Mach number [6]. This is illustrated in Figure 12.7. The significance of the acceleration factor is highlighted for a constant CAS climb segment, where the slope of the curve is shown as dH/dV in Figure 12.7. This derivative is seen to reduce with height up to the *crossover* height (which is defined in Section 17.2.2). The acceleration factor *increases* as the altitude increases for a constant CAS climb—this is evident from the fact that the true airspeed (V) increases (as mentioned earlier in Section 12.4.1) and dV/dH, which is the reciprocal of the slope of the function indicated in Figure 12.7, also increases during the climb.

To determine the climb angle using Equation 12.21, the acceleration factor (f_{acc}) needs to be evaluated for the appropriate climb–speed schedule and atmospheric conditions; this can be accomplished for ISA conditions using the equations presented in Table 12.1 (the derivations of which are given in Appendix D.3). Alternatively, the acceleration factor can be estimated from

Table 12.1 Climb/descent acceleration factors

The acceleration factor for ISA conditions is given by $f_{acc} = 0.7M^2\psi$

where the value of ψ depends on the climb/descent–speed schedule, as follows:

For constant Mach no.:	*For constant EAS:*	*For constant CAS:*
$\psi = -\zeta$	$\psi = 1 - \zeta$	$\psi = \dfrac{[1 + 0.2M^2]^{3.5} - 1}{0.7M^2[1 + 0.2M^2]^{2.5}} - \zeta$

And, where the value of ζ depends on height, as follows:

In the troposphere:	*In the stratosphere:*
$\zeta = 0.190263$	$\zeta = 0$

Notes:

(a) *Sources:* References [2] and [6].

(b) See Appendix D.3 for derivations.

(c) A temperature correction is needed for off-standard atmospheres. For the troposphere, the following correction applies: $\zeta = 0.190263(T_{std}/T)$, where T_{std} is the temperature in the ISA and T is the ambient (i.e., actual) temperature. No correction is needed for the stratosphere.

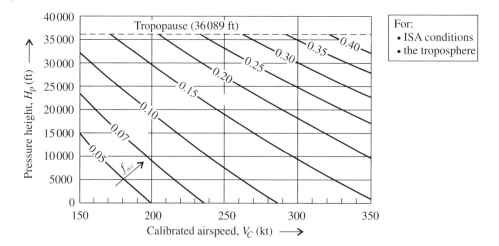

Figure 12.8 Acceleration factor (f_{acc}) for constant CAS climb/descent in the troposphere for ISA conditions.

charts such as the one shown Figure 12.8, which is valid for a constant CAS climb (or descent) below the tropopause. Note that the pressure height scale (H_p) is identical to geopotential height (H) in the ISA, but this is not the case in off-standard atmospheres (see Section 4.3). A temperature correction is thus required for off-standard atmospheres.

Frequently, it is necessary to determine the *climb gradient* rather than the climb angle (γ). The climb gradient is the change in height divided by the horizontal distance covered, which is, of course, equal to tan γ. The climb gradient for small angles (i.e., less than ~15°), can be calculated from the following equation, which follows directly from Equation 12.21:

$$\tan \gamma \cong \sin \gamma = \frac{\dfrac{F_N}{W} - \dfrac{D}{W}}{1 + f_{acc}} \tag{12.22}$$

12.4.3 Climb Angle and Climb Gradient for Constant-Speed Climb

The equations derived in Section 12.4.2 to describe the angle of climb and climb gradient can be simplified by considering a steady-speed (or steady-state) climb. For flight at constant TAS, the rate of change of speed with respect to height (i.e., dV/dH) is zero. An expression for the angle of climb for constant TAS flight can be deduced from Equation 12.21, that is,

$$\sin \gamma = \frac{F_N - D}{W} \tag{12.23}$$

Equation 12.23 illustrates an important result: for a particular airplane weight, the climb angle depends on the excess of thrust to drag. An alternative way of writing Equation 12.23—which is convenient for certain applications—is to introduce the lift-to-drag ratio (E) and to use the small angle approximations sin $\gamma \cong \gamma$ and cos $\gamma \cong 1$:

$$\gamma \cong \frac{F_N}{W} - \frac{D}{L} \cos \gamma \cong \frac{F_N}{W} - \frac{1}{E} \tag{12.24}$$

where γ is the still air climb angle measured in radians.

Equation 12.23 provides the basis for an intuitive graphical method that can be used to estimate an airplane's climb angle at different flight conditions. Plots of F_N and D, for a selected airplane

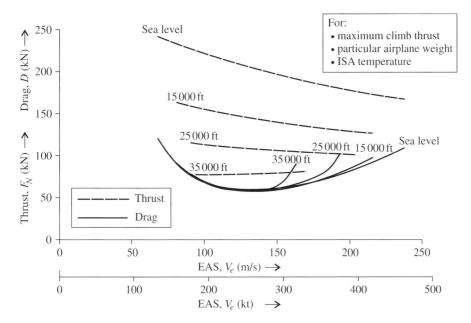

Figure 12.9 Maximum climb thrust and airplane drag plotted as functions of EAS for selected altitudes (example for a mid-size, twin-engine airliner).

weight (W), are superimposed on the same graph as a function of speed. The angle of climb (γ) at a particular speed is proportional to the difference between the two curves (i.e., the excess of thrust to drag). This is illustrated in Figure 12.9, where climb thrust and drag are illustrated as functions of EAS for several selected altitudes. Note that the drag curves for various altitudes, when constructed as a function of EAS, collapse—approximately—onto a single curve, with the notable exception that the EAS at which the drag increases due to compressibility is significantly lower at cruise altitudes than at altitudes below *ca.* 15 000 ft and that the drag rise is more rapid (the drag divergence Mach number occurs at a much slower EAS at cruise than at sea level).

The airspeed that will yield the greatest angle of climb is of interest for flight conditions where adequate clearance from terrain or other obstacles must be maintained. This is of particular importance when considering an engine failure during takeoff (see Section 10.7) or *en route* in mountainous terrain (see Section 17.4). The maximum angle-of-climb speed, at a particular altitude, can be identified by translating downwards the F_N curve until it intersects the associated D curve at only one point. Note that this point corresponds to a speed close to the minimum drag speed ($V_{e_{md}}$), but that its exact point on the graph depends on the shape of the associated thrust curve. An idealization that is often used assumes that engine thrust is invariant with forward speed. With this assumption, the thrust function at any given altitude would be represented by a horizontal line on the chart, and, consequently, the best angle-of-climb speed at all altitudes would be the minimum drag speed.

12.4.4 Rate of Climb

The rate of climb, abbreviated as ROC (or alternatively as RoC or R/C), is the vertical speed of the airplane; herein it is given the symbol V_v. By convention, the rate of climb is measured with respect to a ground datum and not the surrounding air mass, which may be rising or falling. This convention has its origin in the operation of a standard flight instrument: the vertical speed indicator (VSI), which provides an indication of the rate of change of height based on static air pressure measurements.

If the air mass is rising or falling—due to an up- or downdraft, for example—then the rate of climb would be equal to the sum of the vertical component of the true airspeed vector and the vertical speed of the air mass, that is,

$$V_v = \frac{dH}{dt} = V \sin \gamma + U_w \tag{12.25}$$

where V_v is the rate of climb (typical units: m/s, ft/s);
V is the true airspeed (typical units: m/s, ft/s); and
U_w is the vertical speed of the air mass, defined as positive upwards (typical units: m/s, ft/s).

Note: V must be converted into appropriate units for use in Equation 12.25 (e.g., from kt to m/s or ft/s).

In the absence of up- or downdrafts (i.e., $U_w = 0$), the rate of climb can be determined by substituting Equation 12.19 into Equation 12.25:

$$V_v = V \sin \gamma = \frac{\left(\dfrac{F_N}{W} - \dfrac{D}{W}\right) V}{1 + \dfrac{V}{g} \dfrac{dV}{dH}} \tag{12.26}$$

As before, it is possible to introduce the acceleration factor (defined by Equation 12.20):

$$V_v = \frac{\left(\dfrac{F_N}{W} - \dfrac{D}{W}\right) V}{1 + f_{acc}} \tag{12.27}$$

The acceleration factor, f_{acc}, depends on the climb–speed schedule, altitude, and temperature (see Table 12.1). For a typical airliner climb–speed schedule, which is conducted at constant CAS segments up to the crossover height, the acceleration factor will increase as the altitude increases (see Figure 12.8).

For the upper part of the climb—that is, above the crossover height—it is preferable to express the rate of climb as a function of flight Mach number rather than TAS. This is accomplished using Equation 6.8:

$$V_v = \frac{\left(\dfrac{F_N}{\delta} - \dfrac{D}{\delta}\right) M a_0 \sqrt{\theta}}{(1 + f_{acc})\left(\dfrac{W}{\delta}\right)} \tag{12.28}$$

The reason for dividing both numerator and denominator by the pressure ratio (δ) in Equation 12.28 is to write the thrust in corrected form (to be consistent with the commonly used presentation of thrust data for turbofan engines).

12.4.5 Rate of Climb for Constant True Airspeed Climb

For a climb at constant TAS (i.e., a steady, unaccelerated climb), the acceleration factor is zero, and Equation 12.27 can be simplified as follows:

$$V_v = \left(\frac{F_N}{W} - \frac{D}{W}\right) V = \frac{(F_N V - DV)}{W} \tag{12.29}$$

The product of thrust and true airspeed, by definition, is equal to *thrust power* (P_T) and the product of drag and true airspeed is *drag power* (P_D). Thrust power can be considered as the *power available* and drag power as the *power required* (to sustain level, unaccelerated flight)—see Section 7.8.1. With these substitutions, Equation 12.29 can be rewritten as

$$V_v = \frac{P_T - P_D}{W} \tag{12.30}$$

It is evident that the steady rate of climb is proportional to the excess of thrust power to drag power. The term on the right-hand side of Equation 12.30—that is, $(P_T - P_D)/W$—is known as *specific excess power* (SEP). Specific excess power is an instantaneous property of an airplane in flight—it is a measure of its ability to climb and/or accelerate (this is discussed later in Section 15.8.4).

12.4.6   Best Rate-of-Climb Speed

The speed that will give the best angle of climb does not give the best rate of climb; instead, it is necessary to fly at a faster speed to achieve the greatest possible rate of climb. The climb speed used by airlines is generally one that has been calculated to reduce the total trip cost (this subject is discussed in Chapter 18). Now, as an airliner is very efficient in cruise, it is usually advantageous to climb at a high rate in order to maximize the time spent cruising. The optimum climb speed that will reduce the trip cost is typically close to the maximum *rate-of-climb speed*. The speed to achieve the maximum rate of climb will be investigated by assuming a steady-state condition (i.e., the acceleration factor is zero).

To illustrate an airplane's rate of climb graphically, in a similar manner to that done for angle of climb, it is necessary to rewrite Equation 12.29 using EAS rather than TAS. This is done by introducing the square root of the relative air density (σ):

$$V_v = (F_N V_e - D V_e)\frac{1}{W\sqrt{\sigma}} \tag{12.31}$$

Thus, in order to assess the rate of climb (at a particular altitude and air temperature), the functions $F_N V_e$ and $D V_e$ are plotted against V_e, as illustrated in Figure 12.10. It is evident from Equation 12.31 that the rate of climb is proportional to the difference between the two curves and inversely proportional to the airplane weight. The maximum rate of climb—identified by the point on the graph where the difference is greatest—is achieved at a speed that is greater than the minimum drag speed.

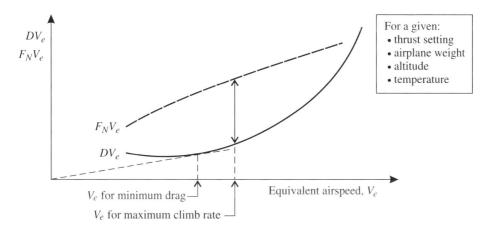

Figure 12.10 Best rate-of-climb speed (EAS).

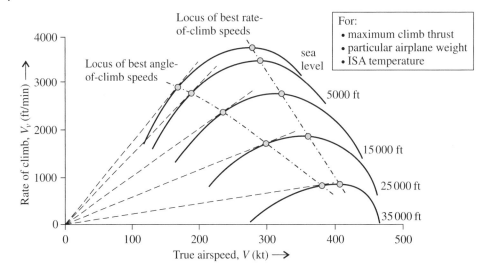

Figure 12.11 Best rate-of-climb and best angle-of-climb speeds for selected altitudes (example for a mid-size, twin-engine airliner).

A plot of climb rate (V_v) versus true airspeed (V) provides a useful mechanism to establish the speeds that will give the greatest *rate of climb* and also the greatest *angle of climb*. This is illustrated in Figure 12.11, where the rate of climb is shown for five selected altitudes for a set of operating conditions (e.g., for a given airplane weight at ISA conditions with maximum climb thrust selected). Note that in this figure the velocity is given as TAS and not EAS, as done earlier. The flight speed that will give the best rate of climb at a particular altitude corresponds to the curve maxima. It is evident from Equation 12.18 that the angle of climb, measured in radians, for a quasi-steady-state condition can be approximated by $\gamma \cong V_v/V$. The flight speed for the best angle of climb at a particular altitude can thus be determined by drawing a line through the origin tangent to the curve. The locus of best rate-of-climb and best angle-of-climb speeds has been constructed through the identified points for each altitude.

An extrapolation of these results leads to an interesting observation: at the maximum altitude at which steady level flight can theoretically be maintained—that is, a flight condition where the rate of climb is reduced to zero and thrust equals drag—the best rate-of-climb and best angle-of-climb speeds coincide. (Service ceilings based on rate of climb are discussed later in Sections 17.3.1 and 20.4.1.)

12.4.7 Summary of Climb Speeds

Figure 12.12 illustrates the climb speeds discussed in this section for a given airplane weight, altitude, and temperature (WAT). It is evident that climbing flight is only possible at speeds where the thrust is greater than the drag. This sets a thrust-limited maximum speed for sustained level flight. Note that this is not the only constraint on maximum level speed—structural and aerodynamic limits associated with high-speed flight are discussed later in Chapter 20.

The maximum angle of climb is achieved at the speed that will maximize the function $F_N - D$ (i.e., excess thrust to drag). This speed can be less than or greater than the minimum drag speed (V_{md}), depending on the shape of the thrust–speed relationship. For certain engine types, the thrust (F_N) is approximately invariant with changes in speed, particularly at cruise altitudes (as illustrated in Figure 12.9). In such cases, the best angle-of-angle speed will be at or very close to the minimum drag speed V_{md}. However, when the thrust lapse with forward speed

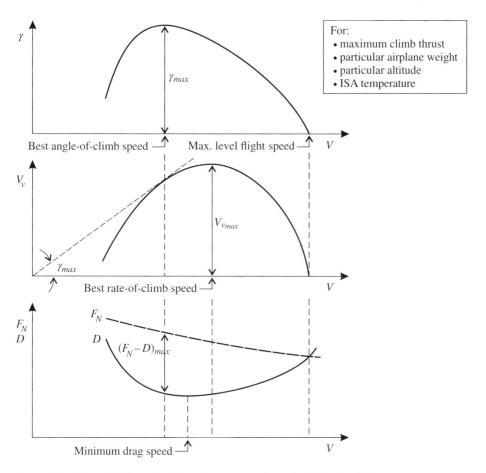

Figure 12.12 Schematic illustrating key reference speeds applicable to the climb.

is significant (as illustrated here in Figure 12.12), the theoretical best angle-of-climb speed will be slower than V_{md}.

The maximum rate of climb is achieved at the speed that will maximize the function $P_T - P_D$ (i.e., excess thrust power to drag power). This speed is always faster than the best angle-of-climb speed.

12.4.8 Time to Climb

The rate of climb (V_v) can be satisfactorily written as the change in geopotential height (H) with respect to time (t)—that is,

$$V_v = \frac{\mathrm{d}H}{\mathrm{d}t} \qquad \text{(from Equation 12.25)}$$

The *time to climb* (t) from height H_1 to height H_2 is thus given by the following integral expression:

$$t = \int_{H_1}^{H_2} \frac{1}{V_v} \, \mathrm{d}H \tag{12.32}$$

where, in the absence of up- or downdrafts, V_v is described by Equation 12.26.

The integration of Equation 12.32 for the general case is not easy because of the manner in which the principal variables (i.e., F_N, D, W, and V) change in the climb. As an illustration, consider an airplane climbing from sea level to its cruising altitude. As the airplane climbs, fuel is consumed and the weight is reduced. The fuel flow (Q), which is equal to the product of the *thrust specific fuel consumption* (TSFC) and the thrust, is not constant. Although the TSFC can be considered to be approximately constant within large climb intervals, the thrust will decrease progressively with increasing altitude. The drag depends on the aircraft weight, air density, and true airspeed.

As mentioned earlier in Section 12.4.2, it is typical for a jet transport airplane to climb initially at a constant CAS and to change later to climbing at constant Mach number. When the airplane climbs at constant CAS, the TAS will increase as the air density reduces. Climbing at constant Mach number implies that there will be a slight decrease in TAS up to the tropopause because the speed of sound will decrease. In the stratosphere, climbing at constant Mach number implies constant TAS.

A numerical approach to determining the *time to climb* and fuel consumed during the climb is outlined below.

Step 1 The climb is divided into n intervals (the height intervals need not be equal). The change of height of the ith interval is designated as ΔH_i.

Step 2 The ROC at the start of the ith interval, designated as V_{v_i}, is determined using Equation 12.27 or Equation 12.28, based on the airplane's weight at the start of the interval (W_i) and the thrust and drag at the height H_i.

Step 3 An estimate of Δt_i (the time to climb through the interval) is then calculated by dividing ΔH_i by V_{v_i}. The fuel burned in the interval is estimated from the product of Q_i and Δt_i. By subtracting this from W_i, an estimate of the weight at the end of the interval (W_{i+1}) is obtained.

Step 4 The ROC at the end of the interval ($V_{v_{i+1}}$) is now determined based on the airplane's estimated weight, thrust, and drag at the height H_{i+1}. A mean value of the ROC for the interval is then calculated from V_{v_i} (determined in step 2) and $V_{v_{i+1}}$.

Step 5 In a repeat of step 3, revised estimates of Δt_i and W_{i+1} are determined using the mean value of the ROC for the interval.

Step 6 The process (steps 2 to 5) is repeated sequentially for each interval. The total time to climb (t) is given by

$$t = \sum_{i=1}^{n} \Delta t_i \qquad (12.33)$$

12.4.9 Distance Covered in the Climb (in Still Air)

The horizontal distance covered during the climb—which is equal to the ground distance in nil wind conditions—depends on the climb–speed schedule and the angle of climb (γ). As γ changes during the climb, this distance has to be calculated by a numerical technique. The calculation is an extension to the numerical computation presented in Section 12.4.8 for the time to climb.

The time that the airplane takes to climb through the ith interval (Δt_i) and the mean true airspeed (V_i) are required information, which can be extracted from the numerical analysis described in Section 12.4.8. The horizontal component of the airspeed vector $\vec{V}$ is equal to $V_i \cos \gamma_i$. Using the trigonometric function $\sin^2 \gamma_i + \cos^2 \gamma_i = 1$ and noting that $\sin \gamma_i = V_{v_i}/V_i$, the horizontal component of the airspeed vector corresponding to the ith interval can be expressed as follows:

$$V_i \cos \gamma_i = V_i \sqrt{1 - \left(\frac{V_{v_i}}{V_i}\right)^2} = \sqrt{(V_i)^2 - (V_{v_i})^2} \tag{12.34}$$

The horizontal still air distance (x) that the airplane covers during the climb is

$$x = \sum_{i=1}^{n} \Delta t_i \, V_i \cos \gamma_i \tag{12.35}$$

and by substituting from Equation 12.34, the distance is given by

$$x = \sum_{i=1}^{n} \Delta t_i \sqrt{(V_i)^2 - (V_{v_i})^2} \tag{12.36}$$

12.4.10   Effect of Altitude on Climb Performance

During the climb, as the altitude increases, the available thrust (at a given thrust setting) will decrease. The drag will also decrease, but—and this is a significant observation—it does so at a lower rate. It is thus evident from Equations 12.23 and 12.29 that both the climb gradient and the climb rate will reduce as the airplane climbs. Practical ceilings for flight operations can be established by considering the greatest altitude at which a set rate of climb (e.g., 300 ft/min) or a set climb gradient can be achieved (see Sections 17.3.1 and 20.4.1).

12.4.11   Effect of Temperature on Climb Performance

When considering the influence of air temperature on climb performance, it is evident that for a given pressure height, an increase in air temperature will correspond to a reduced air density. The result of increased temperature on the climb performance will thus be the same as that for an increase in height—that is, the climb gradient and the climb rate will reduce with an increase in air temperature.

Temperature also impacts the climb performance in another, more subtle, way as it influences the height scale used to measure the rate of climb. It is only in ISA conditions that pressure height equals geopotential height (see Section 5.2.3). Consider an airplane climbing to a fixed pressure height, say corresponding to the cruise altitude. For temperatures above ISA, for example, the geopotential height (and also the geometric height) will be greater than the corresponding pressure height. In this case, the airplane has to climb to a greater geopotential height, requiring a greater energy input. This would imply a lower rate of climb, when measured as a change in pressure height per unit time.

This influence of temperature on the rate of climb can be analyzed by considering the pressure–geopotential height relationship for both standard conditions (designated as "std") and actual conditions (designated as "act"), as shown in Figure 12.13. Corresponding to a time interval (Δt), there is a change in pressure (Δp), which is detected by the altimeter. With the altimeter set at STD (i.e., 1013 hPa or 29.92 inHg), the indicated height increment corresponds to ΔH_{std}. The actual height increment (ΔH_{act}), however, is greater in the case of hotter-than-standard conditions.

Equation 4.23 is recalled:

$$\frac{dp}{dH} = -\frac{pg_0}{RT} \qquad \text{(Equation 4.23)}$$

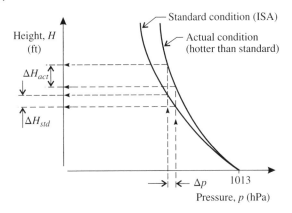

Figure 12.13 Effect of temperature on rate of climb measurement.

For small changes, this can be written as follows:

$$\Delta p = -\frac{p\,g_0}{RT}\Delta H \tag{12.37}$$

Thus $$\Delta p = -\frac{p_{act}\,g_0}{RT_{act}}\Delta H_{act} = -\frac{p_{std}\,g_0}{RT_{std}}\Delta H_{std} \tag{12.38}$$

The actual air pressure (p_{act}) in the climb—as sensed by the altimeter—will be the same as the standard pressure (p_{std}). Furthermore, the change in pressure (Δp) will also be the same; thus the actual change in geopotential height is given by the following expression:

$$\Delta H_{act} = \frac{T_{act}}{T_{std}}\Delta H_{std} \tag{12.39}$$

If Equation 12.39 is divided by the time interval (Δt), the required expression for the rate of climb is obtained:

$$\left(\frac{\Delta H}{\Delta t}\right)_{act} = \left(\frac{T_{act}}{T_{std}}\right)\left(\frac{\Delta H}{\Delta t}\right)_{std} \tag{12.40}$$

or $$(V_v)_{act} = \left(\frac{T_{act}}{T_{std}}\right)(V_v)_{std} = \left(\frac{T_{std} + \Delta T}{T_{std}}\right)(V_v)_{std} \tag{12.41}$$

To illustrate the use of this equation, consider the rate of climb at sea level under ISA $+ 10\,°C$ conditions:

$$(V_v)_{act} = \left(\frac{288.15 + 10}{288.15}\right)(V_v)_{std} = 1.035(V_v)_{std}$$

What this means is that if the flight instruments indicate a rate of climb of 1000 ft/min, then under these conditions the actual rate of climb would be 1035 ft/min. To achieve this requires a greater energy input (i.e., increased thrust) than would be required for a rate of climb of 1000 ft/min under standard conditions. Conversely, if the performance of the airplane were dictated by a given thrust, then the indicated rate of climb would be reduced in hot conditions.

12.4.12   Effect of Airplane Weight on Climb Performance

The airplane's weight directly impacts the rate of climb, as seen from Equation 12.26. It is intuitively obvious that a heavier airplane will have a slower rate of climb, at the same thrust, than a lighter airplane. With other factors constant (i.e., thrust, atmospheric conditions), the best rate-of-climb speed increases with increasing weight; however, the rate of climb will reduce.

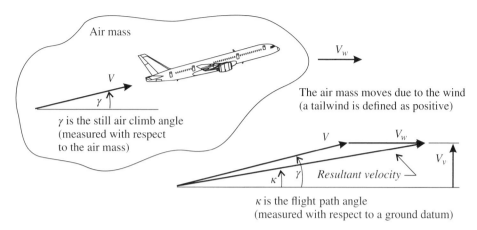

Figure 12.14 Effect of wind on climb performance.

12.4.13 Effect of Uniform Wind on Climb Performance

The air mass within which an airplane operates can move both horizontally and vertically—in fact, this is the norm, rather than the exception. For the purpose of airplane performance analysis, it is convenient to treat the movement of air as two separate velocity components: a horizontal component (which is generally understood as *wind*) and a vertical component (which is understood as an *up-* or *downdraft*). The effects of uniform wind on airplane performance are discussed below; the effects of wind gradients are discussed in Section 12.4.14 and up/downdrafts in Section 12.4.15.

In Figure 12.14, an airplane is shown climbing at a constant true airspeed (V) in a large air mass that is moving with a constant wind speed (V_w). Note that it is customary to treat a tailwind as *positive* for *en route* performance analyses (i.e., climb, cruise, or descent). The still air climb angle (γ), in this case, will not equal the flight path angle (κ). The speed of the airplane with respect to the ground will also be different from the true airspeed (V), because of the wind.

It should be noted that the flight path angle is reduced when the airplane has a tailwind; similarly, a headwind will increase the flight path angle. However, the rate of climb in the presence of a uniform wind is unchanged. This means that the time required for an airplane to climb to a specified height—and the associated fuel needed for the climb—is independent of wind speed or direction. The ground distance covered to reach the specified height, however, does depend on the strength and direction of the wind.

12.4.14 Effect of Wind Gradients on Climb Performance

The conclusion reached in Section 12.4.13, that the rate of climb is unchanged in the presence of wind, applies to conditions where the wind vector is constant with time. If the wind speed changes with altitude, as is often the case, then the airplane's rate of climb at constant TAS will not be exactly constant. A changing headwind or tailwind can arise due to a change in wind speed or wind direction with height—this is called a *wind gradient*. It can also arise when the airplane changes heading in a climb (in a uniform wind condition).

If the headwind component were to increase (or the tailwind component reduce), then both the rate of climb and the angle of climb would increase. The reverse is true for a climb into a reducing headwind or increasing tailwind (e.g., when an airplane ascends into a tailwind jet stream). These statements appear to be inconsistent with the vector diagram given in Figure 12.14, but the situation is not the same, as explained below.

A changing horizontal wind speed encountered during a climb is a common occurrence—in most cases the rate of change of wind speed with height is small and the airplane's performance may be satisfactorily analyzed by ignoring this effect (as described in Section 12.4.13). However, a rapid change in wind speed (e.g., a wind shear) can significantly impact an airplane's climb performance. Low altitude wind shears are a well-recognized hazard to aircraft taking off or landing (see Section 4.5.3).

An intuitive understanding of this situation can be obtained by considering the airplane climbing (with wings level) from a layer of air moving at a constant speed into a higher layer, which has a uniform, but greater headwind (or slower tailwind). Due to its inertia, the airplane will have essentially the same inertial speed (i.e., with respect to the Earth) as it crosses the boundary into the upper layer. The airplane's true airspeed will thus *increase*, and this will produce a small increase in lift and drag. The extra lift will cause the airplane's flight path to steepen (i.e., curve upwards). With no change in thrust, the extra drag (together with a slightly greater component of weight acting along the flight path) will cause the airplane to slow down a little. In addition, the extra lift will produce a small nose-down pitching moment due to the airplane's longitudinal static stability (see Appendix E), which will reduce the angle of attack (moving it towards the original value). The original airspeed will then be regained on a flight path parallel to, but displaced vertically above, the original flight path. This all happens very quickly, of course, with the ground speed stabilizing at a lower value (the reduction being equal to the change in wind speed).

This is the typical response characteristic of an airplane encountering a horizontal wind shear. For an airplane climbing through several such layers of air (with progressively increasing headwind), the cumulative effect will be an increase in the rate of climb. The reverse will be true for a decrease in headwind or an increase in tailwind. Equation 12.26 describes the rate of climb in the absence of a wind gradient. The following expression for the rate of climb (V_v), which includes a change of wind speed with height, is derived in Appendix D.4:

$$V_v = \frac{\left(\dfrac{F_N - D}{W}\right) V}{1 + \dfrac{V}{g}\dfrac{dV}{dH} + \dfrac{V}{g}\dfrac{dV_w}{dH}} \tag{12.42}$$

where dV_w/dH is the wind gradient (typical units: m/s, ft/s), and V_w is defined as positive for a tailwind.

Similarly, the climb angle (which was discussed earlier in Section 12.4.2) in the presence of a wind gradient is given by

$$\sin \gamma = \frac{V_v}{V} = \frac{\left(\dfrac{F_N - D}{W}\right)}{1 + \dfrac{V}{g}\dfrac{dV}{dH} + \dfrac{V}{g}\dfrac{dV_w}{dH}} \tag{12.43}$$

It is interesting to note that flight testing, which is seldom conducted under perfect conditions, has to account for wind gradient effects. It is possible to correct test results to standard conditions (i.e., with zero wind gradient) using the wind gradient factor given in Equations 12.42 and 12.43. A popular and pragmatic approach, however, which eliminates the need for this correction, requires the airplane to be flown (in light winds) at 90° to the wind direction and the results averaged with a subsequent climb conducted in the opposite direction.

12.4.15   Effect of Up- or Downdrafts on Climb Performance

When an airplane flies into an up- or downdraft, both the flight path angle and the rate of climb will change. As the airplane enters an updraft, it will pitch up; the increased angle of attack will result in a sudden increase in lift. The airplane will momentarily accelerate upwards (the lift being greater than the airplane's weight), and it will climb at a new, increased flight path angle. The increased angle of attack—for a statically stable airplane (see Appendix E)—will result in an aerodynamic moment (about the airplane's center of gravity) that will tend to reduce the angle of attack, restoring the original trim condition. When the airplane is fully immersed in the upward moving air current (see Figure 12.15), the rate of climb will be given by

$$V_v = V \sin \gamma + U_w \tag{12.44}$$

where U_w is the speed of the updraft (typical units: m/s, ft/s),

and the flight path angle (κ) will be given by

$$\tan \kappa = \tan \gamma + \frac{U_w}{V \cos \gamma} \tag{12.45}$$

or $\kappa \cong 1 + \dfrac{U_w}{V}$ for κ in radians   (12.46)

The characteristic nature of vertical movements of air is very different from that of horizontal air movements. Their influence on the airplane's flight path are almost always transitory and, unlike the effect of the wind, will not normally have a sustained influence on an airplane's flight path (i.e., the duration will be relatively short). Furthermore, it is common that the direction oscillates between up and down as the airplane flies through gusts, which would mean that the effect on a time-averaged flight path is usually negligible. However, vertical wind shears, where the speed (or direction) of the up- or downdraft changes suddenly, pose a threat to aircraft taking off (or landing) as the rate of climb (or descent) can change very quickly (see Section 4.5.3).

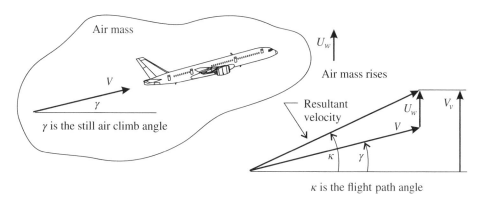

Figure 12.15 Effect of steady updraft on climb performance.

12.4.16   Effect of an Engine Failure on Climb Performance

In this section, consideration is given to the climb performance—that is, the climb angle and the rate of climb—of a multiple engine airplane, after it has suffered an engine failure. An airplane's steady-state (i.e., constant speed) climb performance is a function of the excess of thrust over drag, for a given airplane weight (see Equations 12.21 and 12.27). Following an engine failure, there will be a lower net thrust and an increase in drag. The standard operating procedure would be for the flight crew to select the maximum continuous thrust (MCT) rating (see Section 8.5.5) in this situation. The airplane's net thrust with one engine inoperative can be expressed as follows:

$$F_{N,oei} = \frac{N-1}{N} F_N \qquad (12.47)$$

where $F_{N,oei}$ is the airplane's one-engine-inoperative (OEI) thrust;
$\quad\quad$ N is the total number of engines installed on the airplane; and
$\quad\quad$ F_N is the all engine operating (AEO) thrust (at the selected thrust setting).

The increase in the airplane's drag is associated with windmilling drag, spillage drag, and yaw (or control) drag, as described in Section 7.5.5. The reduction in thrust and increase in drag, following an engine failure, will significantly impact the airplane's climb performance. The climb angle (γ) can be determined from Equation 12.21, with substitutions from Equations 7.29 and 12.47:

$$\sin\gamma = \frac{\left(\frac{N-1}{N}\right)\frac{F_N}{W} - \left(\frac{D + \Delta D_{wm} + \Delta D_{spill} + \Delta D_{\psi}}{W}\right)}{1 + f_{acc}} \qquad (12.48)$$

where ΔD_{wm} is the drag increment due to the windmilling of the engine;
$\quad\quad$ ΔD_{spill} is the drag increment due to the engine spillage effect; and
$\quad\quad$ ΔD_{ψ} is the yaw (or control) drag.

Similarly, an expression for the rate of climb can be derived from Equation 12.27, that is,

$$V_v = \frac{\left(\frac{N-1}{N}\right)\frac{F_N V}{W} - \left(\frac{D + \Delta D_{wm} + \Delta D_{spill} + \Delta D_{\psi}}{W}\right)V}{1 + f_{acc}} \qquad (12.49)$$

where V is the true airspeed.

12.5   Performance in Descending Flight

12.5.1   Angle of Descent and Descent Gradient

The equations developed in Section 12.4.2 for the *angle of climb* (i.e., Equation 12.21) and the *climb gradient* (Equation 12.22) are equally valid for the descent; the only difference being that γ is negative. If the true airspeed in the descent is constant or nearly constant, then $f_{acc} = 0$ and Equation 12.23 can be used:

$$\sin\gamma = \frac{F_N - D}{W} \qquad \text{(Equation 12.23)}$$

As $\gamma < 0$ in a descent, it is evident from Equation 12.23 that the descent angle will depend on the excess drag over thrust.

Typically, descent angles are small (not considering emergency descents), which means that small angle approximations can be used without significant loss of accuracy. Equation 12.24 can be applied to determine the descent angle—that is,

$$\gamma \cong \frac{F_N}{W} - \frac{D}{L}\cos\gamma \cong \frac{F_N}{W} - \frac{1}{E} \qquad \text{(Equation 12.24)}$$

where γ is measured in radians (positive for a climb).

12.5.2   Rate of Descent

The *rate of descent*, which is abbreviated as RoD (or alternatively as RoD or R/D) can be determined from Equation 12.27, derived in Section 12.4.4 for the *rate of climb* (V_v). Note that, by definition, the rate of descent is positive when V_v is negative—this will occur when the drag (D) is greater than the thrust (F_N). For a steady-speed descent (i.e., constant TAS), Equation 12.29 applies—that is,

$$V_v = \left(\frac{F_N}{W} - \frac{D}{W}\right)V = \frac{(F_N V - DV)}{W} \qquad \text{(Equation 12.29)}$$

The airspeed that will result in the minimum rate of descent can be obtained from a plot of V_v against V. In Figure 12.16, this has been done for four thrust settings, expressed as a percentage of the maximum climb thrust (MCLT). Note that the locus of points corresponding to the maximum value of V_v for reducing thrust settings occurs at a reducing true airspeed, but that the speed corresponding to the maximum angle of climb condition increases. When the thrust setting is such that the maximum rate of climb is zero (i.e., 32% MCLT for this illustration), the optimum speed for the rate of climb/descent and that for the climb/descent angle cross over—in fact, the optimum speed is the only speed that will enable a constant height to be maintained

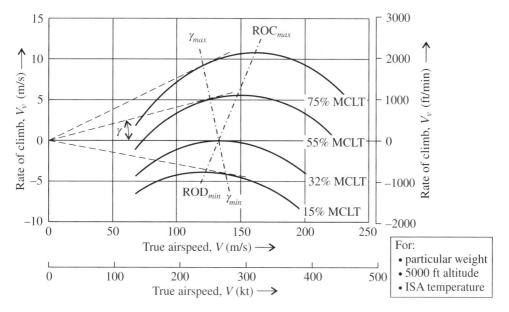

Figure 12.16 Steady-state rate of climb and angle of climb for various thrust settings, indicated as a percentage of the maximum climb thrust, MCLT (example for a mid-size, twin-engine airliner).

at these flight conditions (i.e., for a particular thrust, weight, altitude, and temperature). When the airplane is descending, the speed for the smallest (shallowest) angle of descent occurs at a faster speed than that which would minimize the rate of descent—this is the reverse situation to that which occurs in a climb.

12.5.3   Time and Distance in Descent

The time that it takes an aircraft to descend from its cruise flight level and the associated distance covered can be calculated using a very similar numerical approach to that described in Sections 12.4.8 and 12.4.9 for the climb. The descent would be divided into n intervals and a mean rate of descent determined for each interval in a stepwise fashion starting at the top of descent (TOD). For each step, it would be necessary to compute the airplane's weight based on the associated fuel flow.

12.5.4   Descent–Speed Schedules

For airline operations it is desirable that the trip cost be minimized. As fuel is a significant part of the flight costs, a standard engine rating (setting) for a descent from the cruise altitude is flight idle, as this rating will result in the lowest fuel burn.

From Equation 12.24 (which was derived for a quasi-steady-state condition), it can be seen that for a given thrust-to-weight ratio, the minimum descent gradient, which yields the greatest still air distance, will be obtained at the minimum drag speed (this will maximize the lift-to-drag ratio). Operating the airplane at a speed that produces the greatest air distance is usually advantageous following an engine failure. Descents at such low speeds, however, can take too long for normal flight operations.

The speed that will result in the minimum *rate of descent* is even slower than the minimum drag speed (see Figure 12.16). At this speed, the time that the airplane takes to descend will be a maximum and the air distance covered will be shorter than that associated with flight at the minimum drag speed. There is thus no benefit in descending at a speed slower than the minimum drag speed.

The descent–speed schedule adopted for routine flight operations is one that will reduce the overall trip cost within any restrictions imposed by ATC (air traffic control) procedures and airplane system limitations (see Section 22.3). Utilizing a cost index (see Section 18.3)—appropriately selected for the particular operation—the descent speed can be determined for the airplane's descent weight. For a low cost index, which reduces fuel usage, a relatively low speed (but still greater than the minimum drag speed) is appropriate. However, for a high cost index, which reduces flight time, a relatively high speed (which increases the rate of descent) is appropriate.

A typical descent–speed schedule for an airliner descending from cruise altitude, written in the conventional notation, is M 0.78/300 kt/250 kt. This implies that the airplane descends at Mach 0.78 until the CAS (or IAS) is equal to 300 kt (the crossover height occurs at 29 314 ft in this example) and then descends at 300 kt to 10 000 ft, at which point the airplane slows down to 250 kt for the remainder of the descent.

The descent–speed schedule for an emergency descent—say, following a failure of the cabin pressurization system (see Section 17.5)—will be at the greatest possible rate of descent. In this case, the correct procedure would be a descent at the maximum operating Mach number (M_{MO}), followed by a descent at the maximum operating speed (V_{MO}) below the crossover height.

12.5.5   Glide Angle for Unpowered Flight

Multiple engine failure on airliners is a very rare occurrence. There have been a few cases where the engines have failed almost simultaneously due to fuel starvation[4] and two cases of four-engine jets suffering total engine failure due to the ingestion of volcanic ash.[5] The better an airplane can glide, the more time the flight crew will have to re-start the engines or find a suitable place to land the airplane. It can be shown that an aerodynamically efficient airplane—that is, one with a high E_{max} value—will glide well. For a steady-speed descent (i.e., constant TAS) with zero thrust, an expression for the still air descent angle can be deduced from Equation 12.23:

$$\sin \gamma = -\frac{D}{W} = -\frac{D}{L} \cos \gamma \qquad (12.50)$$

or $\quad \gamma \cong \tan \gamma = -\frac{D}{L} = -\frac{1}{E} \qquad$ (where γ is measured in radians) $\qquad (12.51)$

For modern airliners, the still air gradient in gliding flight is very small, so that $\tan \gamma \cong \gamma$ is a very good approximation.

Equation 12.51 shows that the *gliding angle* of an airplane (measured in radians) is equal to D/L within the accuracy of the basic assumptions. The smallest gliding angle will be achieved when $1/E$ is a minimum—that is, at the flight condition of maximum *lift-to-drag* ratio (E_{max}). The smallest glide angle will ensure that the greatest horizontal distance will be covered for a given change in height. The maximum range for unpowered flight in still air will therefore be achieved at the minimum drag speed (V_{md}).

12.6   Further Reading

The topic of mechanics of flight is described in most publications concerning the performance of fixed-wing aircraft—textbooks that address the topic include the following:

Anderson, J.D., *Aircraft performance and design*, McGraw-Hill, New York, NY, 1999.

Asselin, M., *An introduction to aircraft performance*, AIAA Education Series, American Institute of Aeronautics and Astronautics, Reston, VA, 1997.

Eshelby, M.E., *Aircraft performance: Theory and practice*, Arnold, London, UK, 2000.

Filippone, A., *Advanced aircraft flight performance*, Cambridge University Press, New York, NY, 2012.

Hale, F.J., *Introduction to aircraft performance, selection, and design*, John Wiley & Sons, New York, NY, 1984.

Hull, D.G., *Fundamentals of airplane flight mechanics*, Springer, Berlin, Germany, 2007.

---

4 Examples include (1) a B767-233 (flight AC 143), which ran out of fuel at FL 350 (July 23, 1983) due to a combination of faulty instrumentation and human error and landed on a disused runway (Gimli Airport, Canada) 39 nm (72 km) away; and (2) an A330-243 (flight TS 236), which ran out of fuel at FL 345 over the Atlantic Ocean (Aug. 24, 2001) due to a fuel leak and landed at Lajes Airport, Azores after gliding for 19 minutes. There were no fatalities in either case, and both airplanes were repaired following the incidents [7].

5 A B747-236B (flight BA 009) ingested volcanic ash following the eruption of Mount Galunggung, Indonesia (June 24, 1982), and a B747-406 (flight KL 867) suffered the same fate following the eruption of Mount Redoubt, Alaska (Dec. 15, 1989). In both cases, the crew were able to re-start the engines after descending out of the ash cloud and land safely [7].

Lan, C.-T.E. and Roskam, J., *Airplane aerodynamics and performance*, DARcorporation, Lawrence, KS, 2010.

Mair, W.A. and Birdsall, D.L., *Aircraft performance*, Aerospace Series, Cambridge University Press, Cambridge, UK, 1992.

Ojha, S.K., *Flight performance of aircraft*, AIAA Education Series, American Institute of Aeronautics and Astronautics, Reston, VA, 1995.

Pamadi, B.N., *Performance, stability, dynamics, and control of airplanes*, AIAA Education Series, American Institute of Aeronautics and Astronautics, Reston, VA, 2004.

Perkins, C.D. and Hage, R.E., *Airplane performance, stability and control*, Wiley, New York, NY, 1966.

Ruijgrok, G.J.J., *Elements of airplane performance*, 2nd ed., VSSD, Delft, the Netherlands, 2009.

Saarlas, M., *Aircraft performance*, John Wiley & Sons, Hoboken, NJ, 2007.

Smetana, F.O., *Flight vehicle performance and aerodynamic control*, AIAA Education Series, American Institute of Aeronautics and Astronautics, Reston, VA, 2001.

Yechout, T.R., *Introduction to aircraft flight mechanics: Performance, static stability, dynamic stability, and classical feedback control*, AIAA Education Series, American Institute of Aeronautics and Astronautics, Reston, VA, 2003.

References

1 De Baudus, L. and Castaigns, P., "Control your speed … in cruise," *Safety first*, Product Safety Department, Airbus S.A.S., Blagnac, France, Iss. 21, pp. 6–21, Jan. 2016.

2 ESDU, "Acceleration factors for climb and descent rates at constant EAS, CAS, M," Data item 81046, Amdt. A, IHS ESDU, 133 Houndsditch, London, UK, June 1992.

3 Blake, W. and Performance Training Group, "Jet transport performance methods," D6-1420, Flight Operations Engineering, Boeing Commercial Airplanes, Seattle, WA, Mar. 2009.

4 Boeing, "Jet transport performance methods," D6-1420, The Boeing Company, Seattle, WA, May 1989.

5 Shevell, R.S., *Fundamentals of flight*, 2nd ed., Prentice Hall, Englewood Cliffs, NJ, 1989.

6 Young, T.M., "Climb and descent of fixed-wing aircraft," in *Encyclopedia of aerospace engineering*, Blockley, R. and Shyy, W. (Eds.), Wiley, Hoboken, NJ, pp. 2589–2600, 2010.

7 ASN, "Aviation safety database," Aviation Safety Network, Flight Safety Foundation, Alexandria, VA, Retrieved Aug. 2, 2012. Available from http://aviation-safety.net/database/.

13

Cruising Flight and Range Performance

13.1 Introduction

Cruise describes that portion of a flight from the point where the airplane has leveled off following a climb to its initial cruising altitude until the point where it commences its descent. The key performance considerations for the cruise relate to economic factors. The airplane's fuel efficiency and, consequently, the distance that the airplane can cover while consuming a given quantity of fuel, is a critically important performance metric for an airliner. The distance that an airplane can fly (i.e., the range)—for a given starting weight and fuel load—depends fundamentally on the efficiency of the engines (quantified by the thrust specific fuel consumption), the aerodynamic efficiency of the airplane, (i.e., lift-to-drag ratio), Mach number, ambient air temperature, and the strength and direction of the wind. There are several factors that can influence these parameters, thus indirectly impacting the airplane's range—these include cruise altitude, center of gravity position, fuel energy content, and the demands placed on the engines for ancillary services (such as the cabin pressurization and air conditioning system).

The *specific air range* (SAR), which is introduced in Section 13.2, is the air distance traveled per unit of fuel mass (or fuel weight) consumed. SAR is a point performance parameter, describing the airplane's cruise efficiency at a particular time or location along the flight path. The analytical integration of the SAR function—to provide closed-form expressions for the range that an airplane can cover for a given fuel load and starting weight—can be undertaken in several ways. Three sets of boundary conditions are considered in Section 13.3, resulting in three different mathematical expressions. The implications and limitations of these range equations are discussed in this chapter. The well-known Bréguet equation, which is widely used to estimate an airplane's range, is presented as the second flight schedule in Section 13.3; further details on its use are given in Section 13.7. A numerical method to evaluate the integral expression that describes the still air range is presented in Section 13.4, along with a computational table that illustrates the procedure.

The flight conditions that enable an airplane to travel a particular distance with the least fuel consumed—that is, the conditions of greatest fuel economy—are of considerable interest to aircraft manufacturers and operators alike. In Section 13.5, the conditions that will maximize an airplane's range are assessed by considering aerodynamic parameters based on the airplane's drag polars. This discussion is continued in Section 13.6, where the engine characteristics are included in the assessment of the optimum flight conditions—cruise speeds and flight altitudes that will minimize trip fuel burn or trip cost are discussed. Range analyses are initially presented for conditions of nil wind. The effect of wind on the optimum cruise conditions (i.e., the best cruise speed and cruise altitude) are considered later, in Section 13.8, together with a description of how winds influence the distance that can be achieved for a given fuel load.

Performance of the Jet Transport Airplane: Analysis Methods, Flight Operations, and Regulations, First Edition.
Trevor M. Young.
© 2018 John Wiley & Sons Ltd. Published 2018 by John Wiley & Sons Ltd.

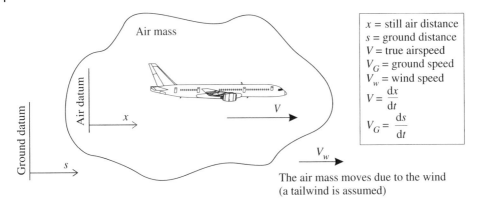

Figure 13.1 Distance and speed definitions for cruise.

13.2   Specific Air Range and Still Air Range Determination

13.2.1   Distance and Speed Definitions

Cruise analysis is based on point (instantaneous) performance attributes, or characteristics, of an airplane in flight. The airplane is assumed to be in a steady-state or quasi-steady-state condition—which essentially means that any changes to the airplane's speed or altitude occur sufficiently slowly that a condition of equilibrium of forces can be assumed. The integration of point performance parameters along the flight path to establish an airplane's range (flight distance) follows as a second step in the analysis process.

The airplane will be assumed to be flying in a straight line, with the wings level, at a constant altitude at a true airspeed V within a large air mass, as shown in Figure 13.1. Relative to the ground, the air mass is assumed to be moving at a steady speed V_w in the direction of flight. Consistent with customary practice, a tailwind is defined as positive since such winds are advantageous for cruising flight.[1]

The air distance—or *still air distance* as it is often called—is defined as x and is measured from a datum that moves with the air mass. The true airspeed (TAS), V, is equal to dx/dt. The ground distance is defined as s, measured from a fixed ground datum; the ground speed, V_G, is equal to ds/dt.

13.2.2   Specific Air Range

The *specific air range* (SAR), which herein is assigned the symbol r_a, is an instantaneous measure of an airplane's efficiency in cruise. Specific air range is also known as *specific range* (SR) or, alternatively, as *fuel mileage*.[2] The term specific air range, however, is more precise and will be used herein. SAR can be defined in two equally correct ways: it is the still air distance traveled

---

1 Note that this convention is not consistent with that adopted for takeoff and landing, where a *headwind* is considered advantageous and is thus defined as positive in Chapters 9–11.

2 Strictly speaking, *specific range* can refer to either *specific air range* (defined in Section 13.2.2) or *specific ground range* (defined in Section 13.8.1). It is thus preferable to avoid the term specific range unless the air/ground distinction is apparent. The term *fuel mileage*, which has the same meaning, is also used widely (with units of nautical miles per lb or nautical miles per 1000 lb). Again, caution is warranted if the air/ground distinction is not apparent.

per unit of fuel *mass* consumed or, alternatively, it is the still air distance traveled per unit of fuel *weight* consumed.[3]

$$r_a = -\frac{dx}{dm_f} \qquad (13.1a)$$

or $\quad r'_a = -\dfrac{dx}{dW_f} \qquad\qquad\qquad\qquad\qquad\qquad\qquad (13.1b)$

where r_a is the SAR defined in terms of fuel mass (typical units: km/kg, nm/kg, nm/slug);

$\quad r'_a$ is the SAR defined in terms of fuel weight (typical units: km/kN, nm/kN, nm/lb);

$\quad x$ is the still air distance (typical units: km, nm);

$\quad m_f$ is the onboard fuel mass (typical units: kg, slug); and

$\quad W_f$ is the onboard fuel weight (typical units: kN, lb).

The reason for the minus sign in Equation 13.1 is that the change in onboard fuel mass, dm_f, (or fuel weight, dW_f) is a negative quantity and SAR is a positive quantity. Commonly used units of SAR are nm/kg, km/kg, and nm/lb.

The rate that fuel is burned in the engines can be expressed as the *mass* of the fuel consumed per unit of time (herein given the symbol Q) or, alternatively, the *weight* of the fuel consumed per unit of time (herein given the symbol Q')—thus,

$$Q = -\frac{dm_f}{dt} \quad \text{(see Equation 8.26a)} \qquad (13.2a)$$

or $\quad Q' = -\dfrac{dW_f}{dt} \quad \text{(see Equation 8.26b)} \qquad\qquad (13.2b)$

The fuel flow Q (or Q') is thus the total fuel consumed by the operating engine(s) per unit of time. These definitions enable the SAR to be written in a more convenient form. By dividing the numerator and denominator in Equation 13.1 by a time increment (dt), the SAR can be expressed as follows:

$$r_a = -\frac{dx}{dm_f} = \frac{\left(\dfrac{dx}{dt}\right)}{\left(-\dfrac{dm_f}{dt}\right)} = \frac{V}{Q} \qquad (13.3a)$$

or $\quad r'_a = -\dfrac{dx}{dW_f} = \dfrac{\left(\dfrac{dx}{dt}\right)}{\left(-\dfrac{dW_f}{dt}\right)} = \dfrac{V}{Q'} \qquad\qquad (13.3b)$

13.2.3 Still Air Range

The range covered by an airplane in still air (i.e., air distance) can be determined from an integral expression based on the definition of SAR. The first step is to make a variable substitution to the description of SAR, as given by Equation 13.1, by noting that the change in *airplane*

---

3 Consistent with the notation adopted in Chapter 8, a prime mark is used in this chapter to designate parameters that are based on fuel *weight* (as apposed to fuel *mass*).

mass in cruise is equal to the change in onboard *fuel* mass. This is true for normal commercial airplane operations. Changes to the airplane's mass that are *not* related to the fuel consumed in the engines—for example, loss of oil, water, or oxygen—are considered negligible. Furthermore, it is assumed for cruise that there is no change to the airplane's mass due to fluids or other items dispensed overboard (e.g., fuel venting). Thus,

$$r_a = -\frac{dx}{dm} \quad \text{or} \quad dx = -r_a \, dm \tag{13.4a}$$

Similarly,

$$r_a' = -\frac{dx}{dW} \quad \text{or} \quad dx = -r_a' \, dW \tag{13.4b}$$

where m is the airplane mass (typical units: kg, slug); and
W is the airplane weight (typical units: N, lb).

Equation 13.4 leads to the following integral expression for the still air range:

$$R = \int_{start}^{end} dx = -\int_{m_1}^{m_2} r_a \, dm = \int_{m_2}^{m_1} r_a \, dm \tag{13.5a}$$

or

$$R = \int_{start}^{end} dx = -\int_{W_1}^{W_2} r_a' \, dW = \int_{W_2}^{W_1} r_a' \, dW \tag{13.5b}$$

where R is the still air range (typical units: km, nm); and
subscript 1 denotes the start-of-cruise condition and subscript 2 the end-of-cruise condition.

Several methods have been developed to evaluate Equation 13.5. A popular approach involves expressing the SAR in terms of the engines' thrust specific fuel consumption (this is explored in Section 13.2.4). An alternative approach is based on the engines' overall efficiency (see Section 13.2.5).

13.2.4 Range Determination Based on Thrust Specific Fuel Consumption

The fuel consumption of a jet engine is often expressed as a *thrust specific fuel consumption* (TSFC), rather than in absolute terms. The TSFC can be defined in two ways, as described in Section 8.4.2.

(1) The TSFC is the *mass* of fuel burned per unit of time, divided by the thrust (this is convenient when working in SI units); herein it is given the symbol c, as defined by Equation 8.30a.
(2) Alternatively, the TSFC is the *weight* of fuel burned per unit of time, divided by the thrust (this is convenient when working in FPS units); herein it is given the symbol c', as defined by Equation 8.30b.

For a jet airplane in steady (i.e., non-accelerating), level (i.e., constant altitude) flight, the total fuel flow (to all engines) can be written as

$$Q = cF_N = cD = c\left(\frac{D}{L}\right)W = \frac{c}{E}mg \tag{13.6a}$$

or

$$Q' = c'F_N = c'D = c'\left(\frac{D}{L}\right)W = \frac{c'}{E}W \tag{13.6b}$$

where c (typical unit: mg N^{-1} s^{-1}) or c' (typical unit: lb lb^{-1} h^{-1}) is the average TSFC for all installed engines;
F_N is the net thrust of all installed engines (typical units: N, lb);

D is the airplane drag (typical units: N, lb);
L is the airplane lift (typical units: N, lb);
E is the lift-to-drag ratio (dimensionless); and
g is the acceleration due to gravity (typical units: m/s^2, ft/s^2).

This expression for fuel flow, when coupled with Equations 13.2 and 13.4, leads to an interesting observation, which can be expressed mathematically as follows:

$$\left(\frac{cg}{E}\right)\Delta t = \frac{-\Delta m}{m} \tag{13.7a}$$

or
$$\left(\frac{c'}{E}\right)\Delta t = \frac{-\Delta W}{W} \tag{13.7b}$$

It is evident from Equation 13.7b that the quotient of TSFC and lift-to-drag ratio in cruise is equal to the relative change in an airplane's weight per unit time. To illustrate this statement, typical values of $c' = 0.59$ lb lb^{-1} h^{-1} and $E = 19$ have been assumed. It can be deduced that the illustrative airplane will consume ~3.1% of its instantaneous (gross) weight per hour during cruise. Obviously, more efficient airplanes will consume a smaller fraction and less efficient types a greater fraction (per flight hour), but, interestingly, the fraction remains relatively constant for a particular airplane during cruise.

Substituting Equation 13.6 into Equation 13.3 enables the SAR to be expressed in a form that is convenient for determining the airplane's still air range, that is,

$$r_a = \frac{VE}{cmg} \tag{13.8a}$$

or
$$r'_a = \frac{VE}{c'W} \tag{13.8b}$$

Alternatively, the SAR can be written as a function of the cruise Mach number (M):

$$r_a = \frac{MaE}{cmg} = \frac{a_0(ME)}{\left(\dfrac{c}{\sqrt{\theta}}\right)(mg)} \tag{13.9a}$$

or
$$r'_a = \frac{MaE}{c'W} = \frac{a_0(ME)}{\left(\dfrac{c'}{\sqrt{\theta}}\right)W} \tag{13.9b}$$

where a is the speed of sound in the ambient air (typical units: m/s, ft/s, kt);
a_0 is the speed of sound at the ISA sea-level datum (typical units: m/s, ft/s, kt); and
θ is the relative temperature (dimensionless).

It is worth noting that Equation 13.9 encapsulates, in an elegant way, the overall efficiency of a jet airplane. By grouping the terms as indicated, it is apparent that the SAR—which is, after all, a measure of the airplane's cruise efficiency—is a function of three terms:

(1) an *aerodynamic term*, given by the product of Mach number and the aerodynamic efficiency of the airplane, that is, $M(L/D)$;
(2) a *powerplant term*, given by the corrected TSFC, that is, $c/\sqrt{\theta}$ (or $c'/\sqrt{\theta}$); and
(3) the airplane's instantaneous weight (W), which—for a given payload and fuel quantity—depends on the airplane's empty weight and hence on the *structural design* of the airplane.

An integral expression for still air range (R) can be obtained by substituting Equation 13.8 into 13.5, that is,

$$R = \frac{1}{g} \int_{m_2}^{m_1} \frac{VE}{cm} \, \mathrm{d}m \tag{13.10a}$$

or $\quad R = \int_{W_2}^{W_1} \frac{VE}{c'W} \, \mathrm{d}W \tag{13.10b}$

Analytical solutions to Equation 13.10, for different boundary conditions, are presented in Section 13.3. A numerical technique to evaluate the integral is presented in Section 13.4.

13.2.5 Range Determination Based on Overall Engine Efficiency

An alternative approach to that described in Section 13.2.4 is based on a *range parameter*. For certain applications—for example, when analyzing airplane concepts with open rotor or ultra-high bypass ratio engines—this approach could be advantageous [1, 2].

The range parameter[4] (r_p) is defined as

$$r_p = \eta_0 E \tag{13.11}$$

where η_0 is the overall engine efficiency (dimensionless).

As described in Section 8.3.5, the overall engine efficiency is the ratio of the net *thrust power* developed by the engines to the rate that *fuel energy* is supplied to the engines. Using mass-based terms, η_0 can be expressed as follows:

$$\eta_0 = \frac{F_N V}{Q H_f} \quad \text{(from Equation 8.24)} \tag{13.12}$$

where H_f is the net heating value of the fuel (i.e., net energy content per unit mass) (typical units: MJ/kg, Btu/lbm).[5]

The significance of the range parameter is that it represents, in non-dimensional form, the product of two fundamental airplane efficiencies: an engine efficiency (i.e., η_0) and an aerodynamic efficiency (i.e., E). For steady, level flight, where thrust equals drag and lift equals weight, the range parameter can be written in terms of the specific air range by substituting Equation 13.12 into Equation 13.11:

$$r_p = \frac{F_N V}{Q H_f} \left(\frac{L}{D} \right) = r_a \left(\frac{F_N}{H_f} \right) \left(\frac{L}{D} \right) = r_a \left(\frac{W}{H_f} \right) \tag{13.13}$$

Using Equations 13.5 and 13.13, the still air range (R) can now be expressed in terms of the overall engine efficiency and the lift-to-drag ratio:

$$R = \int_{m_2}^{m_1} \frac{H_f r_p}{W} \, \mathrm{d}m = \frac{H_f}{g} \int_{m_2}^{m_1} \frac{\eta_0 E}{m} \, \mathrm{d}m \tag{13.14a}$$

or $\quad R = \frac{H_f}{g} \int_{W_2}^{W_1} \frac{\eta_0 E}{W} \, \mathrm{d}W \tag{13.14b}$

---

4 Caution: The term *range parameter* can be ambiguous—it is used by different authors in similar, but not identical, mathematical definitions pertaining to range performance (see also Section 13.7.1). The definition adopted herein is consistent with that used by Torenbeek [1, 2].

5 The properties of turbine-engine aviation fuel are described in Section 22.7.

An analytical solution to the range integral given by Equation 13.14 is presented later in Section 13.7.1, based on the assumption that r_p remains constant during the cruise. It is also possible to derive other solutions corresponding to different boundary conditions—these are not presented herein, but they follow a similar mathematical approach to that described in Section 13.3 (see Torenbeek [1, 2] for further details).

13.3 Analytical Integration

13.3.1 Flight Schedules for Cruise Range Estimation

Closed-form mathematical solutions that describe an airplane's range can be determined by assuming that the TSFC does not change during cruise. This is a reasonable approximation that enables relatively simple equations to be derived, as the parameter can be taken out of the integral expression given by Equation 13.10. It should be remembered, however, that the TSFC of a turbofan engine does vary during cruise, but by assuming a mean constant value (denoted as $\bar{c}$ or $\bar{c}'$), mathematical solutions can be found that are sufficiently accurate for many applications.

The determination of the still air range can be accomplished in several different ways. The approach adopted herein is a two-step one. First, boundary conditions (or constraints) are established that facilitate the integration of Equation 13.10. Thereafter, the implications of these constraints in terms of actually operating the airplane are considered. Three flight schedules that can be represented by closed-form mathematical formulae are presented—the boundary conditions for these solutions are as follows:

(1) cruise at constant *altitude* (H_p) and constant *lift coefficient* (C_L);
(2) cruise at constant *true airspeed* (V) and constant *lift coefficient* (C_L); and
(3) cruise at constant *altitude* (H_p) and constant *true airspeed* (V).

For the sake of convenience, these solutions are herein referred to as the first, second, and third flight schedule.[6] Note that the boundary condition of constant C_L (i.e., constant angle of attack) implies that E is constant, as E is a function of C_L (see Equation 7.40).

13.3.2 First Flight Schedule

The still air range (R) for an airplane flown at constant altitude and constant lift coefficient is given by the following expression, which is obtained by integrating Equation 13.10 (the derivation is given in Appendix D, Section D.5.2):

$$R = \frac{E}{\bar{c}g} \int_{m_2}^{m_1} \frac{V}{m}\,dm = \frac{1}{\bar{c}g}\sqrt{\frac{8g}{\rho S}\left(\frac{C_L}{C_D^2}\right)}\left(\sqrt{m_1} - \sqrt{m_2}\right) \tag{13.15a}$$

$$\text{or} \quad R = \frac{E}{\bar{c}'} \int_{W_2}^{W_1} \frac{V}{W}\,dW = \frac{1}{\bar{c}'}\sqrt{\frac{8}{\rho S}\left(\frac{C_L}{C_D^2}\right)}\left(\sqrt{W_1} - \sqrt{W_2}\right) \tag{13.15b}$$

where $\bar{c}$ and $\bar{c}'$ are mean TSFC values determined for the cruise (see Equation 13.40).

---

6 The terminology used herein—by which the cruise schedules are labeled first, second, and third—is arbitrary and does not reflect any standard convention. The adopted approach is similar to that presented by Hale [3]. The derivation of the solutions to the range integral (Equation 13.10) for the three schedules is contained in Appendix D.5.

An equivalent range formulation can be developed for these boundary conditions in terms of the lift-to-drag ratio (E) and the start-of-cruise true airspeed (V_1):

$$R = \frac{2EV_1}{\bar{c}g}\left(1 - \sqrt{\frac{m_2}{m_1}}\right) = \frac{2EV_1}{\bar{c}g}\left(1 - \sqrt{1 - \varsigma}\right) \tag{13.16a}$$

or

$$R = \frac{2EV_1}{\bar{c}'}\left(1 - \sqrt{\frac{W_2}{W_1}}\right) = \frac{2EV_1}{\bar{c}'}\left(1 - \sqrt{1 - \varsigma}\right) \tag{13.16b}$$

where ς is the cruise fuel mass (or weight) fraction, defined as

$$\varsigma = \frac{m_1 - m_2}{m_1} = \frac{W_1 - W_2}{W_1} \tag{13.17}$$

The implication of this flight schedule in terms of the operation of the airplane can be assessed by considering the lift coefficient for straight and level flight (i.e., when the lift is equal to the weight), as described by Equation 7.4:

$$C_L = \frac{L}{qS} = \frac{2W}{\rho_0 a_0^2 \delta M^2 S} \quad \text{(from Equation 7.4, where } n_z = 1)$$

where q is the dynamic pressure (typical units: N/m^2, lb/ft^2);
  S is the wing reference area (typical units: m^2, ft^2);
  ρ_0 is the air density at the ISA sea-level datum (typical units: kg/m^3, slug/ft^3);
  δ is the relative air pressure (dimensionless); and
  n_z is the load factor.

If the airplane flies at a constant altitude (δ is thus constant), the Mach number (M) must decrease with $\sqrt{W}$, as the lift coefficient must be held at a constant value (to comply with the set boundary conditions). In addition, the engine thrust (F_N) must decrease with W, as

$$F_N = D = \left(\frac{D}{L}\right)W = \frac{W}{E} \tag{13.18}$$

and E is assumed to be constant for this flight schedule.

The flight crew will continually have to reduce the engine thrust setting to maintain a reducing TAS (and Mach number) as the fuel is burned. Although the air distance (range) that an airplane can cover for a given fuel quantity and start-of-cruise weight can be estimated using Equations 13.15 or 13.16, the undesirable consequences of operating the airplane under these constraints limits the usefulness of the resulting solutions.

13.3.3 Second Flight Schedule

The second flight schedule involves what is known as a *cruise-climb* as altitude is not held constant, and, as described later, the airplane gains height in the cruise. The integration of Equation 13.10 is straightforward as both V and E are constant (see Appendix D, Section D.5.3), resulting in the following expression:

$$R = \frac{VE}{\bar{c}g} \int_{m_2}^{m_1} \frac{1}{m}\, dm = \frac{VE}{\bar{c}g} \ln\left(\frac{m_1}{m_2}\right) = \frac{VE}{\bar{c}g} \ln\left(\frac{1}{1-\varsigma}\right) \tag{13.19a}$$

or $\quad R = \dfrac{VE}{\bar{c}'} \displaystyle\int_{W_2}^{W_1} \frac{1}{W}\, dW = \dfrac{VE}{\bar{c}'} \ln\left(\dfrac{W_1}{W_2}\right) = \dfrac{VE}{\bar{c}'} \ln\left(\dfrac{1}{1-\varsigma}\right) \tag{13.19b}$

Equation 13.19 is widely known as the Bréguet range equation or simply the Bréguet equation.[7] The implication of this flight schedule on the operation of the airplane can be deduced from the following expansion of the lift coefficient definition (see Equations 7.4 and 12.9):

$$C_L = \frac{L}{qS} = \frac{2W\cos\gamma}{\rho_0 \sigma V^2 S} \cong \left(\frac{W}{\sigma}\right)\left[\frac{2}{\rho_0 V^2 S}\right] \tag{13.20}$$

where γ is the climb angle; and
 σ is the relative air density (dimensionless).

The climb angle for a cruise-climb is a very small quantity (as illustrated later), and, to an excellent approximation, $\cos\gamma \cong 1$. Now, as C_L and V are to be held constant, it is evident from Equation 13.20 that the airplane must be flown in a way that will ensure that the ratio W/σ remains constant. This is possible if the airplane is allowed to climb very slowly so that the relative density (σ) decreases in direct proportion to the decrease in airplane weight (W). At the same time, as the thrust of a turbofan engine reduces with altitude (as a first approximation, it is often assumed that thrust is proportional to σ in the stratosphere), it will automatically decrease as altitude increases, without the flight crew altering the thrust setting. The flight crew must, therefore, simply maintain a constant TAS, allowing the airplane to "drift" upwards as the flight progresses and the airplane's weight reduces. Note that in the stratosphere, a constant TAS implies a constant Mach number, as the speed of sound in air is constant.

An equivalent approach to that described above results in the same conclusion in regard to the operational implications of this flight schedule in the stratosphere (where temperature is constant). For the sake of completeness, this is also described. Using Equations 7.4 and 12.9, the lift coefficient can be written as

$$C_L = \frac{L}{qS} = \frac{2W\cos\gamma}{\rho_0 a_0^2 \delta M^2 S} \cong \left(\frac{W}{\delta}\right)\left[\frac{2}{\rho_0 a_0^2 M^2 S}\right] \tag{13.21}$$

From this expression, it is evident that the ratio of W/δ must be kept constant for C_L and M to be constant. This approach again illustrates the need for the airplane to climb—albeit very slowly—as it gets lighter when cruising in the stratosphere.

In a cruise-climb, the flight parameters of the starting condition (given the subscript 1) and those of the final condition (subscript 2) are interrelated. The equations that follow, linking these conditions, are useful for elementary problem solving. As the climb angle is small and the lift coefficient is constant, it follows from Equation 13.20 that

$$\frac{W_1}{\sigma_1} = \frac{W_2}{\sigma_2} \tag{13.22}$$

---

7 Named after Louis Charles Bréguet (1880–1955), the French engineer and pioneering aircraft builder. The constant velocity, constant lift coefficient (or constant angle of attack) solution to the range equation for piston–propeller airplanes has been known since at least 1918 [4]. The precise origin of its development, however, is uncertain as there is no seminal publication, from that time, identifying Bréguet with its development [5].

Furthermore, from Equation 13.18, it is apparent that

$$E = \frac{W_1}{F_{N_1}} = \frac{W_2}{F_{N_2}} \qquad (13.23)$$

These two expressions can be combined into a single equation:

$$\frac{W_1}{W_2} = \frac{\sigma_1}{\sigma_2} = \frac{F_{N_1}}{F_{N_2}} \qquad (13.24)$$

As an illustration of the use of Equation 13.22, an airplane conducting a 1000 ft cruise-climb in the ISA stratosphere is considered (the issue of air traffic control permitting such an altitude change is set aside for this illustration). The 1000 ft climb would correspond to a reduction in air density of 4.69%, which would imply the same reduction in the airplane's weight. If, as an approximation, it were assumed that the engines consume 3.1% of the airplane's weight per hour as burned fuel (see Section 13.2.4), then the cruise would last about 1.5 hr. Assuming a cruise speed of Mach 0.82 (i.e., a TAS of 470 kt), an air distance of ~705 nm would be covered. For this illustrative example, the climb gradient would thus be ~0.02%.

The Bréguet solution is the most widely used method for estimating an airplane's range due to its simplicity and accuracy. Several considerations related to the derivation of Equation 13.19 are described below; additional aspects and further uses of the Bréguet solution are discussed later in Section 13.7.

First, it can be noted that during a cruise-climb, an airplane will gain potential energy by virtue of the increase in altitude—this potential energy increase can be considered as energy expended by the airplane that has not contributed toward the predicted range, suggesting that the Bréguet solution under-predicts an airplane's true range potential. At the end of the cruise, this potential energy could, in theory, be converted to a small range increment through the airplane gliding, at zero thrust (and zero fuel expenditure), back down to its initial cruise altitude. For example, the 1000 ft climb, used in the illustration given above, would yield for an airplane with a lift-to-drag ratio of 19 an additional cruise distance of ~3 nm (gliding performance is discussed in Section 12.5.5).

Secondly, it can be noted that there is an apparent anomaly in the derivation of Equation 13.19. The initial conditions established for the cruise analysis were for straight and level (i.e., constant altitude) flight, where thrust equals drag and lift equals weight (see Equation 13.6). The requirement for the airplane to climb clearly invalidates this statement: there is a need for an increase in thrust to maintain constant speed and also an increase in lift (which results in an increase in drag). However, the climb angle in a cruise-climb is very small (less than 0.05°); consequently, the increase in thrust (and fuel) required to maintain the climb is also very small.

Finally, it should be noted that the errors introduced by the aforementioned assumptions and approximations are insignificant in comparison to the impact of the fundamental assumption of constant TSFC, which is necessary to derive Equation 13.19 (see also Section 13.7).

13.3.4 Third Flight Schedule

This flight schedule assumes constant altitude (H_p) and constant true airspeed (V). Note that a constant TAS implies a constant Mach number for flights at constant altitude. The integration of Equation 13.10 with the constraints of constant H_p and constant V is possible if it is assumed that the drag can be modeled by the parabolic drag polar (see Equation 7.14). The resulting expressions (which are derived in Appendix D, Section D.5.4) are given below:

$$R = \frac{2E_{max}V}{g\bar{c}} \arctan\left\{ \frac{\sqrt{B_3}(m_1 - m_2)}{B_3 + m_1 m_2} \right\} \tag{13.25a}$$

or $$R = \frac{2E_{max}V}{\bar{c}'} \arctan\left\{ \frac{g\sqrt{B_3}(W_1 - W_2)}{g^2 B_3 + W_1 W_2} \right\} \tag{13.25b}$$

where $B_3 = \left(\frac{C_{D_0}}{K}\right)\left(\frac{\rho V^2 S}{2g}\right)^2$

The maximum lift-to-drag ratio based on the parabolic drag polar is given by

$$E_{max} = \frac{1}{2}\sqrt{\frac{1}{KC_{D_0}}} \quad \text{(Equation 7.43)}$$

where C_{D_0} is the zero-lift drag coefficient (dimensionless); and
K is the lift-dependent drag factor (dimensionless).

Alternatively, Equation 13.25 can be expressed in terms of the lift coefficient and lift-to-drag ratio evaluated at the start of cruise (i.e., C_{L_1} and E_1):

$$R = \frac{2E_{max}V}{g\bar{c}} \arctan\left\{ \frac{\varsigma E_1}{2E_{max}(1 - KC_{L_1}\varsigma E_1)} \right\} \tag{13.26a}$$

or $$R = \frac{2E_{max}V}{\bar{c}'} \arctan\left\{ \frac{\varsigma E_1}{2E_{max}(1 - KC_{L_1}\varsigma E_1)} \right\} \tag{13.26b}$$

The limitations of the parabolic drag representation are described in Section 7.4. The main issue here is that the airplane's low-speed drag polar should not be used for cruise performance analysis. Appropriate values of K and C_{D_0} for the cruise condition need to be established to determine E_{max} and B_3, for example.

During cruise, under the boundary conditions for this flight schedule, the lift coefficient will reduce with time (as illustrated in Figure 7.5) and, as shown in Figure 7.8, this means that the drag coefficient will reduce. Hence, the drag will also reduce, implying that the flight crew must continuously reduce the thrust to maintain a constant TAS. This can be seen from the following expression, which equates the net thrust to the airplane's drag in cruise:

$$F_N = D = \frac{1}{2}\rho V^2 S C_D \tag{13.27}$$

13.3.5 Summary of Range Expressions and Concluding Remarks

It is apparent from the discussions presented in Sections 13.3.2–13.3.4 that analytical models describing the air distance (range) that an airplane can cover for a given quantity of fuel and a given start-of-cruise airplane weight can be derived when (1) two of the three governing parameters (i.e., altitude, lift coefficient, and TAS) are assumed to be invariant during cruise, and (2) TSFC is assumed to be constant. A summary of the range expressions for the three flight schedules is given in Table 13.1.

Table 13.1 Summary of range expressions based on constant TSFC

Mass flow basis	Weight flow basis

Range integral:[a],[b]

$$R = \int_{m_2}^{m_1} r_a \, dm = \int_{m_2}^{m_1} \frac{V}{Q} \, dm \qquad\qquad R = \int_{W_2}^{W_1} r_a' \, dW = \int_{W_2}^{W_1} \frac{V}{Q'} \, dW$$

Fuel flow:

$$Q = \frac{cgm}{E} \qquad\qquad Q' = \frac{c'W}{E}$$

Range equation for schedule 1 (i.e., flight at constant altitude and lift coefficient):[a],[b],[c]

$$R = \frac{1}{\bar{c}g}\sqrt{\frac{8}{\rho S}\left(\frac{C_L}{C_D^2}\right)}\left(\sqrt{gm_1}-\sqrt{gm_2}\right) \qquad R = \frac{1}{\bar{c}'}\sqrt{\frac{8}{\rho S}\left(\frac{C_L}{C_D^2}\right)}\left(\sqrt{W_1}-\sqrt{W_2}\right)$$

or   or

$$R = \frac{2EV_1}{\bar{c}g}\left[1-\sqrt{\frac{m_2}{m_1}}\right] = \frac{2EV_1}{\bar{c}g}\left[1-\sqrt{1-\varsigma}\right] \qquad R = \frac{2EV_1}{\bar{c}'}\left[1-\sqrt{\frac{W_2}{W_1}}\right] = \frac{2EV_1}{\bar{c}'}\left[1-\sqrt{1-\varsigma}\right]$$

Range equation for schedule 2 (i.e., flight at constant true airspeed and lift coefficient):[a],[b],[c],[d]

$$R = \frac{EV}{\bar{c}g}\ln\left(\frac{m_1}{m_2}\right) = \frac{EV}{\bar{c}g}\ln\left(\frac{1}{1-\varsigma}\right) \qquad R = \frac{EV}{\bar{c}'}\ln\left(\frac{W_1}{W_2}\right) = \frac{EV}{\bar{c}'}\ln\left(\frac{1}{1-\varsigma}\right)$$

Range equation for schedule 3 (i.e., flight at constant altitude and true airspeed):[a],[b],[c],[e]

$$R = \frac{2E_{max}V}{\bar{c}g}\arctan\left\{\frac{\sqrt{B_3}\,(m_1-m_2)}{B_3+m_1m_2}\right\} \qquad R = \frac{2E_{max}V}{\bar{c}'}\arctan\left\{\frac{g\sqrt{B_3}\,(W_1-W_2)}{g^2B_3+W_1W_2}\right\}$$

or   or

$$R = \frac{2E_{max}V}{g\bar{c}}\arctan\left\{\frac{\varsigma E_1}{2E_{max}\left(1-KC_{L_1}\varsigma E_1\right)}\right\} \qquad R = \frac{2E_{max}V}{\bar{c}'}\arctan\left\{\frac{\varsigma E_1}{2E_{max}\left(1-KC_{L_1}\varsigma E_1\right)}\right\}$$

Notes:

(a) Subscript 1 denotes the start-of-cruise condition, defining the mass (m_1), weight (W_1), true airspeed (V_1), lift coefficient (C_{L_1}), and lift-to-drag ratio (E_1).
(b) Subscript 2 denotes the end-of-cruise condition, defining the mass (m_2) and weight (W_2).
(c) The fuel mass (or weight) ratio is defined as $\varsigma = \dfrac{m_1-m_2}{m_1} = \dfrac{W_1-W_2}{W_1}$
(d) The solution to schedule 2, which is widely known as the Bréguet equation, is sometimes simplified by introducing a range factor (Equation 13.35).
(e) The solution to schedule 3 is based on the parabolic drag polar (Equation 7.14), thus,

$$E_{max} = \frac{1}{2}\sqrt{\frac{1}{KC_{D_0}}} \quad \text{and } B_3 = \left(\frac{C_{D_0}}{K}\right)\left(\frac{\rho V^2 S}{2g}\right)^2$$

Flight at constant C_L is clearly problematic when the altitude has to be held constant (which is required for flight schedule 1) as the true airspeed must reduce during the cruise. In the case of the second flight schedule, C_L can be maintained constant through the use of the cruise-climb technique. The significance of the second flight schedule is that it results in a longer cruise distance—for a given set of starting conditions (i.e., weight, altitude, and fuel load)—than the other two solutions, due to the fuel efficiency of the cruise-climb technique. The practical use of the cruise-climb, however, is restricted by air traffic control (ATC), which does not normally allow commercial aircraft to operate in this fashion.[8] Instead, air traffic is normally required

---

8 The Concorde was permitted to fly cruise-climbs at altitudes above those used by subsonic transport airplanes [6].

to operate at set flight levels (as described in Section 5.4). Consequently, long distance flights resort to periodic step climbs during the cruise to maintain proximity to the optimum altitude for the airplane's weight (see Section 13.6.5). The third flight schedule represents a more realistic scenario as altitude and true airspeed are held constant (which in the stratosphere implies that Mach number is constant). However, the use of the parabolic drag polar limits the accuracy of the resulting expression for airplanes that cruise at high subsonic speeds. Furthermore, commercial flights are usually conducted at either an *economy cruise* speed (defined in Section 13.6.2) or a *long range cruise* speed (defined in Section 13.6.3), and neither of these speeds represents a constant TAS condition as they vary as a function of the airplane's instantaneous (gross) weight.

The analytical solutions given in Table 13.1 are intrinsically useful as they identify the parameters that influence an airplane's cruise performance. The boundary conditions and assumptions associated with the derivations, however, should be kept in mind when using these expressions to estimate the range that an airplane can achieve for a given fuel load. Finally, it should be noted that there is a fundamental shortcoming associated with these classic range expressions: no explicit mention is made of the cruise altitude. Thus, the expressions cannot be used in isolation to establish the optimum conditions that will yield the greatest possible range. Cruise efficiency and optimum cruise altitudes are discussed later, in Section 13.5.

13.4 Numerical Integration

13.4.1 Integrated Range Method

The limitations of the analytical approaches to calculating the range that an airplane can cover for a given fuel load are mentioned in Section 13.3.5. When greater accuracy is needed—for example, when computing distances for actual flight operations—a numerical integration technique must be employed. A major advantage of this method, which is sometimes called the *integrated range* method, is that variations in the fundamental parameters (e.g., speed, altitude, fuel flow or TSFC, and air temperature) during the cruise can be accommodated.

The numerical integration of the specific air range can be undertaken in several ways. The basic approach is to divide the cruise into *n* segments, as illustrated in Figure 13.2. In the method described herein, the stations are defined by the airplane's mass (or weight) and not by distance.

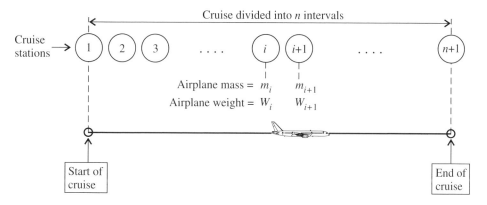

Figure 13.2 Cruise stations (the flight path is shown straight for illustration only).

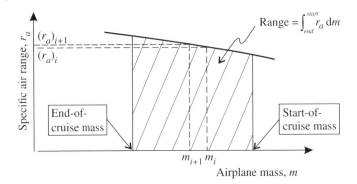

Figure 13.3 Integrated range method (based on airplane mass).

The SAR is now computed at each station; the process depends on the available data. For ease of reference, the various expressions for SAR, derived earlier, are summarized below:

$$r_a = \frac{V}{Q} = \frac{Ma}{Q} = \frac{Ma_0\sqrt{\theta}}{Q} = \frac{a_0\,(ME)}{\left(\dfrac{c}{\sqrt{\theta}}\right)mg} = \frac{H_f\eta_0 E}{mg} \tag{13.28a}$$

or $$r'_a = \frac{V}{Q'} = \frac{Ma}{Q'} = \frac{Ma_0\sqrt{\theta}}{Q'} = \frac{a_0\,(ME)}{\left(\dfrac{c'}{\sqrt{\theta}}\right)W} = \frac{H_f\eta_0 E}{Wg} \tag{13.28b}$$

Figure 13.3 illustrates the integration process on a graph of SAR plotted as a function of airplane mass (alternatively, SAR can be plotted as a function of airplane weight). Note that the range is equal to the area under the graph between the point representing the end of the cruise (i.e., the lowest airplane mass or weight) and the point representing the start of the cruise (i.e., the greatest airplane mass or weight).

The still air range (R) can be calculated by means of a numerical integration of Equation 3.10. For example, if the trapezoidal rule were employed, then the summation would be as follows:

$$R = \sum_{i=1}^{n} \left(\frac{r_{a_i} + r_{a_{i+1}}}{2} \right) (m_i - m_{i+1}) \tag{13.29a}$$

or $$R = \sum_{i=1}^{n} \left(\frac{r'_{a_i} + r'_{a_{i+1}}}{2} \right) (W_i - W_{i+1}) \tag{13.29b}$$

where r_{a_i} is the mass-based SAR and r'_{a_i} is the weight-based SAR at the ith station;
and
m_i is the airplane mass and W_i is the airplane weight at the ith station.

13.4.2 Example of Numerical Computation

An example of an integrated range calculation is presented in Table 13.2 for a mid-size, twin-engine airliner. Five weight intervals were selected for this illustration, with a fuel usage per interval of 2500 lb. The airplane's weight is determined at each station (recorded in column 2). The computational sequence now runs from left to right across the spreadsheet table, with the

Table 13.2 Example of numerical range calculation for a mid-size, twin-engine airliner[a],[b]

Station	W (1000 lb)	W/δ (1000 lb)	C_L	C_D	D/δ (1000 lb)	Q' (lb/h)	r'_a (nm/lb)	Interval	Distance (nm)	Time (h)
1	240.0	1020	0.551	0.0331	61.2	9010	0.0512			
2	237.5	1009	0.546	0.0326	60.3	8878	0.0519	1–2	129	0.280
3	235.0	999	0.540	0.0322	59.6	8763	0.0526	2–3	131	0.283
4	232.5	988	0.534	0.0318	58.8	8653	0.0533	3–4	132	0.287
5	230.0	977	0.528	0.0314	58.1	8543	0.0540	4–5	134	0.291
6	227.5	967	0.523	0.0310	57.3	8433	0.0547	5–6	136	0.295
								Totals:	**662**	**1.436**

Notes:

(a) Conditions: start-of-cruise weight 240 000 lb, altitude 35 000 ft, ISA temperature, and Mach 0.80.
(b) The following information can be ascertained for ISA conditions at this altitude: $\theta = 0.7594$, $\delta = 0.2353$, and $a = 576.4$ kt (see Appendix A).

determination at each station of the following parameters: W/δ, C_L (from Equations 7.1 and 6.12), C_D (from a drag polar—see Figure 7.8), D/δ (from Equations 7.5 and 6.12), Q', and finally r'_a (from Equation 13.28b). The fuel flow, in this example, was based on corrected fuel flow data recorded as a function of corrected thrust (F_N/δ), which in cruise would equate to D/δ. For each interval, the distance traveled is now calculated (column 10) and the results summed to give the total distance (see Equation 13.29b). In this example, the interval flight times have also been determined, based on the computed interval distance and the airplane's speed (column 11), and summed to give the total flight time.

13.4.3 Integrated Range Chart

Figure 13.4 illustrates a convenient representation of an airplane's range data—this is often referred to as an *integrated range chart*. The cruise distance (a) and cruise time (b) corresponding to a pair of airplane weights representing the start-of-cruise and end-of-cruise can be read

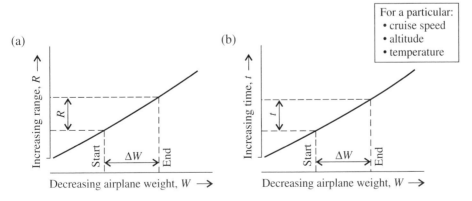

Figure 13.4 Illustration of integrated range chart, which provides a rapid method for estimating an airplane's (a) range and (b) cruise time.

directly off such charts. Alternatively, if the range is known, the end-of-cruise weight can be estimated for a given start-of-cruise weight or vice versa. This information is typically included in an airplane's Flight Crew Operating Manual (FCOM) in a tabulated format.

13.5 Cruise Optimization Based on Aerodynamic Parameters

13.5.1 Introduction to Cruise Optimization

The flight conditions that will give the greatest range for a given fuel quantity are of considerable economic interest to operators. The determination of the best cruise speed and altitude is a subject that has been studied many times—both experimentally and theoretically (see references [1–3, 7–18], for example). In the latter case, it has previously been noted [1, 2] that conclusions reached by different authors were not always in agreement due to the fact that a single unconstrained optimum condition does not exist for airplanes that cruise at subcritical speeds (i.e., below the critical Mach number). This topic will be explored within this section.

One widely used theoretical model is based on the flight condition that yields a maximum value of the ratio $C_L^{1/2}/C_D$. This solution is described in Section 13.5.2, below. The result, however, has limited value for jet transport aircraft, as it only applies to airplanes that cruise at subcritical speeds. Nonetheless, it provides a useful baseline for the more sophisticated techniques that follow.

An airplane will achieve the greatest range possible if, at all times during the cruise, it is operated at the condition that yields the maximum SAR (r_a) evaluated for the airplane's instantaneous weight. It is evident from Equation 13.28 that, for a given airplane weight, $r_a \propto ME/c$. The problem of cruise optimization can thus be expressed as the determination of the optimal combination of altitude and Mach number that will maximize the function ME/c for each value of airplane weight in the cruise.

Considerable insight into the optimum cruise conditions for a jet airplane can be obtained by studying the parameter ME and by assuming that TSFC is invariant over the range of speeds associated with cruise. As the product of Mach number and lift-to-drag ratio (aerodynamic efficiency) depends only on the airplane's aerodynamic characteristics, which can be fully described by the airplane's drag polars, the optimization problem is greatly simplified. This very useful—albeit idealized—approach is described in Sections 13.5.3–13.5.6, below. The discussion on cruise optimization continues in Section 13.6, where the best cruise speed and cruise altitude are established numerically for a reference airplane using all terms contained within the definition of SAR (i.e., including a TSFC or fuel flow term).

13.5.2 Flight Condition for Maximum Range: $(C_L^{1/2}/C_D)_{max}$ Solution

This solution can be derived by considering the cruise altitude to be fixed (i.e., constrained) and by assuming the TSFC to be constant. The objective is to maximize the distance flown per unit of fuel mass (or weight) consumed—that is, to maximize SAR. For steady (i.e., non-accelerating), constant-altitude flight, the SAR (which is given by Equations 13.3 and 13.6) can be written as follows:

$$r_a = \frac{V}{Q} = \frac{V}{cD} = \frac{VE}{cmg} \tag{13.30a}$$

or $$r_a' = \frac{V}{Q'} = \frac{V}{c'D} = \frac{VE}{c'W} \tag{13.30b}$$

Ignoring any variation in the TSFC, it can be deduced that the range will be a maximum when the product of V and E is a maximum. One interpretation of this statement is obtained by expressing the drag in terms of the drag coefficient and then substituting the weight for the lift, that is,

$$VE = \frac{VC_L}{C_D} = \frac{V}{C_D}\left(\frac{2L}{\rho V^2 S}\right) = \frac{\sqrt{2L}}{C_D\sqrt{\rho V^2 S}}\left(\frac{\sqrt{2L}}{\sqrt{\rho S}}\right) = \frac{\sqrt{C_L}}{C_D}\sqrt{\frac{2W}{\rho S}} \tag{13.31}$$

From this equation, it can be concluded that for a given airplane weight (W) and cruising altitude (which will determine ρ in the ISA), VE is a maximum when $C_L^{1/2}/C_D$ is a maximum. It is also evident from Equations 13.30 and 13.31 that V/D is a maximum at this condition. In other words, under these conditions, the cruise speed that will maximize the ratio of true airspeed to drag will give the greatest SAR.

The solution—as described up to this point—is generally applicable as there has been no mention yet of a drag polar. It is when an idealized drag model is used to describe the speed for maximum range—usually as a multiple of the minimum drag speed (V_{md})—that the problems and misunderstandings arise. In Section 7.9.3, a parabolic drag polar was used to show that the flight condition for minimum D/V, when altitude is constant, occurs at $1.32\,V_{md}$ (see Equation 7.64). Furthermore, it was shown that the corresponding lift coefficient is $C_{L_{md}}/\sqrt{3}$ (see Equation 7.65). This lift coefficient results in a maximum value of the ratio $C_L^{1/2}/C_D$.

From Equation 7.64, it would appear that for flight at a fixed altitude, this optimum condition occurs when the airplane is flown at a speed ~32% faster than the minimum drag speed. However—and this is an important observation—the derivation of this result neglected to account for the possible effects of compressibility on the drag polar. The result given by Equation 7.64 is only correct if the parabolic drag polar that was used provides an acceptable approximation to the airplane's drag data over the range of speeds from V_{md} up to the predicted optimum cruise speed. However, this is unlikely to be valid for any jet-powered airplane. In fact, it can be demonstrated, using representative drag polars for jet transport airplanes, that the unconstrained flight condition that yields the optimum cruise speed occurs within the transonic Mach region of drag rise (transonic drag rise is described in Section 7.4.4).

13.5.3 Range Optimization Based on the *ME* (i.e., *ML/D*) Parameter

A more fundamental approach to that presented in Section 13.5.2—in terms of identifying the flight conditions that will yield the greatest range—is possible by considering the SAR as defined by Equation 13.9. Ignoring any variation in the corrected TSFC, it is evident that the condition that will yield the greatest range is that which maximizes the parameter *ME*. Figure 13.5 illustrates the variation of E and *ME*, plotted as a function of Mach number, for a typical cruise lift coefficient, taking into account compressibility effects (the example is for a mid-size, twin-engine airliner and is based on the same drag data used to construct Figures 7.18 and 7.19).

It is apparent that the speed associated with the greatest value of *ME* is significantly faster than that which gives the best E value. The relationship between these speeds depends on the precise way in which compressibility affects the high-speed drag polars. For airplanes that cruise at transonic speeds, the relationship between the two speeds—that is, the best E speed and the best *ME* speed—depends on the airplane type and cannot be represented by a fixed ratio as suggested by the analysis presented in Section 13.5.2.

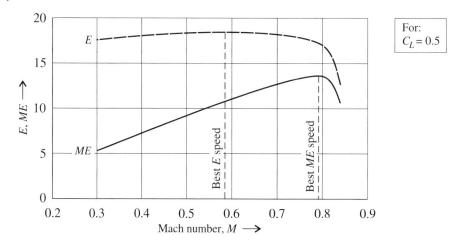

Figure 13.5 Lift-to-drag ratio (*E*) and Mach number times lift-to-drag ratio (*ME*) versus Mach number (example for a mid-size, twin-engine airliner).

During a constant-altitude cruise, the lift coefficient would progressively decrease (as illustrated in Figure 7.5). To represent fully an airplane's aerodynamic characteristics across the range of C_L values associated with a typical cruise requires several graphs of the type shown in Figure 13.5. An alternative approach is illustrated in Figure 13.6 (based on the twin-engine airliner data used to construct Figure 13.5). In this representation, the aerodynamic parameter *ME* is described as a function of two independent variables: lift coefficient and Mach number. The data are presented as isolines (contours) of constant *ME*.

In a similar approach to that adopted when constructing Figure 7.19, partially constrained optima are superimposed on the constant *ME* lines. In Figure 13.6, line I is the locus of

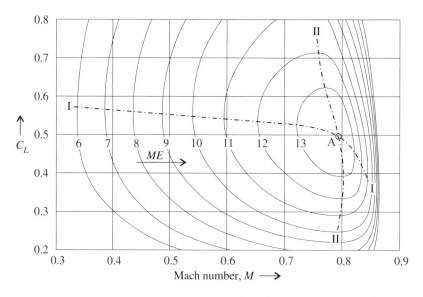

Figure 13.6 Isolines (contours) of constant *ME*, indicated as a function of lift coefficient and Mach number (example for a mid-size, twin-engine airliner).

maximum *ME* for a given Mach number and line II is the locus of maximum *ME* for a given lift coefficient. The point on the chart where the two lines intersect (identified as "A") represents an *unconstrained optimum*. It is seen that the optimum flight condition, which yields the greatest value of *ME* for this airplane, corresponds to a lift coefficient of ~0.50 and a Mach number of ~0.79.

It is worth noting that the existence of an unconstrained optimum value of *ME* on a plane of C_L versus *M* only exists when the drag polars used in its construction display a drag rise due to compressibility effects. For airplanes that cruise at subcritical speeds, the *ME* lines do not form closed loops on charts of this type, and consequently an unconstrained optimum condition does not exist for these aircraft [1, 2]. This is the reason why different solutions to the problem of cruise optimization can be derived for airplanes that cruise at subcritical speeds, depending on whether altitude or speed is constrained. However, jet transport airplanes cruise at speeds within the drag rise region [7]; consequently, unconstrained optima exist (as illustrated in Figure 13.6).

13.5.4 Altitude-Constrained Optimum Based on the *ME* (i.e., *ML/D*) Parameter

Representing the aerodynamic parameter *ME* in the format shown in Figure 13.6 identifies an optimum operating condition for the airplane in terms of C_L and *M*; this is very useful information. What is lacking on the chart, however, is altitude information. This can be introduced using the definition of the lift coefficient (see Equation 7.4), which, for level flight where the airplane's lift is equal to its weight, can be rearranged as follows:

$$C_L M^2 = \frac{2}{\rho_0 a_0^2 S} \left(\frac{W}{\delta} \right) \tag{13.32}$$

Equation 13.32 simply indicates that $C_L M^2 \propto W/\delta$ in cruise. Figure 13.6 has been reproduced as Figure 13.7 with lines of constant $C_L M^2$ superimposed on the *ME* isolines. By interpolation,

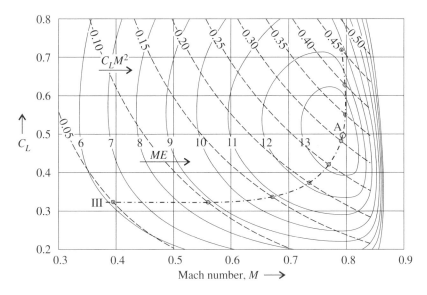

Figure 13.7 Isolines (contours) of constant *ME*, indicated as a function of lift coefficient and Mach number with altitude constraints (example for a mid-size, twin-engine airliner).

it can be seen that the $C_L M^2$ line that would run through the region where *ME* is greatest (i.e., point "A") has a value of ~0.31 for this airplane type. The relative air pressure (δ) can now be determined for a given airplane weight (W) using Equation 13.32, enabling the optimum cruise altitude to be established. It is unequivocally apparent that the ratio W/δ must be kept constant by flying a cruise-climb for the airplane to maintain its optimum efficiency as it gets lighter during cruise.

The penalty of operating at an off-optimum condition can clearly be seen in Figure 13.7. When an altitude constraint is imposed—for example, when the airplane has to operate at an ATC-specified flight level that is above or below the optimum altitude—then it is of interest to know the best cruise speed for the airplane's instantaneous (gross) weight. This condition can be found by tracing along the line of constant $C_L M^2$ determined for the appropriate value of W/δ until the point on the chart is reached where *ME* is greatest. The locus of partially constrained optima, where altitude is set, is indicated on the chart by means of the line identified as III. It is interesting to note that at Mach numbers of less than ~0.58 (for this illustrative example), the line becomes almost straight (i.e., invariant with speed), corresponding to a lift coefficient of ~0.32. The significance of this observation (as noted in reference [1]) is that at Mach numbers less than the drag rise Mach number, the lift coefficient for the partially constrained optima, where altitude is set, is equal to $C_{L_{md}}/\sqrt{3}$. As might be expected, this corresponds to the maximum range solution presented earlier in Section 13.5.2.

13.5.5 Thrust-Constrained Optimum Based on the *ME* (i.e., *ML/D*) Parameter

All jet engines suffer a lapse in thrust with increasing altitude (see Section 8.6). Consequently, the *thrust available*—that is, the thrust produced by all operating engines at the appropriate engine rating (i.e., maximum cruise thrust, as used in normal flight situations, or maximum continuous thrust, as used in emergency situations)—can limit an airplane's cruise speed under certain flight conditions. To maintain constant speed in cruise, at a particular flight condition (characterized by airplane weight, cruising altitude, and air temperature), the thrust available must match the *thrust required*, which is equal to the airplane's drag in cruise.

It is instructive to illustrate graphically how the airplane's available thrust, in some cases, can impose a constraint on the optimum flight condition established in Section 13.5.3. As the airplane is in a state of equilibrium in cruise (and thrust is equal to drag), the definition of the airplane's drag coefficient (C_D) can be used to establish the following relationship:

$$C_D M^2 = \frac{2}{\rho_0 a_0^2 S} \left(\frac{F_N}{\delta} \right) \tag{13.33}$$

where F_N is the thrust required (typical units: N, lb).

The thrust required in cruise, when expressed in corrected form (i.e., F_N/δ), is thus directly proportional to the parameter $C_D M^2$. A similar approach to that adopted in Section 13.5.4 can be used to interpret this statement. This time, Figure 13.6 has been reproduced with lines of constant $C_D M^2$ superimposed on the *ME* isolines—this is presented as Figure 13.8. For an airplane to be able to sustain constant-speed level flight, at a particular combination of C_L and M, the thrust available must exceed the thrust required, which is indicated on the chart by the appropriate $C_D M^2$ value. For example, if the available corrected thrust equates to a $C_D M^2$ value that exceeds ~0.019, then the unconstrained optimum (i.e., point "A") could be achieved. In other words, the available engine thrust would not restrict the airplane from operating at the point of greatest efficiency. However, if the thrust were much lower (e.g., when the airplane has to cruise with one engine inoperative), then the most efficient operating condition would

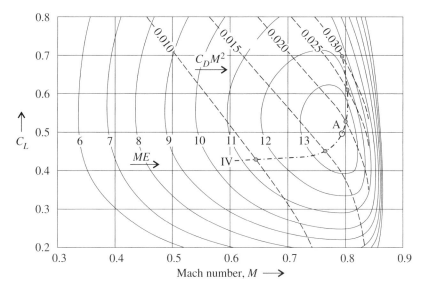

Figure 13.8 Isolines (contours) of constant *ME*, indicated as a function of lift coefficient and Mach number with thrust constraints (example for a mid-size, twin-engine airliner).

correspond to the intersection of the appropriate $C_D M^2$ line and the thrust-constrained optima line identified on the chart as IV.

The thrust produced by the engines, in certain scenarios, can thus limit the airplane to a lower altitude than that which corresponds to the unconstrained optimum. The maximum altitude that an airplane can operate at is known as a ceiling—the procedures for establishing maximum operational altitudes must consider several factors (which are discussed later in Sections 17.3 and 20.4). One of the key requirements is that the thrust available—at the appropriate thrust rating—is sufficient for the airplane to have the capability to climb at a defined rate (e.g., 300 ft/min) when operating at the particular altitude (ceiling) and selected cruise speed. This requirement means that in actual flight operations, the maximum available thrust cannot just equal the thrust required to maintain a constant speed in cruise, but must *exceed* this value to ensure that the airplane has the potential to climb.

13.5.6 Limitations of Optima Based on the *ME* (i.e., *ML/D*) Parameter

In Sections 13.5.3–13.5.5, the assumption of constant TSFC was made, and this facilitated optimum cruise conditions to be established based on the *ME* parameter, which is a function of the airplane's drag polars. It must, however, be remembered that at any particular airplane weight, SAR will be a maximum when *ME/c* is a maximum. At a given altitude, the TSFC (c), in cruise, tends to increase with increasing Mach number (as described in Section 8.7). The manner in which TSFC increases depends on the engine type and on the flight conditions (e.g., altitude, temperature, and net thrust).

Results obtained using the techniques presented in Sections 13.5.3–13.5.5 should thus be regarded as a good approximation of the actual performance characteristics of an airplane in cruise. The true unconstrained optimum, based on SAR, will occur at a slightly lower Mach number (approximately Mach 0.01–0.02 lower for the illustrative airplane data used in this chapter) than that predicted with the assumption of constant TSFC, as done in Section 13.5.3.

13.6 Best Cruise Speeds and Cruise Altitudes

13.6.1 Maximum Range Cruise Speed

The maximum range will be achieved by operating the airplane, throughout the cruise, at the condition that yields the maximum SAR. The flight speed that results in the greatest SAR is known as the *maximum (max.) range cruise* (MRC) speed. The MRC speed for an airplane depends on its gross weight, the cruising altitude, and the ambient air temperature (WAT). Operating at the MRC speed minimizes the fuel needed to cover a given cruise distance (in still air). When operating at a set altitude, the MRC speed will decrease a little during the cruise as the airplane gets lighter—this is illustrated in Figure 13.9.

13.6.2 Economy Cruise Speed

In practice, airlines normally fly a little faster than the MRC speed, sacrificing a small increase in fuel usage, to obtain a shorter cruise time. This makes sense as the trip cost can be reduced. The *direct operating cost* (see Section 18.2) associated with a particular flight can be expressed as the sum of three terms: a fixed cost, fuel cost, and time-dependent cost. Time-dependent costs—which can include all or part of the crew, maintenance, and lease costs—are reduced by flying faster. The cruise speed that would result in the lowest operating cost is thus faster than the MRC speed (which is the speed that would result in the lowest fuel cost).

The cruise speed that would result in the lowest operating cost is called the *economy cruise speed*, which is usually abbreviated as the ECON speed. The calculation of the ECON speed is based on the airline's individual cost structure, which is quantified in terms of a *cost index* (see Section 18.3) determined for the particular airplane type and route.

13.6.3 Long Range Cruise Speed

A simplified approach to that described in Section 13.6.2, which is widely used and does not require knowledge of the airline's costs, is based on the *long range cruise* (LRC) speed. The LRC

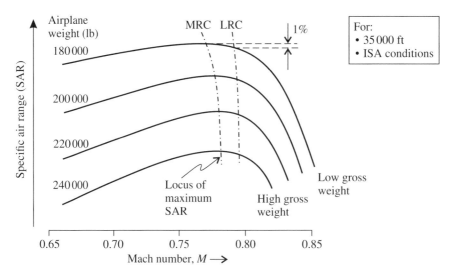

Figure 13.9 Maximum range cruise (MRC) and long range cruise (LRC) speeds (example for a mid-size, twin-engine airliner).

speed is determined by allowing a 1% reduction from the peak SAR and selecting the faster of the two possible speeds [19], as illustrated in Figure 13.9.

As LRC speeds depend solely on the airplane's performance characteristics—that is, without considering costs associated with an individual operator—LRC speeds can be determined by the airplane manufacturer. In fact, this is done as a matter of routine. LRC data are published in such documents as the Flight Crew Operating Manual (FCOM)—for example: Mach number, indicated airspeed, true airspeed, engine setting, and fuel flow would typically be tabulated for selected values of airplane weight, altitude, and temperature (WAT).

13.6.4 Comparison of Cruise Speeds

Figure 13.10 illustrates how the fixed cost, time cost, fuel cost, and total cost vary with Mach number in cruise. The Mach numbers for the three reference cruise speeds (i.e., MRC, LRC, and ECON) are indicated as M_{MRC}, M_{LRC}, and M_{ECON}, respectively. M_{MRC} corresponds to the minimum fuel cost and M_{ECON} to the minimum total cost. M_{LRC} is typically 2–5% faster than M_{MRC} for jet transport airplanes [20, 21]. M_{ECON} is always faster than M_{MRC} when time-dependent costs are included, and, for most commercial passenger operations, M_{ECON} is a little slower than M_{LRC} (as illustrated in Figure 13.10). Note that there is little value in flying slower than M_{MRC} as both fuel-dependent and time-dependent costs increase with reducing speed. The airplane's certified maximum operating Mach number (M_{MO}) sets an upper limit to the allowable cruise speed.

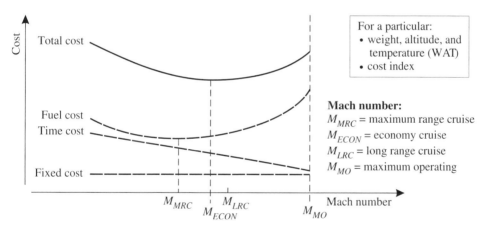

Figure 13.10 Characteristic (reference) Mach numbers for cruise.

13.6.5 Optimum Altitude

An airplane's *optimum altitude* is that altitude which would result in the greatest SAR for a particular airplane weight, air temperature, and Mach number. Optimum altitude data are determined by flight testing and published in the airplane's FCOM for selected speeds (e.g., MRC, LRC) and temperatures (e.g., ISA, ISA + 10 °C, ISA + 20 °C). A common feature of such data—for many airplane types—is that the optimum altitude for a given cruise Mach number tends to increase in an almost linear fashion with reducing airplane weight.

The most efficient cruise—that is, a cruise that will consume the least fuel—corresponds to the airplane flying a cruise-climb along the locus of points representing the altitude optima. Due to flight level restrictions (imposed by ATC), jet transport airplanes routinely fly a stepped approximation of the cruise-climb, climbing to higher cruise altitudes as fuel is burned. This

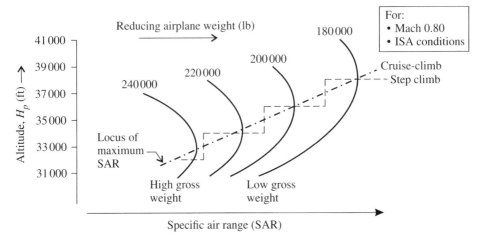

Figure 13.11 Step-climb approximation to the cruise-climb (example for a mid-size, twin-engine airliner).

is illustrated in Figure 13.11 (practical implications of step climbs are discussed later, in Section 17.3.4).

It is apparent from the earlier discussion that at a particular airplane weight there is an optimum combination of altitude and Mach number that results in the greatest SAR. This is illustrated in Figure 13.12—isolines of equal SAR have been constructed for varying cruise altitude and Mach number. Line I is the locus of maximum SAR for a given Mach number and line III is the locus of maximum SAR for a given altitude (i.e., the MRC speed). For this airplane (a medium-range, twin-engine airliner), at the arbitrarily selected weight of 210 000 lb, the greatest SAR corresponds to an altitude of 34 700 ft and a Mach number of 0.777 (i.e., the point where the two partially constrained optima represented by lines I and III intersect). The impact on the SAR of operating at a higher or lower altitude or a faster or slower speed can be seen in

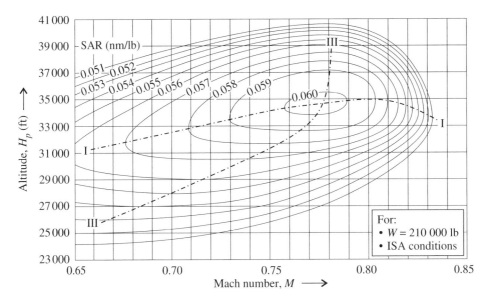

Figure 13.12 Isolines (contours) of constant specific air range, indicated as a function of altitude and Mach number (example for a mid-size, twin-engine airliner).

this illustration. For example, the SAR penalty at Mach 0.79 for flying 2000 ft below the optimum altitude of 35 000 ft is ~1.7%. Note that the relationship is not linear: the SAR penalty for operating 4000 ft below the optimum altitude is ~6.7%.

13.7 Further Details on the Use of the Bréguet Range Equation

13.7.1 General Formulations

In Section 13.3.3, the range integral (Equation 13.5) was evaluated with the assumption that TAS, C_L, and TSFC all remain constant during the cruise. The resulting expression—the Bréguet equation—expressed range as a function of TAS. Alternatively, cruise speed can be expressed in terms of Mach number. From Equation 13.19, it follows that

$$R = \int_{m_2}^{m_1} r_a \, dm = \frac{VE}{\bar{c}g} \ln\left(\frac{m_1}{m_2}\right) = \frac{a_0 \sqrt{\theta}}{\bar{c}g} \left(\frac{ML}{D}\right) \ln\left(\frac{m_1}{m_2}\right) \tag{13.34a}$$

or

$$R = \int_{W_2}^{W_1} r'_a \, dW = \frac{VE}{\bar{c}'} \ln\left(\frac{W_1}{W_2}\right) = \frac{a_0 \sqrt{\theta}}{\bar{c}'} \left(\frac{ML}{D}\right) \ln\left(\frac{W_1}{W_2}\right) \tag{13.34b}$$

Any consistent set of units can be used to calculate the range using these equations. For example, if V has units of knots and c' units of lb lb^{-1} h^{-1}, then the range predicted by Equation 13.34b will be in nautical miles.

For certain applications, it is convenient to make use of a *range factor* (F_R), which is equal to the product of specific air range r_a and airplane mass or, alternatively, the product of specific air range r'_a and airplane weight. The range factor can be written as a function of TSFC or, alternatively, overall engine efficiency and fuel net heating value (see Equations 13.12 and 13.6):

$$F_R = \frac{VE}{\bar{c}g} = \frac{VE}{\bar{c}'} = \frac{H_f \eta_0 E}{g} \tag{13.35}$$

From Equations 13.34 and 13.35, the range can be expressed as follows:

$$R = F_R \ln\left(\frac{m_1}{m_2}\right) = F_R \ln\left(\frac{W_1}{W_2}\right) \tag{13.36}$$

where F_R is the range factor (typical units: km, nm).

The use of the range factor is preferred by certain authors and organizations (e.g., as illustrated in references [7, 22]) for particular performance applications. Boeing [7] indicates that the range factor is "almost constant for maximum range or long range cruise in a climbing cruise operation." For other types of cruise, an average value of the range factor can be used. Establishing a range factor[9] for a particular airplane type simplifies repetitive range calculations.

The range factor F_R is linked to the abovementioned *range parameter*,[10] defined by Equation 13.11 in Section 13.2.5, through the fuel net heating value (and gravitational acceleration).

---

9 The range factor (F_R) has also been used as a metric to quantify airplane cruise performance as it combines key aerodynamic and propulsion efficiencies at a particular cruise speed. This has been done by several authors—e.g., Martinez-Val, *et al.* [23] used estimated range factors (computed from published airplane data) to describe, from a performance perspective, the evolution of jet airliners over time.

10 Caution: The term *range parameter* can be ambiguous—it is used by certain authors to describe what is herein called the range factor (e.g., references [24, 25]; the latter report, however, omits *g* in the definition provided and is thus dimensionally inconsistent).

An important distinction is that the range parameter r_p is dimensionless, whereas the range factor F_R has units of distance.

An alternative approach to the constant-TAS, constant-C_L solution to the range integral described in Section 13.3.3 is based on the assumption that the range parameter r_p remains constant during the cruise. From Equation 13.14, it can be seen that the range equation can be expressed as follows:

$$R = \int_{m_2}^{m_1} \frac{H_f r_p}{mg} \, dm = \frac{H_f r_p}{g} \int_{m_2}^{m_1} \frac{1}{m} \, dm = \frac{H_f \eta_0 E}{g} \ln\left(\frac{m_1}{m_2}\right) \tag{13.37a}$$

$$\text{or} \quad R = \int_{W_2}^{W_1} \frac{H_f r_p}{Wg} \, dW = \frac{H_f r_p}{g} \int_{W_2}^{W_1} \frac{1}{W} \, dW = \frac{H_f \eta_0 E}{g} \ln\left(\frac{W_1}{W_2}\right) \tag{13.37b}$$

H_f is the net heating value of the fuel (i.e., net energy content per unit mass). Assuming a typical value of H_f of 43.28 MJ/kg (see Table 22.3) and the standard value of g, it follows that $H_f/g = 4413\,\text{km} = 2383\,\text{nm}$. Equation 13.37 thus provides a rapid, alternative means to estimate an airplane's range based on the overall engine efficiency (η_0) and lift-to-drag ratio (E). For example, if η_0 equals 0.34 and E equals 20, then F_R equals 30 010 km (16 200 nm)—from which the range can be computed for given start- and end-of-cruise airplane masses (or weights).

13.7.2 Constant-Altitude Cruise and the Use of Mean Values

The Bréguet range equation is widely used because of its simplicity and also because the errors introduced by the assumptions are often not significant. The flight schedule that yielded the Bréguet solution considered C_L to be constant; this can be accomplished by flying a cruise-climb. There is, however, another interpretation of Equation 13.19 (or Equation 13.34): the lift-to-drag ratio represents a mean value, which can be evaluated for any flight profile, including a constant-altitude cruise. For this scenario, the mean lift-to-drag ratio ($\bar{E}$) can be defined as follows:

$$\bar{E} = \frac{1}{m_{f_{used}}} \int_{m_2}^{m_1} E \, dm \quad \text{or} \quad \bar{E} = \frac{1}{W_{f_{used}}} \int_{W_2}^{W_1} E \, dW \tag{13.38}$$

where $m_{f_{used}}$ (or $W_{f_{used}}$) is the fuel mass (or weight) consumed and is given by the difference between the start and end conditions of the cruise, that is,

$$m_{f_{used}} = m_1 - m_2 \quad \text{or} \quad W_{f_{used}} = W_1 - W_2 \tag{13.39}$$

Similarly, the mean thrust specific fuel consumption ($\bar{c}$), which has been used in the analytical solutions presented in Section 13.3, can be defined as follows:

$$\bar{c} = \frac{1}{m_{f_{used}}} \int_{m_2}^{m_1} c \, dm \quad \text{or} \quad \bar{c}' = \frac{1}{W_{f_{used}}} \int_{W_2}^{W_1} c' \, dW \tag{13.40}$$

With this notation, Equation 13.19 can be rewritten as

$$R = \frac{V\bar{E}}{\bar{c}g} \ln\left(\frac{m_1}{m_2}\right) \tag{13.41a}$$

$$\text{or} \quad R = \frac{V\bar{E}}{\bar{c}'} \ln\left(\frac{W_1}{W_2}\right) \tag{13.41b}$$

A computer model, based on representative performance data (for a mid-size, twin-engine airliner), was used to assess the variation of the TSFC and lift-to-drag ratio during a typical constant-altitude, constant Mach number cruise. The results are shown in Figure 13.13. It can

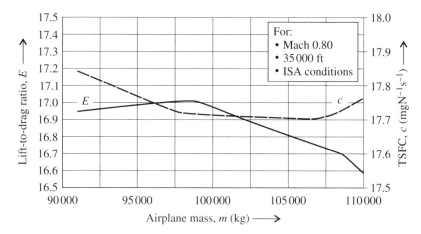

Figure 13.13 Variation of lift-to-drag ratio (E) and TSFC (c) during a constant-altitude, constant-speed cruise for a mid-size, twin-engine airliner.

be seen that the changes are non-linear with respect to the change in airplane mass, but that the variations are relatively small. To investigate the accuracy of the Bréguet equation, compared to the range determined using a numerical integration technique (described in Section 13.4) for a constant-altitude cruise, the airplane's weight, lift-to-drag ratio, and TSFC were determined at the start and at the end of the cruise and at 19 intermediate points. Using E and TSFC averages—based on the 21 data points—Equation 13.41 yielded a distance that was within 0.1% of that given by the integrated range method. This indicates that the Bréguet equation can provide accurate results even when the airplane is not flying a cruise-climb, provided that representative mean values of E and TSFC are used.

13.7.3   Fuel Required for Specified Range

Equation 13.41 enables the range (R) to be calculated for a given fuel mass (or weight) and a specified start-of-cruise mass (or weight) or end-of-cruise mass (or weight). For certain analyses—for example, when considering the impact of a change to an airplane's operating empty weight (OEW) on the trip fuel—it is useful to express the required cruise fuel in terms of the cruise distance, not the other way around. This relationship can be written in terms of either the start condition (subscript 1) or the end condition (subscript 2); both formulations are presented below.

Equation 13.41 can be rewritten as follows:

$$\frac{R\bar{c}g}{V\bar{E}} = \ln\left(\frac{m_1}{m_2}\right) \quad \text{and} \quad \frac{R\bar{c}'}{V\bar{E}} = \ln\left(\frac{W_1}{W_2}\right)$$

hence $\quad \dfrac{m_1}{m_2} = e^{\left(\frac{R\bar{c}g}{V\bar{E}}\right)} = \dfrac{W_1}{W_2} = e^{\left(\frac{R\bar{c}'}{V\bar{E}}\right)}$ $\hfill (13.42)$

The fuel mass consumed in the cruise ($m_{f_{used}}$) or fuel weight consumed ($W_{f_{used}}$) can be deduced from Equations 13.42 and 13.39:

Thus $\quad m_{f_{used}} = \left(\dfrac{m_1}{m_2} - 1\right)m_2 = \left(e^{\left(\frac{R\bar{c}g}{V\bar{E}}\right)} - 1\right)m_2$ $\hfill (13.43a)$

or $\quad W_{f_{used}} = \left(\dfrac{W_1}{W_2} - 1\right)W_2 = \left(e^{\left(\frac{R\bar{c}'}{V\bar{E}}\right)} - 1\right)W_2$ $\hfill (13.43b)$

In Equation 13.43, the mass (or weight) of the fuel is expressed as a function of the airplane's end-of-cruise mass (or weight). A similar expression in terms of the start-of-cruise mass (or weight) can also be derived from Equations 13.42 and 13.39, that is,

$$m_{f_{used}} = \left(1 - \frac{m_2}{m_1}\right) m_1 = \left(1 - e^{-\left(\frac{R\bar{c}g}{V\bar{E}}\right)}\right) m_1 \qquad (13.44a)$$

or $\quad W_{f_{used}} = \left(1 - \frac{W_2}{W_1}\right) W_1 = \left(1 - e^{-\left(\frac{R\bar{c}'}{V\bar{E}}\right)}\right) W_1 \qquad (13.44b)$

Note that the exponent in Equations 13.43 and 13.44 can be simplified by introducing the range factor (F_R), as defined by Equation 13.35.

13.8 Influence of Wind on Cruise Performance

13.8.1 Specific Ground Range

The *still air range* that an airplane can achieve for a given fuel load (under specified flight conditions) is a critically important performance metric for any airplane. For planning flight operations, however, what is of real interest is the *ground range* (R_g). The difference between the two measures is due to the influence of wind. The *specific ground range* (SGR), which is given the symbol r_g, is defined as the ground distance covered (s) per unit of fuel mass (or weight) consumed, that is,

$$r_g = -\frac{ds}{dm_f} \qquad (13.45a)$$

or $\quad r_g' = -\frac{ds}{dW_f} \qquad (13.45b)$

In a similar approach to that described in Section 13.2.2, the specific ground range can be expressed in terms of the airplane's ground speed (V_G) and fuel flow, that is,

$$r_g = -\frac{ds}{dm_f} = \frac{\left(\dfrac{ds}{dt}\right)}{\left(-\dfrac{dm_f}{dt}\right)} = \frac{V_G}{Q} \qquad (13.46a)$$

or $\quad r_g' = -\frac{ds}{dW_f} = \dfrac{\left(\dfrac{ds}{dt}\right)}{\left(-\dfrac{dW_f}{dt}\right)} = \frac{V_G}{Q'} \qquad (13.46b)$

As described in Section 6.3.2, the ground speed can be determined from a vector summation of the true airspeed (V) and the wind speed (V_w). Equation 6.4, which is a valid approximation when the drift angle is small (see Figure 6.2), is repeated below:

$$V_G \cong V + V_w \cos \upsilon \quad \text{(Equation 6.4)}$$

where υ is the wind direction with respect to the track (see Figure 6.2).

Hence $r_g \cong \dfrac{V + V_w \cos \upsilon}{Q} = r_a + \dfrac{V_w \cos \upsilon}{Q} = r_a \left(1 + \dfrac{V_w \cos \upsilon}{V}\right) \qquad (13.47a)$

or $\quad r_g' \cong \dfrac{V + V_w \cos \upsilon}{Q'} = r_a' + \dfrac{V_w \cos \upsilon}{Q'} = r_a' \left(1 + \dfrac{V_w \cos \upsilon}{V}\right) \qquad (13.47b)$

Note that $v = 0°$ means that there is no crosswind component (i.e., there is a straight tailwind); similarly, when $v = 180°$ there is a straight headwind.

13.8.2 Ground Range

If the drift angle is small and the airplane's speed and wind component $V_w \cos v$ are constant during the cruise (or cruise segment), then it follows that the ground range (R_g) is given by

$$R_g = \int_{start}^{end} ds = -\int_{m_1}^{m_2} r_g \, dm = \left(1 + \frac{V_w \cos v}{V}\right) \int_{m_2}^{m_1} r_a \, dm \tag{13.48a}$$

or
$$R_g = \int_{start}^{end} ds = -\int_{W_1}^{W_2} r_g' \, dW = \left(1 + \frac{V_w \cos v}{V}\right) \int_{W_2}^{W_1} r_a' \, dW \tag{13.48b}$$

Thus $R_g = \left(1 + \dfrac{V_w}{V} \cos v\right) R$ (13.49)

where R is the still air range (described in Sections 13.2–13.4).

In an operational context, the requirement is to determine the air distance to be flown for a given ground distance. By only considering a straight tailwind (i.e., $v = 0°$) or straight headwind (i.e., $v = 180°$), Equation 13.49 can be written as follows:

$$\frac{R}{R_g} = \frac{V}{V + V_w} \tag{13.50}$$

where V_w is positive for a tailwind and negative for a headwind.

Figure 13.14 illustrates the effect of a tailwind/headwind on the ratio of air distance (R) to ground distance (R_g), computed using Equation 13.50.

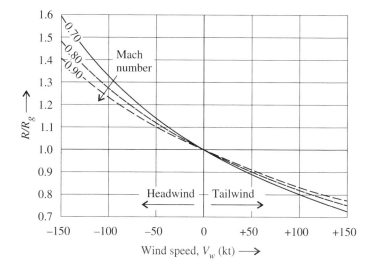

Figure 13.14 Effect of wind: air distance to ground distance ratio versus wind speed.

13.8.3   Maximum Ground Range Speed

Optimizing cruise such that the maximum ground range can be achieved for a given fuel quantity is of significant interest to operators; descriptions of how wind influences cruise performance is given in references [7, 18, 19, 26, 27], for example.

The optimum (reference) cruise speeds presented earlier in Section 13.6 (i.e., MRC, LRC, and ECON speeds) did not consider the influence of wind. It is evident from Equation 13.47 that the condition for maximum specific ground range will be different from that established for maximum specific air range. For constant-altitude, constant-speed flight, it can be deduced from Equations 13.46 and 8.30 that

$$r_g = \frac{V_G}{cF_N} = \frac{V_G}{cD} \tag{13.51a}$$

or $$r'_g = \frac{V_G}{c'F_N} = \frac{V_G}{c'D} \tag{13.51b}$$

The conditions that will maximize SGR will now be investigated. If the TSFC is considered constant, this is equivalent to minimizing the function D/V_G. A graphical explanation as to how winds influence this function is presented in Figure 13.15. The airplane's drag (D) is shown as a function of TAS in Figure 13.15a, and the best D/V speed identified. In Figure 13.15b, three Cartesian coordinate systems (axes)—identified as *tailwind*, *nil wind*, and *headwind*—have been constructed. As true airspeed is equal to ground speed in nil wind conditions, no

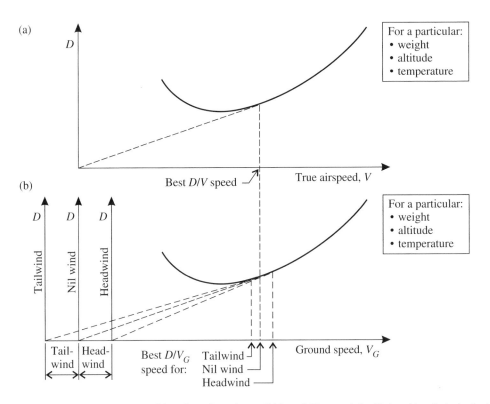

Figure 13.15 (a) Best D/V speed for nil wind condition; (b) Best D/V_G speeds for flight with tailwind, nil wind, and headwind.

error is introduced by labeling the abscissa *ground speed*. The origin of the tailwind axes is displaced to the left of the nil wind origin by the magnitude of the tailwind considered; similarly, the origin of the headwind axes is displaced to the right. Note that the same drag curve applies to all three sets of axes (as drag is a function of airspeed). Lines are now constructed tangent to the drag curve through the three origins to establish the conditions that will yield the greatest D/V_G ratio.

It is evident, from the illustration presented in Figure 13.15, that winds alter the TAS at which an airplane must fly to achieve the greatest specific ground range. To operate at the maximum ground range cruise speed, the flight crew needs to slow down (i.e., reduce the TAS) when there is a strong tailwind and speed up when there is a strong headwind. Although this deduction was made with the assumption of constant TSFC, the same conclusion can be reached using numerical techniques that incorporate TSFC changes with Mach number.

13.8.4   Optimum Cruise Speeds Based on Specific Ground Range

In Section 13.8.3, it was argued that the MRC speed, when based on specific ground range, increases in the presence of a headwind and reduces when there is a tailwind. This is illustrated in Figure 13.16 (the data for the nil wind condition are the same as those presented in Figure 13.9 at the selected airplane weight). The LRC and ECON speeds are influenced by winds in the same way.

13.8.5   Wind Gradients and Wind–Altitude Trades

Altitude optima that will yield the greatest SAR are discussed in Section 13.6.5. Provided that the wind speed and direction are the same at all cruise altitudes, the optimum altitude data determined for nil wind conditions (i.e., based on SAR) also apply to day-to-day operations when winds are present. Strong *wind gradients*, however, often occur at cruise altitudes, particularly in the vicinity of a jet stream (see Section 4.5.6). Wind speeds in such conditions can be significantly different at different flight levels.

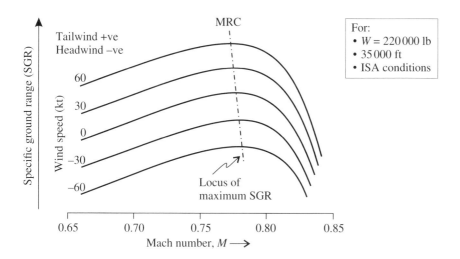

Figure 13.16 Influence of tailwind or headwind on specific ground range (SGR) and maximum range cruise (MRC) speed (example for a mid-size, twin-engine airliner).

It is frequently the case (particularly on eastbound or westbound routes at the mid-latitudes) that a favorable wind condition—that is, a reduced headwind or increased tailwind component—will exist at a lower or higher flight level compared to the still air optimum altitude. It is evident from Figure 13.12 that the SAR is negatively impacted when operating at a flight level below or above the optimum altitude (for a particular airplane weight). However, when there is a favorable wind gradient and the difference in wind speeds between the still air optimum altitude and the selected lower/higher flight level is sufficiently large, it can be economically beneficial to operate at the lower/higher flight level, as this will increase the SGR. If the favorable wind condition occurs at a flight level that is above the still air optimum altitude, then two further considerations need to be taken into account by the flight crew: (1) the higher flight level cannot exceed the airplane's operational ceiling (see Sections 17.3 and 20.4), and (2) the fuel saving must be sufficient to compensate for the additional fuel required to climb to the higher flight level.

Charts for conducting wind–altitude trades are published in the airplane's FCOM; these charts provide flight crews with data to select the flight level that will result in the best SGR. The Airbus A320 has been selected as an example to illustrate this aspect of cruise performance: if a favorable wind difference of greater than 23 kt exists between FL 350 and FL 310, then the airplane at a weight of 150 000 lb (68 000 kg) cruising at Mach 0.78 would consume less fuel (per unit of ground distance) when cruising at the lower altitude of 31 000 ft compared to cruising at the optimum flight level of 35 000 ft [19].

References

1 Torenbeek, E., "Cruise performance and range prediction reconsidered," *Progress in Aerospace Sciences*, Vol. 33, Iss. 5–6, pp. 285–321, 1997.
2 Torenbeek, E., *Advanced aircraft design: Conceptual design, analysis and optimization of subsonic civil airplanes*, John Wiley & Sons, Chichester, UK, 2013.
3 Hale, F.J., *Introduction to aircraft performance, selection, and design*, John Wiley & Sons, New York, NY, 1984.
4 Coffin, J.G., "A study of airplane ranges and useful loads," NACA Report No. 69, National Advisory Committee for Aeronautics, Washington, DC, 1920.
5 Cavcar, M., "Bréguet range equation?," *Journal of Aircraft*, Vol. 43, Iss. 5, pp. 1542–1544, 2006.
6 Orlebar, C., *The Concorde story*, 6th ed., Osprey Publishing, Wellingborough, UK, 2004.
7 Boeing, "Jet transport performance methods," D6-1420, The Boeing Company, Seattle, WA, May 1989.
8 Page, R.K., "Performance calculation for jet-propelled aircraft," *Journal of the Royal Aeronautical Society*, Vol. 51, pp. 440–450, 1947.
9 Jonas, J., "Jet airplane range considerations," *Journal of the Aeronautical Sciences*, Vol. 14, Iss. 2, pp. 124–128, 1947.
10 Miele, A., *Flight mechanics, Vol. 1: Theory of flight paths*, Addison-Wesley Pub., London, UK, 1962.
11 Peckham, D.H., "Range performance in cruising flight," TR 73164, Royal Aircraft Establishment, Farnborough, UK, Mar. 1974.
12 ESDU, "Introduction to estimation of range and endurance," Data item 73018, Amdt. A, IHS ESDU, 133 Houndsditch, London, UK, Oct. 1980.
13 ESDU, "Approximate methods for estimation of cruise range and endurance: Aeroplanes with turbo-jet and turbo-fan engines," Data item 73019, Amdt. C, IHS ESDU, 133 Houndsditch, London, UK, May 1982.

14 Torenbeek, E. and Wittenberg, H., "Generalized maximum specific range performance," *Journal of Aircraft*, Vol. 20, Iss. 7, pp. 617–622, 1983.

15 Martinez-Val, R. and Perez, E., "Optimum cruise lift coefficient in initial design of jet aircraft," *Journal of Aircraft*, Vol. 29, Iss. 4, pp. 712–714, 1992.

16 Miller, L.E., "Optimal cruise performance," *Journal of Aircraft*, Vol. 30, Iss. 3, pp. 403–405, 1993.

17 Cavcar, A. and Cavcar, M., "Approximate solutions of range for constant altitude–constant high subsonic speed flight of transport aircraft," *Aerospace Science and Technology*, Vol. 8, Iss. 6, pp. 557–567, 2004.

18 Blake, W. and Performance Training Group, "Jet transport performance methods," D6-1420, Flight Operations Engineering, Boeing Commercial Airplanes, Seattle, WA, Mar. 2009.

19 Airbus, "Getting to grips with aircraft performance," Flight Operations Support and Line Assistance, Airbus S.A.S., Blagnac, France, Jan. 2002.

20 Roberson, W., Root, R., and Adams, D., "Fuel conservation strategies: Cruise flight," *Aero*, The Boeing Company, Seattle, WA, Vol. Q 04, pp. 23–27, 2007.

21 Seto, L., "Cost index," Flight Operations Engineering, Boeing Commercial Airplanes, Seattle, WA, Sept. 2009.

22 Gallagher, G.L., Higgins, L.B., Khinoo, L.A., and Pierce, P.W., "Fixed wing performance: Flight test manual," USNTPS-FTM-No. 108, United States Naval Test Pilot School, Patuxent River, MD, Sept. 30, 1992.

23 Martinez-Val, R., Palacin, J.F., and Perez, E., "The evolution of jet airliners explained through the range equation," *IMechE Part G: Journal of Aerospace Engineering*, Vol. 222, Iss. 6, pp. 915–919, 2008.

24 Green, J.E. and Jupp, J.A., "CAEP/9-agreed certification requirement for the aeroplane CO_2 emissions standard: A comment on ICAO cir. 337," *Aeronautical Journal*, Vol. 120, Iss. 1226, pp. 693–723, 2016.

25 Greener by Design Science and Technology Sub-Group, "Air travel – greener by design: Mitigating the environmental impact of aviation: Opportunities and priorities," Green, J.E. (Ed.), The Royal Aeronautical Society, London, UK, July 2005.

26 Hale, F.J. and Steiger, A.R., "Effects of wind on aircraft cruise performance," *Journal of Aircraft*, Vol. 16, Iss. 6, pp. 382–387, 1979.

27 Rivas, D., Lopez-Garcia, O., Esteban, S., and Gallo, E., "An analysis of maximum range cruise including wind effects," *Aerospace Science and Technology*, Vol. 14, Iss. 1, pp. 38–48, 2010.

14

Holding Flight and Endurance Performance

14.1 Introduction

In certain flight situations, it is necessary for an airplane to remain in the air for as long as possible on a given fuel load—for example, when on surveillance or other reconnaissance duties. It is desirable, in such circumstances, that the airplane operates at the speed for lowest fuel consumption per unit of time. The greatest time that an airplane can remain aloft—for a given fuel load—is known as its *endurance* and the corresponding speed is known as the *endurance speed.*

A related flight condition is a *hold*. The purpose of a holding maneuver is to keep the airplane within a specified airspace for a period of time before continuing with the mission. The most frequent reason for holding is traffic congestion at the destination airport, when aircraft are delayed before receiving air traffic control (ATC) clearance to continue with the descent to land. Such delays are usually caused by bad weather, but they can also occur when traffic is rerouted to an alternate airport or when a runway is closed. Again, it is desirable that the airplane operates at a speed that reduces the fuel consumed per unit of time for the operating conditions (i.e., airplane weight, altitude, and air temperature). In both scenarios, the distance covered is of no consequence. The optimum speeds for surveillance missions or holding will be shown to be much slower than the optimum speeds for cruise (for comparable conditions).

The holding/endurance equations, which are derived in Sections 14.2 and 14.3 for various boundary conditions (or schedules), provide a means of estimating the airplane's holding/endurance time as a function of the airplane's starting weight, lift-to-drag ratio, thrust specific fuel consumption (TSFC), and available fuel load. Although the main use of these equations is to calculate either (1) the endurance for a given fuel load or (2) the fuel required for the airplane to maintain a holding/reconnaissance position for a given time, the equations are generic and can be applied to other situations where it is required to determine the flight time (e.g., in cruise). A numerical approach to determining the holding/endurance time is presented in Section 14.4, along with a computational table that illustrates the procedure. The optimum speed and altitude to fly to minimize fuel burn per unit time (i.e., best endurance speed) is considered in Section 14.5. The chapter closes with a short discussion on holding procedures and holding speeds (Section 14.6).

Performance of the Jet Transport Airplane: Analysis Methods, Flight Operations, and Regulations, First Edition.
Trevor M. Young.
© 2018 John Wiley & Sons Ltd. Published 2018 by John Wiley & Sons Ltd.

14.2 Basic Equation for Holding/Endurance

The rate of change of airplane mass (or weight) for commercial airplane operations relates to the fuel mass (or weight) burned per unit of time (t), as described in Section 13.2. The relationship can be expressed as follows:

$$-\frac{dm}{dt} = -\frac{dm_f}{dt} = Q \tag{14.1a}$$

or $\qquad -\dfrac{dW}{dt} = -\dfrac{dW_f}{dt} = Q' \tag{14.1b}$

where m is the instantaneous airplane mass (typical units: kg, slug);
$\qquad$ W is the instantaneous airplane weight (typical units: N, lb);
$\qquad$ m_f is the instantaneous onboard fuel mass (typical units: kg, slug);
$\qquad$ W_f is the instantaneous onboard fuel weight (typical units: N, lb);
$\qquad$ Q is the fuel mass flow (typical units: kg/h, slug/h); and
$\qquad$ Q' is the fuel weight flow (typical units: N/h, lb/h).

If the initial airplane mass is m_1 (weight is W_1) and the final mass is m_2 (weight is W_2), the flight time (t) will be given by

$$t = -\int_{m_1}^{m_2} \frac{1}{Q}\, dm = \int_{m_2}^{m_1} \frac{1}{Q}\, dm \tag{14.2a}$$

or $\qquad t = -\displaystyle\int_{W_1}^{W_2} \frac{1}{Q'}\, dW = \int_{W_2}^{W_1} \frac{1}{Q'}\, dW \tag{14.2b}$

The relationship between the parameter $1/Q$ and flight time (see Equation 14.2) is analogous to that of r_a (specific air range) and range (see Equation 13.5). The parameter $1/Q$ (or $1/Q'$) is known as *specific endurance*; it is the flight time per unit of fuel mass (or fuel weight) consumed and typically has units of h/kg (or h/lb). Specific endurance is a point performance parameter (i.e., applicable to a particular time or location along the flight path) describing the time that an airplane can remain airborne for a unit quantity of fuel.

The integral expression given by Equation 14.2 can be evaluated in a similar manner to that undertaken for the range analysis (see Sections 13.3 and 13.4). The fuel flow, which is described by Equation 13.6, is repeated below:

$$Q = cF_N = cD = c\left(\frac{D}{L}\right) W = \frac{c}{E}mg \qquad \text{(Equation 13.6a)}$$

or $\qquad Q' = c'F_N = c'D = c'\left(\dfrac{D}{L}\right) W = \dfrac{c'}{E}W \qquad \text{(Equation 13.6b)}$

where c is the TSFC based on mass flow (typical unit: mg $N^{-1}s^{-1}$);
$\qquad$ c' is the TSFC based on weight flow (typical unit: lb lb^{-1} h^{-1});
$\qquad$ F_N is the net thrust (typical units: N, lb);
$\qquad$ D is the airplane drag (typical units: N, lb);
$\qquad$ L is the airplane lift (typical units: N, lb);
$\qquad$ E is the lift-to-drag ratio (dimensionless); and
$\qquad$ g is the acceleration due to gravity (typical units: m/s^2, ft/s^2).

Equation 13.6 is substituted into Equation 14.2 to give the following expression for the airplane's endurance (or flight time):

$$t = \int_{m_2}^{m_1} \frac{1}{Q} \, dm = \int_{m_2}^{m_1} \frac{E}{cgm} \, dm \tag{14.3a}$$

or

$$t = \int_{W_2}^{W_1} \frac{1}{Q'} \, dW = \int_{W_2}^{W_1} \frac{E}{c'W} \, dW \tag{14.3b}$$

The range integral, given by Equation 13.10, has three variables, that is, true airspeed, lift-to-drag ratio, and TSFC. The endurance integral (Equation 14.3) only has two variables; although, it should be noted that both E and c (or c') are functions of true airspeed (TAS). Nonetheless, the evaluation of Equation 14.3 can be undertaken in a very similar manner to that presented in Chapter 13 for range determination. Analytical solutions to the integral expression are presented in Section 14.3 below, and a procedure for numerically evaluating the endurance integral is presented in Section 14.4.

14.3   Analytical Integration

14.3.1   Flight Schedules for Holding/Endurance

To facilitate the development of closed-form mathematical solutions, it is assumed that the TSFC can be satisfactorily represented by a constant mean value (i.e., $\bar{c}$ or $\bar{c}'$). This assumption allows the integral expression given by Equation 14.3 to be simplified as follows:

$$t = \frac{1}{\bar{c}g} \int_{m_2}^{m_1} \frac{E}{m} \, dm \tag{14.4a}$$

or

$$t = \frac{1}{\bar{c}'} \int_{W_2}^{W_1} \frac{E}{W} \, dW \tag{14.4b}$$

It is possible to evaluate Equation 14.4 in different ways, depending on the boundary conditions that are established. The flight schedules used to evaluate the range in Section 13.3 are repeated below:

(1) flight at constant *altitude* (H_p) and constant *lift coefficient* (C_L);
(2) flight at constant *true airspeed* (V) and constant *lift coefficient* (C_L); and
(3) flight at constant *altitude* (H_p) and constant *true airspeed* (V).

Unlike the constant C_L range formulations, described in Sections 13.3.2 and 13.3.3, in this case a single solution is possible for constant C_L flight irrespective of whether a constraint is placed on altitude or true airspeed.

14.3.2   First and Second Flight Schedules

As flight at constant C_L implies that a constant lift-to-drag ratio is maintained, the integral can be further simplified. The integration of Equation 14.4 to give the flight time (t) is straightforward:

$$t = \frac{E}{\bar{c}g} \int_{m_2}^{m_1} \frac{1}{m} \, dm = \frac{E}{\bar{c}g} \ln\left(\frac{m_1}{m_2}\right) = \frac{E}{\bar{c}g} \ln\left(\frac{1}{1-\varsigma}\right) \tag{14.5a}$$

$$\text{or} \qquad t = \frac{E}{\bar{c}'} \int_{W_2}^{W_1} \frac{1}{W} \, dW = \frac{E}{\bar{c}'} \ln\left(\frac{W_1}{W_2}\right) = \frac{E}{\bar{c}'} \ln\left(\frac{1}{1-\varsigma}\right) \qquad (14.5\text{b})$$

where ς is the cruise fuel mass (or weight) fraction, defined as

$$\varsigma = \frac{m_1 - m_2}{m_1} = \frac{W_1 - W_2}{W_1} \qquad \text{(Equation 13.17)}$$

The implications of the constant C_L boundary condition can be considered for two flight scenarios: (1) constant altitude and (2) cruise-climb. The lift coefficient for straight and level flight (i.e., when the lift is equal to the weight) is described by Equation 7.4, that is,

$$C_L = \frac{L}{qS} = \frac{2W}{\rho_0 \sigma V^2 S} = \frac{2W}{\rho_0 a_0^2 \delta M^2 S} \qquad \text{(from Equation 7.4, where } n_z = 1\text{)}$$

where q is the dynamic pressure (typical units: N/m^2, lb/ft^2);

S is the wing reference area (typical units: m^2, ft^2);

ρ_0 is the air density at the ISA sea-level datum (typical units: kg/m^3, slug/ft^3);

σ is the relative air density (dimensionless);

a_0 is the speed of sound at the ISA sea-level datum (typical units: m/s, ft/s, kt);

δ is the relative air pressure (dimensionless);

M is the Mach number (dimensionless); and

n_z is the load factor.

In normal flight operations, a hold would be conducted at constant altitude under the direction of ATC (see Section 14.6). In this situation (when the air density is constant), the true airspeed would need to reduce in proportion to $\sqrt{W}$ as fuel is burned in order to maintain a constant C_L.

In certain situations, an endurance flight can be conducted without adhering to a fixed altitude—for example, a military aircraft undertaking a surveillance mission. In this case, a cruise-climb (see Section 13.3.3) may be possible. It is evident from Equation 7.4 that if C_L is to remain unchanged at constant true airspeed, then the altitude must increase such that the ratio W/σ is maintained constant. Alternatively, flight at constant Mach number requires that the ratio W/δ be held constant through a very slow, but steady, increase in altitude.

14.3.3 Third Flight Schedule

The third flight schedule is at constant altitude (H_p) and constant true airspeed (V). Under these conditions, Equation 14.4 can be evaluated to give the endurance/holding time if it is assumed that the airplane's drag can be adequately represented by the parabolic drag polar (see Equation 7.14). The solution is practically identical to the range solution evaluated for these boundary conditions (see Appendix D, Section D.5.4). The flight time can be deduced directly from Equation 13.25, as TAS is constant:

$$t = \frac{2E_{max}}{g\bar{c}} \arctan\left(\frac{\sqrt{B_3}(m_1 - m_2)}{B_3 + m_1 m_2}\right) \qquad (14.6\text{a})$$

$$\text{or} \qquad t = \frac{2E_{max}}{\bar{c}'} \arctan\left(\frac{g\sqrt{B_3}(W_1 - W_2)}{g^2 B_3 + W_1 W_2}\right) \qquad (14.6\text{b})$$

$$\text{where } B_3 = \left(\frac{C_{D_0}}{K}\right)\left(\frac{\rho V^2 S}{2g}\right)^2$$

And where the maximum lift-to-drag ratio based on the parabolic drag polar is given by Equation 7.43:

$$E_{max} = \frac{1}{2}\sqrt{\frac{1}{KC_{D_0}}} \qquad \text{(Equation 7.43)}$$

where C_{D_0} is the zero-lift drag coefficient (dimensionless); and
K is the lift-dependent drag factor (dimensionless).

Alternatively, the flight time can be expressed in terms of the starting lift coefficient (C_{L_1}) and the starting lift-to-drag ratio (E_1). From Equation 13.26 it can be deduced that

$$t = \frac{2E_{max}}{g\bar{c}} \arctan\left\{\frac{\varsigma E_1}{2E_{max}(1 - KC_{L_1}\varsigma E_1)}\right\} \tag{14.7a}$$

or $$t = \frac{2E_{max}}{\bar{c}'} \arctan\left\{\frac{\varsigma E_1}{2E_{max}\left(1 - KC_{L_1}\varsigma E_1\right)}\right\} \tag{14.7b}$$

A summary of the analytical endurance/holding time expressions is given in Table 14.1.

Table 14.1 Summary of endurance/holding time expressions based on constant TSFC

Mass flow basis	Weight flow basis
Governing equation:[a],[b]	
$$t = \int_{m_2}^{m_1} \frac{1}{Q}\,dm$$	$$t = \int_{W_2}^{W_1} \frac{1}{Q'}\,dW$$
Fuel flow:	
$$Q = \frac{cgm}{E}$$	$$Q' = \frac{c'W}{E}$$
Holding/endurance equation for schedule 1 or 2 (i.e., flight at constant lift coefficient):[a],[b],[c]	
$$t = \frac{E}{\bar{c}g}\ln\left(\frac{m_1}{m_2}\right) = \frac{E}{\bar{c}g}\ln\left(\frac{1}{1-\varsigma}\right)$$	$$t = \frac{E}{\bar{c}'}\ln\left(\frac{W_1}{W_2}\right) = \frac{E}{\bar{c}'}\ln\left(\frac{1}{1-\varsigma}\right)$$
Holding/endurance equation for schedule 3 (i.e., flight at constant altitude and constant true airspeed):[a],[b],[c],[d]	
$$t = \frac{2E_{max}}{g\bar{c}}\arctan\left(\frac{\sqrt{B_3}\,(m_1 - m_2)}{B_3 + m_1 m_2}\right)$$	$$t = \frac{2E_{max}}{\bar{c}'}\arctan\left(\frac{g\sqrt{B_3}\,(W_1 - W_2)}{g^2 B_3 + W_1 W_2}\right)$$
or	or
$$t = \frac{2E_{max}}{g\bar{c}}\arctan\left\{\frac{\varsigma E_1}{2E_{max}\left(1 - KC_{L_1}\varsigma E_1\right)}\right\}$$	$$t = \frac{2E_{max}}{\bar{c}'}\arctan\left\{\frac{\varsigma E_1}{2E_{max}\left(1 - KC_{L_1}\varsigma E_1\right)}\right\}$$

Notes:

(a) Subscript 1 denotes the start condition, defining the mass (m_1), weight (W_1), true airspeed (V_1), lift coefficient (C_{L_1}), and lift-to-drag ratio (E_1).

(b) Subscript 2 denotes the end condition, defining the mass (m_2) and weight (W_2).

(c) The fuel mass (or weight) ratio is defined as $\varsigma = \dfrac{m_1 - m_2}{m_1} = \dfrac{W_1 - W_2}{W_1}$

(d) The solution to schedule 3 is based on the parabolic drag polar (Equation 7.14), thus,

$$E_{max} = \frac{1}{2}\sqrt{\frac{1}{KC_{D_0}}} \quad \text{and } B_3 = \left(\frac{C_{D_0}}{K}\right)\left(\frac{\rho V^2 S}{2g}\right)^2$$

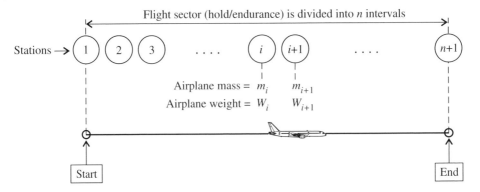

Figure 14.1 Holding/endurance stations (the flight path is shown straight for illustration only).

14.4 Numerical Integration

14.4.1 Integration Method

A similar numerical technique to that used in Section 13.4 to evaluate the range integral can be employed to evaluate Equation 14.2 to give the holding time (or endurance) when fuel flow data is available. The method requires that the airplane's mass (or weight) change be divided into n intervals, as illustrated in Figure 14.1.

The instantaneous mass (or weight) of the airplane is determined at each station. Based on the airplane's drag, the thrust is then determined from which the instantaneous fuel flow (Q) can be established at each station. The evaluation technique of the integral expression can be illustrated using a graph of $1/Q$ versus airplane mass (or $1/Q'$ versus airplane weight), as shown in Figure 14.2. The holding/endurance time is calculated by a numerical integration of the function described by Equation 14.2, which is represented by the area under the graph between the final mass (or weight) and starting mass (or weight). If, for example, the trapezoidal rule were employed, then the summation giving the holding/endurance time would be as follows:

$$t = \sum_{i=1}^{n} \left(\frac{1}{Q_i} + \frac{1}{Q_{i+1}} \right) \left(\frac{m_i - m_{i+1}}{2} \right) \tag{14.8a}$$

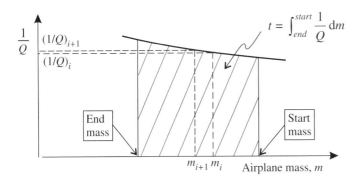

Figure 14.2 Numerical integration to give holding/endurance time.

or $\quad t = \sum_{i=1}^{n} \left(\dfrac{1}{Q_i'} + \dfrac{1}{Q_{i+1}'} \right) \left(\dfrac{W_i - W_{i+1}}{2} \right)$ $\hspace{4cm}$ (14.8b)

where Q_i is the fuel mass flow and Q_i' is the fuel weight flow at the ith station; and
m_i is the airplane mass and W_i is the airplane weight at the ith station.

This method is generally applicable and is not restricted to flights at constant height or constant airspeed, for example (as is required for the analytical solutions described in Section 14.3).

14.4.2   Example of Numerical Computation

An example of the numerical computation of the endurance for a mid-size, twin-engine airliner is presented in Table 14.2. The method is very similar to that presented in Section 13.4.2 for calculating range. For this illustration, four weight intervals (i.e., five stations) were selected, with a fuel usage per interval of 1500 lb. The airplane's weight at each station is recorded in column 2. The computational sequence now runs from left to right across the spreadsheet table, with the determination at each station of the following parameters: W/δ, C_L (Equations 7.1 and 6.12), C_D (from a drag polar—see Figure 7.8), D/δ (Equations 7.5 and 6.12), and Q'. The fuel flow, for this particular example, was based on corrected fuel flow data tabulated as a function of corrected thrust (F_N/δ), which in level, non-accelerating flight equates to D/δ. Finally, the time for each interval is calculated (based on Q' and the airplane weight change for the interval) and the results summed to give the endurance time (see Equation 14.8b).

Table 14.2 Example of endurance calculation for a mid-size, twin-engine airliner[a],[b]

Station	*W* (1000 lb)	*W/δ* (1000 lb)	C_L	C_D	*D/δ* (1000 lb)	*Q'* (lb/h)	Interval	Time (h)
1	200.0	290.8	0.697	0.0397	16.6	6553		
2	198.5	288.6	0.692	0.0394	16.4	6507	1–2	0.230
3	197.0	286.5	0.686	0.0391	16.3	6461	2–3	0.231
4	195.5	284.3	0.681	0.0387	16.2	6415	3–4	0.233
5	194.0	282.1	0.676	0.0384	16.0	6369	4–5	0.235

Total: 0.929

Notes:

(a) Conditions: starting weight 200 000 lb, altitude 10 000 ft, ISA temperature, and Mach 0.38 (210 KCAS).
(b) The following information can be ascertained for ISA conditions at this altitude: $\theta = 0.9312$, $\delta = 0.6877$, and $a = 638.3$ kt (see Appendix A).

14.5   Flight Conditions for Maximum Endurance

14.5.1   Minimum Drag Speed Based on the Parabolic Drag Polar

The maximum endurance will be achieved when the airplane operates at the condition that gives the least rate of fuel consumption. It is often assumed, in the general literature, that TSFC is constant over the associated speed range. With this simplification, it is evident from

Equation 13.6 that the optimum condition occurs when the lift-to-drag ratio is a maximum; consequently the airplane must fly at the minimum drag speed (i.e., the best lift-to-drag ratio speed) to achieve the greatest endurance.

Based on the parabolic drag polar—which generally provides a reasonably accurate representation of an airplane's drag at the relatively low speeds used for holding (or for reconnaissance missions)—the minimum drag speed is given in equivalent airspeed (EAS) by the following equation:

$$V_{e_{md}} = \left[\frac{2W}{\rho_0 S} \right]^{1/2} \left[\frac{K}{C_{D_0}} \right]^{1/4} \qquad \text{(from Equation 7.46)}$$

The corresponding true airspeed is

$$V_{md} = \left[\frac{2W}{\sigma \rho_0 S} \right]^{1/2} \left[\frac{K}{C_{D_0}} \right]^{1/4} \tag{14.9}$$

and the minimum drag Mach number is

$$M_{md} = \left[\frac{2}{a_0^2 \rho_0 S} \left(\frac{W}{\delta} \right) \right]^{1/2} \left[\frac{K}{C_{D_0}} \right]^{1/4} \tag{14.10}$$

With the assumption of constant TSFC, the optimum endurance speed—for a particular airplane where the drag polar can be characterized in terms of the lift-dependent drag factor K and the zero-lift drag coefficient C_{D_0}—is seen to be a function of airplane weight, altitude, and air temperature (WAT).

14.5.2 Minimum Drag Speed Based on Actual Drag Polars

An analysis of the lift-to-drag ratio (E) for the drag polars illustrated in Figure 7.8 (for a mid-size, twin-engine airliner) is presented in Section 7.6.2. Figure 7.19, which contains isolines (contours) of constant E on a plane of lift coefficient versus Mach number, provides a clear graphical illustration of how an airplane's lift-to-drag ratio varies as a function of C_L and M. The minimum-drag lift coefficient ($C_{L_{md}}$) is seen to decrease gradually with increasing Mach number until a relatively high speed is reached (which is in the region of the drag divergence Mach number—see Figure 7.11), at which point $C_{L_{md}}$ starts to reduce rapidly with increasing Mach number. The limitations of representing an airplane's drag relationship over this speed range by a single parabolic drag polar (which is what was assumed in Section 14.5.1) is evident from this chart.

14.5.3 Optimum Speed for Endurance

In Sections 14.5.1 and 14.5.2, the assumption of constant TSFC was used to establish an estimate of the optimum endurance speed. For this particular application, it should be noted that the assumption of constant TSFC is a simplification that results in a systematic bias in the results obtained. The TSFC of a turbine engine *increases* monotonically with increasing Mach number in the vicinity of the optimum endurance speed. Consequently, the speed for minimum fuel burn—at a given airplane weight, altitude, and air temperature (WAT)—is slower than the corresponding minimum drag speed. This can be seen in Figure 14.3, which illustrates, for a mid-size, twin-engine airplane, the fuel flow and drag variation with Mach number at 10 000 ft

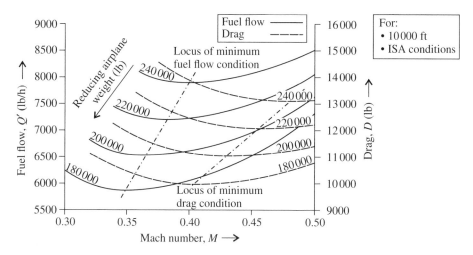

Figure 14.3 Fuel flow and airplane drag as a function of Mach number (example for a mid-size, twin-engine airliner).

for four selected airplane weights. The influence of airplane weight can also be seen in the figure: the speed for minimum fuel burn decreases with reducing airplane weight (at a fixed altitude).

The optimum endurance speed, which will result in the lowest rate of fuel burn (for a given airplane weight, altitude, and temperature), can be established through flight-test data analysis. There are, however, several other considerations that need to be taken into account when determining practical holding speeds for flight operations—these are discussed in Section 14.6.3.

14.5.4 Optimum Altitude for Endurance

For a given Mach number, there is an optimum altitude for minimum fuel burn—this is illustrated in Figure 14.4. This is a similar result to that presented for the cruise in Section 13.6.5. It is evident that the maximum endurance will be achieved by flying a cruise-climb along the locus of points representing the altitude optima. As noted previously in Chapter 13, flight level restrictions imposed by air traffic control (ATC) limit the practical use of the cruise-climb technique. To comply with ATC restrictions, pilots may execute a step climb—climbing to higher

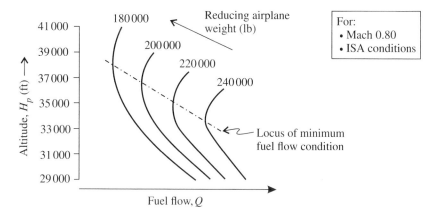

Figure 14.4 Fuel flow variation as a function of altitude (example for a mid-size, twin-engine airliner).

flight levels as fuel is consumed—to maintain proximity to the airplane's optimum altitude (as described in Section 13.6.5).

Holding operations, however, are routinely conducted at ATC specified altitudes—for example, in a holding pattern (see Section 14.6.1).

14.6 Holding Operations

14.6.1 Holding Procedures

A holding procedure is a predetermined flight maneuver designed to keep an airplane in a specified airspace awaiting further ATC clearance. Terminal holding procedures are designed to act as a buffer, absorbing the congestion that can occur due to excessive air traffic arriving with the intention of landing at a particular airport. Holding aircraft, flying the same holding pattern, are typically separated vertically by 1000 ft intervals—this is known as a *holding stack*. Arriving aircraft will usually be directed by ATC to join at the top of the stack. These aircraft, circling in the stack, may occasionally be joined by an airplane that has executed a missed approach (see Section 11.8). Aircraft typically leave the stack from the bottom to make an approach to land, which then allows the stacked aircraft to descend one level. The whole process is controlled by ATC.

Holding aircraft are required to remain in a designated area in the vicinity of the airport. The basic holding area is an airspace that is adequate for pilots to execute a standard holding pattern (see Section 14.6.2), taking into account possible airspeed variations, wind effects, and timing errors. An additional buffer area extends 5 nm beyond the boundaries of the holding area [1]. A minimum holding altitude is established for each holding area, in compliance with regulatory obstacle clearance requirements—for example: 300 m (984 ft) clearance above obstacles, increasing to 600 m (1969 ft) over high terrain or mountainous areas, where the additional clearance is provided to accommodate possible effects of air turbulence [1].

14.6.2 Holding Patterns

Details on holding patterns and operational procedures for holding are given in ICAO Doc. 8168 Vol. 1 Flight Procedures [1] and in the FAA Aeronautical Information Manual (AIM) [2], for example. The conventional holding pattern that pilots are required to fly resembles a racetrack, with two straight legs and two semi-circles, which has a fixed ground location at the end of the inbound leg, as illustrated in Figure 14.5. Right-hand turns are undertaken, where possible. In still air, the outbound and inbound legs are on reciprocal headings with 180° turns executed at the end of each leg. In the presence of wind, pilots must compensate accordingly by increasing or decreasing the bank angle in the turns and by adjusting the airplane's heading

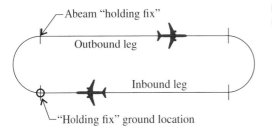

Figure 14.5 Right-hand "racetrack" holding pattern (in still air).

for each leg (the influence of wind on circular flight paths is described later in Section 15.3.6). Modern airplanes equipped with an advanced autopilot and Flight Management System (FMS) are capable of flying holding patterns automatically.

The time for each outbound and inbound leg is normally 1 minute at or below 14 000 ft altitude and $1\frac{1}{2}$ minutes above 14 000 ft; these times can be increased with ATC permission [3]. For example, requests for 2 minute legs are not uncommon. If rate one turns (see Section 15.3.2) are executed, then one complete circuit with standard 1 minute legs takes 4 minutes.

14.6.3 Holding Speeds

ICAO publish maximum holding speeds (in indicated airspeed), based on the airplane's altitude. The speeds are (1) 230 kt (in normal conditions) increasing to 280 kt (in turbulent conditions) for operations up to 14 000 ft; (2) 240 kt (in normal conditions) increasing to 280 kt (in turbulent conditions) for altitudes above 14 000 ft up to 20 000 ft; and (3) 265 kt (in normal conditions) increasing to 280 kt or 0.83 Mach, whichever is less, (in turbulent conditions) above 20 000 ft up to 34 000 ft [1].

It is desirable to minimize fuel consumption in a hold. In Section 14.5.3, the optimum endurance speed was identified as the speed at which to fly to minimize fuel burn. There are several reasons, however, why practical holding speeds used in day-to-day flight operations are often faster than the optimum endurance speed.

The speed for minimum fuel flow is a relatively slow speed, and this can result in undesirable handling qualities or a high workload for the flight crew or even an unacceptable stall margin. As shown in Figure 14.3, the minimum fuel-flow speed is slower than the minimum drag speed. Depending on the thrust characteristics of the engine, the airplane may not be speed stable at such slow speeds (the concept of speed stability is discussed in Section 12.3.4). Consequently, minimum fuel flow will not normally be achieved in typical levels of atmosphere turbulence without the assistance afforded by an autothrottle/autothrust system (see Section 8.5.2). For aircraft without this function, the speed chosen for holding will thus be increased.

Another factor that can result in an increase in the holding speed above the optimum endurance speed relates to the need for the airplane to repeatedly bank and turn. To maintain height in a banked turn, a small increase in thrust is required to compensate for the increase in drag (see Section 15.3.4). The fuel flow will thus increase as the airplane turns at the end of each leg (typically, a jet transport airplane will consume about 5% more fuel when in a standard racetrack holding pattern than when flying straight and level in comparable conditions). Additionally, an airplane's stall speed increases in a turn (see Section 20.2.4); maintaining an adequate margin from stall in the clean configuration, particularly in turbulent conditions, may also necessitate an increase in speed.

Recommended holding speeds for a particular airplane type are determined by the manufacturer and published in the FCOM (Flight Crew Operations/Operating Manual). The speeds are indicated as a function of airplane weight and altitude, and are usually provided as indicated airspeed (IAS). It is interesting to note that the standard holding speed for Airbus aircraft, in the clean configuration, is the *green dot* speed [4], which is an operational reference speed that closely approximates the airplane's minimum drag speed (i.e., best lift-to-drag ratio speed).[1] The fuel penalty of flying at the minimum drag speed compared to the minimum fuel-flow speed is modest, as the fuel flow versus speed relationship tends to be relatively flat over this speed range.

---

1 The green dot speed (which is described in Section 17.4.6) is calculated by the Flight Management System (FMS) on Airbus aircraft, based on the airplane's instantaneous weight and altitude. It is indicated by means of a green circle on the vertical airspeed tape of each Primary Flight Display (PFD) in the cockpit.

14.6.4 Holding with One Engine Inoperative

An interesting feature regarding holding, which is described by Blake, *et al.* [5], concerns aircraft holding with one engine inoperative. The example is given of a twin-engine, single-aisle airplane operating at Mach 0.4 at 10 000 ft. The fuel flow to the "good" engine in the one-engine-inoperative (OEI) case is *less* than the total fuel flow to both engines in the all-engines-operating (AEO) case (this feature is not observed for operations at higher altitudes). What is noteworthy is that there are two factors that work in opposing directions in this scenario. On the one hand, the airplane will have more drag in the OEI condition (see Section 7.5.5); for the B757 example described by Blake, *et al.* [5], this drag increase was ~10% of the total airplane drag. On the other hand, the "good" engine in the OEI case operates at a thrust value that has a significantly better fuel efficiency; this trend can be seen in Figure 8.7. For the AEO case, the engines are operating at low thrust, which will be associated with a very high TSFC. For the OEI case, the "good" engine will be operating at a thrust value that is more than double that of the AEO case, and it will thus have a significantly lower TSFC. The product of thrust and TSFC—which is equal to fuel flow—can thus be less in the OEI case compared to the AEO case.

References

1 ICAO, "Procedures for air navigation services: Aircraft operations, Vol. 1, Flight procedures," 5th ed., Doc. 8168, International Civil Aviation Organization, Montréal, Canada, 2006.
2 FAA, "Aeronautical information manual: Official guide to basic flight information and ATC procedures," Federal Aviation Administration, Washington, DC, Dec. 10, 2015. Electronic version available from www.faa.gov/air_traffic/.
3 Åkerlind, O. and Örtlund, H., *Instrument flight procedures and aircraft performance*, Håkan Örtlund Produktion, Molkom, Sweden, 2006.
4 Airbus, "Getting to grips with aircraft performance," Flight Operations Support and Line Assistance, Airbus S.A.S., Blagnac, France, Jan. 2002.
5 Blake, W. and Performance Training Group, "Jet transport performance methods," D6-1420, Flight Operations Engineering, Boeing Commercial Airplanes, Seattle, WA, Mar. 2009.

15

Mechanics of Maneuvering Flight

15.1 Introduction

Methods for analyzing the performance of an airplane undergoing maneuvers, which result from deliberate deflections of the primary flight controls, are presented in this chapter. A feature of maneuvering flight is that the airplane experiences a significant angular velocity about one or more of the body axes (the body axes coordinate system is fixed with respect to the airplane, as described in Section 3.3.4). Rotations about the X_b, Y_b, and Z_b axes are known as *roll, pitch*, and *yaw*, respectively (see Figure 3.7).

Turning maneuvers—in which the airplane initially rolls about the X_b axis and then moves along a circular flight path with an angle of bank—can be executed in several different ways, depending on how the flight controls are manipulated (Section 15.2). For example, the angle of bank could be held constant or progressively increased or decreased. The turn could be at constant height or the height could increase or decrease. A *coordinated turn* is one in which the ailerons and rudder are appropriately deflected such that the airplane neither skids outwards nor slips inwards towards the center of the turning circle. Equations that describe the motion of the airplane are derived for an airplane performing a constant-height coordinated turn (Section 15.3), a climbing or descending turn (Section 15.4), or an uncoordinated turn (Section 15.5).

By definition, a sustained, level turn is one in which the airplane maintains both speed and altitude in the turn. Such maneuvers are routinely executed when the airplane is in cruise or in a holding pattern, for example. Under certain flight conditions—defined by such factors as the airplane's configuration, its weight, altitude, air temperature, and thrust setting—the airplane's maneuvering capability can be restricted by aerodynamic, structural, or powerplant limitations (Section 15.6). For example, if there is insufficient thrust to maintain a steady, constant-height turn at a particular flight condition, then the airplane's speed and/or height will reduce.

Pitching maneuvers describe the motion of the airplane when it experiences an angular velocity about the Y_b axis (i.e., pitch axis). A pull-up maneuver is one in which the airplane has a positive angular velocity about the Y_b axis, whereas a push-over maneuver has a negative angular velocity about the Y_b axis. The mechanics of such maneuvers are discussed in Section 15.7.

The total energy concept is a powerful technique for evaluating an airplane's capability to climb and/or accelerate along its flight path or execute a maneuver involving a climb and/or acceleration—this concept is introduced in Section 15.8.

Performance of the Jet Transport Airplane: Analysis Methods, Flight Operations, and Regulations, First Edition.
Trevor M. Young.
© 2018 John Wiley & Sons Ltd. Published 2018 by John Wiley & Sons Ltd.

15.2 Turning Maneuvers

15.2.1 Turning Maneuvers: Types of Turns

In turning flight, the airplane is considered to be flying along a circular flight path. A level (i.e., constant height) turn is illustrated in Figure 15.1. The radius of the turning circle is designated by the letter r, and the bank angle, which is the angle from the vertical to the plane of symmetry of the airplane, by the Greek capital letter phi (Φ). The *rate of turn* (or turn rate) of the airplane about the center of the circular flight path is designated by the Greek capital letter omega (Ω). The centripetal acceleration, which acts towards the center of the circle, is denoted as a_c. As height is maintained in the turn, the vertical component of the net lift (L) is equal in magnitude to the airplane's weight (W).

The precise manner in which the turn is executed depends on how the pilot manipulates the flight controls (which are described in Section 3.8) and on the thrust setting. For example, the turn may be at constant height or the airplane could be climbing or descending; it may be at constant true airspeed (TAS) or the speed could be increasing or decreasing. A *coordinated turn* is a correctly banked turn, in which the airplane neither slips inwards towards the center of the turn (i.e., over-banked) nor skids outwards from the center of the turn (i.e., under-banked). The pilot would deflect the ailerons to roll the airplane (by rotating a yoke or moving a sidestick laterally) to achieve the desired bank angle and press a rudder pedal to yaw the airplane appropriately and then adjust the position of the ailerons, if needed. When the aileron and rudder deflections are coordinated—that is, appropriately selected for the speed and bank angle—there is no net sideforce (F_Y) acting on the airplane. This is not the case for an *uncoordinated turn*: a net sideforce acting towards the center of the turn results in a skidding turn (Figure 15.2a), whereas a net force acting outwards results in a slipping turn (Figure 15.2b). A turn with a large net sideforce is uncomfortable for the crew and their passengers as they are forced sideways in their seats. For this reason, coordinated turns are routinely flown. In fact, uncoordinated turns are seldom intentionally performed on jet transport airplanes.

The simplest case to analyze is the coordinated, constant-height, constant-speed turn—this is discussed in Section 15.3. A straightforward extension to this theory enables a climbing or descending turn at constant speed to be analyzed (see Section 15.4). Such turns are frequently encountered during routine maneuvering of the airplane in flight. For this class of airplane, the rate of change of speed in a turn is usually small and the assumption of constant true airspeed is acceptable in most cases. Uncoordinated turns are discussed in Section 15.5.

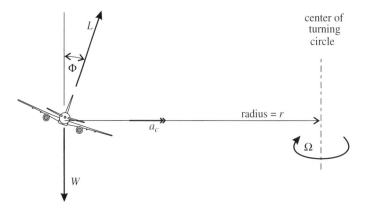

Figure 15.1 Steady (i.e., constant speed), level (i.e., constant height) turn.

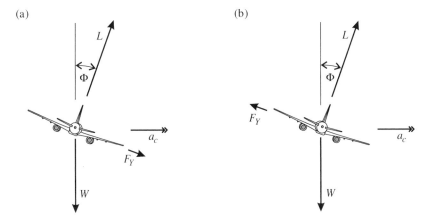

Figure 15.2 Uncoordinated turns (as seen from behind the airplane), illustrating (a) skidding turn and (b) slipping turn.

15.2.2 Turn Coordinator

It is useful, at this point in the discussion on turning performance, to digress and consider the operation of the turn coordinator—a standard component on an airplane's instrument panel. For the sake of illustration, it is easier to consider the traditional type of instrument found on many general aviation aircraft rather than the equivalent display that is integrated into the Primary Flight Display (PFD) of a modern jet airplane.

A turn coordinator is actually two instruments built into one unit (Figure 15.3). The upper part is a gyroscopic instrument, which indicates the *rate of turn* of the airplane. The indicator resembles a little symbolic airplane—seen from behind. There are four white tick marks on the outside of the display: two level tick marks and two angled tick marks. When the wings of the symbolic airplane are aligned with the two level tick marks, the rate of turn is zero; when the wings are tilted, the instrument indicates both the direction and the rate of turn. The lower angled tick marks indicate a turn rate of 3°/s. In other words, when the pilot lines up the symbolic airplane with an angled tick mark, the airplane will execute a turn that can achieve a 360° heading change in two minutes (it is for this reason that "2 MIN" is written on the instrument). Turns conducted at 3°/s are called *rate one* or *standard rate* turns (see Section 15.3.2).

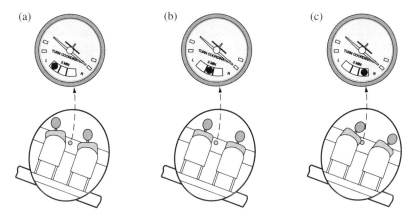

Figure 15.3 Illustration of (a) Skidding turn, (b) Coordinated turn, and (c) Slipping turn.

The lower part of the turn coordinator is essentially an inclinometer. This is a very simple instrument, comprising a curved glass tube, filled with a liquid, and a small ball. When the aircraft is stationary, the ball, in response to its weight, will rest at the lowest point in the tube. During flight, however, the ball will be displaced laterally (i.e., with respect to the center of the instrument) in response to an applied lateral force. Generally speaking, the ball will rest in the central, lowest portion of the tube when the airplane is in straight, level flight. (Should the airplane, however, possess little directional stability or suffer from inherent sideslip, then the ball may not be in the middle in level flight.) In maneuvering flight, the ball will move away from the center if a net sideforce acts on the aircraft. The pilot can drive the ball back towards the center by reducing the net sideforce through an appropriate manipulation of the flight controls. In a coordinated turn, the ball will be at or near the middle of the instrument; in a skidding turn the ball will be displaced outwards with respect to the turn, and in a slipping turn the ball will be displaced inwards with respect to the turn (as illustrated in Figure 15.3). The crew and passengers respond in the same way—in a coordinated turn, they will be pushed directly into their seats, but in an uncoordinated turn, they will also be pushed sideways.

It is part of normal general aviation flying that pilots provide a corrective action (i.e., control input) to keep the ball near the middle—this can be achieved by applying the appropriate rudder deflection for the flight condition (i.e., speed and bank angle). In pilot training, students are instructed to "step on the ball"—this means that the corrective action is to press the rudder pedal on the same side as that to which the ball is deflected.

15.2.3 Load Factor

The load factor n_z is the ratio of net lift (L) of an airplane to its weight (W), as described in Section 7.2.5 and defined by Equation 7.3:

$$n_z = \frac{L}{W} \qquad \text{(Equation 7.3)}$$

In straight, level (constant height) flight with zero bank angle, $n_z = 1$. In any maneuver that results in the lift being greater than the weight—for example, a level turn—the load factor is greater than one.

This load factor is often called the *normal load factor*. By definition, the net lift is the component of the net aerodynamic force that acts normal to the flight path vector in the plane of symmetry of the airplane. As the lift acts parallel to the Z_a axis, defined by the *wind axes* (which is another name for the flight path axes—see Section 3.3.5), the subscript ZW is sometimes used to identify this load factor.[1] The load factor is defined as positive for an aerodynamic force that acts upwards with respect to the airplane (i.e., in the negative Z_a direction).

In a broader context, it should be noted that a load factor is simply a non-dimensional aerodynamic force, where the non-dimensionalizing parameter is the airplane's instantaneous (gross) weight.[2] As forces are vectors, load factors have a defined direction—that is, they are also vectors, although they are seldom considered as such. Other coordinate systems can also

---

1 This convention by which the load factor normal to the flight path is identified by the subscript ZW is used in FAR/CS 25 [1, 2] (see Section 20.5.2).

2 An alternative convention (which is not used herein) defines *load factor* as the ratio of net aerodynamic force to airplane mass (not weight). For this less widely used convention, load factors have units of g (i.e., multiples of gravitational acceleration).

be used to define an aerodynamic load factor—for example, the ratio of the component of aerodynamic force acting normal to the airplane's longitudinal axis (usually defined parallel to the cabin floor) to the airplane's weight is the load factor customarily used for the determination of flight loads for structural design purposes (see Sections 20.6 and 20.7).

The crew and passengers onboard an aircraft performing a maneuver perceive the load factor as a change to their weight. When $n_z = 1$, they perceive their weight as typical, that is, unchanged from the usual condition on Earth where all mass items are subjected to the usual gravitational acceleration (g). However, when $n_z = 2$, they feel as if the acceleration due to gravity were twice its usual value. When $n_z = 0$, they perceive that they are weightless, and when $n_z = -1$, they feel as if they are hanging upside down.

Occupants of an airplane executing a turning or pull-up maneuver thus sense a load factor with respect to their usual reference condition, which is the 1-g state they normally experience on the ground. Pilots colloquially speak about "pulling gees" in a turn (many experienced pilots are able to estimate rather accurately the load factor from the physiological sensation that arises during maneuvers). This perception of load factor by the crew (and passengers) lends itself to the notion of measuring a load factor in "gees" (i.e., multiplies of g). For the most widely used convention, however, a load factor is a dimensionless aerodynamic force that acts on the airplane—and its occupants—such as to cause acceleration in the direction of the force.

15.3 Level Coordinated Turns

15.3.1 Forces and Load Factor in a Level Coordinated Turn

The lift and weight vectors which act on an airplane executing a steady-speed, level coordinated turn are shown in Figure 15.4. To maintain a constant height in the turn, it is necessary that the component of lift acting vertically must equal the airplane's weight. The component of lift acting towards the center of the turn provides the required centripetal force to accelerate the airplane in a circular flight path of radius r. In this analysis, it is assumed that the thrust axis is approximately aligned with the instantaneous flight direction; the thrust and drag forces are thus approximately collinear and are not shown in Figure 15.4.

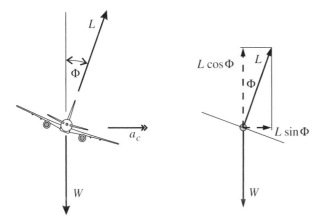

Figure 15.4 Forces acting on the airplane during a steady, level coordinated turn.

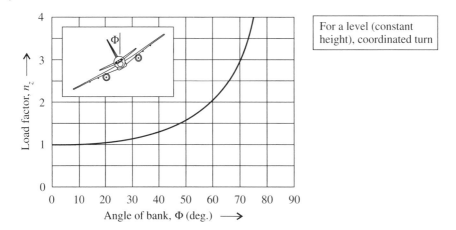

Figure 15.5 Load factor as a function of angle of bank (for a level, coordinated turn).

Resolving the forces shown in Figure 15.4 in the vertical and horizontal directions yields the following equations:

$$L \cos \Phi = W \tag{15.1}$$

and $\quad L \sin \Phi = ma_c = \dfrac{W}{g} a_c$ $\tag{15.2}$

where L is the lift (typical units: N, lb);
$\quad m$ is the airplane mass (typical units: kg, slug);
$\quad W$ is the airplane weight (typical units: N, lb);
$\quad g$ is the gravitational acceleration (typical units: m/s², ft/s²); and
$\quad a_c$ is the centripetal acceleration (typical units: m/s², ft/s²).

In a level turn, the load factor can be written in terms of the bank angle—this can be seen by rewriting Equation 15.1:

$$n_z = \frac{L}{W} = \frac{1}{\cos \Phi} \tag{15.3}$$

It is evident that as the angle of bank (Φ) increases, the load factor increases—this is illustrated in Figure 15.5. Initially, the increase is modest, but at bank angles greater than about 40°, the load factor is seen to increase rapidly with increasing bank angle. Note that the maximum angle of bank in a steady, level turn that an airplane can safely endure will be restricted by the maximum allowable load factor—a critically important structural design limit (this is described later in Section 15.6.3). The angle of bank and corresponding load factor can also be limited by aerodynamic or thrust considerations at a given airplane weight, altitude, and temperature (WAT).

15.3.2 Rate of Turn

The airplane's rate of turn (or turn rate) is its angular velocity about the center of the turning circle. An expression for the turn rate can be obtained by rewriting Equation 2.29:

$$\Omega = \frac{V}{r} \tag{15.4}$$

where Ω is the rate of turn (typical unit: rad/s);
$\quad V$ is the true airspeed (typical units: m/s, ft/s); and
$\quad r$ is the radius (typical units: m, ft).

Note that the airspeed, V, needs to be expressed in units of meters per second (or feet per second) and not knots if the radius has units of meters (or feet).

The centripetal acceleration, which relates to the rate of turn (see Equation 2.31), can be expressed as follows:

$$a_c = V\Omega = \frac{V^2}{r} \tag{15.5}$$

A useful expression for the rate of turn can be obtained by dividing Equation 15.2 by Equation 15.1, that is,

$$\tan \Phi = \frac{V\Omega}{g} \tag{15.6}$$

Equation 15.6 can be simplified by introducing the load factor, which can be related to the angle of bank using the following trigonometric relationship:

$$\tan^2 \Phi = \sec^2 \Phi - 1 = \frac{1}{\cos^2 \Phi} - 1 = n_z^2 - 1 \tag{15.7}$$

Thus $\Omega = \dfrac{g}{V}\sqrt{n_z^2 - 1}$ \hfill (15.8)

It can be deduced from Equation 15.8 that for a given airspeed V, the maximum rate of turn will correspond to the maximum allowable load factor, provided that there is sufficient thrust available to maintain the turn. Conversely, for a given load factor, the theoretical maximum rate of turn will occur at the slowest possible flight speed.

The turn rate is frequently expressed as the time required for the airplane to execute a 180° turn (i.e., π rad). The required time can be determined from Equation 15.4 and related to the true airspeed and angle of bank by means of Equation 15.6.

$$t_\pi = \frac{\pi}{\Omega} = \frac{\pi V}{g \tan \Phi} \tag{15.9}$$

where t_π is the time required for a 180° turn (typical unit: s).

In an operational context, a rate one turn (or standard rate turn) will result in the airplane completing a 180° turn in 1 minute. This translates to a rate of turn, or heading change, of 3°/s. Rate half and rate two turns are defined in a similar manner (Table 15.1). Pilots often execute rate one turns—either following standard operating procedures or when conducting certain maneuvers under ATC (air traffic control) instruction—for example, when in a holding pattern (see Section 14.6).

As airspeed increases, it is necessary for pilots to use a greater angle of bank to maintain the same rate of turn. At high speed, rate one turns result in excessive bank angles. It is thus typical at high speed (e.g., during cruise) that turns are executed with the pilot holding a fixed bank angle—in which case, the turn rate would vary depending on the airspeed.

Table 15.1 Rate-defined turns

	Time for 180° turn (s)	Turn rate (°/s)	Turn rate (rad/s)
Rate 1/2:	120	1.5	$\pi/120$
Rate 1:	60	3	$\pi/60$
Rate 2:	30	6	$\pi/30$

15.3.3 Turn Radius

An expression that describes the radius of turn (r) can be obtained by substituting Equation 15.8 into Equation 15.4. The resulting mathematical expression can be written in several ways: the radius can be expressed as a function of load factor (Equation 15.10) or bank angle (Equation 15.11); additionally, the speed can be written as true airspeed (V), equivalent airspeed (V_e), or Mach number (M), as shown below:

$$r = \frac{V^2}{g\sqrt{n_z^2 - 1}} = \frac{V_e^2}{g\sigma\sqrt{n_z^2 - 1}} = \frac{M^2 a^2}{g\sqrt{n_z^2 - 1}} = \frac{M^2 a_0^2 \theta}{g\sqrt{n_z^2 - 1}} \tag{15.10}$$

or $$r = \frac{V^2}{g\tan\Phi} = \frac{V_e^2}{g\sigma\tan\Phi} = \frac{M^2 a^2}{g\tan\Phi} = \frac{M^2 a_0^2 \theta}{g\tan\Phi} \tag{15.11}$$

where σ is the relative air density (dimensionless);
θ is the relative air temperature (dimensionless);
a is the speed of sound (typical units: m/s, ft/s); and
a_0 is the ISA sea-level speed of sound (typical units: m/s, ft/s).

Note that airspeed (i.e., V or V_e) and speed of sound (i.e., a or a_0) need to be expressed in units of m/s or ft/s to match the typical units of gravitational acceleration (i.e., m/s^2 or ft/s^2).

It is evident that the turn rate (given by Equation 15.8) is a function of the airplane's true airspeed for a particular load factor. A given turn rate and airspeed will correspond to a unique turn radius (see Equation 15.4). This observation enables a single graph to be constructed—which is applicable to all airplane types—linking the turn rate, true airspeed, load factor, and turn radius. This is illustrated in Figure 15.6. Such charts are often annotated to illustrate limiting turn performance for a particular airplane type. For example, a lift boundary can be superimposed on the chart to denote the turn performance available at the maximum lift coefficient.

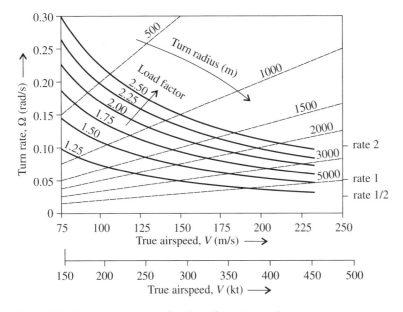

Figure 15.6 Turn parameters as a function of true airspeed.

The conditions for the minimum radius of turn are the same as those for the maximum rate of turn, which was discussed earlier in Section 15.3.2. For a given airspeed, the load factor (n_z) must be the maximum permissible, and for a given load factor, the true airspeed (V) must be the minimum permissible. The radius of turn—and rate of turn—for a sustained turn can also be limited by the maximum available thrust (see Section 15.6.5).

The radius of turn is an important performance parameter when assessing flight paths that must comply with regulatory obstacle clearance requirements (see Section 10.7) after takeoff or following a missed approach (see Section 11.8). The one-engine-inoperative obstacle clearance requirements must be considered during routine flight planning; this can be particularly challenging for flight operations into or out of airports with mountains or tall buildings in the vicinity. In many cases, such flight paths will involve a turn, which will need to be evaluated for the low-thrust, high-drag condition associated with one-engine-inoperative flight (see Section 7.5.5). Another important factor to note when planning such operations is the influence of air density. Climb schedules are defined in calibrated airspeed (CAS). Below 10 000 ft, at speeds less than about 250 kt, the difference between calibrated airspeed and equivalent airspeed (EAS) is small (see Figure 6.4). This observation enables the influence of air density on the radius of the turn (at speeds associated with the takeoff climb-out) to be seen by rewriting Equation 15.11 as follows:

$$r = \frac{V_e^2}{\sigma g \tan \Phi} \cong \frac{V_c^2}{\sigma g \tan \Phi} = \frac{\theta V_c^2}{\delta g \tan \Phi} \tag{15.12}$$

where V_c is the calibrated airspeed (typical units: m/s, ft/s); and
  σ, θ, and δ are the relative density, temperature, and pressure of the air (dimensionless), respectively.

It is evident that at a given CAS and bank angle, the radius of the turn will increase as the air density reduces. The impact of this is seen by pilots operating out of *hot and high* airports, where combinations of high ambient temperature and low atmospheric pressure are commonplace.

15.3.4 Lift, Drag, and Thrust in a Level Turn

It can be deduced from Figure 15.4 that the vertical component of lift must balance the airplane's weight in order for the airplane to maintain a constant height in a turn. When a pilot (of a conventional airplane with positive maneuver stability) enters a constant-speed turn, he/she will need to pull the stick/yoke back a little—thus increasing the angle of attack—to prevent the airplane from losing height in the turn. The lift coefficient in the turn, at the new angle of attack, can be related to the lift coefficient for wings-level, constant-height flight (C_{L_1}) under equivalent conditions (i.e., at the same weight, altitude, temperature, and airspeed), that is,

$$(C_L)_{turn} = \frac{L}{qS} = \frac{n_z W}{qS} = n_z C_{L_1} \tag{15.13}$$

where q is the dynamic pressure (typical units: N/m^2, lb/ft^2);
  S is wing reference area (typical units: m^2, ft^2); and
  C_{L_1} is the equivalent 1-g lift coefficient (i.e., at the same weight, altitude, temperature, and airspeed).

In a banked turn, where the lift coefficient increases as a result of the pilot increasing the angle of attack a little, the drag coefficient—and hence the drag—will also increase. The drag coefficient in the turn can be determined from the lift coefficient using the airplane's drag polar data (see

Section 7.4). If, for the sake of illustration, the parabolic drag polar (see Equation 7.14) were assumed to approximate the airplane's drag–lift relationship, then the drag coefficient in the turn will be given by

$$(C_D)_{turn} = C_{D_0} + K\left[(C_L)_{turn}\right]^2 = C_{D_0} + K\left(n_z C_{L_1}\right)^2 \tag{15.14}$$

where C_{D_0} is the zero-lift drag coefficient (dimensionless); and
K is the lift-dependent drag factor (dimensionless).

The amount by which the drag increases in the turn is thus

$$\Delta D_{turn} = qS\left\{(C_D)_{turn} - (C_D)_{level}\right\} = \frac{1}{2}\rho V^2 SK\left(n_z^2 - 1\right)\left(C_{L_1}\right)^2 \tag{15.15}$$

To maintain constant height when entering a constant-speed banked turn, the pilot would therefore have to increase the thrust a little to account for the increased drag. If the thrust did not increase, the airplane would lose height.

15.3.5 Minimum Drag Speed in a Turn

When sustained turn performance is limited by the available thrust, the rate of turn can be optimized by flying at the speed that minimizes the drag for the aircraft when turning—that speed, measured in EAS, is $(V_{e_{md}})_{turn}$. The minimum drag speed in a turn, however, is not the same as the minimum drag speed for 1-g level flight.

The minimum drag speed in a turn can be determined using the same approach as that used in Section 7.7.2 to calculate $V_{e_{md}}$. Although the final result can be concluded intuitively, the derivation, which assumes that the airplane's drag can be modeled by the parabolic drag polar (see Section 7.4.3), is given as a further illustration of the use of the fundamental equations. It follows from Equation 15.14 that the drag in a turn is

$$D = qS\left(C_{D_0} + Kn_z^2 C_{L_1}{}^2\right) = qS\left[C_{D_0} + Kn_z^2\left(\frac{W}{qS}\right)^2\right] = qSC_{D_0} + \frac{Kn_z^2 W^2}{qS} \tag{15.16}$$

thus $$D = \left(\frac{C_{D_0}\rho_0 S}{2}\right)V_e^2 + \left(\frac{2Kn_z^2 W^2}{\rho_0 S}\right)\frac{1}{V_e^2} \tag{15.17}$$

At this point, it is convenient to adopt the notation introduced in Section 7.7.2 and to express the drag in the turn, as follows:

$$D = A_1 V_e^2 + B_1 n_z^2 V_e^{-2} \tag{15.18}$$

where $A_1 = \dfrac{C_{D_0}\rho_0 S}{2}$ and $B_1 = \dfrac{2KW^2}{\rho_0 S}$ (see Equation 7.49)

For the minimum drag condition, Equation 15.18 is differentiated with respect to V_e and the resultant set equal to zero, that is,

$$\frac{\mathrm{d}D}{\mathrm{d}V_e} = 2A_1 V_e - 2B_1 n_z^2 V_e^{-3} = 0 \qquad \text{(for a minimum value)}$$

thus $$\left(V_{e_{md}}\right)_{turn} = \sqrt{n_z}\left(\frac{B_1}{A_1}\right)^{1/4} = \sqrt{n_z}\,V_{e_{md}} \tag{15.19}$$

where $V_{e_{md}}$ is defined by Equation 7.46.

The final expression, which describes the minimum drag speed (in EAS) in a turn, based on the parabolic drag polar, can be deduced from Equations 15.18 and 15.19:

$$\left(V_{e_{md}}\right)_{turn} = \left[\frac{2n_z W}{\rho_0 S}\sqrt{\frac{K}{C_{D_0}}}\right]^{1/2} \tag{15.20}$$

The minimum drag speed thus increases as the load factor increases, and the load factor (as shown earlier in Section 15.3.1) is directly related to the bank angle. In other words, the optimum speed, from a drag consideration, increases as the bank angle increases.

It was shown in Section 12.5.5 that the airspeed at which an airplane that has suffered a total engine failure must operate, in order to achieve the greatest still air distance, is the minimum drag speed (V_{md}). In that earlier chapter, however, turning maneuvers were not considered. It is apparent from Equation 15.20 that, in order to achieve the greatest still air gliding distance, the pilot should, ideally, fly a little faster if the flight path has steeply banked turns.

15.3.6   Impact of Wind on Turning Flight Paths

In Section 15.3.3, the turn radius was discussed and described in terms of the airplane's true airspeed and bank angle (or load factor) for constant-height maneuvers. This discussion, however, ignored the influence of wind—the resulting turn radius thus describes the airplane's motion with respect to the air mass and not with respect to the ground.

If the pilot executes a coordinated turn with a constant true airspeed and bank angle, the still air turn radius can be computed using Equation 15.11. To determine the airplane's flight path over the ground (i.e., its track), it is necessary to correct the computed still air flight path for the wind. The track of an airplane flying a circular flight path at constant bank angle under steady wind conditions is illustrated in Figure 15.7a. Whereas this may be acceptable in many situations, it is sometimes required that the pilot maintains a track of constant radius (e.g., in a holding pattern). In such cases, he/she needs to vary the bank angle in the turn to account for drift due to the wind; this is illustrated in Figure 15.7b.

The required bank angle needed to achieve a circular track can be described in terms of the ground speed, by rewriting Equation 15.11:

$$\Phi = \tan^{-1}\left(\frac{V_G^2}{gr}\right) \tag{15.21}$$

where V_G is the airplane's ground speed (units: m/s, ft/s).

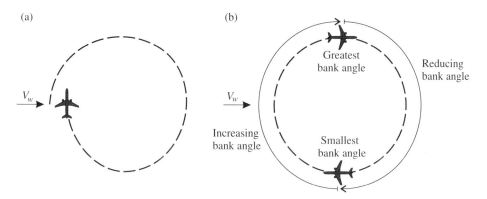

Figure 15.7 (a) Turn with constant bank angle; (b) Turn with varying bank angle (to achieve a circular track).

The magnitude of the ground speed, required for the evaluation of Equation 15.21, can be determined by a vector summation of the true airspeed ($\vec{V}$) and the wind speed ($\vec{V}_w$) vectors (as illustrated in Figure 6.2).

It is evident from Figure 15.7b that the bank angle is greatest when the airplane has a direct tailwind (i.e., highest ground speed) and smallest when the airplane has a direct headwind (i.e., lowest ground speed). A further complexity for the pilot flying a circular track is that the changing bank angle will mean that the drag is not constant; to maintain constant airspeed and height, the thrust will also have to change continuously (as described in Section 15.3.4). The thrust will be greatest when the bank angle is greatest.

Compensating for the wind in this way will ensure that the airplane circles about a fixed ground point. The same principle applies when pilots are required to execute a racetrack pattern hold over a designated location (see Section 14.6.2).

15.4 Climbing or Descending Turns

15.4.1 Climb Angle and Climb Gradient in a Turn

In general, the angle of climb (γ) and the climb gradient ($\tan\gamma$)—see Section 12.4.2—depend on the excess thrust (i.e., the net thrust minus the drag) and on the airplane's acceleration along the flight path (i.e., dV/dt). As described in Section 15.3.4, the airplane's drag increases as it enters into a turn. This means that the airplane will have a reduced capability to climb in a turn compared to an equivalent wings-level climb at the same thrust setting. This reduction in climb gradient or *gradient decrement* is principally a function of the bank angle selected for the turn, but it also depends on the airplane's configuration (e.g., flap position). The climb angle will be further reduced if the airplane increases its speed while turning. These two factors are discussed in this section.

An airplane's *climb potential* can be assessed by considering the maximum climb angle that can be achieved during an accelerated wings-level climb. An appropriate expression that describes the climb angle can be determined from Equation 12.17, that is,

$$\sin\gamma_{wl} = \frac{F_N - D_{wl}}{W} - \frac{1}{g}\left(\frac{dV}{dt}\right) = \frac{F_N}{W} - \frac{D_{wl}}{W} - \frac{1}{g}\left(\frac{dV}{dt}\right) \tag{15.22}$$

where F_N is the net thrust (typical units: N, lb);
D is the drag (typical units: N, lb);
dV/dt is the airplane's acceleration (typical units: m/s², ft/s²); and
where the subscript *wl* denotes a wings-level attitude.

Note that the acceleration term in Equation 15.22 is seen to reduce the climb angle. This can be an important consideration when evaluating takeoff climb obstacle clearance for the one-engine-inoperative condition (see Section 10.7), for example.

The influence of a turn on an airplane's climb performance (at constant thrust setting) can be assessed by considering an additional drag term (ΔD_{tn}), which is inserted into Equation 15.22:

$$\sin\gamma_{tn} = \frac{F_N}{W} - \left(\frac{D_{wl}}{W} + \frac{\Delta D_{tn}}{W}\right) - \frac{1}{g}\left(\frac{dV}{dt}\right) \tag{15.23}$$

where the subscript *tn* implies a banked attitude (i.e., a turn).

As the climb angles under consideration in this analysis are small, it is possible to substitute the weight in Equation 15.23 with the airplane's lift. For convenience, the lift and drag terms are now expressed as coefficients:

$$\sin \gamma_{tn} \cong \frac{F_N}{W} - \frac{C_{D_{wl}}}{C_{L_{wl}}} - \frac{\Delta C_{D_{tn}}}{C_{L_{wl}}} - \frac{1}{g}\left(\frac{dV}{dt}\right) \tag{15.24}$$

Equation 15.24 can be used to conduct trades (trade studies) in which the impact of selected bank angles and accelerations on the airplane's ability to climb in a turn, with one engine inoperative, can be assessed.

15.4.2 Reduction in Climb Angle (or Climb Gradient) in a Steady-Speed Turn

When planning a departure from an airport with obstacles beneath the intended flight path, compliance with the regulatory obstacle clearance requirements [3, 4] must be possible following an engine failure on takeoff (see Section 10.7). In such analyses, consideration should be given to the reduced climb angle (or climb gradient) which occurs if the planned flight path includes a turn.

The change in climb angle ($\Delta \gamma$) in a turn, compared to the equivalent wings-level case, can be expressed as follows:

$$\Delta \gamma = \gamma_{tn} - \gamma_{wl} \tag{15.25}$$

> where the subscript *tn* denotes a banked attitude (turn) and the subscript *wl* a wings-level attitude.

The change in climb gradient ($\Delta \tan \gamma$) is thus

$$\Delta \tan \gamma \cong \sin \Delta \gamma = \sin \gamma_{tn} - \sin \gamma_{wl} \tag{15.26}$$

In the case of a steady-speed climb, $dV/dt = 0$ and Equation 15.23 can be simplified:

$$\sin \gamma_{tn} = \frac{F_N}{W} - \left(\frac{D_{wl}}{W} + \frac{\Delta D_{tn}}{W}\right) = \sin \gamma_{wl} - \frac{\Delta D_{tn}}{W} \tag{15.27}$$

thus $\quad \sin \Delta \gamma = -\dfrac{\Delta D_{tn}}{W} \tag{15.28}$

Equation 15.28 can also be expressed in terms of the lift and drag coefficients, that is,

$$\sin \Delta \gamma \cong -\frac{\Delta C_{D_{tn}}}{C_{L_{wl}}} = -\frac{C_{D_{tn}} - C_{D_{wl}}}{C_{L_{wl}}} \tag{15.29}$$

The reduction in climb angle can thus be determined using the airplane's drag polar and knowledge of the wings-level lift coefficient. Note that the drag coefficient in the turn $C_{D_{tn}}$ would correspond to the airplane's lift coefficient in the turn, that is, $n_z C_{L_{wl}}$.

Alternatively, if the drag polar can be represented by the parabolic drag polar, then the drag increment ΔD_{tn} can be expressed directly in terms of the bank angle. Using Equations 15.15 and 15.7, it can be shown that

$$\Delta D_{tn} = qSK\left(n_z^2 - 1\right)\left(C_{L_1}\right)^2 = qSK \tan^2 \Phi C_{L_1}\left(\frac{W}{qS}\right) = KWC_{L_1} \tan^2 \Phi \tag{15.30}$$

> where C_{L_1} is the 1-*g* lift coefficient at the same airplane weight, altitude, air temperature, and speed.

Substituting Equation 15.30 into 15.28 yields an expression that describes the change in the climb angle, from which the gradient decrement can be deduced:

$$\sin \Delta \gamma = -KC_{L_1} \tan^2 \Phi \tag{15.31}$$

15.4.3   Reduction in Rate of Climb in a Steady-Speed Turn

In the same way that the angle of climb is reduced in a turn due to the increase in drag, so too will the *rate of climb* be lower in a turn compared to the equivalent wings-level case (at the same thrust setting). This can be assessed by assuming that the parabolic drag polar provides a reasonable representation of the airplane's drag and by noting that the rate of climb is equal to $V \sin \gamma$ (see Section 12.4.4). The reduction in climb rate (i.e., compared to the wings-level case) for a steady-speed turning climb at constant thrust can thus be deduced from Equation 15.31:

$$\Delta V_v = (\Delta \sin \gamma)V = -KC_{L_1} V \tan^2 \Phi \tag{15.32}$$

where V_v is the rate of climb (the units will be the same as those selected for V).

15.5   Level Uncoordinated Turns

15.5.1   Forces and Load Factor in a Level Uncoordinated Turn

In Section 15.3, level coordinated turns were considered. In this section, the theory has been extended to consider *uncoordinated* turns, in which a net sideforce (F_Y) acts on the airplane. In Figure 15.8, the sideforce is shown to act inwards (i.e., towards the center of the turning circle)—this would correspond to a skidding turn, as described in Section 15.2.1. As before, the thrust vector is assumed to be approximately aligned with the flight direction (and will not be considered in this simplified analysis). For the component of lift to balance both the weight and the vertical component of the sideforce, the lift must be greater than that required for the equivalent level coordinated turn.

In a repeat of the analysis presented in Section 15.3, the forces acting on the airplane are again resolved in the vertical direction:

$$L \cos \Phi - F_Y \sin \Phi = W \tag{15.33}$$

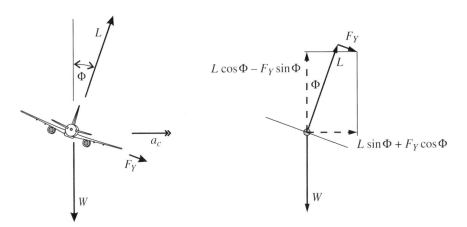

Figure 15.8 Forces acting on the airplane in a level uncoordinated turn.

and in the horizontal direction:

$$L \sin \Phi + F_Y \cos \Phi = ma_c = \frac{W}{g} a_c \tag{15.34}$$

Equation 15.34 is now rearranged and the lift substituted using Equation 15.33:

$$\frac{a_c}{g} = \tan \Phi + \frac{F_Y}{W} \frac{\sin^2 \Phi}{\cos \Phi} + \frac{F_Y}{W} \cos \Phi = \tan \Phi + \frac{F_Y}{W} \left(\frac{\sin^2 \Phi + \cos^2 \Phi}{\cos \Phi} \right)$$

thus $\quad \dfrac{a_c}{g} = \tan \Phi + \dfrac{F_Y}{W \cos \Phi} \tag{15.35}$

This expression can be simplified by introducing the sideforce load factor, which is defined as follows:

$$n_Y = \frac{F_Y}{W} \tag{15.36}$$

thus $\quad \dfrac{a_c}{g} = \tan \Phi + \dfrac{n_Y}{\cos \Phi} \tag{15.37}$

The normal load factor n_z can be obtained directly from Equation 15.33:

$$n_z = \frac{L}{W} = \frac{1}{\cos \Phi} + n_Y \tan \Phi \tag{15.38}$$

By comparing Equation 15.38 to Equation 15.3, it is seen that a net sideforce can increase or decrease the normal load factor n_z, depending on its direction.

15.5.2   Turn Rate and Turn Radius

In an uncoordinated turn executed at a given speed (V) and bank angle (Φ), the presence of a net sideforce alters the rate of turn, the time required to turn through 180°, and the radius of turn. An expression for the rate of turn can be deduced from Equations 15.5 and 15.37:

$$\Omega = \frac{g}{V} \left(\tan \Phi + \frac{n_Y}{\cos \Phi} \right) \tag{15.39}$$

The time (t) required to execute a 180° coordinated turn was given by Equation 15.9. In the case of an uncoordinated turn, the time is not exactly the same:

$$t_\pi = \frac{\pi}{\Omega} = \frac{\pi V}{g} \left(\frac{\cos \Phi}{\sin \Phi + n_Y} \right) \tag{15.40}$$

where t_π is the time required for a 180° turn (typical unit: s).

The radius of turn is obtained by substituting Equation 15.39 into Equation 15.4, that is,

$$r = \frac{V^2}{g} \left(\frac{\cos \Phi}{\sin \Phi + n_Y} \right) \tag{15.41}$$

It is evident that when the net sideforce is zero, Equations 15.39, 15.40, and 15.41 reduce to the forms given for the coordinated turn (i.e., Equations 15.6, 15.9, and 15.10).

From a pilot's perspective, what happens in a skidding turn (shown in Figure 15.8) is that the rate of turn is too great for the angle of bank. The corrective action for a skid—in order to achieve a coordinated turn—would be for the pilot to apply a rudder input (see Section 15.2.2) so as to reduce the turn rate and increase the bank angle. The reverse is true for a slipping turn—that is, the pilot should decrease the bank angle.

A cursory study of Equations 15.39 to 15.41 appears to indicate that a net sideforce acting inwards (i.e., towards the turning center) will augment turning performance by increasing the turn rate and reducing the turn radius. However, this analysis has not considered the impact of the maneuver on the airplane's drag. Uncoordinated turns are, in fact, comparatively inefficient. As mentioned earlier (in Section 15.3.4), the action of banking the airplane increases its drag through an increase in the required lift, which necessitates an increase in thrust (and fuel burn) if speed and height are to be maintained in the turn. Control surface deflections also contribute to the airplane's overall drag. A deliberate sideslip, such as that associated with an uncoordinated turn, will increase the drag still further, resulting in reduced turn performance despite the presence of a component of sideforce acting towards the center of the turn.

15.6 Limits and Constraints in Turning Maneuvers

15.6.1 Limiting Factors in Turns

For a sustained, level turn, the airplane must maintain both speed and height. If the pilot were to increase the angle of bank progressively, tightening the turn, the required lift and thrust would progressively increase. This could, theoretically, continue until a limiting condition was reached. The limit could be imposed by one of three factors: (1) the maximum lift coefficient that can be achieved, (2) the maximum load factor that can be tolerated (i.e., a structural design limit), or (3) the maximum available thrust (at a particular thrust rating). These factors are discussed in Sections 15.6.2–15.6.5.

Depending on the flight conditions (i.e., airplane configuration, weight, altitude, air temperature, and thrust setting), it may not always be possible to maintain speed in a level turn. If the pilot allows the airplane to slow down in the turn, while maintaining altitude, the maximum angle of bank (and load factor) will only be limited by the maximum lift coefficient and the structural strength of the airplane. The same argument applies to a turn at constant speed, but where height is lost. Such turns, in which speed and height are not sustained, are usually called *instantaneous* turns. If the speed and/or height were permitted to drop, the resulting instantaneous turn rate could be higher than the comparative sustained turn rate (unless the bank angle was limited for the reasons mentioned above).

15.6.2 Maximum Lift Coefficient

The lift coefficient for an airplane in a level turn is described by Equation 7.4, which can be written as follows:

$$C_L = \frac{2n_z W}{\rho V^2 S} = \frac{2n_z W}{\rho_0 a_0^2 \delta M^2 S} \qquad \text{(from Equation 7.4)}$$

In the above-mentioned scenario of a progressively increasing angle of bank (and load factor) in a turn, it is apparent that C_L must progressively increase. A limiting condition will be reached when the lift coefficient is equal to the maximum lift coefficient (C_{Lmax})—the magnitude of which depends on the airplane's configuration (i.e., clean or with high lift devices deployed).

The turn rate corresponding to this limiting condition can be determined by substituting the load factor in Equation 15.8 using Equation 7.4. Once the rate of turn is known, the radius can be found using Equation 15.4. Turn rate and turn radius can be expressed as functions of TAS (as performed earlier) or Mach number (the latter option has been selected as the resulting expression will be used to generate a maneuver envelope in Section 15.6.4).

$$\text{Thus} \quad \Omega = \frac{g}{aM} \sqrt{\left(\frac{C_{L_{max}} \rho_0 a_0^2 S}{2W} \right)^2 \delta^2 M^4 - 1} \tag{15.42}$$

$$\text{and} \quad r = \frac{a^2 M^2}{g \sqrt{\left(\frac{C_{L_{max}} \rho_0 a_0^2 S}{2W} \right)^2 \delta^2 M^4 - 1}} \tag{15.43}$$

15.6.3   Structural Design Limits

Normal operation of the airplane should not result in the certified maximum design load factor being exceeded. If the limit load factors are marginally exceeded, structural damage may result, requiring repair or replacement. However, should the limits be substantially exceeded, structural failure will occur. The flight envelope of load factor versus airspeed, within which an airplane can be safely operated, is discussed later in Section 20.7. The positive limit load factor for airplanes certified to FAR/CS 25 [1, 2] varies from 2.5 to a maximum of 3.8, depending on the airplane type. It is evident from Figure 15.5 that high load factors would not be encountered in normal flight operations when the pilot executes typical, routine maneuvers; however, extreme bank angles and high load factors can be encountered in emergencies.

To prevent structural damage, the flight control systems on modern fly-by-wire (FBW) airplanes incorporate envelope protection features—for example, Airbus limits the bank angle on jet transport airplanes to 67°. The corresponding load factor, obtained using Equation 15.3, is 2.56. Although Boeing's design philosophy in this respect is not the same (i.e., the software does not prevent the pilot from exceeding the design limit), the maximum allowable load factors are comparable.

The allowable turn rate and radius of turn that corresponds to a given load factor were derived earlier in Sections 15.3.2 and 15.3.3. Equations 15.8 and 15.10 are repeated with the load factor set equal to the limit load factor ($n_{z,L}$) and the speed expressed as Mach number.

$$\Omega = \frac{g}{aM} \sqrt{n_{z,L}^2 - 1} \tag{15.44}$$

$$\text{and} \quad r = \frac{a^2 M^2}{g \sqrt{n_{z,L}^2 - 1}} \tag{15.45}$$

15.6.4   Maneuver Envelope

A maneuver envelope within which the airplane can be safely operated can be constructed on a chart of turn rate versus airspeed (see Figure 15.9), based on the aerodynamic and structural limits discussed earlier (in Sections 15.6.1–15.6.3). The envelope is bounded on the left by the stall condition, at the top by the limit load factor, and on the right by the certified maximum operating speed (see Section 20.5.3). The characteristic shape of this envelope leads to it sometimes being called a *doghouse* plot. The point where the stall line meets the limit load factor line is known as the *corner speed*.

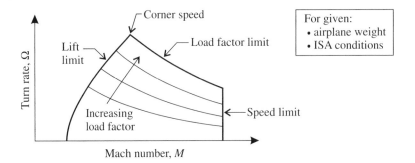

Figure 15.9 Turn rate versus Mach number (doghouse plot).

15.6.5 Available Thrust and Turning Limits

In a sustained level turn, the thrust is equal to the drag, which, for the sake of illustration, has been modeled using the parabolic drag polar (see Section 7.4.3). The *required thrust* is given by

$$F_N = D = qSC_{D_0} + K\frac{(n_z W)^2}{qS} \tag{15.46}$$

and $\quad q = \frac{1}{2}\rho V^2 = \frac{1}{2}\rho_0 V_e^2 = \frac{1}{2}\rho_0 a_0^2 \delta M^2 \tag{15.47}$

If the required thrust cannot be generated by the engine(s), at a particular flight condition, the airplane would not be able to sustain both height and speed in the turn. In other words, the available thrust can limit the bank angle and hence the load factor. Equation 15.46 can be rearranged to express the limiting load factor in terms of the *available thrust*:

$$n_z = \sqrt{\frac{qS}{WK}\left(\frac{F_N}{W} - \frac{qSC_{D_0}}{W}\right)} \tag{15.48}$$

A flight condition where this limit can manifest is cruise at an altitude close to the airplane's ceiling (see Section 17.3). The thrust available is limited by the maximum cruise thrust (MCRT) rating. If the bank angle exceeds the angle at which thrust required equals thrust available, then the airplane will lose height if speed is maintained.

15.6.6 Operational Limits and Constraints

The takeoff and landing phases of a flight, for reasons of safety, are subject to a variety of regulatory constraints, as discussed in Chapters 9–11. Maximum bank angles are specified for certain operating conditions. For example, normal takeoff operations restrict the bank angle to 0° below 50 ft, then to 15° at the takeoff safety speed V_2 up to a height of 400 ft, and to 25° at the final takeoff speed V_{FTO} (i.e., above 400 ft) [4]. A bank angle limit of 30° during approach is also imposed for normal operations [5].

Operators can also define bank angle limits, which are stipulated in the airplane's Operations/Operating Manual (OM). For normal flight operations of commercial passenger aircraft, a bank angle limit of 25° to 30° is typically adhered to for reasons of passenger comfort. On Airbus airplanes, for example, the flight control software imposes a bank angle constraint of 33°—pilots are required to hold a constant sidestick input to maintain a bank angle greater than 33°. Airplane specific limits are sometimes defined for flight with one engine inoperative.

During cruise, the airplane is usually flown by the autopilot in normal flight operations.[3] Heading changes, which result in the airplane banking and turning, are commanded by the flight crew by entering the desired heading into the autopilot. Bank angle limits can be selected by the flight crew or set to auto (automatic). Manual flying at high altitude, in which the pilot directly controls the airplane by yoke/stick inputs, essentially only occurs during flight testing and in emergencies. The flight characteristics of jet transport airplanes at high altitude are known to be quite different compared to their characteristics at low altitude [6]. A more detailed treatment of flight limitations, which includes a discussion on high-speed buffet, is presented in Chapter 20.

15.7   Pitching Maneuvers

15.7.1   Introduction

In a pitching maneuver, the airplane will experience an angular velocity about the Y_b axis (see Figure 3.7), and the airplane will move along an approximately circular flight path, which is usually in a vertical or near-vertical plane. A *pull-up* maneuver (which results in an increased load factor) is commanded by the pilot pulling the stick/yoke backwards, thus deflecting the elevator upwards, whereas a *push-over* maneuver (which results in a reduced load factor) is commanded by the pilot pushing the stick/yoke forwards. Pitching maneuvers can be executed in a manner that holds the wings in a level attitude (i.e., symmetrical flight) or, alternatively, a pitching maneuver could simultaneously involve a roll. A rolling pull-up would be commanded by the pilot pulling the stick/yoke backwards and simultaneously moving the sidestick to one side (or turning the yoke), thus deflecting the ailerons.

15.7.2   Symmetrical Pitching Maneuvers

The discussion presented in this section is restricted to symmetrical pitching maneuvers. Equations are presented that describe the aircraft's centripetal acceleration, flight path radius, and turn rate [7].

The forces that act on the airplane are illustrated in Figure 15.10. The flight path angle (γ)—measured from a horizontal datum—defines the angle of the lift vector, which passes through

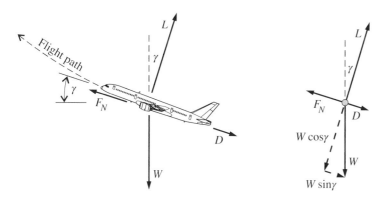

Figure 15.10 Symmetrical pitching maneuver.

---

3 The use of autopilot is mandatory for airplanes operating within RVSM (Reduced Vertical Separation Minimum) airspace (see Section 5.4.4), which extends from FL 290 to FL 410 and covers a large part of the world's airspace [6].

the center of curvature of the flight path. The magnitude of the lift will depend on the amount that the stick/yoke has been pulled back. The thrust and drag vectors are assumed to act tangential to the curved flight path.

The forces are initially considered in the tangential direction, and by the application of Newton's second law it follows that

$$F_N - D - W \sin \gamma = m \frac{dV}{dt} \qquad (15.49)$$

If the rate of change of speed during the maneuver is negligible, then Equation 15.49 can be simplified:

$$F_N = D + W \sin \gamma \qquad (15.50)$$

It is apparent from Equations 15.49 and 15.50 that an increase in thrust is required to maintain a constant speed as the flight path angle increases.

The forces are now considered in the normal direction, and an expression for the centripetal acceleration determined:

$$L - W \cos \gamma = m a_c = W \frac{a_c}{g} \qquad (15.51)$$

Thus $\quad \dfrac{a_c}{g} = n_z - \cos \gamma \qquad (15.52)$

where n_z is the load factor normal to the flight path (dimensionless); and
a_c is the centripetal acceleration (typical units: m/s^2, ft/s^2).

This result enables an expression for the instantaneous radius (r) of the flight path to be determined. Equation 15.52 is substituted into Equation 15.5:

$$r = \frac{V^2}{g(n_z - \cos \gamma)} \qquad (15.53)$$

During a pull-up maneuver, the flight path angle (γ) will be zero at the lowest point of the curved flight path. However, as the maneuver progresses and γ increases, the centripetal acceleration will increase and the radius will reduce. This is due to the changing orientation of the weight vector with respect to the lift vector. One implication of this deduction is that the flight path of an airplane performing a constant-load-factor, constant-speed pull-up maneuver will not be exactly circular.

The turn rate can be deduced from Equation 15.4, that is,

$$\Omega = \frac{g(n_z - \cos \gamma)}{V} \qquad (15.54)$$

where Ω is the turn rate (typical unit: rad/s).

As with the radius of turn, the turn rate is not exactly constant during the maneuver. Conversely, for the pilot to execute a constant-speed pitching maneuver, in which the radius of turn and the rate of turn remain constant, he/she would need to increase progressively the load factor by progressively pulling the stick/yoke back during the maneuver.

15.8 Total Energy

15.8.1 Introduction

The total energy concept is a useful approach to the study of non-steady-state maneuvers. The total energy is the sum of the airplane's potential energy, which is a function of height (for a fixed airplane mass), and its kinetic energy, which is a function of true airspeed (for a fixed airplane mass). Maneuvers executed under conditions of constant total energy (i.e., unchanged with respect to time), involve an *exchange* of potential energy and kinetic energy—which essentially means an exchange between height and speed. The rate of change of an airplane's total energy, however, depends on the power output of its engine(s) and this represents a measure of the airplane's capability to climb or accelerate or conduct a maneuver involving a climb and an acceleration.

The total energy concept has proven to be particularly useful in evaluating the performance of aircraft types where maneuverability is paramount (e.g., tactical air superiority aircraft and aerobatic aircraft). Nonetheless, it is equally valid for other airplane types, and can be useful when assessing an airplane's maneuvering capability in a low thrust condition (e.g., following an engine failure).

15.8.2   Total Energy Expression

The total energy "possessed" by an airplane at any given instant is equal to the sum of its potential and kinetic energies, that is,

$$E_t = E_p + E_k \tag{15.55}$$

where E_t is the total energy (typical units: J, ft lb);
E_p is the potential energy (typical units: J, ft lb); and
E_k is the kinetic energy (typical units: J, ft lb).

The potential energy (see Section 2.4.9) of the airplane can be described as follows:

$$E_p = m \int_{h_0}^{h} g \, dh \qquad \text{(Equation 2.43)}$$

where m is the airplane mass (typical units: kg, slug);
g is the gravitational acceleration (typical units: m/s^2, ft/s^2);
h_0 is the datum height; and
h is the geometric height (typical units: m, ft).

By selecting the height datum as sea level and introducing geopotential height (see Equation 5.1), the airplane's potential energy can be expressed in a simpler way:

$$E_p = m \int_{0}^{h} g \, dh = mg_0 H \tag{15.56}$$

where g_0 is the standard value of g (typical units: m/s^2, ft/s^2), and
H is the geopotential height (typical units: m, ft).

The kinetic energy of the airplane (see Section 2.4.8) may be described as a function of the true airspeed:

$$E_k = \frac{1}{2}mV^2 \qquad \text{(Equation 2.42)}$$

Substituting Equations 15.56 and 2.42 into Equation 15.55 yields the required expression for the total energy:

$$E_t = mg_0 H + \frac{1}{2}mV^2 \tag{15.57}$$

15.8.3 Specific Energy (Energy Height)

To facilitate comparisons between different airplanes, the traditional approach is to divide the total energy (E_t) by the airplane weight, which is taken as the product of airplane mass and the standard value of gravitational acceleration. The resulting parameter is called the *specific energy*:

$$E_s = \frac{E_t}{W} = H + \frac{V^2}{2g_0} \tag{15.58}$$

where E_s is the specific energy (typical units: m, ft); and
W is the airplane's weight (typical units: N, lb).

Note that *specific energy* has the same units as height; for this reason it is also referred to as *energy height* (H_e)—the two terms have the same meaning and both are used in practice. Energy height is the height that an airplane could theoretically reach if *all* of the kinetic energy were converted into potential energy.

Specific energy can be regarded as an inherent performance measure of an airplane, which depends solely on its geopotential height (H) and its true airspeed (V). Using Equation 15.58, lines of constant specific energy (contours) can be constructed on a graph of H (as the ordinate) and V (as the abscissa), as shown in Figure 15.11. Note that this figure is not airplane specific—it applies to all aircraft, irrespective of type or weight. Along any line of constant specific energy, height and speed can theoretically be traded, or exchanged, without restriction. However, for an airplane to move from one energy level to a higher level requires an increase in its total energy (i.e., an energy input)—and the ability of an airplane to do this, at a given height and speed, depends on the performance characteristics of the individual airplane. This is discussed further in Section 15.8.4, after the concept of specific excess power is introduced.

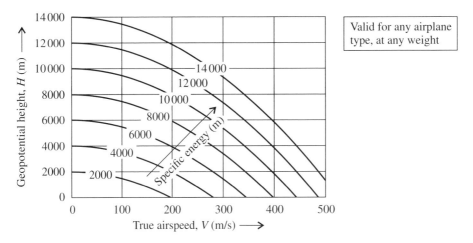

Figure 15.11 Constant specific energy (energy height) curves (contours).

15.8.4 Specific Excess Power

A useful performance parameter that characterizes an airplane's ability to increase its specific energy, enabling it to climb and/or accelerate, is known as *specific excess power* (SEP). This parameter (performance metric) is best introduced by considering the rate of change of specific energy with respect to time. Differentiating Equation 15.58 yields the following expression:

$$\frac{dE_s}{dt} = \frac{dH}{dt} + \frac{V}{g_0}\frac{dV}{dt} \tag{15.59}$$

The next step is to consider the two sides of this equation separately. On the left-hand side is the time rate of change of specific energy. This is a measure of the ability of the airplane to change its energy level (i.e., total energy) at a particular speed and height within a given time. As work is done when a body is moved to a higher energy level, and as power is the time rate of doing work, it is seen that the left-hand side of Equation 15.59 is measure of *specific power* (i.e., power per unit of weight). Interestingly, specific power can be written with units of speed.

The right-hand side of Equation 15.59 relates to an expression that was derived earlier, in Section 12.4.4. Equation 12.26, which describes the rate of climb, may be rearranged as follows:

$$\left(\frac{F_N}{W} - \frac{D}{W}\right)V = \frac{dH}{dt}\left(1 + \frac{V}{g_0}\frac{dV}{dH}\right) \tag{15.60}$$

Thus $\quad \left(\dfrac{F_N V - DV}{W}\right) = \dfrac{dH}{dt} + \dfrac{V}{g_0}\dfrac{dV}{dt} \tag{15.61}$

The product of thrust and velocity is *thrust power* and the product of drag and velocity is *drag power*. The term on the left-hand side of Equation 15.61 thus represents the excess of thrust power over drag power divided by weight—for this reason, it is called *specific excess power* (SEP). The specific excess power is identical to the time rate of change of specific energy. Combining Equation 15.61 with Equation 15.59 yields the following expression for specific excess power:

$$P_s = \frac{dE_s}{dt} = \left(\frac{F_N V - DV}{W}\right) = \frac{dH}{dt} + \frac{V}{g_0}\frac{dV}{dt} \tag{15.62}$$

where P_s is the specific excess power (typical units: m/s, ft/s).

The SEP depends on the airplane's thrust (which is a function of the engine thrust setting, altitude, air temperature, and flight speed), drag (which is also a function of the altitude, air temperature, and flight speed, as well the load factor), and weight.

If thrust and drag data were available for an airplane, Equation 15.62 could be used to determine values of P_s for varying height (H) and speed (V), as illustrated in Figure 15.12. Similarly, flight tests can be used to provide SEP data. The measurement of P_s is not complicated: the airplane's acceleration at constant height is measured across the required speed range, and the SEP deduced from the quantity $(V/g_0)(dV/dt)$ at selected speeds (see Equation 15.64, below).

An appreciation for SEP as a performance metric can be obtained by considering two flight conditions: constant-true-airspeed climb and constant-height acceleration.

(1) In a constant-true-airspeed climb, the rate of climb will equal the airplane's total SEP at the point of interest, that is,

$$P_s = \frac{dH}{dt} = \left(\frac{F_N - D}{W}\right)V \tag{15.63}$$

The maximum rate of climb will correspond to the condition for maximum SEP.

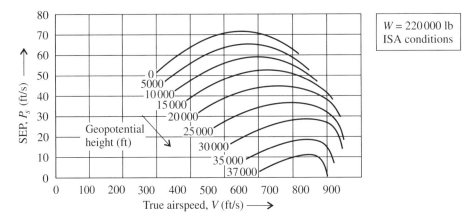

Figure 15.12 SEP versus TAS relationships for a mid-size, twin-engine jet airplane.

(2) For flight at constant height, the airplane's acceleration will relate to the SEP, as follows:

$$P_s = \frac{V}{g_0}\frac{dV}{dt} = \left(\frac{F_N - D}{W}\right)V \tag{15.64}$$

However, in this case, the maximum acceleration is given by the condition for maximum excess thrust (i.e., when $F_N - D$ is a maximum) and not maximum SEP.

15.8.5 Total Energy Performance Graphs (Charts)

The application of the total energy approach usually involves the generation of charts of the two key measures: specific energy (E_s) and specific excess power (P_s). One convenient representation involves superimposing lines of constant P_s on an H–V chart that has lines of constant E_s—as illustrated in Figure 15.13. Along any line of constant energy, height and speed may be exchanged without "using" excess power. Moving from one energy level to a higher

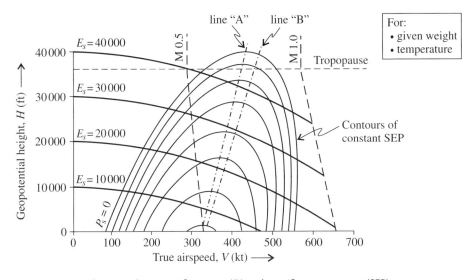

Figure 15.13 Total energy chart: specific energy (E_s) and specific excess power (SEP).

level requires excess power. The ability of the airplane to do this (at a given height and speed) depends on its SEP—and this in turn depends on the available engine thrust (at a given engine setting) and the airplane's weight and drag.

The line of zero SEP in Figure 15.13 represents a critical energy boundary—within this envelope, the airplane has *positive* SEP and this can be used to increase its potential energy or kinetic energy, or both. Outside this envelope, this cannot happen. The $P_s = 0$ boundary describes the performance potential of the airplane in a certain condition; if the weight reduces, the envelope will expand; similarly, if the drag (or load factor) increases, the envelope will contract.

Total energy charts can be used to determine the climb schedule for the fastest climb, for example, as the maximum rate of climb corresponds to the condition of maximum SEP. The locus of points of tangency between contours of constant SEP and lines of constant height (line "A" in Figure 15.13) represents the climb speed schedule for the fastest climb (i.e., minimum time to climb).

The maximum *rate of climb* schedule provides the optimum path for increasing potential energy (E_p); however, no consideration is given to the kinetic energy (E_k) in determining the climb schedule, and E_k can increase, decrease, or even remain constant, depending on the shape of the SEP contours. The climb schedule that results in the maximum rate of increase of *total energy* corresponds to a slightly faster true airspeed climb than in the previous case—and this is given by the locus of points of tangency between contours of constant SEP and lines of constant E_s (line "B" in Figure 15.13). For the twin-engine jet airplane used for this illustration, the speeds were seen to be about 20 kt faster than those that correspond to the maximum rate of climb.

Another application of SEP concerns the turning performance of an airplane. To sustain a steady turn, without losing either height or speed, requires an increase in thrust (compared to level flight). The limiting condition (with the engine at a fixed thrust setting) can be represented on SEP charts with load factor and turn radius data also indicated.

Total energy charts, with contours of E_s and P_s, indicate the potential for an airplane to execute a maneuver. It should be remembered, however, that other factors, such as aerodynamic limits (e.g., stall) and structural limits (e.g., maximum load factor) will restrict the envelope within which the airplane may be safely operated (see Chapter 20). These limits may be superimposed on a total energy chart, if required.

References

1 FAA, *Airworthiness standards: Transport category airplanes*, Federal Aviation Regulation Part 25, Amdt. 25-143, Federal Aviation Administration, Washington, DC, June 24, 2016. Latest revision available from www.ecfr.gov/ under e-CFR (Electronic Code of Federal Regulations) Title 14.

2 EASA, *Certification specifications and acceptable means of compliance for large aeroplanes*, CS-25, Amdt. 18, European Aviation Safety Agency, Cologne, Germany, June 23, 2016. Latest revision available from www.easa.europa.eu/ under Certification Specification.

3 FAA, *Operating requirements: Domestic, flag, and supplemental operations*, Federal Aviation Regulation Part 121, Amdt. 121-374, Federal Aviation Administration, Washington, DC, May 24, 2016. Latest revision available from www.ecfr.gov/ under e-CFR (Electronic Code of Federal Regulations) Title 14.

4 European Commission, *Commercial air transport operations (Part-CAT)*, Annex IV to Commission Regulation (EU) No. 965/2012, Brussels, Belgium, Oct. 5, 2012. Published in *Official Journal of the European Union*, Vol. L 296, Oct. 25, 2012, and reproduced by EASA.

5 EASA, "Acceptable means of compliance (AMC) and guidance material (GM) to Part-CAT," Iss. 2, Amdt. 5, European Aviation Safety Agency, Cologne, Germany, Jan. 27, 2016. Available from www.easa.europa.eu/.

6 Rosay, J., "High-altitude manual flying," *Safety first*, Product Safety Department, Airbus S.A.S., Blagnac, France, Iss. 20, pp. 37–45, July 2015.

7 Young, T.M., "Maneuver performance of fixed-wing transport aircraft," in *Encyclopedia of Aerospace Engineering*, Blockley, R. and Shyy, W. (Eds.), Wiley, Hoboken, NJ, pp. 2601–2612, 2010.

16

Trip Fuel Requirements and Estimation

16.1 Introduction

Specific requirements exist for the determination of the fuel to be uploaded prior to departure. These requirements depend on the operator, type of airplane, planned route to the destination, and the availability of an alternate airport—which will be required if, for any reason, the airplane is unable to land at the destination airport. Operators of public transport aircraft are required, by law, to ensure that all flight operations are conducted in a manner that meets a minimum standard of safety that has been deemed appropriate for the operation; fuel planning is a critical aspect of this process. The requirements pertaining to the planning (and operation) of commercial flights are described in statutory documents published by the state in which jurisdiction for the flight resides—these requirements, to a large extent, reflect a national implementation of the standards and recommendations of ICAO [1]. Although the overarching objective is the same: to ensure that there is adequate fuel onboard for the safe conduct of the mission—taking into account foreseeable contingencies and emergencies—differences exist between the requirements issued by different authorities, such as the FAA [2, 3] and EASA [4, 5]. For example, the contingency fuel for international flights calculated according to FAA rules would be sufficient for the airplane to fly an additional 10% of the trip *time* [2], whereas the corresponding EASA requirement is based on a fixed percentage (e.g., 5%) of the trip *fuel* [5].

An overview of fuel planning requirements for commercial flights is given in this chapter according to ICAO (Section 16.2), FAA (Section 16.3), and EASA (Section 16.4). A computational procedure to determine the takeoff weight and trip fuel in accordance with fuel planning requirements is outlined in Section 16.5. These regulations are also used by aircraft manufacturers to determine an airplane's payload–range performance characteristics (Section 16.6). Such illustrative "brochure" information is based on a set of mission rules, which define the fuel allowances, flight profile, and associated parameters needed to calculate the airplane's range (maximum trip distance) for varying payload.

Techniques that can be used to establish an estimate of trip fuel for non-operational applications (e.g., environmental studies) are described in Sections 16.7 and 16.8. This is followed by discussions on several loosely connected topics regarding trip fuel, including factored great circle distances connecting the departure and destination airports (Section 16.9), transporting fuel (Section 16.10), and reclearance (Section 16.11). The influence of wind, temperature, and barometric pressure changes during cruise are discussed in Section 16.12; mention is also made of the reference conditions used to establish published cruise data—for example, as recorded in the Flight Crew Operations/Operating Manual (FCOM). The chapter closes with the derivation of a mathematical expression that can be used to estimate changes in cruise fuel for small

Performance of the Jet Transport Airplane: Analysis Methods, Flight Operations, and Regulations, First Edition.
Trevor M. Young.
© 2018 John Wiley & Sons Ltd. Published 2018 by John Wiley & Sons Ltd.

changes to an airplane's lift-to-drag ratio, thrust specific fuel consumption (TSFC), or weight, or a combination thereof (Section 16.13).

The definitions of standard airplane weights (such as operating empty weight and zero fuel weight)—which are important when determining trip fuel or establishing payload–range charts, for example—are given in Section 19.2.

16.2 ICAO Requirements

General fuel requirements are stipulated in ICAO Annex 6, section 4.3.6.3 [1]. When an alternate aerodrome (airport) is specified, airplanes with turbine engines must have sufficient fuel (see Figure 16.1):

(A) to fly to and execute an approach, and a missed approach, at the aerodrome to which the flight is planned; and thereafter
(B) to fly to the alternate aerodrome specified in the flight plan; and then
(C) to fly for 30 minutes at holding speed at 1500 ft above the alternate aerodrome under standard temperature conditions, and approach and land; and
(D) to have an additional amount of fuel sufficient to provide for the increased consumption on the occurrence of any of the potential contingencies specified by the operator to the satisfaction of the state of the operator.

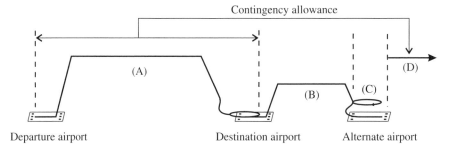

Figure 16.1 ICAO fuel requirements (ICAO Annex 6 [1]).

16.3 FAA Requirements

16.3.1 Overview

The requirements for fuel planning are described in different parts of the FAA regulations [2, 3, 6], depending on the type of airplane, operation, and jurisdiction (e.g., domestic or international). The requirements given in FAR 121 (Operating Requirements: Domestic, Flag, and Supplemental Operations) [2] are outlined below. Three cases are described in this section:

(1) US domestic operations;
(2) international operations where an alternate airport is specified; and
(3) international operations where an alternate airport is not specified ("island" reserves).

The computed fuel must consider realistic operational scenarios—for example, the following has to be taken into account: (1) wind and other weather conditions forecast; (2) anticipated traffic delays; (3) one instrument approach and possible missed approach at destination; and (4) any other conditions that may delay landing of the aircraft (FAR 121.647 [2]).

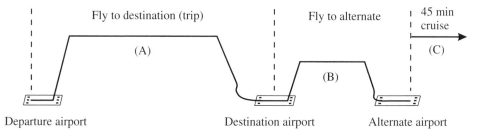

Figure 16.2 FAA domestic fuel requirements (FAR 121.639 [2]).

16.3.2 US Domestic Operations

FAR 121.639 [2] applies to flights within the 48 contiguous states of the USA and the District of Columbia. It is a requirement that no person may dispatch or take off an airplane unless it has enough fuel (Figure 16.2):

(A) to fly to the airport to which it is dispatched; thereafter,
(B) to fly to and land at the most distant alternate airport (where required) for the airport to which dispatched; and thereafter,
(C) to fly for 45 minutes at normal cruising fuel consumption (only applicable to transport category airplanes).

16.3.3 FAA International Operations

FAR 121.645 [2] applies to US flag and supplemental operations of turbine-engine powered airplanes (not including turbo-propeller airplanes). It is a requirement that no person may release for flight or take off an airplane unless it has enough fuel (Figure 16.3):

(A) to fly to and land at the airport to which it is released; after that,
(B) to fly for a period of 10% of the total time required to fly from the airport of departure to, and land at, the airport to which it was released; after that,
(C) to fly to and land at the most distant alternate airport specified in the flight release; and after that,
(D) to fly for 30 minutes at holding speed at 1500 ft above the alternate airport under standard temperature conditions.

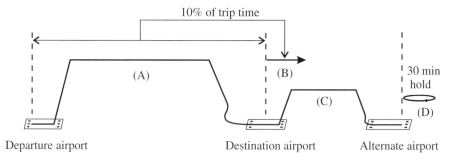

Figure 16.3 FAA international fuel requirements (FAR 121.645 [2]).

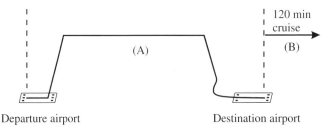

Figure 16.4 FAA "island" reserves fuel requirements (FAR 121.645 [2]).

16.3.4 FAA "Island" Reserves

When an alternate airport is not specified, it is a requirement (FAR 121.645 [2]) that no person may release for flight or take off a turbine-engine powered airplane (not including a turbo-propeller airplane) unless, considering wind and other weather conditions expected, it has enough fuel (Figure 16.4):

(A) to fly to that airport; and thereafter,
(B) to fly for at least two hours at normal cruising fuel consumption.

16.4 EASA Requirements

16.4.1 Overview

The key requirement regarding fuel planning is given in OPS Part CAT.OP.MPA.150 [4], which states that the airplane must carry sufficient fuel for the planned operation and reserves to cover deviations from that operation. This is amplified in AMC1 CAT.OP.MPA.150(b) [5], which describes four different procedures: (1) basic (standard), (2) reduced contingency fuel (RCF), (3) predetermined point (PDP), and (4) isolated aerodrome. An extract from the Acceptable Means of Compliance (AMC) and guidance material (GM) to Part-CAT [5], describing the basic policy, is contained in Appendix F.2.

The amount of fuel that the pilot requires at the start of the mission, to be compliant with the regulations, can be divided into discrete elements—each of which is associated with a particular part (segment) of the mission. This is presented schematically in Figure 16.5 for the EASA basic (standard) fuel planning policy.

16.4.2 Contingency Fuel

The extract contained in Appendix F.2 provides precise details of the manner in which the contingency fuel is to be calculated. This reserve is necessary as it is not always possible to estimate the magnitude of all the factors that have an influence on the fuel consumption, especially as the planning is to be done before the flight (usually at the departure airport) based on *forecast* conditions. Thus, assumptions are made about the weather and traffic conditions *en route* and at the destination airport. Consequently, as mentioned in GM1 CAT.OP.MPA.150(c)(3)(i) [5], contingency fuel is carried to compensate for

(1) deviations of an individual airplane from the expected fuel consumption data;
(2) deviations from the forecast meteorological conditions; and
(3) deviations from planned routings and/or cruising levels/altitudes.

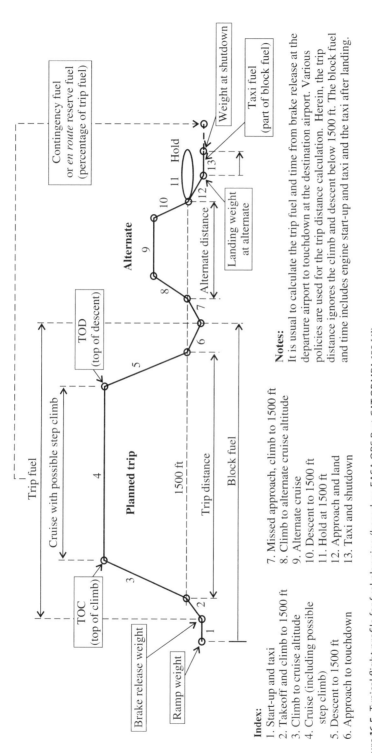

Figure 16.5 Typical flight profile for fuel planning (based on EASA OPS Part CAT.OP.MPA.150 [4]).

Index:
1. Start-up and taxi
2. Takeoff and climb to 1500 ft
3. Climb to cruise altitude
4. Cruise (including possible step climb)
5. Descent to 1500 ft
6. Approach to touchdown
7. Missed approach, climb to 1500 ft
8. Climb to alternate cruise altitude
9. Alternate cruise
10. Descent to 1500 ft
11. Hold at 1500 ft
12. Approach and land
13. Taxi and shutdown

Notes:
It is usual to calculate the trip fuel and time from brake release at the departure airport to touchdown at the destination airport. Various policies are used for the trip distance calculation. Herein, the trip distance ignores the climb and descent below 1500 ft. The block fuel and time includes engine start-up and taxi and the taxi after landing.

Table 16.1 Mission definitions

Term	Definition
Trip fuel/time	This is the fuel/time calculated from *brake release* at the departure airport to *touchdown* at the destination airport.
Trip distance	The trip distance is measured from the departure airport to destination airport. In the absence of wind, this is equal to the ground distance. It typically ignores the climb and descent below 1500 ft.
Range	For mission analyses, the range is equal to the trip distance. In the case of a "cruise only" assessment, the range is the horizontal distance covered in the cruise.
Block fuel/time	The block fuel/time equals the trip fuel/time plus the fuel/time required for the engine start-up and taxi and the taxi after landing.
Reserve fuel	This is the *en route* contingency fuel plus the fuel required for the alternate leg plus the final reserve (or holding fuel). It is permissible for the taxi-in fuel to be taken from the reserve fuel.

16.4.3 Mission Profile

The fuel planning process is aided by establishing a mission profile, which is divided into a number of segments. The process is illustrated in Figure 16.5 for the EASA OPS Part-CAT basic (standard) fuel planning requirements (see Appendix F.2). The requirements (regulations) are often called the *mission rules*. A summary of key definitions is given in Table 16.1. For each segment, the required fuel is determined and summed. The controlling parameter in such analyses is the airplane's weight, which has to be determined for the start of each segment.

This analysis approach is typically used to determine the required fuel for a given trip distance (and mission rules); alternatively the trip distance can be determined for a given fuel load. In Figure 16.5, the climb and descent distances below 1500 ft are not included in the trip distance computation, but the time taken for these segments *is* included. The logic behind this stems from the regulations, which indicate that the trip fuel should take into account *expected* departure and arrival routing [4]. In the absence of actual route details, it is prudent not to take credit for these distances (this is a moot point, anyway, as the distance covered below 1500 ft is usually negligible in comparison to the trip distance).

16.5 Trip Fuel Computational Procedure

Typically, a mission analysis involves establishing the takeoff weight (TOW) and required fuel for a specified trip distance and payload, knowing (1) the airplane's operating empty weight (OEW), (2) principal performance characteristics (e.g., drag polars, climb thrust, corrected fuel flow data), (3) mission profile (e.g., climb and descent speed schedules, cruise speeds, and cruise altitudes), (4) atmospheric conditions (e.g., temperature, wind), and (5) reserve fuel policy (see Sections 16.2–6.4). The computational procedure is non-trivial and requires iteration (as illustrated in Figure 16.6). The TOW is equal to the operating empty weight (which is known) plus the payload (which is also known) plus the takeoff fuel load. The fuel quantity, however, is not an independent variable, but is a function of the TOW.

The minimum fuel at takeoff would comprise the trip fuel and the reserve fuel. Extra fuel may also be uploaded—for example, when transporting fuel (see Section 16.10) or at the discretion of the pilot. Computer programs developed to calculate the trip fuel and reserve

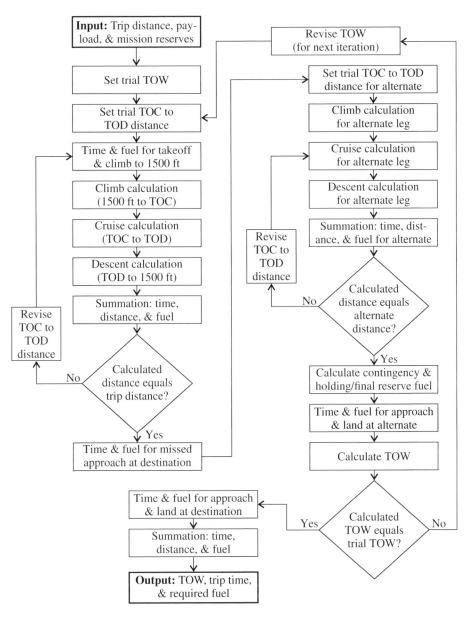

Figure 16.6 Flowchart of a computational sequence to determine the fuel required and takeoff weight, for given trip distance, payload, and fuel reserves.

fuel—and hence the TOW—would usually be constructed with several subroutines, which are able to compute the time, distance, and fuel associated with each mission segment (as identified in Figure 16.5, for example). Computational subroutines would be required for climb, cruise, descent, and hold segments and are based on first-principles (i.e., physics based) numerical procedures, as described in Section 12.4.8 for the climb (the same procedure would apply to the descent) and in Section 13.4 for the cruise, for example. For mission segments that do not warrant a first-principles computation (e.g., the approach and land), lookup tables containing flight-test-derived data may be used together with an interpolation algorithm, usually based on airplane mass (or weight). The choice of whether to compute the fuel and time using a first-principles technique or to utilize a lookup table for these minor segments depends on the application and level of accuracy needed.

The computational logic, implemented in the computer software, can be structured in several different ways. One of the simplest methods (albeit not the most computationally efficient approach) starts by assuming a trial TOW (first iteration) and works sequentially through the mission, calculating the fuel for each segment, which allows the airplane weight at the start of the next segment to be determined. On completion of the mission, the computed fuel and payload is added to the OEW to establish a value for the TOW. If this value differs significantly from the trial TOW, the process loops back and starts again with the next iteration of TOW. The flowchart in Figure 16.6, which illustrates such a computational sequence, was developed to calculate the trip fuel, reserve fuel, trip time, and TOW for user-specified input data concerning trip distance, payload, and fuel reserve policy. The logic adopted in this example used an iterative approach to determine the cruise fuel and cruise time (for both the trip and alternate legs), based on the start-of-cruise airplane weight and a trial cruise distance measured from the top of climb (TOC) to the top of descent (TOD).

16.6 Payload–Range Performance

16.6.1 Basic Payload–Range Diagram

A payload–range diagram (chart) is an airplane performance envelope that depicts the maximum range (the abscissa) that an airplane can achieve with varying payload (the ordinate), under a defined set of conditions, or *mission rules* (see Section 16.6.2, below). The *range*, in this context, is the calculated trip distance corresponding to a particular payload (which, for a commercial airliner, comprises passengers, baggage, and cargo).

A typical payload–range diagram for an airliner is illustrated in Figure 16.7. With the maximum allowable payload, the amount of fuel that can be taken on board will usually not be limited by the size of the fuel tanks, but rather by the allowable TOW, which is a structural design limit (see Section 19.2). Under standard conditions, the maximum allowable TOW will be the maximum (design) takeoff weight (MTOW), and the airplane will have certain nominal range—this point is identified as the "maximum range with maximum payload" in Figure 16.7. If this range is inadequate for the planned flight, then it will be necessary to reduce the payload in order to take on more fuel (but without exceeding the MTOW). Progressively longer stage lengths can

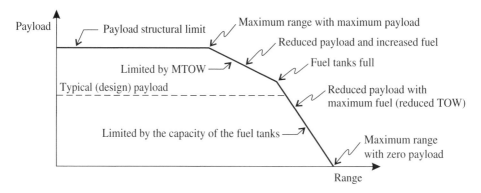

Figure 16.7 Typical payload–range diagram for an airliner.

be achieved by trading payload for fuel. When the point is reached that the fuel tanks are full, the only way that the range can be increased is by further reducing the payload. The greatest possible range corresponds to a zero-payload condition—this is sometimes called the *ferry* range.

16.6.2   Mission Rules

The range indicated on a payload–range chart is determined for a specified mission profile (see Sections 16.2–16.4), which includes the basic mission (i.e., takeoff, climb, cruise, descent, approach, and land at the destination airfield) with an adequate fuel provision for contingencies and diversion requirements (to the alternate airfield). The mission rules are typically based on one of the fuel planning requirements (scenarios) established by the authorities for flight operations, as described in Sections 16.2–16.4. One of the considerations in determining the fuel to be uploaded is an allowance for a possible diversion. For actual flight operations, the distance from the destination airport to the specified alternate airport would be known. For the purpose of generating a payload–range diagram, an arbitrary assumption of the alternate distance is made—this is typically 100 nm for short-haul aircraft and 200 nm for long-haul aircraft. Although the computed range (as indicated on a payload–range chart) does not include the diversion distance, the diversion fuel is included in the computed takeoff weight. The assumed diversion distance thus indirectly influences the computed range.

The *design payload* condition is identified in Figure 16.7. This represents a nominal payload, based on a defined number of passengers, as selected by the airplane manufacturer. Although standardized weights for passengers are specified by the regulatory authorities for operational purposes (see Section 19.6.5), manufacturers are at liberty to select average passenger weights for design or marketing purposes. Typical average weights for passengers, including their check-in and carry-on baggage, vary from about 90 kg for charter or economy-class passengers to about 115 kg for scheduled first-class passengers. Manufacturers have used similar average values in their published payload–range documentation—for example, Boeing used 90.7 kg (200 lb) per passenger for the B737; Airbus also used the same value for short/medium-range airplanes and 95 kg (209.4 lb) for certain long-haul types; and Bombardier used 102 kg (225 lb) per passenger for the C Series. Corresponding to the design payload, the range is determined and shown on the chart. This could be limited by the size of the fuel tank (as shown in Figure 16.7) or by the aircraft's MTOW.

16.6.3   Alternative Form of the Payload–Range Diagram

Further insight into this important representation of performance capability of a jet transport airplane is obtained by noting the following two relationships, which link the operating empty weight (OEW), maximum zero fuel weight (MZFW), and takeoff weight (TOW):

$$MZFW = OEW + \text{Max. payload} \qquad (16.1)$$
$$TOW = MZFW + \text{Fuel} \qquad (16.2)$$

This can be shown graphically using an alternative format (see Figure 16.8), which has the aircraft's weight as the ordinate and the payload–range envelope shown within the diagram (shaded). From point A to point B, the payload is constant, but the TOW increases as additional fuel is required for the increasing range. From point B to point C, payload is traded for fuel (here the TOW is equal to the MTOW). This progressive increase in range, resulting from the increase in fuel, is possible until the maximum fuel limit (i.e., fuel capacity limit) is reached. From point C to point D, the fuel is limited by the size of the fuel tanks, but the payload (and hence the TOW) reduces as the range increases.

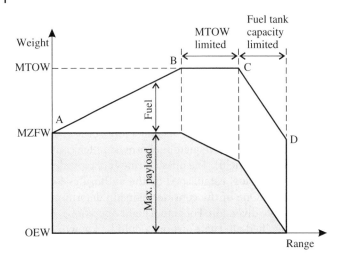

Figure 16.8 Alternative form of the payload–range diagram.

This form of payload–range chart is useful for assessing the impact on these parameters due to changes to the aircraft's MTOW (see Section 16.6.4), OEW (see Section 16.6.5), or fuel capacity (see Section 16.6.6), for example.

16.6.4 Impact of MTOW Increase on the Payload–Range Diagram

Following initial certification, it is not uncommon for a manufacturer to receive approval for an increase in MTOW. This typically follows a detailed engineering assessment, which may include additional flight testing. This is usually as a consequence of the engineers acquiring a better understanding of the aircraft's structure, the applied loads, and the manner in which the loads are distributed through the airframe. It may also be due to modifications to the structure or other systems (e.g., braking system or gust load alleviation system).

The impact of an increase in MTOW on the airplane's payload–range performance is illustrated by the shaded area in Figure 16.9 (note that in this example the MZFW and OEW are unchanged). Within this important part of the envelope, a higher payload, an increase in range, or a combination thereof can now be accommodated.

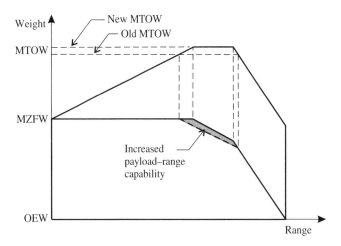

Figure 16.9 Impact of MTOW increase on the payload–range diagram.

16.6.5 Impact of OEW Increase on the Payload–Range Diagram

The OEW of an airplane depends on the customer's selected operational items—and also on how successful the operator is in controlling weight "growth" following the airplane's entry into service. The impact of an increase in OEW is seen in Figure 16.10. As the MZFW is a design weight limit, any increase in OEW means a corresponding reduction in the maximum allowable payload. For operations where the maximum payload is not required, the increase in OEW results in a reduction in maximum range.

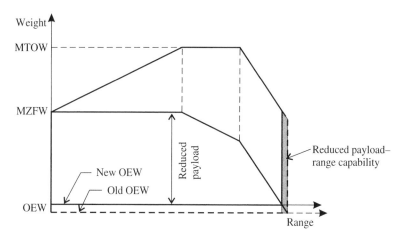

Figure 16.10 Impact of OEW increase on the payload–range diagram.

16.6.6 Impact of Fuel Capacity Increase on the Payload–Range Diagram

Another change to the aircraft that can impact the payload–range chart is an increase in fuel capacity through the installation of additional fuel tanks. In some cases, this may reduce the airplane's freight-carrying capability. This is frequently done for corporate (VIP) versions, for example, where there is a reduced requirement for cargo. Such changes would increase the OEW. However, if the increase in fuel capacity was considered in isolation, then a simple extension to the right-hand boundary on the payload–range chart results—as shown in Figure 16.11.

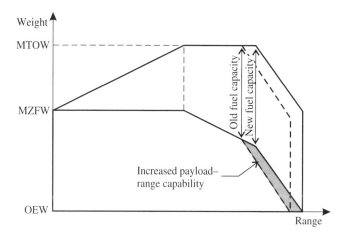

Figure 16.11 Impact of fuel capacity increase on the payload–range diagram.

16.7   Trip Fuel Breakdown and Fuel Fractions

16.7.1   Typical Fuel Breakdown by Mission Sector

The proportion of fuel consumed in each sector of a mission—when expressed as a fraction of the total trip fuel—depends on the airplane type and its particular performance characteristics. The breakdown of trip fuel by sector is strongly influenced by trip distance. This is illustrated in Figure 16.12 for a typical medium-range, single-aisle airliner and in Figure 16.13 for a typical long-range, widebody airliner.

There are several notable features in these two figures. For example, it can be seen in Figure 16.12 that the amounts of fuel consumed in the climb (sector 2) and in the cruise (sector 3),

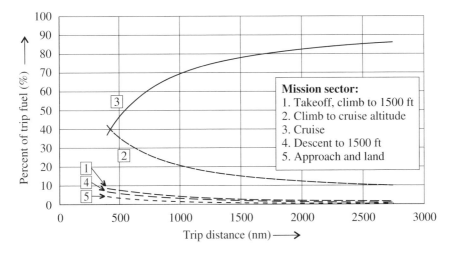

Figure 16.12   Typical trip fuel breakdown for a medium-range, single-aisle airliner.

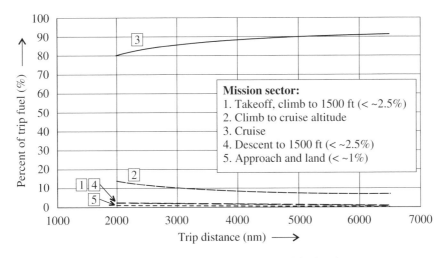

Figure 16.13   Typical trip fuel breakdown for a long-range, widebody airliner.

for this airplane type, are similar for a trip distance of ~400 nm. Furthermore, for this short trip distance, the other sectors of takeoff (1), descent (4), and approach and land (5) account for small, but not insignificant, portions of the total trip fuel. As expected, with increasing trip distance the cruise is responsible for a progressively greater portion of the trip fuel. For trip distances of ~2000 nm (for both airplane types considered), the cruise sector is responsible for more than 80% of the trip fuel. The sum of the smaller elements—that is, takeoff (1), descent (4), and approach and land (5)—is seen to be about 6% for the widebody airliner (Figure 16.13) at ~2000 nm.

16.7.2   Fuel Fractions

It is useful in certain conceptual design activities or performance applications to express the fuel consumed in each leg (or sector) of a mission as a ratio of the aircraft's weight at the end of the leg to its weight at the start of the leg. These ratios are called *fuel fractions*. Note that the term fuel fraction can be misleading, as it is actually a ratio of two *airplane* weights, determined at different points in the mission. Thus, if W_i is the airplane's weight at the start of leg i and W_{i+1} is the weight at the end of that leg, then the ratio W_{i+1}/W_i would be the fuel fraction for leg i. Fuel fractions have numerical values of less than 1.

If the mission is broken into n legs, then the ratio of the final weight to the starting weight for the mission is given by the following expression:

$$\left(\frac{W_{n+1}}{W_1}\right) = \left(\frac{W_2}{W_1}\right)\left(\frac{W_3}{W_2}\right)\left(\frac{W_4}{W_3}\right)\cdots\left(\frac{W_{i+1}}{W_i}\right)\cdots\left(\frac{W_{n+1}}{W_n}\right) \tag{16.3}$$

This is illustrated in Figure 16.14 for a simple mission comprising seven legs. W_1 is the start-up (ramp) weight and W_8 is the weight at shutdown. The value of Equation 16.3 lies in the fact that the fuel used for certain legs is approximately proportional to the aircraft's instantaneous weight. As a consequence, the fuel fractions for these legs do not vary much from one airplane type to another and from one mission to another, enabling a rapid estimation to be made of the total mission fuel. Typical fuel fractions for jet transport aircraft are given in Table 16.2. Once the fuel fractions for all legs are known, it is possible, using Equation 16.3, to estimate the ratio of the shutdown weight to the ramp weight (i.e., W_8/W_1 for the example given in Figure 16.14).

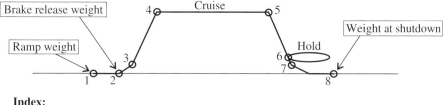

Index:

1–2 Start-up and taxi	4–5 Cruise	7–8 Approach, land,
2–3 Takeoff and climb-out	5–6 Descent	taxi, and shutdown
3–4 Climb to cruise altitude	6–7 Hold	

Figure 16.14 Illustration of mission legs (sectors).

Table 16.2 Typical fuel fractions for jet transport airplanes

Leg	Segment	Fuel fraction	
1–2	engine start and taxi	0.998	
2–3	takeoff and climb-out	0.996	
3–4	climb to cruise altitude[a]	0.980	for low cost index
		0.978	for high cost index
4–5	cruise[b]	calculate using Equation 13.19	
5–6	descent[c]	0.997	for low cost index
		0.998	for high cost index
6–7	hold[d]	calculate using Equation 14.5	
7–8	approach, land, taxi, and shutdown	0.997	

Notes:

(a) The indicated values for leg 3–4 are averages for five airliners (with a takeoff mass ranging from 75 000 kg to 250 000 kg).
(b) The fuel fraction can be calculated using the Bréguet range equation (Equation 13.19), where the TSFC and lift-to-drag ratio are mean values for the cruise.
(c) The indicated values for leg 5–6 are averages for five airliners (with a descent mass ranging from *ca.* 60 000 kg to 180 000 kg).
(d) The fuel fraction can be calculated using endurance Equation 14.5, where the TSFC and lift-to-drag ratio are mean values for the hold.

The mission fuel ($W_{f_{used}}$) can be deduced from the fuel fractions and the airplane ramp weight (W_1) or shutdown weight (W_{n+1}), that is,

$$W_{f_{used}} = W_1 - W_{n+1} \tag{16.4}$$

Thus $$W_{f_{used}} = \left[1 - \left(\frac{W_{n+1}}{W_1} \right) \right] W_1 \tag{16.5}$$

or $$W_{f_{used}} = \left[\left(\frac{W_1}{W_{n+1}} \right) - 1 \right] W_{n+1} \tag{16.6}$$

where W_{n+1}/W_1 is given by Equation 16.3.

16.8 Trip Fuel Estimation

16.8.1 Overhead Approximation

A frequently used approach for estimating the trip fuel for a mission is to assume that the airplane starts the cruise overhead the departure airfield (at the cruise height) and flies to a point overhead the destination airfield. The trip fuel is then estimated using the Bréguet or similar range equation (see Section 13.3), without the complexity of having to determine the climb and descent fuel quantities. From the point of brake-release to the top of climb (TOC) this approximation underestimates the required fuel, but from the top of descent (TOD) to the destination airfield, the approximation typically overestimates the required fuel—the descent error, however, does not completely compensate for the climb error. Furthermore, the estimated trip time is also in error: the time that an airplane takes to climb and descend is obviously greater than the time required to cover the equivalent distance in cruise.

It was mentioned earlier (see Section 16.7.1) that the proportion of fuel consumed in the cruise, expressed as a percentage of the trip fuel, depends on the trip distance. Based on the data presented in Figures 16.12 and 16.13, it can be seen that the cruise fuel is only ~46% of the trip fuel for a range of 500 nm, but this rises quickly to ~81% for a range of 2000 nm and to ~91% for a range of 6000 nm. The implication of this observation is that the overhead approximation will produce increasingly more accurate results as the trip distance is increased. For example, without correction, the approximation is likely to underestimate the actual fuel by more than 10% for a trip of 1000 nm; however, when the range is increased to 2500 nm, the error drops to slightly less than 5%.

In a study conducted by Randle, *et al.* [7], Flight Data Recorder (FDR) information—supplied by a European airline for six airplane types—was analyzed. After correcting for the influence of wind, they concluded that the difference in the fuel actually burned during the climb (i.e., from takeoff to TOC) and the fuel that would be burned during cruising flight over the same ground distance was, on average, 1.5% of the airplane takeoff weight. When they compared the fuel actually burned in the descent (i.e., from TOD to landing) to that which would be burned in cruise over the same ground distance, the difference, on average, was −0.1% of the airplane takeoff weight. Significantly, the descent error results displayed more scatter than that seen for the climb. This variability makes sense as actual flights almost always incorporate a series of turns as the airplane approaches the airport—not only do such maneuvers consume more fuel than an idealized straight-flight-path descent, but it is also likely that considerable flight-to-flight variations exist in such data.

16.8.2   Lost Energy Corrections to the Overhead Approximation

If the range is estimated using the overhead approximation, as shown in Figure 16.15, it can be corrected using the following equation:

$$R_a = R_{a,est} - \Delta R_a \tag{16.7}$$

where R_a is the still air range;
$\quad\quad R_{a,est}$ is the estimated still air range based on the overhead approximation; and
$\quad\quad \Delta R_a$ is the amount by which the range was overestimated.

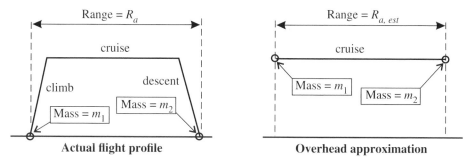

Figure 16.15 Range estimation using the overhead approximation.

ESDU 74018 [8] provides the following expression for estimating ΔR_a:

$$\Delta R_a = k_1 \, \Delta H_{e,cl} \left(\frac{L}{D}\right)_{cr} \tag{16.8}$$

where k_1 is a dimensionless empirical factor;
$\Delta H_{e,cl}$ is the change in energy height in the climb (given by Equation 16.9);
and
$(L/D)_{cr}$ is the cruise lift-to-drag ratio.

The change in energy height can be derived from Equation 15.58

$$\Delta H_{e,cl} = \Delta H_{cl} + \frac{\Delta V^2}{2g_0} \tag{16.9}$$

where ΔH_{cl} is the change in geopotential height in the climb;
ΔV is the change in true airspeed (which in this case is equal to the cruise TAS); and
g_0 is the standard value of gravitational acceleration.

It is suggested in ESDU 74018 [8] that the value of k_1 is set equal to ~0.8. However, an investigation conducted by the author—using the two aircraft models that produced the results shown in Figures 16.12 and 16.13—concluded that a value of $k_1 = 1.0$ consistently produced a more accurate estimation of the range correction ΔR_a.

Cautionary note: Equations 16.7 to 16.9 must be dimensionally consistent (the constituent terms are defined above, but without units). Therefore, if the change in height, ΔH_{cl}, in Equation 16.9 is expressed in feet (or meters), then the change in TAS, ΔV, must be in ft/s (or m/s). Also, if the estimated range, $R_{a,est}$, in Equation 16.7 is given in nm (or km) then the range correction ΔR_a must also be in nm (or km). This means that the change in energy height, $\Delta H_{e,cl}$, must be converted from feet (or meters) to the usual units of range (i.e., nm or km) before being used in Equation 16.8.

The trip time (t) will be underestimated by the overhead approximation; a similar approach to that described above for the range correction may be adopted:

$$t = t_{est} + \Delta t = \frac{R_{a,est}}{V_{cr}} + \Delta t \tag{16.10}$$

where t_{est} is the estimated trip time based on the overhead approximation;
Δt is the amount by which the time was underestimated; and
V_{cr} is the cruise TAS.

and $$\Delta t = k_2 \frac{\Delta R_a}{V_{cr}} \tag{16.11}$$

where k_2 is a dimensionless empirical factor.

There is, however, no fundamental basis for relating the time correction (Δt) to the range correction (ΔR_a); nonetheless, a reasonable approximation has been shown to be possible using Equation 16.10. The magnitude of the empirical factor k_2 will depend on the climb and descent speed schedules. ESDU 74018 [8] indicates that the value could be between 0.6 and 1.2 and suggests a typical value of ~0.8.

16.8.3   Estimating Fuel for a Specified Range

Often the problem at hand is not to calculate the range, but rather to calculate the fuel required (W_f) for a specified trip distance (R). In this case, either Equation 13.43 or Equation 13.44—both of which were derived from the Bréguet range equation in Section 13.7.3—can be used.

If the overhead approximation (see Sections 16.8.1 and 16.8.2) is used, then a correction for the climb and descent is needed. The additional fuel can be estimated using the following equation:

$$\Delta m_f = k_1 \frac{\Delta H_{e,cl}}{r_a}\left(\frac{L}{D}\right)_{cr} = k_1 \Delta H_{e,cl}\left(\frac{Q}{V_{cr}}\right)\left(\frac{L}{D}\right)_{cr} \tag{16.12a}$$

$$\text{or}\qquad \Delta W_f = k_1 \frac{\Delta H_{e,cl}}{r_a'}\left(\frac{L}{D}\right)_{cr} = k_1 \Delta H_{e,cl}\left(\frac{Q'}{r_a'}\right)\left(\frac{L}{D}\right)_{cr} \tag{16.12b}$$

where k_1 is a dimensionless empirical factor (see Section 16.8.2);
   $\Delta H_{e,cl}$ is the change in energy height in the climb (see Equation 16.9);
   r_a (or r_a') is the specific air range based on mass (or weight); and
   Q (or Q') is fuel flow based on mass (or weight).

Cautionary note: The change in energy height, $\Delta H_{e,cl}$, as determined by Equation 16.9, must be converted from the units of height (i.e., ft or m) to the usual units of range (i.e., nm or km) for use in Equation 16.12.

16.8.4   Step-Climb Correction

The purpose of a step climb in cruise is to reduce the overall trip fuel, by operating the airplane at a more efficient flight level (see Section 13.6.5). Steps of 2000 ft or 4000 ft are typically executed during the cruise as the airplane's weight reduces (see Section 17.3.4).

The mass of the additional fuel (Δm_f)—that is, in addition to the fuel needed to cover the climb distance in level flight—required to climb from one flight level to the next available flight level is given by

$$\Delta m_f = m_2 - m_3 \qquad \text{or} \quad \Delta W_f = W_2 - W_3 \tag{16.13}$$

where m_2 and m_3 are defined in Figure 16.16.

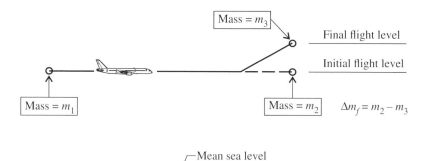

Figure 16.16 Step-climb correction.

The additional fuel can be estimated using Equation 16.12, modified with an empirical factor that accounts for the climb only and assumes a constant TAS:

$$\Delta m_f = k_3 \frac{\Delta H_{cl}}{r_a} \left(\frac{L}{D} \right)_{cr} \tag{16.14a}$$

or $$\Delta W_f = k_3 \frac{\Delta H_{cl}}{r_a'} \left(\frac{L}{D} \right)_{cr} \tag{16.14b}$$

where k_3 is a dimensionless empirical factor, which, it is suggested in ESDU 74018 [8], is equal to ~1.4 for a climb; and
ΔH_{cl} is the change in geopotential height for a constant TAS climb.

16.9 Estimating Trip Distances (To Be Flown)

16.9.1 Factored Great Circle Distance

Theoretical studies of the required trip fuel for a particular route (or city pair) are often based on a trip distance that is computed by factoring the great circle distance (GCD) linking the departure and destination airports. This factor (i.e., percentage increase) depends on the specific route and on several operational issues, as described below.

The GCD between the departure and destination airports can easily be established (see Section 6.2.4)—this represents a theoretical minimum distance along the Earth's surface. Statistical data of actual distances flown—that is, the length of the flight path over the ground—by airplanes conducting routine flight operations can be established for many of the world's airspaces by studying radar tracks. These tracks, of course, are longer than the corresponding GCD. Actual flight operations follow designated terminal departure and arrival procedures, and the *en route* track can be subject to a wide range of influences (such as the requirement to follow a defined airway).

There are often several viable tracks for a single route connecting two airports. The tracks that are available to operators are influenced by many factors—for example: geographical considerations (e.g., mountains), national boundaries (e.g., over-fly rights can be restricted), special-use airspaces (e.g., restricted or partially restricted for military use), security threats (e.g., war zones), and air traffic control (ATC) constraints. Furthermore, the selection of a particular track may also take into account prevailing winds (e.g., the location of jet streams) and the airplane's capabilities, such as whether or not the airplane is approved for ETOPS (i.e., Extended Operations—see Section 17.6). Clark [9] highlighted the issue by considering the Paris–Hong Kong route; he noted that there were approximately 20 different tracks for this particular route, and that the longest track was 24% greater than the GCD—this, however, is an extreme example.

16.9.2 Horizontal *En Route* Flight Efficiency

ICAO [10] defines several key performance indicators for global air navigation. Certain of these indicators are metrics relating to the operational efficiency of air traffic management (ATM) in the world's airspaces. Relevant to this discussion on fuel usage is the "horizontal *en route* flight efficiency" indicator, which expresses the excess distance—that is, the actual distance traveled less the reference distance (i.e., GCD)—as a percentage of the reference distance.

The Federal Aviation Administration and EUROCONTROL, for example, monitor this parameter for their respective airspaces (i.e., United States and Europe[1])—a typical excess *en route* distance, based on actual distances traveled (i.e., from radar data) is 3% [11]. The distances recorded on "filed" flight plans tend to be slightly longer than actual distances flown—in this case, the average excess *en route* distance according to filed flight plans was ~3.5% in the United States and ~5% in Europe [11]. Note that there has been a steady downward trend in this performance indicator, over the past few decades, as more direct flight paths become possible.

16.10   Transporting (Tankering) Fuel

16.10.1   Purpose

Transporting (tankering) fuel is undertaken for a number of practical reasons—for example:

(1) the fuel is significantly cheaper at the departure airport than at the destination airport;
(2) the planned turn-around time is too short for refueling; or
(3) the fuel supply at the destination airport is unreliable.

Carrying extra fuel that is not required for the mission means that the aircraft has a greater takeoff weight, and consequently the trip fuel will be higher (another way of looking at this is to treat tankered fuel as payload). There is thus an economic penalty, in terms of increased fuel burn, which has to be offset to make the exercise worthwhile. The amount of fuel that an aircraft can tanker, for a given mission, will be constrained by one of the following aircraft limits:

(1) MTOW limit (for a given payload);
(2) MLW (maximum landing weight) limit; or
(3) maximum fuel capacity.

16.10.2   Analysis of Transporting Fuel

The financial benefit or penalty of transporting the additional fuel depends directly on the fuel price differential between the airports and the increase in the trip fuel arising from flying a heavier-than-required airplane. The analysis is facilitated by introducing a *fuel transport index* (K_f), which is defined as follows:

$$K_f = \frac{\Delta W_{to}}{\Delta W_l} \tag{16.15}$$

where ΔW_{to} is the change in takeoff weight (due to the additional fuel); and
ΔW_l is the change in landing weight (due to the additional fuel).

Consider, for the sake of illustration, a particular airplane that on a particular route has a fuel transport index of 1.27. This means that for every 1000 kg of fuel that is transported (tankered) to the destination airport, an additional 1270 kg of fuel must be uploaded at the departure airport.

---

1 Europe, in this context, is defined as the geographical area where the air navigation services are provided by the European Union (EU) member states plus those states outside the EU that are members of EUROCONTROL.

The financial benefit or cost incurred in transporting fuel can be determined from the fuel prices at the two airports. The additional fuel cost at the departure airport is

$$\text{Additional cost} = \Delta W_{to} C_{F_{dep}} \tag{16.16}$$

where $C_{F_{dep}}$ is the cost of fuel per unit weight at the departure airport.

The value of the tankered fuel at the destination airport is

$$\text{Fuel value} = \Delta W_l\, C_{F_{arr}} = \frac{\Delta W_{to}}{K_f} C_{F_{arr}} \tag{16.17}$$

where $C_{F_{arr}}$ is the unit cost of fuel at the destination (arrival) airport.

If there is an increase in trip time (Δt) associated with the heavier load, this should also be taken into account:

$$\text{Change in time cost} = \Delta t\, C_T \tag{16.18}$$

where C_T is the time-dependent cost per unit of trip time.

The net change in the cost of the trip is given by

$$\Delta C_{trip} = \Delta W_{to}\, C_{F_{dep}} - \frac{\Delta W_{to}}{K_f}\, C_{F_{arr}} + \Delta t\, C_T \tag{16.19}$$

Now, if ΔC_{trip} is negative, then there is a financial benefit in transporting the fuel. This will be true when the fuel price at the destination airport is given by the following relationship:

$$C_{F_{arr}} > K_f C_{F_{dep}} + \frac{K_f}{\Delta W_{to}} \Delta t\, C_T \tag{16.20}$$

If the mission can be executed with no change in trip time, then it is profitable to carry additional fuel when the fuel price ratio between the destination and departure airports is greater than the fuel transport index—this condition can be written mathematically as follows:

$$\frac{C_{F_{arr}}}{C_{F_{dep}}} > K_f \tag{16.21}$$

The above analysis, however, is simplistic and does not consider the knock-on effects of operating a heavier airplane. There are several less visible, but nonetheless significant, cost implications that arise when tankering fuel: (1) the increased takeoff weights mean that the engines work harder and use up more of their "life" due to higher engine operating temperatures and rotational speeds; (2) landing gear components will be subject to higher loads resulting in greater tire wear and increased maintenance costs; and (3) flap components will suffer increased wear and tear due to higher takeoff and landing speeds. These additional costs need to be included for a comprehensive analysis of the financial gain/loss associated with fuel tankering.

16.11 Reclearance

16.11.1 Introduction

Reclearance (or *redispatch* as it sometimes called) is an operational procedure that can be used to reduce the amount of reserve fuel that has to be uploaded—and, consequently, extend the airplane's range for a given payload. The procedure takes into account the manner in which contingency fuel is specified (see Sections 16.2–16.4). The weight of the reserve fuel contributes to the fuel consumed during flight; hence, if an acceptable operational technique can be employed to reduce the reserve fuel, there will be a reduction in the trip fuel. This can be particularly advantageous for long-haul missions, where the fuel is a large portion of the takeoff weight. The procedure thus permits an increase in payload to be carried on suitable routes. The procedure can also be used on missions where the aircraft's fuel tank capacity limit would prevent it from reaching the desired destination under normal fuel planning rules.

16.11.2 Mission Profile for Reclearance

The concept of reclearance can be illustrated by considering two mission profiles, as shown in Figure 16.17. The upper diagram is a simplified mission profile for a normal mission from the departure airport A to the intended destination B, with C identified as the alternate airport. The lower diagram uses a phantom destination D (which is located near the intended track) and a phantom alternate airport E for the purpose of flight planning. Note that both D and E are actual, fully operational airports. The mission fuel is calculated on this basis—that is, the trip fuel from A to D, plus the *en route* contingency, plus the diversion fuel to E. The *en route* contingency fuel is calculated in accordance with a regulatory fuel policy (see Sections 16.2–16.4)—for example, it could be based on 5% of the trip fuel required to travel from A to D.

When the aircraft gets to the reclearance (redispatch) point, P, the crew will assess the status of the onboard fuel, and, if it is adequate, will change their destination to B (with C as alternate). The onboard fuel at the reclearance point must be sufficient for the trip from P to B, plus the *en route* contingency (which will now be based on 5% of the trip fuel for P to B, in this example), plus diversion fuel to C. If the onboard fuel at P is lower than that which was planned (e.g., due to an inaccurate wind forecast or an *en route* deviation), then the crew would be required to land and refuel at D, before proceeding to B.

The reclearance point, P, is typically at about 90% of the trip distance (under ideal conditions). The viability of an airline actually using the 90% location as the reclearance point depends on

(a) Mission profile for flight to intended destination

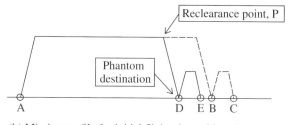

(b) Mission profile for initial flight plan, with reclearance

Figure 16.17 Mission profiles for reclearance procedure.

the availability of suitable airports along the route that can be specified on the flight plan[2] as the initial (phantom) destination and alternate.

16.12 Factors That Can Impact Cruise Fuel

16.12.1 Book Level and Baseline Level Airplane Performance

The *book level* performance of an airplane is established by the aircraft manufacturer and represents a fleet average for new aircraft [12]. This performance level is derived from flight testing (see Section 23.8.1). Book value data are stored in airplane performance databases and is used by performance software to predict the performance of production aircraft. Published performance tables and charts in Flight Crew Operations/Operating Manuals (FCOMs) and in the Quick Reference Handbooks, as well as most Computerized Flight Plan systems, are based on the book level data [12].

The performance levels of an individual airplane can be above or below the book value, due to normal scatter that occurs with complex mechanical products. The *baseline performance* level, which applies to an individual airplane, is normally established at the time that the particular airplane enters service. The baseline level is used as a reference to monitor individual airplane performance in service. An increase in fuel consumption following the airplane's entry into service is normal—this is typically due to an increase in excrescence drag (see Section 7.5.7) and/or engine deterioration (see Section 8.8.5).

Airbus [12] illustrates the trend of cruise performance deterioration with time for the A320 family. The typical in-service performance degradation—that is, change in specific air range—following entry into service is reported to be as follows: 2.0% (± 1%) below book level one year after delivery, 3.5% (± 1%) below book level two years after delivery, and 4.0% (± 1%) below book level three years after delivery [12]. The rate of deterioration is known to decrease with time. Airline maintenance practices can have a big influence on performance degradation.

16.12.2 Reference Conditions for Cruise Performance

The cruise performance data published by an airplane manufacturer for a specific airplane—that is, fuel flow and specific air range (SAR), presented as a function of Mach number, airplane weight, altitude, and air temperature—are based on a typical, reference set of conditions. Small discrepancies between an airplane's actual fuel flow and SAR and the book level (published) data can thus be expected if the actual conditions differ from the reference conditions.

The air distance that an airplane can travel per unit of fuel mass (or fuel weight) consumed (i.e., SAR) is influenced by several principal factors—these are described in Section 13.2. There are also a few secondary factors that can influence SAR. Airplane manufacturers base published cruise data, which are intended for flight operations (e.g., as recorded in the FCOM), on a set of reference conditions. Either flight testing is undertaken at these conditions, or the test data are adjusted afterwards. For flight operations, manufacturers provide a means to correct the baseline data, where necessary, to account for factors that have a significant influence on the fuel consumption.

---

2 Operators are required to submit a flight plan to the relevant authority before departure—a process commonly known as "filing a flight plan." The flight plan contains key information about the airplane (e.g., type, number of persons on board) and the planned flight, including the departure and destination airports, alternate airport, envisaged route, and estimated departure time.

Center of Gravity

The position of the airplane's center of gravity (CG) has an influence on an airplane's drag during cruise, as mentioned in Section 7.5.3. The magnitude of the trim drag associated with the horizontal tailplane is a function of the relative position of the CG with respect to the mean aerodynamic chord; this drag component increases as the CG moves forwards (see Sections 19.3–19.5). Published cruise performance data are based on a reference center of gravity position.

Ancillary Services

The engines provide power for ancillary services onboard the airplane, and this reduces the power available for propulsion. The power is taken as bleed air from the compressors (for most airplane types) and as electricity supplied by generators driven by gearboxes coupled to the engines (see Section 8.8.3 and 8.8.4). An increase in system demand—for example, when additional air-conditioning packs are switched on or when hot air de-icing is used—will negatively impact cruise fuel efficiency. Published performance data are usually based on a typical demand, with correction factors provided for additional demand.

Fuel Net Heating Value

The thrust produced by an airplane's engines for a given fuel flow rate and a given set of flight conditions depends on the characteristics of the fuel—specifically, the net heating value (energy content). The SAR is directly proportional to the fuel net heating value for a given airplane weight, lift-to-drag-ratio, and overall engine efficiency (see Section 13.2.5). The net heating value of approved aviation fuel (see Section 22.7) can vary within a narrow range (but it must meet a minimum value). Flight-test data are thus adjusted to account for this variation. Airplane manufacturers use a reference value when establishing published performance data.

Airplane Geographic Location, Height, and Flight Direction

An airplane of a particular mass will have a slightly different weight depending on its precise location, as gravitational acceleration varies with geographical position and height (see Section 2.3.4). Additionally, the flight direction also has an influence on the airplane's weight due to the centrifugal force resulting from the curvilinear motion of the aircraft in cruise—this opposes the local gravitational acceleration and diminishes its apparent effect (see Section 2.3.5). The lift required to support the airplane's weight in cruise will thus be influenced by the airplane's geographic location, height, and the direction of motion—and this, in turn, will have a small influence on the airplane's drag in cruise. These factors are taken into account during flight testing, and published data are based on a specific reference condition—for example, a north/south flight direction at 45° latitude [13].

16.12.3 Wind

The manner in which wind influences cruise performance is described in Section 13.8. It is apparent that strong winds can have a significant influence on the fuel needed for a mission. The air distance that an aircraft covers within a time interval is generally different from the ground distance (see Section 6.2). The term *equivalent still air distance* (ESAD) is used to describe the air distance equivalent to a particular ground distance under a prevailing wind condition (see Section 6.2.2). Encountering a headwind, for example, will mean that the ESAD is greater than the ground distance; hence, if the airspeed is unchanged, the flight time will be longer and the fuel usage will be greater.

The computation of the required trip fuel for actual flight operations is, of course, based on current and forecast wind conditions (see Section 4.5.9). Operators also utilize wind data when assessing the suitability of operating a particular airplane on a specific route. The speed and

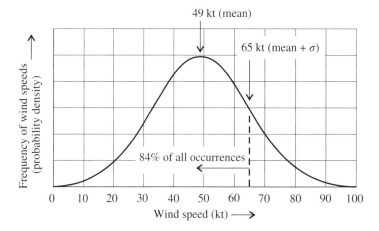

Figure 16.18 Example of idealized wind speed variation based on a normal distribution.

direction of the prevailing winds on a particular route can vary considerably from day to day, although averaged values display clear seasonal patterns. Historical wind data are useful when assessing the trip fuel for (a) an average condition or, (b) an extreme condition. The average wind strength and direction is useful in establishing average or fleet performance characteristics for a particular route. Analysis of an extreme case would likely be undertaken for a different purpose—for example, to determine whether or not the payload–range characteristics of a particular airplane suit the route in question.

Statistical wind data at cruise altitudes—for example, recorded over the past 30 years for city pairs of interest to the airline industry—are primarily used for route planning, comparative airplane performance studies, and economic assessments. A useful representation of historical wind data assumes that the headwind/tailwind variation follows a normal distribution [9]. In the example given in Figure 16.18, the mean wind speed is 49 kt and the standard deviation is 16 kt. Thus, 84% of all occurrences will experience a headwind/tailwind of no greater than 65 kt (the value of 84% probability corresponds to the mean plus one standard deviation).

16.12.4 Barometric Pressure Gradient

Cruise flight levels are determined by pressure altitude, based on the ISA reference pressure of 1013 hPa (29.92 inHg). However, over long distances, barometric pressures can change quite considerably. This means that an airplane flying at a constant pressure altitude (i.e., at a constant flight level) could, in fact, be steadily climbing or descending when measured in terms of geopotential height.

An airplane flying towards a high-pressure zone would increase its geopotential height while maintaining a constant pressure altitude. As a steadily increasing geopotential height implies an increase in potential energy, extra fuel will be expended during the cruise; this negatively impacts the SAR. The reverse is true for an airplane flying towards a low-pressure zone: the geopotential height would reduce, which would have a beneficial effect on the SAR. Flight test procedures for new aircraft take into account the impact of pressure gradients on fuel burn; where practical, the SAR is determined by flying along isobars rather than across isobars. Incidentally, as the wind direction is always parallel to the isobars, this approach means that the test airplane will be flying with a zero drift angle (see Section 6.3.2). The drift angle, in fact, can be used as a means to determine a SAR correction factor for the purpose of monitoring the cruise fuel consumption of an in-service airplane [12].

16.12.5   Temperature Variations

Temperature variations from ISA are known to have a minor influence on SAR, and hence on the range that can be achieved for a given fuel load. To investigate this statement, it is necessary to consider how temperature variations affect both true airspeed (V) and fuel flow (Q) during cruise, as SAR is equal to V/Q, as defined by Equation 13.3. True airspeed (TAS) is directly proportional to $M\theta^{0.5}$, where M is the Mach number and θ is the relative air temperature (see Equation 6.8). If the air temperature were to increase and the Mach number remain constant, then the TAS would increase. Establishing the link between a temperature change and fuel flow for constant Mach cruise is more complicated—as described below.

A temperature increase (e.g., from ISA to ISA + 20 °C) results in a small, but not insignificant, increase in an airplane's drag. The adjustment can be undertaken through the application of a drag coefficient increment, known as a Reynolds number correction (see Section 7.5.8). In a steady-state cruise condition, such an increase in drag would require an increase in thrust to maintain the cruise speed. The fuel flow characteristics of a jet engine, which are discussed in Section 8.4, show that the corrected fuel flow (Q_{cor}) increases—almost linearly—with increasing thrust (see Figure 8.8). An increase in thrust (at a given Mach number and altitude) thus results in an increase in Q_{cor}. Now, the actual fuel flow (Q) for a given Mach number and altitude is directly proportional to $Q_{cor}\theta^x$, where the exponent x is an empirical constant specific to a particular engine type (a typical value is 0.5–0.67). Consequently, an increase in air temperature (under these conditions of constant Mach number and altitude) will increase Q as both Q_{cor} and θ increase.

It can thus be concluded that an increase in air temperature will result in an increase in both the numerator and denominator used to compute SAR; consequently, the SAR may not change appreciably. An example presented by Blake, *et al.* [13], for the B757-200 (at a weight of 220 000 lb) cruising at Mach 0.80 at FL 350, shows that a temperature increase of 20 °C would cause a TAS increase of 4.5% and a fuel flow increase of 6.2%, resulting in a SAR decrease of only 1.7%.

16.13   Impact of Small Changes on Cruise Fuel

16.13.1   Introduction

A mathematical expression for small changes to the key parameters that influence cruise fuel, developed by the author [14, 15], is presented in this section. This expression enables a relative change in cruise fuel to be determined for theoretical or actual changes to an airplane's drag, lift, TSFC, or weight (or a combination of these factors).

The approach has a number of applications—for example, for the analysis of active drag control systems [15]. These techniques are capable of reducing the skin friction drag by delaying the natural transition of the boundary layer from laminar to turbulent. There are a number of proven methods to achieve this; one such technique, which is known as hybrid laminar flow control (HLFC), employs pumps to suck modest amounts of air through perforated skin panels on the airframe surface at or near to the leading edge of a wing, empennage, or engine nacelle.[3] The system would reduce the airplane's drag, but it would also impose a penalty on the TSFC—due to the requirement to extract additional power from the engines to run the pumps—and it

---

3 Suction can also be achieved by connecting the system to a low pressure region on the airplane's surface. The first production airliner to feature HLFC was the B787-9, which employed the technique on the empennage [16]. HLFC has been studied for several decades and experimental installations have previously been tested on a B757 wing and A320 vertical tailplane, for example.

would add weight to the airplane. The equation derived in Section 16.13.2 facilitates an estimation of the impact on cruise fuel burn resulting from the installation of such a system.

Another application of this expression, presented herein, concerns the impact on cruise fuel burn arising from a variation in the net heating value of the fuel—this is explored in Section 22.7.4.

16.13.2 Mathematical Expression for Small Changes

The cruise fuel mass (m_f) that is consumed covering the range R at a true airspeed V is a function of three airplane-related parameters: end-of-cruise mass (m_2), mean TSFC ($\bar{c}$), and mean lift-to-drag ratio ($\bar{E}$). A linear expression for the change in fuel mass (δm_f), due to small changes to the end-of-cruise mass (δm_2), the mean TSFC ($\delta\bar{c}$), or the mean lift-to-drag ratio ($\delta\bar{E}$), can be derived if it is assumed that m_2, $\bar{E}$, and $\bar{c}$ are mutually independent variables in the cruise:

$$\delta m_f = \left(\frac{\partial m_f}{\partial m_2}\right)\delta m_2 + \left(\frac{\partial m_f}{\partial \bar{c}}\right)\delta\bar{c} + \left(\frac{\partial m_f}{\partial \bar{E}}\right)\delta\bar{E} \tag{16.22}$$

The parameters m_2, $\bar{E}$, and $\bar{c}$ are not, in fact, truly independent, and the assumption of independence introduces a small inaccuracy that for most applications is negligible. The next step is to determine mathematical expressions for the three partial derivatives in Equation 16.22. This may be accomplished using Equation 13.43a:

$$m_f = \left(e^{\left(\frac{R\bar{c}g}{V\bar{E}}\right)} - 1\right)m_2 \qquad \text{(Equation 13.43a)}$$

Thus $$\frac{\partial m_f}{\partial m_2} = e^{\left(\frac{R\bar{c}g}{V\bar{E}}\right)} - 1 \tag{16.23}$$

$$\frac{\partial m_f}{\partial \bar{c}} = e^{\left(\frac{R\bar{c}g}{V\bar{E}}\right)}\left(\frac{Rg}{V\bar{E}}\right)m_2 \tag{16.24}$$

and $$\frac{\partial m_f}{\partial \bar{E}} = e^{\left(\frac{R\bar{c}g}{V\bar{E}}\right)}\left(\frac{-R\bar{c}g}{V\bar{E}^2}\right)m_2 \tag{16.25}$$

Substituting the partial derivatives (given by Equations 16.23, 16.24, and 16.25) into Equation 16.22 and dividing through by m_2 yields the following expression:

$$\frac{\delta m_f}{m_2} = \left(e^{\frac{R\bar{c}g}{V\bar{E}}} - 1\right)\frac{\delta m_2}{m_2} + \left(e^{\frac{R\bar{c}g}{V\bar{E}}}\frac{R\bar{c}g}{V\bar{E}}\right)\frac{\delta\bar{c}}{\bar{c}} - \left(e^{\frac{R\bar{c}g}{V\bar{E}}}\frac{R\bar{c}g}{V\bar{E}}\right)\frac{\delta\bar{E}}{\bar{E}} \tag{16.26}$$

Although this equation is useful in its in own right—for the purpose of conducting sensitivity studies—it usually easier to appreciate a relative change in fuel, that is, $\delta m_f/m_f$. Using Equation 13.43a, m_2 is substituted to give the final expression:

$$\frac{\delta m_f}{m_f} = \frac{\delta m_2}{m_2} + \left(\frac{e^\beta \beta}{e^\beta - 1}\right)\frac{\delta\bar{c}}{\bar{c}} - \left(\frac{e^\beta \beta}{e^\beta - 1}\right)\frac{\delta\bar{E}}{\bar{E}} \tag{16.27a}$$

where $\beta = \dfrac{R\bar{c}g}{V\bar{E}}$

Writing the result in this way is convenient as the changes to m_2, $\bar{E}$, and $\bar{c}$ are expressed as nondimensional ratios. The impact on the cruise fuel can thus be determined from relative changes to these terms.

Of course, Equation 16.27a can also be written in terms of the aircraft's weight and the weight-based TSFC. If W_f is the cruise fuel weight (typical unit: lb) and W_2 is the end-of-cruise airplane weight (typical unit: lb), then

$$\frac{\delta W_f}{W_f} = \frac{\delta W_2}{W_2} + \left(\frac{e^\beta \, \beta}{e^\beta - 1} \right) \frac{\delta \bar{c}'}{\bar{c}'} - \left(\frac{e^\beta \, \beta}{e^\beta - 1} \right) \frac{\delta \bar{E}}{\bar{E}} \tag{16.27b}$$

where $\beta = \dfrac{R\bar{c}'}{V\bar{E}}$

and R is the cruise range (typical unit: nm);
      $\bar{c}'$ is the mean TSFC (typical unit: lb lb^{-1} h^{-1});
      V is the true airspeed (typical unit: kt); and
      $\bar{E}$ is the mean lift-to-drag ratio (dimensionless).

References

1 ICAO, "Operation of aircraft: Part 1, International commercial air transport – aeroplanes," Annex 6 to the Convention on International Civil Aviation, 9th ed., International Civil Aviation Organization, Montréal, Canada, July 2010.

2 FAA, *Operating requirements: Domestic, flag, and supplemental operations*, Federal Aviation Regulation Part 121, Amdt. 121-374, Federal Aviation Administration, Washington, DC, May 24, 2016. Latest revision available from www.ecfr.gov/ under e-CFR (Electronic Code of Federal Regulations) Title 14.

3 FAA, *General operating and flight rules*, Federal Aviation Regulation Part 91, Amdt. 91-336A, Federal Aviation Administration, Washington, DC, Mar. 4, 2015. Latest revision available from www.ecfr.gov/ under e-CFR (Electronic Code of Federal Regulations) Title 14.

4 European Commission, *Commercial air transport operations (Part-CAT)*, Annex IV to Commission Regulation (EU) No. 965/2012, Brussels, Belgium, Oct. 5, 2012. Published in *Official Journal of the European Union*, Vol. L 296, Oct. 25, 2012, and reproduced by EASA.

5 EASA, "Acceptable means of compliance (AMC) and guidance material (GM) to Part-CAT," Iss. 2, Amdt. 5, European Aviation Safety Agency, Cologne, Germany, Jan. 27, 2016. Available from www.easa.europa.eu/.

6 FAA, *Operating requirements: Commuter and on demand operations and rules governing persons on board such aircraft*, Federal Aviation Regulation Part 135, Amdt. 135-133, Federal Aviation Administration, Washington, DC, May 24, 2016. Latest revision available from www.ecfr.gov/ under e-CFR (Electronic Code of Federal Regulations) Title 14.

7 Randle, W.E., Hall, C.A., and Vera-Morales, M., "Improved range equation based on aircraft flight data," *Journal of Aircraft*, Vol. 48, Iss. 4, pp. 1291–1298, 2011.

8 ESDU, "Lost range, fuel and time due to climb and descent: Aircraft with turbo-jet and turbo-fan engines," Data item 74018, Amdt. A, IHS ESDU, 133 Houndsditch, London, UK, Aug. 1977.

9 Clark, P., *Buying the big jets: Fleet planning for airlines*, 2nd ed., Ashgate, Aldershot, Hampshire, UK, 2007.

10 ICAO, "2016–2030 Global air navigation plan," 5th ed., Doc. 9750, International Civil Aviation Organization, Montréal, Canada, 2016.

11 FAA and EUROCONTROL, "Comparison of air traffic management-related operational performance: U.S./Europe," Federal Aviation Administration, Washington, DC and EUROCONTROL, Brussels, Belgium, June 2014.

12 Airbus, "Getting to grips with aircraft performance monitoring," Flight Operations Support and Line Assistance, Airbus S.A.S., Blagnac, France, Dec. 2002.

13 Blake, W. and Performance Training Group, "Jet transport performance methods," D6-1420, Flight Operations Engineering, Boeing Commercial Airplanes, Seattle, WA, Mar. 2009.

14 Young, T.M., "Fuel sensitivity analyses for active drag reduction systems," *Aeronautical Journal*, Vol. 108, Iss. 1082, pp. 215–221, 2004.

15 Young, T.M., "Fuel-sensitivity analyses for jet and piston-propeller airplanes," *Journal of Aircraft*, Vol. 45, Iss. 2, pp. 715–719, 2008.

16 Kingsley-Jones, M., "Farnborough: Aero secrets of Boeing's new Dreamliner," Flightglobal, Sutton, UK, July 18, 2014. Available from www.flightglobal.com/news/articles/farnborough-aero-secrets-of-boeings-new-dreamline-401784/.

17

En Route Operations and Limitations

17.1   Introduction

From the time that the airplane leaves the immediate vicinity of the departure airfield until such time as it approaches the destination airfield it is considered to be *en route*. Specific operational limits and restrictions can be imposed by a variety of factors on this portion of a flight—several of the important considerations are discussed in this chapter.

The airspeeds that the crew will select for the departure climb, for example, are often based on predetermined values that are calculated to reduce the total trip cost. The operator, however, does not have complete freedom in this respect as flight operations need to comply with air traffic management requirements. Airspeed and climb thrust selection, which influence the airplane's rate of climb, are discussed in Section 17.2.

The top of climb (TOC) marks the start of the cruise. The selection of an initial flight level—to commence cruise—is based on the airplane's performance capability at the associated weight, altitude, and air temperature (WAT) and on the available flight levels for the route in question. It is important that the selected flight level is close to the optimum altitude—that is, the altitude at which the airplane is most efficient in cruise. It is also important to ensure that the airplane has adequate climb potential at this heavy weight at the start of the cruise (this is discussed in Section 17.3) and that it has a sufficient margin from the buffet onset speed (discussed later in Section 20.3).

When planning routes over mountainous regions, it is a requirement that consideration be given to the possibility of an engine failure (or two engines failing for airplanes with three or more engines), as there will be insufficient thrust to maintain the planned cruise altitude. Vital information for flight crews conducting such operations is the greatest altitude at which the airplane can safely operate after suffering an engine failure (or multiple engine failures in the case of airplanes with three or more engines) and a defined drift-down flight path that provides adequate terrain clearance (see Section 17.4). Another failure case that must be considered when planning flights over high mountains is rapid cabin de-pressurization. In this case, the emergency descent flight profile will be dictated by the available supplemental oxygen carried onboard the airplane (see Section 17.5).

Route planning for twin-engine airplanes over routes that will take the airplane to points that are farther than 60 minutes flying time from an adequate airport are subject to a set of specific requirements (for airplanes with three or more engines, the flying time is 180 minutes)—these *extended operations* are discussed in Section 17.6.

Continuous descent operations—in which airplanes descend continuously from cruise in a low power, low drag configuration prior to the final approach to landing—can provide economic

Performance of the Jet Transport Airplane: Analysis Methods, Flight Operations, and Regulations, First Edition.
Trevor M. Young.

and environmental benefits; details on the operational procedure and its benefits are outlined in Section 17.7.

17.2 Climb to Initial Cruise Altitude (*En Route* Climb)

17.2.1 Climb–Speed Schedules

The climb–speed schedules used by airlines are usually established to reduce the total trip cost, within the constraints of air traffic control (ATC) requirements. The optimum climb speed, in fact, is very close to the maximum rate-of-climb speed (see Section 12.4.6). The true airspeed (TAS) for the best rate of climb varies with altitude, but corresponds closely to a constant calibrated airspeed (CAS) climb.

Air traffic control frequently imposes operational limits on departing (or arriving) aircraft. For example, a speed restriction that applies to many of the world's airspaces is a limit of 250 kt indicated airspeed (IAS) below 10 000 ft (airspace classification is discussed in Section 5.4.4). When the height of the transition altitude is lower than 10 000 ft, flight level 100 is used in place of the 10 000 ft limit [1]. Although the 250 kt speed limit is not imposed universally, it is nonetheless widely used; furthermore, many airlines adopt this speed limit as a standard operating procedure. In a typical climb, the pilot will increase speed upon reaching 10 000 ft, and then continue climbing at a constant CAS. Depending on the airplane type, climb–speed schedules are established either in terms of IAS (aircraft with pneumatic speed indicators) or CAS (aircraft with electronic speed indicators) [2].

A typical climb–speed schedule would be written as 250 kt / 290 kt / M 0.78, for example (see Figure 17.1). This implies that the airplane climbs at a CAS of 250 kt until 10 000 ft is reached. The airplane then accelerates to 290 kt and climbs until Mach 0.78 is reached—this occurs at the *crossover height*[1] (see Section 17.2.2), and, for this example, it corresponds to a pressure height of 30 875 ft. Thereafter, a constant Mach number is held for the remainder of the climb.

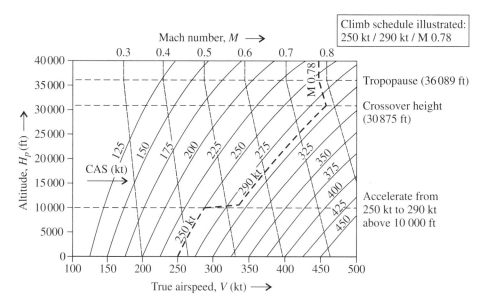

Figure 17.1 Typical airliner climb–speed schedule (ISA conditions).

---

1 The crossover height is sometimes referred to as the *transition height*; this practice is to be discouraged, though, as the term transition height is used in a different context in flight operations (see Section 5.4.2).

It is instructive to consider a constant CAS climb in more detail. The effect of a steadily increasing TAS coupled with a progressive reduction in the speed of sound means that the Mach number will increase rapidly during the climb even though the CAS remains constant—for example: 290 kt CAS corresponds to M 0.52 under ISA (International Standard Atmosphere) conditions at FL 100, but M 0.78 at FL 310. The influence of flow compressibility (see Section 3.5) thus can thus become significant before the airplane has reached its cruise altitude. To avoid a substantial increase in drag during the upper part of the climb, it is usually necessary to hold the Mach number below a defined value—which is M 0.78 in the example illustrated in Figure 17.1. It is typical that this Mach number is usually held constant until the cruise height is reached.

Climbing at constant Mach number implies that there will be a slight decrease in TAS up to the tropopause (as shown in Figure 17.1), as the speed of sound decreases with height at these altitudes. The idealization adopted in the ISA is that the air temperature in the lower stratosphere is a constant $-56.5\,°C$. Thus, climbing at a constant Mach number above the tropopause—which is at a height of 36 089 ft in the ISA—means that the TAS is constant. In reality, the height of the tropopause and the air temperature in the lower stratosphere varies with geographical location (see Section 4.4.3). Nonetheless, the air temperature can be assumed to be invariant in the lower stratosphere at a given location.

17.2.2 Determination of the Crossover Height

Determination of the crossover height is based on finding the pressure height at which a given CAS matches the Mach number selected for the upper climb segment. Under ISA conditions, the pressure height (H_p) in the troposphere can be related to the relative pressure (δ) using Equation 4.21:

$$H_p = \frac{T_0}{L}\left[\delta^{\frac{-RL}{g_0}} - 1\right] \tag{17.1}$$

and from Equation 6.22, noting that $\gamma = 1.4$, the relative pressure at the crossover height is given by

$$\delta = \frac{\left[0.2\left(\frac{V_c}{a_0}\right)^2 + 1\right]^{3.5} - 1}{[0.2\,M^2 + 1]^{3.5} - 1} \tag{17.2}$$

where T_0 is the ISA sea-level datum temperature (typical unit: K);
  L is the ISA lapse rate (typical units: K/m, K/ft);
  R is the specific gas constant (typical units: $m^2\ s^{-2}\ K^{-1}$, $ft^2\ s^{-2}\ K^{-1}$);
  g_0 is the standard value of gravitational acceleration (typical units: m/s^2, ft/s^2);
  V_c is the CAS (typical units: m/s, kt, km/h);
  a_0 is the ISA speed of sound at sea level (typical units: m/s, kt, km/h); and
  M is the Mach number (for the upper climb segment).

17.2.3 Constant Rate-of-Climb Schedule

Climbing at the best rate-of-climb (ROC) speed (see Section 12.4.6) results in a progressively reducing rate of climb. However, sometimes pilots wish to hold the rate of climb constant. The autopilot will then control the airplane's speed and thrust to achieve the required rate of climb.

When the airplane's specific excess power (which reduces with increased height) has decayed to a level that is not able to support the required rate of climb, the airplane is said to have reached its *ceiling* (or maximum altitude) for that rate of climb. This is, in fact, a practical method for defining an airplane's ceiling (as discussed later in Section 17.3.1).

17.2.4   Derated Climb Thrust Settings

The maximum thrust that is available for the climb corresponds to the engine's maximum climb thrust (MCLT) rating (see Section 8.5.5). Modern engines, however, usually have one or more derated climb thrust settings, which operators can choose to use. If the airplane's rate of climb with MCLT at its maximum takeoff weight (MTOW) is considered to be adequate, then at a reduced takeoff weight it would be possible to climb at a comparable rate of climb with reduced thrust. This is particularly evident for long-haul airplanes operating on short sectors (i.e., with a much reduced fuel load).

The use of a derated climb thrust setting results in an increase in the climb time, an increase in the climb distance, and more fuel being consumed in the climb compared to a MCLT climb. These increases, however, are partially offset when considering the whole trip as there will be a reduction in the cruise distance and the cruise fuel. The net increase in trip time and trip fuel can be very small (as indicated in a study conducted by James and O'Dell [3] using the B777-200/Trent 892 airplane as a reference).

The advantage for the operator is that the use of derated climb thrust has a bearing on engine maintenance costs. The number of cycles flown and the environment in which the airplane operates are key factors that influence the time between engine removals and the cost of refurbishment. Also critical is the manner in which the airplane is flown. In particular, the engines' shaft speeds and core temperatures during takeoff and climb are important. The benefit of a derated thrust climb is lower gas temperatures in the engines' turbines. As engine wear and damage are directly related to operating temperatures, significant life cycle cost and reliability benefits can be achieved through the regular use of this procedure.

Engine manufacturers have implemented climb thrust derates in various ways. A typical climb thrust derate does not apply as a fixed percentage reduction in thrust throughout the climb. Rather, a fixed percentage derate is maintained to a certain height (e.g., 10 000 ft), and thereafter the derate is reduced linearly to zero at a defined height (which is often called the "washout" height)—as illustrated in the example given in Figure 17.2. This means that as the

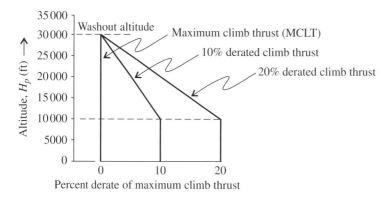

Figure 17.2 Illustration of two climb thrust derate options.

airplane approaches its initial cruise altitude, it will be operating at 100% of the maximum climb thrust (MCLT).

17.3 Cruise Altitude Selection

17.3.1 Ceiling Definitions

The thrust produced by an airplane's engines decays with increasing altitude (as described in Section 8.6.5). However, at typical cruising altitudes, an airplane's drag for a given weight and Mach number does not change appreciably with increasing altitude. Consequently, there exists a maximum altitude or *ceiling* beyond which it will not be possible for an airplane to maintain steady, level (i.e., constant height) flight, as the drag would exceed the thrust available. This *absolute* ceiling depends principally on the engine thrust rating (i.e., maximum climb thrust or maximum cruise thrust), airplane weight, and air temperature.

Operating an airplane at its absolute ceiling, however, is not practical for several reasons. An airplane's rate of climb, at any fixed Mach number, diminishes with increasing altitude (see Section 12.4.4). A plot of altitude versus time (climb duration) will show the airplane approaching the absolute ceiling asymptotically. It thus takes a long time for the airplane to get close to its absolute ceiling. Secondly, steady flight at the theoretical maximum altitude—for a given airplane weight, thrust rating, and air temperature—is only possible at a single speed. Furthermore, there is no margin for an increase in load factor due to atmospheric turbulence or maneuvers. It can be concluded that operating an airplane in the vicinity of its absolute ceiling is problematic and difficult to sustain. (Further information on altitude limits is given in Section 20.4.1.)

An alternative—and more practical—method to define an airplane's ceiling is in terms of a set rate of climb (e.g., 100 ft/min or 300 ft/min) that is achievable at a particular airplane weight, thrust rating, air temperature, and Mach number. This is called a *service ceiling*. For example, the 300 ft/min service ceiling is the greatest altitude at which a rate of climb of 300 ft/min (1.53 m/s) can be achieved. Another way of looking at this is to consider that an airplane cruising at this altitude has the potential to climb at 300 ft/min—this is known as a *residual* rate of climb. As rate of climb is inversely proportional to airplane weight (see Equation 12.27), the service ceiling would increase as the airplane's weight reduces during cruise.

A ceiling can also be defined in terms of the *climb gradient* that can be achieved for a particular set of conditions (i.e., airplane weight, thrust rating, and air temperature). This approach is used when determining the airplane's net flight path with one engine inoperative (or, in the case of airplanes with three or more engines, with two engines inoperative). In determining these maximum flight altitudes, the thrust of the operating engines should be taken as the maximum continuous thrust (MCT). Flight planning over mountainous terrain, where these ceilings are of particular concern, is discussed later in Section 17.4.

17.3.2 Initial Cruise Altitude

As mentioned in Section 17.3.1, the greatest altitude at which a set rate of climb (ROC) can be achieved for a given airplane weight requires the airplane to be operated at a particular speed (which will be close to the airplane's minimum drag speed). In an operational context, however, what an operator needs to know is the maximum altitude that is achievable at an appropriate

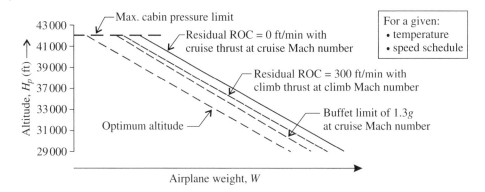

Figure 17.3 Illustration of initial cruise altitude capability.

climb or cruise Mach number for a given airplane weight and air temperature. The methodology used to establish these altitude limits is based on the airplane's rate of climb performance (which is discussed in Section 12.4.4).

The selection of an *initial* cruise altitude for flight operations is based on several considerations—for example, the limiting factor for a given airplane weight and air temperature could be

(1) the altitude at which a residual ROC of 300 ft/min can be achieved at the selected climb speed with maximum climb thrust;
(2) the altitude at which the airplane would have zero residual ROC at the selected cruise speed with maximum cruise thrust;
(3) the altitude that provides adequate margin from buffet onset for the selected cruise speed (see Section 20.3); or
(4) the altitude corresponding to the maximum cabin pressure limit.

The influence of airplane weight on altitude limits is illustrated in Figure 17.3 (note that the relative position of the lines shown in the figure is for illustration only). The greatest usable flight level (see Section 5.4.3) below the most restrictive altitude can be determined by the flight crew using such charts—which are contained in the Flight Crew Operating Manual (FCOM)—based on the airplane's weight and the planned route (specifically the flight direction). Once usable flight levels have been established, an *initial flight level* (i.e., the flight level at which the cruise would commence) can be selected in the proximity of the optimum altitude (i.e., the altitude that would give the greatest specific air range, as described in Section 13.6.5).

17.3.3 Cruise Altitudes for Short Stages

For short stage (trip) lengths, the cruise flight level can be restricted by the climb and descent distances. In such cases, it may not be possible to reach the optimum cruise altitude. Furthermore, there is little practical value in climbing to an initial cruise altitude and then immediately commencing the descent. A widely adopted strategy is based on the cruise sector not being less than 5 minutes [4]. Short stage cruise altitudes, based on the sector length for a given airplane type, are published in the FCOM.

17.3.4   Cruise Step-Climb Schedule

An airplane is most efficient in cruise at its optimum altitude (i.e., the altitude at which the specific air range is a maximum). The specific air range penalty of flying 2000 ft below the optimum altitude is ~1–3%, depending on the airplane type [4, 5]. As described in Section 13.6.5, the optimum altitude increases during the cruise, as the airplane's weight reduces. The ideal cruise procedure would be for the airplane to "drift" slowly upwards as fuel is burned; however, this is not normally possible as standard flight operations restrict the cruise to fixed flight levels (which are described in Section 5.4.3). As a result, a stepped climb is flown, in which periodic changes to a higher flight level (FL) take place during the cruise. The fuel saved by operating at the higher altitude must, of course, be greater than the additional fuel required for the step climb (see Section 16.8.4). Each step involves an increase in altitude of either 2000 ft or 4000 ft, depending on the next available flight level. With Reduced Vertical Separation Minimum (see Section 5.4.3), 2000 ft steps are permitted below FL 410, which means that the flight profile can remain closer to the optimum than would be possible with 4000 ft steps.

Various stepped flight profiles can be flown to approximate the steadily increasing optimum altitude, with the step climbs scheduled as a function of airplane weight. The amount of fuel that has to be consumed before a change in flight level is warranted depends on the airplane's specific air range at the flight levels of interest (see Section 13.6). This is a complex calculation. A reasonable approximation, however, is possible by ignoring changes to the thrust specific fuel consumption (TSFC); in this case, the optimum altitude versus weight function can be established by assuming that W/δ remains constant (see Section 13.5). Using this approach, it can be deduced that an airplane operating in the stratosphere in ISA conditions, which commences its cruise at the optimum altitude, will be 2000 ft below its optimum altitude when its weight has decreased by 9.2% and 4000 ft below its optimum altitude when its weight has decreased by 17.5%. This can be determined from ISA pressure data (see Appendix A) or deduced through the application of Equation 4.16.

As mentioned earlier, the point during the cruise at which a step climb is initiated depends on the airplane's weight (for a given Mach number). Tabulated data are published in the FCOM for selected flight level changes for various temperature conditions (e.g., ISA, ISA + 10 °C, ISA + 20 °C), sometimes with corrections for the air conditioning or anti-icing systems. A pragmatic approach to scheduling step climbs is illustrated in Figure 17.4. The climb is initiated when the

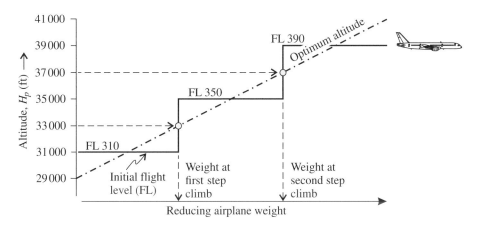

Figure 17.4 Flight profile illustrating step-climb schedule during cruise.

optimum altitude, corresponding to the airplane's weight, is mid-way between the lower (i.e., current) flight level and the higher (i.e., next) flight level.

17.4 *En Route* Engine Failure

17.4.1 *En Route* Obstacle Clearance

Specific regulatory requirements exist in regard to obstacle clearance for the *en route* portion of the trip [6–9]. Following an engine failure, the airplane's maximum cruise altitude will be reduced (an engine failure would clearly have a greater impact on the performance capabilities of twin-engine airplanes, compared to airplanes with three or more engines). It is imperative that the flight planning takes the reduced airplane performance into account to ensure that the flight path is always higher, by a sufficient margin, than any mountain or obstacle that needs to be crossed. For certain airplane types, scheduled to fly over mountainous terrain, this requirement can be difficult to meet and this may limit the takeoff weight. A regulatory provision that improves the situation for operators, without reducing safety margins, is based on a safe *drift-down* flight path that the flight crew may take following an engine failure (see Sections 17.4.4–17.4.6).

Engine failure is recognized as a critically important safety issue. The regulations stipulate that following an engine failure, all airplanes must have a minimum climb performance capability at defined altitudes associated with the planned flight (discussed in Section 17.4.2). In addition to meeting these requirements, airplanes with three or more engines must comply with specific requirements with two engines inoperative (discussed in Section 17.4.3).

17.4.2 One-Engine-Inoperative Performance Requirements

The applicable requirements are specified in FAR 121.191 [6] and OPS Part CAT.POL.A.215 [7]. Based on the one-engine-inoperative (OEI) *en route* net flight path data (as given in the Airplane Flight Manual, or AFM), the airplane must be able to meet condition 1 and either condition 2 or condition 3, as described below, taking into account the expected ambient temperatures:

(1) The airplane must have a positive net flight path slope[2] at 1500 ft above the landing airport.
(2) The airplane must be able to maintain a positive net flight path slope at an altitude of at least 1000 ft above all terrain and obstructions within a defined distance[3] on each side of the intended track.
(3) The net flight path must allow the airplane to continue flight from the cruising altitude to a suitable airport where a landing can be made clearing all terrain and obstructions within a defined distance on each side of the intended track by at least 2000 ft vertically. It must be assumed that the engine fails at the most critical point *en route* and allowance must be made for adverse winds.

The performance calculations that need to be conducted to show compliance with these requirements must take into account the difference between the gross and net flight paths (Figure 17.5). This is identically defined in FAR 25.123 [8] and CS-25.123 [9], as follows: the one-engine-inoperative net flight path data equate to the actual (i.e., gross) climb performance

---

2 In this context, a positive slope can be taken as a climb gradient that is not less than zero.
3 FAR 121.191 [6] defines this distance as five statute miles and CAT.POL.A.215 [7] as five nautical miles on either side of the intended track.

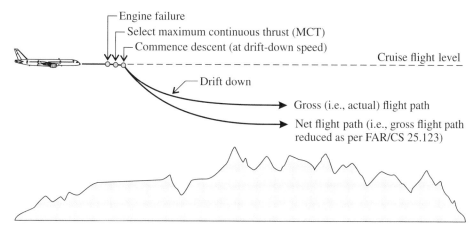

Figure 17.5 Gross and net flight paths following an engine failure.

diminished by a gradient of climb of 1.1% for two-engine airplanes, 1.4% for three-engine airplanes, and 1.6% for four-engine airplanes.

The latter two requirements effectively mean that if the airplane, at the given takeoff weight (and assuming normal fuel consumption), cannot meet condition 2 of being able to maintain a level net flight path with one engine inoperative, 1000 feet above all terrain and obstructions *en route*, then it must meet condition 3 in which the net flight path during descent must clear all terrain and obstructions by at least 2000 feet.

Specific requirements also exist regarding fuel planning for this scenario—for example, fuel jettisoning is permitted, provided that it is a safe procedure and there is enough fuel to continue to a suitable airport with adequate reserve fuel in compliance with the regulations.

17.4.3 Two-Engine-Inoperative Performance Requirements

For airplanes with three of more engines, the requirements for two engines failing *en route* are described in FAR 121.193 [6] and OPS Part CAT.POL.A.220 [7]. If there is a place along the intended track that is more than 90 minutes from a suitable airport (with all engines operating at cruise thrust), then conditions 1 and 3, as described in Section 17.4.2, must be met with two engines inoperative. The engines are assumed to fail simultaneously at the most critical point *en route*. In this case, the net flight path data equate to the gross climb performance diminished by a gradient of climb of 0.3% for three-engine airplanes and 0.5% for four-engine airplanes (as defined in FAR/CS 25.123 [8, 9]).

17.4.4 Route Planning and Drift-Down Flight Paths

Route planning for crossing mountainous terrain[4] can be based on a drift-down flight path. As described in Section 17.4.2 (condition 3), operators are permitted to plan flights in such a way that following the failure of an engine (or two engines in the case of airplanes with three or more engines), the flight crew may follow a defined route, provided that the net flight path

---

4 Mountainous regions where terrain clearance is an issue for route planning for modern airplanes include the Andes, Himalayas, and Hindu Kush, and, for older airplane types, the Rockies and Alps [10].

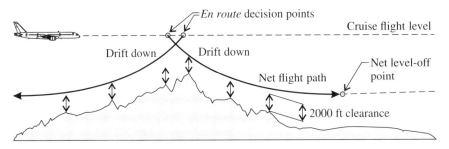

Figure 17.6 Drift-down flight paths.

clears all obstacles on the defined corridor by at least 2000 ft. This provision in the regulations permits certain airplane types to take off with a higher gross weight than would otherwise be possible.

Figure 17.6 illustrates one possible scenario for an airplane crossing a high mountain range. If an engine failure occurs before the airplane gets to the first *decision point*, the flight crew will not be able to cross the mountain peak and must divert from the planned route along an alternative escape route. Failure of the engine(s) after the second *decision point* means that the flight crew must continue to cross the mountain range and cannot turn back. In determining the drift-down flight paths, consideration for the additional loss of height associated with any required maneuvering (e.g., a "turnback") must be taken into account (climbing/descending turns are discussed in Section 15.4). After the airplane has leveled off, it will begin to climb slowly as its weight reduces due to fuel usage.

17.4.5 Performance Analysis Following Single or Multiple Engine Failure(s)

In Chapter 12, methods for analyzing climb and descent performance of the airplane are described; key aspects for the drift-down descent are reviewed below.

The flight path gradient ($\tan \gamma$) in still air conditions, with one or more engines inoperative, can be determined by expressing Equation 12.22 as follows:

$$\tan \gamma \cong \frac{\dfrac{F_N}{W} - \dfrac{D}{W}}{1 + f_{acc}} \qquad (17.3)$$

where γ is the flight path angle in still air (defined as positive upwards);
F_N is the thrust from the remaining engines (typical units: N, lb);
W is the airplane weight (typical units: N, lb);
D is the drag (typical units: N, lb); and
f_{acc} is the acceleration factor (dimensionless) (see Table 12.1).

In this situation, the lift will be approximately equal to the weight; hence, the lift-to-drag ratio can be introduced, as follows:

$$\tan \gamma \cong \frac{\dfrac{F_N}{W} - \dfrac{1}{E}}{1 + f_{acc}} = \frac{\dfrac{F_N}{W} - \dfrac{C_D}{C_L}}{1 + f_{acc}} \qquad (17.4)$$

where E is the lift-to-drag ratio; and
C_L and C_D are the lift and drag coefficients, respectively.

When evaluating Equation 17.4, the following factors are important:

(1) The flight crew are permitted to use the maximum continuous thrust (MCT) setting for the remaining engine(s).
(2) The airplane's drag coefficient must be corrected for the inoperative engine(s)—that is, to account for windmilling drag, spillage drag, and yaw (or control) drag due to rudder deflection (see Section 7.5.5).

The prediction of a drift-down flight path (or profile) is based on the flight crew undertaking a specific series of actions (in accordance with relevant operating procedures). Following an engine failure (or multiple engine failures), the flight crew would select MCT for the remaining engine(s). They would then allow the airplane to decelerate while maintaining the cruise altitude until the drift-down target speed (see Section 17.4.6, below) is reached, before commencing the drift-down descent.

The computational method to determine the drift-down profile is based on the equations presented in Sections 12.4 and 12.5. A stepwise computation is undertaken for selected height intervals (e.g., 1000 ft or 2000 ft), starting at the cruise altitude. Representative values for the fuel flow, rate of descent, and descent gradient are determined for the first interval, based on the airplane's weight and speed. This will enable the time for the airplane to descend through the interval and the horizontal distance corresponding to the net flight path to be calculated. The fuel flow will enable the airplane's weight at the end of the interval to be established, permitting the calculation to start afresh for the next interval. The computation continues until the net level-off altitude (see Sections 17.4.2 and 17.4.3) is reached.

Note that winds will influence the airplane's trajectory, but will not alter the net level-off altitude. Non-ISA conditions are another important factor. An increase in air temperature, for example, will (1) reduce the thrust of the engine (or engines) and (2) increase the true airspeed for a set Mach number. This will reduce the net level-off altitude. Strong headwinds and/or hot conditions (e.g., ISA + 20 °C) can thus result in reduced terrain clearance.

17.4.6   Drift-Down Target Speed

The best speed to execute a drift-down procedure following the failure of an engine (or engines) in cruise is that which will minimize the angle of descent (see Section 12.5). It can be deduced from Equation 17.4 that the airplane needs to operate at the speed that will yield the highest lift-to-drag ratio, that is, the minimum drag speed (see also Sections 7.6 and 7.7). Flight crews are provided with target speeds to fly in such situations (the information is contained in the airplane's FCOM).

On Airbus aircraft, the target speed (or reference speed), known as the *green dot* speed,[5] is identified by a small green circle on the vertical speed tape of the Primary Flight Display (this is particularly useful when the pilot is operating the airplane manually). The green dot speed approximates the best lift-to-drag ratio speed for the airplane in the clean configuration, within an acceptable margin for flight operations. It is determined by a simple formula (established for each airplane type) based on the airplane's instantaneous mass and altitude. An example of a

---

5 Flight crews operating Airbus aircraft are instructed to fly at the green dot speed in certain, specific situations—for example, following an engine failure in cruise (as discussed in this section); in the final takeoff segment following an engine failure (see Section 10.6.1); when in a holding pattern (see Section 14.6.3); or when it is required to climb to a particular altitude over the shortest distance (see Section 12.4.3) [11]. It is also recommended that flight crews do not fly below the green dot speed in cruise (in the clean configuration) [12].

green dot speed formula, which applies to the A330-300 airplane [13], is given below:

- When $H_p \leq 20\,000$ ft, the green dot speed is

$$V_{c_{gd}} = 0.6\frac{m}{1000} + 107 \qquad \text{(all engines operating), or} \tag{17.5}$$

$$V_{c_{gd}} = 0.6\frac{m}{1000} + 97 \qquad \text{(at least one engine inoperative)} \tag{17.6}$$

where $V_{c_{gd}}$ is the green dot speed in knots calibrated airspeed (KCAS); and m is the instantaneous airplane mass in kilograms.

- When $H_p > 20\,000$ ft, 1 kt is added to the computed speed (given by Equation 17.5 or 17.6) for every 1000 ft altitude increase above 20 000 ft.

17.5 *En Route* Cabin Pressurization Failure

17.5.1 *En Route* Obstacle Clearance

Route planning of flights that will overfly mountainous terrain must consider the possibility of a loss of cabin pressure (the cabin pressurization system is discussed in Section 22.3). The emergency descent profile from cruise to an altitude at which crew members and passengers can breathe normally must consider the available supplemental oxygen carried on board the airplane (see Section 22.3.4) and the performance capabilities of the airplane. As the supply of oxygen is limited, it is vital that the airplane descends as quickly as possible—this is usually conducted at M_{MO}/V_{MO} (i.e., the maximum operating Mach number/maximum operating speed for the airplane). If necessary, air brakes can be deployed to increase the rate of descent.

The planning of such flights over mountainous terrain is based on an emergency descent profile (or schedule) that assumes a total pressurization system failure at the most critical point on the intended route and an emergency flight path that ensures that a height of no less than 2000 ft above all obstacles is maintained.[6] An example of an emergency descent profile for an airplane equipped with 22-minute chemical oxygen-generating canisters is illustrated in Figure 17.7. The profile represents the maximum flight levels and durations that can be flown, as dictated by the installed oxygen system, in compliance with the regulations (see Section 22.3.4). The illustrative profile is for a stepped descent from the cruise altitude to an altitude not exceeding FL 140 within 22 minutes, followed by a cruise not exceeding 30 minutes before descending to an altitude not exceeding FL 100 for the remainder of the flight [11].

Unlike the engine failure case (discussed in Section 17.4), the gross flight path following a loss of cabin pressure is not reduced to a net flight path when determining the flight profile. As it is assumed that all engines are operating normally in this case, it is conceivable that the flight crew can increase the altitude again if needed, and, therefore, the standard minimum obstacle clearance of 2000 ft (as prescribed in the airworthiness regulations) applies.

----

6 FAR 121.657 [6] requires that all aircraft operating in mountainous areas maintain a height of at least 2000 ft above the highest obstacle within a horizontal distance of five statute miles from the center of the intended track. For EASA operators, GM1 CAT.OP.MPA.145(a) [14] provides examples of acceptable methods for calculating minimum flight altitudes, which are in compliance with the general requirements of CAT.OP.MPA.145 [7].

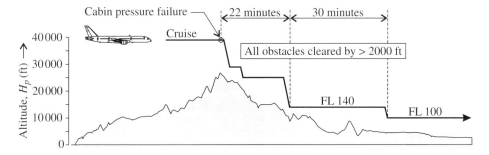

Figure 17.7 Example of a flight profile following cabin de-pressurization for an airplane equipped with a 22-minute chemical oxygen-generating system.

17.5.2 Route Planning Considerations

Route planning must consider both failure cases—that is, engine failure and cabin pressurization failure. Viable descent profiles and critical points—in compliance with the regulations—need to be established for both cases. The airworthiness regulations, however, do not require that consideration be given to both failures occurring simultaneously. It is considered preferential, whenever it is possible, to define the same critical points and the same escape routes, whatever the failure case, to reduce flight crew workload [11]. This strategy can potentially reduce crew reaction time and the risk of mistakes being made. The planned escape route would thus be based on the most penalizing, or restrictive, descent profile.

17.6 Extended Operations

17.6.1 Introduction

The term *Extended Operations* has historically been used to describe the operation of twin-engine jet airplanes over routes that include points that are farther than 60 minutes flying time (under standard conditions in still air) from an adequate airport at the airplane's one-engine-inoperative (OEI) cruise speed. This distance is known as the threshold—it corresponds to approximately 420–450 nm for a jet transport airplane. Extended Operations are thus operations that extend beyond that threshold (illustrated by the route A–C in Figure 17.8). Such operations became known as ETOPS, an acronym for Extended Twin Operations (or Extended Range Operations with Two-Engine Airplanes). Prior to the introduction of the initial ETOPS rules in the mid-1980s, many oceanic routes were restricted to airplanes with three or four engines. The rules have been revised several times since then by the various regulatory authorities (not always in a consistent manner, though).

The FAA revised their ETOPS Advisory Circular in 2008, introducing new requirements for long range and polar operations [15]. The new regulations also applied the basic ETOPS philosophy to airplanes with three or more engines (for the first time) for routes that include points that are farther than 180 minutes flying time from an adequate airport (at the airplane's OEI cruise speed, under standard conditions in still air). The acronym ETOPS was redefined by the FAA to simply mean Extended Operations. EASA, however, continued to use the term ETOPS for twinjet airplanes and later adopted the acronym LROPS (Long Range Operations) to describe Extended Operations of three- and four-engine airplanes. In 2011, ICAO introduced

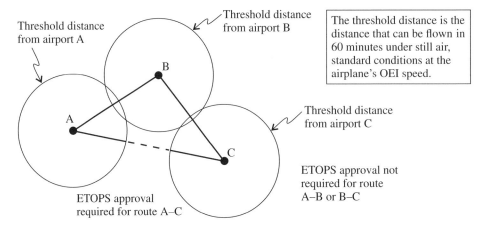

Threshold distance from airport A

Threshold distance from airport B

The threshold distance is the distance that can be flown in 60 minutes under still air, standard conditions at the airplane's OEI speed.

Threshold distance from airport C

ETOPS approval not required for route A–B or B–C

ETOPS approval required for route A–C

Figure 17.8 Schematic illustrating the 60 minute threshold for Extended Operations (ETOPS).

the more descriptive term E̲xtended D̲iversion T̲ime O̲perations (EDTO) to describe Extended Operations of any airplane type, irrespective of the number of engines fitted.

The 60 minute limit, or *rule* as it is usually called, for twin-engine airplane operations is specified in FAR 121.161 [6] and EASA Part-SPA (section SPA.ETOPS.105) [16]; it essentially states that a series of conditions needs to be met before an operator can legally operate an airplane beyond this threshold. The 60 minute rule has a long history—it dates back to the early 1950s and was essentially based on the reliability of piston engines at that time. Turbine engines, however, are much more reliable than piston engines. Responding to requests from airplane manufacturers and long-haul operators, the regulatory authorities (e.g., FAA, CAA, DGAC) in the mid-1980s published (e.g., in FAA AC 120-42 [17]) acceptable means for operators to get the necessary permission—that is, ETOPS approval—to deviate from the 60 minute rule.[7] The adopted approach allowed operators to extend their approved ETOPS time incrementally up to a maximum value, which was initially set as 120 minutes by the FAA (with the possibility of an additional 15% decided on a case-by-case basis). As experience was gained in operating twin-engine airplanes over ETOPS routes, and as the reliability of the new twinjet airplanes at that time (e.g., B767, A310, A300) was demonstrated, so operators could apply for and receive permission to increase the approved ETOPS time. In the late 1980s, the maximum ETOPS time was increased to 180 minutes (again, with the provision for an additional 15% on a case-by-case basis). This permitted more direct and new routes to be operated by twinjet airplanes. More recently, the concept of unlimited ETOPS has been considered, with permission being granted for over 5 hour ETOPS.[8] Not all operators require such capabilities, as the vast majority of the world's long-haul commercial routes can be operated with 180 minute ETOPS approval.

The ETOPS (twin operation) concept is predicated on the extremely improbable likelihood of both engines failing due to independent causes on modern, well-maintained, and professionally operated twin-engine airliners (systems reliability is discussed later in Section 22.2). The requirements for ETOPS approval are outlined in Section 17.6.2, which follows.

---

7 The first ETOPS operations, which took place under a 90 minute ruling, were undertaken by TWA in February 1985 using a B767-200 [18].

8 FAA approval for 330 minute ETOPS for the Boeing B787 was granted in May 2014 [19]. EASA, in October 2014, approved the A350-900 for 180 minute ETOPS in the basic specification and included provisions for 300 minute and 370 minute ETOPS, depending on individual operator selection [20].

17.6.2 Approval for Extended Operations

ETOPS (twin operations) approval is based on (1) the airplane manufacturer demonstrating compliance with a set of requirements [8, 9], and (2) the operator meeting a further set of requirements [6, 7, 15, 21]. In terms of the design and manufacture of the airplane, this imposes stringent requirements regarding the reliability of critical systems. The propulsion system (the most important element affecting ETOPS), for example, must be sufficiently reliable so as to ensure that the probability of a dual engine failure, from independent causes, is lower than a defined limit. For 180 minute ETOPS approval for a twinjet, this translates to an in-flight shutdown (IFSD) rate target, which should not be exceeded in service, of 0.03 per 1000 engine hours [15]. The electrical power supply, similarly, must be designed with a high degree of redundancy, so as to ensure that electrical power is always available for the basic airplane functions (e.g., flight instruments, communication, and navigation equipment). The reliability of the identified systems must be demonstrated by in-service experience (and not estimated by simulation). More than 100 000 engine hours are required to conduct a valid statistical analysis of the reliability of a new engine type and 50 000 for a derivative engine [21].

The second element of the ETOPS approval concerns the operator. Operators must ensure that their airplanes are appropriately equipped (e.g., concerning such items as supplemental oxygen). In addition, operational approval requires that operators meet specific criteria regarding maintenance practices, crew training, in-service experience, operational readiness, and service documentation.

17.6.3 ETOPS (Twin Operations) Tracks and Area of Operations

Designated alternate airfields on ETOPS flight plans must satisfy specific criteria. Airfields that meet these conditions—which relate to technical support, navigation aids, firefighting, and so forth—are denoted as *adequate airfields*. The airplane's track must then stay within the maximum authorized diversion time (e.g., 120 minutes, 180 minutes) from an adequate airfield.

Based on the aircraft's OEI performance capability (taking into account the drift-down from the cruise altitude), estimated aircraft weight (at that point in the mission), and the defined diversion time, an ETOPS area of operation can be determined. The planned track for the mission must remain within this defined "area of operations." This is illustrated in Figure 17.9 for a 120 minute approval. A direct route from A to E is possible, while remaining within the 120 minute, single-engine operating speed circles. Without this approval, a longer route would

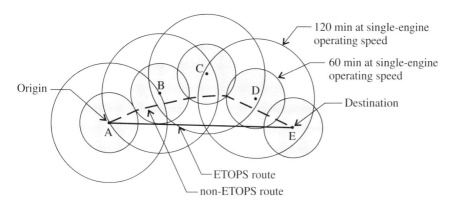

Figure 17.9 ETOPS and non-ETOPS routes for a twinjet with 120 minute ETOPS approval.

have to be flown—with the airplane remaining within the smaller circles (defined by 60 minute flying time from the alternate airfields B, C, and D).

17.6.4 Fuel Planning for ETOPS (Twin Operations)

The minimum required block fuel is determined as the greater of the block fuel determined using a standard fuel planning procedure (see Sections 16.2–16.4) and specific ETOPS requirements, which are described in FAR 121.646 [6] and EASA AMC 20-6 [21]. ETOPS fuel planning assumes a failure at a critical point (i.e., the least favorable point *en route*), which results in the airplane diverting to an alternate airport. It is required that the airplane carry sufficient fuel, taking into account forecast wind and weather, to fly to an ETOPS alternate airport assuming the greater of that required to accommodate:

(1) a rapid decompression at the most critical point, followed by descent to a safe altitude in compliance with stipulated oxygen supply requirements; or
(2) flight at the approved one-engine-inoperative cruise speed assuming a rapid decompression and a simultaneous engine failure at the most critical point, followed by descent to a safe altitude in compliance with stipulated oxygen supply requirements; or
(3) flight at the approved one-engine-inoperative cruise speed assuming an engine failure at the most critical point, followed by descent to the one-engine-inoperative cruise altitude.

For each scenario, the diversion fuel is computed taking into account the associated flight profile, the operator's fuel policies, and the ETOPS critical reserves, as specified in the regulations [6, 21].

17.7 Continuous Descent Operations

A *continuous descent operation* (CDO) is one in which an arriving airplane "descends continuously, to the greatest possible extent, by employing minimum engine thrust, ideally in a low drag configuration, prior to the final approach fix/final approach point (FAF/FAP)"[9] [22]. The adoption of CDO requires appropriate operational procedures and airspace allocation, as well as ATC facilitation.[10]

Ideally, a CDO would start at the top of descent (TOD), following cruise, and would have no segments of level flight during the descent (as illustrated in Figure 17.10). This can, however, be difficult to achieve completely with all flights. Continuous descent operations result in reduced fuel burn and, consequently, reduced exhaust emissions. An airplane that has leveled off—at, say, 2000 ft above the airport elevation—will have a fuel flow approximately four to five times higher than when it is descending with the engines set to idle [24]. An important benefit of CDO is a reduction in the noise generated by arriving airplanes, which can impact communities as far as 20 nm away from the airport (see also Section 21.4.3). Other benefits of CDO indicated by ICAO [22] include (1) more consistent descent flight paths and stabilized approach paths,

---

9 The *final approach fix* (FAF) and *final approach point* (FAP) are specified points that identify the commencement of the final approach segment; the FAF applies to a non-precision instrument approach and the FAP to a precision instrument approach. The FAF and FAP points are often coincident.
10 CDO will be facilitated under ATC procedures developed in the US under the NextGEN (Next Generation Air Transportation) system [23] and in Europe under the SESAR (Single European Sky Air Traffic Management Research) program.

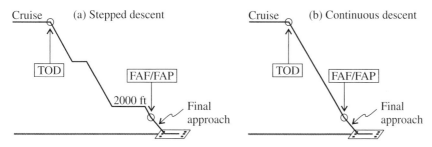

Figure 17.10 (a) Typical stepped descent; (b) Ideal continuous descent.

(2) reductions in pilot and air traffic controller workload, and (3) the potential to reduce the incidence of controlled flight into terrain (CFIT).

References

1 ICAO, "Air traffic services," Annex 11 to the Convention on International Civil Aviation, 13th ed., International Civil Aviation Organization, Montréal, Canada, July 2001.

2 Boeing, "Jet transport performance methods," D6-1420, The Boeing Company, Seattle, WA, May 1989.

3 James, W. and O'Dell, P., "Derated climb performance in large civil aircraft," *Boeing Performance and Flight Operations Engineering Conference*, Seattle, WA, Sept. 2005, Article 6.

4 Airbus, "Getting to grips with fuel economy," Iss. 4, Flight Operations Support and Line Assistance, Airbus S.A.S., Blagnac, France, Oct. 2004.

5 Blake, W. and Performance Training Group, "Jet transport performance methods," D6-1420, Flight Operations Engineering, Boeing Commercial Airplanes, Seattle, WA, Mar. 2009.

6 FAA, *Operating requirements: Domestic, flag, and supplemental operations*, Federal Aviation Regulation Part 121, Amdt. 121-374, Federal Aviation Administration, Washington, DC, May 24, 2016. Latest revision available from www.ecfr.gov/ under e-CFR (Electronic Code of Federal Regulations) Title 14.

7 European Commission, *Commercial air transport operations (Part-CAT)*, Annex IV to Commission Regulation (EU) No. 965/2012, Brussels, Belgium, Oct. 5, 2012. Published in *Official Journal of the European Union*, Vol. L 296, Oct. 25, 2012, and reproduced by EASA.

8 FAA, *Airworthiness standards: Transport category airplanes*, Federal Aviation Regulation Part 25, Amdt. 25-143, Federal Aviation Administration, Washington, DC, June 24, 2016. Latest revision available from www.ecfr.gov/ under e-CFR (Electronic Code of Federal Regulations) Title 14.

9 EASA, *Certification specifications and acceptable means of compliance for large aeroplanes*, CS-25, Amdt. 18, European Aviation Safety Agency, Cologne, Germany, June 23, 2016. Latest revision available from www.easa.europa.eu/ under Certification Specification.

10 Chiles, P., "What lies below," *AeroSafety World*, Flight Safety Foundation, Alexandria, VA, Vol. December, pp. 30–35, 2007.

11 Airbus, "Getting to grips with aircraft performance," Flight Operations Support and Line Assistance, Airbus S.A.S., Blagnac, France, Jan. 2002.

12 De Baudus, L. and Castaigns, P., "Control your speed… in cruise," *Safety first*, Product Safety Department, Airbus S.A.S., Blagnac, France, Iss. 21, pp. 6–21, Jan. 2016.

13 Airbus, "A330 & A340 flight crew training manual," Airbus S.A.S., Blagnac, France, July 2004.

14 EASA, "Acceptable means of compliance (AMC) and guidance material (GM) to Part-CAT," Iss. 2, Amdt. 5, European Aviation Safety Agency, Cologne, Germany, Jan. 27, 2016. Available from www.easa.europa.eu/.

15 FAA, "Extended operations (ETOPS and polar operations)," Advisory Circular 120-42B, Federal Aviation Administration, Washington, DC, June 13, 2008.

16 European Commission, *Special approvals (Part-SPA)*, Annex V to Commission Regulation (EU) No. 965/2012, Brussels, Belgium, Oct. 5, 2012. Published in *Official Journal of the European Union*, Vol. L 296, Oct. 25, 2012, and reproduced by EASA.

17 FAA, "Extended range operation with two-engine airplanes (ETOPS)," Advisory Circular 120-42, Federal Aviation Administration, Washington, DC, June 6, 1985.

18 Airbus, "Getting to grips with ETOPS," Iss. V, Flight Operations Support and Line Assistance, Airbus Industrie, Blagnac, France, Oct. 1998.

19 Boeing, "Press release 28 May 2014: Boeing receives 330-minute ETOPS certification for 787s," Boeing Commercial Airplanes, Seattle, WA, Retrieved July 22, 2015. Available from boeing.mediaroom.com/.

20 Airbus, "Press release 15 Oct. 2014: EASA certifies A350 XWB for up to 370 minute ETOPS," Airbus S.A.S., Blagnac, France, Retrieved July 22, 2015. Available from www.airbus.com/newsroom/.

21 EASA, "Extended range operation with two-engine aeroplanes ETOPS certification and operation," Acceptable Means of Compliance, AMC 20-6 rev. 2, European Aviation Safety Agency, Cologne, Germany, Dec. 16, 2010. Available from www.easa.europa.eu/.

22 ICAO, "Continuous descent operations (CDO) manual," Doc. 9931, International Civil Aviation Organization, Montréal, Canada, 2010.

23 FAA, "NextGEN update: 2014," Federal Aviation Administration, Washington, DC, Aug. 2014. Available from www.faa.gov/nextgen.

24 Da Silva, S., "Continuous descent operations (CDO)," *Workshop on Preparations for ANConf/ 12 − ASBU Methodology*, Nairobi, Kenya, Aug. 13–17, 2012.

18

Cost Considerations

18.1 Introduction

The *cost* incurred when an airplane conducts a mission (flight) is usually taken to mean the financial cost associated with the transportation of passengers and/or freight (cargo) over the mission distance. An introduction to airplane operating costs—a subject that is of critical importance to operators—is presented in Section 18.2. An alternative way of looking at the cost of air transportation from a societal perspective is to consider the *energy* cost of the operation—this is discussed later in Section 18.4. Clearly, the two topics are closely linked as the fuel consumed is a big part of the direct operating cost (DOC) of the mission, and the consumed fuel quantity can, of course, be used to determine the fuel energy expended.

Of interest to the general topic of airplane performance is the *efficiency* with which the operation is conducted—this impacts both the financial cost and energy cost of the mission, although not in precisely the same way. Transportation efficiency is a complex subject—a wide range of different efficiency metrics are used for different purposes. Incorporated into these metrics are fundamental airplane efficiencies. Important contributions come from (1) the aerodynamic efficiency, which can be defined as the lift-to-drag ratio; (2) the engine efficiency, which can be quantified by the thrust specific fuel consumption; and (3) the airplane structural design, which influences the ratio of the airplane's empty weight to its maximum takeoff weight. The specific air range (SAR)—which incorporates aerodynamic, engine, and structural contributions—is a useful overall efficiency metric for minimizing costs for the cruise segment of a mission (as discussed in Section 18.3). Operational efficiencies are generally assessed by considering either the financial or energy cost per unit of output. Several output measures are used for such assessments. One widely used measure is *available seat kilometer*, abbreviated as ASK (by definition, 1 ASK is one passenger seat, available for sale, transported one kilometer).

An introduction to several topics relevant to the study of the financial or energy cost of air transportation is given in this chapter. Airplane operating costs are discussed, albeit at an elementary level, in Section 18.2. The brief discussion includes a summary of commonly used economic terms and their abbreviations. The definition and use of the *cost index*, as it impacts the performance and operation of jet transport airplanes, are presented in Section 18.3. Energy-based metrics, such as *energy intensity*, can be used to compare the performance of different aircraft types (or different modes of transport)—this is discussed in Section 18.4. An extension of this topic of energy cost is the *environmental* cost of a mission. An assessment of the

Performance of the Jet Transport Airplane: Analysis Methods, Flight Operations, and Regulations, First Edition.
Trevor M. Young.

overall impact of an air transport operation on the environment must also consider noise and emissions—these two topics are addressed later in Chapter 21.

18.2 Airplane Operating Costs

18.2.1 Cost Accounting and Financial Data

Airlines, regulatory authorities, and inter-governmental agencies employ accounting procedures and cost allocation methods that are not entirely consistent and uniform across the aviation industry. Despite similar terms being used for cost elements, differences exist in the precise categorization of costs. As a consequence, comparisons of operating costs can be misleading; alternative accounting practices, rather than actual cost differences, can sometimes explain identified differences in operating costs when comparing airlines, airplane types, or regional sectors.

A small number of standardized reporting procedures exist, which are regularly used to derive financial statistics for the industry. For US airlines, there is a statutory requirement for airlines to provide detailed financial information in a specific format—that is, compliant with the US Department of Transportation (DOT) Air Carrier Financial Report Form 41 [1]. International operators supply financial data to the International Civil Aviation Organization (ICAO) using the Air Transport Reporting Form EF [2]—this information provides the basis for ICAO's annual *Digest of Statistics, Series F (Financial Data: Commercial Air Carriers)* [3]. The International Air Transport Association (IATA) also collects a limited amount of financial data from participating airlines, which it then analyzes (principally for member airlines). These databases and economic reports provide a rich source of information for industry financial analysts.

Two points should be noted regarding the use of such data: (1) cost categorization schemes used by different organizations are not entirely consistent, and (2) cost data are categorized according to financial accounting practices. Re-categorization of cost elements is usually required for airplane operating cost analyses (see reference [4], for example).

At the highest level, financial reports separate out non-operating costs (and revenues) from operating costs (and revenues). Operating costs and revenues relate to the operation of the air service—and, as such, exclude such financial items as gains and losses accruing from sale of assets, profit or losses of affiliated companies, repayment of loans, government subsidies, and so forth.

18.2.2 Financial Models for Operating Costs

Data on airline operating costs are valuable to airplane manufacturers—for example, the data can provide an essential input to the decision-making process associated with the adoption of new technologies (e.g., active drag reduction concepts). They also serve to inform researchers on how changes to an airplane's performance parameters, such as speed and cruise altitude, can influence the airplane's operating costs.

These data have facilitated the development of financial models to represent the costs of operating commercial jet transport airplanes. A widely used direct operating cost model was developed by the American Air Transport Association (ATA) in 1944, with a major revision introduced in 1967 to accommodate turbine engines [5]. Periodically, other models have been devised. Notable developments include the direct and indirect operating cost models produced by the Douglas Aircraft Company, under contract to NASA [6], and the model developed by Harris [7] using US DOT Form 41 data, also under NASA contract.

18.2.3   Direct and Indirect Operating Costs

The direct operating costs (DOC) include all expenses associated with operating the airplane.[1] The remaining costs incurred by the operator are often grouped under the heading *indirect operating costs* (IOC). Other categorization schemes are also used—for example, in place of the IOC category, costs not regarded as *aircraft operating costs* can be classed as either *ground operating costs* or *system operating costs* [4].

Various different headings and categories are used to define DOC and IOC costs. The illustrative categorization given in Table 18.1 follows the scheme presented by Doganis [8].

DOC elements have, on average, contributed approximately half of airlines' total operating expenses [4]. The DOC portion, however, tends to vary with time, driven most significantly by the cost of jet fuel—for example, the DOC portion has exceeded 60% of airlines' total operating expenses during times when jet fuel has been very expensive.

Table 18.1 Traditional categorization of airline operating costs[a]

Direct Operating Costs (DOC)	Indirect Operating Costs (IOC)
Costs associated with	*Costs associated with*
• Flight operations—e.g., flight crew salaries and expenses, fuel and oil, airport and *en route* charges[b], aircraft insurance, and rental/lease of flight equipment/crews[c] • Maintenance and overhaul—e.g., engineering staff costs, spare parts consumed, and maintenance administration (could be IOC) • Depreciation and amortization—e.g., flight equipment, ground equipment and property (could be IOC), extra depreciation (in excess of historic cost depreciation), and amortization of development costs and crew training	• Station and ground expenses—e.g., ground staff; buildings, equipment, and transport; handling fees paid to others • Passenger services—e.g., cabin crew salaries and expenses (could be DOC), other passenger service costs, and passenger insurance • Ticketing, sales, and promotion • General and administrative duties • Other operating costs

Notes:

(a) *Source*: Based on Doganis [8].
(b) ICAO classifies airport and *en route* charges as IOC under "station and ground expenses" [8].
(c) US practice is to classify rentals under "depreciation" [8].

18.2.4   Fuel Cost Share

In 2001, fuel represented approximately 14% of an airline's operating costs, based on IATA surveys of major global passenger airlines; as a point of comparison, this was roughly half of the labor cost incurred [9]. The price of jet fuel, which closely tracks the price of crude oil,

---

1 The term *direct operating costs* is a traditional, industry-standard term. Some analysts, however, prefer *aircraft operating costs* or *flight operating costs* as this cost category includes elements that economists consider to be "indirect."

rose in subsequent years to unprecedented levels in 2008—this resulted in the fuel cost share increasing to approximately 32% of the total operating cost [9]. Jet fuel had become the largest single cost item for most airlines. High volatility in the oil market followed, with a steep decline in the price of jet fuel and then a progressive increase. For the financial years 2012 and 2013, IATA reported that fuel and oil represented, as an average for surveyed airlines, 35% and 33% of total operating costs, respectively [10, 11].

Significant regional differences exist in the fuel cost share of an airline's operating costs—for example, in developing economies (such as in the Asia–Pacific region), fuel accounts for a higher share of operating costs as labor costs are comparatively lower. Differences can also be noted between individual airlines for several reasons—among the main factors are labor costs, the fuel efficiency of the airline's fleet, fuel taxes and levies, and fuel hedging (a strategy by which forward purchases of fuel are made at fixed prices).

18.2.5 Airplane Economic Measures

There are several domain-specific parameters that are used for the assessment of the economic performance of an airplane (or airline)—commonly used terms are defined in Table 18.2.

The standard units of measure of operational output (or capacity) of air transportation are *available seat kilometer* (ASK) and *available seat mile* (ASM). These measures are fundamental units of "production" for an airplane when in service. A unit is defined as one seat, available for sale, flown one kilometer (or one mile). A Boeing 787-8, for example, with a seating capacity of 210 flying 1000 km would "generate" 210 000 ASK. Note that the ASK (or ASM) parameter measures the capacity of an air service, but indicates nothing about the actual number of passengers transported. The standard measures of traffic are *revenue passenger kilometer* (RPK) and *revenue passenger mile* (RPM). One RPK represents one revenue (i.e., fee paying) passenger flown one kilometer. Measures for freight (cargo) are defined in a similar way (see Table 18.2).

Table 18.2 Summary of commonly used airplane economic terms and abbreviations

Term/abbreviation	Definition
Available seat kilometer (ASK)	1 seat flown 1 kilometer
Available seat mile (ASM)	1 seat flown 1 statute mile
Available tonne kilometer (ATK)	1 (available) tonne (i.e., 1000 kg) flown 1 kilometer
Available ton mile (ATM)	1 (available) ton (i.e., 2000 lbm) flown 1 statute mile
Revenue passenger kilometer (RPK)	1 revenue passenger flown 1 kilometer
Revenue passenger mile (RPM)	1 revenue passenger flown 1 statute mile
Revenue tonne kilometer (RTK)	1 revenue tonne (i.e., 1000 kg) flown 1 kilometer
Revenue ton mile (RTM)	1 revenue ton (i.e., 2000 lbm) flown 1 statute mile
Cost per ASK (CASK)	Cost per available seat kilometer
Cost per ASM (CASM)	Cost per available seat mile
Revenue per ASK (RASK)	Revenue per available seat kilometer
Revenue per ASM (RASM)	Revenue per available seat mile

The passenger *load factor*[2] (LF) for a single flight is the ratio of revenue passengers to airplane capacity (i.e., the total number of passenger seats). The average load factor (ALF) for a number of flights—or an entire airline or an industry sector measured over a period of time—can be determined from the ratio of RPK to ASK (or, alternatively, from the ratio of RPM to ASM). This ratio is more accurately called the *average network load factor* or the *system load factor*. High average load factors can indicate an efficient usage of the fleet by the operator.

ASK and ASM are used in the determination of unit operating costs for passenger transportation. The *cost per ASK* (CASK) and *cost per ASM* (CASM) are standard measures employed in analyses of the economic performance of airplanes or airlines. CASK (or CASM) is typically expressed in US$ per seat-kilometer (or seat-mile) offered. These measures can be used for cost comparisons between different airplane types or different airlines. A lower CASK (or CASM) means an airplane has lower operating costs for the same output. Comparisons, however, should be made with care. All other things being equal, an airplane operating on routes with a longer average stage length (mission distance) will have a lower CASK, because the fixed costs will account for a smaller portion of its total costs. For this reason, comparative analyses are usually adjusted to a common stage length. *Revenue per ASK* (RASK) and *revenue per ASM* (RASM) are widely used unit revenue measures. RASK (or RASM) is typically expressed in US$ per seat-kilometer (or seat-mile) offered. These measures are used to compare different airlines or the same airline across different sectors or seasons, for example. Generally speaking, the higher the RASK, the more profitable the operation will be (for a given CASK).

The output measures of ASK, ATK, RPK, and RTK (or, alternatively, ASM, ATM, RPM, and RTM) can also be used to assess transportation efficiencies of an airplane type, an airline, or a network, for example. Standard efficiency metrics for passenger air travel include fuel mass (or weight) consumed per ASK or RPK (or, alternatively, per ASM or RPM). Increasingly, assessments of transportation efficiencies consider the energy consumed per unit of output—efficiency metrics are discussed later in Section 18.4.

18.3 Cost Index

18.3.1 Cruise Speeds

With the exception of very short stage lengths, the selected cruise speed can have a significant influence on the total cost of a mission (see Section 13.6). The economy cruise (ECON) speed is the speed at which the sum of the fixed and variable costs that are incurred during the cruise are a minimum. This speed is typically a little faster than the maximum range cruise (MRC) speed, which corresponds to the condition for maximum specific air range. Note that flight at the MRC speed represents an optimum condition in terms of minimum fuel burn, whereas flight at the ECON speed represents an optimum condition in terms of minimum cost.

The determination of the ECON speed for a particular flight is a complex process that requires that the operator establish the relative magnitude of the sum of the time-related costs to the cost of fuel for that flight. This ratio is known as the *cost index*; it is described in more detail in Section 18.3.2, below. The economy cruise Mach number (M_{ECON}) for the flight, based on the selected cost index, is established through a mathematical optimization of a defined cost function (which is described in Section 18.3.4).

---

2 Although the terms have the same name, there is no connection between the *passenger* load factor (described in this chapter) and the *structural design* load factor (described in Sections 20.6 and 20.7).

18.3.2 Cost Index Definition

The *cost index* (CI) is the ratio of the time-dependent cost per unit of trip time to the unit cost of fuel. Any currency unit can be used in the calculation—herein, US dollars have been used. Although fuel is purchased by volume, for this application it is required that the unit fuel cost be expressed in terms of *mass* or *weight*. The cost index can be defined as

$$C_I = \frac{C_T}{C_F} \tag{18.1a}$$

where C_I is the cost index (mass based) (typical units: kg/min, kg/h);
C_T is the time-dependent cost per unit of trip time (typical units: $/h, $/min); and
C_F is the cost of fuel per unit mass (typical unit: $/kg).

Alternatively, where the preference is to use fuel weight, the cost index can be defined as

$$C_I' = \frac{C_T}{C_F'} \tag{18.1b}$$

where C_I' is the cost index (weight based) (typical units: lb/min, lb/h); and
C_F' is the cost of fuel per unit weight (typical unit: $/lb).

Evaluating Equation 18.1 is straightforward; collating the necessary cost data, however, is non-trivial. These costs are unique to the operator and to the airplane. Even airplanes of the same type within a fleet can have slightly different operating costs. The cost of fuel, which varies with geographical location and fluctuates on a daily basis, should ideally reflect the price paid for the fuel consumed on that particular flight. Separating the direct operating costs (DOC) into time-dependent and fixed cost components and then collating the necessary financial data is a difficult and time-consuming exercise. Fixed costs would usually include such items as airport fees, navigation fees, handling and dispatch charges, commissions and insurance, and so forth. Time-dependent costs typically include all or part of the costs associated with passenger services, airplane maintenance, flight crew, cabin crew, and leases.

The influence of varying fuel cost or time-related cost elements on the cost index is illustrated in Figure 18.1. Note that a low cost index is associated with a high fuel cost and/or a low

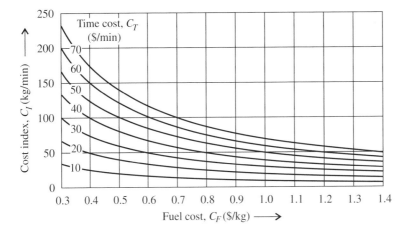

Figure 18.1 Cost index versus fuel cost for varying time-related cost.

time-related cost; conversely, a high cost index is associated with low fuel cost and/or a high time-related cost.

18.3.3 Cost Index and ECON Mach Number for Still Air Cruise

A relationship between the cost index and the ECON Mach number (M_{ECON}) can be derived for an airplane operating in still air conditions (i.e., not considering the impact of wind). For the sake of conciseness, only the mass-based derivation (with units of kg) is presented herein. A weight-based analysis (with units of lb, for example) can be undertaken in an identical manner. Furthermore, to avoid introducing unit conversion factors, time is exclusively expressed in hours and distance in nautical miles in this derivation; thus, C_T will have units of \$/h, C_I units of kg/h, and airspeed units of kt in the ensuing equations.

The total cost (C_{tot}) for an airplane to fly a small cruise distance x is the sum of the fixed cost (C_{fix}) and the variable cost (C_{var}), that is,

$$C_{tot} = C_{fix} + C_{var} \tag{18.2}$$

The conditions that yield the lowest total cost correspond to those that yield the lowest variable cost, which is strongly dependent on the cruise speed. The variable cost component, which is the sum of the time-dependent cost and the fuel cost, can be expressed as follows:

$$C_{var} = C_T t + C_F m_{f_{used}} \tag{18.3}$$

where t is the time to fly the still air distance x (unit: h); and

$m_{f_{used}}$ is the mass of the fuel burned flying the still air distance x (unit: kg).

Equation 18.3 can be used to derive an expression for the ECON Mach number in terms of the cost index and the specific air range (SAR):

$$C_{var} = C_T \frac{x}{V} + C_F \frac{x}{r_a} = \frac{C_T x}{a_0 \sqrt{\theta} M} + \frac{C_F x}{r_a} \tag{18.4}$$

where x is the still air distance (unit: nm);

V is the true airspeed (unit: kt);

r_a is the specific air range (unit: nm/kg);

a_0 is the speed of sound at the ISA sea-level datum (unit: kt);

θ is the relative temperature of the ambient air (dimensionless); and

M is the Mach number (dimensionless).

To establish a minimum condition, C_{var} is now differentiated with respect to M and the resultant set equal to zero:[3]

$$\frac{\partial C_{var}}{\partial M} = \frac{-C_T x}{a_0 \sqrt{\theta} M^2} - \frac{C_F x}{r_a^2} \left(\frac{\partial r_a}{\partial M} \right) \tag{18.5}$$

Now $\dfrac{\partial C_{var}}{\partial M} = 0$ when $M = M_{ECON}$

thus $$\frac{C_T x}{a_0 \sqrt{\theta} (M_{ECON})^2} = \frac{-C_F x}{r_a^2} \left(\frac{\partial r_a}{\partial M} \right)_{M=M_{ECON}} \tag{18.6}$$

---

3 Note that the relative temperature is constant for a given altitude; it is also constant in the stratosphere.

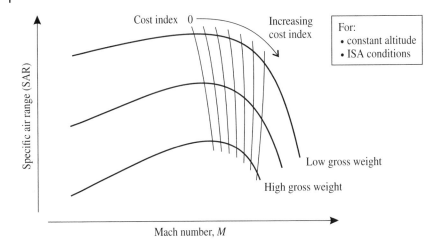

Figure 18.2 Cost index shown on a plot of specific air range (SAR) versus Mach number.

Equation 18.6 can be rearranged, as follows, to introduce the cost index:

$$C_I = \frac{C_T}{C_F} = \frac{-a_0\sqrt{\theta}\left(M_{ECON}\right)^2}{r_a^2}\left(\frac{\partial r_a}{\partial M}\right)_{M=M_{ECON}}$$ (18.7)

where C_I is the cost index (mass based) (unit: kg/h).

Note that when $\partial r_a/\partial M = 0$, $C_I = 0$. This corresponds to the MRC condition. Specific air range data, in the format illustrated in Figure 13.9, can be used in conjunction with Equation 18.7 to generate a cross-plot on which the cost index is indicated (see Figure 18.2).

18.3.4 Economy Cruise Cost Function

The determination of the ECON speed in the general case, which accounts for the influence of wind, requires that a numerical optimization based on an economy cruise cost function be undertaken [12]. In the analysis, the airplane is considered to cover a small ground distance s at a ground speed V_G. As was done in Section 18.3.3, only mass-based equations (with units of kg) are presented in this section. Equations in terms of weight (with units of lb) can be derived in an identical manner.

Using Equation 18.3, the variable cost component can be expressed as follows:

$$C_{var} = C_F\frac{C_T}{C_F}t + C_F m_{f_{used}} = sC_F\left(C_I\frac{t}{s} + \frac{m_{f_{used}}}{s}\right)$$ (18.8)

where s is the ground distance (unit: nm);
  t is the time to fly the distance s (unit: h);
  $m_{f_{used}}$ is the mass of the fuel burned flying the distance s (unit: kg); and
  C_I is the cost index (mass based) (unit: kg/h).

The economy cruise cost function, herein denoted by the Greek letter τ, is defined as

$$\tau = \frac{C_{var}}{sC_F}$$ (18.9)

Substituting Equation 18.8 into Equation 18.9 yields the expression

$$\tau = \frac{C_I t}{s} + \frac{m_{f_{used}}}{s}$$ (18.10)

which can be expressed in the required format by introducing the ground speed:

$$\tau = \frac{C_I}{V_G} + \frac{Q}{V_G} = \frac{1}{V_G}\left(C_I + Q\right)$$ (18.11)

or $\quad \tau = \dfrac{C_I}{V_G} + \dfrac{1}{r_g}$ (18.12)

where τ is economy cruise cost function (mass based) (unit: kg/nm);
V_G is the ground speed (i.e., $V_G = s/t$) (unit: kt);
Q is the fuel flow rate (i.e., $Q = m_{f_{used}}/t$) (unit: kg/h); and
r_g is the specific ground range (unit: nm/kg).

The economy cruise Mach number (M_{ECON}) is defined as the speed at which the economy cruise cost function τ is a minimum [12]. In other words, flight at M_{ECON} satisfies the condition $\partial \tau / \partial M = 0$. The computation is undertaken in an iterative process, in which a computer algorithm searches for the optimum speed and altitude considering all viable combinations of the associated variables (viz., weight, speed, altitude, air temperature, wind, and cost index).

The manner in which the cost index influences τ is evident in Equation 18.11: the fuel flow and the cost index are added (note that the terms have the same units). The cost index can thus be considered as being equivalent to an incremental fuel flow, as regards cost optimization.

If the time-dependent cost component is neglected, the cost index is zero and the optimization of the economy cruise cost function yields a cruise speed equal to the maximum range cruise (MRC) speed corrected for the influence of wind. With increasing cost index, M_{ECON} will increase. The influence of wind can be seen in Equation 18.11. A headwind, which will reduce V_G for a given airspeed, affects τ in a similar way to an increase in the cost index. The presence of a headwind will thus increase M_{ECON} and a tailwind will have the opposite effect. At typical cruise conditions, the change in Mach number will be approximately +0.05 for a 50 kt headwind and −0.05 for a 50 kt tailwind [13].

18.3.5 Flight Operations and the Flight Management System

A cost index can be entered into the Flight Management System (FMS) on modern airliners by the flight crew, via a multifunctional Control Display Unit (CDU). The FMS iteratively solves for the speed that minimizes the cost function for the selected cost index and instructs the autopilot to fly the airplane at the optimum speed.

The cost index, as defined by Equation 18.1, can be expressed with several different units (e.g., kg/min, kg/h, lb/h, or 100 lb/h). For ease of use, commercially available Flight Management Systems employ a scaled value of the cost index that varies between zero and a defined upper limit. Avionics manufacturers have adopted different ranges for their systems—for example, the ranges 0–100, 0–200, 0–500, 0–999, and 0–9999 have all been used [14]. Scaled cost index values, as employed in flight operations, are thus treated as dimensionless coefficients.

It is instructive to consider initially the two extreme cases for the cost index. When $C_I = 0$, this means that the time-dependent cost is effectively zero. The resulting flight schedule will be one that minimizes fuel burn at the expense of an increase in cruise time. In Section 13.6.1, it was shown that flight at the MRC speed results in the maximum specific air range and, consequently, the lowest fuel usage. The selection of zero cost index by the crew will thus result in the FMS commanding wind-corrected MRC speeds.

When C_I = maximum, the unit time-dependent cost dominates the unit fuel cost; consequently, the cruise will be flown at the maximum permissible cruise speed, which cannot exceed M_{MO} (maximum operating Mach number). In this case, the flight time is minimized, but the fuel usage will be comparatively high.

The optimum cruise speed (for a given set of financial and flight conditions), which will result in the minimum total flight cost, is dependent on the relative magnitudes of the fuel- and time-dependent cost components and also on the rate of change of the specific ground range with Mach number. The ECON speed is commanded by the FMS when the flight crew selects an appropriate cost index. The cost index that will minimize the total trip cost can be significantly different for different operators—even for the same route—as airlines often purchase fuel at different prices and they usually have different operating costs. The ECON speed is always greater than the MRC speed when time-dependent costs are accounted for. Note that both the fuel- and time-dependent costs would theoretically increase for an airplane operating at speeds slower than the MRC.

Long range cruise (LRC) speeds, introduced in Section 13.6.3, do not consider operator-specific costs. An LRC speed is the speed that offers 99% of the maximum possible specific air range and is typically 2–5% faster than the corresponding MRC speed [15, 16]. LRC speeds are determined by airplane manufacturers, as a matter of routine, and published in the Flight Crew Operating Manual (FCOM) for varying weight, altitude, and air temperature (WAT) conditions. It is reported [16] that the LRC speed is almost "universally higher than the speed that will result from using the CI selected by most carriers."

Many small airlines do not have the resources to calculate CI values for their operations. There is, however, a technique that can be employed by operators in such cases. Operators are provided with charts, by the airplane manufacturer, to determine an equivalent cost index for the LRC speed associated with a given airplane weight and altitude. The flight crew can then select a CI that corresponds to the appropriate LRC speed. An advantage of this approach, over simply flying the cruise segment at the LRC speeds given in the FCOM, is that winds are considered by the FMS when computing the ECON speed.

18.3.6 Cost Index and Climb Performance

In order to assess the impact of changing the cost index on an airplane's climb performance, it is useful to consider the time taken and costs incurred for the airplane to reach the top of climb (TOC) and to then cruise to a fixed point (typically 150 to 200 nm from the departure airport). This point is identified as "X" in Figure 18.3. The fuel consumed and the time required to reach point X is illustrated in Figure 18.4. The airspeed and hence the rate of climb (which will determine the position of the TOC) depends on the selected cost index.

For $C_I = 0$, the airspeed is relatively slow and the TOC is reached in the shortest possible distance. It was shown earlier in Section 12.4.3 that the *climb gradient* is steepest when the climb takes place at a speed close to the minimum drag speed (V_{md})—which is clearly good for reducing fuel burn. Flight at V_{md}, however, results in a relatively slow rate of climb, which increases the time-dependent costs.

By increasing the airspeed for the climb, the time to reach point X is reduced. A climb conducted at the airspeed that will give the maximum *rate of climb* will result in the shortest climb time (see Section 12.4.6). A high cost index thus results in a small increase in fuel consumption, but a reduction in the time required to reach point X (Figure 18.4).

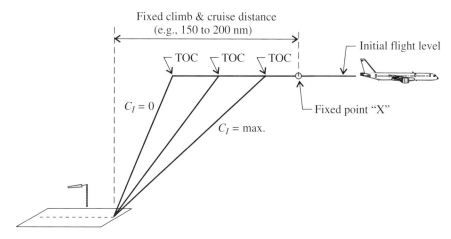

Figure 18.3 Influence of cost index on the climb profile.

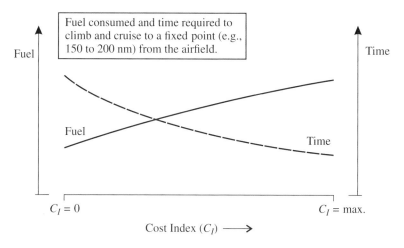

Figure 18.4 Influence of cost index on the fuel and time required to climb and cruise to a fixed point.

Figure 18.3 provides a useful illustration of the influence of cost index on the climb profile; it is, however, an idealization as it ignores possible air traffic control (ATC) restrictions (e.g., a 250 kt speed limit below 10 000 ft). Thus, the airspeed corresponding to the selected cost index would usually only affect the climb from 10 000 ft to the initial cruise altitude.

The difference in time (or fuel consumption) between climbing to point X with a low cost index compared to a high cost index is small—typically less than $1\frac{1}{2}$ minutes [13]. Airbus [13] indicate that, as a general conclusion, optimizing the climb based on a low cost index is only worthwhile if minimizing the time to reach cruising altitude is essential due to competition for flight levels or if so required by ATC. The cost difference between undertaking a climb at a low cost index versus a high cost index is seen to be very small.

18.3.7 Cost Index and Descent Performance

The influence of the cost index on the descent is similar to its influence on the climb: it dictates the descent speed and this determines the position of the TOD (top of descent). For $C_I = 0$,

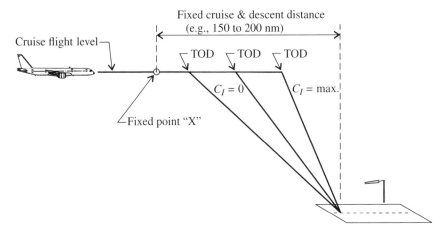

Figure 18.5 Influence of cost index on descent profile.

the descent speed is relatively slow and the TOD is farthest from the airport (Figure 18.5). It is shown in Section 12.5.4 that the smallest *descent gradient* (i.e., shallowest descent) results from flight at the minimum drag speed (V_{md}). However, this can be an unacceptably slow flight speed for normal operations as it protracts the descent unnecessarily.

By increasing the airspeed in the descent (at a given thrust setting), the gradient is increased and the TOD is located closer to the airport. Optimizing the flight using a high cost index will increase the target descent airspeed and therefore reduce the time required to fly from a given point X to the airport (Figure 18.5).

The influence of the cost index on the airplane's performance is greater in the case of the descent compared to the climb. The difference in time between descending using a low and a high cost index is about $2-3\frac{1}{2}$ minutes [13]. As mentioned earlier (in Section 17.2.1), ATC frequently restrict airplane speeds to less than 250 kt below 10 000 ft; thus the cost index usually only affects the descent from the cruise flight level to 10 000 ft.

18.4 Unit Energy Cost

18.4.1 Metrics for Unit Energy Cost (Energy Efficiency)

The concepts described in this section consider the *energy cost* rather than the financial cost of a mission. Energy-based metrics consider the required energy for the mission per unit of output. This broad definition leads to several alternative measures being used to describe the efficiency by which people or goods are transported. There is no single metric that is well established in the literature for such studies, and different authors have considered different permutations of the basic theme (sometimes using different terminology for the same thing). Such metrics are needed to assess the energy required for an aircraft to fly a mission (or for a fleet of aircraft to conduct several missions) or when comparing different modes of transportation (e.g., regional aircraft versus high-speed trains). Energy-based measures are also useful when evaluating the environmental impact of air travel.

The customary units of energy for such studies are MJ or Btu. As the net heating value (i.e., the net energy content per unit mass) of jet fuel does not vary significantly (see Section 22.7.2), it is acceptable to substitute fuel mass (or weight) for fuel energy content when comparisons between turbine-powered aircraft burning kerosene are made. However, it is usually more appropriate to consider the fuel energy and not the fuel mass (or weight) when different modes of transportation are compared or when studies involving unconventional fuel types (e.g., hydrogen) are conducted.

Several output measures that are used in the assessment of air transportation systems were described earlier in Section 18.2.5. The four measures of interest, which fall into two groups, are (1) available seat kilometer (ASK) and available tonne kilometer (ATK), and (2) revenue passenger kilometer (RPK) and revenue tonne kilometer (RTK). Alternatively, these measures can be expressed "per mile" rather than "per kilometer." There is a subtle, but important, difference between these two sets of output measures: ASK and ATK are measures of performance capability of an airplane for a mission or missions over a given time period, whereas RPK and RTK are measures of the realized output of a mission or missions. In economic terms, ASK and ATK measure the capacity for sales, whereas RPK and RTK measure actual sales. This means that there are, theoretically speaking, eight principal ways to define the unit energy cost of a mission—that is, required energy or required fuel mass (or weight) divided by one of the four output measures. In reality, however, not all of these get used.

The unit mission fuel, when based on either ASK or ATK, provides a simple, direct way to compare the energy efficiencies of different airplanes conducting comparable missions or to compare the efficiencies of the same airplane conducting different missions (this is discussed in Section 18.4.2, below). However, to compare or assess the energy efficiencies of a fleet of aircraft or an entire transportation system, or when considering environmental impacts, a specific (or unit) energy metric is more appropriate (these metrics are discussed later in Sections 18.4.3 and 18.4.4).

18.4.2 Unit Mission Fuel

When comparing energy efficiencies of two or more aircraft, metrics based on the fuel used per unit distance flown can be misleading. By normalizing the data to account for the payload (i.e., passengers and/or freight carried), more valid comparisons can be made. Such comparisons of airplane performance capability can be based on the fuel mass (or weight) consumed per ASK or per ATK (or alternatively, per available seat mile or available ton mile).

As an illustration, the comparison conducted by Sutkus, *et al.* [17] is reviewed. They considered the case of two widebody passenger transport airplanes of comparable technology level: the MD-11ER and B747-400. The fuel required per unit distance for the MD-11ER was observed to be about 46% lower than that required for the B747-400 across the mission distances considered. This was largely driven by the fact that the MD-11ER is a lighter airplane. However, when normalized to take into account passenger numbers, the data are similar, as shown in Figure 18.6. The amounts of fuel required per ASK for the two aircraft types are within 6% of each other for the mission distances considered [17].

Another interesting observation that is evident in Figure 18.6 is that the fuel required per ASK reduces—initially rather rapidly—as the trip distance increases; it then levels off at a trip distance of approximately 4200 km (2268 nm) and gradually increases. This trend, in fact, has nothing to do with the selected airplane types, but is a general characteristic of long-range air travel. The observed increase in required fuel per ASK (beyond the minimum point) is a result of the aircraft having to burn fuel to transport the required mission fuel—and this phenomenon gets worse as the trip distance increases. There have been many studies addressing the question of how long-haul air travel can be made more energy efficient. One option is to divide long trips into two (or more) stages. Clearly, a fuel penalty is incurred by incorporating a refueling stop; nonetheless, in considering progressively longer trip distances, a point is reached when there

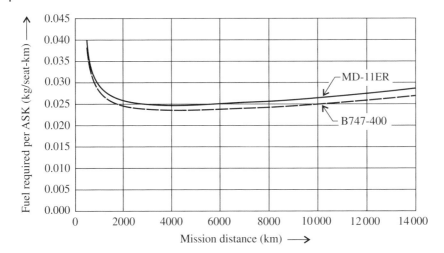

Figure 18.6 Fuel required per ASK versus mission distance for the MD-11ER and B747-400 airplane types. *Source*: Adapted from [17].

is a net fuel—and energy—saving. Multi-stage trips are not popular with passengers, however, but it has been shown that significant fuel and energy savings are possible through this "staging" approach to servicing long-haul routes [18–22]. In one study, Poll [21] concluded that using the same airplane and carrying the same payload (in the same cabin configuration), fuel savings are possible for trips greater than about 6000 km (2700 nm), with the benefit increasing almost linearly with increasing trip distance. Additional savings are possible if the mission is flown by an airplane optimized to the shorter stage lengths; in this case, it is claimed that the savings begin when the trip distance exceeds about 5000 km (3240 nm) [21]. A similar conclusion was reached by Linke, *et al.* [22].

Another idea that has been postulated as a mechanism to reduce the fuel required to service long-haul routes is air-to-air refueling—a technology that has been employed for many years in military aviation, but not for civil aircraft operations [23, 24].

18.4.3 Energy Usage and Energy Intensity

Two energy-based metrics that can be used to evaluate the energy cost or energy efficiency of an airplane (or air transportation system) are described in this section (the terminology used follows that adopted by Lee, *et al.* [25] and Babikian, *et al.* [26]). The application considered is passenger transportation. The concept, however, can also be applied to freight carried by cargo airplanes and also to combined passenger and freight payloads carried by airliners (this is discussed later in Section 18.4.4). For the sake of brevity, only SI units are used in Sections 18.4.3 and 18.4.4—expressions for other units can be derived in an identical manner.

The total fuel energy required for a mission can be expressed as follows:

$$E_f = m_{f_{used}} H_f \tag{18.13}$$

where E_f is the total (i.e., mission) fuel energy (typical unit: MJ).
$m_{f_{used}}$ is the total (i.e., mission) fuel mass (typical unit: kg); and
H_f is the net heating value of the fuel (typical unit: MJ/kg).

For comparative assessments, it is necessary to express the total fuel energy as a specific quantity by dividing by a measurable output parameter, such as ASK or RPK. The *energy usage* (or *specific energy usage*) is defined as[4]

$$E_U = \frac{E_f}{\text{ASK}} \tag{18.14}$$

and the *energy intensity*[5] (or *specific energy intensity*) as

$$E_I = \frac{E_f}{\text{RPK}} \tag{18.15}$$

where E_U is the *energy usage* (typical unit: MJ/seat-km); and
E_I is the *energy intensity* (typical unit: MJ/seat-km).

Linking the two measures is the passenger load factor, which is equal to the ratio of RPK to ASK (see Section 18.2.5):

$$E_I = \frac{E_f}{\text{ASK}} \left(\frac{\text{ASK}}{\text{RPK}} \right) = \frac{E_U}{\alpha} \tag{18.16}$$

where α is the passenger load factor (dimensionless).

The energy usage (E_U), which gives the same information as the unit mission fuel (described in Section 18.4.2), is essentially an airplane design parameter as it is dependent on airplane efficiency measures (e.g., aerodynamic, engine, and structural). The energy intensity (E_I), on the other hand, combines the airplane design efficiencies with a measure of operational efficiency—namely, the passenger load factor. The energy intensity can be used to compare directly the energy needs of difference modes of transport—for example, air transport versus rail or road.[6] Note that both metrics are based on the mission (i.e., total) energy required, which includes airplane ground operations (e.g., engine start-up, taxi, and shutdown). Comparisons between airplanes that have very different mission profiles can thus be misleading. Short-haul aircraft, for example, tend to have a relatively poor unit energy measure compared to long-haul aircraft, as they spend a greater portion of the mission time on the ground.

It is instructive to connect mathematically the various measures of efficiency used to describe aircraft performance. As mentioned earlier, the lift-to-drag ratio is a measure of aerodynamic efficiency, the thrust specific fuel consumption (TSFC) is a measure of engine efficiency, and the passenger load factor is a measure of operational efficiency. The energy intensity, which is a function of the mission (or trip) distance, can be linked to the aforementioned efficiency measures. By combining Equations 18.13, 18.14, and 18.16, the energy intensity can be written as follows:

$$E_I = \frac{m_{f_{used}} H_f}{\alpha R N_{seat}} \tag{18.17}$$

where R is the mission (or trip) distance (typical unit: km); and
N_{seat} is the total number of passenger seats (i.e., the airplane's seating capacity).

---

4 The representation of technical terms or parameters by an abbreviation, rather than by a symbol, in mathematical formulae is generally discouraged (see Section 2.2.2). In this chapter, an exception has been made—consistent with customary usage—where well-known economic abbreviations (e.g., ASK, RPK) are used in equations.
5 Caution: Certain authors (e.g., [27]) consider *energy intensity* as the energy consumed per ASK and not per RPK (as herein defined).
6 US data for different transportation modes are periodically compiled by the Oak Ridge National Laboratory Center for Transportation Analysis [28].

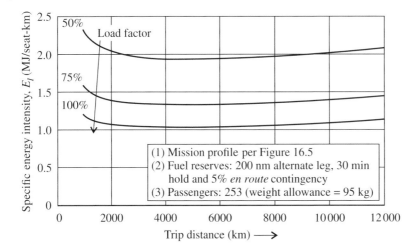

Figure 18.7 Calculated energy intensity versus trip distance for selected load factors for a 253 passenger mid- to long-range widebody twinjet.

The net heating value of the fuel (H_f) is a property of the fuel, with only small variations evident in approved aviation fuel (see Section 22.7.2). The passenger load factor (α) is a measure of capacity utilization—it describes the extent to which the operator fills the aircraft with passengers. Fleet average load factors vary from one airline to another and from one sector to another (e.g., a low-cost, no-frills service is likely to have a higher average load factor than a dedicated business service). Average load factors on a worldwide basis reached record levels of 78–80% for the period 2010–2015 [29]. Although system load factors increased by about 5% per decade from the mid-1960s onwards, this rate of increase could not continue unabated; it is considered likely that global average load factors will fluctuate by a few percent around the 80% level for single-aisle and around the 76% level for widebody services, depending on economic cycles [30].

The mission fuel ($m_{f_{used}}$), which is included in Equation 18.17, can be established using the techniques described in Chapter 16. Figure 18.7 illustrates the results of such an exercise, for which the energy intensity (E_I) of a mid- to long-range widebody twinjet was calculated (using a customized computer model) for three passenger load factors (i.e., 100%, 75%, and 50%). Note that a 100% passenger load factor means that the energy intensity (E_I) is identical to the energy usage (E_U). Figure 18.7 shows a similar pattern to that seen in Figure 18.6—the trend lines reduce to a low point at about 4500 km (2430 nm) and then gradually increase.

For approximate calculations of the mission fuel for mid- to long-range missions, the so-called overhead approximation (see Section 16.8.1) can be used in conjunction with Equation 13.43a, which, for ease of reference, is repeated below.

$$m_{f_{used}} = \left(e^{\left(\frac{R\bar{c}g}{V\bar{E}} \right)} - 1 \right) m_2 \qquad \text{(Equation 13.43a)}$$

where $\bar{c}$ is the mean thrust specific fuel consumption (typical unit: kg N^{-1} h^{-1});
g is the acceleration due to gravity (typical unit: m/s^2);
V is the true airspeed (typical unit: km/h);
$\bar{E}$ is the mean lift-to-drag ratio (dimensionless); and
m_2 is the end-of-mission airplane mass (typical unit: kg).

Caution: Equations 18.17 and 13.43a are both dimensionally consistent; however, care is needed to ensure that appropriate units are used, particularly with regard to distance, speed, and TSFC (typical SI units are used herein for the sake of illustration).

The end-of-mission airplane mass (m_2) can be considered to be the sum of the airplane's operating empty mass plus any additional operational items (e.g., catering supplies) plus the payload plus the reserve fuel (airplane mass/weight definitions are given in Section 19.2.2 and descriptions of reserve fuel requirements are outlined in Sections 16.2–16.4). This brings into the discussion on efficiency several other factors for consideration—for example, the required mission fuel, and hence the energy intensity, would be reduced if

(1) the mass of the empty airplane were reduced (e.g., by combating in-service "weight growth" through regular maintenance);
(2) superfluous onboard items were removed (e.g., by eliminating the sale of duty-free items); or
(3) unnecessary fuel, that is "extra" fuel beyond that needed for the mission, were reduced.

18.4.4 Payload Fuel Energy Intensity

The energy intensity and energy usage metrics discussed in Section 18.4.3 are useful tools for assessing the energy consumed to transport passengers. The metrics, however, do not include cargo (freight), which can account for a significant portion of the payload carried by passenger jet transport airplanes. Payload comprises passengers, their baggage (i.e., baggage carried onboard and stowed in a hold), and cargo (including mail). The cargo carried on passenger aircraft in the cargo holds (located below the passenger deck) is often termed *belly cargo* (or *belly freight*). Note that passenger baggage is not regarded as cargo.

To quantify the efficiency of an airplane or fleet of aircraft or air transportation system for certain applications such as environmental impact assessments, RTK is a superior measure of output compared to ASK or RPK. A productivity measure (i.e., a measure of realized output) that has been used for such studies is the total payload mass (or weight) transported multiplied by the great circle distance measured between the departure and arrival airports. Basing an energy efficiency metric on the shortest distance between the two airports makes sense as any variation from the direct route (e.g., to avail of favorable winds or to comply with air navigation requirements) is included in the assessment as it impacts the fuel (and energy) required for the mission. The efficiency metric can thus be written as

$$E_{I,pl} = \frac{E_f}{m_{pl}\, d_{gc}} \tag{18.18}$$

where $E_{I,pl}$ is the *energy intensity* based on the total payload (typical unit: kJ kg^{-1} km^{-1});
m_{pl} is the total payload mass (typical unit: kg); and
d_{gc} is the great circle distance measured between the departure and arrival airports (typical unit: km).

This metric ($E_{I,pl}$), which has been called *payload fuel energy intensity* (PFEI) [31–33], or minor variations of the definition given by Equation 18.18, has been employed to quantify how efficiently fuel energy is being used to transport payload in commercial aviation on an average or fleet-wide basis [33, 34] and also as the objective function in airplane design optimization studies directed at minimizing carbon dioxide production [31, 32, 35, 36].

References

1 Durso, J.C., "An introduction to DOT Form 41 web resources for airline financial analysis," Rubel School of Business, Bellarmine University, Louisville, KY, Dec. 2007.

2 ICAO, "Statistical air transport reporting forms," International Civil Aviation Organization, Montréal, Canada, Retrieved July 2016. Available from www.icao.int/.

3 ICAO, "ICAO Data," International Civil Aviation Organization, Montréal, Canada, Retrieved Dec. 2013. Available from www.icaodata.com/.

4 Belobaba, P., "Airline operating costs and measures of productivity," in *The global airline industry*, 2nd ed., Belobaba, P., Odoni, A., and Barnhart, C. (Eds.), John Wiley & Sons, Chichester, UK, 2016.

5 ATA, "Standard method of estimating comparative direct operating costs of turbine powered transport airplanes," Air Transport Association of America, Dec. 1967.

6 Douglas Aircraft Company, "Study of short-haul aircraft operating economics," NASA CR-137685, Vol. 1 and NASA CR-137686, Vol. II, Douglas Aircraft Company, Sept. 1975.

7 Harris, F.D., "An economic model of U.S. airline operating expenses," NASA CR-2005-213476, National Aeronautics and Space Administration, Ames Research Center, Moffett Field, CA, Dec. 2005.

8 Doganis, R., *Flying off course: The economics of international airlines*, 3rd ed., Routledge, London, UK, 2002.

9 IATA, "IATA economic briefing: Airline fuel and labor cost share," International Air Transport Association, Montréal, Canada, Feb. 2010.

10 IATA, "2012 airline operational cost task force report," International Air Transport Association, Montréal, Canada, Sept. 2013.

11 IATA, "Airline cost management group (ACMG) FY 2013, enhanced report," International Air Transport Association, Montréal, Canada, Sept. 2014.

12 Boeing, "Jet transport performance methods," D6-1420, The Boeing Company, Seattle, WA, May 1989.

13 Airbus, "Getting to grips with the cost index," Iss. II, Flight Operations Support and Line Assistance, Airbus Industrie, Blagnac, France, May 1998.

14 Roberson, W., "Fuel conservation strategies: Cost index explained," *Aero*, The Boeing Company, Seattle, WA, Vol. Q 02, pp. 26–28, 2007.

15 Seto, L., "Cost index," Flight Operations Engineering, Boeing Commercial Airplanes, Seattle, WA, Sept. 2009.

16 Roberson, W., Root, R., and Adams, D., "Fuel conservation strategies: Cruise flight," *Aero*, The Boeing Company, Seattle, WA, Vol. Q 04, pp. 23–27, 2007.

17 Sutkus, D.J., Jr, Baughcum, S.L., and DuBois, D.P., "Scheduled civil aircraft emission inventories for 1999: Database development and analysis," NASA CR-2001-211216, National Aeronautics and Space Administration, USA, Oct. 2001.

18 Nangia, R.K., "Efficiency parameters for modern commercial aircraft," *Aeronautical Journal*, Vol. 110, Iss. 1110, pp. 495–510, 2006.

19 Hahn, A., "Staging airliner service," *7th AIAA Aviation Technology, Integration and Operations (ATIO) Conference*, Belfast, Northern Ireland, Sept. 18–20, 2007, AIAA 2007-7759, American Institute of Aeronautics and Astronautics.

20 Creemers, W.L.H. and Slingerland, R., "Impact of intermediate stops on long-range jet-transport design," *7th AIAA Aviation Technology, Integration and Operations (ATIO) Conference*, Belfast, Northern Ireland, Sept. 18–20, 2007, AIAA 2007-7849, American Institute of Aeronautics and Astronautics.

21 Poll, D.I.A., "On the effect of stage length on the efficiency of air transport," *Aeronautical Journal*, Vol. 115, Iss. 1167, pp. 273–283, 2011.

22 Linke, F., Langhans, S., and Gollnick, V., "Global fuel analysis of intermediate stop operations on long-haul routes," *11th AIAA Aviation Technology, Integration, and Operations (ATIO) Conference*, Virginia Beach, VA, Sept. 20–22, 2011, AIAA 2011-6884, American Institute of Aeronautics and Astronautics.

23 Nangia, R.K., "Greener civil aviation using air-to-air refuelling—relating aircraft design efficiency and tanker offload efficiency," *Aeronautical Journal*, Vol. 111, Iss. 1123, pp. 589–592, 2007.

24 Spencer, R.J., "Predicting the certification basis for airliner air-to-air refueling," *Aeronautical Journal*, Vol. 119, Iss. 1220, pp. 1175–1192, 2015.

25 Lee, J.J., Lukachko, S.P., Waitz, I.A., and Schafer, A., "Historical and future trends in aircraft performance, cost, and emissions," *Annual Review of Energy and the Environment*, Vol. 26, pp. 167–200, 2001.

26 Babikian, R., Lukachko, S.P., and Waitz, I.A., "The historical fuel efficiency characteristics of regional aircraft from technological, operational, and cost perspectives," *Journal of Air Transport Management*, Vol. 8, Iss. 6, pp. 389–400, 2002.

27 Peeters, P.M., Middel, J., and Hoolhorst, A., "Fuel efficiency of commercial aircraft: An overview of historical and future trends," NLR-CR-2005-669, National Aerospace Laboratory, the Netherlands, Nov. 2005.

28 Davis, S.C., Diegel, S.W., and Boundy, R.G., "Transportation energy data book," 34th ed., Center for Transportation Analysis, Oak Ridge National Laboratory, Oak Ridge, TN, Aug. 2015.

29 Boeing, "Current market outlook 2016–2035," Market Analysis, Boeing Commercial Airplanes, Seattle, WA, July 2016.

30 Forsberg, D., "World fleet forecast: 2014–2033," Avolon, Dublin, Ireland, Sept. 2014.

31 Greitzer, E.M., Bonnefoy, P.A., De la Rosa Blanco, E., *et al.*, "N+3 aircraft concept designs and trade studies," Vol. 1, NASA CR-2010-216794/VOL1, National Aeronautics and Space Administration, Glenn Research Center, Cleveland, OH, Dec. 2010.

32 Drela, M., "Design drivers of energy-efficient transport aircraft," *SAE International Journal of Aerospace*, Vol. 4, Iss. 2, pp. 602–618, 2011.

33 Hileman, J.I., De la Rosa Blanco, E., Bonnefoy, P.A., and Carter, N.A., "The carbon dioxide challenge facing aviation," *Progress in Aerospace Sciences*, Vol. 63, pp. 84–95, 2013.

34 Hileman, J.I., Katz, J.B., Mantilla, J.G., and Fleming, G., "Payload fuel energy efficiency as a metric for aviation environmental performance," *26th International Congress of the Aeronautical Sciences (ICAS)*, Anchorage, AK, Sept. 14–19, 2008, ICAS 546.

35 Poll, D.I.A., "The optimum aeroplane and beyond," *Aeronautical Journal*, Vol. 113, Iss. 1140, pp. 151–164, 2009.

36 Poll, D.I.A., "On the application of light weight materials to improve aircraft fuel burn—reduce weight or improve aerodynamic efficiency?," *Aeronautical Journal*, Vol. 118, Iss. 1206, pp. 903–934, 2014.

19

Weight, Balance, and Trim

19.1 Introduction

An airplane's weight is a critically important performance parameter—consideration is given to an airplane's takeoff weight right at the very start of the design process, and weight calculations and analyses follow every step of the design and manufacture of a new airplane. Airplane weights are recorded as part of type certification, and individual aircraft are weighed before delivery. Depending on the production series and equipment fitted, a single aircraft type can have many weight variants[1]—for example, there are 13 weight variants for the A380-800 and over 25 for the A330-200 [1, 2]. Weight growth during the life of an airplane is unavoidable, arising from repairs, modifications, contamination, paint, and so forth. It is thus standard practice for operators to keep records of each airplane's empty weight—determined strictly according to a set of rules stipulated by the regulatory authorities [3, 4].

Definitions of commonly used weight terms are provided in Section 19.2; this is followed by a discussion on several performance-related topics including center of gravity (CG) determination and CG fore and aft limits (Section 19.3), longitudinal static stability (Section 19.4), and CG control using fuel transfer (Section 19.5). In Section 19.6, the process for determining the payload weight and airplane CG for dispatch is described. The impacts of airplane excess weight and CG location on airplane performance are outlined in Section 19.7. Longitudinal static stability is presented at an introductory level in this chapter with the emphasis on how static stability and stabilizer trim are related. The derivation of the fundamental equations that describe the longitudinal static stability of an airplane is given in Appendix E.

19.2 Airplane Weight Definitions

19.2.1 Weight/Mass Distinction

The distinction between the terms *weight* and *mass* (as discussed in Section 2.3) is important in aircraft engineering. To be entirely consistent—and technically correct—at all times, however, especially when discussing flight operations, can be difficult. FAA documents, for example, generally refer to the aircraft's *weight* (e.g., FAA AC 120-27 [3]), whereas many comparable EASA documents refer to the aircraft's *mass* (e.g., EASA OPS Part-CAT [5]). Similarly, operators are split in their approaches: some use aircraft mass, but the majority prefer aircraft weight. The

---

1 Useful sources of airplane weight information (and other aircraft data) are the Airplane/Aircraft Characteristics for Airport Planning Manuals, which are published individually per aircraft type (see Section 23.7.9).

Performance of the Jet Transport Airplane: Analysis Methods, Flight Operations, and Regulations, First Edition.
Trevor M. Young.
© 2018 John Wiley & Sons Ltd. Published 2018 by John Wiley & Sons Ltd.

consequence is duplicated terminology and two sets of abbreviations: one for weight and one for mass. As there is widespread acceptance of the term *weight* in an operational context (e.g., zero fuel weight, certified weight, weight and balance), the strict weight/mass distinction employed in discussions concerning flight mechanics has been relaxed in this chapter.

The customary unit of measurement for airplane *weight* in USC units is the pound (lb) and that for airplane *mass* in SI units is the kilogram (kg)—these units or multiples of these units (such as the metric ton) are widely encountered in aviation literature and in documentation for flight operations (see Section 23.7). It is also commonplace to see the kilogram unit assigned to an aircraft weight (e.g., takeoff weight). Strictly speaking, this is incorrect, but in this context it seldom results in a misunderstanding.

19.2.2 Airplane Weight Definitions and Breakdown

In some respects, the subject of aircraft weights is akin to financial accounting—with individual elements, or components, assigned to specific categories. The regulatory authorities are not entirely prescriptive regarding the airplane weight definitions used for flight operations. Slightly different weight accounting procedures are used by different manufacturers and operators— subtle differences thus exist between what is included (or excluded) in various airplane weight definitions. The definitions given below—and illustrated in Figure 19.1—are typical (based on references [6–9]).

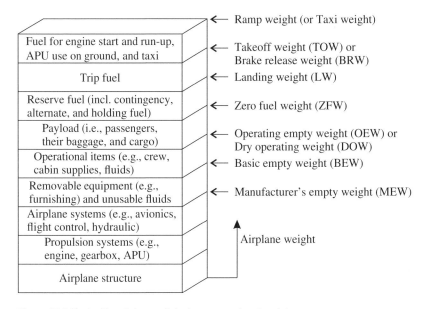

Figure 19.1 Typical breakdown of airplane operational weights.

Manufacturer's Empty Weight
The manufacturer's empty weight (MEW) is the weight of the airframe structure plus the powerplant, systems (e.g., avionics, hydraulics), certain furnishings (e.g., passenger seats),[2] seller-furnished emergency equipment, and fixed equipment that is considered to be an integral part

---

2 Accounting practices differ—seats and other furnishings are often allocated to the basic empty weight.

of the airplane. It is essentially a "dry" weight, including only those fluids contained in closed-circuit systems (e.g., hydraulic systems).

Basic Empty Weight

The basic empty weight (BEW) is equal to the manufacturer's empty weight plus the weight of certain fluids and removable items that do not normally vary for aircraft of the same type (often referred to as *standard* items). These items include (1) unusable fuel[3] and engine oil, and that portion of other liquids (e.g., drinking and washing water) that is unusable; (2) galley structures and associated equipment (e.g., ovens); (3) furnishings such as separation walls, floor coverings, and seats (if not already included in the MEW); and (4) basic emergency equipment (e.g., emergency oxygen equipment, first aid kits). Operators receiving new airplanes would typically base the BEW on the airplane's *delivery empty weight* as recorded by the manufacturer in the airplane's Weight and Balance Manual (WBM).

Operating Empty Weight/Dry Operating Weight

The operating (or operational) empty weight (OEW), which many operators call the dry operating weight (DOW), is the total weight of the airplane ready for a specific operation (mission), excluding usable fuel and payload.[4] It is the basic empty weight plus the weight of the personnel and the equipment and supplies necessary for a particular mission (operation). These *operational* items include the flight and cabin crews (plus their manuals and baggage), catering supplies and usable water (for drinking and washing), cabin service items (e.g., blankets, newspapers, and other literature), replenishable fluids and chemicals (e.g., for toilets), engine and other oils, additional emergency equipment (e.g., life rafts, which are required for overwater routes), tool kits, cargo containers and pallets, and so forth. Operating empty weights are thus specific to each particular airplane (not airplane type) for the mission being flown.

Zero Fuel Weight

The zero fuel weight (ZFW) is the operating empty weight plus the weight of the payload (traffic load), which comprises the passengers, their baggage, and the cargo (freight).

Takeoff Weight

The takeoff weight (TOW) is the zero fuel weight plus the weight of the fuel at takeoff. As the takeoff weight is the weight of the airplane at the start of the takeoff run, it is often called the brake release weight (BRW).

Ramp Weight

The ramp weight (or taxi weight) is the takeoff weight plus the weight of the fuel required for the auxiliary power unit (APU) usage prior to takeoff, the engine start and run-up, and taxi to the start of the takeoff run.

Landing Weight

The landing weight (LW) is the airplane weight at touchdown at the destination airport. From a flight planning perspective, it is equal to the zero fuel weight plus the weight of the reserve

---

3 A relatively small amount of fuel gets trapped in the fuel tanks due to the shape of the tanks and the position of the fuel outlets. *Unusable fuel* is defined the fuel remaining in the airplane after an engine "run-out" test has been conducted. Unusable fuel is regarded as "dead weight" carried by the airplane.
4 EASA refer to this as the aircraft's dry operating mass [4, 5].

fuel (including the *en route* contingency fuel, alternate fuel, final reserve or holding fuel, and any additional fuel required to comply with operational regulations or company policy).

19.2.3 Structural-Limited Design Weights

The structural capability of an airplane dictates the absolute weight limits applicable to an airplane for various conditions (or stages of a mission)—these maximum weights are known as structural-limited weights. The limits are applicable to individual aircraft, and depend on the associated production standards (and hence when it was built). In some cases, earlier manufactured aircraft can have their weight limits increased through approved modifications.

Maximum (Design) Taxi Weight
The maximum (design) taxi weight (MTW) defines the upper limit allowed for the movement of the airplane on the ground (by taxiing or towing). It is a structural limit, dictated by ground loads on the landing gear.

Maximum (Design) Takeoff Weight
The maximum (design) takeoff weight (MTOW) defines the upper limit to the allowable takeoff weight (TOW). It is a structural limit, dictated by either flight loads, braking system limitations, or landing impact considerations, which are based on a vertical impact of 6 ft/s (1.83 m/s) (as required by FAR/CS 25.473 [10, 11]).

Maximum (Design) Landing Weight
The maximum (design) landing weight (MLW) defines the upper limit to the allowable landing weight (LW). It is a structural limit, dictated by either braking system limitations or landing impact considerations, which are based on a vertical impact of 10 ft/s (3.05 m/s) (as required by FAR/CS 25.473 [10, 11]).

Maximum (Design) Zero Fuel Weight
The maximum (design) zero fuel weight (MZFW) defines the upper limit to the airplane's operational weight before usable fuel is loaded. When fuel is consumed during flight, the airplane's weight (and the required lift) reduces; however, the wing root bending moment can *increase* due to the reducing bending moment relief provided by the fuel. The MZFW is a structural limit, dictated by the maximum allowable wing root bending moment. Increasing the MZFW increases the average wing root bending moment, which can result in long-term structural fatigue issues.

19.2.4 Certified and Operational Weights

Certified Weights
The maximum weights that an operator can legally use are those recorded in the Airplane Flight Manual (and also in the Weight and Balance Manual)—these weight limits are known as *certified weights*, which, for operational or financial reasons, can be less than the corresponding design structural limit weights (described in Section 19.2.3). As several operating costs are linked to an airplane's certified takeoff and landing weights, the certified weights are selected by the operator when the airplane is purchased—for this reason, certified weights are often called the *purchased weights*. Certified weights are thus applicable to individual aircraft (not aircraft type), and can be increased up to the corresponding structural limit weights (usually at a cost) at a later time.

Operational Weights

The maximum takeoff weight that an operator can use in service cannot exceed the certified maximum takeoff weight for that specific airplane. During routine flight operations, the allowable takeoff weight can be limited by several other factors, which include (1) the airplane performance capability for the given atmospheric and weather conditions taking into account such constraints as the takeoff distance available (TODA), minimum climb gradient requirements, and obstacle clearance; (2) noise restrictions; (3) runway loading limits; and (4) center of gravity limits. Similarly, the maximum landing weight that an operator can use in service cannot exceed the certified maximum landing weight; it can also be limited by similar operational factors as those mentioned for the takeoff.

19.3   Center of Gravity

19.3.1   CG Determination

An airplane's center of gravity (CG) is measured from a defined datum (e.g., the front pressure bulkhead or a selected point in the vicinity of the nose of the airplane). With the airplane correctly leveled, distances from the datum are designated by X (as shown in Figure 19.2). The term *station* is often used to define an X location measured with respect to the datum (which is then called station zero).

The operating empty weight and the corresponding CG position are determined from the results of routine weighing of the airplane (see Section 19.6.3)—this information is the baseline for calculating the weight and CG position of the airplane before takeoff, and at any stage in

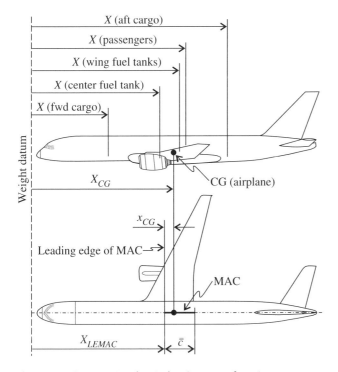

Figure 19.2 Determining the airplane's center of gravity.

flight. This is done by taking into account the weight and CG position of (1) the fuel in the various tanks, (2) the passengers and their carry-on baggage, (3) the check-in baggage, and (4) the cargo. The CG position can be calculated using the following equation:

$$X_{CG} = \frac{W_{OE}X_{OE} + \sum W_i X_i}{W_{OE} + \sum W_i} \tag{19.1a}$$

or

$$X_{CG} = \frac{m_{OE}X_{OE} + \sum m_i X_i}{m_{OE} + \sum m_i} \tag{19.1b}$$

where X_{CG} is the distance to the airplane's CG (typical units: m, ft, in);

W_{OE} (or m_{OE}) is the operating empty weight (or mass) (typical units: N, lb, or kg);

X_{OE} is the distance to the airplane's CG in the operating empty condition (typical units: m, ft, in);

W_i (or m_i) is the weight (or mass) of the ith component (e.g., fuel, passengers, baggage, cargo) (typical units: N, lb, or kg); and

X_i is the distance to the CG of the ith component (typical units: m, ft, in).

For most applications, the airplane's CG position is described as a percentage of a reference chord (RC); it is customary practice to use the mean aerodynamic chord (MAC) for this purpose. Converting the X_{CG} value to a fraction of the MAC can be done using the following equation:

$$\frac{x_{CG}}{\bar{\bar{c}}} = \frac{X_{CG} - X_{LEMAC}}{\bar{\bar{c}}} \tag{19.2}$$

where x_{CG}, X_{CG}, and X_{LEMAC} are defined in Figure 19.2; and

$\bar{\bar{c}}$ is the mean aerodynamic chord.

An airplane's CG position depends on the weight and distribution of the payload and fuel. During flight, as fuel is consumed and as the passengers and crew move about the cabin, the CG will move—both vertically and horizontally. The vertical movement is small and is generally of little consequence, but the horizontal movement is important and, for safety reasons, the CG must remain within specified fore and aft limits for each phase of flight. The lateral position of the CG must also remain within defined limits—this is a consideration for freight loading.

19.3.2 Load and Balance Diagram/CG Envelope

The terms *load and balance diagram* and *CG envelope* are used interchangeably. The usual format of a load and balance diagram, as used for engineering work, has the airplane's weight (or mass) as the ordinate with the airplane's CG position (as the abscissa) shown as a percentage of a reference chord (which is normally the MAC). The airplane's fore and aft CG limits (discussed later in Section 19.3.3) are shown on the diagram together with weight limits, thus defining an envelope within which the airplane can be safely operated.

A convenient format to present this information for flight operational purposes is on a graph of weight (or mass) versus *moment* (about a CG datum). The principle is illustrated in Figure 19.3, in which the weight (or mass) changes and their corresponding moment changes are shown as vectors. The reference point (or reference station) in this example is the 25% MAC point. The advantage of representing fuel or other weight changes (e.g., loading of freight) in this way is that the vectors can be added in any order. This approach also makes it easier for operators to determine the CG position of their airplane prior to dispatch as the moments are

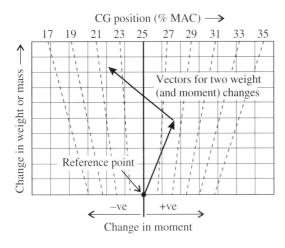

Figure 19.3 Customary method for representing CG changes for flight operations.

simply added or subtracted using a preprepared Load and Trim sheet. It does, however, mean that the CG position is represented by sloping lines on the graph (except, of course, for the reference line). Due to their distinctive appearance, such diagrams are often called *fan charts*. The choice of reference point is arbitrary: it can be selected by the operator to suit the particular airplane and loading requirements (e.g., seating configuration).

The fore and aft CG limits for the various phases of the flight and the limiting weights (e.g., MTOW, MLW, MZFW) are superimposed on the graph to form an operational envelope—this is illustrated in Figure 19.4. Note that for this example (which is based on the A330-200), the takeoff and landing limits are more restrictive than those applicable to flight. This is typical for airplanes of this class. It is vital that the airplane be loaded in a manner that ensures that the CG remains within the limits during all stages of the mission.

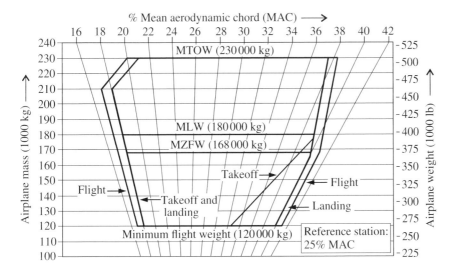

Figure 19.4 Example of load and balance diagram for a widebody airliner (based on A330-200).
Source: Adapted from [12].

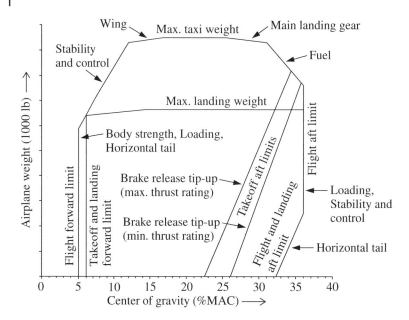

Figure 19.5 Example of CG limits for a single-aisle airliner (based on B737-800). *Source*: Adapted from [13].

19.3.3 Fore and Aft CG Limits on the Load and Balance Diagram

The fore and aft CG limits (see Figure 19.5)—established for airplane certification—are set by the manufacturer and based on a number of structural and aerodynamic design considerations. The forward limit is determined by considering the loads imposed on the airframe—specifically the wing, horizontal tailplane, aft fuselage, and nose landing gear (and the associated attachments)—when the CG is progressively moved forwards. The strength of these components, when loaded under the appropriate design conditions, can thus limit the extent to which the CG can be moved forwards. The aerodynamic factors that can impose a forward CG limit concern the effectiveness of the horizontal tailplane in generating the required pitching moments. If the CG is too far forward, full back stick/yoke movement may be insufficient to generate the pitching moment necessary to rotate (or "un-stick") the airplane for takeoff, or to flare it for landing—these considerations can thus set the forward CG limit.

The aft limits are similarly determined by considering the loads on the airframe when the CG is progressively moved rearwards. In this case, the position and strength of the main landing gear can impose a limit on the aft CG location. With the CG at the aft limit, the load on the nose gear must be adequate to steer the airplane when taxiing or during takeoff and landing. Also, the airplane must not tip up when full thrust is applied at brake release. When establishing the CG aft limit, consideration must also be given to the aircraft's longitudinal static stability and control characteristics (see Section 19.4).

An airplane's static stability depends on its *static margin*, which is the relative position of the CG with regard to the *neutral point* (see Section 19.4.3). If the CG were moved rearwards (by relocating cargo, for example) the static margin would be reduced. A benefit of reducing the static margin is a reduction in the horizontal tailplane air load required to trim the airplane in flight. This, in turn, results in a reduction in the trim drag (see Section 7.5.3). However, if the CG were allowed to move farther back than the certified aft limit, this could create handling difficulties for the crew. In an extreme case, the airplane would not be controllable in certain flight conditions.

19.3.4   Alternate Forward CG for Takeoff

When determining the maximum takeoff weight that an airplane can utilize for a given runway and set of operational conditions (see Section 10.3), the conservative assumption is made that the CG is located at the certified forward limit, as this condition is most restrictive in respect of takeoff performance [14]. Certain operators, however, do not require the entire range of CG values that the airplane is capable of for their particular operations. This offers a mechanism to increase the airplane's takeoff weight—and thus increase the payload and/or the fuel load—when the takeoff weight is field-length limited. In such cases, manufacturers can provide one or more certified *alternate forward CG limit* on the airplane's load and balance charts. These limit lines would be shown aft of the normal forward CG limit for takeoff (say 5% to 15% MAC rearwards) on the load and balance diagrams contained in the Airplane Flight Manual (AFM) and Weight and Balance Manual.

The reason why this mechanism works to increase the takeoff weight is linked to the influence of the CG position on the airplane's lift curve (see Section 7.2.4). For a given takeoff attitude, a slightly greater lift coefficient will be generated when the CG is located at an aft position compared to a forward position. (The takeoff attitude for many airplane types is limited by the airplane's geometry and the requirement that the tail does not strike the ground on rotation.) Furthermore, the airplane's stall speed reduces when the CG is located at an aft position (see Section 19.7.3). Thus, for a given takeoff weight, the takeoff speeds V_R and V_2 (see Section 10.2.2) will be slightly slower for the aft CG case, and the calculated field length will be reduced. Alternatively—and this is what is routinely done—the takeoff weight is increased, and the takeoff calculations rerun until the operation is again field-length limited.

An aft CG also benefits the airplane's climb performance as the lift-to-drag ratio is improved. An aft CG in flight is associated with reduced trim drag for a given weight and flight conditions (see Sections 7.5.3 and 19.4). The benefit is seen in an increase in the maximum climb gradient and maximum rate of climb that can be achieved for a given takeoff thrust setting and airplane weight (see Sections 12.4.3–12.4.6). Consequently, the use of an alternate forward CG for takeoff can have the effect of increasing the obstacle-clearance-limited weight (see Section 10.7), and in certain cases the climb-limited-takeoff weight (see Section 10.6) can also be increased [14].

19.4   Longitudinal Static Stability and Stabilizer Trim

19.4.1   Operation of the Trimmable Horizontal Stabilizer

The horizontal tailplane of a conventional jet transport airplane has two aerodynamic control surfaces: the stabilizer and the elevator, both of which can be used by the pilot to generate pitching moments about the airplane's CG and thus control the airplane about the Y_b axis (see Figure 3.7). The principal function of the elevator is to maneuver the airplane by generating out-of-balance pitching moments about the CG, whereas the horizontal stabilizer is primarily used for longitudinal balance (i.e., trim). The pitch trim system maintains pitch equilibrium in flight, by changing the angle of attack of the stabilizer to counteract moment changes induced by, for example, a change to the airplane's configuration (e.g., flap position), an increase or decrease in speed, or a change to the airplane's weight (e.g., due to fuel usage) or CG location (e.g., due to fuel transfer). Additionally, the system enables the pilot to maintain a pitch input without applying a force to the stick/yoke. It is a common feature on modern airplanes that the horizontal stabilizer automatically follows sustained position changes of the elevator, as demanded by the pilot; the design of such features ensures that the "follow-up" stabilizer movement is slow and that it lags the pilot's input.

During critical phases of the flight, it is important that the trim is correctly set. If an airplane is not in trim and the pilot inadvertently releases the stick/yoke, the airplane will pitch up or down (the pitch rate being dependent on how far the airplane is out of trim). One such critical phase is takeoff. It is standard procedure that flight crews set the horizontal stabilizer position prior to takeoff based on the airplane's actual takeoff weight and CG location. This ensures that the airplane will be in trim—or nearly in trim—during the climb-out after liftoff. The trim setting is typically selected to produce an in-trim condition at the one-engine-operative climb-out speed, V_2. It also means that the pilot's control input to rotate the airplane will be very similar for every takeoff; this is an important consideration in preventing "tail strikes" (see Section 9.5.1).

19.4.2 Equilibrium and the Basic Trim Equation

An airplane is said to be in a state of equilibrium when opposing forces are balanced: the sum of the forces acting on the airplane is zero and the sum of the moments acting about the center of gravity is also zero. From Newton's first law of motion (see Section 2.4.1), it can be deduced that an airplane, in such a condition, will not accelerate in translation or rotation. Equilibrium in pitch (i.e., about the Y_b axis) can be achieved through various combinations of elevator and horizontal stabilizer positions. The airplane is said to be *trimmed* (or in a state of *trim*) if equilibrium is achieved by positioning the stabilizer such that the elevator is in the null (mid) position. Note that trim implies equilibrium, but equilibrium does not necessarily imply trim.

Taking the sum of the moments about the Y_b axis (which passes through the CG) for an airplane that is in equilibrium leads to an interesting—and useful—relationship, which links the stabilizer angular setting with the airplane's weight and CG position. The stabilizer position required to balance the airplane also depends on (1) elevator position, (2) flight speed, (3) air density, and (4) airplane configuration (specifically, the position of the high lift devices and landing gear). This approximate relationship can be expressed, in coefficient form, as follows:[5]

$$C_{M_0} - K_N C_L - \bar{V}'(a_1\delta_s + a_2\delta_e) = 0 \qquad \text{(see Equation E.46)} \qquad (19.3)$$

where C_{M_0} is the pitching moment coefficient about the aerodynamic center of the wing plus body (fuselage)—see Equation 19.4;

K_N is the static margin—see Section 19.4.3;

C_L is the lift coefficient of the airplane;

$\bar{V}'$ is the tailplane volume coefficient—see Equation 19.5;

δ_s is the stabilizer angular position (typical unit: rad)—see Figure 19.6;

δ_e is the elevator angular position (typical unit: rad)—see Figure 19.6;

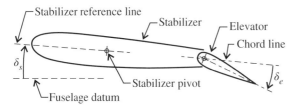

Figure 19.6 Definition of horizontal tailplane angles (shown positive).

---

5 The derivation of Equation 19.3 is presented in Appendix E.

a_1 is the linear derivative of the tailplane lift coefficient with respect to the tailplane angle of attack (typical unit: rad^{-1})—see Equation 19.7; and

a_2 is the linear derivative of the tailplane lift coefficient with respect to the elevator angle (typical unit: rad^{-1})—see Equation 19.7.

The pitching moment coefficient is defined as

$$C_{M_0} = \frac{M_0}{qS\bar{\bar{c}}} = \frac{2M_0}{\rho V^2 S\bar{\bar{c}}} \tag{19.4}$$

where M_0 is the pitching moment about the aerodynamic center of the wing plus body (typical units: N m, ft lb);

q is the dynamic pressure (typical units: N/m^2, lb/ft^2);

ρ is the air density (typical units: kg/m^3, slug/ft^3);

V is the true airspeed (typical units: m/s, ft/s);

S is the wing reference area (typical units: m^2, ft^2); and

$\bar{\bar{c}}$ is the wing mean aerodynamic chord (typical units: m, ft).

The tailplane volume coefficient is defined as

$$\bar{V}' = \frac{\ell_T' S_T}{\bar{\bar{c}} S} \tag{19.5}$$

where ℓ_T' is the aerodynamic tail arm, measured from the aerodynamic center of the wing plus body to the aerodynamic center of the horizontal tailplane (typical units: m, ft); and

S_T is the horizontal tailplane reference area (typical units: m^2, ft^2).

The tailplane lift coefficient is defined as

$$C_{L_T} = \frac{L_T}{qS_T} = \frac{2L_T}{\rho V^2 S_T} \tag{19.6}$$

where L_T is the horizontal tailplane lift force (typical units: N, lb).

The tailplane angle of attack derivative a_1 and the elevator derivative a_2 relate to the ability of the horizontal tailplane to generate a lift force. The tailplane lift coefficient can be written as a linear function of the tailplane angle of attack and elevator angle, as follows:

$$C_{L_T} = a_1 \alpha_T + a_2 \delta_e \tag{19.7}$$

where α_T is the tailplane angle of attack (typical unit: rad).

An important outcome of this analysis of longitudinal static stability is the relationship between the stabilizer angle and the elevator angle, which, it can be deduced, are interdependent (when the airplane is in equilibrium). This can be seen by rearranging Equation 19.3:

$$a_1 \delta_s + a_2 \delta_e = \frac{C_{M_0}}{\bar{V}'} - \frac{K_N C_L}{\bar{V}'} \tag{19.8}$$

It is instructive to consider how the parameters on the right-hand side of Equation 19.8 will vary during flight. The pitching moment coefficient C_{M_0} is essentially constant—this can be deduced from the definition of the aerodynamic center. In practice, the magnitude of C_{M_0} does change a little with varying angle of attack and Reynolds number. The pitching moment will change when the configuration of the airplane changes (e.g., when the flaps are deployed). The tailplane volume coefficient is a non-dimensional ratio of airplane geometric parameters (it has a constant value for a particular airplane type).

The airplane's configuration and flight condition can change in several ways—for example, the crew can deploy the flaps or lower the undercarriage or change the flight speed or altitude. The CG position can also change in flight. Any of these changes will alter the balance of the airplane, requiring a corresponding correction to either or both tailplane angles δ_s and δ_e.

It is theoretically possible for the airplane to be in a state of pitch equilibrium with any combination of stabilizer angle and elevator angle that satisfies Equation 19.8. This, of course, applies within the design limits of the flight controls. In practice, however, the pilot (or autopilot if engaged) will move the stabilizer to the specific angle that reduces the elevator deflection to zero—this process is known as *trimming* the control stick/yoke input (see Section 3.8.6).

19.4.3 Longitudinal Static Stability Margin

The *longitudinal static stability margin* (K_N), which is usually shortened to *static margin*, depends on the rate of change of pitching moment coefficient about the CG with respect to changing lift coefficient. It is also equal to the ratio of the distance (or margin) between the CG and the neutral point (NP) to the mean aerodynamic chord (Figure 19.7), that is,

$$K_N = h_N - h \tag{19.9}$$

where h_N is the position of the NP expressed as a fraction of the MAC; and
$\quad\quad h$ is the position of the CG expressed as a fraction of the MAC.

For an airplane to possess *positive* static stability (i.e., $K_N > 0$), the CG must be ahead of the neutral point, as illustrated in Figure 19.7. The neutral point is the *center of lift* of the whole aircraft. If the CG were located at the neutral point, there would be no change in the aircraft's pitching moment as the angle of attack changes—which, from airplane stability considerations, is not a desirable condition.

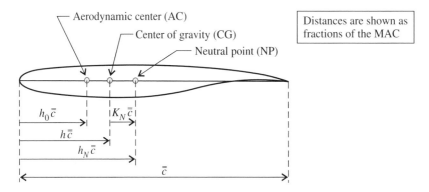

Figure 19.7 Reference distances for longitudinal static stability.

19.4.4 Stabilizer Position for Trim

Equation 19.8 can be rearranged to express the stabilizer angle required to trim the airplane in terms of the static margin and lift coefficient:

$$\delta_s = -\frac{a_2}{a_1}\delta_e + \frac{C_{M_0}}{a_1\bar{V}'} - \frac{K_N}{a_1\bar{V}'}C_L \tag{19.10}$$

or $\quad\quad \delta_s = \dfrac{C_{M_0}}{a_1\bar{V}'} - \dfrac{K_N}{a_1\bar{V}'}C_L \quad\quad$ when $\delta_e = 0$ $\tag{19.11}$

Two substitutions are now made: the static margin is written in terms of the CG position (see Equation 19.9), and the lift coefficient is written in terms of the airplane's weight (the load factor is equal to 1 in this derivation) and the flight Mach number (see Equation 7.4).

$$\delta_s = \frac{C_{M_0}}{a_1 \bar{V}'} - \frac{(h_N - h)}{a_1 \bar{V}'}\left(\frac{2W}{\rho_0 a_0^2 \delta M^2 S}\right) \qquad \text{when } \delta_e = 0 \qquad (19.12)$$

where W is the airplane weight (typical units: N, lb);

ρ_0 is the air density at the ISA sea-level datum (typical units: kg/m^3, slug/ft^3);

a_0 is the speed of sound in air at the ISA sea-level datum (typical units: m/s, ft/s);

δ is the relative air pressure (dimensionless); and

M is the flight Mach number (dimensionless).

Equation 19.12 shows that when the airplane is in a state of trim, for a given Mach number, altitude, and configuration, the stabilizer position depends on the following two parameters:

(1) the CG position (h), which will change as cargo or passengers are rearranged or as fuel is consumed during flight; and

(2) the airplane weight (W), which will change as fuel is consumed during flight.

This relationship provides a mechanism—available on many airplane types—by which the computed gross weight and corresponding CG location can be corroborated independently using a stabilizer position measurement for a particular flight condition and airplane configuration. The ZFW and the corresponding CG location would be entered into the Flight Management System (FMS) before takeoff. During the flight, real-time fuel quantity (from the fuel quantity indicators) and fuel flow (from the FADEC) data would be used to compute the airplane's weight and CG location.

19.4.5 Speed Stability and Longitudinal Static Margin

For a conventional airplane, the longitudinal static margin of the airplane (K_N) directly relates to the control force (to be applied by the pilot on the control stick/yoke) required to hold the airplane at an airspeed other than the trimmed airspeed with the thrust levers set at the trimmed position. For airplanes with positive static stability (i.e., $K_N > 0$), a push control force is required to maintain a speed greater than the trimmed speed and a pull force is required when the airspeed is less than the trimmed speed. The CG position has a significant influence on the required control force (to re-trim at the new speed). If the CG were to be progressively moved rearwards, the control force per knot (of speed change) would reduce—it would reach zero when the CG is at the neutral point and then reverse direction. This parameter of required control force per unit of speed (above or below the trimmed speed) is of fundamental importance in control system design, and minimum values are specified by the regulatory authorities; the average gradient cannot be less than 1 lb for each 6 kt speed change (FAR/CS 25.173 [10, 11]).

19.4.6 Stability Augmentation

Fly-by-wire flight control systems, as employed on modern airliners, permit an airplane to have less longitudinal static stability than earlier generation aircraft—this is often referred to as *relaxed static stability* (RSS). The benefit of reducing the static margin, for a jet transport airplane, is improved fuel efficiency. There are several ways of implementing stability augmentation in flight control systems, and manufacturers have adopted different design philosophies to reduce the undesirable aspects associated with flying with an aft CG position. There is an important provision in the regulations in this regard: FAR/CS 25.672 [10, 11] require that the handling qualities must not be impaired below a level needed for continued safe flight and landing following a failure of the stability augmentation system. This requirement sets a practical limit for manufacturers in terms of how far aft the CG can be permitted to go.

19.5 Center of Gravity Control

19.5.1 Aerodynamic Advantage of CG Control

Several widebody airplane types (e.g., A300, A330, and A340) have fuel tanks in the horizontal tailplane that can be used to control the CG position. The purpose of this design feature is to reduce the trim drag associated with the horizontal tailplane in cruise.

During cruising flight, a state of trim (equilibrium) can be established by an appropriate stabilizer deflection—the value being dependent on airspeed, weight, altitude, temperature, and CG location. For a given lift coefficient, the stabilizer deflection is a linear function of the static margin, which is a relative measure of the distance from the CG to the neutral point (see Equation 19.10). By pumping fuel from the wing (or center) tanks back to a tank in the horizontal stabilizer, the CG is moved rearwards, bringing it closer to the neutral point and reducing the deflection required. A reduced stabilizer deflection is thus needed to balance the air loads and this has the desirable effect of reducing the drag associated with this balancing force. Furthermore, by reducing the magnitude of the tailplane lift (L_T), the lift acting on the wing and body (L_{wb}) is also reduced by a small amount (for the same airplane weight) resulting in a small reduction in the lift-induced drag acting on the wing–body (Figure 19.8).

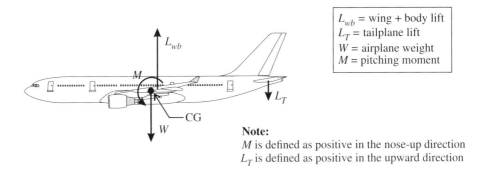

L_{wb} = wing + body lift
L_T = tailplane lift
W = airplane weight
M = pitching moment

Note:
M is defined as positive in the nose-up direction
L_T is defined as positive in the upward direction

Figure 19.8 Trim drag reduction (with aft CG).

19.5.2 Fuel Management System for CG Control

The fuel management system onboard the airplane types mentioned earlier in Section 19.5.1 sequences the transfer of fuel—between the *trim tank* in the horizontal tailplane and the wing tanks—to keep the CG within a narrow band close to the aft CG limit. This CG control band typically has a "width" of 0.5% MAC, and the rear boundary of the band is set approximately 2% forward of the certified flight aft CG limit [9], as illustrated in Figure 19.9.

The trim tank, which in certain cases is permitted to hold fuel at takeoff, receives fuel from the wing tank (or center tank, if installed) during the upper part of the climb (say above FL 200).

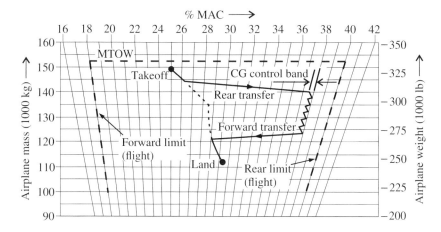

Figure 19.9 Principle of CG control using fuel transfers.

The rearward transfer of fuel moves the airplane's CG backwards. The amount of fuel that is transferred is calculated to bring the CG to the forward boundary of the control band. Normal fuel usage from the wing (or center) tanks results in a progressive aft movement of the CG, and when the CG reaches the rear boundary of the control band, a forward transfer of fuel is automatically instructed to bring the CG back to the forward boundary of the control band. Periodic transfers of fuel "packages" then take place, as needed, during the cruise. Finally, all remaining fuel in the trim tank is transferred forwards to the wing tanks at the end of the cruise or during the initial part of the descent. In normal operations, the trim tank is empty during the landing phase.

Such a technique deliberately "relaxes" the longitudinal static stability of the aircraft in order to reduce cruise fuel burn. This approach is sanctioned by the regulatory authorities as the cruise is a benign flight condition usually requiring only gentle maneuvering.

19.6   Operational Weights and Dispatch Procedures

19.6.1   Operational Loading Procedures

The FAA (in Advisory Circular 120-27 [3]) and EASA (in OPS Part CAT.POL.MAB.100 [5] and in the corresponding AMC and GM [4]) publish material that defines the procedures to be used by operators to calculate the airplane's weight and CG position prior to dispatch. Accurate information—regarding the OEW, the cargo, and the passengers and their baggage—is needed for the correct completion of *Load and Trim* sheets. As errors can lead to performance discrepancies (due to an underestimation of the passengers' weights, for example) or, in extreme circumstances, to a reduction in safety margins (due to an incorrect CG calculation, for example), this work is subject to strict operational procedures.

The baseline data consist of the OEW (or DOW) and the corresponding CG position, which is determined with the airplane in a defined configuration (e.g., flaps and slats in the clean position, landing gear extended, thrust reversers retracted, and water in potable water tanks).

It is a requirement that prior to dispatch, a Load and Trim sheet is prepared. Operators usually customize the format of these sheets. Preprepared tables and charts provide the user with a quick method to determine the weight and CG position of the airplane at the takeoff and zero fuel conditions (see Section 19.6.2). The CG positions can then be checked to ensure compliance with operational limits. Finally, the horizontal stabilizer trim setting for takeoff is determined based on the airplane's weight, CG position, flap setting, and engine thrust level.

The CG limits shown on a Load and Trim sheet are more conservative than those indicated on the certified CG envelope. The amount of conservatism depends on the airplane type and the intended operation. This allows for such operational factors as the extension/retraction of the flaps, slats, or landing gear, the in-flight movement of passengers and crew, fuel density variation, differences of passenger seat locations compared to their assumed positions, and so forth.

19.6.2 CG Checks for Dispatch

For routine flight operations, the method described earlier in Section 19.3.1 to determine an airplane's CG position is not the most appropriate; the moments are large numbers and this can lead to computational errors. A widely used method that overcomes this problem is based on the summation of non-dimensional moment *indices* rather than true *moments*. For the sake of brevity, the equations are only presented in terms of weight—equivalent equations for *mass* can be defined in an identical manner.

The baseline index for the OEW condition (I_{OEW}) is equal to the CG moment about a reference point (e.g., 25% MAC), non-dimensionalized, plus a constant (H), that is,

$$I_{OEW} = \frac{W_{OE}(X_{OE} - X_{ref})}{C} + H \tag{19.13}$$

where W_{OE} is the aircraft's operating empty weight (typical units: N, lb);
X_{OE} is the distance to the aircraft's CG in the OEW condition (typical units: m, in);
X_{ref} is the distance to the reference point (typical units: m, in);
C is a non-dimensionalizing constant (typical units: N m, lb in); and
H is a constant that is selected such that the aircraft's moment index is always positive (dimensionless).

The constants C and H are specific to an airplane type. The weight items that are added to the OEW are similarly expressed as index increments (ΔI), which can be positive or negative (depending on whether they are located aft or forward of the reference point):

$$\Delta I_i = \frac{W_i(X_i - X_{ref})}{C} \tag{19.14}$$

where W_i is the weight of the item of interest (e.g., fuel, passengers, baggage) (typical units: N, lb);
X_i is the distance to the CG of the item of interest (typical units: m, in).

The index (I) for the required weight conditions (e.g., TOW and ZFW) is obtained by the summation of the OEW index and the appropriate index increments, that is,

$$I = I_{OEW} + \sum \Delta I_i \tag{19.15}$$

The dispatch process is straightforward: the index increments are added to the baseline index (i.e., I_{OEW}) using preprepared Load and Trim sheets and the result plotted against the corresponding airplane weight on a customized fan chart. In this way, the airplane's CG positions for the critical cases (i.e., takeoff and zero fuel) can be checked to ensure that the airplane's CG remains within the established operational limits.

19.6.3   Operating Empty Weight (OEW)

The OEW of an airplane (defined earlier in Section 19.2.2) generally increases in service due to modifications (e.g., the installation of new equipment), repairs to the structure or furnishings, and the accumulation of dirt and water (moisture is absorbed by the cabin insulation, for example). Periodic weighing (every 3 or 4 years) is thus required to track the weight growth and any resulting CG shift (FAR 125.91 [15], OPS Part CAT.POL.MAB.100 [5]).

For the purpose of dispatch (see Section 19.6.2), operators can treat each airplane individually, or a group of aircraft as being part of a fleet, in which case a single set of values (i.e., OEW and corresponding CG position) can be used. In the latter case, each airplane in the fleet must meet the deviation requirements stipulated by the regulatory authorities—that is, for the OEW, a deviation of no more than $\pm 0.5\%$ MLW is permitted, and for the CG position, a deviation of no more than $\pm 0.5\%$ of the reference chord is permitted [3, 4].

19.6.4   Standard Crew Weights

Crew weights (with bags) can be determined by operators from surveys of their staff or from published standard values—for example, as shown in Table 19.1 (extracted from FAA AC 120-27E [3]). Comparable EASA information is given in AMC2 CAT.POL.MAB.100(d) [4].

Table 19.1 FAA standard weights (with SI equivalent mass) for crew members

Crew member	Average weight without bags and kit	Average weight with bags and kit
Flight crew member	190 lb (86.2 kg)	240 lb (109 kg)
Flight attendant (average)	170 lb (77.1 kg)	210 lb (95.3 kg)
Male flight attendant	180 lb (81.7 kg)	220 lb (99.8 kg)
Female flight attendant	160 lb (72.6 kg)	200 lb (90.7 kg)

Source: FAA AC 120-27E [3].

19.6.5   Passenger and Baggage Weights

The weight of the passengers and their baggage can be determined from actual weighing prior to boarding[6] or from published standard weight values applicable to the passenger profile for the particular flight. Details of procedures and standard weights are given in FAA AC 120-27 [3] and in AMC1 CAT.POL.MAB.100(e) [4]. The following factors are considered:

(1) the male/female ratio;
(2) the type of flight (e.g., domestic or international; chartered or scheduled); and
(3) the season (e.g., a 5 lb allowance is made for winter clothing, in the case of the FAA).

The FAA standard weights[7] are given in Table 19.2. Standard weights, however, should not be used when the passengers belong to a group that does not reflect the general population in this

6 Weighing of passengers is rare for commercial flight operations using jet transport airplanes, but is undertaken in certain circumstances.
7 The standard weights given in FAA AC 120-27 [3] have been progressively revised upwards, as the average weight of passengers has crept up over the past few decades (and will be revised again, if required).

Table 19.2 FAA standard weights (with SI equivalent mass) for passengers, including hand baggage

Season	Average adult	Male adult	Female adult	Children (2–12 yrs)
Summer	190 lb (86.2 kg)	200 lb (90.7 kg)	179 lb (81.2 kg)	82 lb (37.2 kg)
Winter	195 lb (88.5 kg)	205 lb (93.0 kg)	184 lb (83.5 kg)	87 lb (39.5 kg)

Notes:

(a) *Source*: FAA AC 120-27E [3].
(b) Weight values include 16 lb (7.3 kg) carry-on baggage per passenger (operators are required to conduct a survey if they believe that this allowance is inappropriate).
(c) The weights of infants (younger than 24 months) are factored into the adult weight values.
(d) Summer in the Northern Hemisphere is nominally indicated as May 1 to October 31 and winter from November 1 to April 30. On routes with no seasonal variation, operators are permitted to use average weights that are appropriate to the climate.

respect—for example, a sports team or a military unit. Actual weights should be used in these situations. It is not only a matter of ensuring that an accurate estimate of the takeoff weight be established through this process, but it is also important that the CG be correctly determined. A large group of children seated together at the rear of the cabin, for example, would result in an incorrect CG computation if standard adult weights were used—and this would result in an incorrect stabilizer trim angle being set by the crew for takeoff.

Airlines that wish to adopt non-standard weights (for specific regional operations, for example) are required to conduct passenger surveys according to regulatory guidelines [3]. Operators that wish to use a standard average weight for check-in baggage are required to use a weight not less than 30 lb (13.6 kg), unless a valid survey supports the use of the lower value.

19.7 Performance Implications

19.7.1 Impact on Fuel Burn of a Weight Increase or the Carriage of Excess Weight

As mentioned earlier, an airplane's OEW increases as the aircraft ages, and this will result in a progressive increase in fuel usage. Similarly, the installation of new equipment (e.g., an in-flight entertainment system) or the carriage of unnecessary items (e.g., life rafts on overland operations) or excessive catering supplies will result in an increase in fuel burn. The relationship between a weight increase and the additional fuel burned during cruise can be estimated using the small change expression derived in Section 16.13. From Equation 16.27 it follows that

$$\frac{\delta W_f}{W_f} = \frac{\delta W_2}{W_2} \tag{19.16}$$

where W_f is fuel weight consumed in cruise;
δW_f is the increase in fuel weight;
W_2 is the airplane end-of-cruise weight; and
δW_2 is the increase in airplane end-of-cruise weight (i.e., weight increment).

From Equation 19.16 it can be deduced that a 1% increase in OEW results in a 1% increase in fuel usage. Note that Equation 19.16 strictly applies to the cruise segment (i.e., ignoring the climb and descent segments); nonetheless, it does provide a reasonable means to estimate the increased mission fuel associated with an increase in OEW.

19.7.2   Impact of CG Location on Drag and Specific Air Range

The relationship between an airplane's trim drag (see Section 7.5.3) and its CG position is dependent on the airplane type. Certain airplane types are relatively insensitive to CG movement (e.g., A320); most types, however, display a clear relationship in which the trim drag (and fuel burn) is reduced with a rearward movement of the airplane's CG. The airplane's weight and cruise altitude are factors that influence the relationship between specific air range (SAR) and CG position; the influence is greater at high weights and high altitudes. The SAR, for example, increases by 1.5% to 2% when the CG is moved from the midpoint to the most rearward position permitted (at high aircraft weights at typical cruise altitudes) on the A300-600, A310, and A340 airplane types [9].

19.7.3   Impact of CG Location on Stall Speed and Takeoff and Landing Performance

The stall speed is directly related to the lift generated by the airplane's wing. A forward movement of the CG position will require an increased wing lift; consequently, the stall speed will be greater for a forward CG position. For example, the stall speed for the A340 airplane increases by 1.5 kt when the CG moves from the 26% MAC location to the forward limit [9].

   The CG position influences takeoff performance, as the operating speeds during takeoff are referenced to the stall speed. As mentioned in Section 19.3.4, the takeoff distance is reduced if the CG is moved rearwards. Furthermore, an aft CG can improve takeoff climb performance. The CG position also influences an airplane's landing performance as the approach speed, which is linked to the stall speed, directly impacts the landing distance (see Section 11.5). The farther aft the CG is, the shorter the landing distance will be; similarly, for a given landing distance, moving the CG rearwards results in an increase in the allowable landing weight.

   Adjustments to the takeoff and landing performance data published in the FCOM are thus needed if the airplane's CG is significantly different from the baseline (reference) position utilized for the establishment of the baseline data.

19.7.4   Lateral Imbalance

A lateral imbalance can occur when the payload is loaded asymmetrically about the lateral (roll) axis; this produces a moment about that axis. A deflection of the ailerons is required in flight (as illustrated in Figure 19.10) to correct the lateral imbalance, and this results in increased drag and, consequently, increased fuel burn. The issue is of greater concern for cargo airplanes, compared to passenger airplanes, as the latter do not have much potential for significant lateral imbalances during normal operations.

Figure 19.10 Aileron deflection required in flight to correct a lateral loading imbalance.

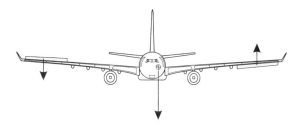

References

1 Airbus, "A380 aircraft characteristics airport and maintenance planning (AC)," Iss. Mar. 30/05, Rev. Dec. 01/16, Customer Services, Airbus S.A.S., Blagnac, France, 2016.

2 Airbus, "A330 Aircraft characteristics airport and maintenance planning (AC)," Iss. Jan. 01/93, Rev. Jan. 01/17, Customer Services, Airbus S.A.S., Blagnac, France, 2017.

3 FAA, "Aircraft weight and balance control," Advisory Circular 120-27E, Federal Aviation Administration, Washington, DC, June 10, 2005. Available from www.faa.gov/.

4 EASA, "Acceptable means of compliance (AMC) and guidance material (GM) to Part-CAT," Iss. 2, Amdt. 5, European Aviation Safety Agency, Cologne, Germany, Jan. 27, 2016. Available from www.easa.europa.eu/.

5 European Commission, *Commercial air transport operations (Part-CAT)*, Annex IV to Commission Regulation (EU) No. 965/2012, Brussels, Belgium, Oct. 5, 2012. Published in *Official Journal of the European Union*, Vol. L 296, Oct. 25, 2012, and reproduced by EASA.

6 Boeing, "Weight definitions and limitations," Training course notes, Weight & Balance, Boeing Commercial Airplanes, Seattle, WA, undated.

7 SAWE, "Introduction to aircraft weight engineering," Society of Allied Weight Engineers, Los Angeles, CA, undated.

8 Airbus, "Getting to grips with aircraft performance," Flight Operations Support and Line Assistance, Airbus S.A.S., Blagnac, France, Jan. 2002.

9 Airbus, "Getting to grips with weight and balance," Flight Operations Support and Line Assistance, Airbus Industrie, Blagnac, Toulouse, France, undated.

10 FAA, *Airworthiness standards: Transport category airplanes*, Federal Aviation Regulation Part 25, Amdt. 25-143, Federal Aviation Administration, Washington, DC, June 24, 2016. Latest revision available from www.ecfr.gov/ under e-CFR (Electronic Code of Federal Regulations) Title 14.

11 EASA, *Certification specifications and acceptable means of compliance for large aeroplanes*, CS-25, Amdt. 18, European Aviation Safety Agency, Cologne, Germany, June 23, 2016. Latest revision available from www.easa.europa.eu/ under Certification Specification.

12 Airbus, "A330 flight deck and systems briefing for pilots," STL 472.755/92, Iss. 4, Airbus Industrie, Blagnac, France, Mar. 1999.

13 Boeing, "Center of gravity limitations," Training course notes, Weight & Balance, Boeing Commercial Airplanes, Seattle, WA, undated.

14 Blake, W. and Performance Training Group, "Jet transport performance methods," D6-1420, Flight Operations Engineering, Boeing Commercial Airplanes, Seattle, WA, Mar. 2009.

15 FAA, *Certification and operations: Airplanes having a seating capacity of 20 or more passengers or a maximum payload capacity of 6,000 pounds or more; and rules governing persons on board such aircraft*, Federal Aviation Regulation Part 125, Amdt. 125-62, Federal Aviation Administration, Washington, DC, July 3, 2013. Latest revision available from www.ecfr.gov/ under e-CFR (Electronic Code of Federal Regulations) Title 14.

20

Limitations and Flight Envelope

20.1 Introduction

Stall is associated with large-scale disruption of the airflow over the upper wing surface, which is caused by an excessive angle of attack, and which results in a significant loss of lift. The stalling phenomenon sets an operational limit on the airplane in the form of a lower speed boundary—which is dictated, to a large extent, by the aerodynamic design of the wing. An airplane, however, does not have a single stall speed—instead, it has a stalling angle of attack for each wing configuration (i.e., the position or setting of the high lift devices), which will give a range of stall speeds depending on the airplane's weight and the air density. The stall speed also depends, to a lesser extent, on the manner in which the airplane is flown. This is discussed in Section 20.2. Preceding the stall, the turbulent airflow from pockets of separating air on the upper wing surface may strike the horizontal tailplane, thus causing a low frequency vibration or shaking of the airframe—this is technically known as *buffeting*. Buffeting also arises at high speed, when shock waves cause flow separation on the wing surface. The high-speed buffet boundaries create an upper speed limit for flight operations (Section 20.3). The combination of aerodynamic, propulsion, and certification limits can be used to establish a safe operating speed envelope, which can be presented as a function of altitude (Section 20.4).

The airworthiness regulations [1, 2] define reference speeds that directly impact flight operations—several key reference speeds are described in Section 20.5. Another important operational envelope is the $V-n$ diagram—a plot of velocity versus load factor. A $V-n$ diagram is an envelope that defines the load factor limits, for maneuvers and atmospheric turbulence, within which an airplane can be safely operated without risking structural damage. The rules for the construction of $V-n$ diagrams are given in the airworthiness regulations FAR 25 [1] and CS-25 [2]—relevant extracts, together with notes on structural design criteria, are presented in Sections 20.6 and 20.7.

20.2 Stall

20.2.1 Airplane Stall

In simple terms, an airplane stalls when the angle of attack of the wing reaches the critical (i.e., initial stall) angle of attack. For a pilot to enter a stall deliberately from level flight, he/she would reduce the airspeed and, to compensate for the reducing lift at the slower speed, simultaneously pull the stick/yoke back, increasing the angle of attack (α)—thus maintaining level flight. At the critical angle of attack—which on some airplane types is reached rather abruptly—the nature

Performance of the Jet Transport Airplane: Analysis Methods, Flight Operations, and Regulations, First Edition.
Trevor M. Young.
© 2018 John Wiley & Sons Ltd. Published 2018 by John Wiley & Sons Ltd.

of the airflow over the wing changes. The flow starts to break away, or separate, from the upper surface, forming strong vortices, which may cause the flow to "double back" and flow forwards along the wing surface in local areas. Separation tends to occur first at one locality and not uniformly across the whole wing. This is an inherent characteristic of the wing geometry, which is influenced by such design features as the wing geometric twist, leading edge sweep, and aerodynamic twist (which results from a change of airfoil section across the wingspan). When the airplane stalls, the pilot typically experiences some buffeting, resulting primarily from the separated airflow striking the tailplane.

An increase in the angle of attack beyond the initial stall angle will result in a substantial area of flow separation on the upper wing surface, a large reduction in lift, and an increase in drag. As the airplane's lift is now less than its weight, it will lose height—in fact, it will accelerate towards the Earth under the influence of gravity. The normal response for a jet transport airplane—when correctly operated—will be a strong nose-down pitching moment. This will quickly reduce the angle of attack, allow the airplane to increase speed, and subsequently restore the airflow over the wing surface. Soon afterwards, adequate lift will be produced to support the airplane's weight. If the stall is asymmetric—that is, with a greater amount of lift being lost on one wing—then one wing will drop more than the other, and the airplane will tend to roll, in addition to its downward motion under gravity. The corrective action by the pilot to arrest this rolling motion will be a rudder input.[1]

An airplane's stall speed (which is usually denoted by the subscript "S") was traditionally taken as the minimum speed at which controllable flight could be maintained. This definition is still used in many aviation sectors (e.g., general aviation); however, since the mid-1980s, a more precise definition has been used for the certification of jet transport airplanes (this is discussed later in Section 20.5.1).

The influence of the airplane's weight and configuration, air density (and hence altitude), and load factor can be seen from the definition of lift coefficient when expressed as a maximum. In flight testing, an airplane's stall speed can be shown to depend also on the thrust and on the rate of deceleration when entering the stall. It is thus misleading to talk about *the* stall speed of an airplane, without specifying the associated conditions. The influences of these factors are discussed in Section 20.2.4.

Knowledge of an airplane's stall speeds are important for safe flight operations (inadvertent stalls without sufficient height for recovery have resulted in many air crashes). Additionally, stall speeds are important as they are used as reference speeds to determine several key operational speeds pertaining to the safety-critical flight phases of takeoff and landing. For this purpose, the airworthiness regulations define a *reference stall speed* (V_{SR})—see Section 20.5.2.

20.2.2   Maximum Lift Coefficient and Airplane Configuration

The lift coefficient corresponding to the minimum flight speed is the maximum lift coefficient ($C_{L_{max}}$), which depends on the airplane configuration, or, more specifically, on the deployed position of the high lift devices. The main purpose of leading and trailing edge high lift devices, such as flaps and slats, is to increase the value of $C_{L_{max}}$ during high angle of attack maneuvers (see Figure 20.1), which are limited to takeoff and landing for routine flight operations.

---

1 The rudder is used to yaw the airplane and generate sideslip, which causes the aircraft to roll due to the dihedral effect. It is necessary for the pilot to control the airplane in this manner as the stalled flow over portions of the wing can make the ailerons ineffective.

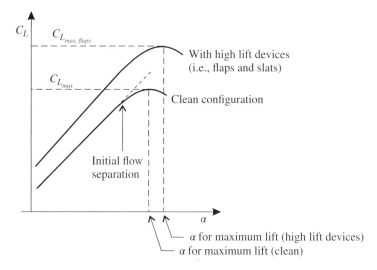

Figure 20.1 Lift curve (lift coefficient versus angle of attack).

20.2.3 Evaluation of Stall Speeds

An airplane's lift coefficient can be expressed as the product of the load factor (n_z) and the airplane weight (W) divided by the product of free-stream dynamic pressure (q) and wing reference area (S), where q can be expressed in terms of true airspeed (TAS), equivalent airspeed (EAS), or Mach number—as described by Equation 7.4. The minimum speed—in TAS (V), EAS (V_e), or Mach number (M)—corresponding to the maximum lift coefficient ($C_{L_{max}}$), can be obtained from the following relationships:

$$C_{L_{max}} = \frac{2n_z W}{\rho V_S^2 S} = \frac{2n_z W}{\rho_0 V_{e_S}^2 S} = \frac{2n_z W}{\rho_0 a_0^2 \delta M_S^2 S} \tag{20.1}$$

where n_z is the load factor normal to the flight path (dimensionless);
    W is the airplane weight (typical units: N, lb);
    ρ is the ambient air density (typical units: kg/m³, slug/ft³);
    V_S is the stall true airspeed (typical units: m/s, ft/s);
    S is the wing reference area (typical units: m², ft²);
    ρ_0 is the air density at the ISA sea-level datum (typical units: kg/m³, slug/ft³);
    V_{e_S} is the stall equivalent airspeed (typical units: m/s, ft/s);
    a_0 is the speed of sound in air at the ISA sea-level datum (typical units: m/s, ft/s);
    δ is the relative air pressure (dimensionless); and
    M_S is the stall Mach number (dimensionless).

In straight and level (constant height) flight, the airplane's lift is equal to its weight, and the load factor is equal to 1. The stall speed in this condition is the lowest speed that is possible in level 1-*g* flight. The subscript 1 (or 1-*g*) is used to designate this stall condition. The corresponding stall speeds can be obtained from Equation 20.1:

$$V_{S_1} = \sqrt{\frac{2W}{\rho S C_{L_{max}}}} \qquad \text{(for 1-}g\text{ flight)} \tag{20.2}$$

$$\text{or} \quad V_{e_{S_1}} = \sqrt{\frac{2W}{\rho_0 S C_{L_{max}}}} \qquad \text{(for 1-}g\text{ flight)} \tag{20.3}$$

$$\text{or} \quad M_{S_1} = \sqrt{\frac{2W}{\rho_0 a_0^2 \delta S C_{L_{max}}}} \qquad \text{(for 1-}g\text{ flight)} \tag{20.4}$$

20.2.4 Factors That Influence Stall Speeds

It is apparent from Section 20.2.2 that an airplane's stall speed depends on its configuration. It also depends on several other factors, including (1) gross weight, (2) load factor, (3) thrust setting, (4) Reynolds number, and (5) center of gravity location. Furthermore, the precise speed at which an airplane stalls has been shown in flight tests to depend also on the airplane's deceleration preceding the stall and on the manner in which the pilot manipulates the flight controls during the stall (this is discussed in later in Section 20.5.1).

It is seen from Equation 20.3 that the 1-g stall speed in EAS, for a given value of $C_{L_{max}}$, is a function of the airplane's gross weight. The stall speed increases with increasing weight (Figure 20.2). This effect can be significant, particularly for long-haul flights, where the landing weight is very much lower than the takeoff weight. A typical ratio of maximum takeoff weight (MTOW) to operating empty weight (OEW) for a jet transport airplane is 2:1. For such an airplane, the stall speed at the MTOW will thus be 1.41 (i.e., $\sqrt{2}$) times the stall speed at the OEW for the same value of $C_{L_{max}}$.

There is also a secondary, much smaller, influence of the airplane's weight on its stall behavior: a higher weight results in greater aeroelastic distortion of the wing, and this tends to reduce the maximum lift coefficient, $C_{L_{max}}$.

The stall speed for an airplane performing a maneuver, such as a pull-up maneuver or a banked turn, will be greater than that given by Equations 20.2 to 20.4. As the airplane is not flying straight and level, but rather accelerating towards the center of a circle, the lift

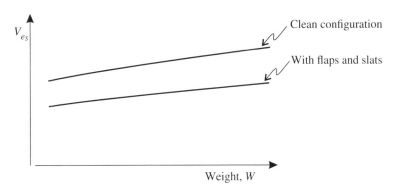

Figure 20.2 Variation of 1-g stall speed with airplane weight and configuration.

cannot equal the weight. The stall speed (EAS) during maneuvers can be determined from Equation 20.1:

$$V_{e_S} = \sqrt{\frac{2n_z W}{\rho_0 S C_{L_{max}}}} = \sqrt{n_z}\sqrt{\frac{2W}{\rho_0 S C_{L_{max}}}} = \sqrt{n_z}V_{e_{S_1}} \tag{20.5}$$

The stall speed during maneuvers (at the same weight) is thus greater than the 1-*g* stall speed by the factor $\sqrt{n_z}$. A 40° angle of bank coordinated turn, for example, will produce a load factor of 1.305 (see Equation 15.3) and a stall speed that is 14% higher than the comparable 1-*g* stall speed. An associated factor is atmospheric turbulence: vertical gusts induce normal accelerations on the airplane (see Section 20.7.6). A strong gust would thus cause an airplane to stall if it were operating at a speed that was only a few knots above the 1-*g* stall speed.

A change in stall characteristics will also occur when the airplane is stalled with the "power on." At high angles of attack, the thrust vector has a significant component normal to the flight path. This component can be regarded as an increment to the total lift, which will result in a small reduction in the stall speed. Certification tests are conducted at idle thrust (see FAR/CS 25.103 [1, 2]); hence, this influence provides a degree of conservatism in published stall data, as most inadvertent stalls occur with "power on."

Mach number also has an influence on an airplane's stall characteristics. Airflow over an airfoil at a high subsonic Mach number separates at a lower C_L than flow at a low Mach number [3, 4]. $C_{L_{max}}$ (for a given configuration) thus reduces with increasing Mach number, reducing the associated stall speed. At sea level, this effect has no operational consequence as flight at high Mach numbers is associated with low angles of attack (i.e., away from the stall). However, at cruise altitudes, the situation is different. If it is assumed, for the sake of illustration, that the airplane climbs from sea level to its cruise altitude at constant calibrated airspeed (CAS), then the Mach number will increase (see Section 6.8.2). Consequently, for a given CAS at cruise altitude, the Mach number is increased and $C_{L_{max}}$ is lower than it would be at sea level. Reference [5] indicates that the stall speed typically reduces by 2–3 kt per 10 000 ft for a jet transport airplane.

Another factor that has an influence on the stall speed is the center of gravity (CG) position (Figure 20.3). The stall speed is slightly lower when the CG is at an aft location compared to when it is at a forward location (when all other factors are considered unchanged). This is due to the reduction in the required *wing* lift that occurs when the CG is moved rearwards. As discussed in Section 7.2.4, the angle of attack in 1-*g* level flight is reduced when the CG is at an aft location compared to when it is at a forward (fwd.) location. The stall angle of attack (shown

Figure 20.3 Effect of CG location on lift coefficient.

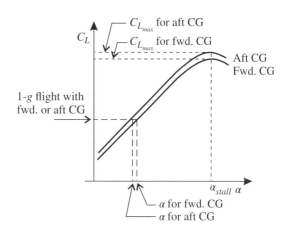

in Figure 20.3 as the angle of attack for maximum lift), however, is independent of CG variation. As $C_{L_{max}}$ is greater for the aft CG location, the stall speed is consequently lower. Certification tests—which are required to determine the highest stall speed—are thus conducted with the most adverse CG position (i.e., the most forward allowable CG location).

20.3 High-Speed Buffet

20.3.1 Introduction to High-Speed Buffet

In a similar way that an airplane's stall speeds can set a lower limit (or boundary) to its safe operational speed range, so high-speed buffet can set an upper limit. High-speed buffet is associated with the creation of shock waves (see Section 3.5.3). Weak shocks (i.e., shock waves that exhibit small losses in total pressure) located near to the point of maximum thickness of the airfoil section are usually not a cause for concern. The formation of strong shock waves aft of the point of maximum thickness, however, will inevitably cause significant flow separation. *Buffet* is said to occur when these shock waves move rapidly backwards and forwards on the upper wing surface, resulting in an unsteady airflow over the wing and a fluctuating lift distribution. When this happens, pulses of unsteady air flow back from the wing, striking the tail. The ensuing vibration and shaking of the airframe emanating from the wing and tail can be felt by the flight crew (and passengers). This is clearly undesirable and is potentially hazardous.

20.3.2 Buffet Onset Speeds

Buffet can occur at any speed: at low to moderate speeds it precedes a stall (see Section 20.2.1), and at high speeds it is associated with shock-induced flow separation. Buffet onset can be correlated to the airplane's lift coefficient (or angle of attack) for each flight Mach number; this is typically done up to the airplane's maximum operating speed. As the speed increases, so the lift coefficient at which buffet occurs reduces. This is illustrated in Figure 20.4 (the indicated buffet line approximates data presented in reference [6] for the B757-200). The data points that are used by manufacturers to establish such a relationship are determined by conducting a series of flight tests, during which time a test airplane is flown to the point of buffet onset by varying the airplane's speed, altitude, and load factor. Speed increases, for example, can be achieved by putting the airplane into a dive; whereas, load factor changes can be achieved by conducting *wind-up* turns (which involve progressively increasing the bank angle, usually at constant speed). Buffet onset is indicated when an accelerometer installed under the pilot's seat measures peak-to-peak accelerations greater than $0.1g$ [7].

As buffet occurs at a particular lift coefficient for a given Mach number, it is possible to establish the flight parameters associated with buffet onset from the following equation, which describes an airplane's lift coefficient in maneuvering flight (see Section 7.2.5):

$$C_L = \left(\frac{n_z W}{\delta} \right) \left(\frac{1}{M^2} \right) \left(\frac{2}{\rho_0 a_0^2 S} \right) \tag{20.6}$$

It is evident from Equation 20.6 that, for a given Mach number (M), the airplane's lift coefficient (C_L) is a function of $n_z W / \delta$. This expression was used to construct the lines that are superimposed on the buffet onset data shown in Figure 20.4. This was accomplished by assuming combinations of airplane weight (W) and altitude (which defines δ) and then calculating C_L for $n_z = 1$. An arbitrarily selected cruise condition, corresponding to the coordinates $C_L = 0.56$

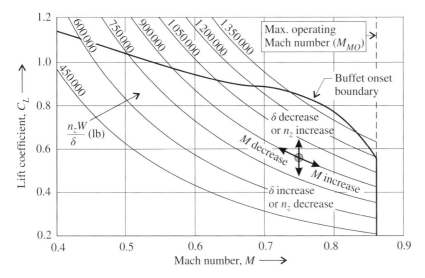

Figure 20.4 Typical buffet onset boundary superimposed on a flight lift coefficient versus Mach number grid for a mid-size, twin-engine airliner.

and $M = 0.75$, is shown on the chart for the sake of illustration. If the airplane's Mach number were to increase or decrease when weight, altitude, and load factor are unchanged, the point of interest would move along the constant $n_z W/\delta$ line, as indicated on the chart. Changes to the parameter $n_z W/\delta$—arising from weight, altitude, or load factor changes (or combinations thereof)—at a constant Mach number would move the point up or down on the chart. This latter observation is particularly important as it relates to the selection of safe operating altitudes for cruise. It is seen that by flying at a higher altitude and/or by conducting a steep turn, at a given Mach number, the margin between buffet onset and the airplane's cruising condition is reduced.

The lowest line on the chart (Figure 20.4), where $n_z W/\delta = $ 450 000 lb, represents several flight conditions (described by combinations of W, n_z, and δ). One such condition corresponds to the airplane operating at a weight of 139 800 lb (where $n_z = 1$) at 29 000 ft (where $\delta = 0.3107$). At this flight condition, the airplane is too light and the altitude too low for high-speed buffet to occur. The upper line, where $n_z W/\delta = 1\,350\,000$ lb, could correspond to a weight of 238 100 lb (where $n_z = 1$) at 41 000 ft (where $\delta = 0.1764$), for example. In this case, the constant $n_z W/\delta$ line intersects the buffet onset boundary at Mach 0.74 (this is called the *low-speed buffet* point) and also at Mach 0.83 (which is the *high-speed buffet* point). This means that at this flight condition—that is, where $n_z W/\delta = 1\,350\,000$ lb—the airplane will be free of buffet only between these speeds.

By considering progressively increasing altitudes, at the same weight, the range of operating speeds at which the airplane would be free from buffet would reduce until, ultimately, only a single operating speed is possible. This point on the chart is colloquially referred to as the *coffin corner*—so called, as the airplane is in an unsafe situation with no easy means of escape for the pilot. The height at which this occurs depends on the design of the airplane. Note that the pilot cannot decelerate from a high-speed buffet condition without immediately encountering low-speed buffet.

For the data presented in Figure 20.4, the coffin corner corresponds to $n_z W/\delta = 1\,416\,000$ lb, where 1-*g* flight is only possible at Mach 0.79. This would correlate to a weight of 238 100 lb (where $n_z = 1$) at 42 000 ft, for example. Note that this is only 1000 ft higher than for the case considered earlier, where the airplane was buffet free from Mach 0.72 to Mach 0.83 at the same weight. Whereas certain military airplanes[2] can fly near to the coffin corner, commercial jet transport airplanes do not produce enough thrust to climb high enough at a weight that would get them into this situation [6].

20.3.3 Altitude Selection

Buffet onset can have a significant operational impact on cruising flight. To facilitate reasonable maneuvering of the airplane, the requirement for a margin of at least 30% between the 1-*g* level-flight lift coefficient and the buffet onset lift coefficient is widely adopted for planning flight operations. This means that an angle of bank of 40° in level flight, which will produce a load factor of 1.3, should be possible without encountering buffet at the selected cruise altitude.[3] Typically, such high bank angles are not used in normal flight operations—this allowance provides for an emergency or a significant overshoot beyond the intended bank angle and/or the possibility of strong atmospheric turbulence (vertical gusts, as mentioned earlier, increase the load factor).

The impact of buffet on cruise altitude selection can be seen in Figure 20.5; in this example, airplane mass, rather than weight, is used. Buffet onset speeds are shown for a range of cruising altitudes for combinations of mass and load factor. For the illustration given in the chart (i.e., mass of 66 500 kg at 35 000 ft altitude), the range of speeds at which the airplane would be

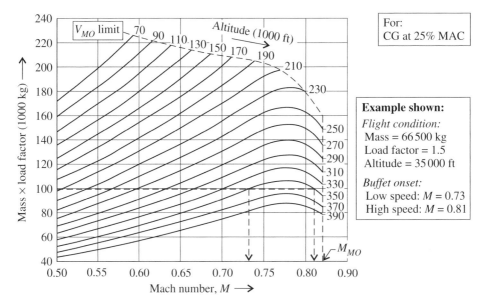

Figure 20.5 Buffet onset speeds (based on data for A319/320/321). *Source*: Adapted from [9].

---

2 Pilots of the U-2 high-altitude reconnaissance airplane regularly operated it in a narrow band of speeds close to the coffin corner, at altitudes as high as 70 000 ft [8].

3 Section AMC 25.251(e) in CS-25 Book 2 [2] provides an acceptable means of compliance to avoid the inadvertent occurrence of buffeting in cruising flight by the requirement that manufacturers "establish the maximum cruise altitude at which it is possible to achieve a positive normal acceleration increment of 0.3*g* without exceeding the buffet onset boundary."

free from buffet with a load factor of 1.5 is Mach 0.73 to Mach 0.81. At 39 000 ft, however, it is evident that the airplane (at this mass) would experience buffet at all speeds when the load factor is equal to or greater than 1.32.

The airplane's CG location has a minor influence on buffet onset; the altitude at which an airplane can conduct a 40° bank turn without encountering buffet is slightly lower in the case of a forward CG compared to an aft CG location. The reason for this can be seen in Figure 20.3: to achieve the same C_L (at 1-*g* flight), a slightly greater angle of attack is needed for a forward CG compared to an aft CG. The higher angle of attack would result in an earlier onset of buffet. For routine flight operations, the data needed to determine safe cruising altitudes are provided in the airplane's Flight Crew Operating Manual (FCOM).

20.4 Altitude–Speed Limitations

20.4.1 Altitude Limits (Ceiling)

An *absolute ceiling* (see also Section 17.3.1) is the maximum altitude at which an airplane theoretically can maintain level flight with the available thrust at the appropriate thrust rating (i.e., maximum climb thrust or maximum cruise thrust). The absolute ceiling depends on the airplane's weight and on the ambient air temperature. Factors that affect engine thrust (e.g., ancillary services, such as air conditioning and anti-icing systems) or airplane drag (e.g., CG position) can also influence the ability of the airplane to maintain a particular altitude. An airplane may be able to fly higher than its absolute ceiling by executing a zoom maneuver (see Section 12.4.1), in which kinetic energy is exchanged for potential energy, but these altitudes cannot be sustained. Referring to Figure 12.9, it can be deduced that the absolute ceiling is achieved when the thrust curve is tangential to the drag curve and steady flight is only possible at one speed. If it is assumed that thrust is independent of speed over the speed range of interest, then the absolute ceiling will be achieved at the minimum drag speed.

A *service ceiling*—which is lower than the absolute ceiling at the same airplane weight and air temperature—is the greatest altitude at which a particular rate of climb (e.g., 100 ft/min or 300 ft/min) can be achieved at a given thrust setting (e.g., maximum climb thrust or maximum cruise thrust). The governing equation that describes an airplane's rate of climb, which was derived in Section 12.4.4 (see Equation 12.28), can be expressed as follows:

$$\frac{F_N - D}{W} = \frac{(1 + f_{acc})V_v}{a_0\sqrt{\theta}M} \tag{20.7}$$

where F_N is the thrust (typical units: N, lb);
$\quad D$ is the drag (typical units: N, lb);
$\quad W$ is the airplane weight (typical units: N, lb);
$\quad f_{acc}$ is the acceleration factor (dimensionless) (see Table 12.1);
$\quad V_v$ is the rate of climb (typical units: m/s, ft/s);
$\quad a_0$ is the speed of sound at the ISA sea-level datum (typical units: m/s, ft/s, kt);
$\quad \theta$ is the relative temperature (dimensionless); and
$\quad M$ is the Mach number (dimensionless).

The maximum altitude at which a set rate of climb (V_v) at a given Mach number (M) can be achieved in the lower stratosphere (where the relative temperature, θ, is constant) depends fundamentally on the manner in which the engines' thrust decays with increasing altitude (see Section 8.6.5). The service ceiling, for many airplane types, increases almost linearly with reducing airplane weight (this is illustrated in Figure 17.3).

20.4.2   Speed Envelope

A speed envelope, presented on a chart of altitude versus airspeed or Mach number, provides a useful graphical representation of the domain in which the airplane can be safely operated. A generic speed envelope for a jet transport airplane is shown in Figure 20.6. The lower speed limit—defined by line A—represents the 1-g stall condition (i.e., V_{S1}). As this is an absolute aerodynamic boundary, rather than a safe operational limit, a second line—defined as B—is shown inside the envelope. Line B provides a maneuver margin from the absolute aerodynamic limits—at low Mach numbers this margin can be set equal to $1.23 \times V_{S1}$, and at high Mach numbers, it can be determined by the 1.3g buffet onset condition [10].

The rate of climb requirement of 300 ft/min can be used to establish a service ceiling for the airplane. As this corresponds to an optimum condition, flight speeds both slower and faster than the optimum speed will result in a reduced altitude—this is represented by line C in Figure 20.6. The 0 ft/min rate of climb limit—shown as line D—represents a theoretical, absolute altitude limit at which the airplane can maintain level flight (in other words, the maximum available thrust will be precisely equal to the airplane's drag at this altitude). The maximum altitude that can be achieved (for a given rate of climb) is not fixed—the ceilings reduce with increasing gross weight or increasing air temperature. Furthermore, the service ceiling and absolute ceiling depend on the performance characteristics of the engines. The maximum certified ceiling (which is set by structural design considerations—see Section 22.3.2), however, is constant for a given airplane type. In the illustration presented in Figure 20.6, the maximum certified ceiling

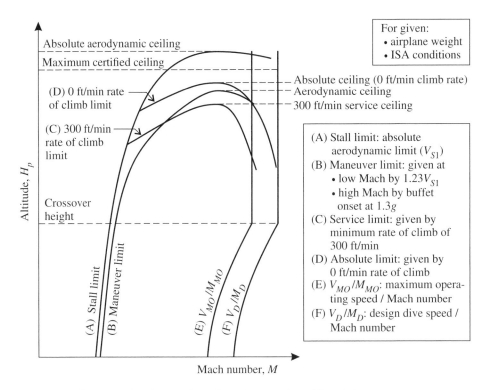

Figure 20.6 Illustrative altitude–speed envelope. *Source*: Adapted from [10].

has been arbitrarily shown above lines B, C, and D, which correspond to the aerodynamic and service ceilings (this can occur when the airplane has a high gross weight).

The high-speed limit, defined by line E, represents the certified maximum operating speed for the airplane. Below the crossover height (see Section 17.2), this limit is defined in terms of calibrated airspeed (V_{MO}) and above the crossover height as Mach number (M_{MO}). Line F represents the *design dive* speed (or design limit speed) for the airplane in calibrated airspeed (V_D) or Mach number (M_D). It is a certification requirement that all aircraft are safe to fly up to V_D/M_D. Specifically, it must be demonstrated that the airplane is controllable and free of flutter up to the declared design dive speed. As operation of the airplane at speeds greater than V_D/M_D can potentially result in structural failure, V_{MO}/M_{MO} is established such that there is an adequate margin (i.e., from V_D/M_D to V_{MO}/M_{MO}) to make it highly improbable that V_D/M_D is inadvertently exceeded. In compliance with certification criteria (see Section 20.7.5), M_D is typically set greater than M_{MO} by a margin of Mach 0.06 or Mach 0.07 [7, 10].

20.5 Key Regulatory Speeds

20.5.1 The 1-*g* Stall Speed (FAR 25 Amendment 25-108)

An important revision to the airworthiness standard for transport category airplanes FAR 25, as regards airplane performance, was implemented with Amendment 25-108 (in 2002). The FAA defined the reference stall speed as a speed not less than the 1-*g* stall speed—instead of the minimum speed obtained in a stalling maneuver, as had previously been used—and, in the process, harmonized the applicable regulations with those adopted in JAR 25, Change 15. Performance engineers now categorize jet transport airplanes into "pre-amendment 25-108" and "post-amendment 25-108" types, depending on their certification basis.[4] A subscript R was introduced to denote the new reference stall speed (i.e., V_{SR})—see Section 20.5.2 for details on the regulations. The absence of the letter R, as in V_S, is generally taken to mean the minimum steady flight speed at which the airplane is controllable.

This action was taken to provide for a consistent, repeatable reference stall speed. For many jet transport airplanes, the minimum speed attained in a stall occurs at a load factor (normal to the flight path) of less than one. Certification flight tests (conducted according to FAR/CS 25.103) require that the airplane decelerates at no more than 1 kt/s from a trimmed, straight and level, 1-*g* flight condition. The minimum speed is attained by the pilot continuing to pull the stick/yoke back deep into the stall. This process, however, can lead to difficulties in flight testing as inconsistent stall speeds can be obtained due to varying pilot technique (which affects the rate and magnitude of the nose-down pitch occurring in the stall). The 1-*g* stall speed, which can be more consistently determined in flight tests, corresponds to the lowest speed at which the aircraft can support its weight in 1-*g* flight.

The major issue at the time, however, was not the determination of the stall speed, *per se*; it was the fact that airworthiness regulations define several safety-critical operating speeds as multiples of the stall speed—for example, the nominal (or reference) approach speed had been defined as 1.3 times the stall speed (determined with the airplane in the approach configuration). New multiplying factors that would provide essentially the same operating speeds regardless of the basis used for determining the reference stall speeds had to be defined. A

---

4 Pre-amendment 25-108 category airplanes include: B707, B727, B737-100/200/Adv/300, B747-100/200/300, B757-200, DC-8, DC-9, and DC-10.
Post-amendment 25-108 category airplanes include: B737-400/.../900, B747-400/8, B757-300, B767-300/400, B777, B717, MD-11, MD-80, and A380 [5], as well as newer types, such as the B787 and A350.

survey of high-speed swept-wing transport-category airplanes indicated that the average load factor at the minimum speed obtained in the stalling maneuver was 0.88 [11]. This means that minimum speeds obtained in stall tests are, on average, equal to 94% of the corresponding 1-*g* stall speeds. The multiplying factors were thus reduced by approximately 6% with the implementation of FAR Amendment 25-108. For example, the regulation defining the nominal (reference) approach speed was changed from 1.3 times the stall speed to 1.23 times the 1-*g* reference stall speed.

An exception, however, was made for airplanes equipped with a "stick pusher" (i.e., a device that pushes the nose of the airplane down near the angle of attack for maximum lift) as unacceptably low operating speeds would have resulted from the use of the new multiplying factors. These airplanes would have been operated at speeds (and angles of attack) closer to the stick pusher activation speed than had previously been the case, thereby reducing safety margins. For these airplane types, the minimum speed obtained in the stalling maneuver was observed to be about 96–97% of the 1-*g* stall speed [11]. Therefore, to maintain equivalency in operating speeds, a supplementary requirement was introduced in the regulations, which stipulates that V_{SR} must not be less than the greater of 2 kt or 2% above the speed at which the device activates.

20.5.2   Reference Stall Speed (V_{SR})

The reference stall speed (V_{SR}) is defined in FAR/CS 25.103 [1, 2], with further details provided in FAA AC 25-7 [12] and AMC 25.103 [2]. It is used as a basis, or *reference*, to determine several operational speeds—for example, the takeoff safety speed (V_2) cannot be less than a fixed multiple of V_{SR} (either 1.08 or 1.13, depending on the airplane type).

The following is a reformatted, abridged extract from FAR/CS 25.103 [1, 2]:

(a) The reference stall speed (V_{SR}) is a calibrated airspeed that may not be less than a 1-*g* stall speed and is expressed mathematically as

$$V_{SR} \geq \frac{V_{C_{L,max}}}{\sqrt{n_{zw}}} \tag{20.8}$$

where $V_{C_{L,max}}$ is the calibrated airspeed obtained when the load-factor-corrected lift coefficient (i.e., $n_{zw} W/qS$) is first a maximum during the maneuver prescribed in paragraph (c) of this section; in addition, when the maneuver is limited by a device that abruptly pushes the nose down at a selected angle of attack (e.g., a stick pusher), $V_{C_{L,max}}$ may not be less than the speed existing at the instant the device operates;

and where: n_{zw} is the load factor normal to the flight path[5] at $V_{C_{L,max}}$; and
q is the dynamic pressure.
(b) $V_{C_{L,max}}$ is determined with:
   (i)   engines idling or, if that resultant thrust causes an appreciable decrease in stall speed, not more than zero thrust at the stall speed;
   (ii)   the airplane in other respects (e.g., flaps, landing gear, and ice accretions) in the condition existing in the test or performance standard in which V_{SR} is being used;
   (iii) the weight used when V_{SR} is being used as a factor to determine compliance with a required performance standard;

---

5 FAR/CS 25 [1, 2] use a convention by which the load factor normal to the flight path is identified by the subscripts *ZW*. Thus, the load factor n_{zw} is identical to n_z as used herein.

(iv) the CG position that results in the highest value of reference stall speed; and

(v) the airplane trimmed for straight flight at a speed not less than $1.13V_{SR}$ and not greater than $1.3V_{SR}$.

(c) Starting from the stabilized trim condition, apply the longitudinal control to decelerate the airplane so that the speed reduction does not exceed one knot per second.

(d) In addition to the requirements of paragraph (a) of this section, when a device that abruptly pushes the nose down at a selected angle of attack (e.g., a stick pusher) is installed, the reference stall speed, V_{SR}, may not be less than 2 kt or 2%, whichever is greater, above the speed at which the device operates.

20.5.3 Operational Limit Speeds

FAR 25 [1] and CS-25 [2] define several limiting speeds that need to be established—during the design and certification of a new airplane—to enable flight operations to be correctly planned and safely executed.[6] Several important speeds are briefly described below.

Flap Extended Speed [FAR/CS 25.1511]

V_{FE} is the maximum speed at which the flaps may be extended. It may not exceed the design flap speed V_F.

Landing Gear Extended Speed [FAR/CS 25.1515]

V_{LE} is the maximum airspeed at which the airplane may be flown with the landing gear fully extended.

Landing Gear Operating Speed [FAR/CS 25.1515]

V_{LO} is the maximum speed at which the landing gear may be operated (i.e., extended or retracted). If the extension speed is not the same as the retraction speed, the two speeds are designated as $V_{LO(EXT)}$ and $V_{LO(RET)}$, respectively.

Maneuvering Speed [FAR/CS 25.1511]

V_A is the design maneuvering speed (see also Section 20.7.5). The maneuvering speed may not exceed V_A.

Maximum Operating Limit Speed [FAR/CS 25.1505]

V_{MO} is the maximum operating speed and M_{MO} is the maximum operating Mach number. V_{MO}/M_{MO} may not be deliberately exceeded in any part of the flight regime (e.g., climb, cruise, descent). V_{MO}/M_{MO} must not be greater than the design cruise speed/Mach number, V_C/M_C, thus ensuring that it is highly improbable that the design dive speed, V_D/M_D, will be inadvertently exceeded in operations. (See also Section 20.7.5.)

Minimum Control Speed in the Air [FAR/CS 25.149]

V_{MC} is the minimum airspeed at which the airplane is deemed to be controllable in the air, while maintaining straight flight after failure of the critical engine. Maximum rudder deflection is required to compensate the large yaw moment resulting from the thrust asymmetry. Banking the airplane (the "dead" engine is raised) reduces the sideslip and permits the speed to be reduced; however, a maximum angle of bank of 5° is permitted for certification. (See also Section 10.2.2.)

---

6 A general list of definitions is provided by the FAA in reference [13] and by EASA in reference [14].

Minimum Control Speed on the Ground [FAR/CS 25.149]

V_{MCG} is the minimum airspeed at which the airplane is deemed to be controllable on the ground, with a lateral excursion of not more than 30 ft after failure of the critical engine. Steering is not used during certification tests; however, in normal operations, steering would be beneficial in controlling the airplane. (See also Section 10.2.2.)

Minimum Control Speed During Approach and Landing [FAR/CS 25.149]

V_{MCL} is the minimum airspeed at which the airplane is deemed to be controllable during approach and landing, maintaining straight flight with a maximum angle of bank of 5° after failure of the critical engine. V_{MCL-2} is the minimum control speed during approach and landing for airplanes with three or more engines, after failure of two engines.

Minimum Unstick Speed [FAR/CS 25.107]

V_{MU} is the airspeed at and above which the airplane can safely lift off the ground and continue the takeoff. The technique adopted to demonstrate this speed during certification requires that the pilot slowly rotate the airplane (while accelerating down the runway) until the rear fuselage strikes the ground (the fuselage is protected). At this angle of attack, the airplane is accelerated until liftoff occurs. Minimum unstick speeds are determined with all engines operating and also with one engine inoperative. (See also Section 10.2.2.)

Reference Stall Speeds [FAR/CS 25.103]

V_{SR} is the reference stall speed (see Section 20.5.2). The configuration of the airplane (e.g., flaps and landing gear positions) corresponds to the condition existing in the test or performance standard in which V_{SR} is being used. V_{SR_0} is the reference stall speed in the landing configuration.

Rough Air Speed and Rough Air Mach Number [FAR/CS 25.1517]

V_{RA}, which is the rough air speed, and M_{RA}, which is the rough air Mach number, are the recommended turbulence penetration speed and Mach number, respectively.

Takeoff V_1 Speed [FAR/CS 25.107]

V_1 is the speed at which action is taken to reject the takeoff (see Section 10.2.3).

Takeoff Safety Speed [FAR/CS 25.107]

V_2 is the takeoff safety speed (see Section 10.2.2).

20.6   Structural Design Loads and Limitations

20.6.1   Design Loads

The loads that an airplane will encounter during normal operation can be broken down into a number of categories—for example: flight loads (due to maneuvers or gusts); ground loads (due to taxiing, takeoff, or landing); powerplant loads (e.g., engine torque and thrust); pressurization of the fuselage; handling and impact loads (e.g., birdstrike); and inertia loads. In the design and certification of new aircraft, it is necessary to take into account all of these factors—and, in service, the airplane should be operated in a manner that does not impose loads on the airframe that exceed the design values. For each structural component, one of these loads will dominate—for example, for the primary structural elements of the wing, this will almost always be the flight loads. Flight loads, or air loads as they are often called, arise due to the aerodynamic forces that act on the airplane during flight. Significant flight loads arise during maneuvers and when flying through strong gusts of upward or downward moving air currents (e.g., in the vicinity of thunderstorms).

Limit loads, which are described in FAR/CS 25.301–307 [1, 2], are defined as the maximum loads that the airplane is expected to encounter in service. The following criteria must be met: (1) the structure shall be capable of supporting limit loads without suffering detrimental permanent deformation, and (2) elastic deformation of the structure due to limit loads shall not interfere with the safe operation of the airplane.

Ultimate loads, which are described in FAR/CS 25.301–307 [1, 2], are equal to the limit loads multiplied by a *factor of safety* (see Section 20.6.2). The airplane shall be capable of supporting ultimate loads without failure for at least three seconds (the three second limit does not apply to dynamic tests simulating actual load conditions). Failure of a structural element can be defined as the inability of the component to satisfactorily perform the task for which it was designed— this can be due to fracture or excessive deformation.

Generally speaking, commercial transport airplanes are designed to a positive limit load factor of no less than +2.5 and a negative limit load factor of −1.0 (see Section 20.7.3). An airplane designed to these values could potentially experience some inelastic structural deformation for maneuvers that exceed +2.5g (or −1.0g in the negative direction) and structural failure for maneuvers that exceed +3.75g (or −1.5g in the negative direction).

20.6.2 Factor of Safety

Unless otherwise specified within the regulations (i.e., FAR/CS 25 [1, 2]), a factor of safety of 1.5 is applied to limit loads. This factor of safety is intended to accommodate a large number of issues associated with the design, manufacture, and operation of aircraft, including the following:

(1) imprecision in the mathematical models used for aerodynamic or stress calculations;
(2) variations in physical properties of structural materials;
(3) variations in fabrication or inspection techniques;
(4) operation of the airplane beyond its flight limits (e.g., in an emergency); and
(5) extreme atmospheric conditions that produce gust velocities greater than those considered in the regulations.

The factor of safety is intended to insure against catastrophic failure from any of the above conditions. It would not be uncommon, however, for a single airplane within a fleet to exceed the design limit loads marginally during a lifetime of service. Although structural damage may occur (resulting in the need to repair or replace components), this should not prove to be serious from an airplane safety standpoint.

20.6.3 Flight Envelope Protection

The flight envelope protection feature of modern fly-by-wire (FBW) flight control systems prevents or restricts a pilot's ability to make control commands that would cause the airplane to exceed structural or aerodynamic design limits. The Airbus A320 was the first jet transport airplane to feature full flight envelope protection. Although the pilot can operate the airplane beyond flight envelope limits in a degraded control law on Airbus FBW airplanes, the flight envelope protection cannot be overridden completely. The Boeing approach is different—the flight crew can override flight envelope limits, for example, by using excessive force on the flight controls.

20.7   *V–n* Diagram (Flight Load Envelope)

20.7.1   Purpose of the *V–n* Diagram

The $V-n$ diagram is a design tool used by engineers to determine the flight loads to which the aircraft's structure has to be designed during the development of a new airplane. Consequently, it defines a set of limits—that is, an envelope—within which the airplane can be safely operated. There are two sources of flight loads (air loads) that need to be considered: *airplane maneuvers* and *atmospheric gusts*. The conventional method to construct a $V-n$ diagram is initially to separate the two factors, constructing one envelope for maneuvers (see Sections 20.7.2–20.7.5) and one for gusts (see Sections 20.7.6–20.7.8). The combined $V-n$ diagram is then constructed so as to encompass the extremes of the two component diagrams (see Section 20.7.9). An overview of the methodology for constructing $V-n$ diagrams is presented in this chapter. The corresponding regulations are given in FAR/CS 25.321–341 [1, 2] with supporting information on gust loads given in FAA Advisory Circular 25.341-1 [15] and CS-25 Book 2 AMC 25.341 [2].

A $V-n$ diagram is a graphical depiction of an airplane's limit load factor as a function of airspeed. By convention, the load factor (n) used to describe flight loads is the ratio of the net aerodynamic force acting normal to the longitudinal axis of the airplane (N) to the weight of the airplane (W). As the angle of attack (α) is usually defined with respect to the same datum, it follows that

$$n = \frac{N}{W} = \frac{L\cos\alpha + D\sin\alpha}{W} \tag{20.9}$$

where N is the net aerodynamic force acting normal to the longitudinal axis of
the airplane (typical units: N, lb); and
L is the net lift and D is the net drag (typical units: N, lb).

Note that this is not the same load factor introduced earlier, in Section 7.2.5, where the load factor n_z was defined normal to the flight path. Differences are negligible, however, for small angles of attack.

The velocity (V), which is shown on the abscissa, is always measured in equivalent airspeed (EAS). This is convenient as it eliminates the need for the loads to be calculated at each altitude of interest. Note that there is an inconsistency regarding the notation used for this discussion on flight loads: the subscript "*e*," which is used elsewhere in this book to denote EAS, has been omitted so that the notation used for flight and gust velocities in Section 20.7 is consistent with that of the regulations (i.e., FAR/CS 25.321–341 [1, 2]).

20.7.2   Flight Maneuvering Envelope

Jet transport airplanes are designed to accommodate symmetrical maneuvers (i.e., pull-up, push-over, and steady turn maneuvers) that generate significant flight loads on the aircraft structure. The flight maneuvering envelope, within which the airplane may be operated without exceeding the design limit loads, is illustrated in Figure 20.7. Note that load factors greater than one are associated with a nose-up pitching maneuver and load factors less than one with a nose-down pitching maneuver.

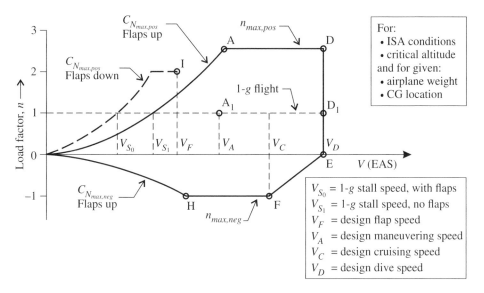

Figure 20.7 Typical maneuver $V-n$ diagram (maneuvering envelope).

The positive and negative limit load factors (i.e., $n_{max,pos}$ and $n_{max,neg}$) that define the upper and lower boundaries of the maneuvering envelope are set by the manufacturer; the selected values, however, must equal or exceed minimum values specified by the regulatory authorities (see Section 20.7.3).

At the left-hand boundary, the load factor is limited by the lift that the airplane is able to generate in positive and negative symmetrical maneuvers—in other words, the airplane will stall at the indicated combinations of load factor and airspeed. The stall speeds that establish the lower speed boundary are functions of the maximum positive and maximum negative normal force coefficients (i.e., $C_{N_{max,pos}}$ and $C_{N_{max,neg}}$)—this is discussed further in Section 20.7.4. Note that level-flight stall (i.e., when $n = 1$) represents a single point on the positive stall boundary. This 1-g stall speed (in the clean configuration, with engines set to idle) is called V_{S_1}. When $n > 1$ (e.g., in a banked turn), the airplane will stall at a higher speed than V_{S_1}, as shown in Figure 20.7. With the flaps extended, the stall boundary moves to the left, and the airplane may be safety operated at slower speeds than V_{S_1}. The 1-g stall speed with flaps fully extended is V_{S_0}.

The right-hand boundary is defined by the design dive speed, V_D (EAS), where the dynamic pressure acting on the airplane is a maximum. Note that the bottom right-hand corner of the envelope has been cropped from point F to point E. Under the regulations this is permissible for this class of airplane as large negative g loads do not occur at such high speeds under normal flight conditions.

20.7.3 Limit Maneuvering Load Factors

The requirements for the upper (i.e., positive g) and lower (i.e., negative g) maneuvering load factor limits are described in FAR/CS 25.337 [1, 2]:

(1) The positive limit maneuvering load factor may not be less than $2.1 + 24\,000/(W + 10\,000)$, where W is the design maximum takeoff weight (unit: lb); additionally, n may not be less than 2.5 and need not be greater than 3.8.

(2) The negative limit maneuvering load factor may not be less than -1.0 at speeds up to V_C and must vary linearly with speed from the value at V_C to zero at V_D.

20.7.4 Stall Boundaries

Stall speeds—for the purpose of constructing $V-n$ diagrams—are based on the maximum *normal force coefficient*, rather than the maximum lift coefficient (which was discussed in Section 20.2). The normal force coefficient (C_N) is defined in Section 3.4.6 as follows:

$$C_N = C_L \cos \alpha + C_D \sin \alpha \qquad \text{(Equation 3.24)}$$

The maximum value of this coefficient ($C_{N_{max}}$) for most applications, however, is not significantly different from $C_{L_{max}}$. Consequently, stall speeds based on the normal force coefficient do not vary significantly from stall speeds calculated using the lift coefficient (see Equation 20.3).

Stall speeds based on the maximum normal force coefficient are described by the following equation:

$$V_S = \sqrt{\frac{2nW}{\rho_0 S C_{N_{max}}}} \qquad \text{(where } V_S \text{ is in EAS)} \tag{20.10}$$

The 1-g stall speed (with flaps retracted) is thus

$$V_{S_1} = \sqrt{\frac{2W}{\rho_0 S C_{N_{max}}}} \qquad \text{(where } V_{S_1} \text{ is in EAS)} \tag{20.11}$$

By rearranging Equation 20.10, it is evident that the positive and negative stall boundaries on a $V-n$ diagram are parabolic functions of airspeed, for a given airplane weight. The load factor corresponding to the maximum positive normal force coefficient ($C_{N_{max,pos}}$) is given by the following equation:

$$n = \frac{\rho_0 S C_{N_{max,pos}}}{2W} V_S^2 \qquad \text{(where } V_S \text{ is in EAS)} \tag{20.12}$$

Similarly, the load factor corresponding to the maximum negative normal force coefficient ($C_{N_{max,pos}}$) is given by

$$n = \frac{\rho_0 S C_{N_{max,neg}}}{2W} V_S^2 \qquad \text{(where } V_S \text{ is in EAS)} \tag{20.13}$$

The speed at the upper left-hand corner of the $V-n$ diagram (sometimes called the *corner speed*) is generally the design maneuvering speed, V_A. Specific requirements for the V_A speed exist in FAR/CS 25 [1, 2] (see Section 20.7.5). The corner point corresponds to the slowest speed at which the maximum load factor (n_{max}) can be achieved. When the V_A speed corresponds to the corner speed, it is evident that

$$V_A = V_{S_1} \sqrt{n_{max,pos}} \qquad \text{(where } V_A \text{ is in EAS)} \tag{20.14}$$

20.7.5 Design Airspeeds

Several important *design speeds* are defined in FAR/CS 25.335 [1, 2]. These speeds, which are measured in EAS, are described in an abridged manner herein (for a complete description, the regulations should be consulted).

Design Cruising Speed (V_C) [FAR/CS 25.335(a)]

The minimum value of V_C must be sufficiently greater than the design speed for maximum gust intensity (V_B) to provide for inadvertent speed increases likely to occur as a result of severe atmospheric turbulence. Except as provided in FAR/CS 25.335(d), V_C may not be less than V_B + 1.32 U_{ref} (where U_{ref} is a reference gust velocity—see Section 20.7.7). However, V_C need not exceed the maximum speed in level flight at maximum continuous power for the corresponding altitude. Additionally, at altitudes where V_D is Mach limited, it is permissible to base V_C on a selected Mach number.

Design Dive Speed (V_D) [FAR/CS 25.335(b)]

V_D or M_D must be at least 1.25 times greater than V_C or M_C, or selected such that the minimum speed margin between V_C or M_C and V_D or M_D is the greater of the following values:

(1) the speed increase that arises when the airplane, initially in a condition of stabilized flight at V_C or M_C, is upset, flown for 20 seconds along a flight path 7.5° below the initial path, and then pulled up at a 0.5g acceleration increment; additional requirements regarding this maneuver exist for airplanes equipped with a high-speed protection function (as described in FAR/CS 25.335(b)(ii) [1, 2]);
(2) a speed margin sufficient to provide for atmospheric variations, instrument errors, and airframe production variations;
(3) the margin at altitudes where M_C is limited by compressibility effects must not be less than 0.07 Mach unless a lower margin is determined using a rational analysis that includes the effects of any automatic systems; in any case, the margin may not be reduced to less than 0.05 Mach.

Design Maneuvering Speed (V_A) [FAR/CS 25.335(c)]

V_A may not be less than $V_{S_1} \sqrt{n}$ where n is the positive limit maneuvering load factor at V_C and V_{S_1} is the 1-g stalling speed with flaps retracted. Both V_A and V_{S_1} must be evaluated at the design weight and altitude under consideration. Additionally, V_A need not be more than V_C or the speed at which the positive $C_{N_{max}}$ curve intersects the positive maneuver load factor line, whichever is less.

The significance of V_A in terms of flight control is that it is a design requirement that the stick/yoke can be fully displaced at this speed so as to generate an extreme nose-up pitching acceleration (full deflections of the ailerons to the stops are also possible at V_A)—hence the term *maneuvering* speed.

Design Speed for Maximum Gust Intensity (V_B) [FAR/CS 25.335(d)]

The minimum value of V_B is given by

$$V_B = V_{S_1} \sqrt{1 + \frac{K_g U_{ref} V_C a}{498 \left(\frac{W}{S} \right)}} \tag{20.15}$$

where V_{S_1} is the 1-g stalling speed in EAS based on the maximum normal force coefficient with flaps retracted (unit: kt);
K_g is the gust alleviation factor—see Equation 20.19;
U_{ref} is the reference gust velocity at V_C in EAS (unit: ft/s)—see Section 20.7.7;
V_C is the design cruise speed in EAS (unit: kt);
a is the slope of the normal force coefficient versus angle of attack curve (unit: rad^{-1}); and
W/S is the wing loading (unit: lb/ft^2).

Additionally, V_B need not be greater than V_C at altitudes where V_C is limited by Mach number. At these altitudes, V_B may be chosen to provide an optimum margin between low- and high-speed buffet boundaries.

Design Wing-Flap Speeds (V_F) [FAR/CS 25.335(e)]

The design wing-flap speed for each wing-flap position must be sufficiently greater than the operating speed recommended for the corresponding stage of flight to allow for probable variations in control of airspeed and for transition from one flap position to another. However, if the airplane is equipped with an automatic flap positioning or load-limiting device, the design flap speeds are those programmed or allowed by the device. In any event, V_F may not be less than (i) 1.6 V_{S_1} with the flaps in the takeoff position at MTOW; (ii) 1.8 V_{S_1} with the flaps in the approach position at MLW (maximum landing weight); and (iii) 1.8 V_{S_0} at MLW (where V_{S_0} is the stall speed with the flaps in the landing position).

20.7.6 Airplane Design for Gust Loads

FAR/CS 25.341(a) [1, 2] requires that loads on each part of the structure be determined by dynamic analysis and that the analysis "must take into account unsteady aerodynamic characteristics and all significant structural degrees of freedom including rigid body motions." An analysis of the time-dependent response of an airplane as it flies through a strong gust is nontrivial. The induced loads depend on several factors, including (1) flight speed, (2) velocity profile of the gust, (3) aerodynamic and structural design features of the airplane (e.g., wing lift versus angle of attack relationship, wing torsional and bending stiffness, and mass distribution), and (4) any onboard control system designed to alleviate the effect of gusts.

As a conventional airplane (with aft tailplane) enters an up-gust, the first effect is to increase the effective wing angle of attack and hence impart an upward acceleration on the airplane. This is accompanied by a pitching acceleration (as the wing lift does not act through the airplane's CG). After a short time lag—dependent upon the airplane's size and speed—the tailplane enters the gust. This results in a small increment to the vertical acceleration, but it also induces a nose-down pitching moment that will tend to reduce the wing's angle of attack and, as a consequence, the vertical acceleration. Finally, the level attitude is restored and the vertical acceleration—and gust-induced loading on the airplane structure—is reduced to zero.

The flexibility of the wing structure significantly affects the gust-induced loading on the airframe. For example, as the airplane enters the gust, the wing will bend upwards and simultaneously twist. Considering a single streamline across the upper wing surface of a conventional swept wing, it is seen that the trailing edge will deflect more than the leading edge—this has the effect of reducing the lift-curve slope and thus reducing the magnitude of gust-induced loads. Furthermore, inertia forces also influence the dynamic response of the wing.

20.7.7 Gust-Induced Loads

An appreciation of the effect of discrete gusts on an airplane can be obtained by considering the airplane in horizontal flight (where $n = 1$) at speed V (EAS) encountering an idealized sharp-edged vertical gust of velocity U (EAS), as illustrated in Figure 20.8. The resulting change in load factor (Δn) can be estimated by considering the change in angle of attack ($\Delta \alpha$) and by assuming quasi-steady conditions:

$$\Delta \alpha = \tan^{-1} \frac{U}{V} \cong \frac{U}{V} \tag{20.16}$$

where $\Delta \alpha$ is the change in angle of attack (unit: rad)

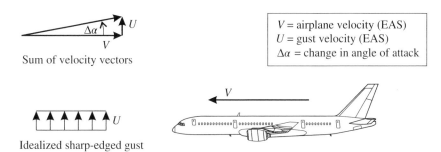

Figure 20.8 Idealized sharp-edged gust encounter.

Using this result, the change in normal force (ΔN) can be approximated as follows:

$$\Delta N = \frac{1}{2}\rho_0 V^2 S(a\Delta\alpha) \cong \frac{1}{2}\rho_0 SUV\,a \tag{20.17}$$

where $a = \mathrm{d}C_N/\mathrm{d}\alpha$

Thus $\quad \Delta n = \dfrac{\Delta N}{W} = \dfrac{\rho_0 UV\,a}{2\left(\dfrac{W}{S}\right)}$ $\tag{20.18}$

Equation 20.18 leads to an interesting observation regarding airplane design: the change in load factor due to the gust (Δn) is directly proportional to the slope of the normal force coefficient curve (a) and inversely proportional to the wing loading (W/S). This explains why airplanes with a low wing loading tend to "bounce around" in gusty conditions and are often uncomfortable to fly in turbulent air.

The assumption in the derivation of Equation 20.18 is that the airplane instantly encounters the gust and that the gust instantly affects the entire airplane. This sharp-edged effect, however, is unrealistic—gusts tend to follow a progressive intensity increase as the airplane penetrates the gust. There is also a time lag in the build-up of lift in response to the gust. These factors substantially reduce the vertical acceleration experienced by the airplane, compared to the idealized sharp-edged assumption. An approach that has historically been used to account for this lack of precision is based on a dimensionless *gust alleviation factor*, K_g, which is defined as follows:

$$K_g = \frac{0.88\mu_g}{5.3 + \mu_g} \tag{20.19}$$

And where the dimensionless *mass parameter*, μ_g, is defined as

$$\mu_g = \frac{2}{\rho\bar{c}ag}\left(\frac{W}{S}\right) \tag{20.20}$$

where ρ is the air density (unit: slug/ft^3);
$\bar{c}$ is the wing mean geometric chord (unit: ft);
a is the slope of the normal force coefficient versus angle of attack curve (unit: rad^{-1}); and
g is the acceleration due to gravity (unit: ft/s^2).

The gust speed U in Equation 20.18 can be replaced by the product of K_g and a *derived* reference gust speed, U_{ref} (EAS). With this idealization, the gust-induced load factor can be expressed as a linear function of airplane speed V (EAS), that is,

$$n = 1 + \Delta n = 1 + \frac{\rho_0 K_g U_{ref} V a}{2\left(\dfrac{W}{S}\right)} \tag{20.21}$$

Equation 20.21 is an expression that is familiar to structural engineers conducting loads analyses on aircraft. It is often written in the following form (e.g., in FAR Part 23 [16]; see also the definition of V_B in Section 20.7.5):

$$n = 1 + \frac{K_g U_{ref} V a}{498\left(\dfrac{W}{S}\right)} \tag{20.22}$$

When Equation 20.21 is expressed in the form given by Equation 20.22, particular units must be used: U_{ref} must be in ft/s, V in kt, a in rad^{-1}, and W/S in lb/ft^2. The constant, which has a value of 498, is evaluated as follows: $498 = 2/(0.002377 \times 1.688)$, where 0.002377 is the ISA sea-level air density in slug/ft^3 and the value 1.688 converts kt to ft/s. Equation 20.21, which was published in 1954 in NACA report 1206 by Pratt and Walker [17], was the basis for airplane gust load analysis for many years (and is still be used for general aviation [16]), but for large jet transport airplanes, this simple approach is no longer adequate.

Current requirements for design limit gust loads are defined in FAR/CS 25.341(a) [1, 2] and are based on an atmospheric model of single discrete extreme turbulence events.[7] The gust model assumes that gusts have a $1 - $ cosine velocity profile normal to the flight direction (Figure 20.9). The gust velocity U (EAS) is defined as a function of the gust penetration distance (s), as follows:

$$U = \frac{U_{ds}}{2}\left(1 - \cos\frac{\pi s}{H}\right) \tag{20.23}$$

where U_{ds} is the design gust velocity in EAS (unit: ft/s)—see Equation 20.24;
s is the gust penetration distance (unit: ft); and
H is a gust gradient distance (unit: ft).

and $\quad U_{ds} = U_{ref} F_g \left(\dfrac{H}{350}\right)^{1/6} \tag{20.24}$

where U_{ref} is the reference gust velocity in EAS (unit: ft/s); and
F_g is the flight profile alleviation factor (dimensionless).

The *gust gradient distance* (H) is that distance over which the gust velocity increases to its maximum value (as illustrated in Figure 20.9). A sufficient number of gust gradient distances in the range 30 to 350 ft (9.1 to 107 m) must be investigated to find the critical response for each load condition.

Reference gust velocities (U_{ref}), which are defined as a function of altitude, are derived effective gust velocities representing gusts expected to be encountered once in 70 000 flight hours (it is assumed that the airplane is flown 100% of the time at the particular altitude). The following

---

7 The history of the evolution of gust load design requirements for airplane certification is described by Fuller [18].

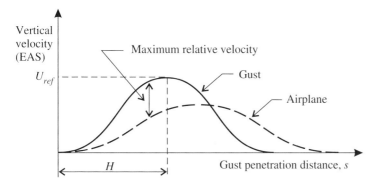

Figure 20.9 Idealized gust and airplane velocities.

values of U_{ref} must be considered: (1) from V_B to V_C, $U_{ref} = 56$ ft/s (EAS) at sea level and U_{ref} is permitted to reduce linearly with altitude to 44 ft/s (EAS) at 15 000 ft and then to 20.86 ft/s (EAS) at 60 000 ft; and (2) at V_D, U_{ref} must be half that which applies at V_C. Positive and negative gust velocities are assumed to have the same magnitude at each altitude.

The *flight profile alleviation factor* (F_g), which has a minimum value at sea level and increases linearly to 1.0 at the airplane's certified maximum altitude, accounts for the probability of the airplane flying at any given altitude. Procedures for determining F_g are given in FAR/CS 25.341(a)(6) [1, 2].

The requirement for manufacturers to determine gust load factors for the gust profile described above at the defined reference speeds for airplane certification is stipulated in FAR/CS 25.341 [1, 2]. Such analyses are complicated and require a sophisticated representation of the airplane's response characteristics to atmospheric turbulence—a description of the methodologies used is beyond the scope of this book (consult FAA Advisory Circular 25.341-1 [15], CS-25 Book 2 AMC 25.341 [2], and references [19–22] for further information on the topic).

20.7.8 Gust Load Diagram

A gust $V-n$ diagram is illustrated in Figure 20.10.

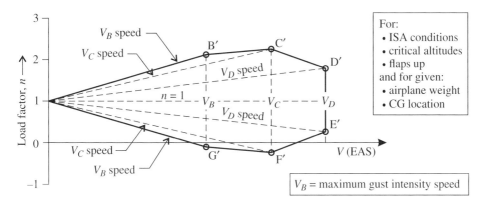

Figure 20.10 Gust $V-n$ diagram (gust load envelope).

In summary, the procedure for constructing the diagram is as follows:

(1) Incremental gust load factors (Δn) for positive (upward) and negative (downward) gusts are determined for the three specified design speeds, that is,
- V_B, the design speed for maximum gust intensity;
- V_C, the design cruising speed; and
- V_D, the design dive speed.

(2) The upper and lower limits (corresponding to positive and negative gusts) for each speed are determined from the equation $n = 1 + \Delta n$. This assumes that the airplane is in a 1-g level-flight condition when the gust is encountered.

(3) Straight lines are drawn between the plotted points to give the gust $V-n$ diagram.

20.7.9   Combined *V–n* Diagram

The combined $V-n$ diagram is obtained by superimposing the gust $V-n$ diagram over the maneuver $V-n$ diagram and considering the extremes of the composite envelope. This is illustrated in Figure 20.11—a combined $V-n$ diagram based on Figures 20.8 and 20.10. For certain airplane designs, the maximum gust load factor will be greater than the limit load factor, $n_{max,pos}$. In such cases, the manufacturer may decide to expand the envelope by setting the limit load factor equal to the peak gust load factor; consequently, the aircraft would be designed to the higher load factor value for all speeds from V_A to V_D.

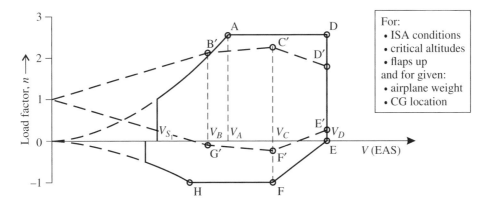

Figure 20.11 Illustrative combined $V-n$ diagram (flight load envelope).

References

1 FAA, *Airworthiness standards: Transport category airplanes*, Federal Aviation Regulation Part 25, Amdt. 25-143, Federal Aviation Administration, Washington, DC, June 24, 2016. Latest revision available from www.ecfr.gov/ under e-CFR (Electronic Code of Federal Regulations) Title 14.

2 EASA, *Certification specifications and acceptable means of compliance for large aeroplanes*, CS-25, Amdt. 18, European Aviation Safety Agency, Cologne, Germany, June 23, 2016. Latest revision available from www.easa.europa.eu/ under Certification Specification.

3 Obert, E., *Aerodynamic design of transport aircraft*, IOS Press, Amsterdam, the Netherlands, 2009.

4 Shevell, R.S., *Fundamentals of flight*, 2nd ed., Prentice Hall, Englewood Cliffs, NJ, 1989.

5 Boeing, "Stall speed, stall warning & limitations on maneuvering flight," Flight Operations Engineering, Boeing Commercial Airplanes, Seattle, WA, 2009.

6 Blake, W. and Performance Training Group, "Jet transport performance methods," D6–1420, Flight Operations Engineering, Boeing Commercial Airplanes, Seattle, WA, Mar. 2009.

7 Rosay, J., "High-altitude manual flying," *Safety first*, Product Safety Department, Airbus S.A.S., Blagnac, France, Iss. 20, pp. 37–45, July 2015.

8 Polmar, N., *Spyplane: The U-2 history declassified*, MBI Publishing Co., Osceola, WI, 2001.

9 Airbus, "Getting to grips with aircraft performance," Flight Operations Support and Line Assistance, Airbus S.A.S., Blagnac, France, Jan. 2002.

10 De Baudus, L. and Castaigns, P., "Control your speed… in cruise," *Safety first*, Product Safety Department, Airbus S.A.S., Blagnac, France, Iss. 21, pp. 6–21, Jan. 2016.

11 FAA, "Proposed revisions to advisory circular 25-7A, Flight test guide for certification of transport category airplanes," Federal Register Vol. 68, No. 1, pp. 149–154, Federal Aviation Administration, Washington, DC, Jan. 2, 2003.

12 FAA, "Flight test guide for certification of transport category airplanes," Advisory Circular 25-7C, Federal Aviation Administration, Washington, DC, Oct. 16, 2012. Available from www.faa.gov/.

13 FAA, *Definitions and abbreviations*, Federal Aviation Regulation Part 1, Amdt. 1-10, Federal Aviation Administration, Washington DC, Mar. 29, 1966. Latest revision available from www.ecfr.gov/ under e-CFR (Electronic Code of Federal Regulations) Title 14.

14 EASA, *Definitions and abbreviations used in certification specifications for products, parts and appliances*, CS-Definitions, Amdt. 2, European Aviation Safety Agency, Cologne, Germany, Dec. 23, 2010. Latest revision available from www.easa.europa.eu/ under Certification Specification.

15 FAA, "Dynamic gust loads," Advisory Circular 25.341-1, Federal Aviation Administration, Washington, DC, Dec. 12, 2014.

16 FAA, *Airworthiness standards: Normal, utility, acrobatic, and commuter category airplanes*, Federal Aviation Regulation Part 23, Amdt. 23-62, Federal Aviation Administration, Washington, DC, Dec. 2, 2011. Latest revision available from www.ecfr.gov/ under e-CFR (Electronic Code of Federal Regulations) Title 14.

17 Pratt, K.G. and Walker, W.G., "A revised gust-load formula and a re-evaluation of V-G data taken on civil transport airplanes from 1933 to 1950," NACA Report No. 1206, National Advisory Committee for Aeronautics, Langley, VA, 1954.

18 Fuller, J.R., "Evolution of airplane gust loads design requirements," *Journal of Aircraft*, Vol. 32, Iss. 2, 1995.

19 Hoblit, F.M., *Gust loads on aircraft: Concepts and applications*, AIAA Education Series, American Institute of Aeronautics and Astronautics, Reston, VA, 1988.

20 Lomax, T.L., *Structural loads analysis for commercial aircraft: Theory and practice*, AIAA Education Series, American Institute of Aeronautics and Astronautics, Reston, VA, 1996.

21 Howe, D., *Aircraft loading and structural layout*, AIAA Education Series, American Institute of Aeronautics and Astronautics, Reston, VA, 2004.

22 ESDU, "An introduction to rigid aeroplane response to gusts and atmospheric turbulence," Data item 04024, Amdt. B, IHS ESDU, 133 Houndsditch, London, UK, June 1, 2010.

21

Noise and Emissions

21.1 Introduction

The key environmental issues regarding the operation of jet transport airplanes—noise and emissions—are discussed in this chapter. Noise (introduced in Section 21.2) may be regarded as an excessive or unwanted sound that causes annoyance to the hearer. Aircraft noise is heard inside the aircraft (by passengers and by the crew) and outside the aircraft by people associated with aircraft operations (e.g., ground crew, airport operations staff, and embarking or disembarking passengers) and, of course, by the general public who live or work near to airports. Standard measures of airplane noise—for certification and operation—are described in this chapter (see Section 21.3), along with the performance implications of noise abatement procedures (Section 21.4).

The impact of jet transport airplane operations on the global atmosphere is of growing concern to aviation regulatory authorities, governments, and environmental scientists worldwide. In Section 21.5, an introduction to the principal emissions and their environmental impact is presented. This is followed by a short discussion in Section 21.6 on mitigation measures, including environmental protection standards, international agreements, and market-based measures.

21.2 Airplane Noise

21.2.1 Impact of Airplane Noise

It has long been recognized that aircraft noise is both an emotive and a subjective issue. It is emotive as aviation enthusiasts are less likely to be irritated by the noise than the general public; nonetheless, most people are aggravated by any extended exposure to aircraft noise. It is subjective as not everyone responds to sound in the same way. Whether a particular sound is aggravating depends on several factors—for example: frequency, intensity, background noise, duration, repetition, and suddenness. It is a combination of these factors and the individual's perception of noise that determines the annoyance level of that sound.

The current approach to the management of aircraft noise is the result of more than 50 years of effort. The International Civil Aviation Organization (ICAO) has played a central role in this process. The approach consists of four principal elements: reducing noise at source (i.e., making airplanes quieter), land-use planning and management (of airports and surrounding areas), noise abatement operational procedures, and operating restrictions (e.g., night-time curfews, quotas on aircraft movements) [1].

Performance of the Jet Transport Airplane: Analysis Methods, Flight Operations, and Regulations, First Edition.
Trevor M. Young.
© 2018 John Wiley & Sons Ltd. Published 2018 by John Wiley & Sons Ltd.

21.2.2 Human Perception of Noise

The human ear is capable of hearing sounds in the frequency range of 20 Hz to about 20 kHz, with peak sensitivity in the range of 2 to 4 kHz and a relatively poor sensitivity in the low frequencies [2]. Naturally, there is a substantial variation in any population group. Humans can discern a very wide range of sound levels from very loud to very quiet—the ratio of the power in a sound wave that causes hearing damage to that corresponding to the faintest sounds that can be heard is of the order of 10^{12}. Interestingly, people perceive the relative loudness of sounds in a manner that resembles a logarithmic scale. For these reasons, the decibel (dB) scale, which is a dimensionless logarithmic ratio, is the standard means of representing sound levels. The *sound pressure level* (SPL) is defined as follows:

$$L_p = 10 \log_{10} \left(\frac{p}{p_0} \right)^2 = 20 \log_{10} \left(\frac{p}{p_0} \right) \tag{21.1}$$

where L_p is the sound pressure level (unit: dB);
p is the sound pressure (unit: N/m^2); and
p_0 is the reference sound pressure of 2×10^{-5} N/m^2.

The numeral 10 in Equation 21.1 converts 1 bel (B) to 10 decibel (dB). The reference sound pressure of 2×10^{-5} N/m^2 is based on the nominal hearing threshold of a young person with perfect hearing—this corresponds to 0 dB. On the upper end of the scale, an impulsive noise exposure with a peak SPL of 140 dB is considered dangerous [2].

Noise levels—say, from several different sources—are added logarithmically, not linearly. The total, for n noise sources, can be computed using the following equation:

$$L_p = 10 \log_{10} \sum_{i=1}^{n} 10^{L_{p,i}/10} \tag{21.2}$$

where $L_{p,i}$ are the sound pressure levels of the component sources (unit: dB).

For example, if two 100 dB noise sources combine, the resultant noise level will be 103 dB. The progressive tightening of jet engine noise regulations (discussed later in Section 21.3) over the past few decades should be viewed in this light: a 3 dB noise reduction may not appear to be a lot, but, in fact, it represents cutting the noise level by almost exactly one-half.

21.2.3 Perceived Noise Level

The human ear displays different sensitivity to different frequencies—as a consequence, the criterion that has been established to assess aircraft noise is based on a metric that takes into account this non-linear hearing response. Experimental data were used to establish relationships between measured sound pressure levels and the human perception of sound. Contours of equal *perceived* noisiness, based on the collective responses of individuals to sounds of varying frequency and intensity, were used to formulate a series of weighting functions, or scales. These relationships are used to convert a measured noise event, consisting of different frequencies, to a *perceived noise level* (PNL) in decibels, which is indicated as PNdB. Basically, the measured event is divided into frequency bands (the audible frequency range is divided into 24, 1/3 octave bands). Each frequency band is weighted according to its perceived noise level using the Noy rating scheme, which allocates a greater weighting to the more sensitive, higher frequencies detectable by the human ear.

A refinement of the PNL measure is the *effective perceived noise level* (EPNL), which has units of EPNdB. The EPNL is the standard measure of aircraft noise for certification; it is also used for certain operational purposes (e.g., local authority restrictions on aircraft movements). The EPNL is obtained from instantaneous PNL values, corrected for spectral irregularities (which take into account pure tones) and the duration of the sound. The use of EPNL enables an objective evaluation of the subjective effect of aircraft noise to be determined from physical measurements of the spectral and temporal sound pressure level variations that are associated with aircraft operations. The procedure for determining EPNL values from measured data is detailed in ICAO Annex 16 Vol. I [3] and also in certain regulatory documents such as FAR Part 36, Appendix A [4].

Another scale that is often used for aircraft noise studies is the *A scale*; the sound level in dB on this scale is written as dB(A) or, alternatively, as dBA. The A scale is one of several standardized weighting scales [5] that are applied to instrument-measured sound levels, for a wide range of applications, to account for the relative loudness of a noise as perceived by the human ear. Airplane design requirements sometimes include noise limits in dB(A), which are supplemental to the certification requirements given in EPNdB.

21.2.4 Sources of Noise

Airplane noise is a mix of two very different types of noise: *tone* noise (which has a single frequency) and *broadband* noise (which has many frequencies). Tone noise can resemble a whistle, whereas broadband noise is more like a hiss (the noise associated with the pressure fluctuations on the surface of a moving vehicle is an example of broadband noise). Rotating assemblies, such as engine spools, produce tone noise (the frequency depends on rotational speed). A noise that has evenly and closely spaced tones sounds like a buzz (which can resemble a circular saw cutting wood). This type of noise is typically heard by passengers seated in the forward part of the cabin ahead of the engines during takeoff when the fan blades reach supersonic speeds. Shock waves form on the blades. As the shocks are not identical, they propagate out of the engine air inlet at varying speeds and merge—the resulting waveform consists of pure tones at multiples of the shaft rotation frequency.

The main contributors to jet engine noise are the exhaust jet, fan, combustor, and turbine. Jet noise in the exhaust—which is a big contributor during takeoff, but a small contributor during landing—results from the shearing action of adjacent airflows. A turbojet engine produces a very high velocity jet and a considerable amount of noise, whereas a high bypass ratio turbofan produces much less noise. By mixing the core and bypass airflows, the velocity of the exhaust gas is significantly reduced. The dramatic reduction in airplane noise levels that occurred during the 1970s and 1980s owed a great deal to the increase in bypass ratio of the new engines introduced during that time.

The engine fan produces a complex noise pattern comprising both tone and broadband components. The noise radiates from the engine after passing forwards through the inlet and rearwards through the bypass duct. Careful design of the fan geometry and the use of sophisticated acoustic linings in the inlet have been very successful in reducing fan noise. The HP and IP turbines and the combustor are traditionally less important sources of noise, but they are becoming more important as the noise from other sources is reduced.

The airframe itself produces a significant amount of noise, and this is becoming a matter of increasing concern as powerplants become progressively quieter. The noise comes from vortex shedding, turbulence interaction, and boundary layer pressure fluctuations. Of particular significance is the noise generated by the flaps, slats, and landing gear, as this is heard during takeoff and landing. Low-frequency noise is generated by airflow over hydraulic lines, wiring, and other small components that are exposed when the undercarriage is extended and high lift devices deployed.

21.3 Noise Regulations and Restrictions

21.3.1 Historical Overview

The ICAO Council formally adopted a noise standard for subsonic jet airplanes in 1971. The standard, which was published as Chapter 2 of ICAO Annex 16 (which now carries the title *Environmental Protection*), became applicable in January 1972. The Federal Aviation Administration (FAA) had introduced similar, but not identical, noise regulations in 1969 for transport category airplanes—these certification requirements were stipulated in the newly introduced FAR Part 36 (Noise Standards: Aircraft Type and Airworthiness Certification). Jet transport airplanes certificated to this standard include the DC-8, DC-9, B747-100 types, and certain models of the A300, for example [6]. A phased timetable for the removal from service of first-generation, noisy jet airplanes was then implemented.

Over the decades that followed, ICAO's Committee on Aircraft Noise (CAN) and its successor, the Committee on Aviation Environmental Protection[1] (CAEP), periodically recommended additions and modifications to the ICAO noise standard. These changes applied to measurement techniques, data analysis procedures, and—most importantly—they introduced more stringent noise limits for the certification of new aircraft.

The initial ICAO noise standard was superseded in 1977. The new standard—which was published as Chapter 3 of ICAO Annex 16—applied to subsonic jet airplanes, including derived versions, for which an application for a type certificate was submitted on or after October 6, 1977 [3]. The FAA, which had tightened its initial certification rules in 1975, introduced the term *stage* in 1977 into FAR Part 36 to differentiate between the initial (i.e., Stage 2) and revised (i.e., Stage 3) standards [7]. Since then, the term *chapter* has been used for ICAO noise standards and the term *stage* for the corresponding FAA regulations.

Examples of Chapter 3/Stage 3 airplane types include the B737-300…/900 series, B757, B767, B777-200/300, A330-300, A340-300, and certain models of the A319/A320/A321 family [6]. The introduction of Chapter 3/Stage 3 requirements was accompanied by agreed timetables—issued in the early 1990s by ICAO, FAA, and European authorities[2]—for the phased withdrawal from service of non-complaint airplane types. In broad terms, this meant that Chapter 2/Stage 2 airplane types could not be operated in affected counties after 1999 (according to the FAA) or 2002 (according to ICAO and the EEC).

In 2001, following a recommendation by the CAEP, the ICAO Council adopted a new noise standard, which is defined in Chapter 4 of ICAO Annex 16 [3]. The standard applies to subsonic jet airplanes, including derived versions, for which the application for a type certificate was submitted on or after January 1, 2006. Examples of airplane types that meet this standard include the B787, A350, and A380. Subsequently, certain manufacturers, for commercial reasons, elected to recertify some of their existing Chapter 3/Stage 3 airplane types to the new Chapter 4/Stage 4 standard.

---

1 The Committee on Aviation Environmental Protection (CAEP) is an ICAO technical committee that reports to the ICAO Council (see Section 23.2). Formal CAEP meetings, which take place every three years, make recommendations to the ICAO Council on matters relating to aircraft noise and emissions. Adopted recommendations are published as SARPs (Standards and Recommended Practices) in ICAO Annex 16.
2 The EEC (European Economic Community), which was the precursor to the European Union, and the ECAC (European Civil Aviation Conference) imposed a different timetable to that of the FAA.

At the ninth meeting of the CAEP (i.e., CAEP/9), in 2013, the committee recommended an even more stringent airplane noise certification standard [8]. This was adopted the following year by the ICAO Council. The new requirements, which are known as Chapter 14, have a phased introduction from 2017 to 2020 depending on airplane weight.[3] The requirements will be embodied in FAR 36 (as Stage 5) by the FAA and in CS-36 by EASA.

21.3.2 Noise Regulations

Noise standards are defined by ICAO in Annex 16, Vol. I [3] and adopted into national legislation by ICAO member states, sometimes with minor variations. Two widely used examples are FAR Part 36 [4] and EASA CS-36 [9], which are the noise certification regulations/standards for jet transport airplanes in the US and Europe, respectively.

To demonstrate compliance with these regulations, noise measurements are taken at ground level at three locations near to the runway (see Figure 21.1): two for takeoff (i.e., the *lateral* and *flyover* locations) and one for landing (i.e., the *approach* location) during a series of prescribed takeoffs and landings. The lateral reference point is 450 m to the side of the runway centerline and positioned where the noise is greatest during takeoff. The flyover reference point is 6500 m from the runway threshold (i.e., the start of the takeoff roll) on the extended runway centerline, and the approach reference point is 2000 m before the runway threshold on the extended runway centerline (the touchdown point is 300 m beyond the threshold). Noise measurements are affected by meteorological conditions; it is thus required that there is zero precipitation during testing, and limits are placed on the air temperature, relative humidity, and maximum wind speed. The measurements are corrected for deviations from the desired flight path, atmospheric absorption, and Doppler shifting.

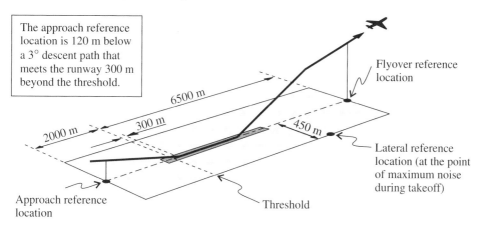

Figure 21.1 Noise measurement locations, as defined by ICAO [3].

The *ICAO Chapter 3* noise limits for two-, three-, and four-engine airplanes are illustrated in Figure 21.2. The following conditions must be met if the measured noise levels are exceeded:

(1) The sum of the exceedance(s) may not be greater than 3 EPNdB.
(2) The exceedance at any single point may not be greater than 2 EPNdB.
(3) The exceedance(s) must be offset by a corresponding amount at another point or points.

---

3 The requirements are known as *Chapter 14*, as chapters 5 to 13 of ICAO Annex 16 Vol. 1 had been allocated to noise standards for other aircraft categories.

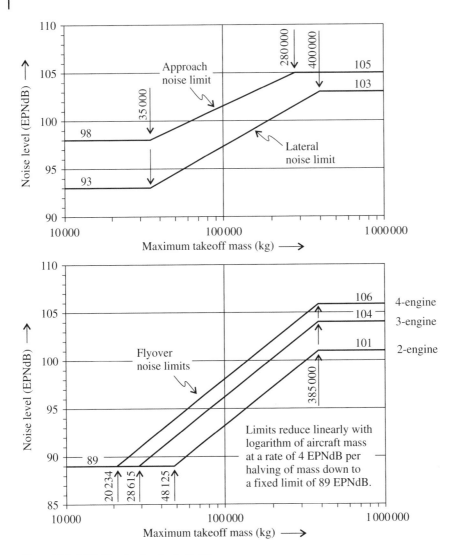

Figure 21.2 ICAO Chapter 3 noise limits [3].

Note that the permissible limits depend on the weight of the aircraft (heavier aircraft are permitted to make more noise within defined upper and lower weight boundaries) and that the specified noise–weight relationships are logarithmic, not linear. In the case of the fly-over measurement, consideration is given to the number of engines: the noise limits for two-engine airplanes are more stringent than those for three-engine airplanes, which, in turn, are more stringent than those for four-engine airplanes. This apparent inconsistency is designed to account for the fact that twin-engine airplanes, with all engines operating, have a higher rate of climb than comparable three- or four-engine airplanes at similar conditions and take-off weight. The one-engine-inoperative takeoff climb requirements for airplane certification (described in Section 10.6) effectively mean that twin-engine airplanes produce more thrust, at similar all-engine-operating takeoff conditions, compared to three- or four-engine airplanes. Consequently, as twin-engine airplanes can climb more quickly, their noise certification limits are set lower.

The *ICAO Chapter 4* noise limits are stricter. With respect to the relevant Chapter 3 noise limits, the following criteria must be met [3]:

(1) The limit at each certification point may not be exceeded (i.e., tradeoffs between points are not permitted).
(2) The cumulative margin at the three certification points must be at least 10 EPNdB (i.e., the sum of the lateral, flyover, and approach noise levels must be at least 10 EPNdB below the summed Chapter 3 limits).
(3) The minimum margin at any two certification points must be at least 2 EPNdB.

The *ICAO Chapter 14* noise limits are stricter still. The requirement under the CAEP recommendation (i.e., CAEP/9) is for a cumulative noise reduction of 7 EPNdB with respect to the ICAO Chapter 4 standard (i.e., 17 EPNdB reduction with respect to Chapter 3) [8].

21.3.3 Local Authority Restrictions

The local noise restrictions enforced at many airports worldwide take several different forms: some authorities enforce more stringent rules than the ICAO requirements,[4] whereas others link landing fees to defined noise levels and fine operators if the limits are exceeded.[5] Widely adopted are nighttime restrictions. Unlike the ICAO rules, it is usual that no alleviation for the weight of the airplane is allowed. This can restrict the operation of certain large airplane types to daylight hours at these airports.

Various techniques are used to monitor airplane noise in the vicinity of airports—for example, the noise of a single event (i.e., an airplane taking off or landing) can be evaluated using one of the following methods:

(1) Measurement of the peak (i.e., greatest) flyover noise level, recorded in dB(A).
(2) The time history of the flyover noise level can be used to produce a single-event-noise-exposure level (SENEL)—this is based on a time-dependent summation of the instantaneous noise levels above a baseline value: typically 65 dB(A).
(3) A noise footprint can be drawn at ground level indicating a noise contour (e.g., for 75 dB(A)) with respect to the runway (often the airport boundaries and neighborhood are also indicated).

The cumulative noise of multiple events can be represented by an *equivalent continuous sound level* measure (i.e., the level of a notional steady sound that, over a given period of time, would have the same A-weighted acoustic energy as the fluctuating noise). It has units of dBA Leq, and is accompanied by a time period (e.g., 12 hours). It provides a single measure of the noise impact on the local community. For example, the size of the 57 dBA Leq noise contour at large airports is monitored by certain authorities.

Another measure that is used to limit the cumulative exposure of local communities to airplane noise is based on a noise quota system. This has been employed by certain international airports that have 24-hour operations, specifically to limit nighttime noise. One such example is the Quota Count (QC) system employed at several UK airports. Each airplane type is assigned an arrival QC value and a departure QC value. The scale, which makes no allowance for airplane

---

4 A database, which is called NoisedB, has been developed by the French DGAC (Directorate General for Civil Aviation), under the aegis of ICAO, as a validated source of certificated airplane noise [10].
5 ICAO provides information and advice for the determination and levying of noise-related charges—e.g., in the Airport Economics Manual [11] and Tariffs for Airports and Air Navigation Services [12]. The latter document is updated annually.

weight or number of engines, extends from 0 to 16 in 0.25 increments based on certification noise measurements. Basically, when the airport's quota has been used up, no more nighttime movements are allowed.

21.4 Noise Abatement and Flight Operations

21.4.1 Noise Abatement Procedures

The objective of a noise abatement procedure is to reduce one or more of the standard noise measures (described earlier in Section 21.3.3), at comparatively low cost, with due regard to all operational safety considerations. Several techniques are employed, as described in Section 21.4.2 for departing aircraft and in Section 21.4.3 for arriving aircraft. The appropriateness of the selected measure depends on many factors including the layout of the airport (e.g., runway orientation), location of noise sensitive buildings (e.g., hospitals and schools), airplane type, and also on the surrounding terrain, airspace, and available navigation aids (both on the ground and in the aircraft).[6]

21.4.2 Departing Aircraft

The airplane's flight profile during the takeoff climb, its weight, and the engine thrust setting all have an impact on the observed noise level on the ground. The influence of these factors can be assessed by constructing noise footprints (illustrated in Figure 21.3). For example, Boeing [14] reported on the potential to reduce the size of the 75 dB(A) footprint by 21% for the B737-600/-700/-800/-900 airplanes series, using an approved thrust reduction technique. This entails reducing the thrust at a height not less than 800 ft AGL (above ground level) and then restoring the normal climb thrust at a selected altitude, typically 3000 ft AGL.

The underlying principle relies on the fact that modern jet airplanes have an excess of climb thrust, that is, above that which is required to satisfy the regulatory one-engine-inoperative climb performance (see Table 10.1). Operators may legitimately reduce (cutback) the engines' thrust to a level that will still ensure that the minimum climb gradient can be met (i.e., with one engine inoperative at the cutback thrust level).[7] For airplanes with avionics systems that can detect an engine failure and automatically increase the thrust on the remaining engine(s), a further cutback in thrust is permitted by the authorities—in such cases the cutback thrust (with one engine inoperative) must be sufficient to maintain a 0% climb gradient. The impact of such a procedure on the noise footprint is illustrated in Figure 21.3.

From an operational point of view, such noise abatement procedures add to a pilot's workload and can increase fuel consumption. Manufacturers have responded by developing onboard systems that automatically reduce and then restore the thrust at specified heights. The economic benefits for airlines of using such procedures depend on a number of factors: first, there is the avoidance of fines, and, secondly, operators can avoid the need to reduce the airplane's takeoff weight (which means a greater payload can potentially be accommodated). Depending on

---

6 The ICAO's noise abatement procedures are described in ICAO Annex 16, Vol. I, Part V [3] and ICAO Doc. 8168 [13].

7 FAA AC 91-53 [15] defines acceptable criteria for thrust cutbacks for various airplane takeoff configurations.

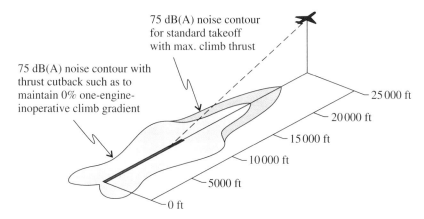

75 dB(A) noise contour
for standard takeoff
with max. climb thrust

75 dB(A) noise contour with
thrust cutback such as to
maintain 0% one-engine-
inoperative climb gradient

25 000 ft
20 000 ft
15 000 ft
10 000 ft
5000 ft
0 ft

Figure 21.3 Impact of reduced thrust on the 75 dB(A) takeoff noise contour (footprint).

the specific details of the operation, these factors will usually offset the small increase in fuel consumed during this non-optimum flight profile.

21.4.3 Arriving Aircraft

Historically, the focus of noise abatement has been on reducing aircraft departure noise, but, in recent years, attention also has turned to reducing arrival noise. The noise generated by an airplane during approach and landing is substantially lower than that which would be generated during takeoff and climb-out; however, it affects a larger area. The optimum flight path that minimizes noise, fuel burn, and engine emissions on approach is a continuous descent operation (CDO) (described in Section 17.7). CDO procedures that have been adopted by airport authorities only allow approaching airplanes to deviate by a small amount from the ideal flight path below a defined height above the airport's elevation. Different airport authorities interpret CDO in slightly different ways—one interpretation, for example, is that level flight cannot exceed 2.5 nm below 5500 ft AAL (above aerodrome level) [16]. Airspace restrictions and certain specific operational arrangements at particular airports, however, can preclude or limit the use of CDO.

An airplane's configuration and speed on approach significantly influences the noise that it generates. The selection of a reduced flap angle (if conditions and operating procedures permit) leads to a reduced airframe noise, a lower engine thrust setting, and reduced engine noise. Whereas such procedures are environmentally beneficial, operations with a reduced landing flap setting can increase the touchdown speed, leading to increased brake wear. Another operational factor that significantly influences measured noise levels is the height at which the landing gear is extended. The operator's Standard Operating Procedures (SOP) will indicate the height above aerodrome level at which the flight crew need to have the airplane in the landing configuration (i.e., with the landing flaps and the landing gear fully deployed)—this is typically 1500 ft AAL, but it can be a little higher. Early deployment of the landing gear without justification generates unnecessary noise.

The standard Instrument Landing System (ILS) glideslope is 3° (see Section 11.7.1). The utilization of a steeper glide path (e.g., 3.25° or 3.5°), when permitted, reduces airplane noise on the ground, as the required engine thrust is reduced and the height of the airplane above the

ground (at each point below the glideslope) is increased. Another acceptable method for mitigating community noise is the use of a displaced threshold (i.e., a threshold not located at the
extremity of the runway). This procedure, however, needs to be carefully balanced against the
potential reduction in safety margins (see ICAO Doc. 8168 [13] for details).

21.5 Airplane Emissions

21.5.1 Introduction

Advances in jet engine and airframe technologies since the 1950s have produced new generations of aircraft that are substantially more fuel-efficient than their predecessors. Jet aircraft
fuel consumption per passenger-km—the most obvious quantifier of aircraft pollution—has
been reduced by over 70%.[8] Smoke from exhausts has been virtually eliminated, and nitrogen
oxide emissions have been drastically cut. Notwithstanding these laudable achievements, aircraft operations contribute to the production of anthropogenic (i.e., that which results from
human activity) greenhouse gases, which lead directly to an increased risk of significant worldwide climate change. Moreover, aircraft are unique in that they operate in the upper troposphere and the lower stratosphere, an atmospheric region where important weather processes
occur (e.g., cloud formation) and where significant amounts of ozone (O_3), an important greenhouse gas (GHG), reside.

Emitted greenhouse gases trap terrestrial radiation, and chemically active gases and particulates alter naturally occurring greenhouse gases in the atmosphere. While the emissions are
known, their long-term effects on the atmosphere are not completely understood, particularly regarding indirect effects (e.g., aircraft-induced cloudiness). Nonetheless, new research
has enabled scientists to quantify the important processes to a reasonable confidence level.
Numerous studies have been conducted to assess the impact of aviation on the global atmosphere.[9] This is a non-trivial task, as the research must consider not only the greenhouse gases
that aircraft engines emit, but also the role that the emissions play in modifying the chemical
and particle microphysical properties of the atmosphere. The studies look at the way in which
these changes influence the delicate balance between the amount of heat entering and leaving
the atmosphere, which is fundamental to an assessment of global warming and climate change.

A standard metric used in many environmental studies—for example, in studies conducted
by the Intergovernmental Panel on Climate Change (IPCC)—is radiative forcing[10] (RF), which
measures the global annual mean radiative imbalance to the Earth's climate system. It can be

---

8 The baseline for this oft-cited comparison—which appears in reference [17], for example—is the de Havilland
Comet 4 (see Section 1.3). The Comet, however, was not a fuel-efficient airplane (in fact, it was substantially less
efficient than the piston-engine airplanes that it was designed to replace) and, in this respect, does not represent the
global fleet in the early days of the jet transport era [18]. For this reason, certain researchers choose to measure the
technological progress made by the aircraft industry in reducing emissions against the state of the art that existed in
the 1960s by using the Boeing 707 or DC-8 as the baseline instead [18–21].
9 A landmark study, *Aviation and the Global Atmosphere* (1999) [17], was carried out for ICAO by the Intergovernmental Panel on Climate Change (IPCC). It was the first comprehensive assessment of the impact of aviation on the
Earth's climate. Periodic updates, with new scientific information, are published by the IPCC (e.g., reference [22]).
Several large collaborative research projects have been sponsored by national governments and by the European
Commission [21, 23, 24]. ICAO publish their own *Environmental Report* [25, 26] on a three-year cycle, which draws
heavily on the research conducted by international agencies.
10 Radiative forcing (RF) is a measure of the change (perturbation) to the energy balance of the Earth–atmosphere
system with respect to the beginning of the industrial era (which is taken as the year 1750 in IPCC studies); it has
units of watts per square meter (W/m^2) [22].

used to make quantitative comparisons of different anthropogenic and natural agents in causing climate change [17, 21, 22, 24]. A positive RF leads to a global mean surface temperature increase (warming effect) and a negative RF to a temperature decrease (cooling effect).

21.5.2 Principal Aircraft Engine Emissions

Kerosene (see Section 22.7.1) is a hydrocarbon fuel (the generic chemical symbol is C_nH_m). The complete or partial combustion of kerosene in an aircraft's engine produces a mix of gases and small particles that are entrained in the exhaust plume. The emitted greenhouse gases (viz., CO_2 and H_2O) have a direct global warming effect; a high degree of confidence exists in the RF calculations for these emissions. Other emissions interact with each other and with the ambient air, producing new chemical species and modifying the composition of the air. Some of these changes indirectly result in a global warming effect; however, there is a significantly higher level of uncertainty associated with these RF calculations.

The principal jet engine emissions are illustrated schematically in Figure 21.4. The emissions typically consist of about 71.5% carbon dioxide (CO_2), 28% water vapor (H_2O), and the remainder of about 0.5% other pollutants—which include nitrogen oxides (NO_x),[11] carbon monoxide (CO), sulfur oxides (SO_x), unburned or partially combusted hydrocarbons (HC), soot, and other

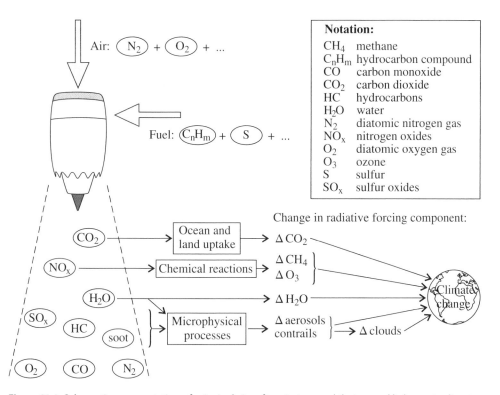

Figure 21.4 Schematic representation of principal aircraft emissions and their causal linkages to climate change. *Source*: Adapted from [21].

---

11 NO_x refers to one or both of the chemical compounds NO and NO_2. Likewise, SO_x refers to one or more of the sulfur–oxygen compounds (e.g., SO_2 and SO_3).

trace compounds [23, 27]. In addition, there may be minute quantities of gases and particles coming from the engine oil, from erosion of engine parts, or from trace amounts of metals that may be present in the fuel [23].

Carbon Dioxide

Carbon dioxide (CO_2) is a product of the *complete* combustion of kerosene in the airplane's engines; essentially, carbon atoms in the fuel join with oxygen atoms in the air in a 1 to 2 ratio. The ejected CO_2 molecules disperse quickly, mixing freely with other atmospheric constituents, contributing to the global build-up of atmospheric carbon dioxide.[12] For aviation, CO_2 yields the largest sustained RF of any contributing agent [22]. The critical factor is not the annual rate of CO_2 emitted into the atmosphere *per se*, but rather the absolute cumulative amount of CO_2 in the atmosphere (CO_2 emitted today will influence the global atmosphere for centuries).

Water Vapor

Water vapor (H_2O) is also a product of the complete combustion of kerosene; it is formed when hydrogen in the fuel combines with oxygen in the air. Emitted water vapor can act as a greenhouse gas (i.e., positive RF) and form persistent contrails in the right conditions (see Section 21.5.3).

Nitrogen Oxides

Nitrogen oxides (NO_x) are formed by thermal dissociation of nitrogen (N_2) and oxygen (O_2) molecules in the air as the gases pass through the combustion chamber. The reaction rates depend on the fuel–air ratio, flame temperature, system pressure, and the residence time (i.e., the time spent by the gases at these conditions) [29]. Note that NO_x is not a greenhouse gas, but it acts as a catalyst to alter the abundance of two principal greenhouse gases: ozone (O_3) and methane (CH_4). The processes by which NO_x affects atmospheric chemistry are complex (and differ according to factors such as season and location). NO_x has three main effects: it increases tropospheric O_3 (positive RF), it reduces ambient CH_4 (negative RF), and it reduces O_3 by a small amount in the longer term (negative RF) [21]. The overall RF effect of NO_x is considered to be positive [21].

Other Trace Species

Numerous trace species are contained in the exhaust plume. Carbon monoxide (CO) and hydrocarbons (HC), which are the result of incomplete combustion of fuel, are considered to be hazardous pollutants and are of concern for air quality near airports. Sulfur oxides (SO_x) are formed when trace quantities of sulfur in the fuel combine with oxygen in the air during combustion. Other sulfur species can also be formed—for example, gaseous sulfuric acid (H_2SO_4), which is considered to be an important precursor of volatile aerosol formation [23]. Jet engine emissions also contain a lot of small solid particles and condensable liquid droplets (i.e., an aerosol), as well as precursor gases that subsequently form aerosols. These particulates can stimulate chemical reactions in the atmosphere, absorb or scatter radiation, and alter natural cloud-forming processes. An important particulate, in terms of its effect on the atmosphere, is soot, which is

---

12 The global increase in atmospheric CO_2 is mainly due to the combustion of fossil fuels, gas flaring, and cement production; other sources are land use changes (e.g., deforestation) and biomass burning [22]. A portion of the CO_2 is absorbed by the terrestrial biosphere and by the oceans (which act as carbon reservoirs). Alarmingly, the amount of CO_2 in the atmosphere increased globally, from all causes, by approximately 120 ppm (parts per million) over *ca.* 260 years from the range 275–285 ppm in the pre-industrial era to 399 ppm (global annual mean) in 2015 [28].

mostly small black carbon particles formed by the incomplete combustion of fuel. Soot has a direct warming effect (positive RF); however, sulfate aerosol is known to have a direct cooling effect (negative RF) [23]. Additionally, aerosols have an important indirect effect—particulates can seed contrails and induce cloudiness (see Section 21.5.3, below).

21.5.3 Contrails and Aviation-Induced Cloudiness

Contrails, which are linear cirrus-type cloud structures (i.e., composed of small ice crystals), are caused by the ejection of water vapor and small particles into cold, ice-supersaturated air. Contrails often dissipate within a few minutes; however—and this is the concern from an environmental viewpoint—they can also persist in a linear form or spread into non-linear structures that closely resemble naturally occurring cirrus (known as contrail-cirrus). It is well known that clouds can reflect incoming solar radiation (cooling the Earth's surface), but they can also trap some of the outgoing long-wave radiation (warming the Earth's surface). The net effect depends on several factors, including the cloud's altitude, thickness, and make-up. Persistent contrails are essentially thin cirrus clouds; the warming effect is expected to dominate, resulting in a net positive RF [17, 21–23].

Aerosols—emitted by the engines and formed from emitted precursor gases—can act as ice nuclei, creating additional cloud cover (sometimes called soot cirrus); aerosols may also modify ambient cirrus cloud properties [21–23]. Accurate modeling of aviation-induced cloudiness (AIC) is exceedingly challenging, and there remains a high level of uncertainty in establishing the RF contribution of AIC [21–23, 26].

21.5.4 Environmental Impact of Aircraft Emissions

Assessments of the global environmental impact of aviation have been undertaken several times since the 1990s (e.g., [17, 21, 24]). Each new assessment builds on previous work and incorporates the results of new research (which often reduce the levels of uncertainty associated with the RF calculations of individual agents). Lee, *et al.* [21] estimated the annual contribution of global aviation to the total radiative forcing associated with all human activities to be about 3.5%, increasing to about 4.9% when the effect of AIC is included (estimate for 2005).

Focusing solely on CO_2 is a much simpler task. The worldwide consumption of fossil fuel by aircraft is a convenient metric—albeit a limited one—for assessing the industry's impact on the environment. Aviation has been responsible for about 2.5% of the total CO_2 emissions produced annually from all anthropogenic sources (for 1995–2005) [21]. Commercial aviation traffic, however, has historically grown at about 5% per year and is predicted to keep expanding at almost that rate for the foreseeable future, based on global RPK growth forecasts.[13] Improvements in aircraft fuel efficiencies due to new technologies and improved air traffic management, however, are expected to lower this annual increase in fuel consumption by at least 1.4% [26]. An aspirational goal to improve annual fuel efficiency by 2% for the international aviation sector was affirmed at the 38th ICAO Assembly (in 2013) [26].

The impact of aircraft emissions is not restricted to the upper atmosphere: about 10% is emitted below 3000 ft (during the takeoff and landing cycle and during ground operations). The pollutants CO and HC are exceptions: about 30% is emitted below 3000 ft, when the aircraft

---

13 Although there are differences in the forecasts of Airbus and Boeing regarding the envisaged demand for different sizes of aircraft, the overall forecasts for RPK (revenue passenger kilometer) growth are similar, as described in Airbus's *Global Market Forecast* [30] and Boeing's *Current Market Outlook* [31].

engines are operating at low combustion efficiencies [27]. Airports have now become one of the largest single sources of localized pollution in certain developed regions of the world [32].

21.5.5 Emission Indices

The usual approach taken to quantify the emissions that are generated by an aircraft's engines is based on the mass of the combusted fuel multiplied by an emission index (EI). Typical EI units are grams emission per kilogram fuel. The amounts of CO_2 and H_2O that are generated depend on the fraction of carbon and hydrogen, respectively, in the fuel. The typical chemical composition of kerosene is determined by analyzing fuel samples taken from airports around the world. This work has led to the establishment of average EI values that can be used to assess the CO_2 and H_2O emissions from aircraft engines on a global basis (Table 21.1).

The presence of sulfur-containing species (e.g., SO_x, H_2SO_4) in the engine exhaust is due to sulfur compounds in the fuel. This is controlled by jet fuel specifications (see Section 22.7.1). In practice, however, actual sulfur quantities can be much lower than the specification limits [33] (a typical EI value is indicated in Table 21.1).

As CO_2 is the dominant aviation emission—and because it is a straightforward calculation to determine the amount of CO_2 produced based on the fuel consumed—it is widely used as a measure of the impact of air travel on the environment. The amount of CO_2 emitted per passenger per flight can be estimated, based on the distance flown coupled with typical fuel consumption data (and a few other factors, such as aircraft type, cabin class, and typical passenger load factor). Several tools exist for this purpose—for example, ICAO's online Carbon Emissions Calculator [26, 34].

Emission indices for NO_x, CO, and HC depend on the physical conditions at which the combustion of the fuel takes place—specifically, the temperatures and pressures in the combustion chamber. A range of EI values are thus needed to model engine emissions during the various phases of a flight, which take into account the atmospheric conditions and the engine thrust setting. The EI value for NO_x is highest at high thrust settings; conversely, EI values for CO and HC are highest at low thrust settings (when combustor temperatures and pressures are lower and the combustion process is less efficient) [35].

Table 21.1 Aviation emission indices for kerosene combustion

Species	Emission index (ranges)[a,b] (grams emission per kilogram fuel)
carbon dioxide (CO_2)	3160
water (H_2O)	1240
nitrogen oxides (NO_x)	14 (12–17)
carbon monoxide (CO)	3 (2–3)
sulfur dioxide (SO_2)	0.8 (0.6–1.0)
hydrocarbons (HC)	0.4 (0.1–0.6)
soot	0.025 (0.01–0.05)

Notes:

(a) *Source*: Lee, *et al.* [23].
(b) The emission indices for NO_x, CO, HC, and soot are 2002 fleet averages [23].

21.6   Mitigating the Effects of Airplane Emissions

21.6.1   Mitigation Measures

ICAO and many other authorities have studied mechanisms to reduce the environmental impact of aircraft emissions—these have included (1) international policies/agreements; (2) engine and airplane certification standards (which promote "green" aircraft technologies); (3) new operational practices; (4) market-based measures (e.g., CO_2 cap and trade policies); and (5) sustainable alternative fuels (see Section 22.7.1).

The role of CO_2 as a radiative forcing agent is well understood. As CO_2 is a direct product of the combustion of kerosene, mechanisms that reduce the amount of fuel burned will reduce CO_2 production—these include the following:

(1) improvements in aircraft efficiency via engine or aerodynamic design improvements or weight reduction;
(2) operational changes (e.g., route optimization, reduced aircraft takeoff weight, reduced holding times); and
(3) fewer or shorter flights (e.g., resulting from improved intermodal transportation systems).

The production of NO_x, CO, HC, and particulates, however, depends on *how* the fuel is combusted—these gaseous and particle emissions are influenced by the design and the operating condition of the engine (e.g., thrust setting). Engine certification standards FAR 34 [36] and CS-34 [37]—which are based on ICAO Annex 16, Vol. II [38]—stipulate upper limits for the production of NO_x, CO, unburned hydrocarbons, and smoke (soot) for the landing/takeoff (LTO) cycle, taking into account such factors as the year of manufacture of the engine and its overall pressure ratio.

The fundamental difficulty in reducing NO_x from an engine design perspective is that the design parameters that are good for fuel efficiency are not necessarily good for controlling NO_x production. For a given standard (or level) of engine technology, an increase in engine core temperatures and pressures will theoretically increase the amount of NO_x emitted, but, all things considered equal, will reduce fuel burn. Complex tradeoffs are thus required. Engine manufacturers have, over the past few decades, developed combustion technologies that have made jet engines more efficient, while simultaneously reducing NO_x emissions [29].

The manner in which the engines are operated can also influence the amount of NO_x produced. For example, if derated or reduced thrust is employed on takeoff (see Section 10.8), less NO_x will be formed as NO_x production is directly related to the engine's EGT (exhaust gas temperature), which is lowered when the thrust is reduced.

The potential also exists to reduce the environmental impact of a flight by changing the route and/or altitude. An altitude change, for example, can be effective in reducing or eliminating persistent contrails and AIC as ice-supersaturated conditions are typically found in narrow vertical layers [39]. Departing from the optimum cruise altitude, however, will increase the rate of fuel consumption (see Sections 13.6 and 17.3). Consequently, in this scenario, non-CO_2 radiative forcing would reduce, but CO_2 radiative forcing (which has a long-term influence) would typically increase. It is theoretically possible to build a climate cost function, which would be analogous to the financial cost function described in Section 18.3. The optimum trajectory would be established through optimization of the cost function. Such a climate cost function would describe the airplane's contribution to climate change at a certain time and location in the atmosphere resulting from H_2O and CO_2 production, ozone formation, methane loss, methane-induced ozone change, contrails, and AIC (see Grewe, *et al.* [40] for further details on such an approach).

21.6.2 ICAO Aircraft Carbon Dioxide Emissions Certification Standard

The CAEP, at its tenth meeting (CAEP/10) in February 2016, recommended a new certification standard—the Aircraft Carbon Dioxide Emissions Certification Standard—as part of ICAO's *basket of measures* to reduce greenhouse gas emissions from the global air transport system [41]. The new standard, to be published as Annex 16, Vol. III, aims to encourage more fuel-efficient technologies in airplane design, and follows a similar approach to that adopted in the current ICAO Annex 16 standards on aircraft noise (Vol. I) and engine emissions for local air quality (Vol. II). The stated intent of these measures is to "equitably reward advances in airplane technologies (e.g., propulsion, aerodynamics, and structures) that contribute to reductions in airplane CO_2 emissions, and differentiate between airplanes with different generations of these technologies" [26].

The certification approach is based on a CO_2 metric value (MV), which is defined as follows [42]:

$$M_V = \left(\frac{1}{r_a}\right) \frac{1}{R_{GF}^{0.24}} \tag{21.3}$$

where M_V is the ICAO metric value (MV);

r_a is the average specific air range (SAR) established at three reference gross weights (unit: km/kg); and

R_{GF} is the ICAO *reference geometric factor* (RGF), which is an airplane size adjustment factor (unit: m^2).

The reciprocal of the SAR is a measure of instantaneous fuel efficiency, which, for a given airplane type (and model), varies primarily with airplane weight, speed, altitude, and air temperature (see Sections 13.2 and 13.6).

The standard [42] calls for the airplane's SAR to be evaluated at optimal combinations of cruise speed and altitude (which can be selected by the manufacturer) at each of three specified airplane gross weights that represent cruise conditions regularly seen in service. The airplane test weights (W) are defined as functions of the airplane's maximum takeoff weight (W_{MTO}), as follows:

(1) high gross weight condition: $W = 0.92\,W_{MTO}$
(2) low gross weight condition: $W = 0.45\,W_{MTO} + 0.63\,W_{MTO}^{0.924}$
(3) mid gross weight condition: average of high and low gross weights.

The three SAR measurements are to be averaged (equally weighted) and used to determine the airplane's MV using Equation 21.3. Reference parameters/conditions for measuring SAR include (1) steady (i.e., un-accelerated), straight, and level flight; (2) true north flight heading; (3) airplane in longitudinal and lateral trim; (4) standard ISA conditions; (5) standard gravitational acceleration; and (6) specified fuel heating value (i.e., 43.217 MJ/kg). Other specified factors that need to be accounted for in the flight testing are (1) center of gravity position; (2) electrical power, mechanical power, and bleed air extraction from the engines; and (3) engine deterioration level [42].

The ICAO reference geometric factor (RGF) is a measurement of fuselage size (an approximation of the pressurized floor area of the cabin). It is defined as the area of a surface bounded by the maximum width of the fuselage outer mold line (OML) projected onto a flat plane parallel with the main deck floor [42]. For airplanes with main and upper decks, it is the sum of the area of a surface bounded by the maximum width of the fuselage OML projected onto a flat plane parallel with the main deck floor and the area of a surface bounded by the maximum width of the fuselage OML at or above the upper deck floor projected onto a flat plane parallel with the upper deck floor. The rear boundary is the rear pressure bulkhead, and the forward boundary is the forward pressure bulkhead except for the cockpit crew zone. The RGF includes all pressurized space on the main or upper deck, including aisles, assist spaces, passageways, stairwells, and areas that can accept cargo and auxiliary fuel containers, but it does not include permanent integrated fuel tanks within the cabin or any unpressurized fairings, nor crew rest/work areas or cargo areas that are not on the main or upper deck [42].

A phased approach to the introduction of MV limits in the new standard is intended. Two tiers of regulatory MV limits have been established for commercial jet transport airplanes in the short term: one for new deliveries of already-certified aircraft types (i.e., in-production aircraft) and a second set of limits for new designs that will begin to be type certified after January 1, 2020 and enter into service after 2024 [26, 43]. The CAEP (CAEP/10) has recommended more stringent MV limits for all new aircraft delivered after January 1, 2028, with a transition period for modified aircraft starting in 2023 [43].

21.6.3   International Agreements on Emissions

The United Nations Framework Convention on Climate Change (UNFCCC) is an international treaty that was agreed at the Earth Summit in Rio de Janeiro in June 1992 [44]; it defines the procedures by which binding international agreements (protocols) on mitigating climate change can be negotiated by the participants (parties). Convention members meet annually at Conferences of the Parties (COPs). The Kyoto Protocol (signed in Kyoto, Japan in 1997 at COP 3) is an international treaty that entered into force in 2005 and which sets binding obligations on industrialized countries to reduce greenhouse gas emissions.[14] The overarching objective was to stabilize greenhouse gas concentrations at a level that was perceived to prevent dangerous long-term damage to the world's climatic system.

Greenhouse gas emissions caused by domestic flights were included in the Kyoto Protocol, but emissions from international air traffic—which does not fall under national jurisdictions— were specifically excluded. It is worth noting that 65% of the CO_2 produced by aircraft worldwide each year is a result of international air travel [26]. Subsequent conferences failed to reach agreement on this issue; instead, after some considerable delay, governments agreed to work through ICAO to determine policies to control emissions associated with international air travel.

In 2013, ICAO announced that a preliminary agreement had been reached with its members to develop global rules that would control airline emissions using market-based measures (MBM) for international travel (i.e., civil aviation flights that depart in one country and arrive in a different country). In 2016, the ICAO Assembly adopted the Carbon Offsetting and Reduction Scheme for International Aviation (CORSIA) [45]. The ICAO agreement[15] aims to achieve carbon neutral growth from 2020 (or a 3-year average around 2020) onwards. Specific rules have been established to determine the CO_2 emissions that would have to be offset by operators utilizing a 2020 baseline [45]. The difference between the emissions in any year after 2020 and the

---

14 Not all industrialized nations are committed to the Kyoto Protocol (e.g., the United States did not ratify the Protocol, and Canada withdrew in 2012) [44].

15 The ICAO agreement was developed in the context of the UNFCCC Paris Agreement (adopted at COP 21 in Paris in 2015)—the most significant international environmental treaty ever adopted (the goal is to hold the increase in global average temperature to well below 2 °C above pre-industrial levels and to pursue efforts to limit the temperature increase to 1.5 °C [44]).

baseline would represent the sector's offsetting requirements for that year. The agreement provides for special circumstances of particular states (countries), in particular developing countries, and has a phased implementation, which is defined as follows: pilot phase (2021–2023), first phase (2024–026), and second phase (2027–2035). Small operators, new entrants, and special operations (e.g., firefighting and search and rescue flights) will be exempt.

References

1 ICAO, "Guidance on the balanced approach to aircraft noise management," 2nd ed., Doc. 9829, International Civil Aviation Organization, Montréal, Canada, 2008.

2 Berger, E.H., Royster, L.H., Royster, J.D., *et al.* (Eds.), *The noise manual*, 5th ed., American Industrial Hygiene Association, Fairfax, VA, 2003.

3 ICAO, "Environmental protection: Vol. I – Aircraft noise," 6th ed., Annex 16 to the Convention on International Civil Aviation, International Civil Aviation Organization, Montréal, Canada, July 2011.

4 FAA, *Noise standards: Aircraft type and airworthiness certification*, Federal Aviation Regulation Part 36, Amdt. 36-30, Federal Aviation Administration, Washington, DC, Mar. 4, 2014. Latest revision available from www.ecfr.gov/ under e-CFR (Electronic Code of Federal Regulations) Title 14.

5 IEC, *Electroacoustics – Sound level meters – Part 1: Specifications*, IEC 61672-1, International Electrotechnical Commission, Geneva, Switzerland, 2013.

6 FAA, "Noise levels for U.S. certificated and foreign aircraft," Advisory Circular 36-1H, Federal Aviation Administration, Washington DC, May 25, 2012.

7 FAA, "Noise standards: Aircraft type and airworthiness certification," Advisory Circular 36-4C, Federal Aviation Administration, Washington, DC, July 15, 2003.

8 ICAO, "Committee on Aviation Environmental Protection, ninth meeting report," Doc. 10012, International Civil Aviation Organization, Montréal, Canada, Feb. 4–15, 2013.

9 EASA, *Certification specifications for aircraft noise*, CS-36, Amdt. 4, European Aviation Safety Agency, Cologne, Germany, Jan. 12, 2016. Latest revision available from www.easa.europa.eu/ under Certification Specification.

10 DGAC, "NoisedB database," Directorate General for Civil Aviation, Paris, France, Retrieved Aug. 2016. Available from noisedb.stac.aviation-civile.gouv.fr./.

11 ICAO, "Airport economics manual," 2nd ed., Doc. 9562, International Civil Aviation Organization, Montréal, Canada, 2006.

12 ICAO, "Tariffs for airports and air navigation services," Doc. 7100, International Civil Aviation Organization, Montréal, Canada, 2016.

13 ICAO, "Procedures for air navigation services: Aircraft operations, Vol. 1, Flight procedures," 5th ed., Doc. 8168, International Civil Aviation Organization, Montréal, Canada, 2006.

14 Friedrich, J., McGregor, D., and Weigold, D., "Quiet climb system," *Aero*, The Boeing Company, Seattle, WA, Vol. Q 01, pp. 26–31, 2003.

15 FAA, "Noise abatement departure profile," Advisory Circular 91-53A, Federal Aviation Administration, Washington, DC, July 22, 1993. Available from www.faa.gov/.

16 CAA, "Managing aviation noise," CAP 1165, Civil Aviation Authority, London, UK, 2014. Available from www.caa.co.uk/.

17 Penner, J.E., Lister, D.H., Griggs, D.J., *et al.* (Eds.), *Aviation and the global atmosphere: A special report of the Intergovernmental Panel on Climate Change*, Cambridge University Press, Cambridge, UK, 1999.

18 Peeters, P.M., Middel, J., and Hoolhorst, A., "Fuel efficiency of commercial aircraft: An overview of historical and future trends," NLR-CR-2005-669, National Aerospace Laboratory, the Netherlands, Nov. 2005.

19 Rutherford, D. and Zeinali, M., "Efficiency trends for new commercial jet aircraft, 1960 to 2008," International Council on Clean Transportation, Washington, DC, Nov. 2009. Available from www.theicct.org/.

20 Kharina, A. and Rutherford, D., "Fuel efficiency trends for new commercial jet aircraft: 1960 to 2014," International Council on Clean Transportation, Washington, DC, Aug. 2015. Available from www.theicct.org/.

21 Lee, D.S., Fahey, D.W., Forster, P.M., *et al.*, "Aviation and global climate change in the 21st century," *Atmospheric Environment*, Vol. 43, Iss. 22/23, pp. 3520–3537, 2009.

22 Forster, P., Ramaswamy, V., Artaxo, P., *et al.*, "Changes in atmospheric constituents and in radiative forcing," in *Climate Change 2007: The Physical Science Basis. Contribution of Working Group I to the Fourth Assessment Report of the Intergovernmental Panel on Climate Change*, Solomon, S., Qin, D., Manning, M., *et al.* (Eds.), Cambridge University Press, Cambridge, UK, 2007.

23 Lee, D.S., Pitari, G., Grewe, V., *et al.*, "Transport impacts on atmosphere and climate: Aviation," *Atmospheric Environment*, Vol. 44, Iss. 37, pp. 4678–4734, 2010.

24 Brasseur, G.P., Gupta, M., Anderson, B.E., *et al.*, "Impact of aviation on climate: FAA's aviation climate change research initiative (ACCRI) Phase II," *Bulletin of the American Meteorological Society*, Vol. 97, Iss. 4, pp. 561–583, 2016.

25 ICAO, "Environmental report 2010: Aviation and climate change," Environment Branch, International Civil Aviation Organization, Montréal, Canada, 2010.

26 ICAO, "Environmental report 2016: Aviation and climate change," Environment Branch, International Civil Aviation Organization, Montréal, Canada, 2016.

27 FAA, "Aviation emissions, impacts & mitigation: A primer," Office of Environment and Energy, Federal Aviation Administration, Washington, DC, Jan. 2015.

28 NOAA, "2015 state of the climate: Carbon dioxide," National Oceanic and Atmospheric Administration, Silver Spring, MD, Aug. 2, 2016, Retrieved Aug. 21, 2016. Available from research.noaa.gov/News/.

29 Faber, J., Greenwood, D., Lee, D., *et al.*, "Lower NO_x at higher altitudes: Policies to reduce the climate impact of aviation NO_x emission," Publication No. 08.7536.32, CE Delft, Oude Delft, the Netherlands, Oct. 2008. Available from www.ce.nl/.

30 Airbus, "Global market forecast: Mapping demand 2016–2035," Airbus S.A.S., Blagnac, France, 2016.

31 Boeing, "Current market outlook 2016–2035," Market Analysis, Boeing Commercial Airplanes, Seattle, WA, July 2016.

32 Marias, K. and Waitz, I.A., "Air transport and the environment," in *The global airline industry*, Belobaba, P., Odoni, A., and Barnhart, C. (Eds.), John Wiley & Sons, Chichester, UK, pp. 405–440, 2009.

33 Hadaller, O.J. and Momenthy, A.M., "The characteristics of future aviation fuels," D6-54940, The Boeing Company, Seattle, WA, 1998.

34 ICAO, "Carbon emissions calculator," International Civil Aviation Organization, Montréal, Canada, Retrieved Aug. 25, 2016. Available from www.icao.int/environmental-protection/CarbonOffset/Pages/default.aspx.

35 Sutkus, D.J., Jr., Baughcum, S.L., and DuBois, D.P., "Scheduled civil aircraft emission inventories for 1999: Database development and analysis," NASA CR-2001-211216, National Aeronautics and Space Administration, USA, Oct. 2001.

36 FAA, *Fuel venting and exhaust emission requirements for turbine engine powered airplanes*, Federal Aviation Regulation Part 34, Amdt. 34-5, Federal Aviation Administration, Washington DC, Dec. 31, 2012. Latest revision available from www.ecfr.gov/ under e-CFR (Electronic Code of Federal Regulations) Title 14.

37 EASA, *Certification specifications and acceptable means of compliance for aircraft engine emissions and fuel venting*, CS-34, Amdt. 2, European Aviation Safety Agency, Cologne, Germany, Jan. 12, 2016. Latest revision available from www.easa.europa.eu/ under Certification Specification.

38 ICAO, "Environmental protection: Vol. II – Aircraft engine emissions," 3rd ed., Annex 16 to the Convention on International Civil Aviation, International Civil Aviation Organization, Montréal, Canada, July 2008.

39 Frömming, C., Ponater, M., Dahlmann, K., *et al.*, "Aviation-induced radiative forcing and surface temperature change in dependency of the emission altitude," *Journal of Geophysical Research: Atmospheres*, Vol. 117, Iss. D19, pp. 2156–2202, 2012.

40 Grewe, V., Frömming, C., Matthes, S., *et al.*, "Aircraft routing with minimal climate impact: The REACT4C climate cost function modelling approach (V1.0)," *Geoscientific Model Development* Vol. 7, pp. 175–201, 2014.

41 ICAO, "Press release: New ICAO aircraft CO2 standard one step closer to final adoption," International Civil Aviation Organization, Montréal, Canada, Feb. 8, 2016. Available from www.icao.int/Newsroom/.

42 ICAO, "CAEP/9 agreed certification requirement for the aeroplane CO_2 emissions standard," Circular 337, International Civil Aviation Organization, Montréal, Canada.

43 ICCT, "Policy update: International Civil Aviation Organization's CO_2 standard for new aircraft," International Council for Clean Transportation, Washington, DC, Feb. 2016.

44 UNFCCC, "UN climate change newsroom," United Nations Framework Convention on Climate Change, Bonn, Germany, Retrieved Aug. 2016. Available from newsroom.unfccc.int/.

45 ICAO, "Assembly 39th session: Report of the executive committee on agenda item 22," A39-WP/530, International Civil Aviation Organization, Montréal, Canada, Oct. 6, 2016.

22

Airplane Systems and Performance

22.1 Introduction

In this chapter, airplane systems that can have a significant impact on the performance of a jet transport airplane or which directly impose a flight limitation, either in normal operation or in emergency conditions following a system failure, are identified and described. The manner in which these systems can influence or restrict an airplane's performance is discussed.

An overview of the regulatory requirements regarding the reliability of airplane systems is given in Section 22.2, as background information to this subject. Selected topics relating to the following systems are then described: cabin pressurization system (Section 22.3), environmental control system (Section 22.4), de-icing and anti-icing systems (Section 22.5), and auxiliary power system (Section 22.6). The powerplant is described in Chapter 8; discussions in this chapter will be limited to the fuel and fuel system (Section 22.7). Specific regulatory requirements pertaining to these systems—as described in FAR 25 [1], CS-25 [2], FAR 121 [3], and OPS Part-CAT [4]—are mentioned.

22.2 Reliability Requirements for Airplane Systems

Specific requirements for the reliability of safety-critical systems are stipulated in FAR/CS 25.1309 [1, 2]. Acceptable means of compliance with these requirements are described by the FAA in AC 1309A [5] and by EASA in AMC 25.1309 [2]. Note that the terminology used by the FAA and EASA is not identical in all respects. In this section, the EASA definitions are provided in the text and also in Table 22.1, with footnotes that describe FAA usage. Failure conditions are classified according to the severity of their effects. The following failure categories, in increasing severity, are defined in AMC 25.1309 [2]: *minor*, *major*, *hazardous*,[1] and *catastrophic* (a description of these terms is given in Table 22.1a). Minor failures, which may occur one or more times during the life of the airplane, do not significantly reduce airplane safety. Airplane systems must be designed so that (1) any major failure condition is *remote*, (2) any hazardous failure condition is *extremely remote*, and (3) any catastrophic failure condition is *extremely improbable* and does not result from a single failure[2] (CS 25.1309 [2]). Qualitative and quantitative descriptions of these probability terms are given in Table 22.1b.

---

1 The FAA's definition in AC 1309A [5] of a major failure condition encompasses the *major* and *hazardous* categories described in AMC 25.1309 [2].

2 Specific exemptions from these requirements are permitted—details are given in CS 25.1309 [2].

Performance of the Jet Transport Airplane: Analysis Methods, Flight Operations, and Regulations, First Edition. Trevor M. Young.
© 2018 John Wiley & Sons Ltd. Published 2018 by John Wiley & Sons Ltd.

Table 22.1a Failure condition classification, description, and probability (based on EASA AMC 25.1309)

Failure condition	Description	Probability of occurrence[b]
Minor failure conditions:	are those that would not significantly reduce airplane safety and which involve crew actions that are well within their capabilities, but may result in a slight reduction in safety margins or functional capabilities, a slight increase in crew workload (such as routine flight plan changes), or some physical discomfort to passengers or cabin crew.	may be *probable.*
Major failure conditions:[c]	are those that would reduce the capability of the airplane or the ability of the crew to cope with adverse operating conditions to the extent that there would be a significant reduction in safety margins or functional capabilities, a significant increase in crew workload or in conditions impairing crew efficiency, discomfort to the flight crew, or physical distress to passengers or cabin crew, possibly including injuries.	must be no more frequent than *remote.*
Hazardous failure conditions:[c]	are those that would reduce the capability of the airplane or the ability of the crew to cope with adverse operating conditions to the extent that there would be: (i) a large reduction in safety margins or functional capabilities; or (ii) physical distress or excessive workload such that the flight crew cannot be relied upon to perform their tasks accurately or completely; or (iii) serious or fatal injury to a relatively small number of the occupants other than the flight crew.	must be no more frequent than *extremely remote.*
Catastrophic failure conditions:	are those that would result in multiple fatalities, usually with the loss of the airplane (note: this was previously defined in the European regulations as a condition that would prevent continued safe flight and landing[d]).	must be *extremely improbable.*

Notes:

(a) *Source*: Based on EASA AMC 25.1309 [2].
(b) See Table 22.1b for failure probability classification.
(c) The FAA treats major and hazardous failure conditions as a single category (called major failure) and uses the term *improbable* to describe the probability of occurrence [5].
(d) This definition is used by the FAA [5] to describe catastrophic failure conditions.

Following a system failure, information concerning unsafe operating conditions, such as a warning indication, must be provided to the crew to enable them to take appropriate corrective action. Systems and controls, including indications and annunciations, must be designed to minimize crew errors that could create additional hazards.

22.3 Cabin Pressurization System

22.3.1 Human Performance and Limitations at Altitude

In an unpressurized airplane (or one in which the cabin pressurization system is inoperative) the ability of humans to perform normal activities diminishes with increasing altitude due to oxygen deprivation. At sea level, a healthy person can extract the oxygen required for normal activities with ease, but at about 8000 ft, the effects of a diminished supply of oxygen to the

Table 22.1b Failure probability classification (based on EASA AMC 25.1309)

Probability condition	Qualitative probability description	Average probability per flight hour
Probable failure conditions:	are those anticipated to occur one or more times during the entire operational life of each airplane.	Greater than of the order of 1×10^{-5}.
Remote failure conditions:[b]	are those unlikely to occur to each airplane during its total life, but which may occur several times when considering the total operational life of a number of airplanes of the type.	Of the order of 1×10^{-5} or less, but greater than of the order of 1×10^{-7}.
Extremely remote failure conditions:[b]	are those not anticipated to occur to each airplane during its total life, but which may occur a few times when considering the total operational life of all airplanes of the type.	Of the order of 1×10^{-7} or less, but greater than of the order of 1×10^{-9}.
Extremely improbable failure conditions:	are those so unlikely that they are not anticipated to occur during the entire operational life of all airplanes of one type.	Of the order of 1×10^{-9} or less.

Notes:

(a) *Source*: Based on EASA AMC 25.1309 [2].

(b) The FAA treats major and hazardous failure conditions as a single category, and uses the term *improbable* to describe the associated probability of occurrence. Improbable failure conditions are those having a probability of occurrence per flight hour of the order of 1×10^{-5} or less, but greater than of the order of 1×10^{-9} [5].

body become apparent for most people. Heavy smokers or elderly or sick people are likely to experience this at lower altitudes.

Although the percentage of oxygen in the air remains approximately constant (at about 21%) from sea level through the altitude range used by jet transport airplanes, the reducing ambient air pressure would reduce the oxygen partial pressure in a person's lungs. As the transfer of oxygen from the air into the blood (traveling through the fine capillaries that surround the lungs) depends on the pressure exerted on the wall membranes of the lungs, the supply of oxygen to the body would reduce with increasing altitude. The relative air pressure (see Figure 4.2) is thus a measure of the relative quantity of oxygen (compared to sea level) that each lungful of air contains; this is three-quarters at 8000 ft (where $\delta = 0.743$), half at 18 000 ft, and one-quarter at 33 700 ft. Early symptoms of hypoxia—a lack of sufficient oxygen to meet the body's needs—are an impairment in night vision and a personality change (not dissimilar to intoxication) [6]. The symptoms progress to impaired judgment, a loss of memory and mental ability, a reduction in muscular coordination, and finally unconsciousness with increasing exposure to rarified air.

Pressurized cabins (or supplemental oxygen systems) are required for humans to operate for any length of time at altitudes above 10 000 ft.[3] In the study of human performance at altitude, the concept of *time of useful consciousness* (TUC) is an important metric. The TUC describes the maximum length of time during which a person can carry out a purposeful activity (or task) at a given altitude in the absence of a pressurization system. Times vary depending on the person's physical condition. For a healthy individual, the TUC is 5–10 minutes at 22 000 ft, 3–5 minutes at 25 000 ft, 2.5–3 minutes at 28 000 ft, and 1–2 minutes at 30 000 ft [6].

For a person breathing 100 percent oxygen, the partial pressure of oxygen in his/her lungs at 34 000 ft altitude is comparable to that of someone breathing air at sea level; however, at 40 000 ft, the partial pressure of oxygen in the lungs reduces and is comparable to someone

---

3 The altitude depends on the physical condition of the individual; 10 000 ft has been selected by the regulatory authorities as the cut-off altitude for flight crews [3, 4]—see Section 22.3.4.

breathing air at 10 000 ft [7]. Therefore, 40 000 ft is the highest altitude at which 100 percent oxygen will provide reasonable protection from the effects of hypoxia for a period of time.

22.3.2 Cabin Altitude in Cruise

The purpose of the cabin pressurization system is to maintain air pressure in the cabin at a level that is comfortable for the crew and passengers. The internal cabin pressure is often expressed as an equivalent *cabin altitude*, where the conversion between pressure and altitude is based on the standard ISA pressure–height relationship (see Section 4.2.4).

The structural design of the fuselage—in terms of its ability to withstand the cyclic loads that it experiences with each flight—sets an upper limit on the pressure differential that can be tolerated between the inside and outside of the fuselage skin. As an airplane ascends, the cabin altitude increases (i.e., the cabin pressure reduces), but not at the same rate at which the airplane climbs. The pressure differential thus increases with altitude. At the airplane's maximum certified cruise altitude, the maximum permissible pressure differential sets a limit on the allowable cabin pressure and this equates to an allowable cabin altitude. The design (lowest) cabin altitude varies depending on the airplane type—for example, it is 8000 ft for the A320 family, B737 NG and B747-400; 7900 ft for the B767 and B777; and 7350 ft for the A330 and A340 for flights longer than 2.5 hr duration (but 8000 ft for flights less than 2.5 hr duration) [8–10]. Airframes of newer airplane types, such as the B787 and A350, are designed to accommodate a cabin altitude of 6000 ft when at maximum cruise altitude [8]—this offers greater passenger comfort than do earlier designs. What this means in terms of structural design is that a B787 airplane cruising at FL 431 has a pressure differential of 652 hPa (9.46 psi) [8]. If the operational limit is exceeded—in an emergency, for example—an overpressure relief valve would open to prevent damage to the airframe.

When an airplane operates at an altitude below its maximum certified cruise altitude, the cabin altitude is set at a lower value—for example, at FL 410 the A340 has a cabin altitude of 7350 ft, but at FL 290, the cabin altitude is 3550 ft [9].

22.3.3 Cabin Pressure Regulation in Climb and Descent

During a normal climb, the cabin altitude will increase (i.e., the cabin pressure will reduce) at a programmed rate (not exceeding 1000 ft/min) to reach the scheduled cabin altitude at the top of climb (TOC). This is illustrated in Figure 22.1 for two selected cruise altitudes: FL 290 and FL 410.

During a normal descent, the cabin altitude will decrease (i.e., the cabin pressure will increase) at a programmed rate (typically 350 ft/min, but not exceeding 750 ft/min) to reach the target pressure just before landing, which is equal to the atmospheric pressure at the destination airport plus 0.1 psi [9]. In some cases, the normal rate of descent of the airplane (with idle thrust) will result in a descent time that is *shorter* than the time needed to repressurize the cabin. In this event, to prevent the airplane landing with the incorrect cabin pressure, a repressurization segment will be included in the descent schedule (Figure 22.2).

22.3.4 Cabin Pressurization System Failure and Supplemental Oxygen Supply

It is a requirement that all pressurized aircraft operating at altitudes greater than 10 000 ft carry supplemental oxygen supply equipment for the crew and passengers in the event of a cabin

Figure 22.1 Example of cabin pressurization—based on information for the A340 [9].

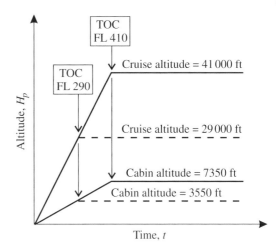

Figure 22.2 Example of descent schedule showing cabin repressurization segment.

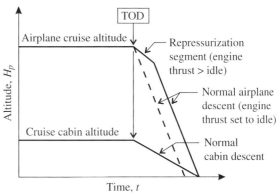

pressurization system failure (see also Section 17.5). Two types of supplemental oxygen supply systems are currently used: gaseous and chemical.[4]

Gaseous systems store compressed oxygen gas in large cylinders. The number and size of gas cylinders are selected by the operator to suit the terrain the airplane will be required to operate over (flights over extensive high terrain will require more oxygen). As the oxygen flow rate and supply pressure depend on the altitude, the time over which oxygen can be supplied depends on the emergency descent profile.

Chemical systems produce oxygen gas by the reaction of chemicals stored in small canisters (located in overhead panels). The mixing of the chemicals is initiated by the mask being pulled downwards; once the flow has started, it cannot be stopped. The units are designed to supply oxygen for a fixed time, which can range from 12 to 22 minutes depending on their design [9, 12]. The oxygen flow rate and supply pressure are not influenced by the altitude. Emergency descent profiles can be determined for aircraft equipped with these systems (as illustrated in Figure 17.7) and published in the Flight Crew Operating Manual (FCOM), for example.

The amount of supplemental oxygen that must be stored onboard the airplane for the flight crew, cabin crew, and passengers is described in FAR 121 (sections 121.329 and 121.333) [3] and in OPS Part-CAT (section CAT.IDE.A.235) [4]. The minimum supplemental oxygen

_____

4 A third system type, known by the acronym OBOGS (On Board Oxygen Generating System), which is qualified for military applications, is being considered by the authorities for use on civil passenger transport aircraft [11].

supply for each flight crew member is two hours; this is described as that quantity of oxygen necessary for the airplane to descend from its maximum certificated operating altitude to 10 000 ft in 10 minutes (at a constant rate of descent) and then fly for 110 minutes at 10 000 ft [3, 4]. For the passengers, there must be sufficient oxygen for (1) all passengers for that part of the flight at pressure altitudes above 15 000 ft; (2) 30% of the passengers for that part of the flight at pressure altitudes above 14 000 ft but not exceeding 15 000 ft; and (3) for 10% of the passengers for that part of the flight that is of more than 30 minutes duration at pressure altitudes above 10 000 ft but not exceeding 14 000 ft [3, 4].

Additionally, there are specific requirements for oxygen for first-aid treatment of a small number of passengers, following an emergency descent, for the rest of the flight (FAR 121.333 [3], OPS Part CAT.IDE.A.230 [4])—this can be achieved through portable oxygen cylinders.

22.4 Environmental Control System

22.4.1 Operation

The environmental control system (ECS) provides a steady supply of air to the cabin for the crew and passengers. For most airplane types, compressed fresh air is extracted from the engines, cooled in air conditioning units (which are often called *packs*), and then passed to a mixing manifold, where it is mixed with filtered recirculated air. In a *no-bleed* system architecture—as used on the B787—air is brought onboard via dedicated inlets and ducted to electrically powered compressors. The mixing ratio on modern airplanes is approximately 50 percent fresh air and 50 percent recirculated air [13, 14]. The air is supplied via overhead outlets to the cabin and an equal amount of air is removed from the cabin.

For health reasons, it is undesirable for atmospheric ozone (which is created by the photochemical conversion of oxygen by solar ultraviolet radiation) to be supplied to the cabin. The ozone in the incoming air is thus converted into oxygen in a catalytic converter prior to the air entering the air conditioning units.

22.4.2 Impact on Airplane Performance

The extraction of bleed air (see Section 8.8.3) from the engines for the ECS has a negative impact on the airplane's fuel consumption. For example, the Pratt & Whitney 4000 engines (5:1 bypass ratio), installed on the B767, suffer a 0.8% increase in fuel burn when the ECS is operating (with 50% recirculation) [13]. The penalty depends, to a large extent, on the bypass ratio of the engines. With low bypass ratio engines, the impact of bleed air extracted for the ECS on the TSFC (thrust specific fuel consumption) is relatively small. For high bypass ratio engines, however, the impact is greater as the bleed air supplied to the ECS represents a much larger percentage of the total core airflow.

The impact of bleed air extraction for air conditioning can be noticed during takeoff on certain airplane types, where a payload penalty can occur in cases when the takeoff is thrust limited. To circumvent this limitation, it is usually permissible for the flight crew to partially or totally turn off the air conditioning packs for the takeoff (the units would be manually or automatically switched on after takeoff). Certain systems (e.g., as installed in the A320) incorporate a feature by which the auxiliary power unit (see Section 22.6) can supply the air conditioning packs during takeoff, thus allowing full engine thrust to be delivered in such cases. Air conditioning packs can also be switched off by the crew in emergencies, and on some airplane types it is acceptable to reduce the volume of air circulated for lightly loaded flights, thus reducing fuel burn.

22.5 De-Icing and Anti-Icing Systems

22.5.1 Airplane Icing

Flight in icing conditions can result in ice accumulating on the protruding parts of the airplane—for example: the nose; the leading edges of the wing, horizontal tailplane, and fin; engine intakes; air inlets; antennae; probes; and exposed hinges. When an airplane collides with droplets of supercooled water (i.e., liquid at a temperature below 0 °C) suspended in the air, the water instantly freezes leading to a build-up of ice. The most severe icing is likely to occur when the outside air temperature (OAT) is in the range 0 °C to −12 °C, although icing can occur at temperatures as low as −40 °C (Figure 22.3). This implies that icing is most likely to occur at altitudes below about 10 000 ft.

Ice accumulations on wing leading edges have the potential to disrupt the airflow severely—reducing lift and increasing drag. The wing's maximum lift coefficient is reduced (due to changes in the wing aerodynamic profile), which implies that the airplane will stall at a lower angle of attack and at a higher stall speed than would otherwise be the case. This situation is exacerbated by the increase in the weight of the airplane due to the ice. The ice can also block Pitot tubes (resulting in a loss of air data) and jam control surfaces. Icing is thus a significant safety concern and steps are taken—in the design and operation of aircraft—to ensure that icing-related risks are minimized.

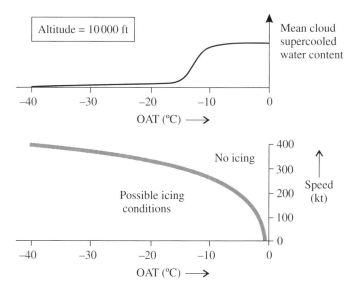

Figure 22.3 Typical icing conditions. *Source*: Adapted from [15].

22.5.2 De-Icing and Anti-Icing Systems

The formation and accumulation of frost, ice, and snow on the surface of an airplane is controlled through the use of de-icing and/or anti-icing systems and procedures. De-icing systems, as the term suggests, remove existing accumulations; whereas, anti-icing systems prevent or delay any further accumulations occurring. Pre-formed ice can, in theory, be removed by several different means—for example, mechanical scrapers, pneumatic boots (which expand to fracture the ice), freezing point depressant (FPD) fluids, or by the application of heat. When the airplane is on the ground, anti-icing is accomplished by treating the aircraft's surface with a glycol-based FPD—this is considered a precautionary measure taken to protect the airplane against the build-up of frost, ice, or snow for a limited time (known as the *holdover* time).

Once airborne, critical components need to be kept free of ice. On most large jet airplanes, hot bleed air systems are used on the wing leading edges and nacelle intakes; electrical heating is used on probes, Pitot tubes, sensors, ports, drains, and so forth. The B787 utilizes an electro-thermal heating system for wing anti-icing and de-icing—heating blankets are bonded to the inside of the slat leading edge [16]. The horizontal and vertical tailplanes are generally not fitted with de-icing systems as current designs can meet certification standards with accumulated ice. De-icing boots are not normally used on jets, but are widely used on turbo-prop airplanes (which have a limited bleed air capability). Liquid anti-/de-icing systems are employed on certain small business jet aircraft.

22.5.3 Impact on Airplane Performance

Bleed air extraction from the engines' compressors (as mentioned in Sections 22.4.2 and 8.8.3) results in a TSFC penalty. The impact of the flight crew switching on the de-icing system during cruise on many airplane types reduces the maximum altitude and sometimes also the optimum altitude (i.e., the altitude that results in the best specific air range). Cruise altitudes (which are discussed in Section 17.3) are published in the FCOM as a function of airplane weight and ambient temperature, with bleed air corrections where applicable.

22.6 Auxiliary Power System

22.6.1 Function

An auxiliary power unit (APU) is a small turbine engine that burns conventional jet fuel (it is usually installed in the rear part of the fuselage). On the ground, the APU can provide power to the electrical systems and bleed air to the air conditioning packs. It can also provide bleed air to start the engines (if a ground unit is not available). On certain airplane types, the APU also operates during takeoff. In flight, the APU can provide back-up power in the event of an engine failure—driving the critical electrical systems and supplying emergency air. As part of the "no-bleed" system architecture adopted for the B787, the APU on this airplane type only provides electrical power.

22.6.2 Fuel Consumption

The APU uses very little fuel in comparison to the total fuel consumed by the engines during a typical mission. The fuel consumption varies depending on the load placed on the unit (and, of course, on the APU type). Typically, an APU installed in a commercial airliner will consume *ca.* 60–80 kg/h (132–176 lb/h) on the ground when running with no load and *ca.* 110–160 kg/h (242–353 lb/h) when supplying power or when starting the engines [17].

As the APU burns much less fuel than an engine, it is prudent for the crew not to start the engines too soon. The extra fuel consumed by running the engines unnecessarily on the ground can result in a significant expense for a fleet over a whole year.

22.7 Fuel and Fuel Systems

22.7.1 Aviation Fuel

Turbine engines are capable of running on a wide range of petroleum-based and synthetic fuels. An introduction to the subject, focusing specifically on the parameters that impact the airplane's performance, is presented in this section.[5] Most of the fuel consumed by aircraft around the world falls into the following three categories:

(1) *Kerosene* is a generic term used to describe a range of petroleum fuel products. It is a hydro-carbon fuel defined by a minimum flash point[6] of 38 °C (100 °F) and an end point[7] of no more than 300 °C (572 °F) [15]. It is the most widely used turbine-engine fuel.
(2) *Aviation gasoline* (AVGAS) is a highly refined hydrocarbon fuel that is blended specifically for use in spark-ignition piston engines designed for aircraft.
(3) *Wide-cut* fuels are essentially hydrocarbon mixtures that span the kerosene and gasoline boiling ranges. They are used in turbine engines in specific instances. The flash point is typically below 0 °C (32 °F), but is not controlled by specification.

Wide-cut fuels have the advantage that a greater amount of fuel can be extracted from a given quantity of crude oil; however, they are inherently more volatile, and are thus less safe from an operational point of view. The four grades of turbine aviation fuel that are widely used for civilian jet operations are listed in Table 22.2, with their respective specification(s). The essential difference between Jet A and Jet A-1 is their freezing points (the freezing point, as defined by ASTM, is the temperature at which the last wax crystal melts on warming, the fuel having previously been cooled with stirring). Jet A-1, which has a freezing point temperature 7 °C lower than that of Jet A, is thus preferred for operations where very cold ambient temperatures are encountered. Jet A, however, is easier to produce as its broader distillation cut permits a refinery to "produce a few percent more Jet A than Jet A-1" [20].

There is also a range of fuels produced specifically for military operators—for example: F-34 and F-35 (codes used by NATO) and fuels prefixed by "JP," whose specifications are controlled by the US military (e.g., JP-5, used widely by the US Navy, and JP-8, used widely by the US Air Force). The main difference between military fuels and commercial fuels is the additives that are used (e.g., to inhibit corrosion or fuel system icing).

Alternative (i.e., non-petroleum) fuels for aviation have been the subject of substantial research interest since *ca.* 2000 [21, 22]—this is primarily driven by two concerns: the supply of crude oil, which is a limited, natural resource, and the environmental impact of aviation. In the-ory, alternative fuels can be produced from a variety of raw materials; however, the extremely demanding requirements for jet propulsion (e.g., energy content, flammability and safety, freez-ing point, storage stability, thermal stability, lubricity, and viscosity), coupled with the generally higher cost of production of synthetic fuels (compared to the current cost of petroleum-based fuels), means that kerosene will remain the principal jet fuel in the medium-term future (i.e., several decades). Biofuels (a term that covers a variety of fuel products manufactured from veg-etable or animal biomass) and synthetic kerosene manufactured using the Fischer-Tropsch (FT)

---

5 Reviews of aviation fuel can be found in references [18–20].
6 The *flash point* is the lowest temperature at which sufficient vapor forms on the surface of the fuel for it to ignite when a flame is applied.
7 The *end point* is the temperature at which all of the fuel will distill into vapor.

process[8] have the potential to replace some petroleum fuel usage in aviation. The first approval[9] for FT-kerosene use in civil aviation was granted in 1999 for a blend containing a maximum of 50% synthetic fuel with conventional kerosene [19]. Since 2009, approvals have been granted for a small number of *drop-in* biofuel processes for civil aviation under the newly developed fuel standard ASTM D7566 [23] (use is also restricted to a maximum of 50% blend with conventional Jet A or Jet A-1 fuel). A drop-in alternative fuel is one that is "fully compatible, mixable, and interchangeable with conventional jet fuel" [22]. Such fuels do not require any adaptation of the aircraft and do not impose any restriction on the use of the aircraft.

Table 22.2 Commonly used grades of jet aviation fuel for civil aircraft[a], [b], [c]

Fuel grade	Specification	Notes
Jet A	ASTM D1655 (Jet A)	This is the principal jet fuel available in the US for civil aviation; it has a maximum freezing point of $-40\,°C$.
Jet A-1	ASTM D1655 (Jet A-1) DEF STAN 91-91 (Jet A-1) JFSCL	This is the most frequently used fuel for civil aviation outside of the US and Russia; it has a maximum freezing point of $-47\,°C$.
TS-1	GOST 10227	This Russian specification covers the jet fuel widely used in Russia, the CIS, and in parts of Eastern Europe; it is similar to Jet A-1.
Jet B	ASTM D1655 (Jet B) CGSB-3.22	This is a wide-cut fuel that has a low flash point; it is not widely used outside of Canada and Alaska.

Notes:

(a) *Sources*: Airbus [15], Hemighaus, *et al.* [20].
(b) The Joint Fueling System Checklist (JFSCL)—which is usually shortened to the *Joint Checklist*—is a standard used by many fuel producers outside of the US, Russia, and China—it combines the most restrictive requirements of ASTM D1655 (Jet A-1) [24] and DEF STAN 91-91 (Jet A-1) [25].
(c) Fuels containing synthesized hydrocarbons approved under ASTM D7566 are considered equivalent to those approved under ASTM D1655 [23].

22.7.2 Fuel Heating Value (Energy Content)

The *fuel heating value* (FHV)—which is also called the *heat of combustion*—is a measure of the fuel's energy content. The amount of heat released when fuel is burned depends on whether or not water formed during combustion condenses to a liquid. A greater amount of heat is released if the water vapor condenses to the liquid phase, giving up its heat of vaporization in the process—this is known as the gross energy content. As jet engines exhaust water as vapor, the heat of vaporization is not considered when determining the *net energy content* of aviation fuel. The *lower heating value* (LHV), which is sometimes called the *lower calorific value* (LCV), is a measure of the net energy content of the fuel—this is the appropriate measure for comparing aviation fuels and for conducting airplane performance analyses.

---

8 The Fischer-Tropsch (FT) process (which is named after Franz Fischer and Hans Tropsch, who developed the process in the 1920s) reacts methane with air, in the presence of a catalyst, to create a carbon monoxide/hydrogen gas mixture, which is then converted into a liquid hydrocarbon fuel.
9 The synthetic FT isoparaffinic kerosene is produced from coal [19].

Table 22.3 Typical properties of turbine-engine aviation fuel

Fuel type	Typical density at 15 °C (60 °F)	Typical heating value	
		per unit mass (i.e., gravimetric)	per unit volume (i.e., volumetric)
Kerosene	0.810 kg/L 6.76 lbm/US gal	43.28 MJ/kg 18 610 Btu/lbm	35.06 MJ/L 125 800 Btu/US gal
Wide-cut	0.762 kg/L 6.36 lbm/US gal	43.54 MJ/kg 18 720 Btu/lbm	33.18 MJ/L 119 000 Btu/US gal

Source: Hemighaus, *et al.* [20].

The net energy content of jet fuel is best expressed in *specific* form, that is, per unit mass or per unit volume.[10] When expressed with respect to unit mass, it is generally called the *gravimetric heating value* or, alternatively, the *specific energy*. The second representation is called the *volumetric heating value* or *energy density*. The typical units for gravimetric heating value are MJ/kg and Btu/lbm; for volumetric heating value, typical units are MJ/L and Btu/US gal. Linking these two parameters is the density of the fuel. The volumetric heating value can be determined from the product of the fuel's gravimetric heating value and its density.[11]

Typical properties (reported by a major fuel supplier) of kerosene and wide-cut fuel are given in Table 22.3. Airplane manufacturers base performance data on a reference fuel. During flight testing, fuel data are collected so that adjustments to airplane flight-test data can be made. Boeing, for example, use a reference value of 18 580 Btu/lbm (43.22 MJ/kg) as a net heating value for the generation of performance data for their airplanes [12].

Notwithstanding the fact that aviation fuel must meet an appropriate specification (see Section 22.7.1) for operational use, the exact properties of aviation fuel will vary from supplier to supplier and from one location to another. This is largely due to varying crude oil composition and differing refining processes. This variation can have an impact—albeit a relatively small one—on the range performance of an airplane, which is a function of the fuel heating value (see Section 13.2.5).

Various experimental methods exist to determine the heating values of combustible substances—and these methods are reported to give slightly different results. To explore the variability of fuel used in airline operations, a set of LHV measurements [26]—acquired by Airbus using a bomb calorimeter (per ASTM D2382[12])—were analyzed and used to construct Figure 22.4. The data illustrates the variation in the gravimetric heating value of typical aviation fuel, based on approximately 120 samples, supplied by airlines operating in different geographical locations worldwide.

---

10 Caution: Because of similar sounding terminology, this parameter can easily be confused with *specific heat capacity* (also called *specific heat*), which is a completely different property of a substance (see Section 2.5.7).

11 For the sake of brevity, the gravimetric/volumetric descriptor is frequently omitted as the context or the units will identify the property of interest (as done in other chapters in this book). Similarly, the qualifier "net"—as in *net heating value* or *net heat of combustion*—is often omitted.

12 Since these tests were conducted, ASTM 2382 (test method for heat of combustion of hydrocarbon fuels by bomb calorimeter) has been withdrawn and replaced by ASTM D4809 [27].

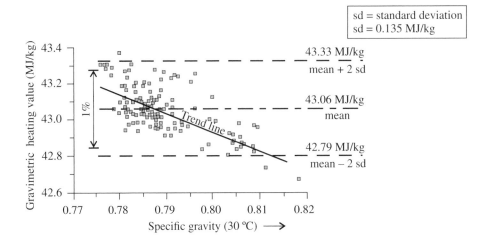

Figure 22.4 Jet fuel gravimetric heating value versus specific gravity. *Source*: Adapted from [26].

It is evident that the limits that represent a 1% variation (i.e., ± 0.5% about the mean) fall within ± 2 standard deviations about the mean. As a point of comparison, it can be noted that the minimum energy content for Jet A and Jet A-1 fuel for approval under ASTM D1655 is a heating value of 42.8 MJ/kg.

An important observation that can be made from Figure 22.4 concerns the relationship between energy content and density. As a general trend, less dense kerosene fuels have relatively high energy content per unit of mass, but relatively low energy content per unit volume. On the other hand, denser fuels tend to have relatively low energy content per unit of mass and relatively high energy content per unit volume. To illustrate this point, isolines (contours) of equal volumetric heating value are shown on a chart of gravimetric heating value versus fuel density (Figure 22.5). Three data points are indicated on the chart—these represent the "typical" kerosene value as reported by Hemighaus, *et al.* [20] (see Table 22.3) and a low and a high density value, based on averages of US fuel [28]. An approximate trend line is shown, illustrating this characteristic of kerosene fuels.

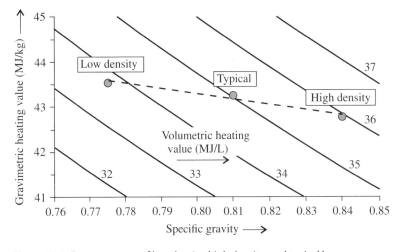

Figure 22.5 Energy content of low density, high density, and typical kerosene.

22.7.3   Influence of Gravimetric Heating Value on Specific Air Range

The heat energy released per unit time due to the continual combustion of fuel in a jet engine is proportional to the product of the fuel mass flow and its gravimetric heating value for a given thermal efficiency—this can be expressed as follows:

$$\dot{E}_h \propto \dot{m}_f H_f \tag{22.1}$$

where $\dot{E}_h$ is the heat energy released per unit time (typical units: kJ/h, Btu/h);
$\quad \dot{m}_f$ is the mass flow rate of the fuel (typical units: kg/h, lbm/h); and
$\quad H_f$ is the gravimetric heating value of the fuel, i.e., the energy content per unit mass (typical units: MJ/kg, Btu/lbm).

To maintain a particular cruise flight condition—that is, corresponding to a specified airplane weight, Mach number, altitude, and air temperature—the engines have to produce a certain thrust, and this requires an appropriate release of heat energy per unit time. If fuel with a reduced gravimetric heating value were to be used, it follows that there would have to be a proportional increase in the mass flow rate to maintain the same thrust (the thrust setting parameter, e.g., N_1 or EPR, is assumed not to change). Changes in engine component efficiencies that arise due to small changes in the gravimetric fuel heating value of kerosene are insignificant.[13]

Of interest is the manner in which a change in net heating value will affect the airplane's cruise performance. The relationship between a change in gravimetric heating value and a change in specific air range (SAR) can be deduced from Equation 13.28a (which applies to a cruise condition where lift is equal to airplane weight and thrust is equal to drag).

$$r_a = \frac{H_f \eta_0 E}{mg} \tag{22.2}$$

where r_a is the specific air range (mass based) (typical unit: km/kg);
$\quad \eta_0$ is the overall engine efficiency (dimensionless);
$\quad E$ is the lift-to-drag ratio in cruise (dimensionless);
$\quad m$ is the airplane mass (typical unit: kg); and
$\quad g$ is the gravitational acceleration (typical unit: m/s^2).

Caution: Unit conversion factors may be required to evaluate Equation 22.2.

Equation 22.2 enables a relationship between a small change in gravimetric heating value (i.e., ΔH_f) and the specific air range (i.e., Δr_a) to be established with respect to a reference flight condition in cruise (i.e., at a particular airplane weight, Mach number, altitude, and air temperature):

$$\frac{\Delta r_a}{r_{a,ref}} = \frac{\Delta H_f}{H_{f,ref}} \tag{22.3}$$

where parameters at the reference condition are denoted by the subscript "*ref.*"

The simple interpretation of Equation 22.3 is that a 1% change in the gravimetric heating value of the fuel will result in a 1% change in the airplane's specific air range. However, as shown in Figure 22.4, a variation of this magnitude is not likely to occur very often for conventional kerosene approved for aviation use.

---

[13] Certain non-petroleum fuels with significantly different compositions (e.g., with different percentages of aromatics) may burn at different rates and consequently the engine's efficiencies may not necessarily be the same when the heating value changes.

22.7.4   Influence of Gravimetric Heating Value on Cruise Fuel

In Section 22.7.3, it was shown that if fuel with a relatively high gravimetric heating value is used, the airplane will have a correspondingly high specific air range, and this means that it will require less fuel (i.e., a reduced fuel mass) to cover a unit distance in cruise. One theoretical approach to quantify this reduction is to consider the cruise sector in isolation (see Section 16.8.1) and to use the "small change expression" derived in Section 16.13.

Before considering the influence of range on the required cruise fuel, it is necessary to relate the relative change in TSFC to the relative change in the fuel gravimetric heating value. It follows from Equation 13.28a that

$$c = \frac{a_0 \sqrt{\theta} M}{\eta_0 H_f} \tag{22.4}$$

where c is the TSFC (based on mass flow rate) (typical unit: mg N^{-1} s^{-1});
$\quad a_0$ is the speed of sound at the ISA sea-level datum (typical unit: m/s);
$\quad \theta$ is the relative temperature of the ambient air (dimensionless); and
$\quad M$ is the flight Mach number (dimensionless).

If the cruise conditions (i.e., airplane weight, Mach number, altitude, and air temperature) remain unchanged, then the relationship between a small change in gravimetric heating value (i.e., ΔH_f) and the change in thrust specific fuel consumption (i.e., Δc) can be established with respect to a reference flight condition (denoted by the subscript "*ref*").

$$\frac{c_{ref} + \Delta c}{c_{ref}} = \frac{H_{f,ref}}{H_{f,ref} + \Delta H_f}$$

$$\frac{\Delta c}{c_{ref}} = \frac{H_{f,ref} - (H_{f,ref} + \Delta H_f)}{H_{f,ref} + \Delta H_f}$$

Thus $\quad \dfrac{\Delta c}{c_{ref}} \cong \dfrac{-\Delta H_f}{H_{f,ref}} \qquad$ as $\quad |\Delta H_f| \ll H_{f,ref}$ $\tag{22.5}$

The change in fuel mass (Δm_f) can be estimated—for a given range (R), true airspeed (V), and mean lift-to-drag ratio ($\bar{E}$)—using the "small change expression" (see Section 16.13). If it is assumed that the lift-to-drag ratio is approximately constant, then it follows from Equation 16.27 that

$$\frac{\Delta m_f}{m_{f,ref}} = \frac{\Delta m_2}{m_2} + \left(\frac{e^\beta \beta}{e^\beta - 1} \right) \frac{\Delta c}{c_{ref}} \tag{22.6}$$

where $\beta = \dfrac{R c_{ref} g}{V \bar{E}}$

and $\quad m_{f,ref}$ is the cruise fuel mass (reference condition) (typical unit: kg);
$\quad m_2$ is the end-of-cruise airplane mass (typical unit: kg);
$\quad c_{ref}$ is the mean TSFC (reference condition) (typical unit: mg N^{-1} s^{-1}); and
$\quad \Delta c$ is the change in TSFC (typical unit: mg N^{-1} s^{-1}) due to ΔH_f.

Substituting from Equation 22.5 into Equation 22.6 yields the required result:

$$\frac{\Delta m_f}{m_{f,ref}} = \frac{\Delta m_2}{m_2} + \left(\frac{e^\beta \beta}{e^\beta - 1} \right) \left(\frac{-\Delta H_f}{H_{f,ref}} \right) \tag{22.7}$$

Equation 22.7 describes the "gearing" effect inherent in this situation, by which the resulting change in the required cruise fuel (when expressed as a percentage) is greater in magnitude than the change in the heating value (when expressed as a percentage)—the magnitude of this "gearing" depends on the cruise range. The change in the required cruise fuel also depends on any change to the end-of-cruise airplane mass. For a fixed operating empty weight (OEW) and a fixed payload, this can arise due to the change in density of the onboard reserve fuel (which contributes to the end-of-cruise airplane mass for a typical mission).

22.7.5 Influence of Fuel Energy Content on Payload–Range

The simplified cruise analysis presented in Section 22.7.4 appears to indicate that fuel with a high H_f is always preferential to fuel with a low H_f. This conclusion, however, is not correct for all situations. The boundary conditions have not been fully considered; furthermore, a full analysis requires the complete mission—not just the cruise—to be considered.

Varying fuel energy content affects different parts of the payload–range envelope (see Section 16.6) in different ways. The upper boundary of the envelope is constrained by the airplane's maximum takeoff weight limit (see Figure 16.7)—the maximum range corresponds to the airplane taking off with the maximum design payload, but the fuel tanks are not full. As the *mass* of the fuel is limited, the greatest range will be associated with fuel with a high gravimetric heating value (in line with the explanation given in Section 22.7.4).

It is a different scenario, however, when determining how far an airplane can fly when the constraint is the capacity of the fuel tanks. As a general trend, less dense fuels (i.e., fuels with a low specific gravity) have relatively high energy content per unit of mass, but relatively low energy content per unit volume; the opposite is true for denser fuels (see Figure 22.5). This mass versus volume distinction was not important in the discussion presented in Sections 22.7.4; however, when the onboard fuel is limited by volume—for example, when the tanks are filled or when a fixed volume of fuel is uploaded—the distinction becomes important. The lower right-hand boundary of a payload–range envelope is constrained by the size of the fuel tanks (see Figure 16.7)—here, the greatest possible range will be associated with fuel with a high volumetric heating value. The problem can be assessed by recalling the basic range expression (see Section 13.2); the range (R) is obtained by integrating the function from the final airplane mass (m_2) to the initial airplane mass (m_1), that is,

$$R = \int_{m_2}^{m_1} \frac{V}{\dot{m}_f}\,dm \qquad \text{(from Equations 13.3 and 13.5)} \tag{22.8}$$

It is evident from Equation 22.1 that fuel with a low H_f will require a proportional increase in the mass flow rate to maintain a given flight condition. This factor, taken in isolation, indicates a reduction in range for low H_f fuel. However, fuel with a high density will increase the start-of-cruise airplane mass (for a given payload mass), and this raises the upper limit of the integral, which increases the range. The net result—that is, a decrease or increase in range—thus depends on the precise amounts by which the gravimetric heating value and the density of the fuel vary from a baseline condition.

Detailed investigations into this topic require a computer model capable of calculating an airplane's range for a given fuel load for a complete mission (as illustrated in Figure 16.5), which also takes into account the mass of the reserve fuel. A study undertaken by the author [29] utilized such a computer model to determine the range for a given fuel load for a medium-range, twin-engine airliner (based on the B757-200). The maximum range associated with the maximum takeoff weight condition for the "typical" kerosene was used as the reference condition and two extreme cases—that is, the "low density" and "high density" kerosene fuels identified in

Figure 22.5—were then considered. Fuel tanks full of "low density" fuel would have relatively low total energy content—and this would translate to a reduction in range compared to the baseline condition established for the "typical" kerosene. The "high density" kerosene produces an interesting situation, as there are two factors that work in opposing directions. The fuel would have a greater heating value per unit volume, but it would also have a greater density and therefore a greater total fuel mass. The increase in the airplane's takeoff mass means a correspondingly greater rate of fuel burn (measured per unit mass); however, the greater available fuel mass can dominate the calculation resulting in an increase in range compared to the baseline "typical" kerosene. But, as indicated above, the outcome depends on the precise characteristics of the fuel.

22.7.6   Low Temperature Operations

Prolonged flight operations at very cold temperatures can cause parts of the fuel to solidify forming wax crystals when the fuel is cooled below its freezing point (aviation fuels are complex mixes of hydrocarbons, and not all parts solidify at the same temperature). Jet A is required to have a freezing point value no greater than −40 °C and Jet A-1 a value no greater than −47 °C (see Section 22.7.1). The freezing point of actual fuels (i.e., consumed in daily flight operations), however, varies and can be considerably lower than the specification limits. Solid particles formed at cold temperatures can cause flow restrictions and blockages in fuel filters, valves, pump inlets, and so forth. This is clearly dangerous, and monitoring of fuel temperatures by flight crews is routine for cold temperature operations (typically ensuring that the fuel temperature is at least 3 °C higher than its freezing point). Aviation fuels often contain small amounts of water, which also freezes (at temperatures higher than the fuel freezing point). Fuel systems in modern aircraft, however, are designed to cope with this occurrence.[14]

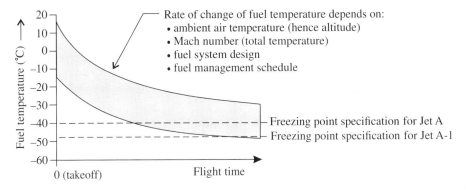

Figure 22.6 Illustration of how fuel temperature reduces with increasing flight time.

During cruise, the temperature of the fuel will progressively decrease—the rate of decrease is a function of the ambient air temperature and several other factors (Figure 22.6). This is an important consideration for the planning and operation of long-haul flights where ambient air temperatures are very low, as can be encountered on polar routes, for example. Selecting a higher Mach number (which will give a greater total air temperature) or a lower flight level

---

14 An unanticipated event resulted in the crash of a B777-200ER airplane (flight BA38) on approach to Heathrow airport in 2008. This was caused by clogging of the fuel/oil heat exchangers by ice accreted from within the fuel (abnormally cold temperatures had been experienced during the 4400 nm flight from Beijing to London).

(with a higher ambient air temperature) improves the situation, as the fuel temperature will not decrease as quickly. Cruising at higher speeds at non-optimum altitudes will, however, consume more fuel.

22.7.7   Fuel Jettison System

Fuel jettison, or fuel dumping as it is also called, is used to reduce an airplane's weight in certain emergency situations, such as when the airplane is required to return to the departure airport or to land at another airport soon after takeoff. This could be the result of mechanical failure, a system warning, an operational abnormality, or a passenger medical emergency, for example. A reduced landing weight reduces the landing speed, which may be desirable for the particular situation.

The landing weight should not be greater than the airplane's certified maximum landing weight (MLW). In an emergency, however, an overweight landing (i.e., above the airplane's MLW) should not pose a risk to the crew or passengers due to the conservatism in the design process. For example, the landing gear is designed to withstand a 6 ft/s impact at the maximum takeoff weight (MTOW). An inspection of the airplane for damage, however, is mandatory following such a landing. Depending on the airplane type, the MLW (see Section 19.2) may be close to the MTOW or there may be a large difference between the two limits (this is usually the case with long-haul airplane types).

Not all airplane types have the ability to jettison fuel. Many short- and medium-range twin-engine airplanes (e.g., B737, B757, A320 family) and regional jets do not have this facility as they meet the relevant airworthiness requirements (i.e., FAR 25.1001 [1]) without it. This is essentially due to the fact that for these airplane types the MLW is not much less than the MTOW. In the event of an emergency—and depending on the nature of the emergency—the crew would reduce the airplane's weight by burning off fuel (e.g., by flying in a holding pattern near to an airport). On certain long-range types (e.g., B747, MD-11, A340) the system is installed in order to meet the airworthiness regulation, and on other types (e.g., B767, B777) it can be installed, although it is not a requirement.

Fuel is pumped overboard through nozzles located near to the wingtips. The fuel jettison rate depends on the airplane type. The B777-200, for example, can dump ~2100 kg/min (when all pumps are operating). As the difference between the certified MTOW and the MLW is 45 360 kg, this means that the maximum "overweight" fuel quantity can be jettisoned in ~21 minutes. Studies reported by Colella [30] have shown that, "in general, fuel jettisoned above 5000 to 6000 ft will completely vaporize before reaching the ground … although there is no restriction on jettisoning at lower altitudes if considered necessary by the flight crew."

References

1 FAA, *Airworthiness standards: Transport category airplanes*, Federal Aviation Regulation Part 25, Amdt. 25-143, Federal Aviation Administration, Washington, DC, June 24, 2016. Latest revision available from www.ecfr.gov/ under e-CFR (Electronic Code of Federal Regulations) Title 14.

2 EASA, *Certification specifications and acceptable means of compliance for large aeroplanes*, CS-25, Amdt. 18, European Aviation Safety Agency, Cologne, Germany, June 23, 2016. Latest revision available from www.easa.europa.eu/ under Certification Specification.

3 FAA, *Operating requirements: Domestic, flag, and supplemental operations*, Federal Aviation Regulation Part 121, Amdt. 121-374, Federal Aviation Administration, Washington, DC,

May 24, 2016. Latest revision available from www.ecfr.gov/ under e-CFR (Electronic Code of Federal Regulations) Title 14.

4 European Commission, *Commercial air transport operations (Part-CAT)*, Annex IV to Commission Regulation (EU) No. 965/2012, Brussels, Belgium, Oct. 5, 2012. Published in *Official Journal of the European Union*, Vol. L 296, Oct. 25, 2012, and reproduced by EASA.

5 FAA, "System design and analysis," Advisory Circular 25.1309-1A, Federal Aviation Administration, Washington, DC, June 21, 1988. Available from www.faa.gov/.

6 Campbell, R.D. and Bagshaw, M., *Human performance and limitations in aviation*, Blackwell Science, Oxford, UK, 1997.

7 FAA, "Pressurization, ventilation and oxygen systems assessment for subsonic flight including high altitude operation," Advisory Circular 25-20, Federal Aviation Administration, Washington, DC, Sept. 10, 1996. Available from www.faa.gov/.

8 Nelson, T., "787 Systems and performance," Flight Operations Engineering, Boeing Commercial Airplane Company, Seattle, WA, Mar. 31, 2009. [Presentation.]

9 Airbus, "Getting to grips with aircraft performance," Flight Operations Support and Line Assistance, Airbus S.A.S., Blagnac, France, Jan. 2002.

10 Airbus, "A330 flight deck and systems briefing for pilots," STL 472.755/92, Iss. 4, Airbus Industrie, Blagnac, France, Mar. 1999.

11 SAE, *On board oxygen generating systems (molecular sieve)*, AIR825/6A, SAE International, Warrendale, PA, Dec. 4, 2015.

12 Blake, W. and Performance Training Group, "Jet transport performance methods," D6-1420, Flight Operations Engineering, Boeing Commercial Airplanes, Seattle, WA, Mar. 2009.

13 Hunt, E.H., Reid, D.H., Space, D.R., and Tilton, F.E., "Commercial airliner environmental control system," *Aerospace Medical Association Annual Meeting*, Anaheim, CA, May 1995.

14 Boeing, "How the environmental control system works on Boeing airplanes," Boeing Commercial Airplanes, Seattle, WA, Retrieved Apr. 2014. Available from www.boeing.com/boeing/commercial/cabinair/environmentcontrol.page.

15 Airbus, "Getting to grips with cold weather operations," Flight Operations Support and Line Assistance, Airbus Industrie, Blagnac, France, Jan. 2000.

16 Sinnett, M., "787 No-bleed systems: Saving fuel and enhancing operational efficiencies," *Aero*, The Boeing Company, Seattle, WA, Vol. Q 04, pp. 6–11, 2007.

17 Airbus, "Getting to grips with fuel economy: Managing flight operations with recommendations on fuel conservation," Iss. 2, Flight Operations Support and Line Assistance, Airbus S.A.S., Blagnac, France, Nov. 2001.

18 Edwards, T., "Advancements in gas turbine fuels from 1943 to 2005," *Journal of Engineering for Gas Turbines and Power*, Vol. 129, Iss. 1, pp. 13–20, 2007.

19 Hemighaus, G., Boval, T., Bosley, C., *et al.*, "Alternative jet fuels: Supplement to aviation fuels technical review," FTR-3/A1, Chevron Products Company, Houston, TX, 2006.

20 Hemighaus, G., Boval, T., Bacha, J., *et al.*, "Aviation fuels technical review," FTR-3, Chevron Products Company, Houston, TX, 2006.

21 Blakey, S., Rye, L., and Wilson, C.W., "Aviation gas turbine alternate fuels: A review," *Proceedings of the Combustion Institute*, Vol. 33, Iss. 2, pp. 2863–2885, 2011.

22 IATA, "IATA 2015 report on alternative fuels," 10th ed., International Air Transport Association Montréal, Canada and Geneva, Switzerland, Dec. 2015.

23 ASTM, *Standard specification for aviation turbine fuel containing synthesized hydrocarbons*, D7566-16B, ASTM International, West Conshohocken, PA, July 2016.

24 ASTM, *Standard specification for aviation turbine fuels*, Standard D1655–13A, ASTM International, West Conshohocken, PA, Dec. 1, 2013.

25 Ministry of Defence, *Turbine fuel, aviation kerosine type, Jet A-1; NATO code: F-35; Joint service designation: AVTUR*, DEF-STAN 91-91, Iss. 7 (Amdt. 2), United Kingdom Ministry of Defence, Glasgow, UK, Dec. 1, 2012.

26 Airbus, "Getting to grips with aircraft performance monitoring," Flight Operations Support and Line Assistance, Airbus S.A.S., Blagnac, France, Dec. 2002.

27 ASTM, *Standard test method for heat of combustion of liquid hydrocarbon fuels by bomb calorimeter (precision method)*, Standard D4809, ASTM International, West Conshohocken, PA, 2006.

28 Speyer, A.A., "Auditing aircraft cruise performance in airline revenue service," *7th Performance and Operations Conference*, Cancun, Mexico, 1992.

29 Young, T.M., "Simplified methods for assessing the impact of fuel energy content on payload-range," *26th International Congress of the Aeronautical Sciences (ICAS) and 8th AIAA ATIO Conference*, Anchorage, AK, Sept. 14–19, 2008, AIAA 2008-8857, American Institute of Aeronautics and Astronautics.

30 Colella, R., "Overweight landing? Fuel Jettison? What to consider," *Aero*, The Boeing Company, Seattle, WA, Vol. Q 03, pp. 15–21, 2007.

23

Authorities, Regulations, and Documentation

23.1 Introduction

Airplane performance is at the center of aviation safety, and the production of accurate and comprehensive documentation that describes the capabilities and limitations of jet transport airplanes is a key element of the safety framework that underpins commercial aviation. Descriptions of the performance requirements and capabilities of jet transport airplanes are prepared by different organizations (e.g., ICAO, regulatory authorities, manufacturers, owners/operators) for different purposes—and this is recorded in a complex, interlinked hierarchy of documentation.

At a high level in this hierarchy are documents that describe the minimum performance capabilities that regulatory authorities stipulate for the certification of new airplane types. Subservient to these regulations are the descriptions of the airplane's demonstrated capabilities as determined by the manufacturer during flight testing, and at a lower level in this hierarchy are documents containing performance data that are used by flight planning personnel and flight crews for day-to-day flight operations.

The International Civil Aviation Organization (Section 23.2) works closely with national aviation authorities (Section 23.3) in regulating and promoting safe and efficient air travel. The two most important regulatory organizations for civil aviation from a global perspective are the United States Federal Aviation Administration and the European Aviation Safety Agency. Important regulations for airplane certification and operation are highlighted in Section 23.4. The responsibilities of safety investigation authorities (air accident investigation authorities) are outlined in Section 23.5. There are also several non-governmental organizations that play an important role in international air travel (Section 23.6). A brief review of the most important documents pertaining to the performance and flight operations of jet transport airplanes is presented in Section 23.7. The generation and exchange of airplane performance data for flight operations are discussed in Section 23.8.

23.2 International Civil Aviation Organization

23.2.1 Introduction

The International Civil Aviation Organization (ICAO), based in Montréal, Canada, is a specialized agency of the United Nations, which is charged with coordinating and regulating international air travel. It has a membership of nearly 200 contracting states (in 2016), including all states (countries) with a significant aviation industry.

Performance of the Jet Transport Airplane: Analysis Methods, Flight Operations, and Regulations, First Edition.
Trevor M. Young.
© 2018 John Wiley & Sons Ltd. Published 2018 by John Wiley & Sons Ltd.

ICAO has a mandate to administer the principles of the Convention on International Civil Aviation—an international agreement signed in Chicago in 1944 (and revised several times since). The sovereign body is the Assembly, which is composed of members of all contracting states—formal meetings of the Assembly are convened at least once every three years. The Council, which is the governing body, is composed of 33 members elected by the Assembly. The Council is responsible for the development and implementation of policy documents, which include Standards and Recommended Practices (SARPs) and Procedures for Air Navigation Services (PANS).

ICAO works in close collaboration with other specialized agencies and organizations, such as the International Air Transport Association (IATA), the International Federation of Air Line Pilots' Associations (IFALPA), and the World Meteorological Organization (WMO).

23.2.2 The Convention on International Civil Aviation

The Convention on International Civil Aviation [1] defines a broad-ranging set of standards relating to the international operations of aircraft. The convention is supported by 19 annexes containing SARPs, which are regularly updated. SARPs do not have a legal standing; countries that adopt these regulations translate the SARPs into their own laws. Contracting states are obliged to publish, in their respective Aeronautical Information Publications (see Section 23.4.6), details of any significant differences between the adopted procedures and the related ICAO procedures.

There is a widespread practice to shorten the title of certain ICAO documents—for example, reference would be made to "ICAO Annex 3," for example, omitting mention of the "Convention on International Civil Aviation," as this is understood. Many of the annexes contain details that directly or indirectly have a bearing on the performance and flight operations of jet transport airplanes—these include: Annex 2 (Rules of the Air); Annex 3 (Meteorological Service for International Air Navigation); Annex 4 (Aeronautical Charts); Annex 5 (Units of Measurement to be Used in Air and Ground Operations); Annex 6 (Operation of Aircraft), specifically Part I (International Commercial Air Transport: Aeroplanes); Annex 8 (Airworthiness of Aircraft); Annex 10 (Aeronautical Telecommunications); Annex 11 (Air Traffic Services); Annex 14 (Aerodromes), specifically Vol. I (Aerodrome Design and Operations); Annex 15 (Aeronautical Information Services); and Annex 16 (Environmental Protection).

23.2.3 Committee on Aviation Environmental Protection

The Committee on Aviation Environmental Protection (CAEP), a technical committee of the ICAO Council, conducts most of the organization's work concerning the environmental impact of aviation, with a specific focus on aircraft noise and emissions. The CAEP was established in 1983, superseding both the Committee on Aircraft Noise (CAN) and the Committee on Aircraft Engine Emissions (CAEE). The CAEP undertakes environmental studies and prepares for the Council recommendations for new policies and SARPs. The CAEP has developed certification standards for aircraft noise (see Section 21.3), local air quality (related to the landing and takeoff cycle), and—most recently—carbon dioxide emissions in cruise (see Section 21.6.2). Formal meetings of the CAEP take place at least once every three years, in a schedule aligned with meetings of the ICAO Assembly.

23.2.4 ICAO Documents

ICAO publishes a wide range of documents (e.g., manuals, guides, and circulars)—many of which are applicable to airplane performance and flight operations, as illustrated by the following sample:

- Manual of the ICAO Standard Atmosphere (ICAO Doc. 7488) [2];
- Procedures for Air Navigation Services: Aircraft Operations, Vol. 1 Flight Procedures (ICAO Doc. 8168) [3];
- Guidance on the Balanced Approach to Aircraft Noise Management (ICAO Doc. 9829) [4];
- Aerodrome Design Manual: Runways (ICAO Doc. 9157) [5]; and
- Air Traffic Management (ICAO Doc. 4444) [6].

23.3 Aviation Authorities

23.3.1 Introduction

National aviation authorities (NAAs)—called civil aviation authorities in many countries—are statutory organizations, established by governments of states, to oversee the approval, regulation, and safe operation of civil aviation within the jurisdiction of the state. The responsibilities of aviation authorities of ICAO contracting states, in respect of the airworthiness of aircraft, is described in ICAO Annex 8 [7] and in the ICAO Airworthiness Manual [8]. Aviation authorities are entrusted by their respective states to maintain the national registry of aircraft, and to issue and renew type certificates and certificates of airworthiness (see Section 23.4.2).

The aviation authorities typically regulate the following activities: design and manufacture of aircraft, maintenance of aircraft, licensing of personnel, licensing of airports and ground-based navigational aids, operation of aircraft, and the use of the country's airspace. Aviation authorities are often referred to as *regulatory authorities* due to their role in developing and/or implementing regulations.

23.3.2 Federal Aviation Administration

The Federal Aviation Administration (FAA), with headquarters in Washington, DC, is an agency of the US Department of Transportation. The FAA was established in 1958, absorbing the responsibilities of the former Civil Aeronautics Administration. It is responsible for overseeing all aspects of civil aviation in the United States. One of the FAA's key activities concerning the airworthiness of aircraft is the development and implementation of Federal Aviation Regulations (FARs) and supporting documentation, such as Advisory Circulars (ACs). The FARs address a diverse range of aviation activities, which includes the certification and operation of jet transport airplanes (see Section 23.4).

23.3.3 European Aviation Safety Agency

The European Aviation Safety Agency (EASA), with headquarters in Cologne, Germany, is responsible for civil aviation safety in the European Union (EU). EASA was created in 2003, and it progressively took over from the Joint Aviation Authorities (JAA) the responsibility for regulating airworthiness and aviation safety issues within EU member states and several associated non-EU European states.[1] The JAA (which was initially called the Joint Airworthiness Authorities) was created in 1970 as an associated body to represent the national civil aviation authorities of a number of European states.[2] Since 2008, EASA has had jurisdiction over new type certificates and other design-related airworthiness approvals [11]. EASA works closely with the national aviation authorities of its members and has taken over many of their functions in the interest of industry standardization (NAAs retain certain legacy responsibilities).

23.3.4 Aviation Authorities Worldwide

Most sovereign states worldwide—either individually or collectively in groups—have established their own aviation authority. The list of national aviation authorities, established in countries with large aviation activities, includes the following: Civil Aviation Safety Authority (CASA, Australia), Agência Nacional de Aviação Civil (ANAC, Brazil), Transport Canada (TC, Canada), Civil Aviation Administration of China (CAAC, People's Republic of China), Direction Générale de l'Aviation Civile (DGAC, France), Luftfahrt-Bundesamt (LBA, Germany), Directorate General of Civil Aviation (DGCA, India), Italian Civil Aviation Authority (ENAC, Italy), Japan Civil Aviation Bureau (JCAB, Japan), Swedish Transport Agency (STA, Sweden), Federal Air Transport Agency (FATA, Russian Federation), Civil Aviation Authority (CAA, United Kingdom), and Federal Aviation Administration (FAA, United States of America).

The Interstate Aviation Committee, denoted by the abbreviation IAC or, alternatively, by the Russian equivalent MAK has its headquarters in Moscow, Russia. The IAC, which was created in 1991 by an intergovernmental agreement between a number of countries of the former USSR (Union of Soviet Socialist Republics), is responsible for civil aviation safety in signatory countries.[3]

Historically, NAAs developed differing rules and regulations. In fact, one of the reasons for the establishment of the International Civil Aviation Organization in 1947 was to resolve such conflicting issues. Harmonization of national regulations has been a slow, but steady activity (which still continues today). As regards the requirements for the certification and operation of jet transport airplanes, enormous progress was made by the European JAA (between *ca.* 1970 and 2006), initially in developing and implementing common regulatory standards in Europe, and later, in conjunction with the FAA, in harmonizing substantial elements of the joint European regulations with those of the United States.

---

1 EU member states (2016) are Austria, Belgium, Bulgaria, Croatia, Cyprus, Czech Republic, Denmark, Estonia, Finland, France, Germany, Greece, Hungary, Ireland, Italy, Latvia, Lithuania, Luxembourg, Malta, Poland, Portugal, Romania, Slovak Republic, Slovenia, Spain, Sweden, the Netherlands, and the United Kingdom (in 2016, the UK voted to leave the EU). In addition, the non-EU countries Iceland, Liechtenstein, Norway, and Switzerland participate in the activities of EASA under a separate agreement [9].

2 The JAA was initiated by the European Civil Aviation Conference (ECAC), which is an intergovernmental organization located in Paris, France. The ECAC was established in 1955 by ICAO and the Council of Europe with the goal of developing safe, efficient, and sustainable air transport systems in Europe. The ECAC has 44 members (2016), including all EASA member states [10]. Harmonizing civil aviation policies and practices among its member states and developing agreed policies with other countries have been key priorities of the ECAC.

3 The countries served by the IAC (2016) are Azerbaijan, Armenia, Belarus, Georgia, Kazakhstan, Kyrgyzstan, Moldova, Russian Federation, Tadjikistan, Turkmenistan, Ukraine, and Uzbekistan [12].

23.4 Regulations, Certification, and Operations

23.4.1 Regulations

Aviation authorities (see Section 23.3) provide the regulatory framework for the approval of new aeronautical products. They are responsible for the issue of *type certificates* for aircraft (see Section 23.4.2). They are also responsible for the approval of organizations (and, in specific cases, personnel) involved in the design, manufacture, maintenance, or operation of aircraft. The authorities, over many years, have established a series of regulations to ensure that all aircraft engaged in public transportation meet a minimum standard of safety that has been deemed appropriate for the intended operation of the aircraft. These regulations—which are variously titled regulations, requirements, specifications, or standards—are informally referred to as *the rules*. Two complementary sets of measures exist—these measures contain specific details regarding the performance of jet transport airplanes. The first set of measures is concerned with the *certification* of new aircraft (see Section 23.4.3); the second set is concerned with the safe *operation* of aircraft (see Section 23.4.4).

In the United States, the regulations pertaining to the initial certification and subsequent operation of jet transport airplanes are published under Chapter 1 of the Aeronautics and Space (Title 14) section of the US Code of Federal Regulations (CFR).[4] An individual regulation, which is understood by the abbreviation FAR (Federal Aviation Regulation), is called a *part*—for example, Part 1 is Definitions and Abbreviations, and Part 3 is General Requirements (there is no Part 2). In Europe, EASA[5] publishes Certification Specifications (i.e., the rules for initial airplane certification) and technical documents for continued airworthiness and safe operation. Most of the topics addressed by EASA in these specifications and documents roughly match those contained in the FARs (although titles, headings, and specific details may differ).[6] Airworthiness regulations carry legal status—in other words, compliance is mandatory and enforceable by law. In the case of EASA, the regulations/specifications are enforceable in EU member states through non-legislative acts (regulations) passed by the European Parliament and in non-EU EASA member states through parallel actions undertaken by respective national governments.

23.4.2 Certificate of Airworthiness and Type Certificate

A *certificate of airworthiness* (C of A) is issued by the national aviation authority of the country in which the airplane is registered. It grants authorization for a particular airplane—as identified by the manufacturer's serial number (MSN)—to operate in that country's airspace, and, through international agreements, in the airspaces of other countries. This is described in Article 31 of the ICAO Convention on International Civil Aviation (see Section 23.2.2), which states that "every aircraft engaged in international navigation shall be provided with a certificate of airworthiness issued or rendered valid by the State in which it is registered." It is a requirement that the certificate of airworthiness be carried on board every aircraft engaged in international

---

4 Current aviation regulations are available as the Electronic Code of Federal Regulations (e-CFR) at the US Government Printing Office (GPO) website www.ecfr.gov/.
5 Current EASA specifications and regulations are available at the EASA website www.easa.europa.eu/.
6 The FAA and EASA publish lists of significant standards differences (SSD) and non-significant standards differences (non-SSD) that exist between their respective regulatory documents—for example, between FAR 25 and CS-25.

air navigation (according to Article 29 of the ICAO Convention on International Civil Aviation). Renewal of the certificate of airworthiness is subject to demonstration of compliance with the requirements for continuing airworthiness, as stipulated by the aviation authority of the country in which the airplane is registered. This includes periodic inspections of the airplane by approved personnel.

Only an airplane for which the manufacturer holds a *type certificate* (TC) and which was produced according to the approved design is eligible for the issue of a certificate of airworthiness. A type certificate is issued by the aviation authority of a country to an airplane manufacturer to certify that the design of the airplane type meets the appropriate airworthiness requirements (i.e., certification standards) of that country. The certification process involves a detailed investigation, which is conducted in accordance with stipulated procedures, into all aspects of the design, manufacture, and performance capabilities of the airplane to demonstrate compliance with the airworthiness requirements (see Section 24.4.3). Flight testing is an essential part of the validation of the airplane's performance. The formal record of this process is a Type Certificate Data Sheet (TCDS)—this document contains a summary of the information required for type certification (e.g., engine make and model, weight limits, and critical performance parameters). Today, FAA and EASA certification is usually undertaken simultaneously; this normally marks the commencement of serial production of the airplane type.

Once a type certificate has been issued, the design cannot be changed without the approval of the aviation authority. Modifications or design changes to the airplane type are approved through the issuing of a *supplemental (or supplementary) type certificate* (STC), once compliance with the authority's requirements has been demonstrated. The STC describes the product modification and identifies any changes to the certification basis.

23.4.3 Requirements for Airplane Certification

The requirements for the issue of a type certificate (see Section 23.4.2) are described in formal documents published by the regulatory authorities—these are variously known as regulations, requirements, standards, or specifications. The set of regulations/specifications for the *certification* of jet transport airplanes by the FAA and EASA include the following key documents:[7]

- FAR 25 (Federal Aviation Regulation Part 25), Airworthiness Standards: Transport Category Airplanes [13];
- EASA CS-25 (EASA Certification Specification 25), Certification Specifications and Acceptable Means of Compliance for Large Aeroplanes: Book 1[8] [14];
- FAR 33 (Federal Aviation Regulation Part 33), Airworthiness Standards: Aircraft Engines [15]; and
- EASA CS-E (EASA Certification Specification E), Certification Specifications and Acceptable Means of Compliance for Engines: Book 1 [16].

These documents are regularly updated, with amendment numbers assigned to indicate changes. The regulatory authorities also publish supplemental information that provides

---

7 The FAA regulations (i.e., FAR 25 and FAR 33) and the corresponding EASA specifications (i.e., CS-25 and CS-E) for airplane and engine certification are almost identical, but there are some differences between the two sets of requirements which manufacturers and operators need to be aware of. For example, the FAA primarily uses USC units, whereas EASA uses a mix of SI and USC units.

8 CS-25 and CS-E are each divided into two books: Book 1 contains the Certification Specification (CS) and Book 2 contains the corresponding Acceptable Means of Compliance (AMC).

guidelines for manufactures to demonstrate compliance with the regulations/specifications. These supplemental documents provide information on methods, procedures, and practices that are acceptable to the authorities. The FAA issues Advisory Circulars (ACs) for such purposes and the European authorities issue Acceptable Means of Compliance (AMC) and Guidance Material (GM). In the latter documents, the associated paragraph number from the standard is preceded by "AMC" or "GM"—for example, AMC 25.101 contains acceptable means of compliance for the requirements given in CS-25.101.

The following are examples of such documents:

- FAA AC 25 is a series of Advisory Circulars covering a wide range of certification-related topics—for example: FAA AC 25-7 is a "Flight Test Guide for Certification for Transport Category Airplanes" [17] and FAA AC 25-13 describes "Reduced and Derated Takeoff Thrust (Power) Procedures" [18].
- EASA CS-25 Book 2 is the companion document to EASA Certification Specification 25, containing the "Acceptable Means of Compliance for Large Aeroplanes" [14].

23.4.4   Requirements for Airplane Operations

The following are key regulatory documents that pertain to the commercial *operation* of jet transport airplanes:

- FAR 91 (Federal Aviation Regulation Part 91), General Operating and Flight Rules [19];
- FAR 121 (Federal Aviation Regulation Part 121), Operating Requirements: Domestic, Flag, and Supplemental Operations [20]; and
- Commercial Air Transport Operations (OPS Part-CAT),[9] published in European Commission Regulation (EU) No. 965/2012 Annex IV [22].

Supporting documents, covering a wide range of topics, are also issued to provide additional information to demonstrate compliance with the regulations/specifications—for example:

- FAA AC 120-27 (Advisory Circular 120-27), Aircraft Weight and Balance Control [23];
- FAA AC 121.195-1A (Advisory Circular 121.195-1A), Operational Landing Distances for Wet runways; Transport Category Airplanes [24]; and
- Acceptable Means of Compliance (AMC) and Guidance Material (GM) to OPS Part-CAT [25].[10]

23.4.5   Continued Airworthiness and Safety Notifications

Aviation authorities utilize a variety of instruments to alert and inform the aviation community of safety issues relating to in-service aircraft. Several key document types are highlighted in this section.

---

9 Following the establishment of EASA, European regulations for commercial air transport operations underwent several revisions. JAR-OPS 1, the Joint Aviation Requirement of the JAA, was initially transposed, with changes, into European law as EC Regulation No. 859/2008 [21] in 2008. These regulations, which became known as EU OPS, were superseded in 2012 by EC Regulation No. 965/2012 [22], which adopted a substantially revised format.
10 European rules on air operations are published in the Official Journal of the European Union as commission regulations (or amendments), with associated AMC and GM published by EASA as "decisions." EASA periodically produce technical publications containing regulations and related AMC and GM in a consolidated format as advisory information [26].

An Airworthiness Directive (AD) is a notification, issued by an aviation authority, to advise owners and operators that a known safety deficiency exists and that it must be corrected as instructed. ADs are issued when (1) an unsafe condition exists in a product (e.g., aircraft, engine, propeller, or appliance) and (2) the condition is likely to exist or develop in other products of the same type design [27]. ADs are mandatory, and failure to comply with the instructions results in the airplane not being considered airworthy.

A Special Airworthiness Information Bulletin (SAIB), as issued by the FAA, is an information tool that alerts, educates, and makes recommendations to the aviation community [28]. SAIBs contain non-regulatory information and guidance that do not meet the criteria for the issuing of an AD. The EASA equivalent of a SAIB is a Safety Information Bulletin (SIB). Critical safety information for operators can also be issued by the FAA using a Safety Alert For Operators (SAFO). A SAFO may contain information alone or a combination of information and recommended (i.e., non-regulatory) action to be undertaken voluntarily by the respective operators identified in each SAFO [29].

A Certification Maintenance Requirement (CMR) is a required scheduled maintenance task established by a regulatory authority during the design certification of the airplane as an operating limitation of the type certificate (TC) or supplemental type certificate (STC). CMRs are a subset of the instructions for continued airworthiness of the airplane [30].

23.4.6   Aeronautical Information Publication (AIP)

Aeronautical Information Publications (AIPs), which are issued by or with the authority of national states (the responsibility is usually delegated to the national aviation authority), contain aeronautical information that is of a lasting nature and is essential to safe air navigation within the state.

The structure and content of these documents follow a standardized format, which is described in ICAO Annex 15 [31]. An AIP typically has three parts: GEN (general), ENR (*en route*), and AD (aerodromes). Charts, providing essential information for flight crews to operate into and out of public airports, for example, are published in the AD section. An AIP also contains a description of the country's airspace (see Section 5.4.4). Revisions to an AIP are published on a defined 28-day cycle, known as the AIRAC (Aeronautical Information Regulation and Control) cycle.

The aviation authorities of some countries also publish parallel information for domestic aviation. In the United States, this is contained in the FAA's Aeronautical Information Manual (AIM) [32].

23.4.7   Instructional or Educational Publications

National aviation authorities also publish a range of handbooks, manuals, leaflets, and so forth that are of an instructional or educational nature. Examples of such documents (which can be a valuable source of information for airplane performance engineering) include the following:

- FAA-H-8083-1A, Aircraft Weight and Balance Handbook [33];
- FAA-H-8083-15B, Instrument Flying Handbook [34];
- FAA-H-8083-16, Instrument Procedures Handbook [35]; and
- FAA-H-8083-25A, Pilot's Handbook of Aeronautical Knowledge [36].

23.5   Safety Investigation Authorities

23.5.1   Introduction

The ICAO Safety Management Manual [37], which provides guidance on the development and implementation of aviation safety programs in member states, introduces the topic of aviation safety as follows:

> While the elimination of aircraft accidents and/or serious incidents remains the ultimate goal, it is recognized that the aviation system cannot be completely free of hazards and associated risks. Human activities or human-built systems cannot be guaranteed to be absolutely free from operational errors and their consequences. Therefore, safety is a dynamic characteristic of the aviation system, whereby safety risks must be continuously mitigated. It is important to note that the acceptability of safety performance is often influenced by domestic and international norms and culture. As long as safety risks are kept under an appropriate level of control, a system as open and dynamic as aviation can still be managed to maintain the appropriate balance between production and protection.
>
> *ICAO Doc. 9859* [37]

In this context, the terms *accident*, *incident*, and *serious incident* have specific meanings [38, 39]. An *accident* is an occurrence associated with the operation of an aircraft in which

(1) a person is fatally or seriously injured (except when the injuries are from natural causes, self-inflicted, or inflicted by other persons); or
(2) the aircraft sustains damage or structural failure which (a) adversely affects the structural strength, performance or flight characteristics of the aircraft; and (b) would normally require major repair or replacement of the affected component (with the exception of engine failure or damage when the damage is limited to the engine, its cowlings, or accessories; or for other damage limited to propellers, wingtips, antennae, tires, brakes, fairings, or small dents or puncture holes in the aircraft skin); or
(3) the aircraft is missing or is completely inaccessible.

An *incident* is an occurrence, other than an accident, associated with the operation of an aircraft that affects or could affect the safety of operation. A *serious incident* is an incident involving circumstances in which there was a high probability of an accident.

23.5.2   Safety Investigations

The European Union's regulation on the investigation and prevention of accidents and incidents in civil aviation [39] describes a *safety investigation* as "a process conducted by a *safety investigation authority* for the purpose of accident and incident prevention which includes the gathering and analysis of information, the drawing of conclusions, including the determination of cause(s) and/or contributing factors and, when appropriate, the making of safety recommendations." This regulation (No. 996/2010) makes the following general points:

- The sole objective of safety investigations should be the prevention of future accidents and incidents without apportioning blame or liability.

- The expeditious holding of safety investigations of civil aviation accidents and incidents improves aviation safety and helps to prevent the occurrence of accidents and incidents.
- Reporting, analysis, and dissemination of findings of safety-related incidents are fundamentally important to improving air safety.

Annex 13 to the Chicago Convention [38] contains internationally accepted Standards and Recommended Practices for investigations into aircraft accidents or serious incidents, as well as the terms of reference and responsibilities of states (countries) associated with such events. The investigation of accidents/serious incidents is conducted under the responsibility of the state where the accident or serious incident occurred or the state in which the aircraft is registered when the location of the accident/serious incident cannot definitely be established as being in the territory of any state. A state may delegate the whole or any part of the investigation to another state by mutual arrangement and consent. States nearest the scene of an accident/serious incident in international waters are required to provide such assistance, if they are able. ICAO recommends that any judicial or administrative proceedings to apportion blame or liability should be separate from any investigation conducted under the provisions of this annex.

23.5.3   Safety Investigation Authorities

Safety investigation authorities—which are widely known as *air accident investigation authorities*—have been established by governments of states in which significant aviation activity takes place to instigate and conduct or oversee safety investigations. Cooperation with aviation authorities (see Section 23.3) is expected; however, the safety investigation authority, under the provisions of ICAO Annex 13, shall have "independence in the conduct of the investigation and have unrestricted authority over its conduct" [38].

The following organizations, which are located in the countries mentioned in Section 23.3.4, are given as examples of the organizations around the world that conduct safety investigations: Australian Transport Safety Bureau; Centro de Investigação e Prevenção de Acidentes Aeronáuticos (Brazil); Transportation Safety Board of Canada; Office of Aviation Safety, Civil Aviation Administration of China; Bureau d'Enquêtes et d'Analyses pour la sécurité de l'aviation civile (France); Federal Bureau of Aircraft Accidents Investigation (Germany); Air Accident Investigation Bureau (India); Agenzia Nazionale per la Sicurezza del Volo (Italy); Japan Transport Safety Board; State Oversight Flight Safety Department, Federal Aviation Authorities of Russia; Swedish Accident Investigation Authority; Air Accidents Investigation Branch, Department of Transport (United Kingdom); National Transportation Safety Board (United States of America).

23.6   Non-Governmental Organizations

23.6.1   Introduction

There are a large number of non-governmental organizations that play an important role in the global air transport industry—these include trade associations, workers' unions, professional societies, and not-for-profit organizations that promote dialog and information exchange. Four organizations are introduced in this section.

23.6.2 International Air Transport Association

The International Air Transport Association (IATA), with its head office in Montréal, Canada, is an industry trade association that was founded in 1945. IATA represents 265 airlines in 117 countries (2016) [40]. Flights by these airlines accounted for over 80% of total worldwide traffic (available seat kilometers). It has a broad range of functions, which include the dissemination of information—in the form of manuals, standards, guidelines, newsletters, and so forth—to the industry. IATA facilitates inter-airline cooperation, promoting safe and efficient air services.

23.6.3 International Federation of Air Line Pilots' Associations

The International Federation of Air Line Pilots' Associations (IFALPA), formed in 1948 and based in Montréal, Canada, represents over 100 000 airline pilots worldwide (2016) [41]. IFALPA specialist committees interact with international organizations on aviation issues of mutual interest and contribute to policy formation. IFALPA has permanent observer status on the ICAO Air Navigation Commission (ANC) and contributes to the creation and adaptation of SARPs.

23.6.4 Flight Safety Foundation

The Flight Safety Foundation (FSF), which was established in 1947 to pursue the continuous improvement of aviation safety and the prevention of accidents, is based in Alexandria, Virginia, USA. The FSF is an independent, not-for-profit international organization that seeks to provide impartial, safety-related guidance and resources to the aviation industry [42].

23.6.5 Air Transport Action Group

The Air Transport Action Group (ATAG), established in 1990, with headquarters in Geneva, Switzerland, is a multi-sectoral, not-for-profit association of transport industry players, comprising airports, airlines, airframe and engine manufacturers, air traffic service providers, airline pilots and air traffic control unions, chambers of commerce, tourism and trade partners, ground transportation, and communication providers [43]. ATAG seeks to promote dialog and information dissemination on cross-industry issues, such as infrastructure and sustainable development.

23.7 Airplane and Flight Crew Documentation

23.7.1 Airplane/Aeroplane Flight Manual

An Airplane/Aeroplane Flight Manual (AFM) is a document prepared by the airplane manufacturer to a stipulated format [44], which is submitted to the appropriate regulatory authority (e.g., FAA, EASA) as part of the certification process [17]. ICAO, in Annex 8 [7], states that "each aircraft shall be provided with a flight manual, placards or other documents stating the approved limitations within which the aircraft is considered airworthy as defined by the appropriate airworthiness requirements and additional instructions and information necessary for the safe operation of the aircraft."

The AFM is based extensively on flight-test data, and is the definitive record of the key performance parameters that have a bearing on airplane safety. Not all performance-related topics are addressed in the AFM—for example, nothing is mentioned about the time required to climb to cruise altitude (this is not an issue for regulation, as it does not have an impact on airplane safety). The document must be formally approved by the appropriate regulatory authority before type certification can be granted. Once approved, it has legal standing.

The manual contains details on the airplane's limitations (e.g., maximum operational weights, airspeed limits), center of gravity limits, normal operational procedures (e.g., flight deck communications, autopilot operation), and non-normal procedures (e.g., concerning an engine fire or ditching). Takeoff and landing are the most regulated elements of a flight—for this reason the AFM contains substantial information on these subjects. The AFM can include appendices and supplements, which may contain details pertaining to a specific airplane model.

Computer software and digital databases have largely replaced paper charts for the representation of airplane performance parameters in the AFM of a modern airplane. For example, the AFM of an Airbus fly-by-wire airplane refers directly to Airbus's OCTOPUS software package; the AFM for a new Boeing airplane[11] refers to Boeing's AFM-DPI (Airplane Flight Manual—Digital Performance Information) software. Guidelines for obtaining approval of a computerized version of an AFM that would replace or supplement parts of a conventional paper AFM are presented in Appendix 1 of FAA AC 25.1581 [44].

A list of the manufacturer's serials numbers (MSNs) of the airplanes to which the AFM applies is mandatory. The manual is periodically revised in line with airplane changes—this may result from configuration changes or compliance with new airworthiness directives (see Section 23.4.5), for example. Each revision requires the approval of the regulatory authority.

Generally speaking, flight crews of jet transport airplanes do not consult the AFM for airplane operational information, making reference instead to the Flight Crew Operations/Operating Manual (FCOM) and the Quick Reference Handbook (QRH). Elements of an FCOM or QRH are often extracted directly from the AFM, and in such cases the content is identical.

23.7.2 Flight Crew Operations/Operating Manual

The Flight Crew Operations/Operating Manual (FCOM)—or simply, the Operations/Operating Manual (OM)—is the primary source of flight performance and operational information used by the flight crew. It is not part of the certification documentation.

The manufacturer's FCOM is prepared for the operator/owner and applies specifically to individual airplanes. It provides the necessary information (e.g., operating limits, procedures, performance data, and system information) for the flight crew to operate the airplane in a safe and efficient manner. The FCOM contains more detailed information on operational procedures than that given in the AFM, but the information may not conflict with the AFM (the AFM is the definitive record). The FCOM is periodically revised to incorporate new information.

Many airlines, especially older legacy national carriers, produce their own operations manuals based on manufacturers' documentation. This process ensures a uniformity of style across the airline's fleet. However, it is a time-consuming process, and many small airlines do not have the resources to customize the manufacturers' documents.

Each manufacturer has its own specific document format for the FCOM—generally speaking, the document is organized into several volumes. The following outline, which is based on an

---

11 Boeing first introduced AFM-DPI into an airplane's AFM with the certification of the B777 and B737NG series in the 1990s.

FCOM for an Airbus airplane, is provided as an illustration of the typical content of such a document:

- Volume 1: general information, flight deck, doors and windows, air conditioning, pressurization, electrical supply, hydraulics, fuel system, flight management and guidance, communication and intercom systems, engine and APU controls, fire protection, emergency equipment, flight controls, flight instruments and displays, indicators, and warning systems.
- Volume 2: flight preparation information, loading procedures and limits (fuel, cargo, weight and balance), takeoff and landing performance, impacts on performance of the landing gear and slats, and flight planning (fuel requirements).
- Volume 3: flight operational procedures, general performance information, airplane operating limits (CG, cabin pressurization, flight controls, engine), emergency conditions, standard operating procedures, in-flight performance (climb, cruise, descent, hold), and one-engine-inoperative procedures and performance.
- Volume 4: flight management and guidance system (FMGS), pilot interface, capabilities, standard flight procedures, and anomalies and abnormal procedures.

23.7.3   Quick Reference Handbook

The Quick Reference Handbook (QRH) is a document prepared by the airplane manufacturer for the operator/owner and applies specifically to individual airplanes. The QRH, which can be a subpart of the FCOM, is intended to be used during flight by the crew to look up procedures or other details pertaining to the airplane. The QRH typically covers normal checklists, emergency/non-normal checklists, in-flight performance, and non-normal maneuvers.

23.7.4   Flight Crew Training Manual

The Flight Crew Training Manual (FCTM) is a support document written for flight crew trainees and their instructors. It supplements the FCOM and QRH, providing additional training-related information concerning normal and emergency procedures.

23.7.5   Flight Planning and Performance Manual

To aid in the flight planning of certain airplane types, manufacturers have produced technical manuals, such as the Flight Planning and Performance Manual (FPPM). An FPPM contains more detailed and complete information for the planning of flight operations than that presented in an FCOM. Such manuals are intended to be used primarily by performance engineers, rather than flight crew. Certain charts (e.g., on takeoff performance) are presented in a simplified format making it easier—and quicker—to use than the corresponding AFM data. For newer airplane types, manufacturers have replaced these manuals with software packages (e.g., Boeing's Performance Engineer's Tool, or PET).

23.7.6   Performance Engineer's Manual

A Performance Engineer's Manual (PEM) is prepared by the airplane manufacturer for the operator/owner. It is intended for use by performance engineers and contains very detailed aerodynamic data (e.g., lift curves, drag polars, buffet boundaries) and engine performance data (e.g., EPR or N_1 values for various thrust settings, fuel flow, and bleed air corrections). It may also contain such *en route* information as specific air range.

23.7.7 Weight and Balance Manual

The Weight and Balance Manual (WBM) is prepared by the airplane manufacturer for the operator/owner. It is a reference document applicable to a particular airplane. It contains weight and balance data (e.g., certified maximum weights and CG limits, stabilizer trim settings), fuel tank information (e.g., volumes, fuel weights and CG locations, unusable fuel), crew and passenger seating information, and cargo loading instructions and limitations. The manual includes, as a supplement, a weighing report reflecting the status of the airplane at the time of delivery. The manual contains all necessary information for the operator to derive operational loading instructions and produce Load and Trim sheets.

23.7.8 Minimum Equipment List and Configuration Deviation List

The Minimum Equipment List (MEL) and Configuration Deviation List (CDL) are complementary documents that describe procedures for the continued safe operation of an airplane with specified items of equipment or parts of the airplane inoperative or missing. The documents are specific to a particular airplane or set of airplanes (applicable airplanes are listed by manufacturer's serial number). In simple terms, the MEL contains a list of items (e.g., onboard equipment) that may be inoperative or missing prior to dispatch. The CDL, which is incorporated into the AFM as an appendix, describes secondary airframe or engine parts (e.g., access doors, fairings) that may be missing prior to dispatch. The MEL and CDL are thus used by the operator/owner to determine the serviceability status of their fleet of aircraft.

The MEL is produced by the operator (or on behalf of the operator) in compliance with the requirements of the national authority. It is derived from the Master Minimum Equipment List (MMEL), which is issued by the regulatory authority, based on manufacturer's information. The MMEL contains information on all variants of a given airplane model within one master list. It is the responsibility of the operator to customize these data to suit their fleet.

Manufacturers provide support for operators to establish acceptable procedures for dispatch under provisions of the MEL or CDR. Boeing, for example, accomplishes this by producing a Dispatch Deviation Guide (DDG) that contains suggested procedures, which operators may use or adapt to suit their specific requirements.

23.7.9 Airplane/Aircraft Characteristics for Airport Planning Manual

An Airplane/Aircraft Characteristics for Airport Planning Manual is a document prepared for each airplane type by the airplane manufacturer to a standardized format, for use by airport operators and airlines for the purpose of airport facilities planning and ground operations. The document contains pertinent descriptions of the airplane and systems (including limiting weights and overall dimensions), certain performance data (e.g., payload–range, takeoff and landing field lengths), ground maneuvering requirements, terminal servicing details, jet wake and noise, and pavement loading limits (i.e., Aircraft Classification Number versus gross weight data).

23.7.10 Supplemental Documents

Supplemental documents concerning flight operations are produced by a wide range of organizations, which include airplane manufacturers, operators, aviation authorities, and independent organizations, such as the Flight Safety Foundation. Airbus, for example, issue

Flight Operations Briefing Notes (FOBNs), which address such topics as compliance with regulations and standards, flying techniques and best practices, operational and human factors, and flight safety. IATA periodically publishes reports and recommendations concerning such issues as fuel and environmental management. These documents provide supplemental information, which does not supersede or override the manufacturer's or operator's official documents.

23.8 Airplane Performance Data

23.8.1 Performance Data Generation

Airplane performance data have traditionally been provided to operators by airplane manufacturers in "hard copy" form—that is, published in various paper manuals (e.g., AFM, Operations Manual, Performance Engineer's Manual). The traditional process of generating performance data is illustrated as method A in Figure 23.1. Reproductions of hard copy manuals as PDF (portable document format) or similar electronic format can be considered to have followed the same process. Progressively, hard copy data have been replaced or supplemented by data in digital format. This has been done in two ways—identified as methods B and C in Figure 23.1.

The root of all performance data used for flight operations lies in flight testing. Physics-based models of airplane behavior, which rely on kinematic equations of motion, are combined with

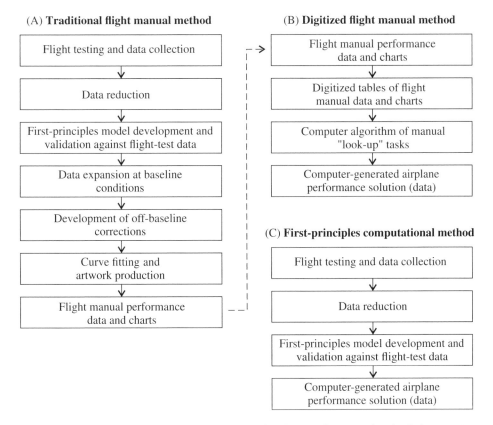

Figure 23.1 Alternative methods for the generation of airplane performance data for flight operations. *Source*: Adapted from [45].

flight-test-derived airplane characteristics (e.g., lift, drag, thrust, and fuel flow) to create what are usually called *first-principles* performance models. These mathematical models would initially be validated against measured flight-test data. The models are then used to expand the data set—obtained at a finite number of test points—across the airplane's operating envelope. The traditional flight manual data presentation relies on meticulously drawn charts of airplane performance characteristics generated using output from the first-principles (expansion) models for selected baseline conditions. A series of corrections to account for off-baseline conditions are then determined and displayed on the charts or recorded in accompanying tables. Throughout the process, a conservative approach is adopted ensuring that the airplane can be operated safely within the indicated performance limits.

For many airplane types, the traditional paper (hard copy) format of data presentation has been supplanted by digitized versions of flight manual data—the process is illustrated by method B in Figure 23.1. Computer algorithms are used to simplify the process of determining the solution to a particular performance problem (e.g., to determine the MTOW for a given set of runway conditions).

The third method (identified as method C in Figure 23.1), which is used for new airplane types, relies on the operator executing a first-principles mathematical model—developed by the manufacturer—to directly compute performance data for a given set of operating conditions. Replacing traditional flight manual charts with computer software that performs direct performance calculations utilizing physics-based models and basic performance data sets—such as drag polars, thrust, and fuel flow—not only streamlines the flight manual development process for manufacturers (as mentioned earlier in Section 23.7.1), but also improves the safety and airplane performance capability available to operators [45]. For example, a higher takeoff weight may be possible using a first-principles model due to the innate conservatism associated with the establishment of paper charts for a traditional hard copy AFM.

Another advantage of digital data is that it can be interfaced with various other computer programs developed by airplane manufacturers, airlines, or third parties. Takeoff performance calculations, for example, can be particularly complex; consequently, most airlines use software developed by the airplane manufacturer. These programs utilize an agreed protocol (see Section 23.8.2, below) for exchanging data with the airline's own computer systems.

23.8.2 SCAP (Standardized Computerized Aircraft Performance)

SCAP (Standardized Computerized Aircraft Performance) is a standard developed by IATA, the Air Transport Association[12] (ATA), and several airplane manufacturers to standardize the interface of computer programs for aircraft performance. There are six SCAP specifications applicable to transport category airplanes: Takeoff, Landing, Climbout, Inflight, Noise, and APM (Aircraft Performance Monitoring). The maintenance and update of SCAP specifications is the responsibility of the IATA SCAP Task Force.

References

1 ICAO, "Convention on international civil aviation," 9th ed., Doc. 7300, International Civil Aviation Organization, Montréal, Canada, 2006.

---

12 The American Air Transport Association (ATA), which was founded in 1936, is the trade organization of the leading US passenger and cargo carriers. It changed its name to Airlines for America (A4A) in 2011.

2 ICAO, "Manual of the ICAO standard atmosphere: Extended to 80 kilometres (262 500 feet)," 3rd ed., Doc. 7488, International Civil Aviation Organization, Montréal, Canada, 1993.

3 ICAO, "Procedures for air navigation services: Aircraft operations, Vol. 1, Flight procedures," 5th ed., Doc. 8168, International Civil Aviation Organization, Montréal, Canada, 2006.

4 ICAO, "Guidance on the balanced approach to aircraft noise management," 2nd ed., Doc. 9829, International Civil Aviation Organization, Montréal, Canada, 2008.

5 ICAO, "Aerodrome design manual: Runways," 3rd ed., Doc. 9157 Part 1, International Civil Aviation Organization, Montréal, Canada, 2006.

6 ICAO, "Air traffic management," 16th ed., Doc. 4444, International Civil Aviation Organization, Montréal, Canada, 2016.

7 ICAO, "Airworthiness of aircraft," Annex 8 to the Convention on International Civil Aviation, 11th ed., International Civil Aviation Organization, Montréal, Canada, July 2010.

8 ICAO, "Airworthiness manual," Doc. 9760, International Civil Aviation Organization, Montréal, Canada, undated.

9 EASA, "Member states," European Aviation Safety Agency, Cologne, Germany, Retrieved Aug. 2016. Available from www.easa.europa.eu/.

10 ECAC, "Home page," European Civil Aviation Conference, Paris, France, Retrieved Aug. 2016. Available from www.ecac-ceac.org/.

11 European Commission, "Regulation (EC) No. 216/2008," Brussels, Belgium, Feb. 20, 2008. Published in *Official Journal of the European Union*, Vol. L 79, Mar. 19, 2008, and reproduced by EASA.

12 IAC, "Interstate Aviation Committee," MAK-IAC, Moscow, Russia, Retrieved Aug. 2016. Available from mak-iac.org/.

13 FAA, *Airworthiness standards: Transport category airplanes*, Federal Aviation Regulation Part 25, Amdt. 25-143, Federal Aviation Administration, Washington, DC, June 24, 2016. Latest revision available from www.ecfr.gov/ under e-CFR (Electronic Code of Federal Regulations) Title 14.

14 EASA, *Certification specifications and acceptable means of compliance for large aeroplanes*, CS-25, Amdt. 18, European Aviation Safety Agency, Cologne, Germany, June 23, 2016. Latest revision available from www.easa.europa.eu/ under Certification Specification.

15 FAA, *Airworthiness standards: Aircraft engines*, Federal Aviation Regulation Part 33, Amdt. 33-34, Federal Aviation Administration, Washington, DC, Nov. 4, 2014. Latest revision available from www.ecfr.gov/ under e-CFR (Electronic Code of Federal Regulations) Title 14.

16 EASA, *Certification specifications and acceptable means of compliance for engines*, CS-E, Amdt. 4, European Aviation Safety Agency, Cologne, Germany, Mar. 12, 2015. Latest revision available from www.easa.europa.eu/ under Certification Specification.

17 FAA, "Flight test guide for certification of transport category airplanes," Advisory Circular 25-7C, Federal Aviation Administration, Washington, DC, Oct. 16, 2012. Available from www.faa.gov/.

18 FAA, "Reduced and derated takeoff thrust (power) procedures," Advisory Circular 25-13, Federal Aviation Administration, Washington, DC, May 4, 1988. Available from www.faa.gov/.

19 FAA, *General operating and flight rules*, Federal Aviation Regulation Part 91, Amdt. 91-336A, Federal Aviation Administration, Washington, DC, Mar. 4, 2015. Latest revision available from www.ecfr.gov/ under e-CFR (Electronic Code of Federal Regulations) Title 14.

20 FAA, *Operating requirements: Domestic, flag, and supplemental operations*, Federal Aviation Regulation Part 121, Amdt. 121-374, Federal Aviation Administration, Washington, DC, May 24, 2016. Latest revision available from www.ecfr.gov/ under e-CFR (Electronic Code of Federal Regulations) Title 14.

21 European Commission, *OPS 1: Commercial air transportation (aeroplanes)*, Annex to Commission Regulation (EC) No. 859/2008, Brussels, Belgium, Aug. 20, 2008. Published in *Official Journal of the European Union*, Vol. L 254, Sept. 20, 2008, and reproduced by EASA.

22 European Commission, *Commercial air transport operations (Part-CAT)*, Annex IV to Commission Regulation (EU) No. 965/2012, Brussels, Belgium, Oct. 5, 2012. Published in *Official Journal of the European Union*, Vol. L 296, Oct. 25, 2012, and reproduced by EASA.

23 FAA, "Aircraft weight and balance control," Advisory Circular 120-27E, Federal Aviation Administration, Washington, DC, June 10, 2005. Available from www.faa.gov/.

24 FAA, "Operational landing distances for wet runways; transport category airplanes," Advisory Circular 121.195-1A, Federal Aviation Administration, Washington, DC, June 19, 1990. Available from www.faa.gov/.

25 EASA, "Acceptable means of compliance (AMC) and guidance material (GM) to Part-CAT," Iss. 2, Amdt. 5, European Aviation Safety Agency, Cologne, Germany, Jan. 27, 2016. Available from www.easa.europa.eu/.

26 EASA, "Air ops: Commission regulation (EU) No. 965/2012 on air operations and related EASA decisions (AMC & GM and CS-FTL.1): Consolidated version," 6th revision, European Aviation Safety Agency, Cologne, Germany, Sept. 2016. Available from www.easa.europa.eu/.

27 FAA, "Airworthiness directives manual," FAA-IR-M-8040.1C, Federal Aviation Administration, Washington DC, May 17, 2010.

28 FAA, "Special airworthiness information bulletin," Order 8110.100B, Federal Aviation Administration, Washington, DC, July 24, 2013. Available from www.faa.gov/.

29 FAA, "Safety alerts for operators," Order 8000.87A, Federal Aviation Administration, Washington, DC, Oct. 24, 2006. Available from www.faa.gov/.

30 FAA, "Certification maintenance requirements," Advisory Circular 25-19A, Federal Aviation Administration, Washington, DC, Oct. 3, 2011.

31 ICAO, "Aeronautical information services," 14th ed., Annex 15 to the Convention on International Civil Aviation, International Civil Aviation Organization, Montréal, Canada, July 2013.

32 FAA, "Aeronautical information manual: Official guide to basic flight information and ATC procedures," Federal Aviation Administration, Washington, DC, Dec. 10, 2015. Electronic version available from www.faa.gov/air_traffic/.

33 FAA, "Aircraft weight and balance handbook," FAA-H-8083-1A, Airman Testing Standards Branch, Federal Aviation Administration, Oklahoma City, OK, 2007.

34 FAA, "Instrument flying handbook," FAA-H-8083-15B, Airman Testing Standards Branch, Federal Aviation Administration, Oklahoma City, OK, 2012.

35 FAA, "Instrument procedures handbook," FAA-H-8083-16, Flight Procedures Standards Branch, Federal Aviation Administration, Oklahoma City, OK, 2014.

36 FAA, "Pilot's handbook of aeronautical knowledge," FAA-H-8083-25A, Airman Testing Standards Branch, Federal Aviation Administration, Oklahoma City, OK, 2008.

37 ICAO, "Safety management manual (SMM)," 3rd ed., Doc. 9859, International Civil Aviation Organization, Montréal, Canada, 2013.

38 ICAO, "Aircraft accident and incident investigation," 10th ed., Annex 13 to the Convention on International Civil Aviation, International Civil Aviation Organization, Montréal, Canada, July 2010.

39 European Commission, "The investigation and prevention of accidents and incidents in civil aviation," Regulation (EU) No. 996/2010, Brussels, Belgium, Oct. 20, 2010. Published in *Official Journal of the European Union*, Vol. L 295/35, Nov. 12, 2010.

40 IATA, "Fact sheet – IATA," International Air Transport Association, Montréal, Canada, June 2016. Available from www.iata.org/pressroom/.

41 IFALPA, "International Federation of Air Line Pilots' Associations: Homepage," Montréal, Canada, Retrieved Oct. 2013. Available from www.ifalpa.org/.

42 FSF, "About the Foundation," Flight Safety Foundation, Alexandria, VA, Retrieved Aug. 2016. Available from flightsafety.org/.

43 ATAG, "About us," Air Transport Action Group, Geneva, Switzerland, Retrieved Dec. 2015. Available from www.atag.org/.

44 FAA, "Airplane flight manual," Advisory Circular 25.1581-1, Federal Aviation Administration, Washington, DC, Oct. 16, 2012.

45 Bays, L.V. and Halpin, K.E., "Improved safety and capability via direct computation of takeoff and landing performance data," *14th Aviation Technology, Integration, and Operations Conference*, Atlanta, GA, June 16–20, 2014, AIAA 2014-2154, American Institute of Aeronautics and Astronautics.

A

International Standard Atmosphere (ISA) Table

Table A.1 International Standard Atmosphere (ISA) (determined as a function of geopotential height, measured in feet)

H (ft)	H (m)	θ	T (K)	T (°C)	δ	p (N/m²)	p (lb/ft²)	σ	ρ (kg/m³)	ρ (slug/ft³)	a (m/s)	a (ft/s)	a (kt)
−2000	−609.6	1.0138	292.11	18.96	1.0744	108866	2273.7	1.0598	1.2983	2.519E-3	342.6	1124	666.0
−1800	−548.6	1.0124	291.72	18.57	1.0668	108092	2257.5	1.0537	1.2908	2.505E-3	342.4	1123	665.6
−1600	−487.7	1.0110	291.32	18.17	1.0592	107322	2241.5	1.0477	1.2834	2.490E-3	342.2	1123	665.1
−1400	−426.7	1.0096	290.92	17.77	1.0516	106557	2225.5	1.0416	1.2760	2.476E-3	341.9	1122	664.7
−1200	−365.8	1.0083	290.53	17.38	1.0441	105797	2209.6	1.0356	1.2686	2.461E-3	341.7	1121	664.2
−1000	−304.8	1.0069	290.13	16.98	1.0367	105041	2193.8	1.0296	1.2612	2.447E-3	341.5	1120	663.7
−800	−243.8	1.0055	289.73	16.58	1.0293	104289	2178.1	1.0236	1.2539	2.433E-3	341.2	1120	663.3
−600	−182.9	1.0041	289.34	16.19	1.0219	103541	2162.5	1.0177	1.2467	2.419E-3	341.0	1119	662.8
−400	−121.9	1.0028	288.94	15.79	1.0145	102798	2147.0	1.0118	1.2394	2.405E-3	340.8	1118	662.4
−200	−61.0	1.0014	288.55	15.40	1.0072	102059	2131.6	1.0059	1.2322	2.391E-3	340.5	1117	661.9
0	0.0	1.0000	288.15	15.00	1.0000	101325	2116.2	1.0000	1.2250	2.377E-3	340.3	1116	661.5
200	61.0	0.9986	287.75	14.60	0.9928	100595	2101.0	0.9942	1.2178	2.363E-3	340.1	1116	661.0
400	121.9	0.9972	287.36	14.21	0.9856	99869	2085.8	0.9883	1.2107	2.349E-3	339.8	1115	660.6
600	182.9	0.9959	286.96	13.81	0.9785	99147	2070.7	0.9826	1.2036	2.335E-3	339.6	1114	660.1
800	243.8	0.9945	286.57	13.42	0.9714	98430	2055.7	0.9768	1.1966	2.322E-3	339.4	1113	659.7
1000	304.8	0.9931	286.17	13.02	0.9644	97717	2040.9	0.9711	1.1896	2.308E-3	339.1	1113	659.2
1200	365.8	0.9917	285.77	12.62	0.9574	97008	2026.0	0.9654	1.1826	2.295E-3	338.9	1112	658.7
1400	426.7	0.9904	285.38	12.23	0.9504	96303	2011.3	0.9597	1.1756	2.281E-3	338.7	1111	658.3
1600	487.7	0.9890	284.98	11.83	0.9435	95602	1996.7	0.9540	1.1687	2.268E-3	338.4	1110	657.8
1800	548.6	0.9876	284.58	11.43	0.9366	94905	1982.1	0.9484	1.1618	2.254E-3	338.2	1110	657.4
2000	609.6	0.9862	284.19	11.04	0.9298	94213	1967.7	0.9428	1.1549	2.241E-3	337.9	1109	656.9
2200	670.6	0.9849	283.79	10.64	0.9230	93525	1953.3	0.9372	1.1481	2.228E-3	337.7	1108	656.5
2400	731.5	0.9835	283.40	10.25	0.9163	92840	1939.0	0.9316	1.1413	2.214E-3	337.5	1107	656.0
2600	792.5	0.9821	283.00	9.85	0.9095	92160	1924.8	0.9261	1.1345	2.201E-3	337.2	1106	655.5
2800	853.4	0.9807	282.60	9.45	0.9029	91484	1910.7	0.9206	1.1277	2.188E-3	337.0	1106	655.1
3000	914.4	0.9794	282.21	9.06	0.8962	90812	1896.6	0.9151	1.1210	2.175E-3	336.8	1105	654.6
3200	975.4	0.9780	281.81	8.66	0.8896	90144	1882.7	0.9097	1.1143	2.162E-3	336.5	1104	654.2
3400	1036.3	0.9766	281.41	8.26	0.8831	89479	1868.8	0.9042	1.1077	2.149E-3	336.3	1103	653.7
3600	1097.3	0.9752	281.02	7.87	0.8766	88819	1855.0	0.8988	1.1011	2.136E-3	336.1	1103	653.2
3800	1158.2	0.9739	280.62	7.47	0.8701	88163	1841.3	0.8934	1.0945	2.124E-3	335.8	1102	652.8

(*continued*)

Performance of the Jet Transport Airplane: Analysis Methods, Flight Operations, and Regulations, First Edition.
Trevor M. Young.
© 2018 John Wiley & Sons Ltd. Published 2018 by John Wiley & Sons Ltd.

Table A.1 (*Continued*)

H (ft)	H (m)	θ	T (K)	T (°C)	δ	p (N/m²)	p (lb/ft²)	σ	ρ (kg/m³)	ρ (slug/ft³)	a (m/s)	a (ft/s)	a (kt)
4000	1219.2	0.9725	280.23	7.08	0.8637	87511	1827.7	0.8881	1.0879	2.111E-3	335.6	1101	652.3
4200	1280.2	0.9711	279.83	6.68	0.8573	86862	1814.2	0.8828	1.0814	2.098E-3	335.3	1100	651.9
4400	1341.1	0.9697	279.43	6.28	0.8509	86218	1800.7	0.8774	1.0749	2.086E-3	335.1	1099	651.4
4600	1402.1	0.9684	279.04	5.89	0.8446	85577	1787.3	0.8722	1.0684	2.073E-3	334.9	1099	650.9
4800	1463.0	0.9670	278.64	5.49	0.8383	84940	1774.0	0.8669	1.0620	2.061E-3	334.6	1098	650.5
5000	1524.0	0.9656	278.24	5.09	0.8320	84307	1760.8	0.8617	1.0555	2.048E-3	334.4	1097	650.0
5200	1585.0	0.9642	277.85	4.70	0.8258	83678	1747.7	0.8565	1.0492	2.036E-3	334.2	1096	649.5
5400	1645.9	0.9629	277.45	4.30	0.8197	83053	1734.6	0.8513	1.0428	2.023E-3	333.9	1096	649.1
5600	1706.9	0.9615	277.06	3.91	0.8135	82431	1721.6	0.8461	1.0365	2.011E-3	333.7	1095	648.6
5800	1767.8	0.9601	276.66	3.51	0.8074	81814	1708.7	0.8410	1.0302	1.999E-3	333.4	1094	648.2
6000	1828.8	0.9587	276.26	3.11	0.8014	81200	1695.9	0.8359	1.0239	1.987E-3	333.2	1093	647.7
6200	1889.8	0.9574	275.87	2.72	0.7954	80589	1683.1	0.8308	1.0177	1.975E-3	333.0	1092	647.2
6400	1950.7	0.9560	275.47	2.32	0.7894	79983	1670.5	0.8257	1.0115	1.963E-3	332.7	1092	646.8
6600	2011.7	0.9546	275.07	1.92	0.7834	79380	1657.9	0.8207	1.0053	1.951E-3	332.5	1091	646.3
6800	2072.6	0.9532	274.68	1.53	0.7775	78781	1645.4	0.8156	0.9992	1.939E-3	332.2	1090	645.8
7000	2133.6	0.9519	274.28	1.13	0.7716	78185	1632.9	0.8106	0.9930	1.927E-3	332.0	1089	645.4
7200	2194.6	0.9505	273.89	0.74	0.7658	77594	1620.6	0.8057	0.9869	1.915E-3	331.8	1088	644.9
7400	2255.5	0.9491	273.49	0.34	0.7600	77005	1608.3	0.8007	0.9809	1.903E-3	331.5	1088	644.4
7600	2316.5	0.9477	273.09	−0.06	0.7542	76421	1596.1	0.7958	0.9749	1.892E-3	331.3	1087	644.0
7800	2377.4	0.9464	272.70	−0.45	0.7485	75840	1583.9	0.7909	0.9688	1.880E-3	331.0	1086	643.5
8000	2438.4	0.9450	272.30	−0.85	0.7428	75262	1571.9	0.7860	0.9629	1.868E-3	330.8	1085	643.0
8200	2499.4	0.9436	271.90	−1.25	0.7371	74689	1559.9	0.7812	0.9569	1.857E-3	330.6	1085	642.6
8400	2560.3	0.9422	271.51	−1.64	0.7315	74118	1548.0	0.7763	0.9510	1.845E-3	330.3	1084	642.1
8600	2621.3	0.9409	271.11	−2.04	0.7259	73551	1536.2	0.7715	0.9451	1.834E-3	330.1	1083	641.6
8800	2682.2	0.9395	270.72	−2.43	0.7203	72988	1524.4	0.7667	0.9392	1.822E-3	329.8	1082	641.2
9000	2743.2	0.9381	270.32	−2.83	0.7148	72428	1512.7	0.7620	0.9334	1.811E-3	329.6	1081	640.7
9200	2804.2	0.9367	269.92	−3.23	0.7093	71872	1501.1	0.7572	0.9276	1.800E-3	329.4	1081	640.2
9400	2865.1	0.9354	269.53	−3.62	0.7039	71319	1489.5	0.7525	0.9218	1.789E-3	329.1	1080	639.7
9600	2926.1	0.9340	269.13	−4.02	0.6984	70770	1478.1	0.7478	0.9161	1.777E-3	328.9	1079	639.3
9800	2987.0	0.9326	268.73	−4.42	0.6931	70224	1466.7	0.7431	0.9103	1.766E-3	328.6	1078	638.8
10000	3048.0	0.9312	268.34	−4.81	0.6877	69682	1455.3	0.7385	0.9046	1.755E-3	328.4	1077	638.3
10200	3109.0	0.9299	267.94	−5.21	0.6824	69143	1444.1	0.7338	0.8990	1.744E-3	328.1	1077	637.9
10400	3169.9	0.9285	267.55	−5.60	0.6771	68607	1432.9	0.7292	0.8933	1.733E-3	327.9	1076	637.4
10600	3230.9	0.9271	267.15	−6.00	0.6718	68074	1421.8	0.7247	0.8877	1.722E-3	327.7	1075	636.9
10800	3291.8	0.9257	266.75	−6.40	0.6666	67545	1410.7	0.7201	0.8821	1.712E-3	327.4	1074	636.4
11000	3352.8	0.9244	266.36	−6.79	0.6614	67020	1399.7	0.7156	0.8766	1.701E-3	327.2	1073	636.0
11200	3413.8	0.9230	265.96	−7.19	0.6563	66497	1388.8	0.7110	0.8710	1.690E-3	326.9	1073	635.5
11400	3474.7	0.9216	265.56	−7.59	0.6512	65978	1378.0	0.7065	0.8655	1.679E-3	326.7	1072	635.0
11600	3535.7	0.9202	265.17	−7.98	0.6461	65463	1367.2	0.7021	0.8600	1.669E-3	326.4	1071	634.6
11800	3596.6	0.9189	264.77	−8.38	0.6410	64950	1356.5	0.6976	0.8546	1.658E-3	326.2	1070	634.1

Table A.1 (*Continued*)

H (ft)	H (m)	θ	T (K)	T (°C)	δ	p (N/m²)	p (lb/ft²)	σ	ρ (kg/m³)	ρ (slug/ft³)	a (m/s)	a (ft/s)	a (kt)
12000	**3657.6**	**0.9175**	**264.38**	**−8.77**	**0.6360**	**64441**	**1345.9**	**0.6932**	**0.8491**	**1.648E-3**	**326.0**	**1069**	**633.6**
12200	3718.6	0.9161	263.98	−9.17	0.6310	63935	1335.3	0.6888	0.8437	1.637E-3	325.7	1069	633.1
12400	3779.5	0.9147	263.58	−9.57	0.6260	63432	1324.8	0.6844	0.8384	1.627E-3	325.5	1068	632.7
12600	3840.5	0.9134	263.19	−9.96	0.6211	62932	1314.4	0.6800	0.8330	1.616E-3	325.2	1067	632.2
12800	3901.4	0.9120	262.79	−10.36	0.6162	62436	1304.0	0.6757	0.8277	1.606E-3	325.0	1066	631.7
13000	**3962.4**	**0.9106**	**262.39**	**−10.76**	**0.6113**	**61943**	**1293.7**	**0.6713**	**0.8224**	**1.596E-3**	**324.7**	**1065**	**631.2**
13200	4023.4	0.9092	262.00	−11.15	0.6065	61453	1283.5	0.6670	0.8171	1.585E-3	324.5	1065	630.7
13400	4084.3	0.9079	261.60	−11.55	0.6017	60966	1273.3	0.6627	0.8119	1.575E-3	324.2	1064	630.3
13600	4145.3	0.9065	261.21	−11.94	0.5969	60482	1263.2	0.6585	0.8066	1.565E-3	324.0	1063	629.8
13800	4206.2	0.9051	260.81	−12.34	0.5922	60001	1253.2	0.6542	0.8014	1.555E-3	323.7	1062	629.3
14000	**4267.2**	**0.9037**	**260.41**	**−12.74**	**0.5875**	**59524**	**1243.2**	**0.6500**	**0.7963**	**1.545E-3**	**323.5**	**1061**	**628.8**
14200	4328.2	0.9024	260.02	−13.13	0.5828	59049	1233.3	0.6458	0.7911	1.535E-3	323.3	1061	628.4
14400	4389.1	0.9010	259.62	−13.53	0.5781	58578	1223.4	0.6416	0.7860	1.525E-3	323.0	1060	627.9
14600	4450.1	0.8996	259.22	−13.93	0.5735	58110	1213.6	0.6375	0.7809	1.515E-3	322.8	1059	627.4
14800	4511.0	0.8982	258.83	−14.32	0.5689	57644	1203.9	0.6334	0.7759	1.505E-3	322.5	1058	626.9
15000	**4572.0**	**0.8969**	**258.43**	**−14.72**	**0.5643**	**57182**	**1194.3**	**0.6292**	**0.7708**	**1.496E-3**	**322.3**	**1057**	**626.4**
15200	4633.0	0.8955	258.04	−15.11	0.5598	56723	1184.7	0.6251	0.7658	1.486E-3	322.0	1057	626.0
15400	4693.9	0.8941	257.64	−15.51	0.5553	56266	1175.1	0.6211	0.7608	1.476E-3	321.8	1056	625.5
15600	4754.9	0.8927	257.24	−15.91	0.5508	55813	1165.7	0.6170	0.7558	1.467E-3	321.5	1055	625.0
15800	4815.8	0.8914	256.85	−16.30	0.5464	55363	1156.3	0.6130	0.7509	1.457E-3	321.3	1054	624.5
16000	**4876.8**	**0.8900**	**256.45**	**−16.70**	**0.5420**	**54915**	**1146.9**	**0.6090**	**0.7460**	**1.447E-3**	**321.0**	**1053**	**624.0**
16200	4937.8	0.8886	256.05	−17.10	0.5376	54471	1137.6	0.6050	0.7411	1.438E-3	320.8	1052	623.6
16400	4998.7	0.8872	255.66	−17.49	0.5332	54029	1128.4	0.6010	0.7362	1.428E-3	320.5	1052	623.1
16600	5059.7	0.8859	255.26	−17.89	0.5289	53590	1119.3	0.5970	0.7314	1.419E-3	320.3	1051	622.6
16800	5120.6	0.8845	254.87	−18.28	0.5246	53155	1110.2	0.5931	0.7266	1.410E-3	320.0	1050	622.1
17000	**5181.6**	**0.8831**	**254.47**	**−18.68**	**0.5203**	**52722**	**1101.1**	**0.5892**	**0.7218**	**1.400E-3**	**319.8**	**1049**	**621.6**
17200	5242.6	0.8817	254.07	−19.08	0.5161	52292	1092.1	0.5853	0.7170	1.391E-3	319.5	1048	621.1
17400	5303.5	0.8804	253.68	−19.47	0.5119	51865	1083.2	0.5814	0.7122	1.382E-3	319.3	1048	620.7
17600	5364.5	0.8790	253.28	−19.87	0.5077	51440	1074.3	0.5776	0.7075	1.373E-3	319.0	1047	620.2
17800	5425.4	0.8776	252.88	−20.27	0.5035	51019	1065.5	0.5737	0.7028	1.364E-3	318.8	1046	619.7
18000	**5486.4**	**0.8762**	**252.49**	**−20.66**	**0.4994**	**50600**	**1056.8**	**0.5699**	**0.6981**	**1.355E-3**	**318.5**	**1045**	**619.2**
18200	5547.4	0.8749	252.09	−21.06	0.4953	50184	1048.1	0.5661	0.6935	1.346E-3	318.3	1044	618.7
18400	5608.3	0.8735	251.70	−21.45	0.4912	49771	1039.5	0.5623	0.6889	1.337E-3	318.0	1043	618.2
18600	5669.3	0.8721	251.30	−21.85	0.4871	49360	1030.9	0.5586	0.6843	1.328E-3	317.8	1043	617.7
18800	5730.2	0.8707	250.90	−22.25	0.4831	48953	1022.4	0.5548	0.6797	1.319E-3	317.5	1042	617.2
19000	**5791.2**	**0.8694**	**250.51**	**−22.64**	**0.4791**	**48548**	**1013.9**	**0.5511**	**0.6751**	**1.310E-3**	**317.3**	**1041**	**616.8**
19200	5852.2	0.8680	250.11	−23.04	0.4752	48145	1005.5	0.5474	0.6706	1.301E-3	317.0	1040	616.3
19400	5913.1	0.8666	249.71	−23.44	0.4712	47746	997.2	0.5437	0.6661	1.292E-3	316.8	1039	615.8
19600	5974.1	0.8652	249.32	−23.83	0.4673	47349	988.9	0.5401	0.6616	1.284E-3	316.5	1039	615.3
19800	6035.0	0.8639	248.92	−24.23	0.4634	46955	980.7	0.5364	0.6571	1.275E-3	316.3	1038	614.8

(*continued*)

Table A.1 (*Continued*)

H (ft)	H (m)	θ	T (K)	T (°C)	δ	p (N/m²)	p (lb/ft²)	σ	ρ (kg/m³)	ρ (slug/ft³)	a (m/s)	a (ft/s)	a (kt)
20000	6096.0	0.8625	248.53	−24.62	0.4595	46563	972.5	0.5328	0.6527	1.266E-3	316.0	1037	614.3
20200	6157.0	0.8611	248.13	−25.02	0.4557	46174	964.4	0.5292	0.6483	1.258E-3	315.8	1036	613.8
20400	6217.9	0.8597	247.73	−25.42	0.4519	45788	956.3	0.5256	0.6439	1.249E-3	315.5	1035	613.3
20600	6278.9	0.8584	247.34	−25.81	0.4481	45405	948.3	0.5220	0.6395	1.241E-3	315.3	1034	612.8
20800	6339.8	0.8570	246.94	−26.21	0.4443	45024	940.3	0.5185	0.6352	1.232E-3	315.0	1034	612.4
21000	6400.8	0.8556	246.54	−26.61	0.4406	44645	932.4	0.5150	0.6308	1.224E-3	314.8	1033	611.9
21200	6461.8	0.8542	246.15	−27.00	0.4369	44269	924.6	0.5115	0.6265	1.216E-3	314.5	1032	611.4
21400	6522.7	0.8529	245.75	−27.40	0.4332	43896	916.8	0.5080	0.6223	1.207E-3	314.3	1031	610.9
21600	6583.7	0.8515	245.36	−27.79	0.4296	43525	909.0	0.5045	0.6180	1.199E-3	314.0	1030	610.4
21800	6644.6	0.8501	244.96	−28.19	0.4259	43157	901.4	0.5010	0.6138	1.191E-3	313.8	1029	609.9
22000	6705.6	0.8487	244.56	−28.59	0.4223	42791	893.7	0.4976	0.6095	1.183E-3	313.5	1029	609.4
22200	6766.6	0.8474	244.17	−28.98	0.4187	42428	886.1	0.4942	0.6053	1.175E-3	313.2	1028	608.9
22400	6827.5	0.8460	243.77	−29.38	0.4152	42068	878.6	0.4908	0.6012	1.166E-3	313.0	1027	608.4
22600	6888.5	0.8446	243.37	−29.78	0.4116	41710	871.1	0.4874	0.5970	1.158E-3	312.7	1026	607.9
22800	6949.4	0.8432	242.98	−30.17	0.4081	41354	863.7	0.4840	0.5929	1.150E-3	312.5	1025	607.4
23000	7010.4	0.8419	242.58	−30.57	0.4046	41001	856.3	0.4807	0.5888	1.142E-3	312.2	1024	606.9
23200	7071.4	0.8405	242.19	−30.96	0.4012	40650	849.0	0.4773	0.5847	1.135E-3	312.0	1024	606.4
23400	7132.3	0.8391	241.79	−31.36	0.3977	40302	841.7	0.4740	0.5807	1.127E-3	311.7	1023	605.9
23600	7193.3	0.8377	241.39	−31.76	0.3943	39956	834.5	0.4707	0.5766	1.119E-3	311.5	1022	605.4
23800	7254.2	0.8364	241.00	−32.15	0.3909	39612	827.3	0.4674	0.5726	1.111E-3	311.2	1021	604.9
24000	7315.2	0.8350	240.60	−32.55	0.3876	39271	820.2	0.4642	0.5686	1.103E-3	311.0	1020	604.4
24200	7376.2	0.8336	240.20	−32.95	0.3842	38932	813.1	0.4609	0.5646	1.096E-3	310.7	1019	603.9
24400	7437.1	0.8322	239.81	−33.34	0.3809	38596	806.1	0.4577	0.5607	1.088E-3	310.4	1019	603.4
24600	7498.1	0.8309	239.41	−33.74	0.3776	38262	799.1	0.4545	0.5567	1.080E-3	310.2	1018	602.9
24800	7559.0	0.8295	239.02	−34.13	0.3743	37930	792.2	0.4513	0.5528	1.073E-3	309.9	1017	602.4
25000	7620.0	0.8281	238.62	−34.53	0.3711	37601	785.3	0.4481	0.5489	1.065E-3	309.7	1016	601.9
25200	7681.0	0.8267	238.22	−34.93	0.3679	37274	778.5	0.4450	0.5451	1.058E-3	309.4	1015	601.4
25400	7741.9	0.8254	237.83	−35.32	0.3647	36949	771.7	0.4418	0.5412	1.050E-3	309.2	1014	600.9
25600	7802.9	0.8240	237.43	−35.72	0.3615	36627	765.0	0.4387	0.5374	1.043E-3	308.9	1013	600.4
25800	7863.8	0.8226	237.04	−36.11	0.3583	36307	758.3	0.4356	0.5336	1.035E-3	308.6	1013	599.9
26000	7924.8	0.8212	236.64	−36.51	0.3552	35989	751.6	0.4325	0.5298	1.028E-3	308.4	1012	599.4
26200	7985.8	0.8199	236.24	−36.91	0.3521	35673	745.0	0.4294	0.5260	1.021E-3	308.1	1011	598.9
26400	8046.7	0.8185	235.85	−37.30	0.3490	35360	738.5	0.4264	0.5223	1.013E-3	307.9	1010	598.4
26600	8107.7	0.8171	235.45	−37.70	0.3459	35049	732.0	0.4233	0.5186	1.006E-3	307.6	1009	597.9
26800	8168.6	0.8157	235.05	−38.10	0.3429	34740	725.6	0.4203	0.5149	9.990E-4	307.3	1008	597.4
27000	8229.6	0.8144	234.66	−38.49	0.3398	34433	719.2	0.4173	0.5112	9.919E-4	307.1	1008	596.9
27200	8290.6	0.8130	234.26	−38.89	0.3368	34129	712.8	0.4143	0.5075	9.848E-4	306.8	1007	596.4
27400	8351.5	0.8116	233.87	−39.28	0.3338	33826	706.5	0.4113	0.5039	9.777E-4	306.6	1006	595.9
27600	8412.5	0.8102	233.47	−39.68	0.3309	33526	700.2	0.4084	0.5003	9.707E-4	306.3	1005	595.4
27800	8473.4	0.8089	233.07	−40.08	0.3279	33228	694.0	0.4054	0.4967	9.637E-4	306.0	1004	594.9

Table A.1 (*Continued*)

H (ft)	H (m)	θ	T (K)	T (°C)	δ	p (N/m²)	p (lb/ft²)	σ	ρ (kg/m³)	ρ (slug/ft³)	a (m/s)	a (ft/s)	a (kt)
28000	**8534.4**	**0.8075**	**232.68**	**−40.47**	**0.3250**	**32932**	**687.8**	**0.4025**	**0.4931**	**9.567E-4**	**305.8**	**1003**	**594.4**
28200	8595.4	0.8061	232.28	−40.87	0.3221	32639	681.7	0.3996	0.4895	9.498E-4	305.5	1002	593.9
28400	8656.3	0.8047	231.88	−41.27	0.3192	32347	675.6	0.3967	0.4860	9.429E-4	305.3	1002	593.4
28600	8717.3	0.8034	231.49	−41.66	0.3164	32058	669.5	0.3938	0.4824	9.361E-4	305.0	1001	592.9
28800	8778.2	0.8020	231.09	−42.06	0.3135	31770	663.5	0.3910	0.4789	9.293E-4	304.7	1000	592.4
29000	**8839.2**	**0.8006**	**230.70**	**−42.45**	**0.3107**	**31485**	**657.6**	**0.3881**	**0.4754**	**9.225E-4**	**304.5**	**999.0**	**591.9**
29200	8900.2	0.7992	230.30	−42.85	0.3079	31202	651.7	0.3853	0.4720	9.158E-4	304.2	998.1	591.4
29400	8961.1	0.7979	229.90	−43.25	0.3052	30921	645.8	0.3825	0.4685	9.091E-4	304.0	997.2	590.9
29600	9022.1	0.7965	229.51	−43.64	0.3024	30642	640.0	0.3797	0.4651	9.025E-4	303.7	996.4	590.3
29800	9083.0	0.7951	229.11	−44.04	0.2997	30365	634.2	0.3769	0.4617	8.958E-4	303.4	995.5	589.8
30000	**9144.0**	**0.7937**	**228.71**	**−44.44**	**0.2970**	**30090**	**628.4**	**0.3741**	**0.4583**	**8.893E-4**	**303.2**	**994.7**	**589.3**
30200	9205.0	0.7924	228.32	−44.83	0.2943	29817	622.7	0.3714	0.4549	8.827E-4	302.9	993.8	588.8
30400	9265.9	0.7910	227.92	−45.23	0.2916	29546	617.1	0.3686	0.4516	8.762E-4	302.6	992.9	588.3
30600	9326.9	0.7896	227.53	−45.62	0.2889	29277	611.5	0.3659	0.4483	8.698E-4	302.4	992.1	587.8
30800	9387.8	0.7882	227.13	−46.02	0.2863	29010	605.9	0.3632	0.4449	8.633E-4	302.1	991.2	587.3
31000	**9448.8**	**0.7869**	**226.73**	**−46.42**	**0.2837**	**28745**	**600.3**	**0.3605**	**0.4417**	**8.569E-4**	**301.9**	**990.3**	**586.8**
31200	9509.8	0.7855	226.34	−46.81	0.2811	28482	594.9	0.3579	0.4384	8.506E-4	301.6	989.5	586.3
31400	9570.7	0.7841	225.94	−47.21	0.2785	28221	589.4	0.3552	0.4351	8.443E-4	301.3	988.6	585.7
31600	9631.7	0.7827	225.54	−47.61	0.2760	27961	584.0	0.3526	0.4319	8.380E-4	301.1	987.7	585.2
31800	9692.6	0.7814	225.15	−48.00	0.2734	27704	578.6	0.3499	0.4287	8.317E-4	300.8	986.9	584.7
32000	**9753.6**	**0.7800**	**224.75**	**−48.40**	**0.2709**	**27449**	**573.3**	**0.3473**	**0.4255**	**8.255E-4**	**300.5**	**986.0**	**584.2**
32200	9814.6	0.7786	224.36	−48.79	0.2684	27195	568.0	0.3447	0.4223	8.194E-4	300.3	985.1	583.7
32400	9875.5	0.7772	223.96	−49.19	0.2659	26944	562.7	0.3421	0.4191	8.132E-4	300.0	984.3	583.2
32600	9936.5	0.7759	223.56	−49.59	0.2635	26694	557.5	0.3396	0.4160	8.071E-4	299.7	983.4	582.6
32800	9997.4	0.7745	223.17	−49.98	0.2610	26447	552.3	0.3370	0.4128	8.010E-4	299.5	982.5	582.1
33000	**10058.4**	**0.7731**	**222.77**	**−50.38**	**0.2586**	**26201**	**547.2**	**0.3345**	**0.4097**	**7.950E-4**	**299.2**	**981.7**	**581.6**
33200	10119.4	0.7717	222.37	−50.78	0.2562	25957	542.1	0.3319	0.4066	7.890E-4	298.9	980.8	581.1
33400	10180.3	0.7704	221.98	−51.17	0.2538	25715	537.1	0.3294	0.4036	7.830E-4	298.7	979.9	580.6
33600	10241.3	0.7690	221.58	−51.57	0.2514	25474	532.0	0.3269	0.4005	7.771E-4	298.4	979.0	580.1
33800	10302.2	0.7676	221.19	−51.96	0.2491	25236	527.1	0.3245	0.3975	7.712E-4	298.1	978.2	579.5
34000	**10363.2**	**0.7662**	**220.79**	**−52.36**	**0.2467**	**24999**	**522.1**	**0.3220**	**0.3944**	**7.653E-4**	**297.9**	**977.3**	**579.0**
34200	10424.2	0.7649	220.39	−52.76	0.2444	24764	517.2	0.3195	0.3914	7.595E-4	297.6	976.4	578.5
34400	10485.1	0.7635	220.00	−53.15	0.2421	24531	512.3	0.3171	0.3885	7.537E-4	297.3	975.5	578.0
34600	10546.1	0.7621	219.60	−53.55	0.2398	24300	507.5	0.3147	0.3855	7.480E-4	297.1	974.6	577.5
34800	10607.0	0.7607	219.20	−53.95	0.2376	24070	502.7	0.3123	0.3825	7.422E-4	296.8	973.8	576.9
35000	**10668.0**	**0.7594**	**218.81**	**−54.34**	**0.2353**	**23842**	**498.0**	**0.3099**	**0.3796**	**7.365E-4**	**296.5**	**972.9**	**576.4**
35200	10729.0	0.7580	218.41	−54.74	0.2331	23616	493.2	0.3075	0.3767	7.309E-4	296.3	972.0	575.9
35400	10789.9	0.7566	218.02	−55.13	0.2309	23392	488.5	0.3051	0.3738	7.253E-4	296.0	971.1	575.4
35600	10850.9	0.7552	217.62	−55.53	0.2287	23169	483.9	0.3028	0.3709	7.197E-4	295.7	970.2	574.9
35800	10911.8	0.7539	217.22	−55.93	0.2265	22948	479.3	0.3004	0.3680	7.141E-4	295.5	969.4	574.3

(*continued*)

Table A.1 (*Continued*)

H (ft)	H (m)	θ	T (K)	T (°C)	δ	p (N/m²)	p (lb/ft²)	σ	ρ (kg/m³)	ρ (slug/ft³)	a (m/s)	a (ft/s)	a (kt)
36000	10972.8	0.7525	216.83	−56.32	0.2243	22729	474.7	0.2981	0.3652	7.086E-4	295.2	968.5	573.8
36200	11033.8	0.7519	216.65	−56.50	0.2222	22512	470.2	0.2955	0.3620	7.024E-4	295.1	968.1	573.6
36400	11094.7	0.7519	216.65	−56.50	0.2200	22297	465.7	0.2927	0.3585	6.956E-4	295.1	968.1	573.6
36600	11155.7	0.7519	216.65	−56.50	0.2179	22083	461.2	0.2899	0.3551	6.890E-4	295.1	968.1	573.6
36800	11216.6	0.7519	216.65	−56.50	0.2159	21872	456.8	0.2871	0.3517	6.824E-4	295.1	968.1	573.6
37000	11277.6	0.7519	216.65	−56.50	0.2138	21663	452.4	0.2844	0.3483	6.759E-4	295.1	968.1	573.6
37200	11338.6	0.7519	216.65	−56.50	0.2117	21455	448.1	0.2816	0.3450	6.694E-4	295.1	968.1	573.6
37400	11399.5	0.7519	216.65	−56.50	0.2097	21250	443.8	0.2789	0.3417	6.630E-4	295.1	968.1	573.6
37600	11460.5	0.7519	216.65	−56.50	0.2077	21047	439.6	0.2763	0.3384	6.567E-4	295.1	968.1	573.6
37800	11521.4	0.7519	216.65	−56.50	0.2057	20846	435.4	0.2736	0.3352	6.504E-4	295.1	968.1	573.6
38000	11582.4	0.7519	216.65	−56.50	0.2038	20646	431.2	0.2710	0.3320	6.442E-4	295.1	968.1	573.6
38200	11643.4	0.7519	216.65	−56.50	0.2018	20449	427.1	0.2684	0.3288	6.380E-4	295.1	968.1	573.6
38400	11704.3	0.7519	216.65	−56.50	0.1999	20253	423.0	0.2658	0.3257	6.319E-4	295.1	968.1	573.6
38600	11765.3	0.7519	216.65	−56.50	0.1980	20059	418.9	0.2633	0.3225	6.258E-4	295.1	968.1	573.6
38800	11826.2	0.7519	216.65	−56.50	0.1961	19867	414.9	0.2608	0.3195	6.199E-4	295.1	968.1	573.6
39000	11887.2	0.7519	216.65	−56.50	0.1942	19677	411.0	0.2583	0.3164	6.139E-4	295.1	968.1	573.6
39200	11948.2	0.7519	216.65	−56.50	0.1923	19489	407.0	0.2558	0.3134	6.081E-4	295.1	968.1	573.6
39400	12009.1	0.7519	216.65	−56.50	0.1905	19303	403.1	0.2534	0.3104	6.022E-4	295.1	968.1	573.6
39600	12070.1	0.7519	216.65	−56.50	0.1887	19118	399.3	0.2509	0.3074	5.965E-4	295.1	968.1	573.6
39800	12131.0	0.7519	216.65	−56.50	0.1869	18935	395.5	0.2485	0.3045	5.908E-4	295.1	968.1	573.6
40000	12192.0	0.7519	216.65	−56.50	0.1851	18754	391.7	0.2462	0.3016	5.851E-4	295.1	968.1	573.6
40200	12253.0	0.7519	216.65	−56.50	0.1833	18574	387.9	0.2438	0.2987	5.795E-4	295.1	968.1	573.6
40400	12313.9	0.7519	216.65	−56.50	0.1816	18397	384.2	0.2415	0.2958	5.740E-4	295.1	968.1	573.6
40600	12374.9	0.7519	216.65	−56.50	0.1798	18221	380.5	0.2392	0.2930	5.685E-4	295.1	968.1	573.6
40800	12435.8	0.7519	216.65	−56.50	0.1781	18046	376.9	0.2369	0.2902	5.630E-4	295.1	968.1	573.6
41000	12496.8	0.7519	216.65	−56.50	0.1764	17874	373.3	0.2346	0.2874	5.577E-4	295.1	968.1	573.6
41200	12557.8	0.7519	216.65	−56.50	0.1747	17703	369.7	0.2324	0.2847	5.523E-4	295.1	968.1	573.6
41400	12618.7	0.7519	216.65	−56.50	0.1730	17533	366.2	0.2302	0.2819	5.470E-4	295.1	968.1	573.6
41600	12679.7	0.7519	216.65	−56.50	0.1714	17366	362.7	0.2279	0.2792	5.418E-4	295.1	968.1	573.6
41800	12740.6	0.7519	216.65	−56.50	0.1697	17200	359.2	0.2258	0.2766	5.366E-4	295.1	968.1	573.6
42000	12801.6	0.7519	216.65	−56.50	0.1681	17035	355.8	0.2236	0.2739	5.315E-4	295.1	968.1	573.6
42200	12862.6	0.7519	216.65	−56.50	0.1665	16872	352.4	0.2215	0.2713	5.264E-4	295.1	968.1	573.6
42400	12923.5	0.7519	216.65	−56.50	0.1649	16711	349.0	0.2194	0.2687	5.214E-4	295.1	968.1	573.6
42600	12984.5	0.7519	216.65	−56.50	0.1633	16551	345.7	0.2173	0.2661	5.164E-4	295.1	968.1	573.6
42800	13045.4	0.7519	216.65	−56.50	0.1618	16393	342.4	0.2152	0.2636	5.114E-4	295.1	968.1	573.6
43000	13106.4	0.7519	216.65	−56.50	0.1602	16236	339.1	0.2131	0.2611	5.066E-4	295.1	968.1	573.6
43200	13167.4	0.7519	216.65	−56.50	0.1587	16080	335.8	0.2111	0.2586	5.017E-4	295.1	968.1	573.6
43400	13228.3	0.7519	216.65	−56.50	0.1572	15927	332.6	0.2091	0.2561	4.969E-4	295.1	968.1	573.6
43600	13289.3	0.7519	216.65	−56.50	0.1557	15774	329.5	0.2071	0.2536	4.922E-4	295.1	968.1	573.6
43800	13350.2	0.7519	216.65	−56.50	0.1542	15623	326.3	0.2051	0.2512	4.874E-4	295.1	968.1	573.6

Table A.1 *(Continued)*

H (ft)	H (m)	θ	T (K)	T (°C)	δ	p (N/m²)	p (lb/ft²)	σ	ρ (kg/m³)	ρ (slug/ft³)	a (m/s)	a (ft/s)	a (kt)
44000	**13411.2**	**0.7519**	**216.65**	**−56.50**	**0.1527**	**15474**	**323.2**	**0.2031**	**0.2488**	**4.828E-4**	**295.1**	**968.1**	**573.6**
44200	13472.2	0.7519	216.65	−56.50	0.1513	15326	320.1	0.2012	0.2464	4.782E-4	295.1	968.1	573.6
44400	13533.1	0.7519	216.65	−56.50	0.1498	15179	317.0	0.1992	0.2441	4.736E-4	295.1	968.1	573.6
44600	13594.1	0.7519	216.65	−56.50	0.1484	15034	314.0	0.1973	0.2417	4.691E-4	295.1	968.1	573.6
44800	13655.0	0.7519	216.65	−56.50	0.1470	14890	311.0	0.1955	0.2394	4.646E-4	295.1	968.1	573.6
45000	**13716.0**	**0.7519**	**216.65**	**−56.50**	**0.1455**	**14748**	**308.0**	**0.1936**	**0.2371**	**4.601E-4**	**295.1**	**968.1**	**573.6**
45200	13777.0	0.7519	216.65	−56.50	0.1442	14607	305.1	0.1917	0.2349	4.557E-4	295.1	968.1	573.6
45400	13837.9	0.7519	216.65	−56.50	0.1428	14467	302.1	0.1899	0.2326	4.514E-4	295.1	968.1	573.6
45600	13898.9	0.7519	216.65	−56.50	0.1414	14328	299.3	0.1881	0.2304	4.470E-4	295.1	968.1	573.6
45800	13959.8	0.7519	216.65	−56.50	0.1401	14191	296.4	0.1863	0.2282	4.428E-4	295.1	968.1	573.6
46000	**14020.8**	**0.7519**	**216.65**	**−56.50**	**0.1387**	**14056**	**293.6**	**0.1845**	**0.2260**	**4.385E-4**	**295.1**	**968.1**	**573.6**
46200	14081.8	0.7519	216.65	−56.50	0.1374	13921	290.7	0.1827	0.2238	4.343E-4	295.1	968.1	573.6
46400	14142.7	0.7519	216.65	−56.50	0.1361	13788	288.0	0.1810	0.2217	4.302E-4	295.1	968.1	573.6
46600	14203.7	0.7519	216.65	−56.50	0.1348	13656	285.2	0.1793	0.2196	4.261E-4	295.1	968.1	573.6
46800	14264.6	0.7519	216.65	−56.50	0.1335	13525	282.5	0.1775	0.2175	4.220E-4	295.1	968.1	573.6
47000	**14325.6**	**0.7519**	**216.65**	**−56.50**	**0.1322**	**13396**	**279.8**	**0.1758**	**0.2154**	**4.180E-4**	**295.1**	**968.1**	**573.6**
47200	14386.6	0.7519	216.65	−56.50	0.1309	13268	277.1	0.1742	0.2133	4.140E-4	295.1	968.1	573.6
47400	14447.5	0.7519	216.65	−56.50	0.1297	13141	274.5	0.1725	0.2113	4.100E-4	295.1	968.1	573.6
47600	14508.5	0.7519	216.65	−56.50	0.1285	13015	271.8	0.1708	0.2093	4.061E-4	295.1	968.1	573.6
47800	14569.4	0.7519	216.65	−56.50	0.1272	12891	269.2	0.1692	0.2073	4.022E-4	295.1	968.1	573.6
48000	**14630.4**	**0.7519**	**216.65**	**−56.50**	**0.1260**	**12767**	**266.7**	**0.1676**	**0.2053**	**3.983E-4**	**295.1**	**968.1**	**573.6**
48200	14691.4	0.7519	216.65	−56.50	0.1248	12645	264.1	0.1660	0.2033	3.945E-4	295.1	968.1	573.6
48400	14752.3	0.7519	216.65	−56.50	0.1236	12524	261.6	0.1644	0.2014	3.908E-4	295.1	968.1	573.6
48600	14813.3	0.7519	216.65	−56.50	0.1224	12404	259.1	0.1628	0.1995	3.870E-4	295.1	968.1	573.6
48800	14874.2	0.7519	216.65	−56.50	0.1213	12286	256.6	0.1613	0.1976	3.833E-4	295.1	968.1	573.6
49000	**14935.2**	**0.7519**	**216.65**	**−56.50**	**0.1201**	**12168**	**254.1**	**0.1597**	**0.1957**	**3.796E-4**	**295.1**	**968.1**	**573.6**
49200	14996.2	0.7519	216.65	−56.50	0.1189	12052	251.7	0.1582	0.1938	3.760E-4	295.1	968.1	573.6
49400	15057.1	0.7519	216.65	−56.50	0.1178	11937	249.3	0.1567	0.1919	3.724E-4	295.1	968.1	573.6
49600	15118.1	0.7519	216.65	−56.50	0.1167	11822	246.9	0.1552	0.1901	3.689E-4	295.1	968.1	573.6
49800	15179.0	0.7519	216.65	−56.50	0.1156	11709	244.6	0.1537	0.1883	3.653E-4	295.1	968.1	573.6
50000	**15240.0**	**0.7519**	**216.65**	**−56.50**	**0.1145**	**11597**	**242.2**	**0.1522**	**0.1865**	**3.618E-4**	**295.1**	**968.1**	**573.6**
50200	15301.0	0.7519	216.65	−56.50	0.1134	11486	239.9	0.1508	0.1847	3.584E-4	295.1	968.1	573.6
50400	15361.9	0.7519	216.65	−56.50	0.1123	11376	237.6	0.1493	0.1829	3.549E-4	295.1	968.1	573.6
50600	15422.9	0.7519	216.65	−56.50	0.1112	11268	235.3	0.1479	0.1812	3.515E-4	295.1	968.1	573.6
50800	15483.8	0.7519	216.65	−56.50	0.1101	11160	233.1	0.1465	0.1794	3.482E-4	295.1	968.1	573.6
51000	**15544.8**	**0.7519**	**216.65**	**−56.50**	**0.1091**	**11053**	**230.8**	**0.1451**	**0.1777**	**3.449E-4**	**295.1**	**968.1**	**573.6**
51200	15605.8	0.7519	216.65	−56.50	0.1080	10947	228.6	0.1437	0.1760	3.416E-4	295.1	968.1	573.6
51400	15666.7	0.7519	216.65	−56.50	0.1070	10843	226.5	0.1423	0.1743	3.383E-4	295.1	968.1	573.6
51600	15727.7	0.7519	216.65	−56.50	0.1060	10739	224.3	0.1410	0.1727	3.350E-4	295.1	968.1	573.6
51800	15788.6	0.7519	216.65	−56.50	0.1050	10636	222.1	0.1396	0.1710	3.318E-4	295.1	968.1	573.6

(continued)

Table A.1 (*Continued*)

H (ft)	H (m)	θ	T (K)	T (°C)	δ	p (N/m²)	p (lb/ft²)	σ	ρ (kg/m³)	ρ (slug/ft³)	a (m/s)	a (ft/s)	a (kt)
52000	15849.6	0.7519	216.65	−56.50	0.1040	10534	220.0	0.1383	0.1694	3.287E-4	295.1	968.1	573.6
52200	15910.6	0.7519	216.65	−56.50	0.1030	10434	217.9	0.1370	0.1678	3.255E-4	295.1	968.1	573.6
52400	15971.5	0.7519	216.65	−56.50	0.1020	10334	215.8	0.1356	0.1662	3.224E-4	295.1	968.1	573.6
52600	16032.5	0.7519	216.65	−56.50	0.1010	10235	213.8	0.1343	0.1646	3.193E-4	295.1	968.1	573.6
52800	16093.4	0.7519	216.65	−56.50	0.1000	10137	211.7	0.1331	0.1630	3.163E-4	295.1	968.1	573.6
53000	16154.4	0.7519	216.65	−56.50	0.0991	10040	209.7	0.1318	0.1614	3.132E-4	295.1	968.1	573.6
53200	16215.4	0.7519	216.65	−56.50	0.0981	9944	207.7	0.1305	0.1599	3.102E-4	295.1	968.1	573.6
53400	16276.3	0.7519	216.65	−56.50	0.0972	9849	205.7	0.1293	0.1584	3.073E-4	295.1	968.1	573.6
53600	16337.3	0.7519	216.65	−56.50	0.0963	9755	203.7	0.1280	0.1569	3.043E-4	295.1	968.1	573.6
53800	16398.2	0.7519	216.65	−56.50	0.0953	9661	201.8	0.1268	0.1554	3.014E-4	295.1	968.1	573.6
54000	16459.2	0.7519	216.65	−56.50	0.0944	9569	199.8	0.1256	0.1539	2.985E-4	295.1	968.1	573.6
54200	16520.2	0.7519	216.65	−56.50	0.0935	9477	197.9	0.1244	0.1524	2.957E-4	295.1	968.1	573.6
54400	16581.1	0.7519	216.65	−56.50	0.0926	9387	196.0	0.1232	0.1509	2.929E-4	295.1	968.1	573.6
54600	16642.1	0.7519	216.65	−56.50	0.0918	9297	194.2	0.1220	0.1495	2.901E-4	295.1	968.1	573.6
54800	16703.0	0.7519	216.65	−56.50	0.0909	9208	192.3	0.1209	0.1481	2.873E-4	295.1	968.1	573.6
55000	16764.0	0.7519	216.65	−56.50	0.0900	9120	190.5	0.1197	0.1466	2.845E-4	295.1	968.1	573.6
55200	16825.0	0.7519	216.65	−56.50	0.0891	9033	188.6	0.1186	0.1452	2.818E-4	295.1	968.1	573.6
55400	16885.9	0.7519	216.65	−56.50	0.0883	8946	186.8	0.1174	0.1439	2.791E-4	295.1	968.1	573.6
55600	16946.9	0.7519	216.65	−56.50	0.0874	8861	185.1	0.1163	0.1425	2.764E-4	295.1	968.1	573.6
55800	17007.8	0.7519	216.65	−56.50	0.0866	8776	183.3	0.1152	0.1411	2.738E-4	295.1	968.1	573.6
56000	17068.8	0.7519	216.65	−56.50	0.0858	8692	181.5	0.1141	0.1398	2.712E-4	295.1	968.1	573.6
56200	17129.8	0.7519	216.65	−56.50	0.0850	8609	179.8	0.1130	0.1384	2.686E-4	295.1	968.1	573.6
56400	17190.7	0.7519	216.65	−56.50	0.0841	8526	178.1	0.1119	0.1371	2.660E-4	295.1	968.1	573.6
56600	17251.7	0.7519	216.65	−56.50	0.0833	8445	176.4	0.1108	0.1358	2.635E-4	295.1	968.1	573.6
56800	17312.6	0.7519	216.65	−56.50	0.0825	8364	174.7	0.1098	0.1345	2.610E-4	295.1	968.1	573.6
57000	17373.6	0.7519	216.65	−56.50	0.0818	8284	173.0	0.1087	0.1332	2.585E-4	295.1	968.1	573.6
57200	17434.6	0.7519	216.65	−56.50	0.0810	8205	171.4	0.1077	0.1319	2.560E-4	295.1	968.1	573.6
57400	17495.5	0.7519	216.65	−56.50	0.0802	8126	169.7	0.1067	0.1307	2.535E-4	295.1	968.1	573.6
57600	17556.5	0.7519	216.65	−56.50	0.0794	8048	168.1	0.1056	0.1294	2.511E-4	295.1	968.1	573.6
57800	17617.4	0.7519	216.65	−56.50	0.0787	7971	166.5	0.1046	0.1282	2.487E-4	295.1	968.1	573.6
58000	17678.4	0.7519	216.65	−56.50	0.0779	7895	164.9	0.1036	0.1270	2.463E-4	295.1	968.1	573.6
58200	17739.4	0.7519	216.65	−56.50	0.0772	7820	163.3	0.1026	0.1257	2.440E-4	295.1	968.1	573.6
58400	17800.3	0.7519	216.65	−56.50	0.0764	7745	161.8	0.1017	0.1245	2.416E-4	295.1	968.1	573.6
58600	17861.3	0.7519	216.65	−56.50	0.0757	7671	160.2	0.1007	0.1233	2.393E-4	295.1	968.1	573.6
58800	17922.2	0.7519	216.65	−56.50	0.0750	7597	158.7	0.0997	0.1222	2.370E-4	295.1	968.1	573.6
59000	17983.2	0.7519	216.65	−56.50	0.0743	7525	157.2	0.0988	0.1210	2.348E-4	295.1	968.1	573.6
59200	18044.2	0.7519	216.65	−56.50	0.0736	7453	155.7	0.0978	0.1198	2.325E-4	295.1	968.1	573.6
59400	18105.1	0.7519	216.65	−56.50	0.0728	7381	154.2	0.0969	0.1187	2.303E-4	295.1	968.1	573.6
59600	18166.1	0.7519	216.65	−56.50	0.0722	7311	152.7	0.0960	0.1176	2.281E-4	295.1	968.1	573.6
59800	18227.0	0.7519	216.65	−56.50	0.0715	7241	151.2	0.0950	0.1164	2.259E-4	295.1	968.1	573.6
60000	18288.0	0.7519	216.65	−56.50	0.0708	7172	149.8	0.0941	0.1153	2.238E-4	295.1	968.1	573.6

B

Units and Conversion Factors

B.1 SI System of Units

The SI system of units (Système international d'unités)[1]—the modern form of the metric system—is based on the meter–kilogram–second (MKS) variant of the original metric system. It has seven base units and two supplementary units (Table B.1). There are also a number of derived units (a selection relevant to the study of engineering mechanics is given in Table B.2). Additionally, there are a few non-SI units that are accepted for use in the SI system (e.g., liter). Prefix symbols (Table B.3) are applied to the base units, with the exception of mass where they are applied to the gram. SI unit symbols are not considered abbreviations and a period (full stop) is thus not needed (e.g., it is incorrect to write "rad."). By convention, the same symbol is used for both singular and plural (e.g., it is incorrect to write "kgs").

The CGS (centimeter–gram–second) system, which is another variant of the metric system dating back to the 1800s, has *base* units of centimeter, gram, and second for length, mass, and time, respectively. CGS units are not accepted for use with the SI system and are considered obsolete in most scientific disciplines (the exception is electromagnetism, where certain CGS units remain popular).

B.2 Other Systems of Units and Conversion Factors

British Imperial units and United States Customary (USC) units (also known as Standard units) are systems of measurement that have a common origin—both were derived from an older system (usually called English units) that was used throughout the British Empire and its region of influence. Units in both systems are precisely defined in terms of SI units by a series of international agreements [5].

A subset of USC units is used in a coherent system of measurement based on the foot, pound, and second as the base units of length, force, and time, respectively. This foot–pound–second (FPS) system considers the pound (lb) as the fundamental unit of force and the slug as the derived unit of mass (see Section 2.3.1 in Chapter 2).

Commonly used conversion factors to SI units are listed in Table B.4.

---

1 For further information on the SI system of units, consult ISO (International Organization for Standardization) Standard 80000 [1], BIPM (International Bureau for Weights and Measures) International System of Units (SI) [2], or NIST (National Institute of Standards and Technology) Special Publication 330 [3]. ISO 31 Quantities and Units (1992) [4] was recently superseded by ISO 80000 (a harmonized ISO and International Electrotechnical Commission standard), which is published in several parts under the general title Quantities and Units (Part 4 is Mechanics).

Performance of the Jet Transport Airplane: Analysis Methods, Flight Operations, and Regulations, First Edition. Trevor M. Young.
© 2018 John Wiley & Sons Ltd. Published 2018 by John Wiley & Sons Ltd.

Table B.1 Base and supplementary SI units

Unit	Symbol	Quantity
ampere	A	electric current
candela	cd	luminous intensity
kelvin	K	thermodynamic temperature
kilogram	kg	mass (see note a)
metre	m	length (see note b)
mole	mol	amount of substance
radian	rad	plane angle (supplementary unit)
second	s	time
steradian	sr	solid angle (supplementary unit)

Notes:

(a) The base unit of mass is the kilogram, which uniquely for a SI base unit has a prefix symbol.
(b) The US spelling is meter.

Table B.2 Selected SI-derived units (relevant to engineering mechanics)

Unit	Symbol	Equivalent	Quantity
degree Celsius	°C		temperature (see notes a and b)
hertz	Hz	s^{-1}	frequency
joule	J	N m	energy *or* work *or* quantity of heat
newton	N	$kg\ m/s^2$	force
pascal	Pa	N/m^2	pressure *or* stress
watt	W	J/s	power

Notes:

(a) The Celsius temperature scale is formally accepted for use with the SI system. The scale is defined such that $0\,°C = 273.15\,K$ (exactly) and that the unit interval is identical to that of the kelvin scale (i.e., a temperature change of 1 °C is equal in magnitude to a change of 1 K).
(b) The Centigrade scale is the precursor thermodynamic temperature scale to the Celsius scale (which is precisely defined in terms of the triple point of water). The Centigrade scale is obsolete and not defined in the SI system.

Table B.3 Abridged list of prefixes for SI units

Factor	Prefix	Symbol	Factor	Prefix	Symbol
10^{12}	tera	T	10^{-1}	deci	d
10^{9}	giga	G	10^{-2}	centi	c
10^{6}	mega	M	10^{-3}	milli	m
10^{3}	kilo	k	10^{-6}	micro	μ
10^{2}	hecto	h	10^{-9}	nano	n
10^{1}	deca *or* deka	da	10^{-12}	pico	p

Notes:

(a) The prefix symbol is applied to the base unit; an exception is made for mass, where the prefix is applied to the gram (e.g., 1 mg, which equals 10^{-3} g or 10^{-6} kg).
(b) Multiple prefixes (e.g., μ kg) are not used.

Table B.4 Conversion factors to SI units

Unit	SI unit
length	
1 in	= 25.4 mm (exactly)
1 ft	= 12 in (exactly) = 0.304 8 m (exactly)
1 yd	= 3 ft (exactly) = 0.914 4 m (exactly)
1 mile	= 5 280 ft (exactly) = 1 609.344 m (exactly)
1 nm	= 1 852 m (exactly) (see notes b and c)
area	
1 in^2	= 645.16 mm^2 (exactly)
1 ft^2	= 0.092 903 04 m^2 (exactly)
1 yd^2	= 0.836 127 36 m^2 (exactly)
1 mile^2	= 2.589 988 km^2
1 nm^2	= 3.429 904 km^2 (exactly)
volume	
1 in^3	= 16.387 064 cm^3 (exactly)
1 ft^3	= 28.316 85 dm^3
1 yd^3	= 0.764 554 9 m^3
1 pt (UK)	= 0.568 261 25 dm^3 (exactly)
1 liq pt (US)	= 0.473 176 5 dm^3
1 gal (UK)	= 277.420 in^3 = 4.546 092 dm^3 (exactly)
1 gal (US)	= 231 in^3 (exactly) = 3.785 412 dm^3
1 barrel (bbl)	= 42 US gal (exactly) = 158.987 3 dm^3
1 L	= 1 dm^3 (exactly) (see note d)
time	
1 min	= 60 s (exactly) (see note e)
1 h	= 3600 s (exactly) (see note e)
speed	
1 ft/s	= 0.304 8 m/s (exactly)
1 mile/h	= 0.447 04 m/s (exactly)
1 km/h	= (1/3.6) m/s (exactly) = 0.277 777 8 m/s
1 kt	= 0.514 444 4 m/s (see notes b and f)
rotational speed	
1 rev/min	= 0.104 719 8 rad/s
acceleration	
1 ft/s^2	= 0.304 8 m/s^2 (exactly)

(*continued*)

Table B.4 (*Continued*)

Unit	SI unit
mass	
1 lbm	= 0.453 592 37 kg (exactly) (see notes g and h)
1 oz	= (1/16) lbm (exactly) = 28.349 52 g
1 ton (UK)	= 2 240 lbm (exactly) = 1 016.047 kg (see note i)
1 ton (US)	= 2 000 lbm (exactly) = 907.184 7 kg (see note j)
1 tonne	= 1 000 kg (exactly) (see note k)
1 slug	= 14.593 90 kg
density	
1 lbm/ft^3	= 16.018 46 kg/m^3
1 slug/ft^3	= 515.378 8 kg/m^3
force	
1 lb	= 4.448 222 N (see note h)
1 kgf	= 9.806 65 N (exactly) (see note l)
pressure	
1 bar	= 10^5 Pa = 100 kPa (exactly)
1 lb/in^2	= 6 894.757 Pa
1 lb/ft^2	= 47.880 26 Pa
1 atm	= 101 325 Pa (exactly)
1 inHg	= 25.4 mm Hg (exactly) = 3 386.389 Pa
1 Torr	= 133.322 4 Pa
work or torque	
1 ft lb	= 1.355 818 J
power	
1 ft lb/s	= 1.355 818 W
1 hp	= 550 ft lb/s (exactly) = 745.699 9 W
temperature	(see note m)
1 °R	= (5/9) K (exactly)
°F to °C	$T_C = (5/9)(T_F - 32)$ (where T_C and T_F are temperature values in °C and °F, respectively)
°C to K	$T_K = T_C + 273.15$ (where T_K and T_C are temperature values in K and °C, respectively)
heat or energy	
1 Btu	= 788.169 3 ft lb = 1 055.056 J (see note n)
1 cal	= 4.186 8 J (exactly) (see note o)
1 Btu/s	= 1.414 85 hp = 1 055.056 W
1 Btu/h	= 0.293 071 1 W
1 Btu/US gal	= 278.716 3 J/L
1 Btu/(lbm °R)	= 4 186.8 J kg^{-1} K^{-1} (exactly)
1 Btu/lbm	= 2 326 J/kg (exactly)

Table B.4 (*Continued*)

Unit	SI unit
dynamic viscosity	
1 cP	= 0.001 Pa s (exactly) (see note p)
1 slug ft^{-1} s^{-1}	= 47.880 26 Pa s = 47.880 26 N s m^{-2}
angle (plane)	
1°	= (π/180) rad = 0.017 453 29 rad

Notes:

(a) *Sources*: References [4, 6].

(b) Historically, several different definitions of a nautical mile were in use in various countries. Progressively, these older definitions were abandoned. Today, the only definition that is in widespread use, and supported by international agreement, considers 1 nm to be exactly equal to 1852 m.

(c) There is no internationally accepted standard abbreviation (symbol) for nautical mile. The popular abbreviation nm conflicts with the SI symbol for nanometer; for that reason, it is not advocated by certain authorities. However, where confusion is unlikely to occur (e.g., in this book), it is widely used. Other abbreviations that are used include NM (used by ICAO [7], for example), M (used by BIPM [2]), nmi, and naut. Mi.

(d) The litre, or alternatively liter (US spelling), is a non-SI unit that is formally accepted for use with the SI. Accepted symbols are L and l (the upper case letter is preferred to the lower case letter due to the possible confusion with the numeral 1).

(e) The unit symbols min and h are formally accepted for use with the SI, but should not be used in conjunction with SI prefixes (see Table B.3)[2]. The abbreviation hr, for hour, is widely used in a non-scientific context.

(f) There is no single abbreviation (symbol) for knot that has universal acceptance. Several authorities, including the ISO [4] and BIPM [2], advocate kn. The abbreviations kt and KT are widely used in aviation literature and in commercial flight operations, as indicated in ICAO Annex 5 [7]. Popular abbreviations linked to the knot are KTAS (knots true airspeed), KCAS (knots calibrated airspeed), KEAS (knots equivalent airspeed), and KIAS (knots indicated airspeed).

(g) The pound-mass (avoirdupois pound) unit (lbm), while widely used in everyday situations in certain English-speaking countries, is not used much for scientific applications. The conversion of mass to a weight equivalent is based on the standard value of gravitational acceleration (see Section 2.3.3).

(h) The convention of writing lbm as the pound-mass unit and lbf as the pound-force unit is widely advocated to avoid confusion in technical publications. The customary use of lb for pound-force in the aerospace industry has been adopted herein, rather than lbf (which is recommended by several authorities, e.g., NIST [6] and ICAO [7]).

(i) A ton as defined in the UK is 2240 lbm; this is often called a long ton.

(j) A ton as defined in the US is 2000 lbm; this is often called a short ton.

(k) The tonne (symbol t), which is also called a metric ton, is a non-SI unit that is formally accepted for use with the SI.

(l) The kilogram-force unit (kgf), which is also known as the kilopond (kp), is widely considered obsolete and is seldom used for scientific applications.

(m) The kelvin unit is defined as the fraction 1/273.16 (exactly) of the thermodynamic temperature of the triple point of water [2, 6]. An interval of 1 °C is equal to an interval of 1 K, and 0 °C = 273.15 K (exactly). An interval of 1 °F is equal to an interval of 1 °R, and 32 °F = 491.67 °R (exactly).

(n) The conversion factor applies to the International Table (IT) British thermal unit (Btu), as adopted by the Fifth International Conference on Properties of Steam (London, July 1956) [4]. Several other British thermal units have been used (which have slightly different conversion factors).

(o) The conversion factor applies to the International Table (IT) calorie (cal), as adopted by the Fifth International Conference on Properties of Steam (London, July 1956) [4]. Other definitions exist (e.g., thermochemical calorie).

(p) Poise (P) and centipoise (cP) are units of dynamic viscosity in the CGS system. Conversion to SI units is as follows: 1 poise (which is equal to 100 cP) is equal to 0.1 Pa s (and 1 Pa s = 1 N s m^{-2} = 1 kg m^{-1} s^{-1}).

C

Coordinate Systems and Conventions

C.1 Introduction

Several right-handed Cartesian coordinate (X, Y, Z) systems are used in the study of airplane flight dynamics. In Chapter 3, the following coordinate systems are defined:

- The *ground axis system* (X_g, Y_g, Z_g): fixed with respect to the Earth (see Section 3.3.2).
- The *Earth axis system* (X_e, Y_e, Z_e): has its origin at the airplane's center of gravity (CG) and moves with the airplane (see Section 3.3.3).
- The *body axis system* (X_b, Y_b, Z_b): fixed with respect to the airplane, the longitudinal axis (X axis) being defined positive in the forward direction (see Section 3.3.4).
- The *flight path, or wind, axis system* (X_a, Y_a, Z_a): has its origin at the body axes origin, the X axis being defined positive towards the incoming relative wind (see Section 3.3.5).

Further details on the coordinate systems and conventions are provided in this appendix. Angles of rotation are defined in Section C.2 and velocity components in Section C.3.

C.2 Angles of Rotation

C.2.1 Roll, Pitch, and Yaw Angles

The roll angle (ϕ), pitch angle (θ), and yaw angle (ψ) describe rotations about the X_b, Y_b, and Z_b axes, respectively[1]—their precise definitions, however, need to be carefully observed. The roll angle is the angle between the Y_b axis and the intersection of the Y_b–Z_b plane with the X_e–Y_e plane (see Figure C.1). The pitch angle is the angle between the X_b axis and its projection on the X_e–Y_e plane. The yaw angle is the angle between the projection of the X_b axis on the X_e–Y_e plane and the X_e axis.

The angles ϕ, θ, and ψ are usually described as Euler angles and are measured in degrees or radians. Euler angles represent a set of sequential rotations of a coordinate system (usually fixed to a rigid body) with respect to a reference coordinate system. In this case, the angles are described by three successive rotations of the body axes with respect to the Earth axes:

(1) Initially, the body axes are aligned with the Earth axes; the body axes are then rotated about the Z_b axis through the angle ψ (Figure C.1a).

---

1 Cautionary note: In the study of flight dynamics, it is commonplace to use the lowercase Greek letters ϕ, θ, and ψ to denote angular perturbations about an equilibrium attitude.

Performance of the Jet Transport Airplane: Analysis Methods, Flight Operations, and Regulations, First Edition.
Trevor M. Young.
© 2018 John Wiley & Sons Ltd. Published 2018 by John Wiley & Sons Ltd.

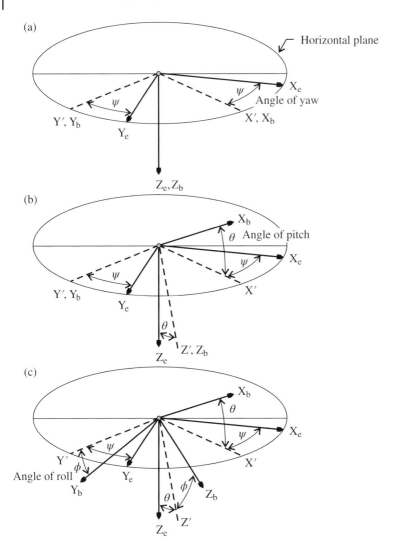

Figure C.1 Yaw, pitch, and roll angles with respect to Earth axes.

(2) The body axes are now rotated about the Y_b axis through the angle θ (Figure C.1b).
(3) Finally, the body axes are rotated about the X_b axis through the angle ϕ (Figure C.1c).

C.2.2 Bank Angle

The bank angle (Φ) is the angle between the Y_b axis and its projection on the X_e–Y_e plane (Figure C.2). When the pitch angle is zero, the bank angle is equal to the roll angle (ϕ); however, in the general case, this is not true. The bank angle can be expressed in terms of the roll angle and the pitch angle.

$$\sin \Phi = \sin \phi \, \cos \theta \tag{C.1}$$

Figure C.2 Bank angle with respect to Earth axes.

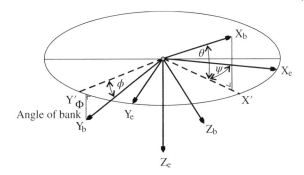

Figure C.3 Azimuth and flight path angles with respect to Earth axes.

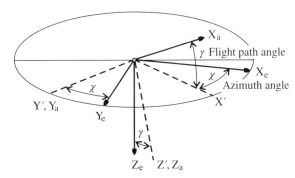

C.2.3 Azimuth Angle and Flight Path Angle

The azimuth angle (χ) is the angle between the projection of the X_a axis on the X_e–Y_e plane and the X_e axis (Figure C.3). Similarly, the flight path angle (γ) is the angle between the X_a axis and its projection on the X_e–Y_e plane (Figure C.3).

C.2.4 Angle of Attack and Angle of Sideslip

The angle of attack (α) is the angle between the projection of the X_a axis on the X_b–Z_b plane and the X_b axis (Figure C.4). The angle of attack links the pitch angle and the flight path angle by the following relationship:

$$\alpha = \theta - \gamma \qquad \text{(C.2)}$$

The sideslip angle (β) is the angle between the X_a axis and its projection on the X_b–Z_b plane (Figure C.4).

Figure C.4 Angle of attack and sideslip angle with respect to flight path axes.

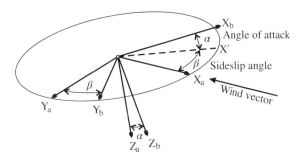

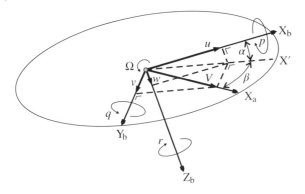

Figure C.5 Velocity components with respect to body axes.

C.3 Velocity Components

A description of the components of the airplane's velocity is given with respect to a fixed Earth axes system, in the absence of winds. The motion of the CG of the airplane, at any point in time, can be described by two vectors: the airspeed (V) and the angular velocity (Ω). It is customary that these linear and angular velocities are resolved into components with respect to the body axes.

The vector V, which by definition acts along the X_a axis, is resolved into components u (along the X_b axis), v (along the Y_b axis), and w (along the Z_b axis).[2] The following expressions can be deduced from Figure C.5, which illustrates these relationships:

$$u = V \cos \beta \, \cos \alpha \tag{C.3}$$

$$v = V \sin \beta \tag{C.4}$$

$$w = V \cos \beta \, \sin \alpha \tag{C.5}$$

It thus follows that

$$V^2 = u^2 + v^2 + w^2 \tag{C.6}$$

The angular velocity (Ω) can be resolved into components p (about the X_b axis), q (about the Y_b axis), and r (about the Z_b axis); it thus follows that

$$\Omega^2 = p^2 + q^2 + r^2 \tag{C.7}$$

The components of angular velocity can be expressed in terms of the time rates of change of the angles ϕ, θ, and ψ. The direction of these velocity vectors can be deduced from Figure C.1: $d\phi/dt$ is directed along the X_b axis, $d\theta/dt$ is directed along the Y' axis, and $d\psi/dt$ is directed along the Z_e axis. The three equations that link the angular velocity components and the time rates of change of the rotation angles are as follows:

$$p = -\frac{d\psi}{dt} \sin \theta + \frac{d\phi}{dt} \tag{C.8}$$

$$q = \frac{d\psi}{dt} \cos \theta \sin \phi + \frac{d\theta}{dt} \cos \phi \tag{C.9}$$

$$r = \frac{d\psi}{dt} \cos \theta \cos \phi - \frac{d\theta}{dt} \sin \phi \tag{C.10}$$

---

2 Cautionary note: In the study of flight dynamics, it is commonplace to use the lowercase letters u, v, and w to denote velocity perturbations about an equilibrium condition.

D

Miscellaneous Derivations

D.1 Introduction

This appendix contains several mathematical derivations that were deemed to be too detailed for inclusion the main body of the book. Derivations are presented relating to fundamental fluid properties (Section D.2), acceleration factors for climb/descent for three flight conditions (Section D.3), wind gradient correction for rate of climb/descent (Section D.4), and range equations corresponding to three different flight schedules (Section D.5).

D.2 Fundamental Fluid Properties

D.2.1 Total Pressure

The derivation of Equation 2.82 (Section 2.5.10), which appears in the appendix as Equation D.5, is given below. The expression for the total pressure (p_t) can be derived by initially expressing the ratio T_t/T in terms of the ratio p_t/p, using Equation 2.66:

$$\frac{T_t}{T} = \left(\frac{p_t}{\rho_t R}\right)\left(\frac{\rho R}{p}\right) = \left(\frac{p_t}{p}\right)\left(\frac{\rho}{\rho_t}\right) \tag{D.1}$$

Equation 2.75 is rearranged with the constant of proportionality represented by the symbol C, as follows:

$$\rho = \left(\frac{p}{C}\right)^{\left(\frac{1}{\gamma}\right)} \tag{D.2}$$

Thus $$\frac{T_t}{T} = \left(\frac{p_t}{p}\right)\left(\frac{p}{C}\right)^{\left(\frac{1}{\gamma}\right)}\left(\frac{p_t}{C}\right)^{\left(\frac{-1}{\gamma}\right)} = \left(\frac{p_t}{p}\right)^{\left(\frac{\gamma-1}{\gamma}\right)} \tag{D.3}$$

Substituting Equation D.3 into Equation 2.79 yields

$$\frac{T_t}{T} = \left(\frac{p_t}{p}\right)^{\left(\frac{\gamma-1}{\gamma}\right)} = \left(1 + \frac{\gamma-1}{2}M^2\right) \tag{D.4}$$

Equation D.4 can be rearranged to give the required expression for total pressure:

$$p_t = p\left(1 + \frac{\gamma-1}{2}M^2\right)^{\left(\frac{\gamma}{\gamma-1}\right)} \tag{D.5}$$

Performance of the Jet Transport Airplane: Analysis Methods, Flight Operations, and Regulations, First Edition.
Trevor M. Young.
© 2018 John Wiley & Sons Ltd. Published 2018 by John Wiley & Sons Ltd.

D.2.2 Bernoulli Equation for Compressible Flow

The derivation of Equation 2.84 (Section 2.5.11), which appears in the appendix as Equation D.8, is given below. Equation D.2 is substituted into Equation 2.61 to give the following differential equation:

$$\mathrm{d}p = -\left(\frac{p}{C}\right)^{\left(\frac{1}{\gamma}\right)} V \mathrm{d}V$$

or $\qquad \left(\dfrac{p}{C}\right)^{\left(\frac{-1}{\gamma}\right)} \mathrm{d}p = -V \mathrm{d}V$ $\qquad\qquad$ (D.6)

Thus $\quad \left(\dfrac{1}{C}\right)^{\left(\frac{-1}{\gamma}\right)} \displaystyle\int p^{\left(\frac{-1}{\gamma}\right)} \mathrm{d}p = -\int V \, \mathrm{d}V$

$$\left(\frac{1}{C}\right)^{\left(\frac{-1}{\gamma}\right)} \frac{p^{\left(\frac{-1}{\gamma}+1\right)}}{\left(\frac{-1}{\gamma}+1\right)} + \frac{1}{2}V^2 = \text{constant}$$

or $\qquad \left(\dfrac{\gamma}{\gamma-1}\right) \left(\dfrac{p}{C}\right)^{\left(\frac{-1}{\gamma}\right)} p + \dfrac{1}{2}V^2 = \text{constant}$ $\qquad\qquad$ (D.7)

By substituting from Equation D.2, the required result is obtained, that is,

$$\left(\frac{\gamma}{\gamma-1}\right)\left(\frac{p}{\rho}\right) + \frac{1}{2}V^2 = \text{constant} \qquad\qquad \text{(D.8)}$$

D.2.3 Binomial Expansion of the Bernoulli Equation

The derivation of Equation 6.17 (Section 6.6.3), which appears in the appendix as Equation D.12, is given below. The total pressure (p_t) can be described by Equation 2.82, that is,

$$\left(\frac{p_t}{p}\right)_{compressible} = \left(1 + \frac{\gamma-1}{2}M^2\right)^{\frac{\gamma}{\gamma-1}} \qquad \text{(Equation 2.82)}$$

The pressure differential $(p_t - p)$ is isolated on the left-hand side of the equation using Equation 2.66:

$$p_t - p = p\left\{\left(1 + \frac{\gamma-1}{2}M^2\right)^{\frac{\gamma}{\gamma-1}} - 1\right\} = \rho RT\left\{\left(1 + \frac{\gamma-1}{2}M^2\right)^{\frac{\gamma}{\gamma-1}} - 1\right\} \qquad \text{(D.9)}$$

The dynamic pressure is now introduced and the velocity-squared term expressed as Mach number using Equations 2.77 and 2.78:

$$p_t - p = \frac{1}{2}\rho V^2 \left(\frac{2}{\gamma M^2}\right)\left\{\left(1 + \frac{\gamma-1}{2}M^2\right)^{\frac{\gamma}{\gamma-1}} - 1\right\} \qquad \text{(D.10)}$$

A binomial expansion can be used to rewrite the right-hand side of Equation D.10:

$$p_t - p = \frac{1}{2}\rho V^2 \left(\frac{2}{\gamma M^2}\right) \left\{ 1 + \left(\frac{\gamma}{\gamma - 1}\right) \left(\frac{\gamma - 1}{2}\right) M^2 + \left(\frac{\gamma}{\gamma - 1}\right) \left(\frac{\gamma}{\gamma - 1} - 1\right) \left(\frac{\gamma - 1}{2}\right)^2 \frac{M^4}{2!} \right.$$

$$+ \left(\frac{\gamma}{\gamma - 1}\right) \left(\frac{\gamma}{\gamma - 1} - 1\right) \left(\frac{\gamma}{\gamma - 1} - 2\right) \left(\frac{\gamma - 1}{2}\right)^3 \frac{M^6}{3!} \qquad \text{(D.11)}$$

$$\left. + \left(\frac{\gamma}{\gamma - 1}\right) \left(\frac{\gamma}{\gamma - 1} - 1\right) \left(\frac{\gamma}{\gamma - 1} - 2\right) \left(\frac{\gamma}{\gamma - 1} - 3\right) \left(\frac{\gamma - 1}{2}\right)^4 \frac{M^8}{4!} + \cdots - 1 \right\}$$

Setting $\gamma = 1.4$ (see Section 2.5.7), Equation D.11 can be reduced to the following form:

$$p_t - p = \frac{1}{2}\rho V^2 \left\{ 1 + \frac{M^2}{4} + \frac{M^4}{40} + \frac{M^6}{1600} + \cdots \right\} \qquad \text{(D.12)}$$

D.3 Acceleration Factors for Climb/Descent

D.3.1 Acceleration Factors in the ISA

The derivation of the equations that describe the climb/descent acceleration factors, as recorded in Table 12.1 (Section 12.4.2), is presented below. The acceleration factor (f_{acc}) depends on the rate of change of airspeed with height, as defined by Equation 12.20:

$$f_{acc} = \frac{V}{g} \frac{dV}{dH} \qquad \text{(Equation 12.20)}$$

Three solutions are developed using the properties of the ISA (see Section 4.2)—these apply to (1) constant Mach number climb/descent, (2) constant EAS climb/descent, and (3) constant CAS climb/descent. In these derivations, height is expressed as geopotential height (H) and g is set equal to the standard sea-level ISA value (g_0).

D.3.2 Constant Mach Number Climb/Descent in the ISA

The true airspeed (V) can be expressed as a function of the Mach number (M) and temperature ratio (θ) by combining Equations 6.8, 2.77, and 4.1:

$$V = Ma = M\sqrt{\gamma RT} = M\sqrt{\gamma R\theta T_0} \qquad \text{(D.13)}$$

As Mach number is constant for this solution, it follows that

$$\frac{dV}{d\theta} = \frac{1}{2}M(\gamma RT_0)^{1/2}\theta^{-1/2} = \frac{M}{2}\sqrt{\frac{\gamma RT_0}{\theta}} \qquad \text{(D.14)}$$

The temperature ratio in the ISA is a function of H (for a given lapse rate), as described by Equation 4.12:

$$\theta = \frac{T}{T_0} = \left(1 + \frac{LH}{T_0}\right) \qquad \text{(Equation 4.12)}$$

hence $\dfrac{d\theta}{dH} = \dfrac{L}{T_0}$ \qquad (D.15)

The acceleration factor, which is defined by Equation 12.20, can be expressed as follows:

$$f_{acc} = \frac{V}{g_0}\frac{dV}{dH} = \frac{M\sqrt{\gamma R\theta T_0}}{g_0}\left(\frac{dV}{d\theta}\right)\left(\frac{d\theta}{dH}\right) \tag{D.16}$$

By substituting from Equations D.14 and D.15, the required result is obtained, that is,

$$f_{acc} = \frac{M\sqrt{\gamma R\theta T_0}}{g_0}\left[\frac{M}{2}\sqrt{\frac{\gamma R T_0}{\theta}}\right]\left(\frac{L}{T_0}\right) = \frac{\gamma M^2}{2}\left(\frac{LR}{g_0}\right) \tag{D.17}$$

Equation D.17 can be evaluated to give the acceleration factors for constant Mach number climb/descent (as recorded in Table 12.1) using the values for the ISA (see Table 4.1). The required data in SI units are $\gamma = 1.4$, $R = 287.05287$ m² s⁻² K⁻¹, $L = -0.0065$ K/m (in the troposphere) and $L = 0$ (in the lower stratosphere), and $g_0 = 9.80665$ m/s².

Hence, for the troposphere

$$f_{acc} = 0.7M^2(-0.1902631) \tag{D.18}$$

and for the lower stratosphere

$$f_{acc} = 0 \tag{D.19}$$

D.3.3 Constant EAS Climb/Descent in the ISA

The true airspeed (V) can be expressed in terms of EAS (V_e) using Equation 6.14:

$$V = V_e\sigma^{-\frac{1}{2}} \tag{D.20}$$

hence $$\frac{dV}{d\sigma} = -\frac{1}{2}V_e\sigma^{-\frac{3}{2}} \tag{D.21}$$

The density ratio (σ) in the ISA is a function of H (for a given lapse rate), as described by Equation 4.14:

$$\sigma = \left[1 + \frac{L}{T_0}H\right]^{\frac{-g_0}{RL}-1} \quad \text{(Equation 4.14)}$$

For the sake of convenience let $A_2 = L/T_0$ and $B_2 = -g_0/RL$,

thus $$\sigma = [1 + A_2H]^{B_2-1} \tag{D.22}$$

and $$\frac{d\sigma}{dH} = (B_2 - 1)(A_2)[1 + A_2H]^{B_2-2} \tag{D.23}$$

The acceleration factor (see Equation 12.20) can be expressed as follows:

$$f_{acc} = \frac{V}{g_0}\frac{dV}{dH} = \frac{V_e}{\sqrt{\sigma}g_0}\left(\frac{dV}{d\sigma}\right)\left(\frac{d\sigma}{dH}\right) \tag{D.24}$$

Substituting from Equations D.21 and D.23, yields

$$f_{acc} = \frac{V_e}{\sqrt{\sigma}g_0}\left[-\frac{1}{2}V_e\sigma^{-\frac{3}{2}}\right](B_2 - 1)(A_2)[1 + A_2H]^{B_2-2}$$

$$= -\frac{V_e^2}{2\sigma^2 g_0}(B_2 - 1)(A_2)[1 + A_2H]^{B_2-2}$$

$$f_{acc} = -\frac{V_e^2}{2g_0}(A_2B_2 - A_2)\frac{[1 + A_2H]^{B_2-2}}{[1 + A_2H]^{2B_2-2}} \quad \text{(substituting from Equation D.22)}$$

$$= \frac{V_e^2}{2g_0}\frac{[A_2 - A_2B_2]}{[1 + A_2H]^{B_2}} \tag{D.25}$$

The EAS is now expressed in terms of Mach number using Equations 6.15 and 2.77, and the density ratio substituted using Equation D.22:

$$V_e = \sqrt{\sigma}M\sqrt{\gamma RT}$$

Thus $V_e^2 = \gamma M^2 RT\left[1 + A_2H\right]^{B_2-1}$ (D.26)

The temperature (T) can be expressed in terms of H (see Equation 4.11), that is,

$$T = T_0 + LH = T_0\left(1 + \frac{LH}{T_0}\right) = T_0(1 + A_2H) \tag{D.27}$$

Substituting from Equations D.25, D.26, and D.27, f_{acc} can be expressed as follows:

$$f_{acc} = \frac{\gamma M^2 RT_0(1 + A_2H)[A_2 - A_2B_2]}{2g_0[1 + A_2H]} = \frac{\gamma M^2 RT_0}{2g_0}[A_2 - A_2B_2]$$

or $\quad f_{acc} = \frac{\gamma M^2 RT_0}{2g_0}\left[\frac{L}{T_0} + \frac{g_0}{RT_0}\right] = \frac{\gamma M^2}{2}\left[\frac{RL}{g_0} + 1\right]$ (D.28)

Equation D.28 can now be evaluated to give the acceleration factors (recorded in Table 12.1) using the standard values for the ISA (see Section D.3.2), that is,

for the troposphere

$$f_{acc} = 0.7M^2(1 - 0.1902631) \tag{D.29}$$

and for the lower stratosphere

$$f_{acc} = 0.7M^2 \tag{D.30}$$

D.3.4 Constant CAS Climb/Descent in the ISA

The pressure ratio (δ) in the ISA is a function of H (for a given lapse rate), as described by Equation 4.13:

$$\delta = \left[1 + \frac{L}{T_0}H\right]^{\frac{-g_0}{RL}} \quad \text{(Equation 4.13)}$$

hence $\dfrac{d\delta}{dH} = \left(\dfrac{-g_0}{RL}\right)\left[1 + \dfrac{L}{T_0}H\right]^{\left(\frac{-g_0}{RL}-1\right)}\left(\dfrac{L}{T_0}\right)$ (D.31)

Substituting the density ratio (see Equation 4.14) into D.31 simplifies the expression:

$$\frac{d\delta}{dH} = \frac{-g_0\sigma}{RT_0} \tag{D.32}$$

Based on the definition of Mach number (see Equations 2.78 and 2.77), it follows that

$$M^2 = \frac{V^2}{\gamma RT} \tag{D.33}$$

hence $\dfrac{dM^2}{dH} = \dfrac{2V}{\gamma RT}\left(\dfrac{dV}{dH}\right) - \dfrac{V^2}{\gamma RT^2}\left(\dfrac{dT}{dH}\right) = \dfrac{2V}{\gamma RT}\left(\dfrac{dV}{dH}\right) - \dfrac{V^2 L}{\gamma RT^2}$ (D.34)

Equation 6.23, which relates CAS to Mach number, can be rewritten as follows:

$$\delta\left\{\left[1 + \frac{\gamma-1}{2}M^2\right]^{\left(\frac{\gamma}{\gamma-1}\right)} - 1\right\} = \left[1 + \frac{\gamma-1}{2}\left(\frac{V_c}{a_0}\right)^2\right]^{\left(\frac{\gamma}{\gamma-1}\right)} - 1 \tag{D.35}$$

Equation D.35 is differentiated with respect to H (note that for this solution $dV_c/dH = 0$):

$$\frac{d\delta}{dH}\left\{\left[1 + \frac{\gamma-1}{2}M^2\right]^{\left(\frac{\gamma}{\gamma-1}\right)} - 1\right\} + \delta\left(\frac{\gamma}{\gamma-1}\right)\left[1 + \frac{\gamma-1}{2}M^2\right]^{\left(\frac{\gamma}{\gamma-1}-1\right)}\left(\frac{\gamma-1}{2}\right)\left(\frac{dM^2}{dH}\right) = 0 \tag{D.36}$$

Equations D.32 and D.34 are now substituted into D.36, and the result rearranged:

$$\delta\left(\frac{\gamma}{2}\right)\left[1 + \frac{\gamma-1}{2}M^2\right]^{\left(\frac{1}{\gamma-1}\right)}\left[\frac{2V}{\gamma RT}\left(\frac{dV}{dH}\right) - \frac{V^2 L}{\gamma RT^2}\right] = \frac{g_0\sigma}{RT_0}\left\{\left[1 + \frac{\gamma-1}{2}M^2\right]^{\left(\frac{\gamma}{\gamma-1}\right)} - 1\right\}$$

or $\quad \dfrac{2V}{\gamma RT}\left(\dfrac{dV}{dH}\right) - \dfrac{V^2 L}{\gamma RT^2} = \left(\dfrac{2g_0\sigma}{RT_0\delta\gamma}\right)\dfrac{\left[1 + \frac{\gamma-1}{2}M^2\right]^{\left(\frac{\gamma}{\gamma-1}\right)} - 1}{\left[1 + \frac{\gamma-1}{2}M^2\right]^{\left(\frac{1}{\gamma-1}\right)}}$ (D.37)

The acceleration factor is isolated on the left-hand side of the equation. The expression is simplified and the TAS written in terms of Mach number using Equation D.33:

$$\frac{V}{g_0}\left(\frac{dV}{dH}\right) = \left(\frac{\sigma T}{\delta T_0}\right)\frac{\left[1 + \frac{\gamma-1}{2}M^2\right]^{\left(\frac{\gamma}{\gamma-1}\right)} - 1}{\left[1 + \frac{\gamma-1}{2}M^2\right]^{\left(\frac{1}{\gamma-1}\right)}} + \frac{M^2\gamma RTL}{2g_0 T}$$

but $\quad \theta = \dfrac{T}{T_0} = \dfrac{\delta}{\sigma} \quad$ (Equation 4.4)

hence $f_{acc} = \dfrac{V}{g_0}\left(\dfrac{dV}{dH}\right) = \dfrac{\left[1 + \frac{\gamma-1}{2}M^2\right]^{\left(\frac{\gamma}{\gamma-1}\right)} - 1}{\left[1 + \frac{\gamma-1}{2}M^2\right]^{\left(\frac{1}{\gamma-1}\right)}} + \left(\dfrac{\gamma RL}{2g_0}\right)M^2$ (D.38)

Equation D.38 can now be evaluated to give the constant CAS acceleration factors (as recorded in Table 12.1) using the standard values for the ISA (see Section D.3.2), that is,

for the troposphere:

$$f_{acc} = \frac{\left[1 + 0.2M^2\right]^{3.5} - 1}{\left[1 + 0.2M^2\right]^{2.5}} - 0.13318M^2 \tag{D.39}$$

and for the lower stratosphere:

$$f_{acc} = \frac{\left[1 + 0.2M^2\right]^{3.5} - 1}{\left[1 + 0.2M^2\right]^{2.5}} \tag{D.40}$$

D.3.5   Acceleration Factors in Off-Standard Atmospheres

In an off-standard atmosphere, pressure height (H_p) does not match geopotential height (H). The temperature gradient, with respect to geopotential height, in an off-standard atmosphere is described by Equation 4.27, that is,

$$\frac{\mathrm{d}T}{\mathrm{d}H} = \frac{\mathrm{d}T}{\mathrm{d}H_p}\left(\frac{T_{std}}{T_{std} + \Delta T}\right) = L\left(\frac{T_{std}}{T}\right) \qquad \text{(Equation 4.27)}$$

This relationship provides a simple mechanism to correct the acceleration factors derived in Sections D.3.2–D.3.4 for an off-standard atmosphere: the ISA lapse rate (L) in Equations D.17 (for constant Mach number climb/descent), D.28 (for constant EAS climb/descent), and D.38 (for constant CAS climb/descent) should be multiplied by the temperature correction T_{std}/T, where T_{std} is the ISA temperature and T is the actual temperature.

D.4   Wind Gradient Correction for Rate of Climb/Descent

The derivation of Equations 12.42 and 12.43, which appear in the appendix as Equations D.45 and D.44, respectively, are given below. The scenario of an airplane climbing into a progressively increasing headwind (or decreasing tailwind)—that is, a wind gradient—is described in Section 12.4.14. The impact of a wind gradient on an airplane's climb rate can be analyzed by considering a comparable scenario with the airplane climbing within a large air mass, where the air mass itself is accelerating in a horizontal direction (see Figure D.1).

To determine the airplane's rate of climb in this situation, it is convenient to write the acceleration of the air mass as follows:

$$\frac{\mathrm{d}V_w}{\mathrm{d}t} = \frac{\mathrm{d}V_w}{\mathrm{d}H}\frac{\mathrm{d}H}{\mathrm{d}t} \tag{D.41}$$

where V_w is the wind speed (a tail wind is defined as positive);
$\mathrm{d}V_w/\mathrm{d}H$ is the rate of change of wind speed with height; and
$\mathrm{d}H/\mathrm{d}t$ is equal to the rate of climb (V_v).

Based on Newton's second law, the sum of the forces along the flight path is equal to the product of the airplane's mass and its acceleration in the direction of the flight path (i.e., the rate of change of TAS). In this case, the air mass is also accelerating; hence, an inertial force—which is

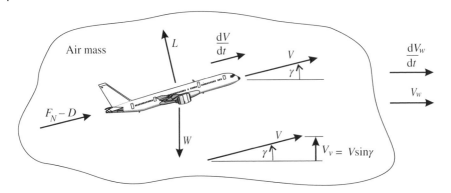

Figure D.1 Airplane climbing in an accelerating air mass.

equal to the product of the airplane's mass and the acceleration of the air mass with respect to the ground—must be taken into account:

$$F_N - D - W \sin\gamma = \left(\frac{W}{g}\right)\left(\frac{dV}{dt}\right) + \left(\frac{W}{g}\right)\left(\frac{dV_w}{dt}\cos\gamma\right)$$

thus $\quad \sin\gamma + \left(\frac{1}{g}\right)\frac{dV}{dH}\frac{dH}{dt} + \left(\frac{1}{g}\right)\frac{dV_w}{dH}\frac{dH}{dt}\cos\gamma = \frac{F_N - D}{W}$ $\qquad$ (D.42)

The rate of climb, as shown in Section 12.4.2, is given by Equation 12.18:

$$\frac{dH}{dt} = V\sin\gamma \qquad \text{(Equation 12.18)}$$

Thus $\quad \sin\gamma + \left(\frac{V}{g}\right)\frac{dV}{dH}\sin\gamma + \left(\frac{V}{g}\right)\frac{dV_w}{dH}\sin\gamma\cos\gamma = \frac{F_N - D}{W}$

and $\quad \sin\gamma = \dfrac{\left(\dfrac{F_N - D}{W}\right)}{1 + \dfrac{V}{g}\dfrac{dV}{dH} + \dfrac{V}{g}\dfrac{dV_w}{dH}\cos\gamma}$ $\qquad$ (D.43)

If the angle of climb is relatively small, then Equation D.43 can be simplified by assuming that $\cos\gamma \cong 1$:

$$\sin\gamma = \dfrac{\left(\dfrac{F_N - D}{W}\right)}{1 + \dfrac{V}{g}\dfrac{dV}{dH} + \dfrac{V}{g}\dfrac{dV_w}{dH}} \qquad (D.44)$$

The rate of climb (V_v) taking into account both airplane acceleration and wind gradient can be deduced from Equation D.44:

$$V_v = V\sin\gamma = \dfrac{\left(\dfrac{F_N - D}{W}\right)V}{1 + \dfrac{V}{g}\dfrac{dV}{dH} + \dfrac{V}{g}\dfrac{dV_w}{dH}} \qquad (D.45)$$

D.5 Still Air Range Equations for Various Flight Schedules

D.5.1 Flight Schedules for Range Equations

The derivations of the range equations described in Section 13.3 for the three flight schedules are given below—that is (1) cruise at constant *altitude* and constant *lift coefficient*, (2) cruise at constant *true airspeed* and constant *lift coefficient*, and (3) cruise at constant *altitude* and constant *true airspeed* (TAS). Note that flight at constant lift coefficient (C_L) implies a constant lift-to-drag ratio (E). In all cases, the thrust specific fuel consumption (TSFC) (c) is assumed constant and equal to a mean value ($\bar{c}$). The parabolic drag polar (see Section 7.4.3) is used in determining the range equation for flight schedule 3.

D.5.2 Constant Altitude and Constant Lift Coefficient

With the assumption of constant TSFC, the still air range (R) is given by

$$R = \int_{m_2}^{m_1} \frac{VE}{\bar{c}mg}\, dm \qquad \text{(Equation 13.10a)}$$

As C_L is constant, E is constant. This observation permits the range equation to be expressed as follows:

$$R = \frac{E}{\bar{c}g} \int_{m_2}^{m_1} \frac{V}{m}\, dm \tag{D.46}$$

For straight (i.e., the airplane is not banking), level (i.e., constant height) flight, the airplane's lift (L) is equal to its weight (W), and hence

$$C_L = \frac{L}{qS} = \frac{2L}{\rho V^2 S} = \frac{2W}{\rho V^2 S} = \frac{2mg}{\rho V^2 S} \tag{D.47}$$

or $$V = \sqrt{\frac{2mg}{\rho S C_L}} \tag{D.48}$$

Noting that ρ and C_L are both constant and substituting Equation D.48 into Equation D.46 enables the required expression for the range to be determined:

$$R = \frac{E}{\bar{c}g} \int_{m_2}^{m_1} \sqrt{\frac{2mg}{\rho S C_L}} \frac{1}{m}\, dm$$

$$R = \frac{E}{\bar{c}g} \sqrt{\frac{2g}{\rho S C_L}} \int_{m_2}^{m_1} m^{-0.5}\, dm = \frac{E}{\bar{c}g} \sqrt{\frac{2g}{\rho S C_L}} \left[2m^{0.5}\right]_{m_2}^{m_1}$$

Thus $$R = \frac{2E}{\bar{c}g} \sqrt{\frac{2g}{\rho S C_L}} \left(\sqrt{m_1} - \sqrt{m_2}\right) \tag{D.49}$$

or $$R = \frac{2E}{\bar{c}g} \sqrt{\frac{2g m_1}{\rho S C_L}} \left(1 - \frac{\sqrt{m_2}}{\sqrt{m_1}}\right) \tag{D.50}$$

If V_1 is the start-of-cruise TAS, then, using Equation D.48, the range can be expressed as

$$R = \frac{2EV_1}{\bar{c}g}\left(1 - \sqrt{\frac{m_2}{m_1}}\right) = \frac{2EV_1}{\bar{c}g}\left(1 - \sqrt{1-\varsigma}\right) \tag{D.51a}$$

or $\quad R = \frac{2EV_1}{\bar{c}'}\left(1 - \sqrt{\frac{W_2}{W_1}}\right) = \frac{2EV_1}{\bar{c}'}\left(1 - \sqrt{1-\varsigma}\right) \tag{D.51b}$

where $\varsigma = \dfrac{m_1 - m_2}{m_1} = \dfrac{W_1 - W_2}{W_1}$ $\quad$ (Equation 13.17)

Alternatively, the range can be written in terms of C_L and C_D. Using Equation D.49 and writing E as C_L/C_D, the range can be expressed as follows:

$$R = \frac{1}{\bar{c}g}\sqrt{\frac{8g}{\rho S}\left(\frac{C_L}{C_D^2}\right)}\left(\sqrt{m_1} - \sqrt{m_2}\right) \tag{D.52a}$$

or $\quad R = \frac{1}{\bar{c}'}\sqrt{\frac{8}{\rho S}\left(\frac{C_L}{C_D^2}\right)}\left(\sqrt{W_1} - \sqrt{W_2}\right) \tag{D.52b}$

Equations D.51 and D.52 appear as Equations 13.16 and 13.15, respectively, in Section 13.3.2.

D.5.3 Constant TAS and Constant Lift Coefficient

The derivation of the range (R) for the second flight schedule is straightforward as both V and E are constant:

$$R = \int_{m_2}^{m_1} \frac{VE}{\bar{c}mg}\,dm \quad \text{(Equation 13.10a)}$$

$$R = \frac{VE}{\bar{c}g}\int_{m_2}^{m_1}\frac{1}{m}\,dm = \frac{VE}{\bar{c}g}[\ln m]_{m_2}^{m_1} = \frac{VE}{\bar{c}g}(\ln m_1 - \ln m_1) \tag{D.53}$$

Thus $\quad R = \frac{VE}{\bar{c}g}\ln\left(\frac{m_1}{m_2}\right) = \frac{VE}{\bar{c}g}\ln\left(\frac{1}{1-\varsigma}\right) \tag{D.54a}$

or $\quad R = \frac{VE}{\bar{c}'}\ln\left(\frac{W_1}{W_2}\right) = \frac{VE}{\bar{c}'}\ln\left(\frac{1}{1-\varsigma}\right) \tag{D.54b}$

Equation D.54 appears as Equation 13.19 in Section 13.3.3.

D.5.4 Constant Altitude and Constant TAS

The range (R) is given in Section 13.2.3 as

$$R = \int_{m_2}^{m_1} r_a\,dm \quad \text{(Equation 13.5a)}$$

In level cruising flight, the thrust (F_N) is equal to the airplane's drag (D). The specific air range (r_a) can be expressed as follows:

$$r_a = \frac{V}{Q} = \frac{V}{\bar{c}F_N} = \frac{V}{\bar{c}D} \tag{D.55}$$

Thus $R = \displaystyle\int_{m_2}^{m_1} \frac{V}{\bar{c}D}\, dm$ (D.56)

Based on the parabolic drag polar (see Section 7.4.3), the drag can be written as

$$D = \frac{1}{2}\rho V^2 S\left(C_{D_0} + KC_L^2\right) = \frac{\rho V^2 S}{2}\left(C_{D_0}\right) + \frac{\rho V^2 SK}{2}\left(\frac{2mg}{\rho V^2 S}\right)^2$$

$$D = \frac{C_{D_0}\rho V^2 S}{2} + \frac{2Km^2 g^2}{\rho V^2 S} \tag{D.57}$$

Thus $\dfrac{V}{\bar{c}D} = \dfrac{V}{\bar{c}\left(\dfrac{C_{D_0}\rho V^2 S}{2} + \dfrac{2Km^2 g^2}{\rho S V^2}\right)} = \dfrac{\left(\dfrac{\rho V^3 S}{2Kg^2\bar{c}}\right)}{\left(\dfrac{C_{D_0}\rho^2 V^4 S^2}{4Kg^2}\right) + m^2}$ (D.58)

Let $\quad A_3 = \dfrac{\rho V^3 S}{2Kg^2\bar{c}} \quad$ and $\quad B_3 = \dfrac{C_{D_0}\rho^2 V^4 S^2}{4Kg^2}$ (D.59)

hence $R = \displaystyle\int_{m_2}^{m_1} \frac{A_3}{B_3 + m^2}\, dm = \frac{A_3}{\sqrt{B_3}}\left[\arctan\frac{m}{\sqrt{B_3}}\right]_{m_2}^{m_1}$

$$R = \frac{A_3}{\sqrt{B_3}}\left(\arctan\frac{m_1}{\sqrt{B_3}} - \arctan\frac{m_2}{\sqrt{B_3}}\right) \tag{D.60}$$

Equation D.60 can be simplified by introducing the angles α and β, where

$$\tan\alpha = \frac{m_1}{\sqrt{B_3}} \quad \text{and} \quad \tan\beta = \frac{m_2}{\sqrt{B_3}}$$

hence $\alpha = \arctan\dfrac{m_1}{\sqrt{B_3}} \quad$ and $\quad \beta = \arctan\dfrac{m_2}{\sqrt{B_3}}$ (D.61)

But $\quad \tan(\alpha - \beta) = \dfrac{\tan\alpha - \tan\beta}{1 + \tan\alpha\tan\beta}$

thus $\quad (\alpha - \beta) = \arctan\left(\dfrac{\tan\alpha - \tan\beta}{1 + \tan\alpha\tan\beta}\right) = \arctan\left\{\dfrac{\dfrac{1}{\sqrt{B_3}}(m_1 - m_2)}{1 + \dfrac{1}{B_3}(m_1 m_2)}\right\}$ (D.62)

Also $\quad \dfrac{A_3}{\sqrt{B_3}} = \left(\dfrac{\rho V^3 S}{2Kg^2\bar{c}}\right)\sqrt{\dfrac{4Kg^2}{C_{D_0}\rho^2 V^4 S^2}} = \sqrt{\dfrac{1}{C_{D_0}K}}\left(\dfrac{V}{g\bar{c}}\right) = \dfrac{2E_{max}V}{g\bar{c}}$ (D.63)

The maximum L/D ratio (E_{max}), as given by Equation 7.43, is introduced into Equation D.63 to simplify the expression. Using Equations D.60, D.61, D.62, and D.63, the range can be written as follows:

$$R = \frac{2E_{max}V}{g\bar{c}} \arctan \left\{ \frac{\dfrac{1}{\sqrt{B_3}}(m_1 - m_2)}{1 + \dfrac{1}{B_3}(m_1 m_2)} \right\} \tag{D.64}$$

thus $\quad R = \dfrac{2E_{max}V}{g\bar{c}} \arctan \left\{ \dfrac{\sqrt{B_3}\,(m_1 - m_2)}{B_3 + m_1 m_2} \right\}$ $\tag{D.65a}$

or $\quad R = \dfrac{2E_{max}V}{\bar{c}'} \arctan \left\{ \dfrac{g\sqrt{B_3}\,(W_1 - W_2)}{g^2 B_3 + W_1 W_2} \right\}$ $\tag{D.65b}$

Equation D.65 appears as Equation 13.25 in Section 13.3.4. Alternatively, the range can be expressed in terms of the start-of-cruise lift coefficient (C_{L_1}) and start-of-cruise lift-to-drag ratio (E_1). This is accomplished by initially substituting the term B_3 in Equation D.64:

$$R = \frac{2E_{max}V}{g\bar{c}} \arctan \left\{ \frac{\sqrt{\dfrac{4K}{C_{D_0}\rho^2 V^4 S^2}}(W_1 - W_2)}{1 + \dfrac{4K}{C_{D_0}\rho^2 V^4 S^2}(W_1 W_2)} \right\}$$

$$R = \frac{2E_{max}V}{g\bar{c}} \arctan \left\{ \frac{\sqrt{KC_{D_0}}\left(\dfrac{2}{C_{D_0}\rho V^2 S}\right)\varsigma W_1}{1 + \dfrac{4K}{C_{D_0}\rho^2 V^4 S^2}(1 - \varsigma)W_1^2} \right\}$$

$$R = \frac{2E_{max}V}{g\bar{c}} \arctan \left\{ \frac{\left(\dfrac{1}{2E_{max}}\right)\varsigma W_1}{\dfrac{C_{D_0}\rho V^2 S}{2} + \dfrac{2KW_1^2}{\rho V^2 S} - \dfrac{2K\varsigma W_1^2}{\rho V^2 S}} \right\}$$

$$R = \frac{2E_{max}V}{g\bar{c}} \arctan \left\{ \frac{\varsigma W_1}{2E_{max}\left(D_1 - KC_{L_1}\varsigma W_1\right)} \right\}$$

$$R = \frac{2E_{max}V}{g\bar{c}} \arctan \left\{ \frac{\varsigma E_1}{2E_{max}\left(1 - KC_{L_1}\varsigma E_1\right)} \right\} \tag{D.66}$$

Equation D.66 appears as Equation 13.26 in Section 13.3.4.

E

Trim and Longitudinal Static Stability

E.1 Introduction

Pitch trim (see Section 19.4) is intricately linked to the topic of *longitudinal static stability*, which is discussed in this appendix. The concept can be introduced by considering an airplane—flying in a wings-level attitude, in equilibrium—that is subjected to a small disturbance in pitch. This disturbance may be the result of a deliberate pilot action (e.g., involving the stick/yoke being abruptly pulled backwards and then returned to its original position), or the result of the airplane striking a gust. The change in angle of attack that arises due to the disturbance results in a change in the aerodynamic forces acting on the airplane in the X–Z plane. If the resulting forces and moments move the airplane towards the original equilibrium state, then it is said to be *statically stable*.

The *static* forces and moments that arise are assumed to be due to the displaced angular *position* of the airplane and exclude any effects caused by rates of change of position. *Dynamic* stability takes into account the forces and moments that arise due to the velocities and accelerations (in translation and rotation). Damping forces (the magnitude of which depend on velocity) will tend to reduce the resulting displacement and, in the case of an oscillatory response, will reduce the amplitude of the motion. Static stability can be described as a necessary but not sufficient condition for dynamic stability.

It is necessary to define a number of angles, derivatives, and coefficients pertaining to the airplane (Section E.2) before the analysis of the longitudinal static stability can be undertaken (Sections E.3 and E.4).

E.2 Definitions and Conventions

E.2.1 Definitions of Angles, Derivatives, and Coefficients

The lift of the complete airplane (L) is the sum of the *wing plus body* contribution (L_{wb}) and the horizontal tailplane contribution (L_T). The subscript *wb* refers to *wing plus body*, but in fact includes contributions from all parts of the airplane less the tail (e.g., engine nacelles), that is,

$$L_{wb} = L - L_T \tag{E.1}$$

Performance of the Jet Transport Airplane: Analysis Methods, Flight Operations, and Regulations, First Edition.
Trevor M. Young.
© 2018 John Wiley & Sons Ltd. Published 2018 by John Wiley & Sons Ltd.

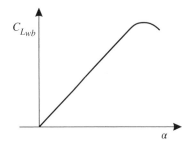

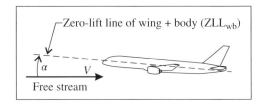

Figure E.1 Wing-plus-body lift.

The coefficient of lift for the wing plus body is written in the conventional way:

$$C_{L_{wb}} = \frac{L_{wb}}{qS} = \frac{2L_{wb}}{\rho V^2 S} \tag{E.2}$$

where L_{wb} is the lift of the wing plus body (typical units: N, lb);
q is the free-stream dynamic pressure (typical units: N/m², lb/ft²);
ρ is the air density (typical units: kg/m³, slug/ft³);
V is the true airspeed (typical units: m/s, ft/s); and
S is the wing reference area (typical units: m², ft²).

For this discussion on static stability, the angle of attack (α) is measured with respect to the *zero-lift line* (ZLL) of the *wing plus body* (Figure E.1). The slope of the $C_{L_{wb}}$ versus α graph within the linear range is given the symbol a.

$$a = \frac{dC_{L_{wb}}}{d\alpha} \tag{E.3}$$

A moment (M) about the Y axis (pitch axis) is called a *pitching moment*; it is defined as positive if it tends to rotate the nose of the airplane upwards.[1] The pitching moment coefficient is defined as follows:

$$C_M = \frac{M}{qS\bar{\bar{c}}} = \frac{2M}{\rho V^2 S \bar{\bar{c}}} \tag{E.4}$$

where M is the pitching moment (typical units: N m, ft lb); and
$\bar{\bar{c}}$ is the wing mean aerodynamic chord (typical units: m, ft).

When a wing produces lift, the airflow behind the wing is deflected downwards (Figure E.2). The flow immediately ahead of the horizontal tailplane is shown as V_T in Figure E.3. The *downwash angle* (ε) measured ahead of the tailplane depends on the amount of lift generated by the wing, which in turn depends on the angle of attack of the wing. The relationship of ε versus α is typically linear (up to the stall), thus:

$$\varepsilon = \frac{d\varepsilon}{d\alpha}\alpha \tag{E.5}$$

The derivative $d\varepsilon/d\alpha$ is assumed to be approximately constant (for a particular airplane for flight speeds up to the critical Mach number).

1 Caution: The symbol M is used for pitching moment (and not Mach number) here.

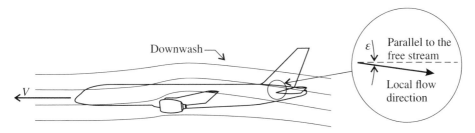

Figure E.2 Downwash angle.

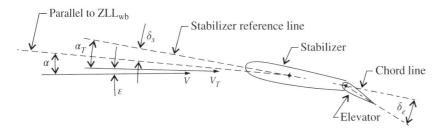

Figure E.3 Notation for horizontal tail (angles are shown positive).

Several angles relating to the tailplane are identified in Figure E.3. The *stabilizer angle* (δ_s) is measured from the wing plus body zero-lift line (ZLL$_{wb}$) to the stabilizer reference line, which is defined such that $C_{L_T} = 0$ when $\alpha_T = 0$ and $\delta_e = 0$. In order that the tail balances the moments generated by the wing and body in normal flight, δ_s will be negative for a statically stable airplane. The *elevator angle* (δ_e) is measured relative to the tailplane chord line and is defined as positive for a downward rotation of the control surface. The *tail angle of attack* (α_T) is the angle of the local airflow measured relative to the tail chord line; it is a function of three angles (see Figure E.3), that is,

$$\alpha_T = \alpha - \varepsilon + \delta_s \tag{E.6}$$

The tail lift coefficient (C_{L_T}) is the tail lift force (L_T), which is defined as positive upwards, divided by the product of the free-stream dynamic pressure (q) and the horizontal tail reference area (S_T), that is,

$$C_{L_T} = \frac{L_T}{qS_T} = \frac{2L_T}{\rho V^2 S_T} \qquad \text{(Equation 19.6)} \tag{E.7}$$

The tail lift coefficient is a function of two angles: α_T and δ_e. The influence of each of these angles on C_{L_T} is approximately linear (up to the region of tailplane stall):

$$C_{L_T} = a_1\alpha_T + a_2\delta_e \qquad \text{(Equation 19.7)} \tag{E.8}$$

where a_1 and a_2 are linear constants (derivatives).

The *tailplane volume coefficient* ($\bar{V}'$), by definition, is a ratio of two volumes, that is,

$$\bar{V}' = \frac{\ell'_T S_T}{\bar{c}S} \qquad \text{(Equation 19.5)} \tag{E.9}$$

where ℓ'_T is the *tail arm* (i.e., the distance from the aerodynamic center of the *wing plus body* to the aerodynamic center of the tail).

The numerator is a measure of the effectiveness of the tailplane, and the denominator is the product of two wing reference parameters.

E.2.2   Pitch Flight Control and Trim

It is instructive to consider a steady (i.e., constant speed), level (i.e., constant height) flight condition in which the pilot's input is required to balance the moments that act about the airplane's center of gravity (CG)—that is, to maintain a state of pitch equilibrium. Pitch control of the airplane can be commanded by the pilot in two ways: (1) by displacing the stick/yoke in the fore/aft direction, or (2) by activating the pitch trim system. Consider, for the sake of illustration, what happens when the pilot pushes the stick/yoke forwards. The trailing edge of the elevator is deflected downwards (i.e., a positive angular displacement); the resulting pressure distribution causes an upward aerodynamic force on the tailplane, which results in a nose-down (i.e., negative) pitching rotation of the airplane about its CG.

The pilot can command stabilizer angle changes by activating the trim switch on the stick/yoke or by rotating the trim wheel—either action will cause the stabilizer angle to change (i.e., a nose-upward or -downward rotation, depending on the direction of the control input). A change in stabilizer angle (δ_s) will change the tail angle of attack (α_T) and hence its contribution to the tailplane lift force (see Equation E.8)—this will reduce or increase the required elevator deflection needed to maintain the airplane in equilibrium (i.e., in trim). If the trim wheel is rotated in one direction, the required stick/yoke force will increase and if the wheel is rotated in the opposite direction, the required force will decrease. A pilot operating an airplane with a trimmable horizontal stabilizer and manual flight controls (as fitted on low-speed general aviation aircraft) will experience nil stick/yoke force (i.e., the force is *trimmed* out) when the elevator is in the neutral position. On fly-by-wire aircraft, the stick/yoke force will not necessarily be zero when the elevator is in the neutral position.

If the flight condition changes—which, in this context, means a change to the lift coefficient (e.g., resulting from a change in airspeed and/or altitude)—then the airplane will no longer be in equilibrium about the pitch axis and the stabilizer angle setting will no longer be correct. A state of equilibrium can be restored on airplanes with manual flight control systems by the pilot appropriately displacing the stick/yoke and/or by means of a trim system input. The *elevator position* changes needed to trim an airplane with a conventional flight control system at different flight conditions (i.e., different C_L values) are a measure of the airplane's *stick-fixed* static stability.

E.3   Conditions for Longitudinal Static Stability

E.3.1   Equilibrium Condition

In Figure E.4, an airplane is shown to be flying at a steady speed and is in a state of equilibrium. Newton's first law leads to the conclusion that the airplane will continue in steady rectilinear flight only if the resultant force acting on it is zero and the resultant moment acting about the CG (M_G) is also zero. In coefficient form, the second condition can be written as follows:

$$C_{M_G} = \frac{2M_G}{\rho V^2 S \bar{\bar{c}}} = 0 \tag{E.10}$$

where C_{M_G} is the pitching moment coefficient about the CG (dimensionless).

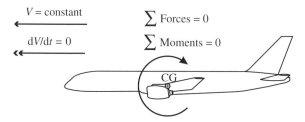

Figure E.4 Equilibrium condition.

E.3.2 Static Stability Criteria

For the airplane to be *statically stable*, it is necessary that following a small disturbance in pitch, a moment must be produced that moves the airplane towards the equilibrium condition. If a discrete disturbance causes an increase in the angle of attack, then the aerodynamic moment that is produced—as a result of the increased angle of attack—must tend to reduce the angle of attack. For a positive change in angle of attack, a negative (i.e., nose-down) moment is therefore required for the airplane to be stable about the trimmed flight condition. Mathematically, this can be written as follows:

$$\left(\frac{dC_{M_G}}{d\alpha}\right)_{C_{M_G}=0} < 0 \tag{E.11}$$

The derivative is now expanded to introduce the lift-curve slope:

$$\left(\frac{dC_{M_G}}{d\alpha}\right)_{C_{M_G}=0} = \left(\frac{dC_L}{d\alpha}\right)\left(\frac{dC_{M_G}}{dC_L}\right)_{C_{M_G}=0} \tag{E.12}$$

As $dC_L/d\alpha > 0$ within the linear region of the lift curve, Equation E.11 can be written as

$$\left(\frac{dC_{M_G}}{dC_L}\right)_{C_{M_G}=0} < 0 \tag{E.13}$$

A graph of C_{M_G} versus C_L for a stable airplane will have a negative slope, as shown by the "initial flight condition" line in Figure E.5. Note that the only point on the line where the airplane will be in equilibrium is where $C_{M_G} = 0$. If the initial disturbance (viz., a vertical gust or discrete control input) increases the angle of attack, C_L will increase. It is evident from the graph that a negative (i.e., nose-down) pitching moment will then be produced, which will tend to rotate the airplane towards the trimmed angle of attack. This relatively straightforward deduction is very important: it provides a simple means of assessing whether or not an airplane will be statically stable. However, it says nothing about the manner in which the airplane tends towards the trimmed flight condition (e.g., it provides no indication as to whether or not the airplane will oscillate about the point of equilibrium).

The slope of the line (i.e., the "initial flight condition" in Figure E.5) depends on the degree (or amount) of stability. The derivative dC_{M_G}/dC_L will be shown later (in Section E.4) to be a direct measure of the static stability of an airplane, which is quantified by a parameter called the *static margin* (K_N). It will be shown that K_N is a function of the CG position and the aerodynamic design of the airplane (in particular, the tailplane). If the airplane is re-trimmed at a new angle of

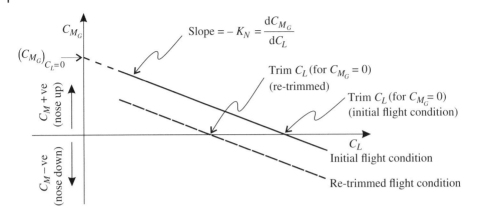

Figure E.5 Stability relationship for a statically stable airplane.

attack (e.g., following a change in airspeed and/or altitude), the basic airplane stability (i.e., the derivative dC_{M_G}/dC_L) does not change. This is illustrated by the "re-trimmed" line in Figure E.5.

The extrapolation of the "initial flight condition" line in Figure E.5 to the theoretical point of $C_L = 0$ corresponds to the moment coefficient $(C_{M_G})_{C_L=0}$. The condition of $C_L = 0$ implies infinite flight speed; alternatively, it may also be considered to be the flight condition corresponding to the angle of attack that will generate zero net lift on the airplane. One way of interpreting this condition is to consider an airplane model in a wind tunnel. If the model is progressively rotated in a nose-down direction, a point will be reached where the upward aerodynamic forces will exactly balance the downward forces. Although the air pressure acting on the model, at this angle of attack, creates no net lift, the pressure distribution is not symmetrical and it will create a moment (which is the result of a pure couple). This means that the moment about the CG, at this particular angle of attack, will equal the moment about any reference point on the model, including the aerodynamic center. The significance of this deduction can be appreciated by considering the definition of the aerodynamic center.

The aerodynamic center (AC) of a wing (or wing plus body), by definition, is that point on the wing chord about which there is zero change of pitching moment with changes in angle of attack. If the wing is working in its linear range below the stall, then there will be no change of pitching moment about the aerodynamic center with change of lift. It also means that the pitching moment about the CG at zero net lift is equal to the pitching moment about the aerodynamic center. In non-dimensional form, this can be written as follows:

$$\left(C_{M_G} \right)_{C_L=0} = C_{M_0} \tag{E.14}$$

where C_{M_0} is the pitching moment coefficient about the aerodynamic center of the wing plus body (fuselage).

E.4 Simplified Trim Equation

E.4.1 Simplifications and Assumptions

A number of simplifications and assumptions are made in this elementary analysis of an airplane's longitudinal static stability to facilitate the derivation of the basic trim equation:

- The airplane is in steady level flight or at an angle of climb or descent small enough for the total lift to be approximately equal to the airplane's weight.
- The thrust and drag forces are assumed to have a negligible influence on the static stability of the airplane.
- The CG is located close to the mean aerodynamic chord (MAC) of the wing.
- The lift forces act in a direction approximately perpendicular to the MAC of the wing (this assumes that the angle of attack is small).
- Compressibility effects are neglected, so that the aerodynamic coefficients are independent of Mach number.
- The influence of aeroelastic effects is negligible.
- The aerodynamic force and moment derivatives are constant (which implies that the airplane is at a speed well above the stall and the controls have linear characteristics).

E.4.2 Summation of Moments About the CG

The principal forces and moments acting on a conventional airplane are shown in Figure E.6. The sum of the moments acting about the CG will be evaluated. The approach adopted here is to determine the contributions from the wing–body combination (Figure E.7), the tail (Figure E.8), and the powerplant (Figure E.9) separately, and then to simplify the ensuing expressions before summing the individual components.

The mean aerodynamic chord ($\bar{\bar{c}}$) of the wing is used as a reference length and its leading edge as a datum (Figure E.7). The positions of critical points are indicated as fractions of the MAC. The weight acts through the CG, which is located a distance $h\bar{\bar{c}}$ from the leading edge of the MAC, and the lift of the *wing plus body* acts through the aerodynamic center of the *wing plus body*, which is located at a distance $h_0\bar{\bar{c}}$.

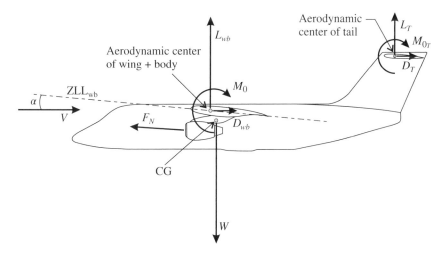

Figure E.6 Forces and moments acting on a conventional airplane.

Sum of the Moments Due to the Wing Plus Body (See Figure E.7)

$$\sum_{wing/body} M = (L_{wb} \cos \alpha)(h - h_0) \bar{\bar{c}} - (L_{wb} \sin \alpha)Z_w + (D_{wb} \cos \alpha)Z_w$$
$$+ (D_{wb} \sin \alpha)(h - h_0)\bar{\bar{c}} + M_0 \tag{E.15}$$

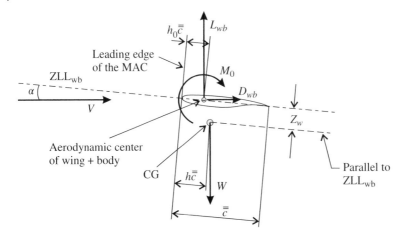

Figure E.7 Airplane stability: wing plus body details.

The following simplifications are introduced:

- The drag terms are assumed to be small in comparison to the lift terms and are thus neglected.
- The angle of attack is assumed to be small, permitting the use of the small angle approximations $\cos \alpha \cong 1$ and $\sin \alpha \cong \alpha$ (for α measured in radians).
- The $(L_{wb} \sin \alpha)Z_w$ term is small in comparison to the other terms and can be neglected.

Hence $\sum_{wing/body} M = L_{wb}(h - h_0)\bar{\bar{c}} + M_0$ (E.16)

Sum of the Moments Due to the Tailplane (See Figure E.8)

$$\sum_{tail} M = -L_T \cos(\alpha - \varepsilon)\ell_T - L_T \sin(\alpha - \varepsilon)Z_T + D_T \cos(\alpha - \varepsilon)Z_t$$
$$- D_T \sin(\alpha - \varepsilon)\ell_T + M_{0_T}$$
(E.17)

The following simplifications are introduced:

- The drag terms are assumed to be small in comparison to the lift terms and are thus neglected.
- The small angle approximations $\cos(\alpha - \varepsilon) \cong 1$ and $\sin(\alpha - \varepsilon) \cong \alpha - \varepsilon$ are used.
- The moment M_{0_T} (i.e., the moment about the tailplane aerodynamic center), which is due to the negative airfoil camber, is small. Furthermore, the tail area for a conventional airplane

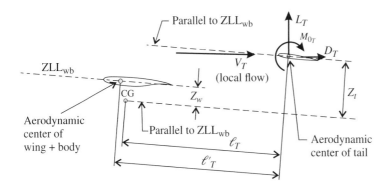

Figure E.8 Airplane stability: tailplane details.

is small compared to the wing area; hence, the term M_{0_T} has been assumed to be negligible when compared to the other terms.

- The $L_T \sin(\alpha - \varepsilon)Z_t$ term is small in comparison to the other remaining term and is thus neglected.

Hence $\sum_{tail} M = -L_T \ell_T$ (E.18)

Moment Due to the Powerplant (See Figure E.9)

$$\sum_{powerplant} M = -(F_N \cos i_p)Z_p \tag{E.19}$$

It is assumed that the angle i_p is small; thus $\cos i_p \cong 1$ and Equation E.19 can be simplified:

$$\sum_{powerplant} M = -F_N Z_p \tag{E.20}$$

The thrust contribution to the stability of the airplane depends on whether the thrust line is above or below the CG. An engine installed below the CG (as illustrated in Figure E.6) will be destabilizing.

Figure E.9 Airplane stability: powerplant details.

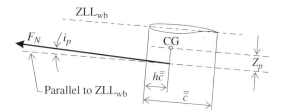

Sum of the Moments for the Whole Airplane

The resulting moment about the CG is obtained by summing the components given in Equations E.16, E.18, and E.20:

$$\sum_{airplane} M = M_G = M_0 + L_{wb}(h - h_0)\bar{\bar{c}} - L_T \ell_T + F_N Z_p \tag{E.21}$$

The magnitude of the thrust component (i.e., $F_N Z_p$) can be small in comparison to the other terms (depending on the position of the engines); the term will be neglected in this elementary analysis of the airplane's static stability. Using Equation E.1, the term L_{wb} can be substituted, enabling Equation E.21 to be written as follows:

$$M_G = M_0 + (h - h_0)\bar{\bar{c}}(L - L_T) - \ell_T L_T = M_0 + (h - h_0)\bar{\bar{c}}L - [\ell_T + (h - h_0)\bar{\bar{c}}]L_T$$

or $\quad M_G = M_0 + (h - h_0)\bar{\bar{c}}L - \ell'_T L_T \tag{E.22}$

To assess the stability criterion defined earlier (see Equation E.15), Equation E.22 has to be written in coefficient form. The expression is non-dimensionalized by dividing by $qS\bar{\bar{c}}$.

$$C_{M_G} = C_{M_0} + (h - h_0)C_L - \frac{\ell'_T S_T}{\bar{\bar{c}} S}C_{L_T}$$

or $\quad C_{M_G} = C_{M_0} + (h - h_0)C_L - \bar{V}' C_{L_T} \quad$ (from Equation E.9) $\tag{E.23}$

If the airplane is in a state of equilibrium, the sum of the moments about the CG is zero.

Thus $\quad C_{M_G} = C_{M_0} + (h - h_0)C_L - \bar{V}' C_{L_T} = 0 \tag{E.24}$

Equation E.24 is the *simplified trim equation* and is a statement that the airplane is in a state of equilibrium at the lift coefficient C_L.

E.4.3   Static Stability and Neutral Point

A criterion for positive static stability can be established by differentiating Equation E.24. As the derivative $dC_{M_0}/dC_L = 0$, by the definition of aerodynamic center, the resulting expression can be written as follows:

$$\frac{dC_{M_G}}{dC_L} = (h - h_0) - \bar{V}' \frac{dC_{L_T}}{dC_L} \qquad (E.25)$$

In Section E.3.2, it was argued that for an airplane to be statically stable, the derivative dC_{M_G}/dC_L must be negative. From Equation E.25 it is evident that if the CG is sufficiently far forward (i.e., h is sufficiently small), then the derivative will be negative and the airplane will be statically stable. Mathematically, the criterion for static stability is that the CG must be sufficiently far forward that

$$h < h_0 + \bar{V}' \frac{dC_{L_T}}{dC_L} \qquad (E.26)$$

If the CG were progressively moved rearwards—in a hypothetical situation—the airplane would become progressively less stable. A limiting condition arises when the CG is located at the *neutral point* (i.e., $h = h_N$). In this situation, the airplane will be *neutrally stable*. Theoretically, the position of h_N can be determined from the following expression:

$$h_N = h_0 + \bar{V}' \frac{dC_{L_T}}{dC_L} \qquad (E.27)$$

With the CG located at the neutral point, it is evident from Equations E.25 and E.27 that the airplane has zero static stability, that is,

$$\frac{dC_{M_G}}{dC_L} = (h_N - h_0) - \bar{V}' \frac{dC_{L_T}}{dC_L} = 0 \qquad (E.28)$$

The neutral point has a close similarity to the aerodynamic center. The neutral point is effectively the aerodynamic center of the complete airplane (in practice the term *neutral point* is usually used for the complete airplane, and the term *aerodynamic center* is used for either the wing, wing plus body, or tail alone). Equation E.27 shows that the effect of the tail is to move the aerodynamic center (of the complete airplane) a distance $\bar{V}' \, dC_{L_T}/dC_L$ aft of the aerodynamic center of the wing plus body.

E.4.4   Tailplane Contribution to the Airplane's Static Stability

The tailplane's contribution to the airplane's longitudinal static stability is equal to $\bar{V}' \, dC_{L_T}/dC_L$. This term will now be evaluated by expanding the derivative into three components and evaluating each one separately:

$$\frac{dC_{L_T}}{dC_L} = \left(\frac{dC_{L_T}}{d\alpha_T} \right) \left(\frac{d\alpha_T}{d\alpha} \right) \left(\frac{d\alpha}{dC_L} \right) \qquad (E.29)$$

The *first* derivative in the sequence (i.e., $dC_{L_T}/d\alpha_T$) is the lift-curve slope of the tailplane with respect to the tail angle of attack. It can be obtained by differentiating Equation E.8. If the elevator is fixed (i.e., the elevator angle does not change within the time frame for which the stability analysis is carried out), then $d\delta_e/d\alpha_T = 0$. Thus

$$\frac{dC_{L_T}}{d\alpha_T} = a_1 \tag{E.30}$$

The *second* derivative in the sequence (i.e., $d\alpha_T/d\alpha$) is evaluated by starting with Equation E.6 and substituting for ε from Equation E.5. As the stabilizer angle does not change within the time frame for which the stability analysis is carried out, this terms drops out:

$$\alpha_T = \alpha \left\{ 1 - \frac{d\varepsilon}{d\alpha} \right\} + \delta_s \tag{E.31}$$

Thus $\quad \dfrac{d\alpha_T}{d\alpha} = 1 - \dfrac{d\varepsilon}{d\alpha}$ \hfill (E.32)

The *third* derivative in the sequence (i.e., $d\alpha/dC_L$) requires a non-dimensional version of Equation E.1. This is obtained by dividing through by qS:

$$C_L = C_{L_{wb}} + \frac{S_T}{S} C_{L_T} \qquad \text{(using Equation E.7)}$$

or $\quad C_L = a\,\alpha + \dfrac{S_T}{S} C_{L_T} \qquad$ (using Equation E.3) \hfill (E.33)

Differentiating Equation E.33 with respect to α yields the following equation:

$$\frac{dC_L}{d\alpha} = a + \left(\frac{S_T}{S} \right) \left(\frac{dC_{L_T}}{d\alpha} \right) = a + \left(\frac{S_T}{S} \right) \left(\frac{dC_{L_T}}{dC_L} \right) \left(\frac{dC_L}{d\alpha} \right)$$

Thus $\quad \dfrac{dC_L}{d\alpha} = \dfrac{a}{1 - \left(\dfrac{S_T}{S} \right) \left(\dfrac{dC_{L_T}}{dC_L} \right)}$

or $\quad \dfrac{d\alpha}{dC_L} = \dfrac{1}{a} \left[1 - \left(\dfrac{S_T}{S} \right) \left(\dfrac{dC_{L_T}}{dC_L} \right) \right]$ \hfill (E.34)

The three derivatives, given by Equations E.30, E.32, and E.34, are now substituted into Equation E.29:

$$\frac{dC_{L_T}}{dC_L} = \left(1 - \frac{d\varepsilon}{d\alpha} \right) \left(\frac{a_1}{a} \right) \left(1 - \frac{S_T}{S} \frac{dC_{L_T}}{dC_L} \right)$$

or $\quad \dfrac{dC_{L_T}}{dC_L} = \dfrac{1}{1 + \dfrac{S_T}{S} \dfrac{a_1}{a} \left(1 - \dfrac{d\varepsilon}{d\alpha} \right)} \left(\dfrac{a_1}{a} \right) \left(1 - \dfrac{d\varepsilon}{d\alpha} \right)$ \hfill (E.35)

The derivative dC_{L_T}/dC_L can be simplified by introducing the term τ, where

$$\tau = \frac{S_T}{S} \frac{a_1}{a} \left(1 - \frac{d\varepsilon}{d\alpha} \right) \tag{E.36}$$

Thus $\quad \dfrac{dC_{L_T}}{dC_L} = \left(\dfrac{1}{1 + \tau} \right) \dfrac{a_1}{a} \left(1 - \dfrac{d\varepsilon}{d\alpha} \right)$ \hfill (E.37)

Equation E.37 is a rigorous expression for dC_{L_T}/dC_L, and can be taken forwards in a static stability analysis of an airplane in this form. This approach usually leads to the introduction of a new tail volume coefficient, which is defined as $\bar{V}'/(1 + \tau)$. The alternative approach (which is adopted herein) is to simplify Equation E.37 by noting that τ is relatively small compared to 1; hence $1/(1 + \tau) \cong 1$. With this approximation, Equation E.37 can be written as follows:

$$\frac{\mathrm{d}C_{L_T}}{\mathrm{d}C_L} \cong \frac{a_1}{a}\left(1 - \frac{\mathrm{d}\varepsilon}{\mathrm{d}\alpha}\right) \tag{E.38}$$

Neglecting τ amounts to neglecting the contribution of the tailplane to the lift-curve slope of the complete airplane. This is seen by combining Equation E.37 with Equation E.34:

$$\frac{\mathrm{d}C_L}{\mathrm{d}\alpha} = (1 + \tau)a \tag{E.39}$$

E.4.5 Static Margin

The static margin (K_N) is the distance from the CG to the neutral point expressed as a fraction of the MAC.

$$K_N = h_N - h \tag{E.40}$$

It was stated earlier (see Section E.3.2) that static stability depends on the slope of the function C_{M_G} plotted against C_L. This can be demonstrated by substituting Equation E.27 into Equation E.40.

$$K_N = \left(h_0 + \bar{V}'\frac{\mathrm{d}C_{L_T}}{\mathrm{d}C_L}\right) - h = -\frac{\mathrm{d}C_{M_G}}{\mathrm{d}C_L} \quad \text{(using Equation E.25)} \tag{E.41}$$

Equation E.40 shows that for positive static stability, the CG must be ahead of the neutral point; that is, the static margin must be positive. Substituting Equation E.38 into Equation E.27 yields the following expression for the neutral point:

$$h_N = h_0 + \bar{V}'\frac{a_1}{a}\left(1 - \frac{\mathrm{d}\varepsilon}{\mathrm{d}\alpha}\right) \tag{E.42}$$

Finally, by combining Equations E.40, E.41, and E.42, the static margin can be written in the following consolidated form:

$$K_N = -\frac{\mathrm{d}C_{M_G}}{\mathrm{d}C_L} = h_N - h = (h_0 - h) + \bar{V}'\frac{a_1}{a}\left(1 - \frac{\mathrm{d}\varepsilon}{\mathrm{d}\alpha}\right) \tag{E.43}$$

E.4.6 Trim Condition

The trim equation, given by Equation E.24, will now be elaborated. The tail angle of attack can be expressed using Equations E.31 and E.3:

$$\alpha_T = \frac{C_{L_{wb}}}{a}\left(1 - \frac{\mathrm{d}\varepsilon}{\mathrm{d}\alpha}\right) + \delta_s \tag{E.44}$$

As the contribution of the tailplane to the lift-curve slope of the complete airplane is comparatively small (see Section E.4.4), C_L can be substituted for $C_{L_{wb}}$ to give the following approximate expression:

$$\alpha_T \cong \frac{C_L}{a}\left(1 - \frac{\mathrm{d}\varepsilon}{\mathrm{d}\alpha}\right) + \delta_s \tag{E.45}$$

The trim equation can be rewritten by substituting Equation E.45 into Equation E.8 and then substituting for the tail lift coefficient in Equation E.24:

$$C_{M_G} = C_{M_0} + (h - h_0)C_L - \bar{V}' \left\{ a_1 \left[\frac{C_L}{a} \left(1 - \frac{d\varepsilon}{d\alpha} \right) + \delta_s \right] + a_2 \delta_e \right\} = 0$$

$$C_{M_0} - C_L \left\{ (h_0 - h) + \bar{V}' \frac{a_1}{a} \left(1 - \frac{d\varepsilon}{d\alpha} \right) \right\} - \bar{V}'(a_1\delta_s + a_2\delta_e) = 0$$

$$C_{M_0} - K_N C_L - \bar{V}'(a_1\delta_s + a_2\delta_e) = 0 \tag{E.46}$$

In conclusion, it can be noted that Equation E.46 is an expansion of the basic trim equation (given by Equation E.24), which is itself a statement that the sum of the moments acting on the airplane about the CG is zero when the airplane is in a state of equilibrium.

F

Regulations (Fuel Policy)

F.1 Introduction

EASA's Acceptable Means of Compliance and Guidance Material to Part-CAT [8] describes four approved fuel planning procedures—these are known as (a) basic procedure, (b) reduced contingency fuel (RCF) procedure, (c) predetermined point (PDP) procedure, and (d) isolated aerodrome procedure.

This appendix contains a reformatted extract from the regulatory document [8] describing the basic procedure (the latest version of these requirements can be obtained from the cited reference).

F.2 Fuel Planning: EASA Basic Procedure

The usable fuel to be on board for departure should be the sum of the following:

(1) *Taxi fuel*, which should not be less than the amount expected to be used prior to takeoff. Local conditions at the departure aerodrome and auxiliary power unit (APU) consumption should be taken into account.
(2) *Trip fuel*, which should include:
    (i) fuel for takeoff and climb from aerodrome elevation to initial cruising level/altitude, taking into account the expected departure routing;
    (ii) fuel from top of climb to top of descent, including any step climb/descent;
    (iii) fuel from top of descent to the point where the approach is initiated, taking into account the expected arrival procedure; and
    (iv) fuel for approach and landing at the destination aerodrome.
(3) *Contingency fuel*, except as provided for in the reduced contingency fuel (RCF) procedure, which should be the higher of:
    (i) either:
        (A) 5% of the planned trip fuel or, in the event of in-flight replanning, 5% of the trip fuel for the remainder of the flight;
        (B) not less than 3% of the planned trip fuel or, in the event of in-flight replanning, 3% of the trip fuel for the remainder of the flight, provided that an *en route* alternate (ERA) aerodrome is available;
        (C) an amount of fuel sufficient for 20 minutes flying time based upon the planned trip fuel consumption, provided that the operator has established a fuel consumption monitoring program for individual airplanes and uses valid data determined by means of such a program for fuel calculation; or

Performance of the Jet Transport Airplane: Analysis Methods, Flight Operations, and Regulations, First Edition. Trevor M. Young.
© 2018 John Wiley & Sons Ltd. Published 2018 by John Wiley & Sons Ltd.

(D) an amount of fuel based on a statistical method that ensures an appropriate statistical coverage of the deviation from the planned to the actual trip fuel (this method is used to monitor the fuel consumption on each city pair/airplane combination, and the operator uses this data for a statistical analysis to calculate contingency fuel for that city pair/airplane combination);

    (ii) or an amount to fly for 5 minutes at holding speed at 1 500 ft (450 m) above the destination aerodrome in standard conditions.

(4) *Alternate fuel*, which should:

    (i) include:

        (A) fuel for a missed approach from the applicable DA/H or MDA/H at the destination aerodrome to missed approach altitude, taking into account the complete missed approach procedure;

        (B) fuel for climb from missed approach altitude to cruising level/altitude, taking into account the expected departure routing;

        (C) fuel for cruise from top of climb to top of descent, taking into account the expected routing;

        (D) fuel for descent from top of descent to the point where the approach is initiated, taking into account the expected arrival procedure; and

        (E) fuel for executing an approach and landing at the destination alternate aerodrome;

    (ii) where two destination alternate aerodromes are required, be sufficient to proceed to the alternate aerodrome that requires the greater amount of alternate fuel.

(5) *Final reserve fuel*, which should be:

    (i) for airplanes with reciprocating engines, fuel to fly for 45 minutes; or

    (ii) for airplanes with turbine engines, fuel to fly for 30 minutes at holding speed at 1 500 ft (450 m) above aerodrome elevation in standard conditions, calculated with the estimated mass on arrival at the destination alternate aerodrome, or the destination aerodrome when no destination alternate aerodrome is required.

(6) The minimum *additional fuel*, which should permit:

    (i) the airplane to descend as necessary and proceed to an adequate alternate aerodrome in the event of engine failure or loss of pressurization, whichever requires the greater amount of fuel based on the assumption that such a failure occurs at the most critical point along the route, and

        (A) hold there for 15 minutes at 1 500 ft (450 m) above aerodrome elevation in standard conditions; and

        (B) make an approach and landing, except that additional fuel is only required if the minimum amount of fuel calculated in accordance with (2) to (5) is not sufficient for such an event; and

    (ii) holding for 15 minutes at 1 500 ft (450 m) above destination aerodrome elevation in standard conditions when a flight is operated without a destination alternate aerodrome.

(7) *Extra fuel*, which should be at the discretion of the commander.

Source: AMC1 CAT.OP.MPA.150(b) Fuel policy [8].

G

Abbreviations and Nomenclature

G.1 Mathematical Notation

A	aspect ratio (Equation 3.3); *or* cross-sectional area
a	speed of sound in air (Equation 2.77); *or* acceleration (Equation 2.17); *or* linear slope of C_L versus α curve; *or* linear slope of C_N versus α curve
a_c	centripetal acceleration (Figure 15.1)
A_{eff}	effective aspect ratio (Equation 7.10)
a_0	speed of sound in air at ISA sea-level datum conditions (Table 4.1b)
a_1	linear derivative of tailplane lift coefficient with respect to the tailplane angle of attack (Equation E.8)
A_1, B_1	drag parameters used with parabolic drag polar (Equation 7.49)
a_2	linear derivative of tailplane lift coefficient with respect to the elevator angle (Equation E.8)
A_2, B_2, C_2	parameters used to simplify mathematical derivations (Equations 9.26 and D.22)
A_3, B_3	parameters used to simplify range derivations (Equation D.59)
b	wingspan (Figure 3.3)
C	non-dimensionalizing constant for moment index (Equation 19.13); constant of proportionality
c	thrust specific fuel consumption (mass flow basis) (Equation 8.30a); *or* airfoil chord length (Figure 3.1)
$\bar{c}$	mean thrust specific fuel consumption (mass flow basis) (Equation 13.40); *or* geometric (standard) mean chord (Equation 3.10)
c'	thrust specific fuel consumption (weight flow basis) (Equation 8.30b)
$\bar{c}'$	mean thrust specific fuel consumption (weight flow basis) (Equation 13.40)
$\bar{\bar{c}}$	mean aerodynamic chord (Equation 3.5)
C_D	drag coefficient (Equation 3.22)
C_d	section (local) drag coefficient (Equation 3.15)
C_{D_g}	drag coefficient in ground effect (Equation 7.31)
C_{D_i}	lift-dependent drag coefficient (Equations 7.6 and 7.7)
C_{D_v}	vortex drag coefficient
C_{D_0}	zero-lift drag coefficient (Equations 7.6 and 7.7)
ΔC_{D_w}	wave drag coefficient (compressibility drag increment)
C_F	fuel cost per unit mass (Equation 18.1a)
C_{fix}	fixed cost component (Equation 18.2)

Performance of the Jet Transport Airplane: Analysis Methods, Flight Operations, and Regulations, First Edition.
Trevor M. Young.
© 2018 John Wiley & Sons Ltd. Published 2018 by John Wiley & Sons Ltd.

C_I	cost index (Equation 18.1a)
C_L	lift coefficient (Equation 3.21)
C_l	section (local) lift coefficient (Equation 3.14)
C_{L_K}	lift coefficient where drag polar changes (Figure 7.13)
C_{L_T}	tailplane lift coefficient (Equation 19.6)
C_{L_α}	lift-curve slope (i.e., $dC_L/d\alpha$) (Figure 7.1b)
C_{L_1}	1-g lift coefficient (Equation 7.2); *or* start-of-cruise lift coefficient (Equation 13.26)
C_M	pitching moment coefficient (Equation 3.25)
C_m	section (local) pitching moment coefficient (Equation 3.18)
C_{M_G}	pitching moment coefficient about the center of gravity (Equation E.10)
C_{M_0}	pitching moment coefficient about the aerodynamic center of the wing plus body (Equation 19.4)
C_N	normal force coefficient (Equation 3.23); *or* yawing moment coefficient (Equation 7.28)
C_p	specific heat at constant pressure (Equation 2.70)
c_r	root chord
C_T	time-dependent cost per unit of trip time (Equation 18.1a)
c_t	tip chord (Figure 3.3)
C_{tot}	total cost (Equation 18.2)
C_v	specific heat at constant volume (Equation 2.70)
C_{var}	variable cost component (Equation 18.3)
c_0	centerline chord (Figure 3.3)
D	drag
d	section (local) drag (Figure 3.11); *or* distance
d_{gc}	great circle distance
d_{rl}	rhumb-line distance
E	lift-to-drag ratio (Equation 7.35)
$\bar{E}$	mean lift-to-drag ratio (Equation 13.38)
e	Oswald factor (Equation 7.11)
E_f	fuel energy
$\dot{E}_h$	heat energy released per unit time (Equation 22.1)
E_I	energy intensity (passenger load) (Equation 18.15)
$E_{I,pl}$	energy intensity (total payload) (Equation 18.18)
E_k	kinetic energy (Equation 2.42)
$\dot{E}_k$	time rate of change of kinetic energy
E_p	potential energy (Equation 2.43)
E_s	specific energy (Equation 15.58)
E_t	total energy (Equation 15.55)
E_U	energy usage (passenger load) (Equation 18.14)
E_1	start-of-cruise lift-to-drag ratio (Equation 13.26)
F	force; *or* thrust
f	fuel–air ratio (Equation 8.9)
f_{acc}	acceleration factor (Equation 12.20)
F_{br}	braking force (Equation 9.56)
F_{fr}	friction force (Equation 9.7)
$\dot{F}_g$	flight profile alleviation factor (Equation 20.24)
F_m	momentum force (Equation 8.6)
F_N	net thrust of jet engine(s)

F_p	pressure force (Equation 8.7)
F_R	rudder force (Figure 7.15); *or* range factor (Equation 13.35)
F_s	specific thrust (Equation 8.25a)
F_Y	sideforce (Figure 15.2)
G	universal gravitational constant (Equation 2.6); *or* gas generator function (Equations 8.43 and 8.44)
g	gravitational acceleration
g_{SL}	gravitational acceleration at sea level (Equation 2.7)
$g_{SL,\phi}$	gravitational acceleration at sea level with centrifugal correction (Equation 2.10)
g_z	gravitational acceleration at height z above sea level (Equation 2.8)
$g_{z,\phi}$	gravitational acceleration at height z with centrifugal correction (Equation 2.13)
$g_{z,\phi,cm}$	gravitational acceleration at height z with centrifugal and airplane motion corrections (Equation 2.15)
g_0	standard value of gravitational acceleration (in the ISA)
H	geopotential height; *or* gust gradient distance (Figure 20.9); *or* constant for moment index (Equation 19.13)
h	height; *or* position of CG expressed as a fraction of the MAC (Figure 19.7)
H_d	density height
H_e	energy height
H_f	net heating value (gravimetric heating value) of fuel
h_N	position of neutral point expressed as a fraction of the MAC (Figure 19.7)
H_p	pressure height (altitude)
h_{sc}	screen height
h_0	datum height; *or* position of aerodynamic center expressed as a fraction of the MAC (Figure 19.7)
I	CG moment index (Equation 19.15)
K	lift-dependent drag factor used with the parabolic drag polar (Equation 7.14)
k	total temperature recovery factor (Equation 2.81); *or* slope of line; *or* correction factor; *or* empirical correlation factor
K_f	fuel transport index (Equation 16.15)
K_g	gust alleviation factor (Equation 20.19)
K_N	static margin (Equation E.43)
K_1, K_2, K_3, K_4	drag coefficient factors (Equations 7.22 and 7.23)
L	lift; *or* temperature gradient with height in the atmosphere (lapse rate) (Equation 4.10); *or* length; *or* rolling moment (Figure 3.9)
l	section (local) lift (Figure 3.11); *or* characteristic length
L_a	adiabatic lapse rate (Equation 4.34)
l_E	engine moment arm (Figure 7.15)
ℓ'_T	aerodynamic tail arm (Figure E.8)
M	Mach number (Equation 2.78); *or* pitching moment (Figure 3.9)
m	mass; *or* section (local) pitching moment
$\dot{m}$	mass flow rate
M_0	pitching moment about the aerodynamic center of the wing plus body
m_0	molar mass of gas (Equation 2.67)
M_C	design cruising Mach number
M_{cr}	critical Mach number
M_D	design dive Mach number

M_{dd}	drag divergence Mach number (Figure 7.10)
m_E	mass of the Earth
M_{ECON}	economy cruise Mach number
m_f	fuel mass
$m_{f_{used}}$	mass of fuel consumed (e.g., in cruise)
M_{LRC}	long range cruise Mach number
M_{MO}	maximum operating Mach number
M_{MRC}	maximum range cruise Mach number
M_V	ICAO metric value (Equation 21.3)
N	rotational speed of engine spool; *or* number of engines; *or* normal aerodynamic force (Equation 20.9); *or* yawing moment (Figure 3.9)
n	load factor normal to the longitudinal axes of the airplane (Equation 20.9); *or* empirical constant; *or* index
n, t	normal and tangential coordinates (directions) (Figure 2.5)
N_{seat}	number of passenger seats on the airplane (seating capacity)
n_Y	sideforce load factor (Equation 15.36)
n_z	load factor normal to the flight path (Equation 7.3)
$n_{z,L}$	limit load factor
n_{zw}	notation used in FAA/EASA regulations for n_z
O	origin of Cartesian coordinates
P	power (Equation 2.39)
p	pressure (Equation 2.49); *or* empirical constant
p, q, r	angular velocity components about X_b, Y_b, Z_b axes (Figure C.5)
P_D	drag power (Equation 7.52)
P_s	specific excess power (Equation 15.62)
P_T	thrust power
p_t	total pressure (Equation 2.82)
Q	fuel flow rate (mass flow basis) (Equation 8.26a)
Q'	fuel flow rate (weight flow basis) (Equation 8.26b)
q	dynamic pressure (Equation 2.64)
Q_{cor}	corrected fuel flow (mass flow basis) (Equation 8.36)
R	range; *or* specific gas constant (Equation 2.66); *or* resultant (net) force (Figure 3.11); *or* reaction force (Equation 9.7)
$\bar{R}$	universal gas constant (Equation 2.67)
R'	gas constant (Equation 2.68)
r	radius
r_a	specific air range (mass flow basis) (Equation 13.3a)
r'_a	specific air range (weight flow basis) (Equation 13.3b)
R_e	Reynolds number (Equation 2.92)
r_E	equivalent radius of the Earth (Equation 6.1)
R_g	ground range
r_g	specific ground range (mass flow basis) (Equation 13.46a)
r'_g	specific ground range (weight flow basis) (Equation 13.46b)
R_{GF}	ICAO reference geometric factor (Equation 21.3)
r_p	range parameter (Equation 13.11)
S	wing reference area (Figure 3.3); *or* Sutherland's empirical constant (Equation 2.91)
s	semispan (Figure 3.3); *or* ground distance; *or* gust penetration distance (Equation 20.23)
S_{exp}	exposed wing area

S_T	horizontal tailplane reference area
S_π	characteristic (reference) area (Equation 3.12)
T	absolute temperature
t	time; *or* airfoil maximum thickness
T_C	temperature value in degree Celsius (Equation 2.2a)
T_F	temperature value in degree Fahrenheit (Equation 2.2b)
T_K	temperature value in kelvin (Equation 2.3a)
T_{KP}	kink point temperature
T_R	temperature value in degree Rankine (Equation 2.3b)
T_t	total (stagnation) temperature (Equation 2.79)
t_π	time for a 180° turn (Equation 15.9)
U	work (Equation 2.36); *or* speed of air/gas; *or* gust velocity (in EAS)
u	flow velocity (Figure 2.11)
u, v, w	velocity components along X_b, Y_b, Z_b axes (Equations C.3, C.4, and C.5)
U_{ds}	design gust velocity (in EAS) (Equation 20.24)
U_{ref}	reference gust velocity (in EAS) (Equation 20.24)
U_w	vertical speed of the air (speed of updraft) (Figure 12.15)
V	true airspeed
$\vec{V}$	TAS velocity vector (Table 6.1)
$\bar{V}'$	tailplane volume coefficient (Equation 19.5)
v	volume
V_A	design maneuvering speed (Figure 20.7)
V_{APP}	approach speed
V_B	design speed for maximum gust intensity (Figure 20.10)
V_{BR}	speed at which the airplane is fully configured for braking
V_C	design cruising speed (Figure 20.7)
V_c	calibrated airspeed (Equation 6.24)
V_D	design dive speed (Figure 20.7)
V_e	equivalent airspeed (Equation 6.14)
V_{EF}	speed at which the critical engine is assumed to fail (during takeoff)
V_{EV}	speed at which an abort event, other than an engine failure, is assumed to occur
V_F	design flap speed (Figure 20.7)
V_{FTO}	final takeoff speed
V_G	ground speed (Equation 9.1)
$\vec{V}_G$	ground velocity vector (Table 6.1)
$V_{G,hp}$	hydroplaning ground speed
V_I	instrument indicated airspeed (Equation 6.30)
V_i	indicated airspeed (Equation 6.30)
V_{LOF}	liftoff speed
V_{MBE}	maximum brake energy speed
V_{MC}	minimum control speed
V_{MCG}	minimum control speed on the ground
V_{MO}	maximum operating speed
V_{MU}	minimum unstick speed
V_R	rotation speed
V_{REF}	reference landing speed (Equation 11.1)
V_S	stall speed
V_{SR}	reference stall speed
V_{SR_0}	reference stall speed in landing configuration

V_{S_0}	1-g stall speed in landing configuration (flaps fully extended)
V_{S_1}	1-g stall speed
V_{TD}	touchdown speed
V_v	rate of climb (Equation 12.25)
V_w	wind speed
$\vec{V}_w$	wind vector (Table 6.1)
V_1	takeoff reference speed (first RTO action speed); *or* start-of-cruise TAS
V_2	takeoff safety speed
W	weight (Equation 2.5)
$\dot{W}$	weight flow rate
w	specific weight (Equation 2.46)
W_f	fuel weight
$W_{f_{used}}$	weight of fuel consumed (e.g., in cruise)
W_{MTO}	maximum takeoff weight
W_{OE}	operating empty weight
X	distance from datum (e.g., weight datum) (Figure 19.2)
x	still air distance (Figure 13.1); *or* distance from datum (e.g., along chord)
$\bar{\bar{x}}_0$	distance from wing apex to the leading edge of the MAC (Figure 3.4)
X_a, Y_a, Z_a	flight path (or velocity or wind) axes Cartesian coordinates
X_b, Y_b, Z_b	body axes Cartesian coordinates (Figure 3.7)
X_e, Y_e, Z_e	Earth axes Cartesian coordinates (Figure 3.6)
X_g, Y_g, Z_g	ground axes Cartesian coordinates (Figure 3.6)
z	height above sea level

Greek letters

α	angle of attack (Figures 3.13, 12.2, and C.4); *or* passenger load factor (Equation 18.16)
β	sideslip angle (Figure C.4); *or* exponent (Equation 16.27)
β_s	Sutherland's empirical constant (Equation 2.91)
Γ	wing dihedral angle (Figure 3.3)
γ	ratio of specific heats of air (Equation 2.71); *or* still air climb/descent angle (flight path angle) (Figures 12.2 and C.3)
γ_G	runway inclination (slope angle) (Figure 9.5)
Δ	change or correction to parameter of interest
δ	relative pressure (of air) (Equation 4.2); *or* planform correction factor (Equation 7.10); *or* boundary layer thickness (Figure 3.19); *or* change to parameter of interest
δ_e	elevator deflection angle (Figure E.3)
δ_s	stabilizer deflection angle (Figure E.3)
δ_t	total pressure ratio (Equation 4.6)
ε	flight path drift angle (Figure 6.2); *or* downwash angle (Figure E.2)
ε_g	geometric wing twist
ς	cruise fuel mass (or weight) fraction (Equation 13.17)
η	non-dimensional semispan coordinate (Equation 3.2)
η_B	braking efficiency (Equation 9.57)
η_0, η_t, η_p	engine efficiency: subscript 0 denotes overall efficiency (Equation 8.16), t thermal efficiency (Equation 8.17), and p propulsive efficiency (Equation 8.18)

θ	relative temperature (of air) (Equation 4.1); *or* pitch angle (Equation 12.5)
θ_t	total temperature ratio (Equation 4.5)
κ	wing semispan fraction (Figure 3.4); *or* flight path angle (Figure 12.14)
κ_A	airfoil technology factor (Equation 7.16)
Λ	sweep angle (a subscript is used to identify the reference line—e.g., *LE* for leading edge) (Figure 3.3)
λ	bypass ratio (Equation 8.3); *or* planform taper ratio (Equation 3.1); *or* ground effect factor (Equation 7.30)
μ	dynamic viscosity (Equation 2.89); *or* coefficient of (rolling) friction (Equation 9.7)
μ_B	airplane braking coefficient (Equation 9.56)
μ_g	mass parameter (Equation 20.20)
ν	kinematic viscosity (Equation 2.90)
ρ	density (Equation 2.45)
σ	relative density (of air) (Equation 4.3)
σ_{SG}	specific gravity (Equation 2.48)
τ	shear stress (Equation 2.89); *or* economy cruise cost function (Equation 18.9); *or* lift-curve slope factor (Equations E.36 and E.39)
υ	wind direction with respect to the track (Figure 6.2)
Φ	bank angle (Figures 15.1 and C.2)
ϕ	roll angle (Figure C.1); *or* geographic latitude (Equation 6.1); *or* lift-dependent drag increment factor (Equation 7.26)
ϕ, θ, ψ	roll, pitch, yaw angular rotations about X_b, Y_b, Z_b axes (Figures 3.7 and C.1)
ϕ_T	thrust offset angle (Figure 7.14)
χ	azimuth angle (Figure C.3); *or* true track angle of the flight path (Equation 2.14)
ψ	yaw angle (Figure C.1); *or* geographic longitude (Equation 6.1)
ψ, ζ	acceleration factor parameters (Table 12.1)
Ω	angular velocity (Equation C.7); *or* rate of turn (Figure 15.1)
ω	rotational (or angular) velocity (Figure 2.3)
ω_E	rotational velocity of the Earth (Figure 2.1)

Subscripts

A	air segment
a	air (gas)
ac	aerodynamic center
acc	accelerated
act	actual
AEO	all engines operating
amb	ambient
an	anemometer
APP	approach
arr	arrival (airport)
avg	average
B	brake
BR, br	braking
CG	center of gravity
cl	climb
con	contaminant

cor	corrected
cp	center of pressure
cr	cruise
dep	departure (airport)
e	exit
EF	engine failure
est	estimated
f	fuel
FTO	final takeoff
G	ground; *or* ground segment
gd	green dot (speed)
H	horizontal
hp	hydroplaning (aquaplaning)
i	induced; *or* indicated
LE	leading edge
LEMAC	leading edge of mean aerodynamic chord
max	maximum
md	minimum drag
min	minimum
mp	minimum drag power
MTO	maximum takeoff weight
MW	main wheel
n	normal direction (coordinate)
NW	nosewheel
oe	operating empty
oei/OEI	one engine inoperative
OEW	operating empty weight
ot	power off-take
p	profile
pl	payload
R	rotate
RL	rotate to liftoff
S	stall
SL	sea level
spill	spillage
std	standard (ISA) condition
T	tail; *or* transition
t	total; *or* tip; *or* tangential direction (coordinate)
test	measured (test) condition
to	takeoff
TR	transition
uc	undercarriage (landing gear)
unacc	unaccelerated
V	vertical; *or* vortex
w	wind; *or* wing; *or* wave (drag)
wb	wing plus body
wl	wings level
wm	windmill
x, y, z	with respect to X, Y, Z coordinates

zl	zero lift
zll	zero-lift line
ψ	yaw (control)
∞	free stream
0	ISA sea level (datum); *or* free stream (ambient); *or* lift independent (drag); *or* centerline (chord); *or* aerodynamic center (moment, distance)
1	start-of-cruise/hold condition; *or* 1-*g* flight condition
2	end-of-cruise/hold condition

Superscript

*	conditions at tropopause; *or* reference condition (e.g., speed)

G.2   Chemical Symbols

CH_4	methane
C_nH_m	hydrocarbon compound
CO	carbon monoxide
CO_2	carbon dioxide
H_2O	water
H_2SO_4	sulfuric acid
N_2	diatomic nitrogen gas
NO_x	nitrogen oxides
O_2	diatomic oxygen gas
O_3	ozone
S	sulfur
SO_x	sulfur oxides

G.3   Abbreviations and Acronyms

AAL	above aerodrome (airfield) level
AC	aerodynamic center; *or* Advisory Circular
ACN	Aircraft Classification Number
AD	Airworthiness Directive
ADC	Air Data Computer
AEO	all engines operating
AFE	above field (airfield) elevation
AFM	Airplane/Aeroplane Flight Manual
AGL	above ground level
AIC	aviation-induced cloudiness
AIM	Aeronautical Information Manual
AIP	Aeronautical Information Publication
AIRAC	Aeronautical Information Regulation And Control
ALAR	Approach and Landing Accident Reduction
ALF	average load factor
ALR	adiabatic lapse rate
AMC	Acceptable Means of Compliance

AMSL	above mean sea level
AOA	angle of attack
APM	Aircraft Performance Monitoring
APU	auxiliary power unit
ARC	Aviation Rulemaking Committee; *or* Aeronautical Research Council (British)
ASDA	accelerate–stop distance available
ASI	airspeed indicator
ASK	available seat kilometer
ASM	available seat mile
ATA	Air Transport Association
ATAG	Air Transport Action Group
ATC	air traffic control
ATIS	Automatic Terminal Information Service
ATK	available tonne (metric ton) kilometer
ATM	assumed temperature method; *or* air traffic management; *or* available ton mile
A4A	Airlines for America
BADA	Base of Aircraft Data
BEW	basic empty weight
BIPM	International Bureau of Weights and Measures
BPR	bypass ratio
BRW	brake release weight
BTV	Brake To Vacate
C of A	certificate of airworthiness
CAA	Civil Aviation Authority
CAAC	Civil Aviation Administration of China
CAEE	Committee on Aircraft Engine Emissions
CAEP	Committee on Aviation Environmental Protection
CAN	Committee on Aircraft Noise
CAS	calibrated airspeed
CASK	cost per ASK
CASM	cost per ASM
CAT	commercial air transport
CDL	Configuration Deviation List
CDO	continuous descent operation
CDU	Control Display Unit
CFIT	controlled flight into terrain
CFR	Code of Federal Regulations
CG	center of gravity
CGS	centimeter–gram–second
CI	cost index
CIS	Commonwealth of Independent States
CL	climb (speed/rating)
CMR	Certification Maintenance Requirement
CODATA	Committee on Data for Science and Technology
COP	Conference of the Parties
CORSIA	Carbon Offsetting and Reduction Scheme for International Aviation
CP	center of pressure
CS	Certification Specification
DA/H	decision altitude/height

DALR	dry adiabatic lapse rate
DDG	Dispatch Deviation Guide
DGAC	Direction Générale de l'Aviation Civile
dim	dimension
DOC	direct operating cost(s)
DOT	Department of Transportation
DOW	dry operating weight
E	east
EAS	equivalent airspeed
EASA	European Aviation Safety Agency
EC	European Commission
ECAC	European Civil Aviation Conference
e-CFR	Electronic Code of Federal Regulations
ECON	economy (cruise speed)
ECS	environmental control system
ECU	Electronic Control Unit
EDTO	Extended Diversion Time Operations
EEC	Electronic Engine Controller
EFH	engine flight hours
EGT	exhaust gas temperature
EI	emission index
ELR	environmental lapse rate
EPNdB	effective perceived noise level in decibel
EPNL	effective perceived noise level
EPR	engine pressure ratio
ERA	*en route* alternate
ESAD	equivalent still air distance
ETOPS	Extended Twin Operations; *or* Extended Operations
EU	European Union
FAA	Federal Aviation Administration
FADEC	Full Authority Digital Engine Control
FAF	final approach fix
FAP	final approach point
FAR	Federal Aviation Regulation
FCOM	Flight Crew Operations/Operating Manual
FCTM	Flight Crew Training Manual
FDR	Flight Data Recorder
FHV	fuel heating value
FL	flight level
FLX	flexible (takeoff speed/rating)
FMC	Flight Management Computer
FMS	Flight Management System
F/O	first officer
FOBN	Flight Operations Briefing Note
FPD	freezing point depressant
FPPM	Flight Planning and Performance Manual
FPS	foot–pound–second
FSF	Flight Safety Foundation
FT	Fischer-Tropsch

GA	go-around
GCD	great circle distance
GDP	gross domestic product
GHG	greenhouse gas
GM	Guidance Material
GMT	Greenwich Mean Time
GPS	Global Positioning System
GS	ground speed; *or* glideslope
HC	hydrocarbons
HDG	heading
HLFC	hybrid laminar flow control
HP	high pressure
HTP	horizontal tailplane
IA	indicated altitude
IAC	Interstate Aviation Committee
IAE	International Aero Engines
IAS	indicated airspeed
IATA	International Air Transport Association
ICAO	International Civil Aviation Organization
IFALPA	International Federation of Air Line Pilots' Associations
IFR	Instrument Flight Rules
IFSD	in-flight shutdown
IGE	in ground effect
ILS	Instrument Landing System
IMC	instrument meteorological conditions
incl.	including
IOC	indirect operating cost(s)
IP	intermediate pressure
IPCC	Intergovernmental Panel on Climate Change
ISA	International Standard Atmosphere
ISO	International Organization for Standardization
IT	International Table
JAA	Joint Aviation Authorities
JAR	Joint Airworthiness Requirement
JFSCL	Joint Fueling System Checklist
KCAS	knots calibrated airspeed
KEAS	knots equivalent airspeed
KIAS	knots indicated airspeed
KT	knots
KTAS	knots true airspeed
LCV	lower calorific value
LDA	landing distance available
LE	leading edge
LEMAC	leading edge of MAC
LF	load factor
LHV	lower heating value
LOC	localizer
LP	low pressure
LRC	long range cruise

LROPS	Long Range Operations
LTO	landing and takeoff (cycle)
LW	landing weight
M	Mach number
MAC	mean aerodynamic chord
MAK	Interstate Aviation Committee (Russian)
max.	maximum
MBM	market-based measures
MCLT	maximum climb thrust
MCRT	maximum cruise thrust
MCT	maximum continuous thrust
MDA/H	minimum descent altitude/height
MEL	Minimum Equipment List
METAR	Meteorological Aerodrome Report
MEW	manufacturer's empty weight
min.	minimum
MKS	meter–kilogram–second
MLW	maximum landing weight
MMEL	Master Minimum Equipment List
MRC	maximum range cruise
MSL	mean sea level
MSN	manufacturer's serial number
MTOW	maximum takeoff weight
MTW	maximum taxi weight
MV	metric value
MZFW	maximum zero fuel weight
N	north
NAA	national aviation authority
NACA	National Advisory Committee for Aeronautics
NASA	National Aeronautics and Space Administration
NGV	nozzle guide vane
NIST	National Institute of Standards and Technology
no.	number
NOAA	National Oceanic and Atmospheric Administration
NP	neutral point
NTSB	National Transportation Safety Board
OAA	obstacle accountability area
OAT	outside air temperature
OBOGS	On Board Oxygen Generating System
OEI	one engine inoperative
OEW	operating/operational empty weight
OGE	out of ground effect
OM	Operations/Operating Manual; *or* outer marker
OML	outer mold line
OPS	operations
OTS	Organized Tracking System
PACOTS	Pacific Organized Tracking System
PANS	Procedures for Air Navigation Services
PCN	Pavement Classification Number

PDP	predetermined point
PEM	Performance Engineer's Manual
PFC	Porous Friction Course
PFD	Primary Flight Display
PFEI	payload fuel energy intensity
PIREP	pilot report
PNL	perceived noise level
ppm	parts per million
QC	Quota Count
QFE	Q code for altimeter pressure setting
QNE	Q code for altimeter pressure setting
QNH	Q code for altimeter pressure setting
QRH	Quick Reference Handbook
RASK	revenue per ASK
RASM	revenue per ASM
RC	reference chord
R/C	rate of climb
RCAM	Runway Condition Assessment Matrix
RCF	reduced contingency fuel
R/D	rate of descent
ref.	reference
RF	radiative forcing
RGF	reference geometric factor
RoC	rate of climb
ROC	rate of climb
RoD	rate of descent
ROD	rate of descent
RPK	revenue passenger kilometer
RPM	revolutions per minute; *or* revenue passenger mile
RSS	relaxed static stability
RTK	revenue tonne (metric ton) kilometer
RTM	revenue ton mile
RTO	rejected takeoff
RTOW	regulatory takeoff weight
RVSM	Reduced Vertical Separation Minimum
S	south
SAFO	Safety Alert For Operators
SAIB	Special Airworthiness Information Bulletin
SALR	saturated adiabatic lapse rate
SAR	specific air range
SARPs	Standards and Recommended Practices
SCAP	Standardized Computerized Aircraft Performance
SENEL	single-event-noise-exposure level
SEP	specific excess power
SFC	specific fuel consumption
SG	specific gravity
SGR	specific ground range
SI	International System of Units (French: Système International d'Unités)
SIB	Safety Information Bulletin

SIGMET	Significant Meteorological Information
SOP	Standard Operating Procedures
SPECI	Aviation Selected Special Weather Report (French translation)
SPL	sound pressure level
SR	specific range
SSD	significant standards differences
STC	supplemental type certificate
STD	standard
STOL	short takeoff and landing
TA	transition altitude
TAF	Terminal Aerodrome Forecast
TALPA	Takeoff And Landing Performance Assessment
TAS	true airspeed
TC	type certificate
TCDS	Type Certificate Data Sheet
TET	turbine entry temperature
TGT	turbine gas temperature
THS	trimmable horizontal stabilizer
TL	transition level
TO	takeoff
T/O	takeoff
TOC	top of climb
TOD	top of descent; *or* takeoff distance
TODA	takeoff distance available
TOGA	takeoff/go-around
TO/GA	takeoff/go-around
TOP	takeoff parameter
TORA	takeoff run available
TOW	takeoff weight
TPR	turbofan power ratio
TR	track
TSFC	thrust specific fuel consumption
TSP	thrust setting parameter
TUC	time of useful consciousness
UK	United Kingdom
UNFCCC	United Nations Framework Convention on Climate Change
US	United States (of America)
USA	United States of America
USC	United States Customary
UTC	Coordinated Universal Time
VFR	Visual Flight Rules
VHF	very high frequency
VMC	visual meteorological conditions
VSI	vertical speed indicator
VTP	vertical tailplane
W	west
WAT	weight, altitude, and temperature
WBM	Weight and Balance Manual
WD	wind direction

WGS	World Geodetic System
WMO	World Meteorological Organization
ZFW	zero fuel weight
ZLL	zero-lift line
ZLL_{wb}	zero-lift line of wing plus body

References

1 ISO, *Quantities and units – Part 1: General*, ISO 80000-1, International Organization for Standardization, Geneva, Switzerland, 2009.
2 BIPM, "The international system of units (SI)," 8th ed., Bureau International des Poids et Mesures (BIPM), Paris, France, 2006.
3 NIST, "The international system of units (SI)," Taylor, B.N. and Thompson, A. (Eds.), Special Publication 330, National Institute of Standards and Technology, Gaithersburg, MD, Mar. 2008.
4 ISO, *Quantities and units*, ISO 31, International Organization for Standardization, Geneva, Switzerland, 1992.
5 Cardarelli, F., *Encyclopaedia of scientific units, weights and measures: Their SI equivalences and origins*, translated by Shields, M.J., Springer-Verlag, London, 2003.
6 NIST, "Guide for the use of the international system of units (SI)," Thompson, A. and Taylor, B.N. (Eds.), Special Publication 811, National Institute of Standards and Technology, Gaithersburg, MD, Mar. 2008.
7 ICAO, "Units of measurement to be used in air and ground operations," 5th ed., Annex 5 to the Convention on International Civil Aviation, International Civil Aviation Organization, Montréal, Canada, July 2010.
8 EASA, "Acceptable means of compliance (AMC) and guidance material (GM) to Part-CAT," Iss. 2, Amdt. 5, European Aviation Safety Agency, Cologne, Germany, Jan. 27, 2016. Available from www.easa.europa.eu/.

Index

Performance of the Jet Transport Airplane: Analysis Methods, Flight Operations, and Regulations, First Edition.
Trevor M. Young.
© 2018 John Wiley & Sons Ltd. Published 2018 by John Wiley & Sons Ltd.